Parasitology

Parasitology

An Integrated Approach

Second Edition

Alan Gunn
School of Biological and Environmental Sciences
Liverpool John Moores University
Liverpool
UK

Sarah J. Pitt
School of Applied Sciences
University of Brighton
Brighton
UK

Contents

Preface

This book is designed mainly for undergraduate students of biological sciences, biomedical sciences, medicine, and veterinary sciences who need to know about protozoan and helminth parasites and understand how they affect their hosts. It would also be useful for postgraduate students who need background information about parasites to support their research and for members of any other professional group who need an insight into the subject for their work.

Protozoan and helminth parasites are fascinating organisms, and examples of their parasitism are found in a broad range of hosts, including plants, invertebrates, and vertebrates. The interaction between a parasite and its host is complex and dynamic. Therefore, studying parasitology is useful for appreciating a range of concepts in biological sciences including growth and reproduction, biochemistry, immunology, and pathology.

Parasites do not live within their hosts in isolation. We feel that it is instructive to recognise how an individual organism might interact with other members of the same species, other species of protozoa and helminths, and other classes of microorganism within a particular host. The effects on the host of harbouring a particular species of parasite are influenced by various host factors, including genetic constitution, immune status, and behaviour. Also, for parasites of humans, consideration of social, religious, and cultural factors is often necessary. We therefore call this book 'Integrated Parasitology' to emphasise how parasites influence, and are influenced by, a complex web of interacting factors. All the material from the first edition has been extensively revised and updated. In addition, many new illustrations are included, and the on-line support material has been expanded.

We divide the book into a series of conventional chapters, but because we wish to show how topics are inter-related, the reader will find certain subjects are picked up, put down, discussed in more detail elsewhere, and then returned to in a later chapter. This is also a good way of learning since it is better to take in bite sized chunks of information and return to them frequently rather than attempt to grasp all aspects of a topic in a single sitting. We first introduce the concept of parasitism and the terms used by parasitologists to describe parasite lifestyles. There then follows a brief consideration of taxonomy after which, in Chapters 3–7, we introduce some of the 'key players' explain their basic biology and how they interact with one another. Within this second edition, we have included more diagrams of parasite life cycles, but the reader is also recommended to consult the excellent online resource available at the DPDx – CDC Parasitology Diagnostic Web Site (http// dpd.cdc.gov/dpdx). There follows a chapter on parasite transmission in which we consider, amongst other topics, not only how parasites exploit other animals as vectors and intermediate hosts but also how they manipulate their host's behaviour to increase the chances of transmission. We provide separate chapters on immunology and pathology, but in truth, it is virtually impossible to separate these topics because they are so inter-dependent. Chapter 12 provides a counterbalance to the bad press that parasites receive. Parasites can not only be used for the treatment of medical conditions but also (in small doses) actually be good for us. Before one can begin to study parasites, one needs to be able to find them and count them. Even if the host is dead, this is not necessarily as simple as it sounds. Correct parasite identification is essential before treatment can be given and to determine whether a control programme is performing effectively. We therefore devote Chapter 13 to parasite diagnosis that encompasses techniques ranging from straightforward light microscopy to advanced molecular biology. We then consider methods of treating and vaccinating against parasites in separate chapters before ending with a chapter on control strategies.

In the spirit of integration, we have provided web-based support material via the publisher's website. This includes the numerous photographs of parasites that we could not include in the book without increasing its size and cost. There are also quizzes based on each chapter, and some ideas for projects that do not require access to complex laboratory facilities or use laboratory animals such as mice, rats, and rabbits.

Alan Gunn and Sarah J. Pitt, 2021

About the Companion Website

This book is accompanied by a companion website:

www.wiley.com/go/gunn/parasitology2

This website includes:

- Figures
- SAQs and answers
- MCQs and answers
- Practical exercises

1

Animal Associations and the Importance of Parasites

<table>
<tr><td colspan="2">CONTENTS</td></tr>
<tr><td>1.1</td><td>Introduction, 1</td></tr>
<tr><td>1.2</td><td>Animal Associations, 1</td></tr>
<tr><td>1.2.1</td><td>Symbiosis, 3</td></tr>
<tr><td>1.2.2</td><td>Commensalism, 5</td></tr>
<tr><td>1.2.3</td><td>Phoresis, 5</td></tr>
<tr><td>1.2.4</td><td>Mutualism, 6</td></tr>
<tr><td>1.2.5</td><td>Parasitism, 6</td></tr>
<tr><td>1.2.6</td><td>Parasitoids, 8</td></tr>
<tr><td>1.2.7</td><td>The Concept of Harm, 10</td></tr>
<tr><td>1.3</td><td>Parasite Hosts, 10</td></tr>
<tr><td>1.4</td><td>Zoonotic Infections, 13</td></tr>
<tr><td>1.5</td><td>The Co-evolution of Parasites and Their Hosts, 13</td></tr>
<tr><td>1.5.1</td><td>The Red Queen's Race Hypothesis, 14</td></tr>
<tr><td>1.5.2</td><td>Parasites in the Fossil Record, 15</td></tr>
<tr><td>1.5.3</td><td>Parasites and the Evolution of Sexual Reproduction, 16</td></tr>
<tr><td>1.6</td><td>Parasitism as a 'Lifestyle': Advantages and Limitations, 17</td></tr>
<tr><td>1.7</td><td>The Economic Cost of Parasitic Diseases, 18</td></tr>
<tr><td>1.7.1</td><td>DALYs: Disability-Adjusted Life Years, 20</td></tr>
<tr><td>1.8</td><td>Why Parasitic Diseases Remain a Problem, 22</td></tr>
</table>

1.1 Introduction

In this chapter, we introduce the concept of parasitism as a lifestyle and explain why it is such a difficult term to define. We also introduce some of the terms commonly used by parasitologists. Like all branches of science, parasitology uses specialist terms such as 'intermediate host', 'definitive host' and 'zoonosis' that one must understood before one can make sense of the literature. We explain why the study of parasites is so important and why parasitic infections will remain a problem in human and veterinary medicine for many years yet to come.

1.2 Animal Associations

All animals are in constant interaction with other organisms. These interactions can be divided into two basic types: intra-specific interactions and inter-specific interactions.

Parasitology: An Integrated Approach, Second Edition. Alan Gunn and Sarah J. Pitt.
© 2022 John Wiley & Sons Ltd. Published 2022 by John Wiley & Sons Ltd.
Companion website: www.wiley.com/go/gunn/parasitology2

Intra-specific interactions are those that occur between organisms of the same species. They range between relatively loose associations such as those between members of a flock of sheep and highly complex interactions such as those seen in colonial invertebrates. For example, the adult (medusa) stage of the Portuguese man o' war 'jellyfish' (*Physalia physalis*) may appear to be a single organism but is actually composed of colonies of genetically identical but polymorphic individuals. These colonies divide labour between themselves in a similar manner to that of organ systems within a non-colonial organism. For example, some colonies are specialised for reproduction, whilst others are specialised for feeding. The term 'jellyfish' is in inverted commas because although *P. physalis* superficially resembles a jellyfish and is a member of the Phylum Cnidaria, it is taxonomically not a true jellyfish. Instead, it belongs to the order Siphonophora within the class Hydrozoa. The true jellyfish belong to the Class Sycphozoa within which there are several orders but in all of these, the medusa stage is a single multicellular organism.

Inter-specific interactions are those that take place between different species of organism (Figure 1.1). As with intra-specific interactions, the degree of association can vary between being extremely loose and highly complex. Odum (1959) classified these interactions on the basis of their effect on population growth using the codes '+' = positive effect, '−' = negative effect, and '0' = no effect. This leads to six possible combinations (00, 0−, 0+ etc.), and these too can be broken down into further subdivisions. Some authors also include a consideration of the direction and extent of any physiological and biochemical interactions between the two organisms. Many terms have been suggested to compartmentalise these interactions (e.g., phoresis, mutualism, predation), but these are merely convenient tags, and they cannot be defined absolutely. This is because there is a huge diversity of organism interactions, and even within a single interaction there are many variables, such as the relative health of the two organisms, that will determine the consequences of the interaction for them both. It is therefore not surprising that there is a multiplicity of definitions in the scientific literature, and it is not unusual for two authors to use different terms for the same type

Figure 1.1   Different species will sometimes co-operate for mutual benefit.

of interaction between species. In this section, we will discuss symbiosis, commensalism, phoresis, mutualism and finally parasitism, with some examples of each.

1.2.1 Symbiosis

The term symbiosis derives from the Greek *συμβίωση* and is usually translated as 'living together'. It was originally used in 1879 by Heinrich Anton de Barry to define a relationship of 'any two organisms living in close association, commonly one living in or on the body of the other'. According to this definition, symbiosis covers an extremely wide range of relationships. Some authors state that both organisms in a symbiotic relationship benefit from the association (i.e., it is [++]) although this is clearly a much more restrictive definition, and it is more appropriately referred to as mutualism. However, some authors consider symbiosis and mutualism are synonymous – this only adds to the confusion. For the purposes of this book, we will keep to de Barry's original definition.

1.2.1.1 Symbionts

Strictly speaking, a 'symbiont' is any organism involved in a symbiotic relationship. However, most scientists tend to restrict the term to an organism that lives within or upon another organism and provides it with some form of benefit – usually nutritional. The association is therefore referred to as a host: symbiont relationship and most symbionts are microorganisms such as bacteria, algae, or protozoa. Where the symbiont occurs within the body of its host, it is referred to as an endosymbiont, whilst those attached to the outside are referred to as ectosymbionts. There are two types of endosymbiont: primary endosymbionts (or p-endosymbionts) and secondary endosymbionts. Primary endosymbionts form obligate relationships with their host and are the product of many millions of years of co-evolution. They are usually contained within specialised cells and are transferred vertically from mother to offspring. Consequently, they undergo co-speciation with their host and form very close host-specific relationships. By contrast, secondary endosymbionts probably represent more recent host: symbiont associations. In the case of insects, these symbionts live within the haemolymph (blood) rather than specialised cells or organs. Secondary endosymbionts tend to be transmitted horizontally and therefore do not show a close host: symbiont relationship. Horizontal transmission occurs when a symbiont (or parasite) is transmitted from one host to another that is not necessarily related to it.

It is uncertain how endosymbionts begin their association with their hosts, but some authors suggest that they arise from pathogens that attenuated over time. The suggestion that a parasite–host relationship tends to start off acrimoniously and then mellows with time is widespread in the literature, and whilst this may sometimes occur it is not a foregone conclusion.

1.2.1.2 The Importance of Symbionts to Blood-feeding Organisms

Although vertebrate blood contains proteins, sugars, and lipids, as well as various micronutrients and minerals, it lacks the complete range of substances most organisms require to sustain life and to reproduce. Consequently, many of the animals, which derive most or all their nutrition from feeding on blood (haematophagy), have symbiotic relationships with bacteria that provide the missing substances, such as the B group of vitamins. The need for supplementary nutrients is particularly acute in blood sucking lice (sub-order Anoplura) because they have lost the ability to lyse (break up) red blood cells, and therefore many nutrients will remain locked within these cells. In many cases, the symbiotic bacteria are held within special cells called mycetocytes that are grouped together to form an organ called a mycetome. Although these terms appear to indicate the involvement of fungi, they originate from a time when scientists could not distinguish between the presence of yeasts and bacteria within cells. Many scientists continue to use the term 'myceto-cyte' regardless of the nature of the symbiont, but others use the term 'bacteriocyte' where it is known that the cells harbour only bacteria.

In blood-feeding leeches belonging to the order Rhynchobdellida (there is a popular misconception that all leeches feed on blood; many of them are predatory), mycetomes surround or connect to the oesophagus. Mycetomes do not form in all blood-feeding leeches, and in the medicinal leech, *Hirudo medicinalis* (Figure 12.1), the symbiotic bacteria live within the lumen of the gut (Graf et al. 2006). The bacteria present in *H. medicinalis* are *Aeromonas veronii*; earlier work on leeches often refers to this bacterium as *Aeromonas hydrophila*. *Aeromonas veronii* also forms associations with other blood-feeding invertebrates, as well as vampire bats, but it can also live independently as a free-living organism. Interestingly, both *H. medicinalis* and *A. veronii* produce antimicrobial peptides that suppress the growth of other microbes in the leech's gut (Tasiemski et al. 2015). This reduces the diversity of the gut microbial flora and emphasises the close relationship between the two organisms. *Aeromonas veronii* is not always beneficial: in humans, it causes wound infections, septicaemia, and gastroenteritis. Blood-feeding leeches are useful in modern medicine, particularly to aid wound drainage following reconstructive surgery, but there is a risk of them facilitating an *Aeromonas* infection in the patient. The infections are often trivial, but they can become serious and lead to abscesses or cellulitis. This is a difficult problem to solve because the symbiotic bacteria are essential for the long-term survival of the leech. One cannot develop a strain or culture of *Aeromonas*-free leeches. However, treating the leeches 1–4 weeks before use with an antibiotic such as ciprofloxacin removes the bacteria without compromising the willingness of the leech to feed (Mumcuoglu et al. 2010). A leech is only used once in reconstructive surgery because of their potential to transmit diseases between patients. Consequently, the long-term survival of antibiotic-treated leeches is not a concern.

The nymphs and adults of the human body louse *Pediculus humanus humanus* (sub-order Anoplura) (Figure 7.11) have a symbiotic relationship with the gamma (γ) proteobacterium *Riesia pediculicola* – also referred to as *Candidatus Riesia pediculicola*. The term '*Candidatus*' is used in prokaryote taxonomy for an organism that may be well characterised from molecular and other studies but cannot be cultured in the laboratory. *Riesia pediculicola* are primary intracellular endosymbionts that provide their hosts with pantothenic acid (vitamin B5) and are essential to the survival of the lice. In nymphs and adult male lice, the symbionts live within a mycetome that some texts refer to as the 'stomach disc'. This is an unfortunate term because lice, like other insects, do not have a stomach in the mammalian sense of the word. Anyway, the mycetome is located on the ventral side of the mid-gut but unlike the leeches mentioned previously, there is no connection between the mycetome and the gut lumen (Perotti et al. 2008). In adult female lice, the bacteria re-locate to the oviducts and the developing eggs. Interestingly, molecular phylogenetic analysis cannot distinguish between the symbiotic bacteria of human head lice (*Pediculus humanus capitis*) and human body lice (*Pediculus humanus humanus*). This adds support to phylogenetic analysis of the lice themselves that indicates that although head lice and body lice occupy different ecological niches and body lice tend to lay their eggs on clothing whilst head lice attach their eggs to hair shafts, they are two morphotypes of the same species rather than two separate species. One suggestion is that body lice evolved from head lice relatively recently in human evolution once we started wearing clothing. The association between *Riesia* and *Pediculus* is between 12.95 and 25 million years old – which makes it one of the youngest host: primary endosymbiont relationships so far recorded (Allen et al. 2009). In common with other primary endosymbionts, *Riesia* has undergone a reduction in genome complexity and lost genes (Moran and Bennett 2014): this is because it has come to rely on its host for the provision of many nutrients and protection from the environment etc. In addition, because its transmission is via the eggs of its host, each louse symbiont population is in reproductive isolation and unable to undergo recombination with other strains of *Riesia* in other lice. This has led to the suggestion that *Riesia* will lack the capacity to develop rapid resistance mechanisms to antibiotics – and because the *Riesia* is essential for the lice, killing the symbiont would result in host mortality.

1.2.2 Commensalism

The term 'commensalism' derives from the Latin *commensalis* and means 'at the same table together'. Most definitions state that one species benefits from the association and the other is unharmed (0+). Including the concept of 'harm' within any definition is seldom a good idea because harm is difficult to measure and varies with the circumstances. Similarly, a 'benefit' may not be immediately apparent, and some associations commonly cited as commensal might involve a degree of benefit to both parties (++) albeit they may not benefit to the same extent. A commensal association may be 'facultative', in which both species can live independently of one another or 'obligatory', in which one of the associates must live in association with its partner. For example, in many warmer parts of the world, the cattle egret (*Bubulcus ibis*) perches on the back of cattle and big game from which it swoops down periodically to capture lizards and insects that are disturbed as its ride moves through the undergrowth. The egret is perfectly capable of living apart from cattle, but it benefits from its mobile vantage point-cum-beater. The egrets probably do not remove many ectoparasites from the cattle and they get their Arabic name *Abu Qerdan* 'father of ticks' from the abundance of ticks associated with their nesting colonies. The cattle, therefore, appear to gain little from the relationship although the egret acts as an early warning system of the approach of predators. African Cape Buffalo (*Syncerus caffer*) have a good sense of smell but a notoriously poor eyesight: they are therefore vulnerable to predators approaching from downwind. The red-billed oxpecker (*Buphagus erythrorhynchus*) is sometimes said to have a similar commensal relationship with cattle, but this is almost certainly false. Unlike cattle egrets, the red-billed oxpecker has an obligatory relationship with cattle and big game and far from removing ticks it feeds primarily on scabs and wound tissue pecked from their host. Their feeding delays wound healing and thereby makes the affected animal vulnerable to infections and infestations with blowfly larvae.

The amoeba, *Entamoeba coli* (not to be confused with the gastro-intestinal bacterium *Escherichia coli,* which is also abbreviated to *E. coli*) is a common commensal that lives in our large intestine. Unlike its pathogenic cousin, *Entamoeba histolytica*, *E. coli* feeds on bacteria and gut contents and does not invade the gut mucosa or consume red blood cells. Therefore, *E. coli* is of little interest *per se*, although a study in Mexico suggested an association between moderate-heavy infestations and childhood obesity (Zavala et al. 2016). The most important feature of *E. coli* is that its morphological similarity to *E. histolytica* means that one must be careful to distinguish between the two species in microscope surveys of faecal samples.

1.2.3 Phoresis

The term 'phoresis' derives from the Greek verb φέρω ('*phero*') meaning to bear/carry. This association involves one species providing shelter, support, or transport for another organism of a different species and may be temporary or permanent. For example, apart from during their first instar, the larvae, and pupae of the blackfly *Simulium neavei* attach themselves to the outer surface of freshwater crabs. The larvae feed by filtering out phytoplankton and detritus from the water and the crabs act as a firm yet mobile substrate on which to attach. An appreciation of this association is important because adult *S. neavei* are vectors of the filarial nematode *Onchocerca volvulus* that causes 'River Blindness'.

1.2.4 Mutualism

Mutualistic (Latin, *mutuus* meaning 'reciprocal') relationships are those in which both species benefit from the association in terms of their growth and survival (++). Some authors further restrict the definition to one in which neither partner can live on its own, whilst others are less prescriptive. The association between *Wolbachia* bacteria and *O. volvulus* is clearly mutualistic. The bacteria live within the cells of the reproductive tissues and hypodermis in the adult female worms and provide them with essential metabolites. In the absence of the bacteria, the worms cannot establish themselves in their host and grow and adult females become infertile. The bacteria are therefore a potential target for the chemotherapy of filarial nematode infections (Jacobs et al. 2019; Taylor et al. 2019).

Whether the relationship between the Cnidarian *Hydra viridissima* and its algal partner *Chlorella* is mutualistic depends upon the strictness of one's definition. *Hydra viridissima* can grow and reproduce in the absence of their algal partner, but it is uncertain whether the strains/species of *Chlorella* associated with *H. viridissima* can survive independently. The algae live within vacuoles in the endodermal cells of the *Hydra* and thereby impart it with its characteristic green coloration. Whether this provides camouflage that is beneficial is uncertain. When the *Hydra* reproduces by budding, its algal partner is passed on to the offspring; the algae are not essential to the budding process, but *H. viridissima* seldom undergoes sexual reproduction if the algae are absent. Experiments in which the algae are removed from the *Hydra* by exposure to high light intensities (Habetha et al. 2003) indicate that the nature of the relationship depends upon the environmental conditions. Like other *Hydra* species, *H. viridissima* obtains its food by capturing prey on tentacles that are armed with nematocysts, whilst the alga carries out photosynthesis and releases the sugars maltose and glucose-6-phosphate that can potentially be used by *H. viridissima*. If there is suitable illumination and plenty of prey for the *Hydra*, the growth of *H. viridissima* with and without algae is similar. This indicates that, under these conditions, the sugars released by the algae have little importance for the *Hydra*. If, however, there is illumination but no food for the *Hydra*, then those lacking algae die after a few weeks, whilst those containing algae shrink but can survive for at least 3 months and commence feeding again if presented with food. Therefore, the symbiotic algae play an important role in the survival of *H. viridissima* whose normal food supply is low/absent. By contrast, if *H. viridissima* are kept in the dark but with plenty of prey available, those lacking algae grow much better than those containing them. Furthermore, the algal population declines by about 60% although they are not lost entirely and the *H. viridissima* remain pale green. This indicates that under these conditions, the algae receive nutrients from the *Hydra* to such an extent that the relationship changes from mutualism to one akin to parasitism.

1.2.5 Parasitism

Parasitism is a surprisingly difficult term to describe, and there are numerous definitions in the literature. We have adopted the definition that: 'parasitism is a close relationship in which one organism, the parasite, is dependent on another organism, the host, feeding at its expense during the whole or part of its life (− +)'. Parasitism is frequently a highly specific relationship that always involves a degree of metabolic dependence of the parasite upon its host and often, though not always, results in measurable harm to the host. The association is usually prolonged, and although it may ultimately result in the death of the host, this is not usually the case. It is therefore distinct from predation in which the predator usually quickly kills and consumes its prey. However, owing to the complexities of animal relationships, there are always 'grey areas' in which any definition starts to become unstuck. This is particularly apparent in the case of animals that feed on blood.

Mosquitoes and tsetse flies are not considered parasites because they only feed for a few seconds or minutes before departing. By contrast, hookworms and crab lice are parasites, because they live in permanent associated with their host. Blood-feeding leeches and lampreys, however, are free-living organisms that attach to their victim for several hours whilst taking a blood meal; some authors consider them parasites, whilst others define their feeding as a type of predation.

From Welcome Guest to Villain: The Derivation of the Term 'Parasite'

The word 'parasite' derives from the Greek παρά ('*para*') meaning 'beside' and σῖτος ('*sitos*') that means 'food'. In Ancient Greece, the term 'parasite' had religious connotations and nothing to do with infectious organisms. According to a stone tablet discovered in the temple of Heracles (Hercules) in Cynosarges, the priest was required to make monthly sacrifices in the presence of parasites who were to be drawn from men of mixed descent. Declining a request to act as a parasite was a punishable offence. (Cynosarges was an area near to the city walls of Athens. In addition to the temple there was also a gymnasium, and it was here that the Cynic philosophers taught.) Subsequently, the word came to mean someone who shared one's food in return for providing amusement and flattery. The '*parasitus ridiculosissimus*' was a popular character in Greek and early Roman comedies and they even had joke books to help them should they run out of witticisms. The greed of the parasite was a constant source of fun for dramatists, and he was often given crude nicknames such as 'little brush – because he swept the table clean'. Double entendres were as popular over 2000 years ago as they are today and the Latin for little brush '*peniculus*' is also a diminutive for a penis (Maltby 1999).

An obligate parasite is one that has no alternative but to develop as a parasite of its host. On the other hand, a facultative parasite can develop as a parasite or a free-living organism depending upon the circumstances. For example, the larvae of the warble fly *Hypoderma bovis* must develop as parasites of cattle and are therefore obligate parasites. By contrast, the larvae of the blowfly *Lucilia sericata* are facultative parasites. This is because if the female fly lays her eggs upon a live sheep, the larvae will feed on living tissue and therefore be parasites. Conversely, if she lays her eggs on a dead sheep, the larvae will feed as free-living detritivores. Similarly, the amoeba *Naegleria fowleri* can live as a free-living organism in ponds and lakes but if it enters the nostrils of someone swimming in the water, then it can become an opportunistic parasite and infect their brain.

As mentioned above, some organisms, such as the human body louse *Pediculus humanus*, are parasitic at all stages of their life cycle, whilst others are only parasitic at one or more stages. For example, the blood fluke *Schistosoma haematobium* parasitises us during its adult stage and snails during two of its larval stages but it also has two non-feeding free-living stages. The act of being a parasite is therefore stage specific. Some estimates suggest that as many as 50% of all known species are parasites at some point in their life cycle. However, this estimate is subject to the caveat that there is no consensus about what constitutes a species, especially among the prokaryotes. The number of known species is also a reflection of the interests of biologists in different groups of animals. For example, the fact that insects account for 72% of all known species is, at least partly, a consequence of them being studied intensively for over 200 years. In one insect order alone, the Hymenoptera (ants, bees, wasps), there are approximately 100,000 parasitoid species. By contrast, fewer people have studied mites and nematodes and the diversity of their parasitic species is probably vastly underestimated. Nevertheless, parasitism is a remarkably common lifestyle and parasites (and their hosts) exist in all the major groups of living organisms including the archaea, bacteria, fungi, plantae, protozoa, invertebrates, and vertebrates.

There is an endless debate as to whether viruses are parasitic organisms. At one level, this would appear to be self-evident since viruses are incapable of maintaining themselves or reproducing except when within their host cell. However, being composed of complex organic molecules and having the capacity to evolve is not necessarily synonymous with being a living entity, especially when those attributes are dependent upon existing within a host cell. We will discuss this topic further in Section 2.2.

1.2.5.1 Intra-specific Parasites

Although most parasitic relationships involve two different species, intra-specific parasitism also occurs. Brood parasitism is a common example of intra-specific parasitism among many birds (Tomás et al. 2017) and some social wasps (Oliveira et al. 2016). It involves a female laying her eggs in the nest of a conspecific (member of the same species) – this means that the costs of rearing, the young will be borne by another individual. Intra-specific parasitism sometimes occurs during sexual reproduction when the male attaches to the female and becomes dependent upon her for the provision of nutrients. For example, in certain deep-sea angler fish belonging to the suborder Ceratioidea, the larval fish develop in the upper 30 m of sea water and then gradually descend to deeper regions as they metamorphose into adults. The adolescent males have a very different morphology to the females: they are much smaller; they have larger eyes, and, in some species, they develop a large nasal organ that presumably helps in their search for females. Furthermore, the males cease feeding and rely upon reserves laid down in their liver during the larval period to fuel their swimming. Upon finding a suitable female, the male grasps onto her using special tooth-like bones that develop at the tips of his jaws (his actual teeth degenerate during metamorphosis). After attaching, the male grows (although he remains much smaller than his consort), and his testes mature. His skin and blood vessels fuse with hers at the site of attachment, and he remains attached for the rest of his life and draws all his nourishment from her. Some authors suggest that the male must find a virgin female. However, although most females carry only a single male, there are records of females with three or more males attached to them. This is presumably an adaptation to life in the deep-sea regions in which the opportunity to locate mates is limited. Nevertheless, this raises questions about how sexual selection occurs because it is unusual in nature for a female to mate with just one male for life, especially if that male is the first one to turn up. This type of relationship is not found in all ceratioid anglerfish; in some species, the males are facultative parasites rather than obligate ones as described in the above scenario, whilst in other species the males are free-living, capable of capturing their own food, and form only temporary attachments to females. Molecular evidence suggests that the development of the parasitic males is a relatively plastic phenomenon among anglerfish and has evolved and subsequently become lost on several occasions (Pietsch 2005).

1.2.6 Parasitoids

The term parasitoid is restricted to certain parasitic insects whose hosts are almost exclusively other insects – although a few species attack certain crustacea, spiders, millipedes, centipedes, and earthworms. Some parasites cause mortality and may even depend on the death of their host to effect transmission to the next stage of their life cycle, but host death is not inevitable. By contrast, parasitoids slowly consume their host's tissues so that the host remains alive until the parasitoid has completed its development. At this point, the host dies either through the loss of vital tissues or through the parasitoid physically eating its way out of its host. Parasitoids are all parasitic during their larval stage, and the adult insect is free living and feeds on nectar, pollen, dead organic matter, or is predatory, depending upon the species. Parasitoids can develop as endoparasites within their host or as ectoparasites attached to the outside but with their mouthparts buried deep within the host's body. The larva has only the one host in or on which it develops and those that are endoparasites tend to exhibit the most host specificity. This lifestyle is therefore distinct from those insects such as warble flies (e.g., *Hypoderma bovis*) and bot flies (e.g., *Gasterophilus intestinalis*), which exhibit a more 'traditional' parasitic way of life that does not inevitably result in the death of the host. Many species of the order Hymenoptera (bees, ants, wasps) are parasitoids, and it is also a common lifestyle among the Diptera (true flies), but it is absent or very rare among the other orders. By contrast, most of the insect orders are hosts to parasitoids. Hyperparasitism is also common in which a parasitoid parasitizes another species of parasitoid. Parasitoids are effective for the control of agricultural pests, particularly within closed environments such as greenhouses. However, they have had limited success as control agents for parasites, their vectors, or intermediate hosts.

The parasitoid lifecycle typically begins with the adult female locating its host and either injecting one or more eggs or attaching them to the host's outer surface. Sometimes she also injects a toxin that temporarily or permanently disables her victim. The host is chosen based on its stage of development, which may be anywhere from the egg to the adult stage.

Parasitoid: Virus Interactions

Some endoparasitic wasps belonging to the families Ichneumonidae and Braconidae have a fascinating relationship with polydnaviruses. The polydnaviruses from these two wasp families are morphologically distinct, and they probably arose from the 'domestication' of two different viruses. However, through convergent evolution they exhibit many biological similarities (Drezen et al. 2017; Strand and Burke 2019).

The viruses replicate within the calyx cells of the wasps' ovaries and are then secreted into the oviducts. Therefore, when a wasp injects her eggs into a suitable host, usually a lepidopteran caterpillar, she also injects the virus. The viruses cannot replicate within the caterpillar, but they do invade several of its cell types. Within these cells, the viruses integrate into the caterpillar's genome and cause the expression of substances that facilitate the establishment of the parasitoid. For example, one of the main immune responses that insects express in response to an invader is encapsulation. Encapsulation depends upon recognition of the invader and then a co-ordinated physiological response: amoeboid-like cells present in the haemolymph surround the invader and then either kill it through the production of toxic chemicals and/or lack of oxygen or physically isolate it and thereby prevent it damaging the host.

If one implants wasp eggs without the virus into a host, then these are rapidly encapsulated and killed. The protective effect of the virus probably results from it causing the caterpillar to express protein tyrosine phosphatase enzymes and thereby interfering with the encapsulation process. Protein tyrosine phosphatases dephosphorylate the tyrosine residues of several regulatory proteins and are therefore closely involved in the regulation of signal transduction. Altering the levels of regulatory proteins makes it impossible for the host to express an effective immune response and therefore the parasitoid egg develops unmolested. The viruses also have other effects on the caterpillar including preventing its further development once it reaches the stage at which the parasitoid is to emerge. Consequently, the polydnaviruses have a mutualist-like relationship with the parasitoid within which they replicate. They are transmitted vertically as an endogenous provirus that integrates into the wasp genome but have a pathogenic relationship with the parasitoid's host, within which it cannot replicate.

It is probable that there are many other examples of symbiotic virus-parasitoid/parasite relationships awaiting discovery. However, not all wasp parasitoids have relationships with viruses and these inject toxins that cause similar disruptions to the host immune response and host development.

1.2.7   The Concept of Harm

The term 'harm' is often employed when describing interactions between organisms but is particularly pertinent to any discussion of host: parasite relationships. Unfortunately, harm is a difficult term to define and is not always easy to measure. For example, parasites are usually much smaller than their host and a single parasite may have such a minor impact that its effect on the physiology and well-being of the host is too trivial to measure. By contrast, a large number of the same parasite could cause serious illness or even death. Similarly, a low parasite burden may have little impact upon a healthy, well-nourished adult host, but the same number of parasites infecting an unhealthy, starving young host may prove fatal. Consequently, harbouring a pathogen (being infected) and expressing the signs and symptoms of being infected (suffering from a disease) are not necessarily synonymous. A common analogy is that a single glass of water will not harm you and may even do you good, but the rapid consumption of a thousand glasses of water would kill you. Does that mean that water is beneficial or poisonous? Clearly, it can be both and, likewise, harm is dependent upon the context in which it is being considered. For human parasites, one should also consider the context and psychological consequences. Among some poor communities, being infected by lice and parasitic worms may be considered an unremarkable fact of everyday life. By contrast, in affluent communities, the very thought of harbouring worms inside the body or being bitten by fleas may cause mental torment far above any physical harm caused. It is therefore not a good idea to make the ability to record measurable harm as a prerequisite for the classification of the relationship between two organisms. Indeed, in certain instances, low levels of parasitic infection may benefit the well-being of the host (Maizels 2020). Nevertheless, many parasite species have the capacity to cause morbidity, that is, a diseased state, and some may cause mortality (death). We discuss the possible beneficial consequences of low parasite burdens in more detail in Chapter 12.

The morbidity that parasite infections induce is often reflected in a reduction in the host's fitness as measured in terms of its growth or reproductive output. This is often attributed to the direct pathogenic effect of the parasite, such as through the loss of blood and the destruction of tissues or competition for resources. For example, many gut helminths act as so-called kleptoparasites (literally, thieving parasites) and compete with their host for nutrients within the gut lumen. However, the situation is far more complicated than this. Although a functional immune system is crucial for an organism to protect itself against pathogens, immune systems are energetically costly and when nutrients are limiting, it must trade these costs against other physiological processes. Ilmonen et al. (2000) demonstrated this by injecting one group of breeding female pied flycatchers (*Ficedula hypoleuca*) with a diphtheria-tetanus vaccine and a control group with a saline solution. The vaccine was not pathogenic and did not induce an infection, but it activated the birds' immune system. They found that birds injected with the vaccine exhibited a lower feeding effort, invested less in self-maintenance and had a lower reproductive output, as determined by fledgling quality and number. The authors therefore concluded that the energetic consequences of activating the immune system can be sufficient to reduce the host's breeding success.

1.3   Parasite Hosts

A parasite host is an organism on or in which the parasite lives and from which it derives its nutrition. The host is usually not related taxonomically to the parasite although this is not always the case (see intra-specific parasites). Most parasites are highly host specific and only infect one host species or a group of closely related species. This is because all hosts represent a unique challenge in terms of the complex adaptations the parasite requires to evolve to identify, invade, and survive within/upon them. Nevertheless, a few parasite species, exploit a wide range of hosts. For example, the protozoan parasite *Toxoplasma gondii* infects, grows, and asexually reproduces in virtually all warm-blooded vertebrates although sexual reproduction only takes place within the small intestine of cats.

Hosts can be divided into classes, depending upon the role they play in the parasite's life cycle. The 'definitive' (or final) host is the one in, or on, which the parasite reaches maturity and undergoes sexual reproduction, whilst the 'intermediate' host is the one in which the parasite undergoes its developmental stage(s). There may be just one or several intermediate hosts and the parasite may or may not undergo asexual reproduction during this time, but it cannot develop into an adult or reproduce sexually. In this way, some parasites exploit their hosts to maximum effect by combining the reproductive power of asexual reproduction in the larval stage with the advantages of sexual reproduction during the adult stage.

Parasites devote more of their energies to reproduction than free-living animals because they do not have to worry about food, shelter, and fluctuations in environmental conditions. This is important because the chances of any offspring locating and establishing themselves within a suitable host are very low. The completion of a parasite's life cycle sometimes depends upon the death of the intermediate host and the subsequent consumption of the larval form by the definitive host. In this situation, the parasite is often very pathogenic in its intermediate host but has relatively minor effects on the definitive host. The intermediate host is not always killed or consumed by the definitive host. For example, after undergoing asexual reproduction in the snail intermediate host, the cercariae of the liver fluke *Fasciola hepatica* physically and chemically bore their way out and swim off to transform into metacercariae attached to aquatic vegetation. The snail survives the damage to its tissues, and the lifecycle is completed when the metacercaria are consumed by the sheep definitive host (see Section 5.2.1.1.1 for more details).

Parasites of Parasites

Viruses infect several parasitic protozoa such as *Leishmania* spp. (Rossi and Fasel 2018) and *Giardia lamblia* (Janssen et al. 2015) but, at the time of writing, there was surprisingly little evidence of their presence in helminths – though this is probably because few scientists have looked for them. Some workers suggest that viruses could be used to combat parasite infections (Hyman et al. 2013), but there is increasing evidence that many of the viruses found in parasitic protozoa contribute to their pathogenicity (Gómez-Arreaza et al. 2017).

Parasites are also infected by prokaryotic (e.g., bacteria) and eukaryotic (e.g., fungi and protozoa) parasites. Those parasites that infect other parasites are known as hyperparasites. For example, the microsporidian *Nosema helminthorum* is parasitic on the tapeworm *Moniezia expansa* that lives within the small intestine of sheep and goats (Canning and Gunn 1984). Sheep become infected by the tapeworm when they accidentally ingest oribatid mites containing the cysticercoids of *M. expansa*. Subsequently, the sheep must consume the infective cysts of *N. helminthorum* and these must then penetrate the tegument (tapeworms lack a gut of their own) of the tapeworm. Within the tapeworm, *N. helminthorum* reproduces and causes numerous raised opaque bleb-like patches but is not especially pathogenic. Related microsporidia affect various other platyhelminth parasites (Canning 1975; Sokolova and Overstreet 2020), but there are remarkably few reports of them infecting parasitic nematodes (e.g., Kudo and Hetherington 1922). The discovery of microsporidia infecting the free-living nematode *Caenorhabditis elegans* has opened the potential of developing a laboratory model for studying both nematode immunity and the biology of microsporidia (Zhang et al. 2016). This is because *C. elegans* is a commonly used model organism whose full genome is known. Several species of microsporidia cause pathogenic infections in humans and domestic animals and a simple laboratory model would prove extremely useful in the development of drug treatments etc.

A paratenic host, also sometimes referred to as a transport host, is one that a parasite enters but within which it cannot undergo further development. Paratenic hosts are not usually essential for a parasite to complete its life cycle although they may provide a useful bridge between the infective stage/intermediate host and definitive host. For example, the definitive hosts of the nematode *Capillaria hepatica* are primarily rodents although it infects several other species of mammals including dogs, cats, and pigs. Human infections are rare but potentially serious. The adult worms reside in the definitive host's liver and their unembryonated eggs remain there until the host dies/ is killed and a scavenger/ predator consumes them (Figure 1.2). The unembryonated eggs pass through the gut of the scavenger/predator and then out with the faeces. This helps disperse the

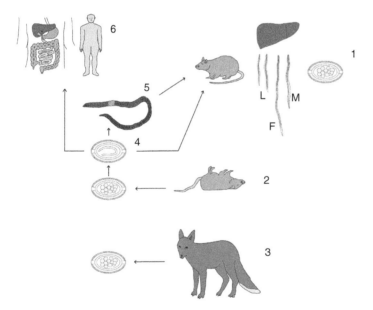

Figure 1.2   Life cycle of the nematode *Capillaria hepatica* illustrating the role of paratenic hosts in the transmission cycle. Drawings not to scale. 1 = A rodent becomes infected when it consumes embryonated eggs. These hatch in the small intestine, the larvae penetrate the gut, enter the circulation, and reach the liver. The larvae (L) moult become adult male (M) and female (F) worms and commence laying eggs. The unembryonated eggs remain in the liver. 2 = When the rodent dies its body decays and the unembryonated eggs enter the soil. If a scavenger eats the body, the unembryonated eggs pass through the gut and are dispersed. 3 = If a fox or other predator eats a live infected rodent, the unembryonated eggs are passed in its faeces. Scavengers and predators therefore act as dispersal hosts. 4 = The eggs embryonate to the infective stage in the soil. A rodent, human, or other susceptible mammal becomes infected when it consumes the infective eggs. 5 = Earthworms that consume infective eggs act as paratenic hosts if they are subsequently eaten by a rodent (or other susceptible mammal such as a pig). 6 = Humans are accidental, dead-end hosts within whose liver the parasites can develop to adulthood and produce eggs.

eggs in the environment. Development of the eggs to the infective stage occurs within the soil and takes several weeks or even months. If the definitive host's body is not consumed, the eggs embryonate to the infective stage, but there will be little dispersal. Earthworms ingest infective embryonated eggs of *C. hepatica* whilst feeding on soil and detritus. Because many rodents consume earthworms, these probably facilitate the transfer of the nematode to its definitive host.

1.4 Zoonotic Infections

A zoonotic infection (zoonosis) is one that is freely transmissible between humans and other vertebrate animals. The transfer of *Plasmodium falciparum* malaria between two people by a mosquito is therefore not zoonosis because a mosquito is not a vertebrate and *P. falciparum* only infects humans. By contrast, a mosquito transmitting *Plasmodium knowlesi* from a monkey to a human would be an example of a zoonosis because the *P. knowlesi* infects both monkeys and humans and we are both vertebrates. A disease that is only transmitted between humans is called an anthroponosis and a good example would be *P. falciparum*.

Many of the most important parasites in human and veterinary medicine are zoonotic infections. For example, pigs are the normal intermediate host of the pork tapeworm *Taenia solium,* and we are its definitive host. Therefore, pigs infect humans, and we infect pigs. Sometimes, humans are just one additional host within a parasite's life cycle. For example, the blood fluke *Schistosoma japonicum* has many definitive hosts apart from humans, including dogs, cattle, pigs, and rats. Consequently, all these definitive hosts can shed eggs that will infect the snail intermediate hosts, and the resultant cercariae can infect all of them.

The transmission of zoonotic parasites is usually heavily influenced by the nature of human: animal relationships. Therefore, they can be both simultaneously theoretically simple and recalcitrant to control. This is because their control often depends upon changing human behaviour, and this depends upon a complex mix of culture, religion, tradition, economics, personality, and politics. For example, theoretically, many zoonotic infections might be halted by simple acts of basic hygiene or the cooking of food. However, people are often unable or unwilling to change the way they live their life for all sorts of reasons. Zoonotic infections should not always be considered from the risks that they pose to us. Sometimes, wild animal populations can be threatened by the diseases that we transmit to them. We will consider specific instances of this throughout the book.

1.5 The Co-evolution of Parasites and Their Hosts

Evolution can be defined as a change in gene frequency between generations, but for this to occur three criteria need to be met. First, there must be genetic variation within the population. If the population is genetically homogeneous, then variation can only occur sporadically through random mutation. The second criterion is that the variation must be heritable: if the variation cannot be passed on to offspring, then it will be lost regardless of the benefits it imparts. The third and final criterion is that the variation must influence the probability of leaving reproductively viable offspring. If the variation is beneficial, then the organism possessing it will leave more offspring; however, unless these are reproductively viable, the variation would be quickly lost from the gene pool. Parasites live in close association with their hosts and the two organisms will co-evolve. The nature of the host: parasite relationship may therefore change with time. For example, provided the three criteria are met, the host will evolve resistance/susceptibility factors depending upon the pressure exerted by the parasite. Although ever greater resistance to infection may appear to be 'ideal', this is unlikely to arise if the energetic cost impacts on the ability to leave viable offspring. At the same time, the parasite will evolve virulence/avirulence factors that promote its own survival.

It is often stated that long-standing parasite: host relationships are less pathogenic than those that have established more recently. This is based on the reasoning that if the parasite kills its host, then it will effectively 'commit suicide' because it will have destroyed its food supply. Consequently, over time, it is to be expected that the parasite will become less harmful to its host – that is, it becomes less virulent. However, this assumption is questionable because a pathogen's virulence often reflects its reproductive success. For example, let us consider two hypothetical strains, A and B, of the same nematode species that lives in the gut of sheep. Strain A is highly virulent and causes the death of the sheep whilst strain B is relatively benign and seldom causes any mortality. At first glance, one might expect that strain B would leave more offspring because its host lives for longer. However, if virulence was linked to the nematode's reproductive output and the eggs were released at a time when they were likely to infect new hosts, then strain A would bequeath more of its genes to subsequent generations. Consequently, the proportion of strain A in the nematode population would increase with time and there would be constant selection for increasing virulence. The sheep and the parasites may eventually be driven to extinction by these changes, but individual animals (and humans) are almost always driven by their own immediate self-interest rather than hypothetical future prospects.

1.5.1 The Red Queen's Race Hypothesis

The scenario described above naturally begs the question of, if this is true, why does life still exist today. This is because, on this basis, parasites and other pathogens should have killed everything off many millions of years ago. The answer is that the scenario is too simplistic and all host: parasite/pathogen relationships involve a complex array of competing factors. Consequently, the evolutionary endpoint of any relationship is case-dependent. Sometimes the parasite becomes more virulent, and sometimes its virulence attenuates to an intermediate level, but one cannot assume that the natural endpoint is a mutually beneficial form of mutualism. Indeed, the relationship between a parasite and its host is often likened to a 'co-evolutionary arms race' in which the parasites attempt to acquire more resources from the host to produce their offspring whilst the host evolves mechanisms for reducing its losses and eliminating the parasite. This has given rise to the ecological theory known as the Red Queen's Race. The name derives the Red Queen in Lewis Carroll's *Alice Through the Looking Glass* who says, "Now, here, you see, it takes all the running you can do, to keep in the same place" (Ladle 1992). One should also bear in mind that a parasite and its host are not co-evolving in isolation. Hosts usually harbour various parasites and other pathogens, and these may influence its response to an infection. Similarly, the parasite may be competing with other infectious agents for the host's resources. For example, experiments using bacteria infected with phage viruses suggest that the presence of numerous pathogens can speed up host evolution (Betts et al. 2018).

Parasites and other pathogens are generally smaller than their hosts are and reproduce faster. Consequently, one might expect them to win any arms race because, potentially, they could evolve adaptations to overcome their host's defences faster than the host could generate new ones. However, hosts that are comparatively long-lived usually have sophisticated immune systems that identify and kill or neutralize new parasite variants. The host is therefore not a constant environment for the parasite. Parasite virulence is also affected by the mode of transmission. Horizontally transmitted parasites, especially those with a wide host range, can 'afford' to be highly virulent because there are lots of potential hosts and if one or more of them dies it has no direct consequences. However, when the parasite is vertically transmitted (e.g., via the eggs of its host or across the placenta) there is a direct link between the effect of the parasite on its host and its own reproductive success. For example, a virulent parasite's genes will not be transmitted; if the parasite is so pathogenic, it kills the host before it can reproduce. Similarly, if it kills the host's eggs while they are *in utero* or reduces the number of host eggs that are produced or survive to become adults and reproduce themselves, then the parasite is compromising its own reproduction. It is therefore to be expected that, as a rule (there will always be exceptions), vertically transmitted parasites should be less pathogenic than those that are transmitted horizontally. There is some support for this hypothesis. For example, two ectoparasites of swifts – a louse and fly – that are vertically transmitted have no effect on nestling growth or fledgling success even when the numbers of these parasites are artificially increased or the birds are stressed (Tompkins et al. 1996). Similarly, in feral pigeons, a vertically transmitted louse has little impact on the birds' health but horizontally transmitted ectoparasitic mites cause so much distress that the birds' reproductive success drops to zero (Clayton and Tompkins 1995).

1.5.2 Parasites in the Fossil Record

Most parasites are soft-bodied organisms, and they lack the hard structural features that facilitate preservation in the fossil record. It is therefore impossible to ascertain whether parasitism has always been a common 'lifestyle' – although this is highly likely. Conway Morris (1981) suggested surveying the commensals, symbionts and parasites of those organisms that have remained apparently unchanged for millions of years (the so-called living fossils) might reveal unusual organisms and provide insights into animal associations. For example, horseshoe crabs (Phylum Chelicerata, Subclass Merostomata) have existed almost unchanged for hundreds of millions of years. There is little published information on their parasites although flatworms of the family Bdellouridae only form associations with them (Riesgo et al. 2017). Despite the paucity of the fossil record, studies to date suggest that many parasite–host relationships persist for millions of years, and that parasite life cycles and morphology remain remarkably constant (Leung 2017)

Copepod ectoparasites that were morphologically similar to those in existence today have been identified attached to fossil teleost fish dating to the Lower Cretaceous Period (145–100.5 million years ago) (Cressey and Boxshall 1989). Evidence of nematode parasites is largely restricted to those infecting insects that became trapped in amber (Poinar 1984). Helminth eggs can be identified in coprolites (fossilised faeces), but while there have been extensive studies on animal and human faeces found in archaeological sites (Camacho et al. 2018), there is less data on coprolites dating back millions of years. As with any faecal analysis, one must not assume that presence indicates parasitism. An organism's presence may result from passage through the gut following accidental consumption (e.g., eggs of a parasite of another animal) or invasion of faeces after its deposition (e.g., eggs of a detritivore). Preservation of animals following rapid mummification under desiccating conditions or freezing in tundra enables the identification of soft-bodied parasites with greater accuracy. For example, nematodes and botfly larvae can be identified from woolly mammoths that died thousands of years ago on the Siberian tundra (Grunin 1973; Kosintsev et al. 2010).

Sometimes one can infer the presence of parasites in fossilised remains from the pathology they cause (Donovan 2015). For example, the pearls found in mussels and oysters often form because of infection by trematode parasites. Pearls thought have been caused by trematode parasites have been identified in fossil mussels dating back to the Triassic era (250–200 million years ago) (Newton 1908). Dinosaurs almost certainly had their full complement of parasites although their evidence is sadly lacking from the fossil record. However, marks found on the bones of the dinosaur *Tyrannosaurus rex* are thought to resemble the pathology caused by the protozoan parasite of birds *Trichomonas gallinae* (Wolff et al. 2009). Similarly, Tweet et al. (2016) found sufficient evidence in the fossilised gut contents of a hadrosaurid dinosaur to describe a vermiform organism that they called *Parvitubilites striatus* that may have been parasitic. Poinar and Poinar (2008) have even suggested that parasites were a major factor in the ultimate extinction of the dinosaurs – although this is not a widely accepted view amongst palaeontologists.

1.5.3 Parasites and the Evolution of Sexual Reproduction

Sex has fascinated biologists (amongst others) for generations. From a logical point of view, sexual reproduction does not make sense because of what is referred to as the two-fold cost of sex. Firstly, the males, who usually constitute in the region of 50% of a population, serve only to inseminate the females and do not reproduce themselves. Furthermore, a lot of time and effort is often employed in searching for a mate and mating can itself be an energetically expensive and potentially dangerous process. By contrast, in an asexually reproducing organism 100% of the population can reproduce. Consequently, an asexually reproducing population is theoretically able to grow faster and respond to changes in the environment (e.g., increased food supply) faster than one that reproduces sexually. The other 'cost' of sexual reproduction is that the gametes are haploid and the process of recombination at meiosis means that an individual can only pass on 50% of its genes to each of its offspring. Consequently, useful genes and gene combinations could be lost in the process of generating new genetic variants. Despite these problems, and several others, most organisms undertake sexual reproduction and therefore it must have some major advantage(s)

There are several theories why so many organisms reproduce sexually (Burke and Bonduriansky 2017). One of the most popular is that of Hamilton et al. (1990) who suggest that sexual reproduction arose as a mechanism by which organisms can limit the problems of parasitic infections. Parasites can potentially reproduce faster than their hosts, and therefore, they will evolve to overcome the most common combination of host resistance alleles. Therefore, hosts with rarer resistance alleles will then be at a competitive advantage and ultimately one of these will become the most common resistance allele combination in the host population. The arms race will continue *ad infinitum* with the parasites adapting to the most common resistance allele combination and the host generating new allele combinations. The process of recombination ensures that (provided the initial gene pool is sufficiently diverse) there will be a constant supply of novel resistance alleles. Furthermore, a resistance allele combination to which parasites have adapted need not be lost from the population because it may prove useful again in the future. By contrast, in an asexually reproducing organism the offspring will have the same resistance allele combinations as their parents, and once parasites have overcome these, then the whole population is vulnerable to infection.

If sexual reproduction arose as means of reducing the depredations of parasites, then one would expect it to be common where parasites are abundant and challenge frequent. By contrast, asexual reproduction should be favoured where parasites are absent, or the level of challenge is low. Although there are several instances of exactly this in the literature, they remain remarkably few. The best-known example is that of the snail *Potamopyrgus antipodarum* that originated in New Zealand and has since spread to many parts of the world. It exists as sexually reproducing populations, asexually reproducing populations, and mixed sexually and asexually reproducing populations. Positive correlations have been described between the extent of parasitism by parasitic flatworms and the frequency of sexual reproduction. Sexual reproduction is rare where flatworm parasite challenge is low, and conversely, it is common where the parasite challenge is high (Lively and Jokela 2002). Another commonly cited example is that of certain minnow populations living in Mexico (Lively 1996). These minnows exist as both asexually reproducing and sexually reproducing populations, but those reproducing sexually tend to have lower parasite burdens (except where inbreeding has resulted in reduced genetic diversity). Most multicellular parasites reproduce sexually themselves, although some combine it with asexually reproducing larval stages, such as schistosomes and the tapeworm *Echinococcus granulosus*. Even some parasitic protozoa, such as the trypanosomes, exhibit something akin to sexual reproduction. This suggests that even endoparasites living in protected environments such as the gut or bloodstream of another animal remain vulnerable to infections. However, although there is experimental evidence that parasitism influences the evolution and maintenance of sexual reproduction (Auld et al. 2016), there are almost certainly many other factors involved. For example, sexual reproduction may help protect against transmissible cancer cells (Thomas et al. 2019).

1.6 Parasitism as a 'Lifestyle': Advantages and Limitations

Provided one can get away with it, stealing something is easier than making it oneself or earning money to purchase it. Therefore, it is unsurprising that so many organisms have adopted a parasitic lifestyle to some extent. If one takes the view that the main purpose of an organism's existence is to transfer as many of its genes as possible into the next generation, then all organisms should maximise their reproductive output. However, an organism must trade the costs of reproduction against other activities such as finding food and then digesting and absorbing it, finding a mate, and protecting itself against competitors, predators, and the environment. By living upon or within a host, a parasite can reduce many of these 'other costs' and thereby devote more of its time and energy to reproduction. Most parasites stay in association with their host for the duration of a life cycle stage, and therefore, having located and infected their host, the need for sensory apparatus and locomotion are reduced because the parasite has access to a guaranteed food source. This guarantee also means that the parasite does not have to extract as much energy as possible from each 'unit of resource'. Instead, it can afford to be wasteful, and many parasites have reduced metabolic pathways. Furthermore, there is no need to lay down metabolic reserves beyond those required for the next life cycle stage. Parasites rarely need well-developed food gathering apparatus and, in some cases, such as the tapeworms, they have dispensed with a mouth and gut altogether, relying on nutrients being absorbed across the body wall.

Because parasites live within or upon their host, they have less need to maintain body surfaces and behaviours that protect them from desiccation, heat, cold because this is done by the host. Similarly, the parasite is to a large extent protected from predators and pathogens, because these must overcome the host's immune system before locating the parasite. Even ectoparasites receive protection to some extent because hosts cannot always distinguish between a predator attempting to take a bite out of them from an animal solely interested in removing a flea or louse.

A parasite will be transported wherever the host goes and therefore the limits of its dispersal depend upon the dispersal powers of its host, coupled with whatever other special needs the parasite must complete its life cycle (e.g., the presence of a suitable vector or environmental conditions). Consequently, a parasite does not have to devote energy to dispersal.

Table 1.1 Summary of advantages and disadvantages associated with the parasite lifestyle.

Advantages	Disadvantages
Once host located, no need for further searching	Extreme host specificity can increase vulnerability to extinction
Food permanently available	
Limited requirement for complicated food capturing mechanisms	Must locate at optimal site on/in host to ensure food/survival
Reduced need for food processing	
Protection from environmental extremes	Must adapt to host's internal physiological environment (internal parasites only)
Protection from predators and diseases	Must overcome host's immune defences
Reduced need for dispersal because host (+ vector) carries the parasite.	Spread limited by host's geographic range
Can devote larger proportion of energy intake to reproductive output than a free-living organism	Transmission can be extremely risky and most offspring die before establishing in a new host

If the benefits of parasitism are so enormous, this therefore begs the question why there are not more highly specialized parasites and why parasitism tends to be extremely common among some groups of organisms but rare among others. For example, there are comparatively few parasitic higher plants, Lepidoptera, or vertebrates.

Any would-be parasite must first overcome the putative host's immune defences and adapt to its internal physiological environment: this involves many physiological modifications, and therefore most parasites are host specific. However, host-specificity places the parasite in a difficult situation because its existence then becomes dependent upon that of its host. Should the host become extinct, then its parasites will follow suit unless they are able to infect other organisms. Furthermore, for the individual parasite, finding hosts is seldom easy. Although many parasites produce huge numbers of offspring, the chances of any one of them managing to locate a suitable host, establishing an infection, and reproducing successfully are extremely small. The advantages and disadvantages of the parasite lifestyle are summarised in Table 1.1.

1.7 The Economic Cost of Parasitic Diseases

The morbidity (illness) and mortality (death) associated with parasitic diseases causes financial losses to both an individual, their family, and to the wider society. These losses divide into the direct costs and indirect costs, and these are used in 'cost-of-illness' studies to prioritise healthcare funding decisions (Onukwugha et al. 2016). The direct costs include factors such as the costs of diagnosis and treatment. They are therefore relatively easy to identify and calculate because they consist of purchase costs and wages. By contrast, the indirect costs are much more wide-ranging and nebulous. For example, they include the costs associated with the infected individual's inability to work or reduced efficiency/productivity. They also include wider and often unappreciated costs that are borne by the family and/or the community. For example, the death of someone results in their family incurring the funeral costs (which can be considerable), as well as debilitating psychological stress that may impair their ability to work. Because most parasites cause chronic infections that persist for months or even years, the indirect costs associated with them often exceed the direct costs. For example, a study in China found that one case of malaria cost $US 239 (1,691.23 Chinese Yuan) of which the direct costs constituted 43% and the indirect costs 57% (Xia et al. 2016). Furthermore, the costs were equivalent to 11% of a household's income. Similarly, in southern India, lymphatic filariasis costs in the region of US$ 811 million per year and cause productivity losses as high as 27% in the weaving sector (Ramaiah et al. 2000). Parasitic diseases that cause disfigurement often results in social exclusion that further traps the sufferer in poverty and mental ill health. People suffering from lymphatic filariasis can become so isolated that they will not venture out to seek freely available treatment at government clinics, let alone to look for paid employment (Wijesinghe et al. 2007). Although it is not a financial calculation, experimental studies indicate that for wild animals living communally, it is also the indirect costs of parasitism that impact most upon the group (Granroth-Wilding et al. 2015).

For domestic animals, there are the direct costs of diagnosis and treatment along with mortalities but the losses that result from lost productivity (e.g., milk yield, live weight gain) and/ or work capacity (e.g., draught oxen, camels, donkeys) are much greater. Unfortunately, the calculation of losses associated with parasites in the agricultural industry is problematic, and there is a lot of variation between individual farms. In addition, published figures can rapidly become out of date through currency fluctuations, changes in farming practices and the value of stock (amongst many other factors). Therefore, we provide just a few figures to illustrate the potential of parasites to cause financial losses. In the United Kingdom, gastrointestinal parasitic infections in lambs are estimated to cost the British sheep industry ~£84 million per year (~USD$ 102.4 million); the costs associated with infections in breeding ewes are not known but the combined figure would obviously be much higher (http://beefandlamb.ahdb.org.uk/wp-content/uploads/2013/04/Economic-Impact-of-Health-Welfare-Final-Rpt-170413.pdf). Brazil is a much larger country with a huge cattle industry, and the financial impact of parasitic diseases is correspondingly massive. They are estimated to cause losses of approximately US$13.96 billion per year; gastrointestinal nematodes are responsible for ~51% of these losses and the tick *Rhipicephalus microplus* a further 23% through direct effects and as a vector of other parasites (Lopes et al. 2015a). In the United States, the protozoan parasite *Neospora caninum* is estimated to cause in the region of US$ 546 million per annum in the dairy industry alone. The losses it causes in agriculture on a worldwide basis could be as high as US$ 2.38 billion per annum (Reichel et al. 2013). There are no figures for the economic cost of *N. caninum* infection in dogs, but many dog owners will spend large sums of money on the welfare of their pets and pedigree dogs can sell for hundreds or even thousands of pounds. Consequently, control of the disease in dogs is of concern to owners, as well as a means of preventing its transmission to cattle.

In developing countries, the economic costs of parasitic diseases of livestock can have consequences for the expansion of agriculture and the ability of populations to feed and clothe themselves. For example, in Pakistan, the increasing demand for milk and milk products has seen the import of high-yielding Holstein-Friesian breeds. Unfortunately, these are particularly susceptible to the tick-borne protozoan parasite *Theileria annulata* (causative agent of Tropical Theileriosis) and the losses it causes can account for 13.8% of a total farm's costs (Rashid et al. 2018). Similarly, in east, central, and southern Africa, East Coast Fever in cattle caused by *Theileria parva* results in annual losses of hundreds of millions of pounds/dollars and is one of the reasons many people in the region remain subsistence farmers (Muhanguzi et al. 2014). Although vaccines against both *T. annulata* and *T. parva* have been available for many years, there are practical problems associated with their use. Consequently, preventing the transmission of infections is mostly through acaricides that kill the tick vectors. However, because tick populations are increasingly resistant to these, there is a fear that the ticks will spread and consequently so will the diseases.

1.7.1 DALYs: Disability-Adjusted Life Years

A common means of measuring the consequences of human disease and other causes of morbidity is to calculate disability-adjusted life years (DALYs). These are derived by summing an estimate of a disease or condition's potential for reducing lifespan and an estimate of the amount of time a person suffering from the disease/cause is disabled (www.who.int/evidence/bod). One DALY is the equivalent of the person losing a year of healthy life.

DALY = Number of years of life lost through premature mortality + Years of life lived with disability

For example, a person committing suicide or dying in a traffic accident would suffer premature death, but there would be little or no disability (assuming they died instantly), whilst a person with malaria may suffer prolonged ill health and ultimately die prematurely years later. DALYs facilitate the comparison of morbidity and mortality factors and thereby help prioritize funding and policy decisions and determine the effectiveness of health initiatives. In some studies, the DALY model is refined to place greater value on the life of a young adult than of a child or older person. This version considers young adults more economically beneficial to society and with a longer productive life in front of them than a child or older person. However, the use of age weighting is contentious and the WHO ceased using this approach in 2010.

The use of DALYs began in 1994 and although the WHO and many other organisations employ them, they have always been controversial. For a detailed consideration of the limitations of DALY calculations, see Parks (2014). The use of DALYs to assess the importance of parasitic diseases is particularly difficult because the estimation of the years of life with disability includes a weighting factor that supposedly accounts for the severity of the disease. This can result in wildly different estimations. For example, although some studies suggest that the global burden of human schistosomiasis is ~3 million DALYs, others have put it as high as 70 million (Hotez et al. 2010). Furthermore, coinfections with several parasite species and parasite–pathogen interactions (e.g., *Leishmania*-HIV) are common and can have major implications for disease progression and outcome.

A comprehensive study of global health metrics by Hay et al. (2017) provides an insight into the relative importance of various causes of mortality and morbidity. Table 1.2 shows a selection of their data. Except for malaria, many parasitic diseases have comparatively small DALYs compared with other sources of morbidity/mortality – this is because they operate within restricted distributions. For example, car accidents are a common source of morbidity and mortality in all countries, and therefore, it is not surprising that they have high DALY values. Similarly, diarrhoeal diseases, sexually transmitted infections, and measles are serious diseases throughout the world – though many people do not realise that in addition to causing morbidity, many can also be fatal. The accuracy of all statistics depends upon the accuracy with which the data are recorded. For developing countries with few resources and those in the grip of armed conflict, this is extremely difficult. Consequently, the literature often includes huge discrepancies about how many people suffer from a disease and how many people die from it. For example, according to Wang et al. (2016), the nematode *Ascaris lumbricoides* was responsible for 2,700 deaths in 2015, but a WHO website suggested that around 60,000 people die of the disease every year (https://www.who.int/water_sanitation_health/diseases-risks/diseases/ascariasis/en/).

Although some workers have attempted to use economic costings for wildlife diseases, it is a controversial approach: how much is a blackbird worth? Indeed, there has been a tendency for parasitologists to view wildlife mainly from the perspective of their potential as reservoirs of disease for human infections or those of our domestic animals (Thompson et al. 2010). This has

Table 1.2 A comparison of global disability adjusted life years (DALYs) and mortality for selected parasites and other factors.

Factor	All-age DALY (million) (year = 2016]	DALY range (year = 2016]	Mortality per annum (year)	Reference for mortality data
Malaria	56.2	45.8–67.9	435,000 (2017)	https://www.who.int/news-room/fact-sheets/detail/malaria
Visceral leishmaniasis	0.71	0.40–1.21	24,200 (2015)	Wang et al. (2016)
Cutaneous/mucocutaneous leishmaniasis	0.27	0.18–0.40	Rarely fatal	
African trypanosomiasis	0.13	0.06–0.22	3,510 (2015)	Wang et al. (2016)
Schistosomiasis	1.86	1.12–3.18	4,400 (2015)	Wang et al. (2016)
Lymphatic filariasis	1.19	0.59–2.11	Rarely fatal	
Ascariasis	1.31	0.88–1.94	2,700 (2015)	Wang et al. (2016)
			60,000 (date not stated, website accessed 2019)	https://www.who.int/water_sanitation_health/diseases-risks/diseases/ascariasis/en/
Hookworm	1.69	1.00–2.65	Rarely fatal	
HIV/AIDS	57.6	54.6–61.0	570,000–1.1 million (2018)	https://www.unaids.org/en/resources/fact-sheet
Measles	5.72	2.15–12.26	73,400 (2015)	Wang et al. (2016)
Ebola	0.0003	0.0002–0.001	5,500 (2015)	Wang et al. (2016)
			33 (2018)	https://www.afro.who.int/health-topics/ebola-virus-disease
Diarrhoeal diseases	74.41	63.4–93.4	1.65 million (2016)	Troeger et al. (2018)
Syphilis	9.42	5.47–14.60	107,000 (2015)	Wang et al. (2016)
Road injuries	71.40	67.52–76.13	1.35 million (2018)	https://www.who.int/violence_injury_prevention/road_safety_status/2018/en/

DALY and DALY range data were derived from Hay et al. (2017). The mortality data were derived from the most recent year available at the time of writing and from various sources.

sometimes led to widespread culling of wildlife. For example, in parts of Africa it was once common practice to kill antelopes and other large game animals to prevent them acting as a reservoir of *Trypanosoma brucei* infection. Similarly, at the time of writing, the practice of culling badgers in the United Kingdom to prevent the spread of TB in cattle is proving hugely controversial and expensive. Its effectiveness is also debateable.

The rate of extinctions amongst animals and plants is proceeding at an alarming rate and with it the realisation that we need to do more to conserve them. This is not just an ethical issue, but it also has economic implications since wildlife tourism is big business in some countries. Any attempt at conserving an organism must consider the diseases it suffers from. In addition to natural infections, wild animals are also afflicted by parasites introduced to their habitat by humans. It would be wrong to consider natural infections as invariably benign and those introduced by humans as invariably malign. For example, until the introduction of the New World screwworm fly (*Cochliomyia hominivorax*) eradication campaign in the USA, one estimate suggested that it killed up to 80% of white-tailed deer fawns in the southern states every year (Fuller 1962). The screwworm fly was present naturally and the eradication campaign was solely to prevent infections in cattle and other domestic animals, but the result was beneficial to wildlife too. More commonly, a parasite colonises a new area through contamination (e.g., in soil or ship ballast water) or through infections of us and our domestic animals. The consequences then depend upon whether the invading species finds other suitable hosts and, if it needs one, a suitable vector or intermediate host. The exposure of any naïve animal (or human) to an agent capable of establishing an infection in them often ends badly and if that agent can complete its life cycle in the area, then the consequences for the local population of new hosts is equally dire. For example, on the Galapagos Islands, the populations of several of the species of Darwin's finches have been devastated following the arrival of the fly *Philornis downsi*. It probably came to the islands in the 1960s among imported fruit and vegetables. The adult flies are free living, but their blood-feeding larvae are ectoparasitic on nestling birds and cause high mortalities (McNew and Clayton 2018). Wildlife tourism brings in hundreds of millions of dollars per year to the Galapagos Islands (https://www.galapagos.org/wp-content/uploads/2012/01/TourismReport2.pdf). Although most people do not visit the Galapagos Islands to spot Darwin's finches, the loss of iconic species such as the Giant Tortoises to introduced parasitic infections would undoubtedly have serious implications for the tourist industry.

1.8 Why Parasitic Diseases Remain a Problem

Whenever a seemingly simple but intractable problem arises, a commonly heard refrain is 'if we can put a man on the moon, why can't we do X, Y, or Z'. As we have seen, parasitic diseases cause suffering to us and to our domestic animals, and the economic costs are enormous. Furthermore, many diseases could be controlled by simple measures such as providing safe drinking water and appropriate waste disposal facilities. So, one might ask, why do parasitic diseases continue to afflict so many people and impact so heavily on agriculture?

As with so many apparently simple questions, the reason parasitic diseases remain a problem does not have a single simple answer and is also tied up with the most exasperating factor of all – human behaviour (Table 1.3). To begin with, human parasitic diseases are predominantly (although not entirely) a problem of poor people who live in insanitary conditions and who do not have a healthy diet. The diseases are therefore most prevalent in developing countries where neither the government nor individual people have money to spare. For example, in 2016 the total healthcare

Table 1.3 Summary of factors contributing to the problems of parasitic diseases.

Poverty

Lack of sanitation

Complacency

Poor nutrition

Lack of health infrastructure

Lack of government interest

Corruption

Urbanization

Social conflict/wars

Movement of non-immune people to regions where they become infected from the resident population.

Movement of infected people to regions where they infect non-immune resident population

Man-made environmental damage

Natural disasters

Lack of effective drugs/ parasite resistance

Increasing resistance of vectors/ intermediate hosts

---

expenditure in Zimbabwe as a percentage of the gross domestic product (GDP) was similar to that of the United Kingdom (9.41% cf 9.76%) and considerably more than that of oil-rich Saudi Arabia (5.74%) (https://data.worldbank.org/indicator/SH.XPD.CHEX.GD.ZS). However, in terms of total health expenditure per capita, the United Kingdom spent US$4192, Saudi Arabia US$1147, and Zimbabwe US $94 (https://knoema.com/atlas/Zimbabwe/Health-expenditure-per-capita). Needless to say, US $94 does not buy many medicines.

We humans are extremely adaptable creatures. Consequently, we can survive harsh environments, oppressive regimes, and cruel exploitation. Unfortunately, this adaptability can degenerate into acceptance and complacency on the parts of both individuals and governments. Because parasitic diseases are so prevalent in developing countries, there is a tendency not to prioritise them: fevers and diarrhoea become an accepted part of everyday life. Furthermore, parasitic diseases tend to cause chronic disease and although the patient may ultimately die, the condition does not capture the attention of the local or world media. For example, Ebola virus is well known in the developed world because of its appalling pathology and images of patients being treated by nurses and doctors dressed in spacesuit-like protective clothing. However, although Ebola virus causes about 70% mortality, the numbers of people who have died of the infection are relatively few. By comparison, Human African Trypanosomiasis (HAT, often referred to as 'sleeping sickness') causes almost 100% mortality if untreated and kills many more people than Ebola (Table 1.2), but it seldom receives a mention in the media. The reason is simple, HAT kills slowly by comparison. Furthermore, the transmission of HAT depends upon tsetse flies, and these have demanding environmental requirements that limit their distribution. Consequently, HAT is only a threat to people living in certain parts of Africa. By contrast, Ebola spreads through close human contact and therefore the virus could conceivably spread anywhere in the world. Consequently, people in distant countries feel threatened even though their risk is incredibly small. The fact that Ebola virus has been touted as a possible biological warfare agent also helps to engender interest in the disease and funds to study and control it.

In addition to being poor, the countries in which parasitic diseases are most problematic are often unstable and suffer high levels of corruption. Consequently, those in control often devote much of their revenue into the trappings of power and military spending: many developing countries spend less than 4% of their GDP on healthcare. This means that even less of not very much is available for the treatment and control of parasitic diseases. The instability of the regimes and conflicts, which can last for decades makes it difficult to provide health services and co-ordinate control strategies. They also lead to the destruction of basic infrastructure and the decline in agricultural and commercial activity – and this contributes to poverty and malnutrition. At its worst, conflicts lead to large numbers of refugees who are frequently housed in squalid campsites, which lack proper sanitation. These displaced people are often in poor health and malnourished, they take their parasites with them wherever they go, and they are highly vulnerable to the local strains of parasites at wherever they arrive. For example, the civil wars in the Central Asian states such as Tajikistan, which occurred after the breakup of the Soviet Union in the early 1990s, displaced people to neighbouring countries including Afghanistan. The most common type of malaria in Tajikistan at that time was caused by *Plasmodium vivax,* whereas in Afghanistan, the more virulent *Plasmodium falciparum* was found, and drug-resistant strains were circulating. Some of the refugees who returned home in the late 1990s were infected with drug-resistant *P. falciparum* and since there was a suitable mosquito vector, this form of malaria was subsequently transmitted among people who had never left Tajikistan (Pitt et al. 1998). Similarly, at the time of writing, the wars in Syria and Yemen had resulted in an almost complete collapse of their health infrastructure. In both Syria and Yemen, leishmaniasis was becoming a serious problem, and the disease was being transmitted to refugee camps in surrounding countries (Al-Salem et al. 2016; Du et al. 2016). Syria also saw a rise of almost 100,000 cases of malaria between 2015 and 2016 (https://www.globalcitizen.org/en/content/malaria-yemen-crisis-increasing-cases/) whilst in the Yemen, control programmes that aimed to eliminate onchocerciasis and lymphatic filariasis by 2015 foundered with no prospect of them resuming (Abdul-Ghani 2016).

Natural disasters, such as cyclones and earthquakes, can lead to similar destruction of infrastructure and refugee problems to those of war. Widespread flooding also provides extensive breeding conditions for mosquitoes and thereby increases the spread of mosquito-borne diseases such as malaria. The destruction of sewage systems and facilities for waste disposal, in conjunction with a warm wet environment, also facilitates the spread of faecal-oral transmitted protozoa and helminths. It is therefore not surprising that widespread flooding in tropical countries usually results in an increase in malaria and water-borne diseases (Boyce et al. 2016; Okaka and Odhiambo 2018).

The damage we cause to the environment can encourage the spread of disease by making conditions more suitable for vectors and intermediate hosts and/or the survival of parasite eggs and cysts. For example, clearance of the rainforests in the Amazon produces open sunlit pools that are ideal breeding grounds for the mosquito vector of malaria *Anopheles darlingi* (Harris et al. 2006). Also, as people move into these clearings to live or work, they come into contact with zoonotic infectious agents that may not be perfectly adapted to living in us but can still cause disease.

The way we live and organise our societies is a major contributor to the spread of parasitic diseases. Throughout the world, there is an increase in urbanization. This means that more people are living close together and the potential for disease transmission between them is therefore high (McMichael 2000). Vector species that can live in an urban environment, such as *Anopheles stephensi* and certain other mosquitoes, therefore pose a particular risk (Takken and Lindsay 2019).

If a high population density combines with inadequate sanitation, then widespread transmission of contaminative diseases is inevitable. In some slums, over 50 households may share a single toilet. Furthermore, this toilet may be 50 m or more from the dwellings. Consequently, urinating and defecating on the bare ground by both children and adults are common in some of these communities. In a study of slum dwellers in Gujarat (western India), 71% of the participants were infected with parasitic protozoa and 26% with helminth infections (Shobha et al. 2013). Not surprisingly, many claimed to suffer from diarrhoea. Similarly, a study of slum children (1–5 years old) in Karachi (Pakistan) found that the prevalence rate of intestinal parasites was 53 and 10% of the children harboured two or more parasite species (Mehraj et al. 2008). Many of these children suffered from stunted growth.

Sometimes, parasites and their vectors spread by less obvious means. For example, the increased use of cars and motorised transport has resulted in large numbers of used tyres entering the ecosystem. Used tyres retain water after it has rained, and they make excellent breeding grounds for some mosquito species. There is a huge international market in used tyres that are loaded onto lorries and ships and moved within and between countries. In the process, mosquitoes are also moved around the world and notorious vectors of disease such as the Asian tiger mosquito *Aedes albopictus* are now established in countries such as Spain where they were formerly absent. *Aedes albopictus* does not transmit parasitic diseases but is an important vector of viruses such as Dengue virus, yellow fever virus, and Zika virus. The adults are not capable of dispersing far by flight, but it has colonized many countries through the transport of its larvae in used tyres. The adult mosquitoes also disperse by unintentionally hitching a ride inside a car or other vehicle (Eritja et al. 2017). It is likely that many other mosquitoes and other vectors disperse in similar fashions. For example, there are several reports of 'airport malaria' in which a person contracts the disease from a mosquito that has been carried from one country to another within a plane (Isaäcson and Frean 2001).

Before the COVID-19 pandemic that began in 2019, people were increasingly mobile and cheap air travel meant that millions of people rapidly moved between countries for leisure and business. In addition, large numbers of people moved long distances as economic migrants and political refugees. The COVID-19 pandemic brought much of this movement to a sudden halt, and at the time of writing, it was uncertain when and to what extent mass movements will return. Anyone who moves to a new environment becomes exposed to diseases to which they have no previous experience, and hence immunity. They are therefore vulnerable to infection. Similarly, those who are already infected (but may not be aware of the fact) carry their diseases with them and could potentially transmit their infections to a non-immune population on arrival. Obviously, when many people are moving there are many opportunities for disease transmission. For domestic animals, it is possible to instigate legislation that governs their movement. For example, a passport scheme can ensure that they have received appropriate vaccinations and/or drugs to remove infections. Similarly, a period of quarantine upon arrival at their destination can be imposed. Except in very authoritarian regimes, this is seldom feasible as a long-term solution for human populations. Although some countries closed their borders and/or imposed strict quarantines on people during the COVID-19 pandemic, this approach cannot be sustained for any length of time because of the economic consequences. Some countries insist that all persons entering their borders have documentation proving they have received certain vaccinations, such as for yellow fever. However, there are few anti-parasite vaccines and even where effective prophylactic medicines are available to treat parasites, such as anti-malarial drugs, it is notoriously difficult to persuade people to take them as prescribed.

Another of the major reasons why parasites remain a problem is the lack of suitable drugs and vaccines to treat them. The development of drugs for use in human medicine takes many years and is extremely expensive. Consequently, the drug companies need to be sure that they will obtain a good rate of return for their investments. See Chapter 14 for more information on the treatment of parasitic diseases. Unfortunately, those who suffer most severely from parasitic diseases are usually poor and cannot afford expensive drugs. Similarly, the development of anti-parasite vaccines is hampered by a combination of cost and the difficulty of generating protective immunity against parasitic infections. These issues are dealt with in detail in Chapter 15.

The control of parasites by targeting their vectors/intermediate hosts is also becoming more problematic. For many years, this approach proved highly effective, and in the 1950s, it was even believed possible that malaria might be eradicated by killing the anopheline mosquito vectors. However, some vectors are exhibiting increasing resistance against a wide range of insecticides and new chemicals are not being developed to replace those in current use. Furthermore, there are mounting concerns for the environmental damage that can result from inappropriate use of insecticides and fears over risks they pose to our health.

2

Taxonomy

CONTENTS

2.1   Introduction

In this chapter, we will provide a very brief introduction to the study of taxonomy. Correct diagnosis is essential for treatment and control of any disease and that requires consensus on the names and terms used in the identification process. Without it, there cannot be effective communication between workers both within and between countries. For example, even within a country, a disease or organism may be known by various common names, and language differences further complicate communication. Therefore, before we begin to consider specific parasites, it is necessary to understand of how the taxonomic system works and its relevance to parasitology.

Those who study the identification of organisms are called taxonomists, and they arrange organisms into a hierarchy of categories to demonstrate their relationship to one another. Phylogeny is the study of the evolutionary relationships between organisms. This is increasingly informed by comparisons of gene sequences in a process called molecular phylogeny in which phylogenetic trees are generated to represent the closeness of relationships.

The Ancient Greek philosopher Heraclitus of Ephesus is accredited with the well-known saying that 'All is flux. Nothing stays still'. This is certainly true of taxonomy, and frequent name changes and taxonomic re-arrangements will be a constant refrain throughout this book. One needs to be aware of these changes in order to compare past reports with those published more recently. For example, an organism might now be known under a different name or what was once described as a single species is now considered to consist of two or more distinct species with different biological characteristics.

Parasitology: An Integrated Approach, Second Edition. Alan Gunn and Sarah J. Pitt.
© 2022 John Wiley & Sons Ltd. Published 2022 by John Wiley & Sons Ltd.
Companion website: www.wiley.com/go/gunn/parasitology2

Over the years, taxonomists have identified numerous organisms and grouped them together in many different arrangements. Primarily, this has been on the basis of their morphology, and this remains a major feature of taxonomy. Increasingly, morphological studies are complemented by molecular phylogeny, and this is having have a major impact on our understanding of animal relationships, confirming some groupings whilst questioning the validity of many others. However, molecular phylogenetics does not always provide clear evidence of the relationships between species. Consequently, there is often a lack of consistency between texts, and there are frequent rearrangements.

There is some debate about how many kingdoms exist although most modern textbooks refer to six: Archaea, Bacteria, Protista, Fungi, Plantae, Animalia. Parasitic species are common in all the kingdoms but traditionally, parasitologists deal almost exclusively with organisms belonging to the kingdoms Protista and Animalia. Although many prokaryotes (archaeans and bacteria) are parasitic, their study falls within a remit of microbiology. Similarly, parasitic fungi fall within the realm of mycology; parasitic plants are reserve of botany (although, these days, many practitioners prefer the title of plant scientist).

2.2   Viruses: A Special (Unresolved) Case

Viruses are not usually considered to be living entities and therefore do not have a kingdom of their own. This, however, is a hotly debated topic. For example, although Moreira and Lopez-Garcia (2009) argue strongly against viruses being living entities, Koonin and Starokadomskyy (2016) consider the very question of whether viruses are alive to be unscientific because the definition of what one means by life is arbitrary. Didier Raoult and his co-workers argue that the giant viruses called nucleocytoplasmic large DNA viruses (NCLDVs) should be considered as an additional distinct domain of living organisms (Boyer et al. 2010). NCLDVs are so large that they can be mistaken for bacteria, and their genomes are typically twice the size of other viruses. The suggestion that a specific group of viruses might be living organisms has generated a great deal of controversy for which no resolution is in sight. Nevertheless, phylogenetic analysis suggests that NCLDVs evolved before modern eukaryotes, that is, before the organisms that are their current hosts. Furthermore, they may have been the source of two DNA-dependent RNA polymerases and a DNA topoisomerase that are found in modern eukaryotes (Guglielmini et al. 2019).

2.3   Taxonomic Hierarchy

Kingdoms are subdivided into units or taxa (singular taxon) such as class, family, genus (Table 2.1). There are no rules about how many species constitute a genus, how many orders constitute a class, or whether families are divided into subfamilies. However, it is essential that a 'taxon' forms a natural grouping. Consequently, research, especially molecular phylogeny, causes taxonomists to re-arrange the hierarchy of individual species and groups of organisms on a regular basis. A class, family or any other category within one group of organisms is therefore not evolutionarily comparable with those in another group.

The International Commission on Zoological Nomenclature (ICZN) provides rules on legal aspects of nomenclature (e.g., precedence). However, it is not unusual for workers to continue using old names that have been superseded or to fail to agree on an accepted single name. For example, the blowflies known as *Lucilia cuprina* and *Lucilia sericata* within the United

Table 2.1 The taxonomic hierarchy with specific reference to the sheep nasal bot fly *Oestrus ovis*.

Taxonomic division	Taxon name	Common name
Kingdom	Animalia	Animals
Subkingdom	Bilateria	
Branch	Protostomia	
Infrakingdom	Ecdysozoa	Moulting invertebrates
Phylum	Arthropoda	
Subphylum	Hexapoda	Insects and related species
Class	Insecta	Insects
Infraclass	Pterygota	Winged insects
Division	Neoptera	
Subdivision	Endopterygota	
Superorder	Panorpita	
Order	Diptera	True flies
Suborder	Cyclorrhapha	Higher flies
Superfamily	Oestroidea	
Family	Oestridae	
Subfamily	Oestrinae	
Genus	*Oestrus*	
Subgenus		
Species	*ovis*, Linnaeus, 1758	
Subspecies		

Not all taxonomists agree on the appropriate division for a grouping (taxon). For example, some workers consider there to be two suborders of Diptera: the Nematocera and the Brachycera and that the term Cyclorrhapha should be considered a division of the Brachycera rather than a suborder. Note that only the genus name and lower taxonomic descriptors are placed in italics.

Kingdom and Europe are often called *Phaenicia cuprina* and *Phaenicia sericata* by workers in the United States.

2.3.1 The Binomen System

All organisms (apart from viruses) have a two-part name, or binomen – hence, the term binomial nomenclature. The two parts consist of the generic (or genus) name and the trivial name (also called the specific epithet or specific name).

The trivial name may be followed by the naming authority, i.e., the name of the person who first described the organism, along with the date the description was published – this is placed in brackets.

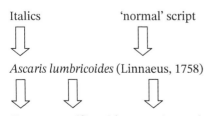

Italics   'normal' script

Ascaris lumbricoides (Linnaeus, 1758)

Genus   specific epithet naming authority
Genus + specific epithet = the binomen, also called the 'species name'.

Surprisingly, there is no universally accepted definition of what is meant by the term 'species. Indeed, there are currently over 20 different definitions. Furthermore, over the course of thousands of years, there is never a single point at which one species becomes two: it is like attempting to identify the day one ceases being a child and becomes an adult. To further complicate matters, some species have distinct forms that are called sub-species, and these are distinguished through the use of trinomens. For example, the human body louse *Pediculus humanus humanus* and the head louse *Pediculus humanus capitis* are usually distinguished as separate sub-species. However, for many years there has been a debate about whether the reported differences in their morphology and behaviour are consistent enough to justify them being considered closely related sub-species or separate species in their own right. Current molecular evidence suggests that they are morphotypes of a single species (Light et al. 2008). Similarly, there are two physiological variants of the mosquito *Culex pipiens*: *Culex pipiens pipiens,* which bites only birds, and *Culex pipiens molestus,* which only bites humans. The two variants of *Culex pipiens* cannot be differentiated morphologically. They can be crossed in the laboratory, but, in the United Kingdom, the populations remain genetically isolated in the wild. Distinguishing between the variants is important because this mosquito can act as a vector for the potentially fatal West Nile Virus, and therefore its biting behaviour has a major impact on whether the disease spreads from birds to humans.

The difficulty of differentiating between species and sub-species can give rise to 'taxonomic inflation' in those groups that are particularly well studied. For example, ant taxonomists tend not to recognize sub-species, so everything is separated at the species level. By contrast, butterfly taxonomists are enthusiastic users of trinomens. Not surprisingly, this often results in ecological surveys revealing a greater species diversity of ants than butterflies.

Some scientists state that the increasing use of the phylogenetic concept of species rather than the older biological species concept is driving taxonomic inflation. The traditional biological species concept operates on the premise that two organisms should be considered different species if either they are incapable of mating or, if they do mate, then their progeny are infertile. Allowances have to be made for the likelihood of gene flow between populations. For example, the fact that tigers and lions can hybridise does not mean that they are the same species, as this would never happen in the wild. By contrast, the phylogenetic species concept is based on the observation that separate populations of organisms often have distinct inheritable differences (for example, a colour pattern or the size of a body part). What constitutes an inheritable difference sufficient to qualify a population as a 'species would depend upon the views of the taxonomist. The increasing use of DNA analysis by taxonomists has undoubtedly contributed to the popularity of the phylogenetic species concept, because it often identifies differences in gene sequences between populations. Some workers have suggested that the adoption of a phylogenetic species concept can result in up to 48% more species than the biological approach for the same group of organisms (Marris 2007). It also causes complications by suggesting some unlikely taxonomic relationships. For example, for many years, microsporidian parasites were classified as protozoa until molecular studies indicated they have a closer relationship to fungi. However, molecular evidence is not always conclusive, and some scientists now consider that the microsporidians should be moved back to the kingdom Protista.

2.4 Kingdom Protista

The kingdom Protista is a loose assemblage of organisms linked together by their shared characteristic of a membrane bound nucleus – which therefore makes them eukaryotes – and their lack of the organizational features found in the kingdoms Fungi, Plantae, and Animalia. Although they are for the most part single-celled, some are colonial. The lack of unifying morphological and molecular features indicates that the kingdom is polyphyletic – that is, it is composed of individuals that arose from various different ancestors. Consequently, there are frequent calls for its division into several separate kingdoms or clades to reflect these differences.

2.5   Kingdom Animalia

Many workers consider there to be two groups of animals: the Parazoa and the Eumetazoa. The position of the Placozoa remains enigmatic with some authors including them amongst the Parazoa, some placing them in the Eumetazoa, and others isolating them into their own independent grouping. For many years, *Trichoplax adhaerens* was the only known placozoan, but at the time of writing, there were three species. Placozoans are small (~1 mm), flat aquatic organisms with a rather amoeboid shape. They glide across benthic surfaces using cilia and absorb algae and detritus across certain cells lining their ventral surface. There are currently no records of them being parasitic or acting as hosts for parasites. However, they harbour rickettsia and bacteria endosymbionts, and it would be surprising if other microbes and viruses did not parasitize them. Interestingly, *T. adhaerens* is currently the only metazoan animal known to express the protein 'apicortin' (Orosz 2018). Apicortins probably help stabilize microtubules and are characteristic of apicomplexan parasites such as malaria and certain free-living algae to which they are distantly related. It is uncertain whether the placozoans acquired the apicortin genes from consuming algae or sharing genes through long distant evolutionary events.

2.5.1   Parazoa

The Parazoa are organisms that lack true tissues. This group contains either only the phylum Porifera, better known as the sponges, or two phyla, the Porifera and the Placozoa depending upon one's taxonomic preferences. All sponges are aquatic and gain their nutrition through filter feeding and, in some cases, in a symbiosis with algae or bacteria. Sponges belonging to the family Clionaida bore into the shells of molluscs and penetrate the calcareous skeleton of corals (Mote et al. 2019). Other sponges will, in their turn, exploit the burrows and also grow over them. Some workers refer to this as parasitism, although this is debatable.

2.5.2   Eumetazoa

The Eumetazoa comprise those animals containing recognisable muscles, nerves, and other tissues. This group contains the vast majority of animal phyla (i.e., all vertebrates and most invertebrates). One can identify two sub-group of eumetazoans on the basis of their embryonic development. These are the diploblasts and the triploblasts. Diploblastic animals are those that form only two germ layers during embryonic development. These layers are the outer ectoderm and the inner endoderm. Only two phyla exhibit this form of development: the Cnidaria (e.g., jellyfish, sea anemones, and hydra) and the Ctenophora (sea gooseberries). Virtually all members of these two phyla are free-living and most are predatory, although there are a small number of exceptions. For example, among the Ctenophora, juveniles belonging to the genus *Lampea* begin life as external parasites of salps. However, as they mature and become larger, the relationship changes to one of predation in which they engulf and consume the salp to which they initially attached. Intriguingly, some ctenophores are hosts to the parasitic planula (larval stage) of the sea anemone

Edwardsiella carnea. Among the Cnidaria, the best-known parasitic species *Polypodium hydriforme* infests the eggs of freshwater fish (e.g., sturgeons). *Polypodium hydriforme* spends most of its life cycle within the eggs in the ovaries of its fish host. When the fish spawns, it leaves and forms a free-living medusiform-like stage (i.e., resembling a small jellyfish) before subsequently infecting another female fish. The genus *Polypodium* is currently considered a sister group to the Myxozoa within the Cnidaria (Okamura and Gruhl 2016). The myxozoans are a highly unusual group of parasitic cnidarians with complex life histories. Most species utilise a fish intermediate host and an annelid or bryozoan definitive host. However, there are few species, such as *Soricimyxum minuti*, that utilise shrews and other mammals as intermediate hosts (Székely et al. 2015). The most economically important species is *Myxobolus cerebralis* that causes 'whirling disease' in wild and farmed salmonid fish, such as trout and salmon. The parasites develop into myxospores within the cartilage of the vertebrae and skull (as well as other organ systems) of infected fish. This results in deformity and a characteristic 'whirling' motion in which the fish appears to chase its tail. The mortality rate can exceed 90%, and after death, the spores enter the surrounding water. The next stage of the life cycle requires ingestion of the myxospores by *Tubifex* spp. worms. Within the worms, the parasite undergoes sporogony to form sporocysts that contain triactinomyxon spores. Completion of the life cycle then occurs through either of two events. It might be through a fish consuming an infected *Tubifex* worm or through the worm releasing the triactinomyxon spores with its faeces into the surrounding water. In the latter circumstance, the spores rapidly penetrate the skin of a suitable fish host upon contact.

Triploblastic animals are those in which a third germ layer, the mesoderm, develops during embryogenesis: the mesoderm forms between the outer ectoderm and the inner endoderm layers. Most invertebrate species and all vertebrate species are triploblastic organisms. One can divide triploblastic animals into three broad categories based on their internal morphology: acoelomates, pseudocoelomates, and coelomates. The acoelomates are those that lack a body space (coelom) other than the gut (e.g., phylum Platyhelminthes: tapeworms, flukes). The pseudocoelomates, also known as the blastocoelomates, have a characteristic pseudocoelom (blastocoelom) between the gut and the body wall (e.g., phylum Nematoda: nematodes). A pseudocoelom is a body cavity that develops temporarily in most metazoan animals during embryonation, but in the pseudocoelomates it persists into adulthood. Coelomate animals (also known as eucoelomates) are those in which a true coelom (body space surrounded by mesoderm) develops between the gut and the body wall (e.g., phylum Annelida [earthworms], phylum Arthropoda [scorpions, crabs, insects], phylum Vertebrata [fish, amphibians, reptiles, birds, mammals]). There are numerous examples of triploblastic invertebrate species that are parasites of other organisms as well many that act as the intermediate hosts or vectors of parasites.

3

Parasitic Protozoa Part A: Phyla Rhizopoda, Metamonada, Apicomplexa

3.1 Introduction

For many years, the protozoa were placed within the Kingdom Animalia, but they now represent a subkingdom within the Kingdom Protista. That is, the protists (single celled eukaryotes) now having a separate kingdom of their own. Thomas Cavalier-Smith and his co-workers have proposed even more radical re-arrangements (Ruggiero et al. 2015). They suggest that several phyla that have always been considered protozoa should be moved to form a further new Kingdom, the Chromista. This would include the apicomplexans, the ciliates, dinoflagellates, and the foraminifera (Cavalier-Smith 2018). We are not adopting their proposals because they are not yet widely accepted but this may change.

Parasitology: An Integrated Approach, Second Edition. Alan Gunn and Sarah J. Pitt.
© 2022 John Wiley & Sons Ltd. Published 2022 by John Wiley & Sons Ltd.
Companion website: www.wiley.com/go/gunn/parasitology2

In this and the next chapter, we summarise the life cycles and biology of some of the most important parasitic protozoa in human and veterinary medicine. Although they consist of just a single cell, parasitic protozoa come in numerous shapes and sizes. Many of them have complicated life cycles involving two or more taxonomically unrelated hosts with reproduction occurring in both. These complexities often contribute to their success as parasites. They exhibit a vast array of immune-avoidance mechanisms, and their pathology is often influenced by their interactions with other microorganisms. Parasitic protozoa live in all the organs of our body and cause diseases ranging from benign to rapidly and incurably fatal. They also exhibit every imaginable means of infecting their hosts from simple contamination to sexual and vector-assisted transmission.

3.2 Phylum Rhizopoda

The Rhizopoda used to be known as the Sarcomastigophora. It is a small phylum consisting of about 200 species and contains the amoebas. Most rhizopods are free-living or commensal, but it includes important parasitic species such as *Entamoeba histolytica*. The name 'Rhizopoda' translates as 'root-like foot' and refers to the process by which the cytoplasm flows within the cell to form projections of the body wall called pseudopodia (false feet) that are used for both movement and acquiring food. By sending out pseudopodia, they can surround and entrap small food particles within membrane bound vesicles (food vacuoles) in a process called phagocytosis. In some species, their movement is aided by one or more flagellae. Although amoebas are sometimes described as 'primitive', ultrastructural, and molecular studies indicate that this is incorrect.

3.2.1 *Entamoeba histolytica*

Entamoeba histolytica is essentially a human parasite. There are records of infections in other primates (Deere et al. 2019), but it is uncertain whether zoonotic transmission occurs in the wild. Its pathogenicity in other primates is uncertain, but in humans it causes potentially fatal amoebic dysentery. Dysentery is a generic term for a serious inflammatory disorder that affects the intestines and results in intense diarrhoea, pain, and fever. It can arise from various causes and amoebic, bacterial, and viral dysentery occur in both temperate and tropical regions. Dysentery has long been known as a 'handmaiden of war', often inflicting more casualties than bullets and bombs. Dysentery epidemics have accompanied nearly every account of war from antiquity to the present day. Wherever large numbers of people (especially if they are malnourished) live in proximity and in squalid conditions, the situation is ripe for an outbreak of dysentery. In recent years, large-scale migrations, conflicts, and deteriorating economic conditions have led to enhanced levels of amoebic dysentery in some countries; tourists travelling on 'exotic adventure holidays' have also occasionally found themselves victim of amoebic dysentery.

Entamoebae and Amoebic Dysentery

In the past, it was often stated that *E. histolytica* infected about 10% of the world's population, but that the majority never expressed disease symptoms. The reason so many apparently infected people remained asymptomatic was put down to different strains of the parasite varying in their virulence. It is now clear that three species of morphologically identical *Entamoeba* are commonly found in our intestines: *Entamoeba histolytica*, *Entamoeba dispar* and *Entamoeba moshkovskii*. These three species are only distinguishable by species-specific antigens or DNA analysis. *Entamoeba histolytica* is notoriously pathogenic, whereas *E. dispar* is generally considered a harmless commensal. However, in hamsters, *E. dispar* damages the intestinal wall and causes liver abscesses, and there are occasional case reports of pathogenic effects in humans. The pathogenic status of *E. moshkovskii* remains uncertain. When faecal surveys distinguish between *Entamoeba* species, *E. dispar* is always by far the commonest species present and *E. moshkovskii* is also relatively common (Calegar et al. 2016). Consequently, it is difficult to draw firm conclusions from literature in which the three species are not differentiated. Nevertheless, strain differences occur between populations of *E. histolytica,* and these reflect the pathology they cause. Many people who are genuinely infected with *E. histolytica* remain asymptomatic, and whilst sometimes this relates to host factors, there are avirulent strains of the parasite (Escueta-de Cadiz et al. 2010). A characteristic feature of virulent strains is that they over-express genes coding for lysine-rich factors and glutamic- and lysine-rich proteins. The function of these genes, referred to as KRiPs and KERPs, respectively, is unknown but one of them, KERP1, is associated facilitating the adhesion of *E. histolytica* to red blood cells and causing liver abscesses (Santi-Rocca et al. 2008). The gene responsible KERP1 also occurs in *Entamoeba nutalli,* which is a virulent intestinal parasite of macaques that also cause liver abscesses. However, this gene has also been identified in *E. dispar* and *E. moshkovskii,* which are generally considered non-pathogenic (Weedall 2020). It is possible that variations in the pathogenicity of species of *Entamoeba* and their various strains may reflect differences in the expression of the gene coding for KERP1 rather than its presence or absence.

Despite the problems of identification, there is no doubt that *E. histolytica* is a major cause of disease in many parts of the world but particularly in developing countries. Every year, many millions develop amoebic dysentery or hepatic amoebiasis. The debilitating symptoms of these conditions can last for months or even years and 40,000–110,000 people are thought to die each year from their infections.

There are two stages in the life cycle of *E. histolytica*: the actively growing and feeding stage called the trophozoite form and the cyst transmission stage (Figures 3.1 and 3.2a,b). One should never refer to protozoa as producing eggs! In common with all other parasitic protozoa (but unlike the free-living amoebae), *E. histolytica* has no contractile vacuole. Although it also lacks mitochondria, it has genes coding for proteins of mitochondrial origin within its nuclear genome. It also has organelles called 'mitosomes' that are double-walled structures that lack DNA. Their function is uncertain, but they probably represent the remnants of mitochondria.

The trophozoite is 12–60 μm in size and has a clear granular outer cytoplasm, a more densely granular inner cytoplasm, and there is an aggregated region of chromatin referred to as a karyosome centrally located within the nucleus. Reproduction takes place asexually by cell division and through cyst formation. The stimuli causing the trophozoites to transform into cysts are uncertain, but it is an essential part of the life cycle. Therefore, targeting cyst formation using drugs might reduce parasite transmission (Mi-ichi et al. 2019). The cysts are 10–15 μm in diameter and (when mature) contain four nuclei and characteristic bar-shaped chromatoidal bodies that serve as a store of nucleoprotein. The cell wall contains chitin that provides protection and enables the cyst to survive in the outside environment for several months under favourable conditions. The output of cysts is enormous, and an infected person may excrete over 10 million cysts per day in their faeces.

Any trophozoites voided with faeces soon die and transmission usually occurs through faecal–oral contamination with infective cysts. Remember the f-words: 'flies, fingers, faeces, food'. Common means of contamination include drinking water (or ice made from contaminated water),

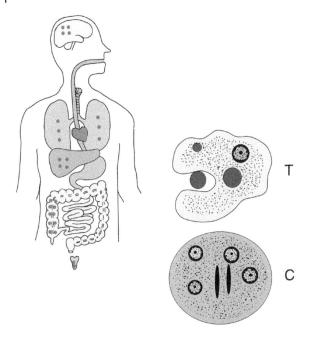

Figure 3.1 Life cycle of *Entamoeba histolytica*. The trophozoite stage (T) has a single spherical nucleus with a central karyosome; the presence of phagocytosed red blood cells is considered diagnostic of *E. histolytica*. There are four nuclei in the mature cyst (C) and two bar-shaped chromatoidal bodies. Infection normally occurs through faecal-oral transmission of the cyst stage although sexual transmission of the trophozoite stage can result in cutaneous infection of the genitalia. The main infection site is the colon although secondary infections can occur in the liver, lungs, and brain. Drawings not to scale.

vegetables grown on land fertilized with human faeces, and insects that move between faeces and our food. Dogs may act as transport hosts when they pick up cysts on their fur and then transfer them to us. Dogs are sometimes coprophagic, but it is uncertain whether the cysts would survive travelling through their digestive system. The safe disposal of human faeces is therefore a crucial factor in reducing the transmission of *E. histolytica*.

After an infective cyst reaches our small intestine, the amoebae emerge and undergo a complicated series of divisions to produce eight trophozoites. Subsequently, peristalsis sweeps the amoebae down to the large intestine (colon) where they multiply in the lumen and may invade the gut wall. Avirulent strains of *E. histolytica* remain in the lumen of the colon and cause humans no harm. Those that are virulent attack and ingest the epithelial cells lining the gut wall and then proceed to spread through underlying layers. In the process of invasion, they cause the formation of flask-shaped ulcers. In serious infections, ulceration and bleeding occur over large areas of the intestine (Figure 3.2c). Consequently, the trophozoites of virulent strains often contain ingested red blood cells within their food vacuoles. This can be useful in laboratory diagnosis. A great deal of water is normally re-absorbed in our colon. Consequently, reducing its functional surface causes a decline in water re-absorption. In severe cases, the reduction in water reabsorption coupled with the loss of blood and fluids leads to emaciation and death from dehydration. The ulceration explains why patients suffering from amoebic dysentery frequently complain of gastric pain. Also, together with loss of fluids it means that patients pass stools that are loose and contain mucus and blood mixed with faecal material. These symptoms are distinct from bacterial dysentery in which there is no cellular exudate.

(a)

(b)

(c)

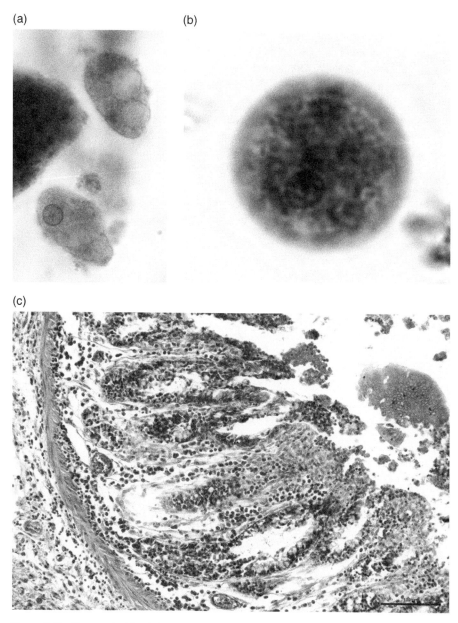

Figure 3.2 *Entamoeba histolytica.* (a) Trophozoite in a histological section through an intestinal ulcer. The cytoplasm is vacuolated in these specimens – this arises if there is a delay in fixing the sample. (b) Cyst. Cysts are spherical and one must focus through the cyst to see all four nuclei; immature cysts have 1-3 nuclei. It can be difficult or impossible to distinguish the chromatoidal bodies in the cysts using light microscopy and their nuclear structure may be lost after prolonged storage. (c) Ulceration of the colon caused by *Entamoeba histolytica.* Note the huge numbers of trophozoites and the destruction of the villi. The parasites have penetrated the lower layers of the gut wall.

The ulcers in the intestine often suffer secondary invasion by bacteria – this extends and deepens the ulcers and leads to increased blood and fluid loss. When the ulcers start to heal, functional tissue is replaced by fibrous scar tissue. This reduces gut elasticity and, if extensive, may impair peristalsis in the colon and even cause a potentially fatal gut blockage.

If the amoebae damage the lining of the blood vessels, they gain entrance to the general circulation and are then swept up in the blood stream. Wherever the amoebae come to rest, they establish secondary ulcers that are potentially life threatening. The liver is the most commonly affected organ (hepatic amoebiasis) although the lungs, brain, and other organs may be invaded. Sometimes these secondary ulcers cause symptoms that are mistaken for those of other diseases, such as cancer. This can result in delays in providing the correct treatment and therefore serious long-term damage. Hepatic amoebiasis is usually characterised by a single liver abscess that develops on the right lobe. Liver abscesses can become extensive and produce a copious purulent exudate (i.e., a thick fluid containing white blood cells, cell debris, and dead and dying cells) that resembles chocolate sauce. Depending on the site of the abscess, it can drain into the peritoneal cavity or the lungs – in which case it may be coughed up. Pulmonary amoebiasis often results from the extension of a pre-existing hepatic infection and therefore most cases afflict the right lobes of the lungs. Cutaneous amoebiasis often afflicts the perianal region and results from an infection spreading from the bowel.

For further details of the biology and pathogenesis of *E. histolytica*, see Nozaki and Bhattacharya (2015).

3.2.2 *Entamoeba dispar*

Entamoeba dispar is morphologically indistinguishable from *E. histolytica* and has the same life cycle, but it is normally considered a harmless commensal. Nevertheless, there is some evidence suggesting it occasionally causes lesions in the intestines and liver (Oliveira et al. 2015). It is much more prevalent than *E. histolytica,* and therefore, it is essential to distinguish between the two species to avoid a false diagnosis of amoebic dysentery and thereby initiating inappropriate treatment. In mixed xenic cultures, *E. dispar* soon outgrows *E. histolytica* – which could cause problems where the amoebas are cultured to confirm an initial diagnosis by microscopy. Whether this reflects better fitness and/or whether *E. dispar* influences the establishment of *E. histolytica* is uncertain.

3.2.3 *Entamoeba moshkovskii*

Originally described from Moscow sewage, subsequent surveys identified *E. moshkovskii* from various types of ponds and sediments around the world. Therefore, unlike *E. histolytica*, it can survive as a free-living organism. Although it can infect humans, the difficulty of distinguishing it from *E. histolytica* and *E. dispar* has undoubtedly led to under-reporting. Although often considered a harmless commensal, there are reports of it causing diarrhoea (Shimokawa et al. 2012). Possibly, more cases of *E. moshkovskii* in association with diarrhoea will be reported once molecular-based diagnostic techniques become widely used.

3.2.4 *Entamoeba gingivalis*

Entamoeba gingivalis is commonly found in swab samples taken from the gingival crevices of our mouths. *E. gingivalis* does not form cysts, and therefore, transmission is probably through kissing or sharing food and eating implements. There are occasional case reports of the recovery of *E. gingivalis* from the vagina (Bradbury et al. 2019) and the lungs (Jian et al. 2008). Although often implicated in periodontitis (an inflammatory disease that affects the gums and the bone surrounding the teeth), it infects both healthy and diseased individuals (Bonner et al. 2018). Part of the problem in determining its association with disease is the wide variation of rates of recovery of *E. gingivalis* from samples. This is probably owing to the collection techniques employed, and the recommendation is to take 5–10 samples from each person.

3.2.5 *Naegleria fowleri*

Although there are over 30 species of *Naegleria*, only one of these, *Naegleria fowleri*, is pathogenic. Like the other members of the genus, *N. fowleri* is a free-living amoeboflagellate. That is, an amoeba that in one of its life cycle stages possesses flagellae. It is a cosmopolitan species normally found in freshwater ponds and lakes, but it lives in numerous wet or moist environments such as swimming pools, humidifier systems, and damp soil (Siddiqui et al. 2016). There are three life cycle stages: the active amoeboid trophozoite, the non-feeding flagellate stage that is produced when the food supply runs low and acts as a dispersal stage, and a cyst stage that forms in response to adverse environmental conditions (Figure 3.3). Free-living *N. fowleri* feed on bacteria that they ingest using special 'feeding cups' on the outer cell membrane, and they are often packed with vacuoles containing microbes. If the flagellate stage enters the nasal cavity, it can transform to the trophozoite stage and become invasive.

The trophozoite is the infective stage, and it gains entry via our nose. We usually become infected by swimming in infected water. However, Mahmood (2015) suggested that a rise in the number of cases in Pakistan might be linked to the practice of 'wudu' (also referred to as 'wuzu' and 'ablution') by muslims before they pray. Wudu involves irrigating the nose with water, and if contaminated water is used, the practitioner could become infected. After entering our nose, the trophozoite migrates from the nasal mucosa along the olfactory nerves through the cribiform plate and thence into the brain where it causes primary amoebic meningoencephalitis. The use of the term 'primary' distinguishes it from encephalitis caused by *E. histolytica* in which invasion of the brain is a secondary consequence of infection in the gut. The trophozoite of *N. fowleri* moves remarkably quickly: the time between initial exposure and first symptoms takes as little as 24 hours and death commonly occurs after 4–10 days. The trophozoites ingest host tissues and red blood cells and cause a serious inflammatory reaction that contributes to the pathology. Highly pathogenic strains of *N. fowleri* kill cells on contact – presumably by secreting toxic substances. The mortality rate is more than 90%. The symptoms of infection are non-specific and often start with neck stiffness followed by headaches, photophobia, confusion, seizures, and the patient then enters a coma from which he/

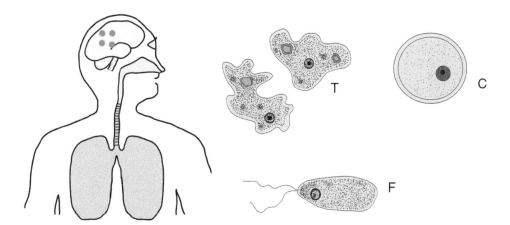

Figure 3.3 Life cycle of *Naegleria fowleri*. This is normally a free-living species found in warm ponds, lakes, and soil. The trophozoite stage (T) has 1–12 suckers that it uses to feed on bacteria. The flagellate stage (F) develops in response to adverse conditions and is the dispersal stage; it forms in response to adverse conditions. The cyst stage (C) develops in response to dehydration and is not infective. Human infections arise when the trophozoite stage enters the nasal cavity. The amoebae penetrate the lining of the nose and reach the brain via the cribiform plate where they cause primary amoebic meningoencephalitis. Drawings not to scale.

she seldom recovers. There are isolated case reports of it causing meningoencephalitis in various domestic and wild animals including cattle (Pimentel et al. 2012) and rhinoceros (Yaw et al. 2019).

3.2.6 *Balamuthia mandrillaris*

Balamuthia mandrillaris is another cosmopolitan amoeba capable of causing fatal encephalitis (*Balamuthia* amoeba encephalitis). Its species name derives from its discovery as the cause of a fatal brain infection of a mandrill baboon at San Diego Wild Animal Park in California in 1986. Although considered free-living, particularly in association with soil, there are many more reports of it causing infections than of its recovery from the environment. There are two life cycle stages – the trophozoite and the cyst stage: which of these is/are the infective stage is uncertain (Figure 3.4). Many amoebae feed on bacteria but while *B. mandrillaris* ingests them, they do not appear to sustain growth. By contrast, at least in cultures, *B. mandrillaris* grows well when fed other species of amoebae or human tissue culture cells. Unlike *N. fowleri*, *B. mandrillaris* tends to cause a chronic disease in humans that may last up to 2 years – although with a similar almost invariable (>98%) fatal outcome. The mode of entry is uncertain but, in several case reports, it appears to have been through puncture wounds in the skin from which it then spread via the blood stream. There are also case reports of infections through organ transplant and nasal lavage using (probably) infected water. *Balamuthia mandrillaris* will infect tissues other than the brain, including the kidneys, pancreas, and the skin. It probably gains access to the brain via the choroid plexus (Jayasekara et al. 2004). Once established in the brain, the amoebae cause a granulomatous reaction and the site of the infection becomes surrounded by macrophages. The pathology is therefore called granulomatous amoebic encephalitis.

Balamuthia mandrillaris is a zoonotic parasite and natural (and often fatal) infections occur in many species of wild and domestic animals (Visvesvara et al. 2007). In common with other free-living amoebae, *B. mandrillaris* often acts as host for various bacteria, including *Legionella pneumophila* (Shadrach et al. 2005). However, its importance as a transport host for microbial infections is uncertain.

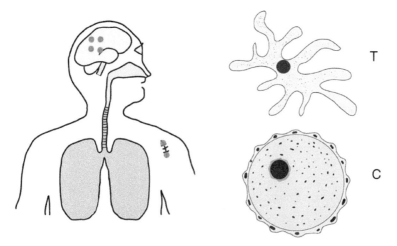

Figure 3.4 Life cycle of *Balamuthia mandrillaris*. This species exists either as a free-living organism in ponds, lakes, and soil or as a parasite. Both the trophozoite (T) and the cyst stage (C) are probably infectious and enter via skin wounds or through the lining of the nose. It damages various organs including the skin, kidneys, lungs, adrenal glands, and pancreas. Invasion of the central nervous system probably occurs via the choroid plexus and results in potentially fatal granulomatous encephalitis. Drawings not to scale.

3.2.7 Genus *Acanthamoeba*

The genus *Acanthamoeba* contains over 20 species, and they are amongst the most common free-living soil amoebae. They also live in various freshwater habitats, and several are opportunistic pathogens. There are two life cycle stages: the active feeding and dividing trophozoite stage and the cyst stage (Figure 3.5). Opportunistic infections arise through skin wounds and possibly via breathing in the cysts. Infections also possibly arise through swimming in infected water. Immunocompromised individuals are at particular risk of infection.

The most serious consequences of infection arise when the amoebae disseminate from their initial entry site via the blood stream and reach the brain. This can result in granulomatous amoebic encephalitis that has a high fatality rate. More commonly, *Acanthamoeba* causes keratitis (inflammation of the cornea) (Roozbahani et al. 2018) when it invades the surface of our cornea. This usually occurs following trauma to the eye and/or if we use contaminated contact lenses. Soft contact lenses are particularly likely to harbour the parasite and transfer it to the wearer's cornea. The infection is extremely painful and if not successfully treated may lead to the loss of the eye. Acanthamoebae are extremely common organisms in the environment, so it is surprising that infections are not reported more frequently than they are. Presumably, this is at least in part owing to an effective host immune response.

Acanthamoeba castellanii is the species most associated with keratitis. It harbours numerous other microorganisms including bacteria such as *L. pneumophila, Chlamydia pneumoniiae* and *Pseudomonas* spp., yeast such as *Cryptococcus neoformans* and various Mimiviridae. The extent to which the amoeba transmits these infections is uncertain, but the associations result in the acquisition and maintenance of microbial virulence factors (Guimaraes et al. 2016). *Cryptococcus neoformans* is not pathogenic in healthy individuals, but it causes potentially fatal meningitis in those who are immunocompromised. Many species of amoebae harbour mimiviridae although the effect they have on their hosts is uncertain. There are isolated case reports of mimiviruses in association with pneumonia (e.g., Saadi et al. 2013). However, the role of mimiviruses in human disease remains controversial (Abrahão et al. 2018).

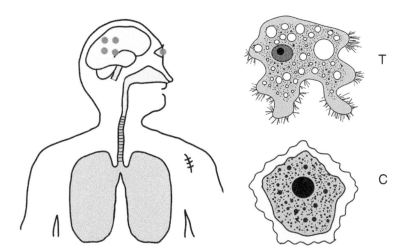

Figure 3.5 Life cycle of *Acanthamoeba* spp. These amoebae are normally free-living in ponds, lakes, and soil. Both the trophozoite (T) and the cyst stage (C) are probably infectious and enter via skin wounds or through the lining of the nose. Infections of the eye also occur. Fatal disease occurs when they invade the brain. Drawings not to scale.

3.3 Phylum Metamonada

This is a large group of anaerobic flagellate protozoa of uncertain composition and most probably derived from various ancestors (i.e., it is polyphyletic). They lack mitochondria although this is almost certainly a derived characteristic rather than them never having possessed them during their evolution. They have groups of four flagellae and/or basal bodies that are often arranged in association with their nucleus to a form a structure called the karyomastigont.

3.3.1 Order Diplomonadida

Most members of the Diplomonadida are parasites, the best known of which belong to the genus *Giardia*. A characteristic feature of diplomonads is the presence of two haploid nuclei each of which is associated with four basal bodies and flagellae. They lack Golgi apparatus and mitochondria, but organelles called mitosomes, that probably represent relict mitochondria, exist in some species. Mitosomes lack DNA and do not undertake oxidative phosphorylation. They probably manufacture iron–sulphur proteins that undertake essential tasks in the cytosol (Tachezy 2019). In common with many other protozoa, the taxonomic position of the diplomonads is under constant revision.

3.3.1.1 Genus *Giardia*

The taxonomy of *Giardia* spp. is an interesting illustration of the insights provided by molecular phylogenetics. Despite the morphological similarities between *Giardia* spp. and other parasitic flagellates (e.g., *Trichomonas* spp., *Trypanosoma* spp.), the giardias represent much more primitive organisms. Within the genus *Giardia*, allocating isolates into species groups is very difficult. The first attempts were based on apparent host specificity since cysts are morphologically identical and attempts at *in vitro* culture to produce trophozoites was often unsuccessful. Subsequently, improvements in culture methods and microscope resolution enabled the examination of trophozoites. On this basis, there are six species of *Giardia*, but only one of them infects humans. For historical reasons, the species infecting humans is called *Giardia duodenalis* in Europe and Australia, the North Americans favour *Giardia lamblia*, while some authors refer to it as *Giardia intestinalis*. These species names are synonyms, and there is no 'correct' one; the only important issue is that locally, clinicians and scientists use the same name.

Numerous mammal species harbour *Giardia* spp. Organisation of the *Giardia* into groups and strains within those groups is based predominantly on genetic sequences (Thompson and Ash 2019). Unfortunately, many of the published studies involve characterisation and genetic sequencing of small numbers of isolates – in some cases only one parasite from one animal. This is a serious limitation, and the area of *Giardia* taxonomy clearly requires further work.

3.3.1.1.1 *Giardia duodenalis*

Giardia duodenalis is one of the commonest human parasites and has prevalences of 4–43% in low-income countries and 1–7% in high-income countries. It has the distinction of being the first parasitic protozoa to be described. This happened in 1681 when the pioneering microscopist Antoine van Leeuwenhoek observed it in a sample of his own diarrhoea. Currently, scientists divide *G. duodenalis* into eight genetic assemblages (A–H) but only two of these, A and B, infect us. In addition to humans, assemblages A and B both parasitise wild and domestic mammals. In some parts of America, the implication of beavers as reservoirs of infection has resulted in giardiasis gaining the popular moniker of 'beaver fever' (Tsui et al. 2018). However, the extent to which *G. duodenalis* is a zoonotic infection cycling between humans and other mammals is uncertain.

The trophozoite stage of *G. duodenalis* is pear-shaped 12–15 μm in length and has four pairs of flagellae (Figure 3.6). Its ventral surface has a concave profile on which there are two depressions referred to as 'adhesive discs' or 'suckers' although they have a supportive function rather than being contractile. A pair of flagella located within the 'ventral groove' work as a 'pump' that removes fluid from underneath the adhesive discs and may facilitate the removal of nutrients from the underlying host mucosa. The oval-shaped cyst stage is 8–12 μm in size and initially contains two nuclei but once they are mature, four nuclei are present along with several prominent axonemes (microtubules that constitute the core of the flagella): the flagellae and adhesive discs are broken down and stored as fragments during the cyst stage. Infected people shed huge numbers of cysts in their faeces – possibly as high as 1×10^8 viable cysts per gram of faeces. Cyst shedding is intermittent, and laboratory confirmation of the diagnosis often requires the patient to produce several faecal samples. Transmission usually occurs through consuming the cysts in food and water, or through touching contaminated surfaces and then transferring the cysts to one's mouth.

Giardia duodenalis normally resides in our duodenum and upper small intestine although sometimes our stomach, ileum, and colon become infected. The parasites attach to the surface epithelium and overlying mucus layer and although they may completely cover the surface of the gut, they do not invade the underlying tissues. Many people become non-symptomatic carriers of the parasite, but some develop an acute form of enteritis, referred to as giardiasis, that manifests as profuse watery diarrhoea. The diarrhoea has a characteristic foul smell because the parasite interferes with the absorption of fats. If undiagnosed and untreated, the infection can become chronic. This is characterised by episodes of abdominal pain and defaecating loose, clay-coloured stools that have a smell reminiscent of bad eggs. The consequence of a long-term infection can be malnourishment due to malabsorption. It can also result in a deficiency in fat-soluble vitamins, such as vitamin K, and hence associated metabolic disorders. Interestingly, some people who suffer

Figure 3.6 Life cycle of *Giardia duodenalis*. The trophozoite stage (T) normally infects the duodenum and upper section of the small intestine where it attaches to the gut lining. The cyst stage (C) is produced in huge, but intermittent, numbers and passed in the faeces. Transmission is via faecal-oral contamination of the cysts. Drawings not to scale.

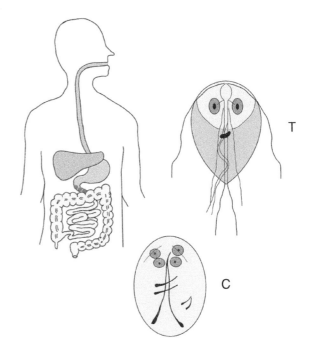

giardiasis develop lactose intolerance, and this may persist even after the infection is cured. Parasite strain differences and our own immune status probably contribute to the reason *G. duodenalis* afflicts some of us more severely. In addition, *Giardia* spp. have complex interactions with the resident host gut microbiome that may lead to protection or exacerbation of an infection. For example, in mice, *Giardia* infection results in a dramatic shift in the balance of aerobic and anaerobic commensal bacteria (Barash et al. 2017). Whether the shift is a direct consequence of the waste products of *Giardia* metabolism or an indirect one resulting from the host's immune response to the parasite is uncertain. Of course, it could also be a consequence of both factors. This interaction with the microbiome has led workers to investigate the possibility of using probiotics or prebiotics to either prevent infections or treat established infections. For example, supplementing the diet of mice with the prebiotic inulin reduces the severity of giardiasis (Shukla et al. 2016).

Giardia duodenalis is classed as a re-emerging infection. This is mainly due to increased incidences in the developed regions of Australasia, North America, and Western Europe. Several factors contribute to these increases. The most obvious is that global travel is now cheap and accessible to many people in high-income countries. Consequently, increasing numbers of them travel to countries with a high prevalence where they become infected. Furthermore, they often transfer the parasite to others on their return home. However, blaming other countries is sometimes unfair. For example, a study of giardiasis in Northwest England found that 75% were acquired within the United Kingdom (Minetti et al. 2015). In addition, apparent rises in the numbers of cases of giardiasis are sometimes a consequence of hospital laboratories introducing new, more sensitive diagnostic techniques (Minetti et al. 2016). Nevertheless, giardiasis is a frequent cause of diarrhoea outbreaks among young children in day care facilities, following contamination of domestic water supplies, and among people living in buildings with inadequate sanitation. *Giardia duodenalis* is also a sexually transmitted infection. It is reportedly common among men who have sex with men and those indulging in risky sexual practices such as oral–anal sex (Escobedo et al. 2014).

3.3.2 Order Trichomonadida

Most species within this group are parasites or endosymbionts within vertebrates and invertebrates. Several species are important parasites of domestic animals (e.g., *Histomonas meleagridis*, *Tritrichomonas foetus*) and humans (e.g., *Trichomonas vaginalis*). The trophozoites are often ovoid or pear-like in shape: the anterior is usually rounded and the posterior pointed, although amoeboid forms occur in some species. The number of flagellae varies between species, but there are often 4–6 emerging at their anterior apex. In addition, one flagellum usually curves backwards so that it runs along the cell wall to form an undulating membrane – this flagellum is therefore described as 'recurrent'. Most species do not form cysts. They have a single nucleus and internally there is a prominent median tube-like organelle called the axostyle.

Trichomonads have genomes ranging from 86 to 177 Mb. This is unusually large for protozoa: by comparison, the genome of *Plasmodium falciparum* is ~23 Mb, whilst that of *Trypanosoma cruzi* is ~34 Mb. The large size is largely a consequence of extensive gene duplication, and this may contribute to their success as parasites and facilitated their ability to infect a variety of host species (Barratt et al. 2016). Gene duplication also occurs in other parasitic protozoa although not usually to the same extent as in the trichomonads. Gene duplication is not, however, an invariable feature of a parasitic lifestyle and some parasitic protozoa have unusually small genomes. For example, in the apicomplexan *Babesia bovis*, it is 8.2 Mb, whilst the microsporidian *Encephalitozoon intestinalis* has a genome of only 2.3 Mb.

3.3.2.1 *Histomonas meleagradis*

This parasite infects a wide range of birds but for some reason it is particularly pathogenic in young turkeys, in which untreated infections are usually fatal. It lives within the lumen of the caecum and the liver parenchyma and causes the disease histomoniasis. Infected birds lose condition, become listless and suffer from anorexia, poor growth and sulphur-yellow diarrhoea. The neck and head often become black – and hence the infection is commonly known as 'blackhead disease'. The pathogenicity often links to concurrent infections with other parasitic protozoa, such as *Coccidia* spp., and pathogenic bacteria such as *Escherichia coli* and *Salmonella typhimurium*.

The morphology of *H. meleagradis* is variable (i.e., it is pleomorphic) and depends upon the organ that it infects and the stage of the disease. For example, the form found free within the lumen of the caecum (and in culture) is amoeboid, 5–30 μm in diameter, with a clear outer ectoplasm a more granular endoplasm and with one or two flagella emerging from close to the nucleus. The invasive form that lives within tissues is also amoeboid, but it is smaller (8–15 μm) and the flagellum is absent.

Histomonas meleagradis has an unusual means of transmission that involves becoming incorporated within the eggs of *Heterakis gallinarum*. *Heterakis gallinarum* is a common nematode parasite that lives in the caecum of many wild and domestic birds. The nematode has a direct life cycle in which its eggs pass out in the faeces of its host and, after embryonation, are ingested by another bird in which they hatch and initiate an infection. Earthworms can act as paratenic hosts for the nematode. Interestingly, both *H. meleagradis* and *H. gallinarum* both require interactions with bacteria to establish themselves in their bird host. The infection and subsequent pathology associated with *H. meleagradis* therefore depends upon a complex interplay between the protozoan, a nematode, and microbial flora (Bilic and Hess 2020).

3.3.2.2 *Trichomonas vaginalis*

Millions of people are infected with *Trichomonas vaginalis* and some estimates suggest that it is the most common non-viral sexually transmitted infection in the world (Kissinger 2015). However, prevalences vary considerably between countries, and it is not especially common in the United Kingdom with around 6,000 cases per year (Field et al. 2018). By contrast, a study of trichomoniasis in four African cities found prevalences ranging from 6.5 to 40% (Buvé et al. 2001). Despite its name, *T. vaginalis*, frequently infects men, but it causes them little harm and they usually clear the infection rapidly. Infected men can transmit the infection to women during sexual intercourse and rectal *T. vaginalis* infections can occur in men-who-have-sex-with-men (Hoffman et al. 2018).

The parasite is 'tear-drop' shaped with five flagellae emerging at the anterior end: four of these flagella are free whilst the fifth curves back to form a short undulating membrane that extends just over half the length of the cell (Figure 3.7). *Trichomonas vaginalis* expresses only one body shape, but its size varies considerably: the length can be 7–32 μm, whilst the width is 5–12 μm. In common with other diplomonads, there are no mitochondria, but they possess a row of hydrogenosomes alongside the axostyle. This arrangement distinguishes *T. vaginalis* from other trichomonad species. Hydrogenosomes probably derive from mitochondria but unlike the more degenerate mitosomes, they make ATP, as well as the hydrogen that gives them their name. Interestingly, although *T. vaginalis* only infects us, molecular evidence indicates that it originated in pigeons and subsequently switched hosts (Peters et al. 2020).

Although *T. vaginalis* infects millions of women every year, most of them remain asymptomatic. The protozoa feed on bacteria and sloughed off epithelial cells present in the reproductive tract. However, in some women the parasites induce a severe inflammatory response that manifests as a

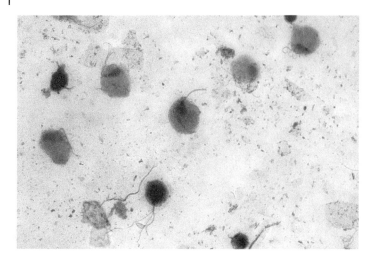

Figure 3.7 *Trichomonas vaginalis.* This protozoan parasite expresses only one body shape, but its size varies considerably: the length can be 7–32 μm, whilst the width is 5–12 μm. Five flagellae emerge at the anterior end: four of these flagella are free whilst the fifth curves back to form a short undulating membrane that extends just over half the length of the cell.

copious frothy white or greenish vaginal discharge. Infection during pregnancy is often associated with poor outcomes such as premature delivery and below-average birth weight for the baby, although whether the parasite induces these effects is uncertain.

3.3.2.3 *Trichomonas tenax*

This is a common human parasite and occurs throughout the world. It is rather like *T. vaginalis* in appearance although it is slightly smaller (5–16 μm long, 2–15 μm wide), has a somewhat shorter undulating membrane, and the hydrogenosomes are arranged differently. Molecular studies indicate a close similarity between the two species and *T. tenax* may be a variant of *T. vaginalis* (Kucknoor et al. 2009). *Trichomonas tenax* is usually restricted to the oral cavity where it feeds on bacteria and tissue debris.

   Trichomonas tenax does not form cysts and cannot survive passage through the digestive tract. Consequently, transmission probably involves contamination via kissing and the sharing of food and eating/drinking utensils. It is less host fastidious than *T. vaginalis,* and in addition to humans, it infects various domestic animals, including cats and dogs (Kellerová and Tachezy 2017). There is therefore the potential for zoonotic transmission between those besotted owners who inexplicably kiss their pets and those who do not follow basic hygiene during food preparation and consumption.

   Although *T. tenax* is associated with periodontal diseases (Marty et al. 2017), it is also common in people with good dental hygiene. This is probably because strains of *T. tenax* differ in their pathogenicity (Benabdelkader et al. 2019). There are occasional reports of bronchopulmonary infections, but these are usually associated with pre-existing pulmonary conditions, such as cancer.

3.3.2.4 *Trichomonas gallinae*

Trichomonas gallinae has a worldwide distribution and is a common parasite of poultry, pigeons, and many other birds. In doves and pigeons, it causes a condition called 'canker' whilst in birds of prey the condition is known as 'frounce'. It lives predominantly in the upper digestive tract and particularly the crop. In contrast, the related species *Tetratrichomonas gallinarum* tends to live in the lower digestive tract, caeca, and sometimes the liver.

The trophozoites of *T. gallinae* are usually ovoid in shape, 7–11 μm in length, and have four free flagella with a fifth recurving to form an undulating membrane (Figure 3.8). There is no cyst stage. Strains differ in their pathogenicity and pigeons, doves, and other members of the colubriformes are more badly affected than most other birds. In addition, raptors such as peregrine falcons (*Falco peregrinus*) and sparrowhawks (*Accipiter nisus*) that consume pigeons often suffer badly from *T. tenax* infections, and this can have consequences for their conservation (Dudek et al. 2018).

Avian trichomoniasis (trichomonosis) is an emerging disease in finches and other passerine birds in the UK, Europe, and Canada (Chi et al. 2013; Forzán et al. 2010). The first report of the problem was from the United Kingdom in 2005. Since then, it has become an ongoing epidemic as migrating birds spread the parasite across the country and throughout Europe. In the United Kingdom, this has resulted in a catastrophic decline in the number of greenfinches (*Chloris chloris*). In Europe, the epidemic is associated with a single clonal strain of *T. tenax* that is distinct from the strain causing extensive bird mortalities in America (Alrefaei et al. 2019).

The pathogenic strains of *T. tenax* induce lesions in the wall of the intestine and the parasites, then spread around the bird's body and cause liver pathology similar to that of *Histomonas*

Figure 3.8 Trichomonad parasites of animals. (a): *Tritrichomonas foetus*; (b): *Trichomonas gallinae*. *Source:* Reproduced from Chandler and Read, (1961), © Wiley-Blackwell.

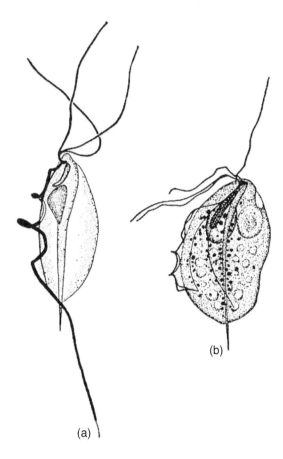

(b)

(a)

meleagridis. Young birds are the worst affected, and whilst adults are often infected, they do not show evidence of disease – although they act as carriers of infection. In badly affected birds, necrotic lesions to the intestine and mouth can extend to the bones.

Pigeons often clash bills during social interactions and until they are 10 days old; young squabs feed by pushing their beaks into their parent's mouth to feed on 'milk' held within the parent's crop. This provides many opportunities for parasite transmission to occur. *Trichomonas tenax* probably spread from pigeons to finches through infected pigeons contaminating garden bird-baths. In addition, in the United Kingdom, there has been a marked rise in the population of wood pigeons, and this has probably increased the levels of contamination. The parasites can survive for up to 1 hour in water, so a single infected bird can potentially infect many others after using a popular birdbath. They can also survive on moist but not dry bird seed, so it is possible that in some circumstances bird feeders may act as an additional source of infection (McBurney et al. 2017).

3.3.2.5   *Tritrichomonas foetus*

This is a sexually transmitted parasite of cattle, and although it is also found as a sexually transmitted infection in other animals (e.g., horses), it is not usually pathogenic in them. The trophozoite of *T. foetus* is a pear-shaped organism 10–25 µm long and 3–15 µm wide with four anterior flagella, three of which are free and one flagellum curves backwards to form an undulating membrane that extends the length of the body and then projects freely from the posterior apex (Figure 3.8). There is no cyst stage although they form pseudocysts in response to iron depletion. It is uncertain whether the pseudocyst stage plays an important role in parasite transmission.

In bulls, the parasites are usually found in the preputial cavity and cause little harm although there is sometimes inflammation that causes painful urination and unwillingness to copulate. In cows, the parasite causes more serious pathology. The infection begins with vaginitis and then spreads to the uterus where they can cause early abortion and permanent sterility. The parasite remains in the lumen and does not penetrate the underlying tissues.

There are increasing reports of *T. foetus* causing large bowel diarrhoea in domestic cats in the UK, USA, and parts of Europe. The parasites isolated from cats are morphologically identical to those from cattle, and there are only very minor differences in their DNA sequences (Yao and Köster 2015). Therefore, the current consensus is that they represent two isolates of a single species. However, in cats, transmission of the parasite occurs through faecal–oral contamination.

3.3.2.6   *Pentatrichomonas hominis*

This parasite usually lives as a harmless commensal in our large intestine and caecum. It has a worldwide distribution and in addition to humans, it colonizes the large intestine of many wild and domestic mammals, including sheep, dogs, pigs, and monkeys. However, the extent to which zoonotic transfer occurs is uncertain.

The trophozoite is pear-shaped, 5–15 µm long, and 7–10 µm wide with four free flagellae at the anterior end (Figure 3.7). A fifth flagellum curves back to form an undulating membrane that extends the length of the body and then projects freely from the posterior apex.

Prevalences tend to be higher in children than in adults. Sometimes it is associated with diarrhoea but whether it causes the condition is not known. Similarly, although there is a higher prevalence of *P. hominis* in patients suffering from gastrointestinal cancer than in healthy patients, whether there is a causative association is uncertain (Zhang et al. 2019).

3.4 Phylum Apicomplexa

The Apicomplexa is one of the largest phyla amongst the protozoa and includes many important parasites of humans and domestic animals (Table 3.1).

As with many other protozoa, the taxonomic arrangements within the Apicomplexa are in a state of flux. All members of the phylum are obligate intracellular parasites of invertebrates and vertebrates. A common feature shared by all apicomplexans is the presence within their invasive stage of a unique intracellular structure called the apical complex that is composed of a group of secretory organelles called the micronemes and rhoptries (Figure 3.9). The apical complex lies at the anterior apex of the cell where it is associated with a region called the oral structure. The secretions of the micronemes and the rhoptries play an important role in the invasion of red blood cells by the malaria parasites (Suarez et al. 2019).

Table 3.1 Representative examples of parasitic protozoa belonging to the phylum Apicomplexa to illustrate the wide variety of hosts and transmission strategies.

Genus	Example	Host	Transmission	Disease
Plasmodium	*Plasmodium falciparum*	Humans	Vector: Anopheline mosquitoes	Malaria
Toxoplasma	*Toxoplasma gondii*	All warm-blooded animals	Contamination, congenital, ingestion of infected flesh	Toxoplasmosis
Neospora	*Neospora caninum*	Dogs, cattle	Contamination, congenital	Neosporosis
Cyclospora	*Cyclospora cyetanensis*	Humans	Contamination	Cyclosporosis
Eimeria	*Eimeria tenella*	Poultry	Contamination	Coccidiosis
Theileria	*Theileria parva*	Cattle	Vector: *Rhipicephalus* ticks	East Coast Fever
Babesia	*Babesia bigemina*	Cattle	Vector: *Rhipicephalus* ticks	Texas Fever
Isospora	*Isospora belli*	Humans	None	Isosporosis
Cryptosporidium	*Cryptosporidium hominis*	Humans	Contamination	Cryptosporidiosis

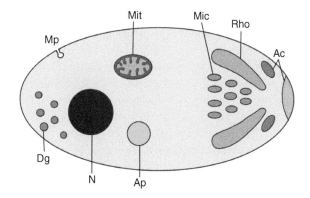

Figure 3.9 Generalized diagram of the invasive stage of an apicomplexan parasite. Abbreviations: Ac: apical complex (conoid + polar ring); Ap: apicoplast; Dg: dense granules; Mic: micronemes; Mit: mitochondrion; Mp: micropore; N: nucleus; Rho: rhoptries.

Plastids in Parasites

Apicomplexans contain a unique organelle called the apicoplast that probably evolved from a chloroplast (plastid) although it does not contain any pigments. Molecular studies indicate that the protozoa currently comprising the Apicomplexa arose from at least three independent transitions of free-living photosynthetic algae to intracellular parasites (Mathur et al. 2019). The apicoplast has four membranes and contains DNA although most of the genes that code for proteins within it have transferred to the nucleus. Interestingly, a protist, *Chromera velia*, that is phylogenetically related to the Apicomplexans contains a functioning plastid that is morphologically similar to that found in the Apicomplexans (Moore et al. 2008). *Chromera velia* is usually free-living, but it can form endosymbiotic relationships with the larvae of certain corals. It is related to the dinoflagellate algae – a group that includes species that combine photosynthesis with other forms of nutrition, including predation. The apicoplast is of interest because it has no equivalent in the parasite's animal hosts and therefore is a potential target for specifically designed chemotherapeutics. The apicoplast produces essential metabolites and the parasites die if exposed to drugs that interfere with its functions (Lim et al. 2016).

Molecular phylogeny suggests that some of the Kinetoplastida may also have contained plastids at an early stage in their evolution, but these have since been lost. Many euglenid protozoa contain chloroplasts (e.g., *Euglena gracilis*), and these are closely related to the Kinetoplastida. However, ultrastructural studies suggest that the euglenids acquired their plastids after the point at which they diverged from the Trypomastigota (Leander 2004).

3.4.1 Genus *Plasmodium*

The genus *Plasmodium* probably evolved hundreds of millions of years ago and long before the arrival of the vertebrates (Escalante and Ayala 1995). There are over 200 species, most of which are parasites of birds. In addition, many species infect reptiles, rodents, and primates. Blood-feeding insects, especially mosquitoes, act as their vectors. Four principal species infect humans: *Plasmodium falciparum, Plasmodium vivax, Plasmodium ovale,* and *Plasmodium malariae.* A fifth species, *Plasmodium knowlesi* normally parasitizes monkeys but sometimes causes fatal human infections. In addition, there are rare case reports of people becoming infected by species of *Plasmodium* that normally parasitize other animals. In most vertebrate hosts, *Plasmodium* parasites are not particularly pathogenic but in humans, they cause the disease malaria. Historical literature usually refers to it as the ague or marsh fever. The word ague derives from the Latin for sharp (*acutus*) because it manifests as a sharp fever (*febris acuta*) whilst living in a marshy district has long been associated with ill health – although this was attributed to the bad air (hence *mal aria*) rather than the anopheline mosquitoes flying in it.

Plasmodium falciparum, P. vivax, P. ovale and *P. malariae* exhibit marked differences in their biology, and molecular evidence suggests that they evolved from separate lineages. That is, they are more closely related to *Plasmodium* species that parasitise other animals than they are to one another. Interestingly, molecular evidence indicates that *P. falciparum* originated in gorillas and humans recently acquired the disease in a single cross-species transmission. Therefore, contrary to previous assumptions, *P. falciparum* does not derive from chimpanzees nor did it originate in primitive human ancestors (Liu et al. 2010). It is also now apparent that *P. vivax* originated in Africa millions of years ago among chimpanzees, gorillas, and ancestral humans (Loy et al. 2017).

Malaria causes a higher human mortality than any other parasitic disease although the situation is improving in many countries. Between 2000 and 2017, there were significant reductions in both the number of cases and the mortality associated with malaria https://www.afro.who.int/health-topics/malaria. In 2000, there were 233 million cases of malaria resulting in 985,000 deaths, but by

2017, the figures had declined to 219 million cases and 435,000 deaths. Although malaria transmission occurs in 91 countries and approximately 50% of the world's population is at risk of contracting the disease, most cases occur in just 15 countries with Nigeria accounting for 27% of them and the Democratic Republic of Congo for 10%. Indeed, approximately 90% of the fatal cases occur in sub-Saharan Africa and involve children less than 5-years-old. Even within a country, there can be big differences in the risks of contracting malaria. In Nigeria (population ~201 million), 76% of people live in malaria prone areas, whilst the remaining 24% live in areas of low transmission. Similarly, in Kenya (population ~47.6 million), there is virtually no risk of the disease in some regions, whilst in others the risk is high, and it is a major cause of childhood mortality. Unfortunately, at the time of writing (2021), efforts to make further reductions appear to have stalled. Regardless of whether it causes fatal disease, wherever malaria is common amongst a community it has serious socioeconomic consequences and contributes to poverty. This is because illness prevents people from working and/or they are less productive. In addition, because the disease is particularly severe in young children, they are unable to attend school and gain the education that would improve their chances in life. Some estimates suggest that the direct costs of malaria are US$12 billion per year and the indirect costs resulting from reduced economic growth substantially more. However, the causes of poverty are complex and whilst malaria undoubtedly causes immense hardship, it is not the main reason that so many people in the developing countries remain poor (Utzinger and Tanner 2013).

How Malaria Has Influenced the Course of History

Malaria has influenced the course of history for thousands of years and remains relevant today. The distinctive symptoms of chronic and repeated infections of malaria can be identified with almost complete certainty from historical descriptions from ancient civilisations of Egypt, Sumeria, China, and India. For example, a Chinese medical text, the *Nei Cheng*, written in approximately 2700 BC refers to epidemics of the 'Mother of Fevers' that is undoubtedly malaria. The authors describe the cyclical fevers and an enlarged spleen that are features of malaria. Similarly, the symptoms of malaria can be identified from the writings of Hippocrates in Ancient Greece (~500 BC). Some authors have suggested that the decline of the Ancient Greek and Roman civilisations was associated with effects of malaria epidemics (Poser and Bruyn 1999). More recently, during WW2, an outbreak of malaria among allied troops in 1943 seriously compromised their attempts to invade Sicily. American troops suffered similar problems with malaria during the Vietnam war in the 1960s. Although malaria caused only about 0.2% of fatalities among the American troops, the debilitating effects reduced the combat strength of some units by up to 50%. The Viet Cong were aware of the problems that malaria caused the American troops and intentionally sabotaged local mosquito and malaria control programmes. They were successful in undermining malaria control over large areas, but consequently the combat strength of their own troops was also severely compromised (Drisdelle 2011).

Today, malaria is considered a tropical disease, but it was once a common disease in many of the temperate regions of the world. Malaria was a major cause of mortality in parts of Italy and Greece as late as the 1920s (Snowdon 2006). Malaria existed in the United Kingdom until the early years of the twentieth century, particularly in the fenland regions (Reiter 2000). The potentially fatal consequences of the ague are alluded to in works by Geoffrey Chaucer (c1343–1400) and William Shakespeare (1564–1616), so the disease was obviously common enough for their audiences to be familiar with it. Samuel Pepys (1633–1703) describes suffering from the ague in his diaries and historical accounts state that Oliver Cromwell (1599–1658) died from an attack of the 'tertian ague'.

3.4.1.1 *Plasmodium* **Life Cycle**

The *Plasmodium* species that infect humans have a complex life cycle that involves both asexual and sexual reproductive stages (Figure 3.10). They are all transmitted by Anopheline mosquitoes and multiplication takes place in both humans and the mosquito vector. Only female mosquitoes feed on blood, as they require nutrients contained within it to produce their eggs. Male mosquitoes feed on nectar and other sugary solutions, and therefore, only female mosquitoes transmit malaria.

We contract malaria when a female mosquito harbouring the sporozoite stage of the parasite bites us. The sporozoites enter our body with her saliva and the blood stream transports them to the liver where they penetrate the hepatocytes (liver cells). Within the hepatocytes, the sporozoites

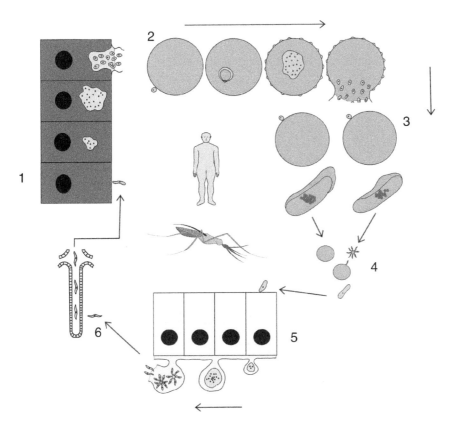

Figure 3.10   Life cycle of *Plasmodium falciparum*. 1: An infected mosquito injects sporozoites into the blood stream and these travel to the liver. The sporozoites invade the hepatocytes and undergo exoerythrocytic schizogony to produce numerous merozoites. 2: Merozoites leave the liver cells, enter the circulation, and infect red blood cells. They transform into ring-shaped trophozoites that develop into schizonts (erythrocytic schizogony), which then form numerous merozoites. The parasites export proteins that re-model the cell membrane of infected red blood cells so that 'knobs' are formed. When an infected cell dies, it releases merozoites that reinfect other red blood cells. 3: At some point, merozoites develop into male and female gametocytes. Those of *P. falciparum* have a characteristic banana shape that deforms the host red blood cell. 4: After ingestion by a mosquito, the male and female gametocytes are released from the red blood cells and fuse to form a zygote. 5: The zygote differentiates into an ookinete that invades a mosquito gut cell within which it forms an oocyst. The oocyst undergoes sporogony to form numerous sporozoites that are released into the haemolymph. 6: The sporozoites penetrate the mosquito salivary glands and are transmitted within mosquito's saliva when it feeds. Drawings not to scale.

change their morphology and multiply asexually by 'exoerythrocytic schizogony' to form thousands of merozoites. The term exoerythrocytic indicates that the reproduction takes place in cells other than the red blood cells. Schizogony is a form of asexual reproduction in which first, the nucleus divides several times and then the parent cell divides to form as many individual merozoites as there are nuclei. Schizogony also occurs in many other apicomplexan parasites. In *P. vivax* and *P. ovale*, not all the parasites immediately undergo schizogony but instead some remain in a quiescent state known as the hypnozoite form. These hypnozoites can remain in the liver for weeks or even years before undergoing further development. They are therefore responsible for the onset of illness/relapses long after the initial infection.

The formation of merozoites destroys their host liver cells, thereby releasing them into the blood stream. The merozoites proceed to invade red blood cells in which they transform into the trophozoite stage – these also reproduce asexually by shizogony. Because these divisions take place in red blood cells, it is referred to as erythrocytic schizogony. The ingestion of host cytoplasm by a trophozoite causes the formation of a large food vacuole. This gives the parasite the appearance of a ring of cytoplasm with the nucleus conspicuously displayed at one edge – hence, the moniker of the 'signet ring stage'. As the trophozoite grows, its food vacuole becomes less noticeable by light microscopy, but pigment granules of hemozoin in the vacuoles become apparent. Hemozoin is an insoluble polymer of haem and is the end product of the parasite's digestion of the host's haemoglobin. The growth of the parasites within an infected red blood cell eventually destroys it and it ruptures. This releases the merozoites, hemozoin and other parasite waste products, and dead cell material. These merozoites then infect other red blood cells and the process of infection, replication, and destruction repeats many times. At some point in this cycle, certain merozoites transform into sexual stages referred to as macrogametocytes (female) and microgametocytes (male). These gametocytes do not develop any further and remain within their host red blood cells until a suitable female anopheline mosquito ingests them.

Shortly after ingestion by a mosquito, the male and female gametocytes swell, leave their host red blood cells, locate one another, and fuse to form a zygote. This is the only diploid stage in the whole *Plasmodium* life cycle. The zygote then differentiates into an ookinete. The ookinete is capable of movement and it bores through the mosquito's gut until it comes to rest at the outer wall of the midgut epithelium where it transforms into an oocyst. Inside the oocyst further rounds of asexual multiplication take place called sporogony that result in the formation of numerous sporozoites. Once it is mature, the oocyst bursts and the sporozoites migrate through the mosquito's haemolymph (blood) to the salivary glands. The next time the mosquito feeds, it injects the sporozoites along with its saliva. Depending upon the species of *Plasmodium* and mosquito, the oocyst stage lasts between 8 and 35 days, thereby making it the longest part of the life cycle. It also means that successful transmission depends heavily on the lifespan of the mosquito. This is because the mosquito must survive long enough after its initial infected blood meal for the sporozoites to form and then reach its salivary glands. Infection stimulates mosquitoes to feed more frequently, thus increasing the chances of transmission. Once infected, a mosquito remains infective for the rest of her life.

3.4.1.2 *Plasmodium falciparum*

This is the most dangerous species because of its capacity to cause potentially fatal cerebral malaria – also sometimes referred to as malignant tertian malaria. Its distribution has become restricted in recent years although it remains very common in tropical and sub-tropical regions such as sub-Saharan Africa, Southeast Asia, and parts of South America. It does not form hypnozoites, but sudden relapses (recrudescence) can occur months or even years after the last fever episode because the parasite persists in the blood at low sub-clinical levels. The merozoites can invade and develop in both young red blood cells (reticulocytes) and mature red blood cells. Consequently, falciperan malaria is often characterised by high levels of parasitaemia.

3.4.1.3 *Plasmodium vivax*

The name *vivax* derives from the lively nature of the trophozoites in our red blood cells (*vivax* is the Latin adjective for vivacious). This stage often has an amoeboid appearance in blood films. *Vivax* malaria was once the most widespread form of the disease and common as far north as Norway and Siberia. However, as European countries developed, they largely eradicated malaria, and *vivax* malaria is now mostly restricted to Asia and the countries bordering the Mediterranean. However, it is still the most common *Plasmodium* species in most countries in which malaria remains endemic. Vivax malaria is rare in West Africa because the merozoites only penetrate red blood cells carrying the Duffy buffer blood group antigens Fy^a and Fy^b – and most West Africans do not express these. In addition, the merozoites cannot penetrate mature red blood cells and therefore they invade developing reticulocytes. Consequently, *P. vivax* cannot form the same high parasitaemias achieved by *P. falciparum*. *Plasmodium vivax* is notorious for causing latent infections in which the hypnozoite stage remains quiescent within the liver and then, after years of apparent good health, the patient suddenly develops malarial fevers. The factors determining the length of the latent period probably relate to genetic differences between the parasites causing the initial infection and/or sudden changes in the host's immune status.

3.4.1.4 *Plasmodium ovale*

This species exists in many parts of the world although it is not as common as *P. vivax* or *P. falciparum,* and its natural distribution is sub-Saharan Africa and the western Pacific. The merozoites only develop in reticulocytes and the parasitaemia tends to be low. *Plasmodium ovale* forms hypnozoites in the liver, and therefore, it can produce long-lasting latent infections. Some workers consider that there are two sub-species of *P. ovale, P. ovale curtisi* and *P. ovale wallikeri,* and these differ in their clinical manifestations.

3.4.1.5 *Plasmodium malariae*

This species has a worldwide distribution, but it is not common anywhere and infections are normally relatively benign. Natural infections are common in chimpanzees, but these seldom live near human dwellings, and they are probably not an important reservoir of infection. Humans and chimpanzees both experience latent *P. malariae* infections that persist for many years. Unlike *P. vivax* and *P. ovale,* the sudden occurrence of malarial symptoms after a period of good health is not a result of hypnozoite activation. Instead, the parasite persists within the blood for years, possibly even for life, at a very low parasitaemia and it is only when some change in the immune status of the host occurs that the numbers increase sufficient to cause fever. Unlike *P. vivax* and *P. ovale,* the merozoites of *P. malariae* preferentially invade older red blood cells.

3.4.1.6 *Plasmodium knowlesi*

This species inhabits many parts of Southeast Asia. Its normal hosts are monkeys such as the long-tailed macaque (*Macaca fascicularis*). It is also capable of infecting humans and is therefore a zoonosis. Human infections are a particular problem in Sarawak and peninsular Malaysia. In blood smears, it is often mistaken for *P. malariae* and therefore considered 'low risk'. Unfortunately, the consequences of misdiagnosis can be fatal. This is because *P. knowlesi* causes serious pathology in a similar manner to *P. falciparum*. Like *P. vivax*, the merozoites of *P. knowlesi* only invade red blood cells carrying the Duffy buffer blood group antigens Fy^a and Fy^b.

3.4.2 Genus *Theileria*

Several species in this genus parasitize wild and domestic animals, but the most well known is *Theileria parva*, which causes East Coast fever (theileriosis) in cattle in sub-Saharan Africa. Other important species include *Theileria annulata* that also parasitizes cattle and *Theileria hirci* that infects sheep and goats – these species occur in parts of North Africa, the middle East, Europe, and Asia. *Theileria* parasitize red blood cells, lymphocytes, and tissue macrophages (histiocytes), and they are common causes of disease and potentially fatal infections in their mammalian hosts. Unlike the genus *Babesia*, the genus *Theileria* does not include zoonotic species. They are all transmitted by Ixodidid ticks (so-called hard ticks) and are therefore examples of tick-borne diseases.

The genera *Theileria* and *Babesia* belong to the order Piroplasmida and hence these parasites are referred to as piroplasmids. The genomes of *Theileria* species exhibit important differences from other apicomplexans (Nene et al. 2016). For example, although the *T. parva* genome is much smaller (36.5%) than that of *P. falciparum,* it contains 76.6% of the number of genes encoding proteins. This means its genes are packed extremely closely together. In addition, some metabolic pathways are abbreviated/absent, which indicates considerable metabolic dependence upon the host.

3.4.2.1 *Theileria* Life Cycle

The *Theileria* life cycle begins when a tick injects infectious sporozoites into the blood stream of a suitable mammal host (Figure 3.11). *Theileria parva* sporozoites are non-motile and unlike those of many other apicomplexans, they do not actively invade the host cell. Instead, they make random contact with T and B-lymphocytes and attach to host cell receptors. The host cell then internalizes the parasite and surrounds it with a vacuole membrane. The parasites do not orientate themselves in relation to the host cell, and they discharge the contents of their rhoptries and micronemes only after entering it. These chemicals cause the vacuole membrane to disperse so that the parasites lie free within the cytoplasm of the lymphocytes. The sporozoites then transform into multinucleate schizonts and induce the host cell to proliferate: the parasites are closely associated with the lymphocyte mitotic apparatus and divide with their host cell so that daughter lymphocytes are also infected. The first generation of schizonts are called 'macroschizonts' and these give rise to 'macromerozoites' that invade other lymphocytes and give rise to either more macroschizonts and macromerozoites or to microschizonts that give rise to micromerozoites. The micromerozoites may invade either lymphocytes or red blood cells. Those invading lymphocytes continue to multiply by schizogony, but those invading red blood cells differentiate into 'piroplasms' that do not divide any further but are infectious to the tick vector. The piroplasms are extremely small, typically 1.5–2 µm long and 0.5–1 µm wide and rod-shaped although oval, comma, and ring-shaped forms also occur. They are a characteristic diagnostic feature of East Coast fever. The digestion of infected red blood cell within the tick gut lumen releases the piroplasms, and they differentiate into either microgametocytes or macrogametocytes. The microgametocytes and macrogametocytes fuse to form a diploid zygote, and this invades the tick gut epithelial cells where it transforms into a motile 'kinete' form. The kinetes make their way through the body to the tick salivary glands where they invade the type III acini cells and undergo sporogony to produce numerous infectious sporozoites.

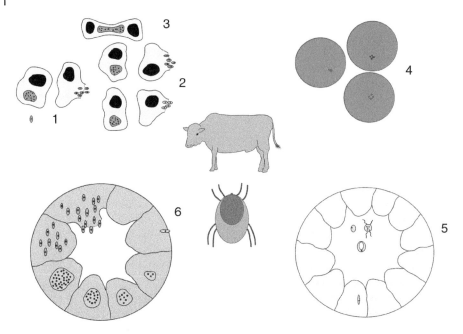

Figure 3.11 Life cycle of *Theileria parva.* 1: An infected tick injects sporozoites into a cow's bloodstream, and these are internalized by T- and B-lymphocytes within which they transform into multinucleate schizonts. 2: Schizogony results in the formation of merozoites that are released when the host cell dies. 3: Infected lymphocytes are induced to divide and the parasites infect each new cell as it forms. The first generation of schizonts (macroschizonts) form 'macromerozoites' that invade other lymphocytes and give rise to either more macroschizonts and macromerozoites or to microschizonts that give rise to micromerozoites. The micromerozoites invade either lymphocytes or red blood cells. 4: Micromerozoites invading red blood cells differentiate into piroplasms that do not divide any further but are infectious to the tick vector. 5: The piroplasms are released within the tick gut lumen and differentiate into either microgametes or macrogametes. The microgametes and macrogametes fuse to form a diploid zygote, and this invades the tick gut epithelial cells where it transforms into a motile kinete. 6: The kinetes make their way through the tick's body to its salivary glands where they undergo sporogony to produce sporozoites. The sporozoites are released into the saliva and are transmitted when the tick feeds. Drawings not to scale.

3.4.2.2 *Theileria parva*

Theileria parva is principally a disease of cattle. Ticks belonging to the genus *Rhipicephalus* (mainly *Rhipicephalus appendiculatus* and *Rhipicephalus zambesiensis*), transmit it, and its distribution is largely determined by the presence of its vectors.

Theileria parva exhibits considerable genotypic diversity. Indeed, a single cow may harbour several distinct genotypes. This complicates vaccine design because there is a lack of cross-protection between different strains of the parasite (Katzer et al. 2010). The virulence of theileriosis varies between regions and probably relates to the ecology of the tick vector. Where the tick life cycle stages do not usually coincide (e.g., adults plus nymphs/nymphs plus larvae/adults plus larvae), the disease tends to be less virulent (Tindih et al. 2010). This is because a highly virulent parasite would kill its cattle host before there was an opportunity for the next generation of ticks to become infected. *Rhipicephalus appendiculatus* and *R. zambesiensis* are typical three host ticks and the larval, nymphal, and adult stages exploit different hosts. After feeding (engorging), the tick drops off to moult or in the case of the adult females to lay their eggs. In subtropical and southern regions of Africa, the ticks are seasonal with one generation per year (i.e., they are unimodal) but in tropical regions where there is high rainfall, up to three generations may occur.

3.4.3 Genus *Babesia*

Some reports state that there are over 100 species of *Babesia*, whilst others suggest that there are somewhat fewer. Most species are tick borne parasites of mammalian erythrocytes although some parasitize bird red blood cells. Some *Babesia* species also parasitize other blood cells, such as lymphocytes and histiocytes. The genus *Babesia* is primarily of economic importance as parasites of cattle, sheep, and other domestic animals (Table 3.2). The most important species are *Babesia bigemina*, *Babesia bovis*, and *Babesia divergens*. *Babesia microti* is primarily a parasite of rodents but it also infects other mammals and is the principal cause of human babesiosis. For example, between 2006 and 2015, the incidence of human infections in New York State, USA, increased from 1.7 cases per 100,000 persons to 4.5 cases per 100,000 persons. It has been isolated from ticks in Europe and human cases are increasingly reported – and not just in immunocompromised individuals. Because ticks are vectors for various bacterial and viral diseases, it is common for those who contract babesiosis to suffer from co-infections such as Lyme disease. Rather confusingly, some literature refers to this parasite as *Theileria microti*. Even more confusingly, there is a parasite referred to as *Babesia* cf. *microti*, *Babesia vulpes* and *Theileria annae* that has a similar distribution and causes severe disease in dogs. The 'cf' denotes that it is uncertain whether it is actually *B. microti*, a sub-species, or a closely related but different species. In addition to tick vectors, transmission of *B. microti*, and *B.* cf. *microti* is also possible across the placenta. Dogs and other canines infected with babesiosis are reportedly capable of directly transmitting the infection in their bites.

3.4.3.1 *Babesia* Life Cycle

The life cycle begins when ticks inject infective sporozoites are into the mammal bloodstream and these then invade the red blood cells (Figure 3.12). Initially, the host cell encloses the sporozoites within a membrane-bound vacuole. However, the parasites escape from this and therefore come

Table 3.2 The distribution, vectors, and host ranges of representative *Babesia* species of medical and veterinary importance.

Species of *Babesia*	Distribution	Mammal host	Tick vector
Babesia Bigemina	Central & South America, North & South Africa, Australia, Asia (not in UK)	Cattle, water buffalo, zebu, deer	Various species of *Rhipicephalus* and also *Ixodes ricinus*
Babesia bovis	Central & South America, North & South Africa, Australia, Asia, Southern Europe (not in UK)	Cattle, deer	Various species of *Rhipicephalus* and also *Ixodes ricinus*
Babesia divergens	Northern Europe, including UK	Cattle	*Ixodes ricinus*
Babesia Microti	North-eastern USA, Europe	Rodents, human infections increasingly reported	*Ixodes dammini, Ixodes scapularis*
Babesia Ovis	Southern Europe, Middle East, Africa	Sheep, goats	Various species of *Rhipicephalus*
Babesia Canis	Southern Europe, Middle East, Africa, Asia, Central, South and North America	Dogs and other caniids	Various species of *Ixodes, Dermacentor variabilis, Haemaphysalis leachi*

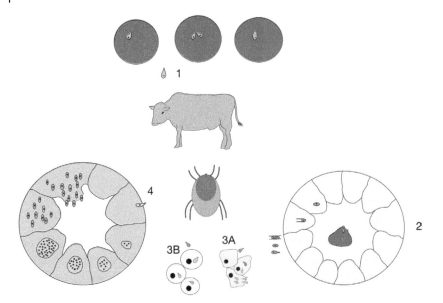

Figure 3.12 Life cycle of *Babesia* spp: 1: An infected tick injects saliva containing sporozoites into the bloodstream and these invade red blood cells. The sporozoites transform into merozoites that divide asexually until they kill their host cell, and they then infect other red blood cells. Some merozoites transform into oval-shaped gamonts, and these develop no further unless a suitable tick vector ingests them. 2: Within a tick's gut, the gamonts invade the intestinal cells and differentiate into gametocytes (ray bodies). Once they are mature, the gametocytes leave the tick gut cells and fuse to produce a zygote. 3A: Zygotes transform into motile primary kinetes that invade other cell types within which they transform and multiply to form secondary motile kinetes. 3B: Primary kinetes that invaded the tick oocytes result in the young tick emerging from its egg already infected, and in these, the parasites ultimately invade their salivary glands. 4: Secondary kinetes that parasitize the salivary glands transform into sporonts, and these give rise to infectious sporozoites. Drawings not to scale.

into intimate contact with the erythrocyte cytoplasm. The sporozoites now transform into the merozoite stage that divides by binary fission and accumulates within the cell and eventually kill it. The merozoites are usually pear shaped (in *Babesia bigemina* they are 4–5 µm long × 2–3 µm wide) and in blood smears, they are seen singly or in pairs. When their host cell dies, the parasites are released, and they re-infect other red blood cells. Mechanical transmission between hosts can occur – for example, via re-used needles and veterinary instruments and blood transfusions. In susceptible hosts, the parasitaemia builds up rapidly and 70–80% of the red blood cells can become infected. Some merozoites transform into oval-shaped gamonts and these develop no further unless a suitable tick vector ingests them.

When a tick ingests blood from an infected host, it digests the *Babesia* merozoites along with the red blood cells they parasitize. The gamonts, however, survive and invade the cells lining the tick's gut. Within these intestinal cells the gamonts differentiate into gametocytes that are referred to as 'ray bodies' (derived from the original German '*strahlenkörper*' that is still used by some authors), which are globular with one or two thorn-like projections. Once they are mature, the ray bodies leave the gut cells and two of them fuse to produce a zygote. The zygote then transforms into a motile kinete that invades several other cell types including muscles, Malpighian tubules, and in female ticks the ovaries and oocytes. Within these cells, they transform and multiply to form numerous secondary motile kinetes. Upon release, the secondary kinetes parasitize other cells including the salivary glands. Within the salivary glands, the kinetes transform into sporonts, and these give rise to numerous infectious sporozoites. Those primary kinetes that invaded the tick oocytes result in the young tick emerging from its egg already infected, and in these the parasites ultimately invade their salivary glands. For one-host ticks, trans-ovarial transmission of the infection is essential if they are to function as effective vectors. This is because they usually spend their whole life on a single animal and therefore cannot transmit infections directly from one host to another. For two- and three-host ticks, transovarial transmission may be of less importance for the parasite. This is because the ticks drop off and re-attach to new hosts between moults. Therefore, the ticks have greater opportunity to spread the infection. Many parasites that grow and develop within blood-feeding vectors damage their vector in the process (e.g., sand-flies and *Leishmania* spp.) but *Babesia microti* facilitates the feeding and improves the survival of its tick host, *Ixodes trianguliceps* (Randolph 1991). Whether similar *Babesia*-tick relationships occur in other *Babesia* species is uncertain.

3.4.3.2 *Babesia bigemina*

The different species of *Babesia* vary in their pathogenicity and distribution but they share many similarities, so we shall discuss only one of them, *Babesia bigemina*, in detail. This species was once a major cause of disease in North America, where it caused 'Texas fever' in infected cattle. Following successful eradication campaigns, *B. bigemina* is of much less economic importance in the USA today, but it remains a serious problem in Central and South America, North and South Africa, Australia, and Asia. It is principally a disease of cattle, although it also infects water buffalo, zebu, and deer. Ticks belonging to the genus *Rhipicephalus* are responsible for most of the transmissions, and these have a widespread distribution in tropical and subtropical countries.

Cattle normally acquire their infection from others that have recovered from the disease but continue to harbour a subclinical infection. The pathology associated with *B. bigemina* is unusual in that adult cattle tend to be more severely affected than the young are. This is particularly the case in cattle not previously exposed to the parasite or the local strain – for example those transported to a new region or country or following exposure to infected ticks newly introduced into the area. High-performance milking cows imported into Africa from temperate regions are at particular risk of succumbing to severe disease symptoms. In Mexico, white-tailed deer (*Odocoileus virginianus texanus*) have high seroprevalence for both *B. bigemina* and *Babesia bovis* but how important these (and other wildlife) might be as reservoirs of infection is uncertain.

A host of variables, including the age and immune status of the host (e.g., previous exposure and vaccination status) and the strain of the parasite influence the pathogenicity. The damage is primarily associated with the loss of function and destruction of the red blood cells. Lysing of infected erythrocytes releases haemoglobin into the blood stream. The destruction of small numbers of red blood cells has little effect and many infections are sub-clinical. However, the rapid destruction of numerous cells in a short period can overload the body's ability to remove the waste material. Consequently, haemoglobin and its breakdown products accumulate resulting in jaundice and their appearance in the urine – and hence the common name for the infection of 'red water fever' – or haemoglobinuria. The loss of functioning red blood cells also gives rise to severe haemolytic anaemia. The infected animal develops a fever and cerebral involvement is possible. The precise mechanism by which *Babesia* induces brain pathology is uncertain although Schetters and Eling (1999) suggest that it might provide a useful animal model for human cerebral malaria.

Human cases of babesiosis are uncommon, and severe disease is usually associated with splenectomised patients or those already suffering from immunodeficiency or AIDS. The importance of retaining the spleen wherever possible means that far fewer splenectomies take place than was once the case. Surgeons perform the operation for a variety of reasons ranging from injury following a car crash to cancer.

3.5 Subclass Coccidiasina

Commonly known as the Coccidia, this is the largest group of the Apicomplexa. All members are intracellular parasites of vertebrates and invertebrates. They usually parasitize the intestinal cells although other cell types may also be infected. Some species have only a single host, whilst others employ two – commonly a vertebrate and an invertebrate, although it may be two vertebrates, one of which feeds on the other. The life cycle usually begins with the invasion of a host cell by a sporozoite stage followed by a cycle of merogony, gametogony, and sporogony. The group used to contain just the *Eimeria*, the *Isospora* group, and the haemogregarines (mainly parasites of red blood cells of amphibians and reptiles) but currently includes *Cryptosporidium, Sarcocystis*, and *Toxoplasma*.

3.5.1 Genus *Eimeria*

There are probably tens of thousands of *Eimeria* species and new ones are described on a regular basis. Unfortunately, there are the usual problems with taxonomy and species identification, so the literature is a bit confusing. The host range encompasses fish, lizards, and mammals, and most *Eimeria* species are host specific or infect a few closely related host species. Several species are of economic importance. For example, estimates for the annual worldwide losses owing to avian coccidiosis in commercially reared chickens and other birds are probably in the region of £500 million (Shirley et al. 2007). In addition, some *Eimeria* species play an important role in wildlife ecology although they are not easy to detect (Jarquín-Díaz et al. 2019). By contrast, there are few reports of *Eimeria* infections in primates and human infections either do not occur or are exceedingly rare. We will only consider one species, *Eimeria tenella* as a representative example.

3.5.1.1 *Eimeria tenella*
This is the commonest and most pathogenic of the seven species of *Eimeria* that infect domestic poultry. Each *Eimeria* species develops in a different region of the bird's digestive tract, and co-infections with two or more species are common. *Eimeria tenella* occurs throughout the world and is responsible for a great deal of economic loss. Although it can cause high flock mortality, the availability of vaccines and anticoccidial drugs coupled with effective hygiene means that most losses result from chronic and subclinical infections causing reduced growth and egg production.

The life cycle is monoxenous (i.e., involves a single host) and begins when a bird ingests an infective oocyst with its food or in its water. The oocyst releases infective sporozoites when it reaches the small intestine. Gut peristalsis moves sporozoites down the intestinal tract with the digesta and once they reach the caecum, they invade the intestinal epithelial cells. In common with the other species of *Eimeria*, a membrane bound parasitophorous vacuole of host cell origin surrounds all the intracellular stages. The incorporation of parasite proteins into the vacuole membrane prevents the fusion of lysosomes or other vesicles. The sporozoites travel through the epithelial cells and emerge into the lamina propria where macrophages promptly ingest them. The macrophages act as a sort of transport host and move the parasites to the glands of Lieberkuhn where they escape and invade the glandular epithelial cells. Within the epithelial cells, the parasites transform into meronts and undergo a form of asexual reproduction called merogony to produce numerous merozoites. This kills the host cell and releases the merozoites to invade other caecal epithelial cells within which they produce another generation of merozoites. The parasites kill these host cells and following their release, these second-generation merozoites invade new epithelial cells. However, at this point, for some reason, the subsequent development can follow one of two paths. Some of the merozoites will give rise to a third generation of merozoites, whilst others undergo gametogony to produce become either macrogametocytes (female) or microgametocytes (male). The microgametocytes leave their host cell and invade those containing a macrogametocyte and fuse with it to effect fertilization. After fertilization, the macrogamete transforms into a zygote and then into an oocyst. The oocyst contains only a single cell – referred to as the sporont. The oocyst leaves the host bird in its faeces but is not infectious at this stage. It now undergoes sporogony in which four sporocysts each of which contains two sporozoites develop. This takes two or more days depending upon the environmental temperature. Therefore, prompt removal of faeces and good farm hygiene can effectively prevent the transmission of disease both within and between rearing sheds.

Serious disease primarily affects young poultry particularly those between 3 and 8 weeks of age but older birds infected for the first time later in life also suffer badly. Infected birds become listless, cease to feed, and huddle together to keep warm. Damage to the caecum results in bleeding into the gut and stains the birds' faeces with blood. The damage allows secondary invasion by bacteria present naturally in the gut, and this extends the lesions and causes further pathology. Acutely infected birds often die from blood loss 5–6 days after infection. In addition to haemorrhages, the gut swells and thickens, so it appears 'sausage-like'. Birds that are still alive 9 days after infection will usually recover: a caseous (cheese-like) plug may form in the lumen of their caecum, which is voided with the faeces.

Recovering birds develop protective, species-specific, cell-mediated immunity to re-infection based on CD4+ and CD8+ T cells found in the lymphoid tissues associated with the gut (Shirley et al. 2007). Co-infections with two or more species of *Eimeria* do not necessarily compromise the development of immunity (Jenkins et al. 2009).

3.5.2 Genus *Isospora*

Taxonomic revisions have split the genus *Isospora* into two: those that do not express a tissue cyst stage in their life cycle remain in the genus *Isospora*, whilst those that do have a tissue cyst stage form the genus *Cystoisospora*. Most of the several hundred species belonging to the genus *Isospora* are parasites of birds. They are all monoxenous and therefore complete their life cycle in a single host – although some may exploit paratenic hosts to effect transmission. Berto et al. (2011) provide a review of both *Isospora* and *Eimeria* that infect birds. *Isospora* do not appear to affect chickens and other poultry, but they can be pathogenic in finches, sparrows, and other passerine birds, as well as mynahs and starlings.

3.5.3 Genus *Cystoisospora*

Members of this genus formerly belonged to the genus *Isospora*. The presence of a tissue cyst stage within their life cycle distinguishes them from parasites of the genus Isospora. They are monoxenous although some species (e.g., those infecting cats) exploit paratenic hosts. The genus includes several species that infect mammals and us. For example, *Cystoisospora canis* causes diarrhoea in puppies, *Cystoisospora suis* causes neonatal piglet diarrhoea, and *Cystoisospora belli* is an important human pathogen.

3.5.3.1 *Cystoisospora (Isospora) belli*

This species occurs throughout the world but is particularly common in tropical and subtropical countries. The first cases were recognised in people returning from the battlefields during the First World War, hence the species name (Latin *bellum* = war). It only infects humans and there do not appear to be any paratenic hosts (Dubey and Almeria 2019). It causes diarrhoea and is frequently associated with persons suffering from AIDS or immunosuppressive illnesses.

The life cycle begins when we ingest oocysts (17–37 × 8–21 µm) containing infective sporozoites (Figure 3.13). In common with other members of the genus, each oocyst contains two sporocysts and each sporocyst contains four sporozoites. The sporozoites emerge in the small intestine and invade the epithelial cells where they transform into merozoites that divide to produce more merozoites that in turn divide in a cycle that repeats many times. This destroys the host cell and releases the parasites so that they can invade new host cells. In serious infections, this results in the loss of large areas of the gut lining and allows secondary invasion by gut microbes. Presumably, the parasites gain access to the circulation because tissue cysts develop in variety of organs distant from the gastrointestinal tract. For example, within the spleen, liver, and lymph nodes. Each cyst contains a single parasite (zoite) (i.e., they are monozoic) that resembles a typical coccidian sporozoite. The formation of tissue cysts usually occurs in immunocompromised patients but their role in pathology and relapses is uncertain. There is a suggestion that in pigs, the formation of tissue cysts by *C. suis* contributes to the maintenance of immunity to the parasite (Shrestha et al. 2015). Nowadays,

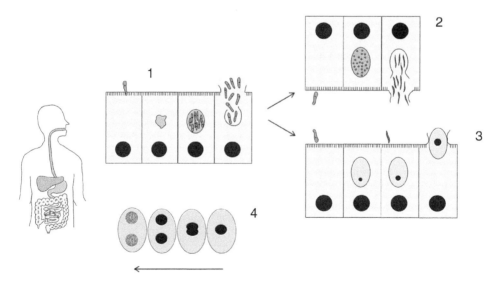

Figure 3.13 Life cycle of *Isospora belli*. 1: Infection commences following ingestion of a sporulated oocyst. The oocyst releases eight sporozoites when it reaches the small intestine and these invade the gut epithelial cells where they transform into merozoites. The merozoites divide asexually to produce more merozoites. 2: Merozoite forming a multinucleate meront that gives rise to microgametocytes (male). 3: Merozoite giving rise to a macrogametocyte (female). Fusion of a male and a female gamete results in the formation of a zygote, and this develops into an oocyst. 4: Sporulation of the oocyst occurs before it is released in the host's faeces. Drawings not to scale.

few animals get the opportunity to eat humans, so the tissue cysts probably play no role in transmission of *C. belli*. The importance of tissue cysts in the transmission of other mammalian *Cystoisospora* species is uncertain.

At some point, the merozoites transform into multinucleate meronts and these give rise to microgametes (male) or macrogametes (female). Unless the microgametes and macrogametes occur within the same cell (which is possible), the microgametes leave their host cell to locate a macrogamete and their fusion results in the formation of a zygote that then develops into an oocyst. The death of the host cell releases the oocyst into the gut lumen and leaves the body with the faeces. The oocyst then undergoes sporulation to produce two sporocysts, each of which contains four sporozoites. Transmission is therefore passive by faecal–oral contamination. Environmental conditions probably heavily influence the time taken for sporulation. For example, in some *Isospora* species, temperatures below 20 °C inhibit sporulation, but it takes less than 16 hours at 30 °C. Nevertheless, transmission is probably through infected food or water rather than direct contact. For example, through not washing one's hands after defaecation or by contacting an article touched by such a person. This is therefore distinct from *Cryptosporidium parvum* in which the oocysts are immediately infective after shedding.

3.5.4 Genus *Cyclospora*

This genus currently contains 13 species that are parasitic in a range of vertebrates (e.g., snakes, rodents), but the most important is *Cyclospora cayetanensis* that only infects humans.

3.5.4.1 *Cyclospora cayetanensis*

First described in 1979, this is an emerging parasitic disease with a cosmopolitan distribution (Almeria et al. 2019). There are regular outbreaks in USA and Canada associated with the importation of fresh berries (Dixon et al. 2016). The life cycle begins with the ingestion of oocysts (8–10 μm diameter) containing infective sporozoites. Once the oocysts reach the duodenum and upper small intestine, they release the sporozoites and these invade the gut epithelial cells. They then transform into type I merozoites that then divide to form type II merozoites. The type II merozoites transform into meronts and these give rise to microgametocytes and macrogametocytes. Fusion of a microgametocyte with a macrogametocyte gives rise to zygote, and this develops into an oocyst that is shed with the faeces. The oocyst then undergoes sporulation to produce two sporocysts each containing two sporozoites. Sporulation takes 7–15 days depending upon the environmental conditions and therefore transmission probably occurs through contamination of food and water rather than direct person-to-person/contaminated object contact.

The symptoms of infection are non-specific and resemble those of many other gastrointestinal diseases. Patients suffer from abdominal pain, watery diarrhoea, flatulence, low-grade fever, anorexia, and weight loss. In endemic regions, the symptoms tend to be worse in young children and in non-endemic regions most people who become infected express symptoms. Persons who are immunosuppressed and AIDS patients are more seriously affected.

3.5.5 Genus *Sarcocystis*

Members of this genus are obligate intracellular parasites with a life cycle that involves two hosts – a herbivore intermediate host in which only asexual multiplication occurs and a carnivore definitive host in which sexual reproduction takes place. Most *Sarcocystis* species are very host-specific and infect a limited number of closely related intermediate/final hosts (Table 3.3). You will not be

Table 3.3 Summary of the most important species of *Sarcocystis* in human and veterinary medicine.

Species of *Sarcocystis*	Synonym	Intermediate host	Definitive host
Sarcocystis bovicanis	*Sarcocystis cruzi*	Cattle	Dogs and other canines
Sarcocystis bovihominis	*Sarcocystis hominis*	Cattle	Humans
Sarcocystis bovifelis	*Sarcocystis hirsuta*	Cattle	Cats and other felines
Sarcocystis Suihominis	*Isospora hominis*	Pigs	Humans and some primates
Sarcocystis ovifelis	*Sarcocystis tenella*	Sheep	Cats and other felines
Sarcocystis hovathi	*Sarcocystis gallinarum*	Chicken	Dogs and other canines

surprised to hear that the taxonomy is confused, and one must take care when using the older literature. For example, some animals once thought to harbour just a single *Sarcocystis* species actually contain several species. In addition, some species are morphologically identical, and others have synonyms (e.g., *Sarcocystis cruzi* is also known as *Sarcocystis bovicanis* – a reflection that the intermediate and definitive hosts are cattle and dogs/ other canines respectively).

The life cycle of *S. bovicanis* is typical of most *Sarcocystis* species. In this species, the definitive host is a dog or other canine, whilst the intermediate hosts are cows and other bovids (Figure 3.14). The life cycle begins when an infected dog sheds free sporocysts or oocysts in its faeces. A cow must then consume the sporocysts/oocysts, and this usually happens through contamination of food or water. When the sporocyts/oocysts reach the cow's small intestine, they release the sporozoites. The sporozoites then invade the gut epithelial cells and make their way to the blood vessels. The bloodstream then distributes them around the body. The parasites invade the endothelial cells of the blood vessels that serve many of the body's organ systems. Within the endothelial cells, the parasites transform into merozoites and undergo four cycles of merogony (asexual reproduction). After each cycle, the newly formed merozoites are released, and these infect new endothelial cells downstream of the initial infection. After the last cycle, the merozoites invade skeletal and cardiac muscle cells. Occasionally, smooth muscle, the brain, and spinal cord are also infected. Within these cells, the merozoites transform into metrocytes or 'mother cells' each of which divides asexually to form a structure called a sarcocysts (Figure 3.15). With repeated asexual division, a sarcocyst steadily becomes larger and larger and in some *Sarcocystis* species may become big enough to be visible to the naked eye. Eventually, the globular metrocytes cease producing new metrocytes and form crescent-shaped bradyzoites. The time taken for this depends upon the species of *Sarcocystis* but can be around 2 months. During this time, the sarcocysts are non-infectious since only bradyzoites can transmit the infection. Completion of the life cycle requires a dog to consume flesh containing the bradyzoites. Digestion of the sarcocyst within the dog's small intestine releases the bradyzoites, and these become motile. The bradyzoites initially invade the gut epithelial cells and then make their way to the *lamina propria* region where they transform into either male or female gametes. After gamete fusion, the parasites undergo sporogony to form oocysts that contain two sporocysts. The oocysts are therefore already sporulated when shed and each contains four sporozoites. The oocysts are shed into the lumen of the gut and passed with the faeces. The oocyst has a thin wall and often ruptures when one is preparing faecal samples for microscopy. Consequently, one normally sees sporocysts (16.3 × 10.8 μm) in faecal samples. *Sarcocystis* seldom causes serious pathology in its definitive hosts.

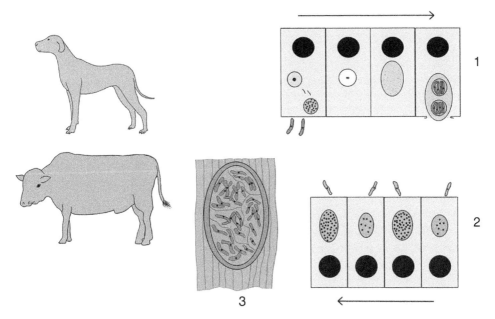

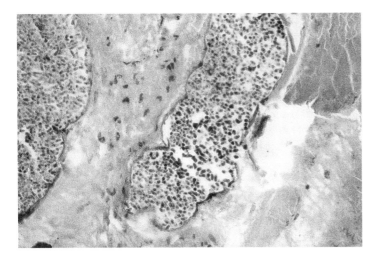

Figure 3.14 Life cycle of *Sarcocystis bovicanis*. 1: Digestion of a sarcocyst within the dog's small intestine, releases bradyzoites that invade the gut epithelial cells and then make their way to the *lamina propria* region where they transform into either male or female gametes. After gamete fusion, the parasites undergo sporogony to form oocysts that contain two sporocysts. The oocysts are shed into the lumen of the dog's gut and pass with the faeces. 2: A cow consumes the sporocysts/oocysts, and these release the sporozoites that invade its gut epithelial cells and make their way to the blood vessels. The parasites invade the endothelial cells of the blood vessels, transform into merozoites, and undergo four cycles of merogony. After each cycle, the newly formed merozoites infect new endothelial cells. 3: After the last cycle, the merozoites invade skeletal and cardiac muscle cells and transform into metrocytes, each of which divides asexually to form a sarcocyst. Eventually, the metrocytes cease producing new metrocytes and form bradyzoites. Completion of the life cycle requires a dog to consume flesh containing the bradyzoites. Drawings not to scale.

Figure 3.15 Transverse section through a sarcocysts of *Sarcocystis muris* in the trachea of a mouse.

Human *Sarcocystis* Infections

Humans are the definitive hosts for *S. bovihominis*, *S. suihominis* and several other species of *Sarcocystis* that we usually acquire from eating raw or poorly cooked meat. As definitive hosts to these parasites, we suffer from intestinal infections. The symptoms are non-specific and typically include nausea and diarrhoea. The infections are usually self-limiting and seldom serious.

We can also act as intermediate hosts for some *Sarcocystis* species. For example, we can act as an intermediate host for *Sarcocystis nesbitti* although in most cases the species responsible is uncertain. The definitive hosts for *S. nesbitti* are probably snakes or other reptiles. Therefore, in common with the other species for which we are intermediate hosts, we are 'dead end' hosts because few animals have the opportunity to eat us. Presumably, we suffer accidental infections with the sporocysts/oocysts through contamination of food or water and the normal intermediate hosts are other species of primates. The symptoms of infection depend upon the site at which the sarcocysts grow and their abundance. Typically, they induce inflammatory responses that result in pain, fever, and swelling at the infected site. There are reports of regular outbreaks of human sarcocystosis amongst tourists visiting parts of Malaysia (Fayer et al. 2015). Whether these link to one or more species of *Sarcocystis* is uncertain.

In intermediate hosts, the consequences of infection vary between species, the level of challenge, and the species of *Sarcocystis* parasitizing them. However, most pathology is usually associated with damage caused to the vascular epithelium during the second stage of merogony. Heavy infections of *S. bovicanis* in cattle can result in widespread haemorrhages afflicting virtually every organ in the body. This results in anaemia, emaciation, and the animal may become anorexic; abortion can occur in breeding cattle. The immune response results in lymphadenopathy and submandibular oedema whilst the hair at the end of the tail is often lost. Most infections in domestic livestock, however, are subclinical and not discovered until the sarcocysts are detected during meat hygiene inspections after the animal is slaughtered.

3.5.6 Genus *Toxoplasma*, *Toxoplasma gondii*

This genus contains only one species, *Toxoplasma gondii*. However, it has a remarkable host range and can probably infect all mammals and birds.

This intracellular parasite was initially described from a desert rodent, the North African *gondi* (*Ctenodactylus gondi*) [sometimes spelled '*gundi*' – vowels are not used in written Arabic, so some words appear in various spellings when translated into English] in Tunisia 1908, but the life cycle was not established until 1969–1970. The name *Toxoplasma* has nothing to do with toxins but derives from the curved shape of the tachyzoite stage of the life cycle (Figure 3.16). '*Toxon*' (Τόξον) is the Greek word for a 'bow or something that is crescent shaped and '*plasma*' (πλάσμα) is Greek for 'creature'. Since these humble beginnings, it has become apparent that not only does *T. gondii* have a cosmopolitan distribution but also it has perhaps the widest host range of any parasite.

The life cycle of *T. gondii* has two parts and three infectious stages (Figure 3.17). The two parts of the life cycle are the sexual cycle that occurs within cats and other felines that are the definitive

Figure 3.16 *Toxoplasma gondii* tachyzoites. This protozoan parasite has an unusually large host range that encompasses most warm-blooded vertebrates although sexual reproduction only occurs in the intestine of cats.

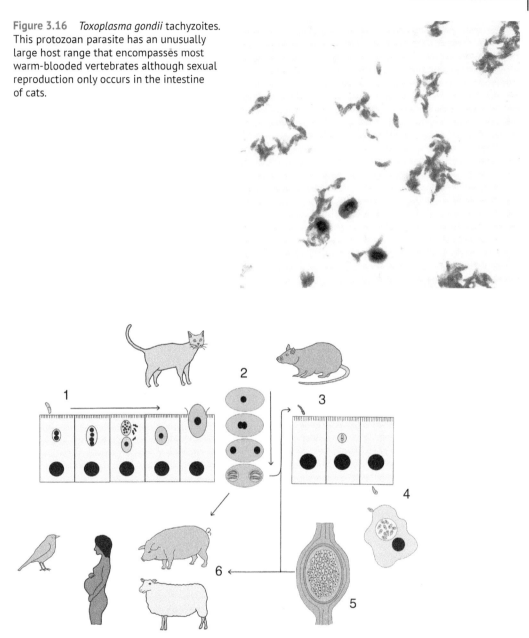

Figure 3.17 Life cycle of *Toxoplasma gondii*: 1: Cats acquire their infection by consuming either sporulated oocysts passed in another cat's faeces or an intermediate host containing the tachyzoites and bradyzoites. Following ingestion, the parasites invade the epithelial cells lining the cat's small intestine and multiply by endodyogeny followed by endopolygeny. Next, comes gametogenesis and the formation of microgametes and macrogametes. Fusion of the gametes results in the formation of a zygote, and this develops into an oocyst that is shed with the faeces whilst it is still unsporulated. 2: Sporogony happens outside the host. 3: When an infectious oocyst reaches the small intestine of an intermediate host, it releases the sporozoites and these then invade the gut epithelial cells. 4: They then leave these cells and invade macrophages and many other cell types within which they transform into tachyzoites which multiply by endodyogeny. 5: After a series of parasite division cycles, they form tissue cysts (zoitocysts) containing bradyzoites. If consumed, the tachyzoites and bradyzoites are infectious to both other intermediate hosts and to cats. 6: Virtually all warm-blooded animals can act as intermediate hosts. Trans-placental transmission in humans can lead to abortion and birth defects. The importance of birds as intermediate hosts in transmission dynamics is uncertain. Drawings not to scale.

hosts and the asexual cycle that occurs in the intermediate hosts – which are virtually any warm-blooded animal, including us. The three infectious stages are the sporulated oocysts that contain the sporozoites, the tachyzoites, and the tissue cysts that contain bradyzoites: all three stages are infectious to both the feline definitive hosts and the intermediate hosts. Cats acquire their infection by consuming either sporulated oocysts passed in another cat's faeces (e.g., through contamination) or an intermediate host containing the tachyzoites and bradyzoites. Following ingestion, the parasites invade the epithelial cells lining the cat's small intestine and undergo a complex series of asexual divisions. The first step is multiplication by endodyogeny – this involves two daughter cells developing within a mother cell and consuming her before separation takes place. After endodyogeny comes reproduction by endopolygeny in which the mother cell produces several daughter cells simultaneously. Next, comes gametogenesis and the formation of microgametes and macrogametes. Fusion of the gametes results in the formation of a zygote, and this develops into an oocyst that is shed with the faeces whilst it is still unsporulated. Sporogony happens outside the host and takes about 2–3 days under ideal conditions. Fresh faeces are therefore not infectious because the oocysts first need to undergo sporogony to form sporocysts. The infectious sporulated oocysts are oval (10–13 × 9–11 μm) and contain two sporocysts each of which contains four sporozoites. The sporocysts can remain infectious in the environment for months and possibly even 1–2 years. The shedding of oocysts starts 3–10 days after initial infection and continues for only 1–2 weeks although some estimates suggest that a cat may shed up to 100 million oocysts in its faeces during that time. After an initial infection, a cat is usually resistant to re-infection. The prolonged survival of the oocysts in the environment therefore compensates for the short period over which an infected cat sheds them in its faeces. In some cats, developmental stages may persist in a dormant state and give rise to another batch of oocysts following their subsequent reactivation. This could happen because of changes in the cat's immune status. Immunity can also wane naturally with time and a cat may suffer another *T. gondii* infection several years after the first challenge.

Within infected cats, *T. gondii* also invades other tissues, including the muscles and nervous tissues where they divide asexually to produce tachyzoites and tissue cysts containing bradyzoites – just as in the intermediate hosts. If the cat is pregnant, *in utero* infection of the developing kittens may occur. Most cats, however, acquire their *T. gondii* infection through preying on mice, rats, and other rodents. As might be expected, the prevalence of infection is higher in stray cats and those that are good mousers. Many bird species, from sparrows and pigeons to ducks and owls, are naturally infected with *T. gondii*, but they are probably not as important as rodents as sources of infection to cats.

For *T. gondii* infection to circulate between cats and rodents, it is essential that the rodents ingest the oocysts. Domestic cats normally bury their faeces and although this could theoretically reduce the chances of contaminative transmission, it could contribute to oocyst survival by reducing exposure to environmental conditions such as desiccation and UV light. The smell of cat faeces repels mice and voles, so the oocysts must survive until the faeces breaks down and disperses in the soil. Dogs sometimes consume cat faeces (Lewin 1999), and therefore one might expect them to be at risk of serious infections. The seroprevalence of *T. gondii* in dogs is often high, especially in strays (e.g., Valenzuela-Moreno et al. 2020), but this is probably due to their scavenging of dead animals and waste food rather than consumption of cat faeces – although the latter habit is unlikely to help.

Contaminated water sources are responsible for some outbreaks of toxoplasmosis in humans and many wild and domestic animals probably become infected in the same way. Presumably, the oocysts wash from the soil into ponds, streams, reservoirs, and they can survive the chlorination of drinking water and sewage treatment. The oocysts also survive for prolonged periods in brackish water and at least a year in seawater (Shapiro et al. 2019).

When an infectious oocyst reaches the small intestine of an intermediate host, it releases the sporozoites, and these then invade the gut epithelial cells. They then leave these cells and invade macrophages and many other cell types within which they transform into tachyzoites, which multiply by endodyogeny. *Toxoplasma gondii* does not invade the anucleate red blood cells of mammals. However, they do parasitize the nucleated red blood cells of birds. Within an infected cell, the parasites multiply until they consume the whole cell and all that is left is the cell membrane. At this point, the host cell is called a pseudocyst. Ultimately, the pseudocyst membrane breaks down, thereby releasing the tachyzoites that invade new host cells and repeat the process. As in other apicomplexan parasites, the tachyzoites actively invade their host cells. Having made initial contact with a suitable cell, the tachyzoites re-orientate themselves and then discharge the contents of their micronemes, rhoptries, and dense granules. This enables the parasite to attach to the host cell surface after which it forces its way inside and becomes enclosed within a parasitophorous vacuole. Because *T. gondii* infects such a wide variety of types of cells and species of animal it presumably identifies host cell receptors that have a widespread distribution in warm-blooded vertebrates. One of the microneme proteins, TgMIC1, has a cell binding motif called 'microneme adhesive repeat' (MAR) that binds to a group of carbohydrates called sialylated oligosaccharides. Sialic acid oligosaccharides are common components of cell surface membranes and play important roles in a variety of carbohydrate-mediated cell surface interactions.

After a series of parasite division cycles, the host's immune system starts to exert an effect, and this stimulates *T. gondii* to form tissue cysts (zoitocysts) containing bradyzoites. These tissue cysts develop predominantly within nervous (e.g., central nervous system and eyes) and skeletal and cardiac muscle tissue, but they may develop elsewhere. The tissue cysts probably persist for life in some intermediate hosts. Prolonged infections may also result from periodical re-activation, transformation into tachyzoites, followed by the formation of new tissue cysts. In addition to infection through consuming oocysts, intermediate hosts may also acquire a *T. gondii* infection through consuming meat containing the tachyzoite and/or bradyzoites. In humans, this occurs through consuming raw or undercooked meat. In these instances, the parasites invade the gut epithelial cells, and the life cycle continues as described above. Human infections may also result from blood transfusions and organ transplants (Robert-Gangneux et al. 2018).

In pregnant mammals, vertical transmission occurs when tachyzoites cross the placenta and infect the developing embryo. There is extensive documentation of transplacental transmission of *T. gondii* infections in humans but the extent to which it occurs in other animals, and its importance in the epidemiology of the parasite is uncertain. Acquisition of primary toxoplasmosis during pregnancy can be a significant cause of congenital infection. Consequently, in some countries, such as France, screening for the infection is a routine part of antenatal care. Although there are sporadic calls for *Toxoplasma* testing for all pregnant women in the United Kingdom, the incidence is too low to make it cost effective. Instead, the emphasis is on identifying and treating individual women. *Toxoplasma gondii* is susceptible to treatment with antibiotics such as spiramycin that are relatively safe to administer during pregnancy and help to reduce the chances of congenital transmission, following the exposure of the mother to infection during the early stages of pregnancy. However, the outcome is often not a happy one; if that foetus is infected it can lead to spontaneous (or therapeutic) abortion or a severely handicapped baby.

Population structure of *Toxoplasma gondii*

The population structure of *T. gondii* is highly clonal (Galal et al. 2019). In Europe and North America, most strains of *T. gondii* belong to one of three distinct genotypes: Types I, II, and III. However, other genetically distinct strains exist in Brazil and French Guiana. There is limited variability within genotypes I, II, and III, and this suggests that they probably arose relatively recently. Furthermore, this polymorphism appears limited to just two alleles called A and E – short for Adam and Eve – on the basis that all extant strains derive from mixtures of two ancestral founder strains. The continued existence of these separate strains probably results from the parasite's ability to maintain asexually reproducing populations, possibly indefinitely, by horizontal and vertical transmission between intermediate hosts and the limited opportunity for genetic mixing during sexual reproduction within the cat definitive host. Because *T. gondii* is haploid, genetic mixing only occurs if a cat has simultaneous infections with two different strains of the parasite. Furthermore, the development of the two strains would have to be synchronous because the production of gametes occurs over a relatively short period. This scenario might occur if the cat consumed a mouse harbouring two or more strains of the parasite or two mice harbouring different strains in a short period. There is limited evidence of intermediate hosts harbouring mixed strain *T. gondii* infections, but there is a need for more work in this area.

3.5.7 Genus *Neospora*

This is a genus of emerging parasites. The first description was in Norway in 1984 when a cyst-forming protozoan parasite was associated with the development of encephalomyelitis and myositis in a litter of puppies. Since then, the parasite responsible has been given the name *Neospora caninum* although Heydorn and Melhorn (2002) consider this to be a '*nomen nudum*' – that is it fails to conform to International Code of Zoological Nomenclature. Despite the taxonomic uncertainty, most workers continue to refer to the parasite as *N. caninum* and we shall do the same. Presumably, up until 1984, the parasite was overlooked because it rapidly became recognised as having a cosmopolitan distribution and being a major cause of abortion in cattle. Another species of *Neospora, Neospora hughesi,* causes myeloencephalitis in horses and donkeys but we currently know very little about its biology.

3.5.7.1 *Neospora caninum*
Neospora caninum is essentially a disease (neosporosis) of dogs and cattle although many other mammals may also be infected. On a worldwide basis, it causes losses of hundreds of millions of pounds per year and is a particular problem in cattle-breeding countries (Liu et al. 2020). It parasitises many wild animals from mice to dugongs although the consequences of infection for wild animals are uncertain.

The life cycle of *N. caninum* resembles that of *T. gondii.* Apart from having different definitive hosts, the two parasites occupy a very similar ecological niche. It is therefore not surprising that competition probably exists between the two species in the intermediate hosts (Sundermann and Estridge 1999). Sexual reproduction in *N. caninum* occurs within dogs and other canines (rather than cats in the case of *T. gondii*) and they are therefore the definitive host. A separate asexual cycle occurs within cattle, particularly dairy cattle, but also within several other mammals, including dogs, in which transmission takes place vertically via the placenta, rodents, shrews, pigs, and camels. Although the parasite will grow in laboratory human cell line cultures, and some reports mention finding low antibody titres to the parasite in human serum samples, there are no confirmed cases of human infection to date (Calero-Bernal et al. 2019).

Dogs normally acquire an infection by consuming meat containing the tissue cysts. When these cysts reach the small intestine, they release bradyzoites that invade the intestinal cells and reproduce by schizogony to form schizonts. The subsequent stages leading to the formation of the oocysts are a bit uncertain. Presumably, the schizonts give rise to microgametes and macrogametes that then fuse and develop into oocysts. The oocysts (10–11 μm) are subsequently shed with the dog's faeces and sporulate in the environment to produce two sporocysts, each containing four sporozoites.

A cow (or other intermediate host) acquires an infection when it ingests a sporulated oocyst – for example, whilst grazing or through contaminated water. The oocyst releases the sporozoites when it reaches the cow's small intestine, and these invade the intestinal epithelium and ultimately find their way to a wide variety of cell types around the body. The majority of parasites locate themselves in the reticulo-endothelial system where they transform into tachyzoites that multiply asexually by endodyogeny until they kill their host cell after which they invade other cells. Healthy cows usually do not exhibit any signs of infection. The parasites induce a strong gamma-interferon-based cell-mediated immune response, and this causes *N. caninum* to form tissue cysts containing bradyzoites. These tissue cysts are approximately 100 μm in size and occur predominantly in nervous tissue, but they may be located elsewhere. If the cow is pregnant at the time of her initial infection, the tachyzoites can cross the placenta and infect the developing embryo. Alternatively, if she becomes pregnant after being infected, the resting tissue cysts are activated because of the normal reduction in cell-mediated immunity that happens during the first trimester of pregnancy. Within the tissue cysts, the bradyzoites transform into tachyzoites that find their way to the developing calf. The consequences for the developing calf depend upon the number of invading tachyzoites and the stage of pregnancy. The earlier in pregnancy that the calf is infected the worse the prognosis. Infection during the first and second trimester is most likely to result in abortion, whilst infection during the third trimester may apparently have no harmful effects – although 80–95% of calves born to infected mothers are infected at birth. Occasionally, a calf infected during the third trimester may abort or be born with neurological symptoms, such as paralysis. Within infected but otherwise healthy calves, the parasite forms dormant tissue cysts. When infected female calves reach adulthood and become pregnant, these cysts are activated and the parasite transfers to the next generation. Although *N. caninum* can cause abortion and birth defects in sheep and goats, there are fewer published reports on this than in cattle.

The colostrum of infected cows may contain *N. caninum* tachyzoites, but these are unlikely to be important for the transmission of the infection. This is because over 80% of calves born to infected mothers are already infected at birth. Similarly, although the parasite is detectable in the semen of infected bulls, it is uncertain whether it is sexually transmissible – and if it is whether it happens sufficiently often to be a significant epidemiological factor.

Because so many calves are born infected, this is probably the principal means by which the infection is maintained within herds. This would occur through the normal farm practice of retaining a proportion of the female calves as replacement heifers (female cows). However, modelling studies suggest that this is not sufficient to maintain the infection over prolonged periods and there must be periodic horizontal transmission from infected dogs. Because farm dogs are often closely associated with cattle herding, dog: cow transmission cycles are probably common.

Within dogs, tissue tachyzoite stages also form and neurological symptoms occur if nervous tissues are infected. In addition, in pregnant bitches, the tachyzoites can invade across the placenta and infect the developing pups. The consequences of congenital infection in dogs can result in the pups being born with serious damage to nervous and skeletal muscle tissue that results in paralysis.

3.5.8 Genus *Cryptosporidium*

The taxonomy of the genus *Cryptosporidium* has undergone several re-arrangements, so one must take care when reading the literature. Although traditionally considered coccidian parasites, albeit somewhat aberrant ones (e.g., they lack an apicoplast), molecular and ultrastructural studies indicate that cryptosporidians have much closer affinities to the gregarines. Consequently, the genus no longer resides amongst the Coccidia and currently lives in a new subclass, the Cryptogregaria (Ryan and Hijawa 2015; Ryan et al. 2016). The gregarines are a group of intracellular parasites that infect a wide range of invertebrates but not vertebrates. For more details on gregarines, see Desportes and Schrével (2013).

There are numerous *Cryptosporidium* species, and they include parasites of amphibia, reptiles, birds, and mammals *Cryptosporidium* (Fayer et al. 2018). Several species are zoonotic (Table 3.4) but humans are predominantly infected by just two species: *Cryptosporidium parvum* and *Cryptosporidium hominis*. Both species express different subtypes that vary in their distribution and host preferences (Nader et al. 2019). In immunocompetent adults, cryptosporidiosis causes watery diarrhoea, abdominal cramps, and a slight fever, but it is usually self-limiting. However, in children under 5 years of age, it can be highly pathogenic, and it is the second most common cause of death from diarrhoea after rotavirus infections. It is also potentially fatal in those who suffer from AIDS or are immunocompromised.

The life cycle of *C. parvum* is typical for the genus. When an oocyst reaches the small intestine or the respiratory system, it releases four sporozoites and these adhere to the surface mucus lying above the host epithelial cell's apical membrane. They then release enzymes that break down the mucus, thereby enabling the sporozoites to contact and bind to the host cell membrane through receptor-ligand interactions. The sporozoites then discharge the contents of their apical organelles. These include chemicals that induce the host cell membrane to form protrusions that surround the parasite and encapsulates it within a parasitophorous vacuole. Although the host cell initially encapsulates *Cryptosporidium*, the membrane that ultimately surrounds the parasitophorous vacuole contains both parasite and host cell components. This parasitophorous vacuole sits just inside the host cell apical membrane and is separate from the host cytoplasm. This location is unusual as

Table 3.4   Selected species of *Cryptosporidium* to indicate their usual host and zoonotic potential.

Species name	Usual host	Zoonotic potential
Cryptosporidium muris	Mousc	Yes
Cryptosporidium parvum	Mouse	Yes
Cryptosporidium meleagridis	Turkey	Yes
Cryptosporidium wrairi	Guinea pig	No
Cryptosporidium felis	Cats	Yes
Cryptosporidium serpentis	Snakes	No
Cryptosporidium baileyi	Chicken	No
Cryptosporidium saurophilum	Skink	No
Cryptosporidium andersoni	Cattle	Yes
Cryptosporidium canis	Dog	Yes
Cryptosporidium hominis	Humans	Yes

it means that the vacuole is effectively extracytosolic. There is a spot, known as the annular ring, where the host apical membrane and the parasitophorous vacuole membrane fuse. In addition, the parasite develops a unique organelle called a 'feeder organelle' through which it derives nutrients from the host cell. Bones et al. (2019) provide more details of the invasion process. Within the parasitophorous vacuole, the sporozoite transforms into a trophozoite and then into a meront that divides asexually by merogony to produce type 1 meronts that contain 8 merozoites and then type 2 meronts that contain four merozoites. Further cycles of merogony can take place or the type 2 meronts may undergo gametogony to produce microgamonts (male) and macrogamonts (female). The microgamonts produce microgametocytes that leave their host cell and search for macrogamonts with which they fuse to form a zygote. About 20% of zygotes form thin-walled oocysts that re-infect the host. The other 80% of zygotes develop into thick-walled oocysts that are shed into the environment where they are immediately infectious although they can survive for prolonged periods. Unlike *Cyclospora*, *Isospora*, and *Sarcocystis*, parasites of the genus *Cryptosporidium* do not form sporocysts.

Most people acquire their *Cryptosporidium* infection through ingesting oocysts transmitted by the faecal–oral contamination. However, infections also occur through inhaling the oocysts and autoinfection. On rare occasions, pulmonary infections occur, and these patients sometimes shed oocysts in their sputum. Cryptosporidiosis outbreaks are often associated with contamination of drinking water and infections of wild and domestic animals probably also occur through this route. The oocysts can survive in seawater and are resistant to chlorination. Consequently, infection can result from ingesting oocysts whilst swimming in recreational pools, lakes, or the sea. Flies and other invertebrates may also pick up and transmit the oocysts to food. Clams, oysters, and other shellfish concentrate *Cryptosporidium* oocysts from the surrounding water, but it is uncertain whether they pose a significant source of human infections. Detection of oocysts in patients' samples or water sources using a light microscope is difficult because they are only ~5 μm long and often present in low numbers.

4

Parasitic Protozoa Part B: Phylum Kinetoplastida; Parasitic Algae and Fungi

4.1   Introduction

In this chapter, we will introduce just a few of the kinetoplastid parasites that are important in human and veterinary medicine. This is a remarkable group of protozoa that includes parasites of plants, invertebrates, and vertebrates. Their transmission strategies range from contamination to vector assisted and sexually transmitted. We do not usually think of algae as parasitic organisms but there are a few that have adopted this lifestyle and some even parasitise us. This should not come as a surprise as several notorious protozoan parasites, such as *Plasmodium*, probably evolved from algal ancestors. The fungi have a Kingdom of their own and are normally considered the preserve of mycologists. However, the microsporidia are something of a special case. Originally classed as protozoa, they fell within the remit of parasitologists. Now, their subsequent reclassification as fungi presents parasitologists with a bit of dilemma. Namely, do they still belong within parasitology textbooks when yeast infections etc. are firmly excluded? At the time of writing, the classification of the Microsporidia is being called into question again, and therefore, we feel justified in including them here.

4.2   Phylum Kinetoplastida

The Kinetoplastida is a large diverse group of protozoa that includes both plant and animal parasites (Table 4.1). Some authors consider the Kinetoplastida to be a phylum, while others refer to it as a class or an order. They are commonly known as the trypanosomes from the genus *Trypanosoma* that includes the causative agents of Human African Trypanosomiasis (HAT) and several other parasites of medical and veterinary importance. The genus name *Trypanosoma* derives from the

Parasitology: An Integrated Approach, Second Edition. Alan Gunn and Sarah J. Pitt.
© 2022 John Wiley & Sons Ltd. Published 2022 by John Wiley & Sons Ltd.
Companion website: www.wiley.com/go/gunn/parasitology2

Table 4.1 Examples of kinetoplastid parasites of medical, veterinary, and agricultural importance and the diseases they cause.

Genus	Example	Host	Vector/transmission	Disease
Leishmania	Leishmania donovani	Humans, dogs, rats	Phlebotomus sandflies	Kala-azar (visceral leishmaniasis)
	Leishmania major	Humans, monkeys, dogs, rodents	Phlebotomus sandflies	Cutaneous leishmaniasis
	Leishmania tropica	Humans, monkeys, dogs, rodents	Phlebotomus sandflies	Cutaneous leishmaniasis
	Leishmania braziliensis	Humans, sloths, monkeys, opossums, and many others	Lutzomyia sandflies	Cutaneous/mucocutaneous leishmaniasis
Trypanosoma	Trypanosoma brucei gambesiense	Humans	Tsetse flies (Glossina spp.)	African trypanosomiasis (sleeping sickness)
	Trypanosoma congolense	Cattle	Tsetse flies (Glossina spp.)	African trypanosomiasis (nagana)
	Trypanosoma equiperdum	Horses	venereal	Dourine
	Trypanosoma cruzi	Humans, dogs, cats, rats, and many others	Triatomid bugs	Chagas disease
Phytomonas	Phytomonas staheli	Coconut palms	Lincus lobuliger (Pentatomid bug)	Hartrot

Greek words *trypano* (τρύπανο) = an auger [a device for boring holes in wood] and *soma* (σώμα) = body that refer to their corkscrew-like locomotion. Magez and Radwanska (2014) provide a comprehensive review of all aspects of trypanosome biology and their transmission.

The Kinetoplastida are characterised by their possession of a flagellum and a unique intracellular structure called a kinetoplast (Figure 4.1). The kinetoplast is a disk of interlocking DNA circles (kDNA) located within a large mitochondrion. The structure of kinetoplast DNA is unlike that found in any other organism and its complex replication involves special proteins (da Silva et al. 2017). It may therefore be possible to design drugs to interfere with the replication of kinetoplast DNA. The position of the mitochondrion is such that the kinetoplast is just underneath the kinetosome that is itself situated underneath the base of the flagellum. The kinetosome (sometimes called the basal body), is a structure found in many organisms and is homologous with the centriole; it is involved in the formation of the flagellum. The Kinetoplastida always have a flagellum that may be long and free, incorporated into the cell surface to form an undulating membrane, or small and enclosed within a pocket. The inner core of the flagellum is the axoneme. Alongside this, and connecting to it, is the paraxial rod that consists of a lattice-like crystalline array of structural proteins. In the promastigote, epimastigote, and trypomastigote stages, the flagellum emerges at the anterior end of the cell and therefore acts as a propeller that pulls the cell along rather than pushing it from behind (Figure 4.2).

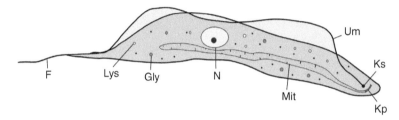

Figure 4.1 Diagram of a typical trypanosome. Kp: kinetoplast; Ks: kinetosome; F: flagellum; Mit: mitochondrion; Gly: glycosome; N: nucleus; Lys: lysosome; Um: undulating membrane.

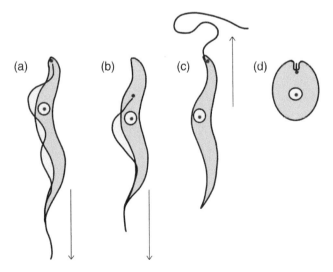

Figure 4.2 Morphological forms of trypanosomes. (a) Trypomastigote. (b) Epimastigote. (c) Promastigote. (d) Amastigote. The flagellum pulls the cell forwards rather than pushes it from behind. Arrows: direction of movement.

The Kinetoplastida have a unique organelle called the glycosome. This may be related to the peroxisomes (which they do not have) found in other organisms. The glycosomes are about 0.25 µm in diameter and contain glycolytic enzymes that are normally present in the cytoplasm of other organisms. The bloodstream forms of trypanosomes are extremely metabolically active and soon die if they run out of glucose to metabolise. The glycosomes are the site of glucose metabolism, and therefore, like the kinetoplast DNA, they are a potential target for antiparasitic drug design.

Many species within the Kinetoplastida are only parasitic in insects. For example, members of the genus *Leptomonas* live in the gut of various insects including blood-feeding reduviid bugs (Kaufer et al. 2017). Because they have only a single host, these species are referred to as monoxenous although there are rare case reports of human infections in patients who are HIV+ve. There are also accounts of *Leptomonas* co-infections with visceral leishmaniasis and post kala-azar dermal leishmaniasis caused by *Leishmania donovani* (Thakur et al. 2020).

Many of the Kinetoplastida alternate between an invertebrate host such as a blood-feeding insect or leech, and a vertebrate host with development occurring in both. Parasites that have more than one type of host are called heteroxenous. Heteroxenous Kinetoplastida species often express two or more morphological forms with one form present in the invertebrate and the other in the vertebrate (Table 4.2). Some members of the Kinetoplastida exhibit sexual reproduction, or something similar (Berry et al. 2019; Gibson and Peacock 2019), but it is uncertain whether it is a widespread phenomenon in the group.

4.2.1 Genus *Leishmania*

Members of the genus *Leishmania* exhibit two distinct morphologies: the amastigote form that occurs in the vertebrate host, and the promastigote form that occurs in the invertebrate vector.

Table 4.2 Morphological forms of Kinetoplastida parasitic in humans and domestic animals.

Morphological form	Description	Example
Amastigote	Kinetoplast and kinetosome above the nucleus, flagellum short and confined in pocket. Cell shape globular	*Leishmania donovani* inside vertebrate macrophage *Trypanosoma cruzi* in human spleen, liver, muscle, and other cell types
Promastigote	Kinetoplast and kinetosome at anterior end of cell, flagellum free, and long. Cell shape elongate	*Leishmania donovani* in sandfly gut
Epimastigote	Kinetoplast and kinetosome close and anterior to the nucleus. There is a short undulating membrane before the flagellum emerges at the anterior of the cell. Cell shape elongate	*Trypanosoma cruzi* in triatomid gut
Trypomastigote	Kinetoplast and kinetosome at posterior end of cell. Flagellum forms an undulating membrane that runs the length of the cell and may continue free when it reaches the anterior end. Cell shape elongate	*Trypansoma cruzi* in human bloodstream

Trypanosomes that are parasitic insects exhibit other morphological forms, such as choanomastigote, opisthomastigote, and paramastigote.

Table 4.3 Taxonomic divisions within the genus *Leishmania*.

Genus	Subgenus	Disease	Example
Leishmania	*Leishmania*	Visceral	*Leishmania donovani* phenetic complex
			Leishmania infantum phenetic complex
		Old World cutaneous	*Leishmania major* phenetic complex
			Leishmania tropica phenetic complex
		New World cutaneous	*Leishmania mexicana* phenetic complex
	Viannia	New World	*Leishmania braziliensis* phenetic complex
	Sauroleishmania	Lizard leishmaniasis	*Leishmania tarentolae*

The vertebrate hosts are mostly mammals, whilst the invertebrates are various species of sandflies. Those species parasitic in reptiles belong to the subgenus *Sauroleishmania* and do not cause zoonotic infections. Perhaps counter-intuitively, molecular evidence indicates that the *Leishmania* evolved in the Neotropical regions during the Mesozoic era as parasites of mammals and those species parasitizing reptiles, the *Sauroleishmania*, subsequently evolved from them (Noyes et al. 2000).

The taxonomy of the genus *Leishmania* is complex, and it is extremely difficult or impossible to distinguish many of them from their morphology using a light microscope. Lainson and Shaw (1987) comprehensively reviewed the genus and molecular and phylogenetic studies have largely supported their proposals for how it should be divided (Table 4.3). However, agreement concerning the status of several species is far from complete, and there remain uncertainties about many aspects of the evolution of the *Leishmania* – see Schönian et al. (2018) for further details. Lainson and Shaw (1987) identified two subgenera: *Leishmania* and *Viannia*. In the subgenus *Leishmania,* the parasites begin their development in the sandfly vector's midgut and then move forward to the pharynx. The sandfly then injects the parasites into the vertebrate host when she feeds. By contrast, in the subgenus *Viannia* the parasites begin development in the vector's hindgut and then move forward to the pharynx.

Those species within the subgenus *Viannia* (e.g., *Leishmania braziliensis, Leishmania peruviana, Leishmania guyanensis, Leishmania panamensis*) are restricted to South America and are primarily responsible for cutaneous disease. Species belonging to the subgenus *Leishmania* (e.g., *Leishmania donovani, Leishmania major, Leishmania infantum, Leishmania tropica, Leishmania mexicana*) have representatives in both the New World and the Old World and include agents of both visceral and cutaneous disease. Because of the difficulties associated with identifying the parasites and the diversity of the pathologies they cause, there is a tendency to refer to *Leishmania* phenetic complexes. That is, a species exhibits various phenotypes that may or may not relate to underlying genotypic differences. In addition, some species hybridize, and this can affect their subsequent transmission and pathology. For example, hybrids between *L. infantum* and *L. major* have greater transmission potential (Volf et al. 2007).

Present day members of the genus *Leishmania* are responsible for a great deal of morbidity and mortality in us and some cause serious disease in domestic and wild animals. There is even a suggestion that ancestral versions of *Leishmania* (with help from some other parasites) killed off the dinosaurs (Poinar and Poinar 2008). However, this theory has not gained a lot of support and most scientists continue to blame a meteorite. In humans, leishmaniasis exists as a complex of diseases caused by various species of *Leishmania*. While it is convenient to group those species of *Leishmania,* which invade organs as 'visceral' and those which affect the outer body surface as 'cutaneous', the distinctions are far from clear. For example, *L. donovani* is normally associated with the development of visceral leishmaniasis, but it can become cutaneous – as in the development of post kala-azar dermal leishmaniasis (see later). Similarly, *L. tropica* usually causes cutaneous leishmaniasis but can infect the viscera. Some 12 million people are infected with cutaneous leishmaniasis, which leaves long-lasting sores, and a further 500,000 with the potentially fatal visceral leishmaniasis of whom up to 80,000 die every year. The incidence of HIV/*Leishmania* co-infections is also increasing and a serious cause for concern.

4.2.1.1 *Leishmania* Life Cycle

In the vertebrate host, *Leishmania* exists in the amastigote form (2.5–5 μm) within mononuclear phagocytes and in particular the macrophages (Figure 4.3). These are a sub-group of the leukocytes (white blood cells) that have an essential role in the immune response in which they phagocytose and destroy foreign organisms. The parasites multiply by binary fission within a phagocyte until the host cell is destroyed after which they are released to be ingested by, and subsequently invade, new phagocytes (Figure 4.4). Transmission is normally by female sandfly vectors: only the female sandfly feeds on blood – the males (Figure 4.5) are harmless nectar feeders. Parenteral transmission, for example via contaminated needles or blood transfusion is also possible. Venereal and transplacental transmission of *L. infantum* sometimes occurs in dogs (Magro et al. 2017), but at the time of writing, there was little information on whether it also occurs in other species of *Leishmania* or in hosts other than dogs.

The sequence of development within the sandfly vector varies slightly between species, but in all cases involves transformation, replication, and subsequent movement to the anterior region of the gut. A sandfly acquires an infection when it ingests infected mononuclear phagocytes during feeding. Like mosquitoes, sandflies are 'batch processors' that take in a large blood meal that is then enclosed within a peritrophic membrane when it enters the midgut. They hold the blood meal within the midgut and digest it after which the products are absorbed. There is therefore usually a

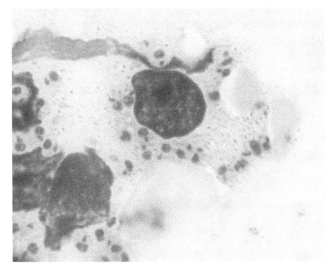

Figure 4.3 Spleen smear showing *Leishmania donovani* amastigotes infecting macrophages. Many of the infected cells have burst and released the parasites. Owing to the small size of the amastigotes, it is extremely difficult to distinguish between *Leishmania* species using a light microscope.

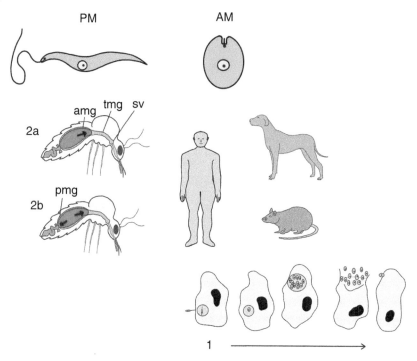

Figure 4.4 Generalized life cycle of *Leishmania* spp. The parasite exists in the promastigote form (PM) in the sandfly vector and the amastigote form (AM) in the mammalian host. Some *Leishmania* species infecting humans are zoonotic and infect other mammals such as dogs and rodents. 1: An infected sandfly injects promastigotes when it feeds, and these are taken up by macrophages and other mononuclear phagocytes. Within these cells, the parasite is enclosed within a phagosome. Here, it transforms into the amastigote form and reproduces asexually. Eventually, the host cell is destroyed, and this releases amastigotes that infect other cells. A sandfly becomes infected when it ingests blood containing infected phagocytes. After reaching the sandfly midgut, the parasites transform to the promastigote stage, reproduce asexually and undergo several morphological changes and migrations within the gut. 2a: In species belonging to the subgenus *Leishmania*, the promastigotes move to the anterior midgut (amg) and then to the stomadeal valve (sv). 2b: In species belonging to the subgenus *Viannia*, the parasites first move the posterior midgut (pmg) and then forward to the stomadeal valve. In both subgenera, the parasites are ejected when the fly attempts to feed. For further details, see text. Drawings not to scale.

gap of several days between blood meals. When the blood meal first reaches the insect's midgut, the amastigotes transform to the procyclic promastigote stage. This occurs in response to the decrease in temperature and increase in pH. The procyclic stage has a relatively short flagellum, is not very motile, and multiplies by binary fission within the blood meal. The promastigotes undergo a series of morphological transformations and multiplications as they move up from the midgut to the region of the stomodeal valve that marks the boundary between the foregut and the midgut. Along the way, some of the parasites attach to the lining of the gut and stomodeal valve using their flagellae, and this attachment is an important part of the life cycle. In those species belonging to the subgenus *Viannia,* most parasites make their way to the pylorus region (hind triangle) at the posterior of the midgut before moving forward. Unlike *Plasmodium* (in which the infective stages invade the salivary glands of the vector mosquito), the metacyclic stages of *Leishmania* do not, as a rule, infect the salivary glands of the sandflies. However, sandfly saliva does play an important role in the transmission process.

Figure 4.5 *Lutzomyia longipalpis* adult male; male sandflies have prominent claspers at the end of their abdomen. Sandflies are small (<4 mm in length), and their wings and body are covered in hairs. Their palps are very long and fold backwards. In common with many other nematoceran flies (e.g., mosquitoes), they have long legs, and their long antennae consist of many joints. Only female sandflies consume blood.

How Harming the Vector Facilitates Transmission

On reaching the stomodeal valve, the *Leishmania* promastigotes secrete a gel-like substance called promastigote secretory gel, the main component of which is filamentous proteophosphoglycan, and some of them transform into the infective metacyclic promastigote stage (this has a long flagellum and is very active). The gel physically blocks the gut, and this together with the vast numbers of parasites severely compromises the fly's ability to feed. Further compounding this, the parasites also produce chitinase enzymes that physically damage the peritrophic membrane and stomodeal valve. Because the insect's ability to ingest food is impaired, it becomes hungry thereby increasing its probing and number of visits to hosts all of which increases the chance of transmission. Physical probing probably does not transfer many parasites but to ingest food the infected fly must first expel some of the promastigote secretory gel. This gel contains numerous infective metacyclic stage parasites, as well as other non-infective stage(s). The secretory gel also facilitates the establishment of the infection in the vertebrate host, so it has a dual role in both the invertebrate and vertebrate host (Giraud et al. 2019).

The transmission mechanism(s) employed by the *Sauroleishmania* remain uncertain. Within the sandfly vector, these species tend to remain in the posterior regions of the gut, and therefore, it is unlikely that transmission occurs when the sandfly feeds. Furthermore, sandflies do not usually defaecate whilst feeding, so it is unlikely that transmission resembles that of *T. cruzi* by triatomid bugs. A third possibility is that the transmission occurs through the lizards consuming infected sandflies.

Much of the work on how *Leishmania* establishes and develops within its mammalian hosts involves mice as model organisms and those few *Leishmania* species capable of being cultured in the laboratory. One should therefore be careful of extrapolating from these studies to the likely behaviour of other *Leishmania* species and infections in other hosts. Mice and humans are both mammals, but one cannot assume that their immune systems react identically to the same infectious agent.

Furthermore, leishmaniasis manifests itself in numerous ways. Therefore, different *Leishmania* species and strains probably exhibit variations in the way they establish themselves and interact with the host immune system. Nevertheless, all species follow a basic pattern of development following their entry into the blood stream that involves morphological and physiological transformations and establishment in the mononuclear phagocytes and in particular the macrophages.

How *Leishmania* Establishes Within Mammalian Phagocytes

After an infected sandfly feeds, mononuclear phagocytes quickly detect and ingest the promastigotes that enter the blood circulation. Initial attachment of a parasite to a phagocyte begins at the tip of the parasite's flagellum via a ligand-receptor process. The principal ligands are phosphoglycans and the zinc metalloprotease enzyme GP63 (glycoprotein 63) on the surface of the promastigote whilst on the phagocyte, a variety of complement receptors are involved in the attachment process. After phagocytosing the *Leishmania* parasite, a phagocyte holds it within a membrane-bound vesicle called a phagosome. Lysomes then fuse with the phagosome and discharge hydrolytic enzymes and microbicidal peptides into it. They also acidify the contents. The structure then becomes to known as a parasitophorous vacuole or phagolysosome. Different species of *Leishmania* cause the formation of different types of phagolysosomes: those produced by *L. mexicana* tend to be large and contain many parasites whilst those formed by *L. donovani* tend to be much smaller.

The combination of a rise in temperature associated with moving to a mammalian host and the drop in pH caused by enclosure within a phagolysosome induce the promastigote to change into the amastigote form. This transformation takes about 1–4 hours following ingestion and is essential if the parasite is to survive the acidic pH and hydrolytic enzymes within the phagolysosome. Transformation includes changes the composition of the cell surface phosphoglycans. Some of these, the glycoinositol phospholipids (GIPLs) directly inhibit the production of nitric oxide (NO) by the phagocyte. The expression of GP63 is downregulated in the amastigote but they continue to express it, and it is important for both the survival of the parasite and as a virulence factor (Olivier et al. 2012). The membrane changes associated with transformation to the amastigote stage means amastigotes invade phagocytes using a different set of ligands and receptors to the promastigote stage. Despite these changes, phagocytes remain capable of destroying amastigotes. Therefore, the reasons why some people develop serious or even fatal infections, whilst in others the infection is resolved, possibly without displaying any symptoms is uncertain.

4.2.1.2 Visceral Leishmaniasis

Classical visceral leishmaniasis is commonly known as kala-azar and causes fevers like those of malaria. Both malaria and kala-azar occur in the same areas, so it is important for doctors to distinguish between them. The name kala-azar (black head) derives from the symptomatic darkening of the forehead and mouth of patients suffering from visceral leishmaniasis. In India, great plagues of visceral leishmaniasis occurred in Assam in the late nineteenth and early twentieth centuries that depopulated whole villages. Serious outbreaks still occur today, and visceral leishmaniasis remains an important cause of morbidity and mortality in over 70 countries around the world. Most cases of visceral leishmaniasis occur in South Asia (~67%), but it is also a big problem in parts of East Africa (e.g., Sudan in the region bordering with Ethiopia). There is also a focus of infection in South America, especially in Brazil.

The clinical picture of visceral leishmaniasis differs geographically. There is, nonetheless, a basic pattern to the course of the disease. The first stage begins when a papule develops at the site of the sandfly bite, but this eventually regresses. Low-grade recurrent fevers then develop anything from 10 days to 2 years or more afterwards, and these persist throughout the course of the disease. Within the spleen, immune-related responses destroy red blood cells, and this causes anaemia. In addition, the liver becomes enlarged (hepatomegaly) as does the spleen (splenomegaly). Enlargement of the spleen results from a combination of hyperplasia induced by the need to produce new mononuclear phagocytes and from infected mononuclear phagocytes filling with parasites. The patient often suffers from diarrhoea and this, together with the fever, leads to anorexia, malnutrition, and dehydration. If the disease is not treated, 90% of those suffering visceral leishmaniasis will die. Recovery can be rapid and complete with or without treatment. However, in many cases the parasite persists and may appear on the skin in raised macules causing the disfiguring condition post kala-azar dermal leishmaniasis.

4.2.1.3 Post Kala-Azar Dermal Leishmaniasis

Post kala-azar dermal leishmaniasis (PKDL) is a condition that is usually associated with *L. donovani* and develops as sequel to visceral leishmaniasis in 2.5–20% of cases (hence 'post-kala-azar'). It may manifest itself anything from immediately afterwards to several years following the condition. It is characterised by the development of nodules and/or macules that can be extensive and cover any area of the body and may be mistaken for leprosy. The nodules are irregular raised masses on the skin surface whilst macules (Latin *macula* = blemish or small spot) are flat discoloured areas on the skin surface. These regions contain numerous amastigotes, and if they occur on exposed parts of the body, they are a ready source of infection for sandflies. The identification and treatment of patients suffering from PKDL is therefore an important part of any control programme. There are marked differences the occurrence and development of PKDL between countries, which suggests that host, and/or parasite factors may be important in whether it develops. For example, most cases of PKDL (~50%) occur in the Sudan and the condition tends to develop more rapidly there than in India.

Some workers consider that the development of PKDL is associated with the incomplete or inefficient treatment of visceral leishmaniasis following treatment with the drugs sodium stibogluconate and pentamidine. However, PKDL may also occur after treatment with miltefosine. In India, Das et al. (2009) found about 20% of PKDL cases occurred in people for whom there was no record of either visceral leishmaniasis or the prescription of the drugs used to treat it. However, these people tested positive for *L. donovani* and therefore carried an asymptomatic infection.

4.2.1.4 Cutaneous Leishmaniasis

Cutaneous leishmaniasis manifests itself in a variety of different forms depending upon the species of *Leishmania*. Clinically, cutaneous leishmaniasis divides into three basic types depending upon how the disease presents. Localized Cutaneous Leishmaniasis (LCL) generally takes the form of a dry ulcer. It develops at the bite site of the sandfly vector and usually heals by itself although this may take some time and leave permanent scarring. This form of disease is common in India, Central Asia, the Middle East, and parts of southern Europe; *L. major* and *L. tropica* are responsible for most infections in these regions. LCL is also common in South America where species such as *L. venezuelensis* and *L. mexicana* are responsible. Diffuse (disseminated) cutaneous Leishmaniasis (DCL) is a rarer and much more serious condition than LCL. It manifests as numerous raised (but not ulcerating) papules and nodules that spread to cover large areas of the body.

The condition is often associated with immune suppression, and there are several reports of HIV co-infection (e.g., Corrêa Soares et al. 2020). Unlike LCL, patients with DCL seldom recover without treatment. In the 'Old World', *Leishmania aethiopica* is the most common cause of DCL whilst in South America *L. mexicana* and *L. amazonensis* are implicated. As mentioned earlier, any one species of *Leishmania* may cause different types of leishmaniasis.

Mucocutaneous leishmaniasis (MCL) arises from the formation of an ulcerative lesion that afflicts the mouth, palate, and nose. As a rule, MCL develops and spreads slowly over a period of years and eventually destroys the affected region. It is most common in South America, particularly Brazil and the Amazon regions of Peru, Ecuador, Colombia, and Argentina where the condition is known as 'espundia'. The discovery of pre-Inca pottery illustrating disfigured faces suggests that the disease pre-dates the European invasion of South America. However, some workers claim that the conquistadors and early Spanish settlers introduced MCL into South America. Espundia has a low mortality (~5%) provided the patient receives medical care. However, in the absence of treatment, mortality but would be undoubtedly higher. Death from MCL commonly results from complications such as aspiration pneumonia although some sufferers suffocate owing to laryngeal closure. *Leishmania braziliensis* is responsible for most cases of MCL although *L. guyensis* is also important. *Leishmania major* and some other species of *Leishmania* can also cause MCL, but these cases are rare. *Leishmania guyanensis* and *L. braziliensis* are naturally infected with *Leishmania* RNA virus 1 (Cantanhêde et al. (2018), and this results in the development of a particularly rapid and aggressive form of MCL (Olivier and Zamboni 2020).

4.2.2 Genus *Trypanosoma*

All members of this genus are parasites of vertebrate animals, and they are almost all transmitted by invertebrate vectors. Asexual reproduction usually takes place in both the vertebrate host and the invertebrate vector (i.e., the parasites are heteroxenous). However, in mechanically transmitted trypanosomes, such as *Trypanosoma evansi*, there is no development outside the vertebrate host. There is evidence of sexual reproduction (or something like it) occurring in the invertebrate host in some trypanosome species. *Trypanosoma equiperdum* is something of an exception to this general lifestyle because it is a sexually transmitted parasite of horses and other equine species. Nevertheless, most of the genus *Trypanosoma* are heteroxenous and two distinct groups are identifiable based on where they develop within their invertebrate host. Those developing within the anterior regions of the gut and transmitted when the vector bites exhibit 'anterior station' development and belong to the 'Salivaria' group. Species developing in the vector's hindgut and transmitted via its faeces exhibit 'posterior station development' and belong to the 'Stercoraria' group. Both groups include species of medical and veterinary importance (Table 4.4; Figure 4.6). It should come as no surprise that the taxonomy of trypanosomes undergoes regular rearrangements. It is highly likely that some well-known species are actually 'sub-species' or 'synonyms', but for the sake of comparison with past literature we have retained the most common species' names.

In several trypanosome species, the kinetoplast is modified or lacking entirely. This has had a dramatic effect on their subsequent spread and evolution. In the case of *Trypanosoma brucei*, the development of the mitochondrion undergoes down regulation within the vertebrate host. However, the mitochondrion is fully functional during the procyclic stage that develops in the tsetse fly vector (Figure 4.7). The partial loss of kinetoplast DNA (dyskinetoplastidy) or its complete loss (akinetoplastidy) means that the mitochondria cannot fully function, and therefore the parasite can only develop within its vertebrate host. Consequently, trypanosomes lacking a kinetoplast and those in which it is only partially functional must rely on mechanical vector transmission

Table 4.4 Examples of *Trypanosoma* species of medical and veterinary importance.

Parasite	Transmission Group	Vector/means of transmission	Host
Trypanosoma brucei brucei	Salivaria	Tsetse flies (*Glossina* spp.)	Ruminants (cause of 'nagana')
Trypanosoma brucei gambiense	Salivaria	Tsetse flies (*Glossina* spp.)	Humans (cause of 'sleeping sickness'). No important animal reservoir of infection
Trypanosoma brucei rhodesiense	Salivaria	Tsetse flies (*Glossina* spp.)	Humans (cause of 'sleeping sickness'). Reservoir of infection in cattle, wild game, lions, hyena etc.
Trypanosoma congolense	Salivaria	Tsetse flies (*Glossina* spp.)	Cattle, pigs, wild game. Cause of 'nagana'
Trypanosoma evansi (*Trypanosoma brucei evansi*)	Salivaria	Tabanid and other biting flies (mechanical transmission)	Horse, cattle, pigs, dogs, rodents. Cause of 'surra'
Trypanosoma equinum (synonym of *T. evansi*)	Salivaria	Tabanid and other biting flies (mechanical transmission)	Horses, donkeys, cattle, dogs Cause of 'mal de caderas'
Trypanosoma equiperdum (*Trypanosoma brucei equiperdum*)	Salivaria	Sexual transmission	Horse, asses. Cause of 'dourine'
Trypanosoma cruzi	Stercoraria	Reduviid bugs (e.g., *Triatoma infestans*)	Humans (cause of Chagas Disease). Reservoir of infection in many domestic and wild animals
Trypanosoma theileri	Stercoraria	Tabanid flies	Cattle

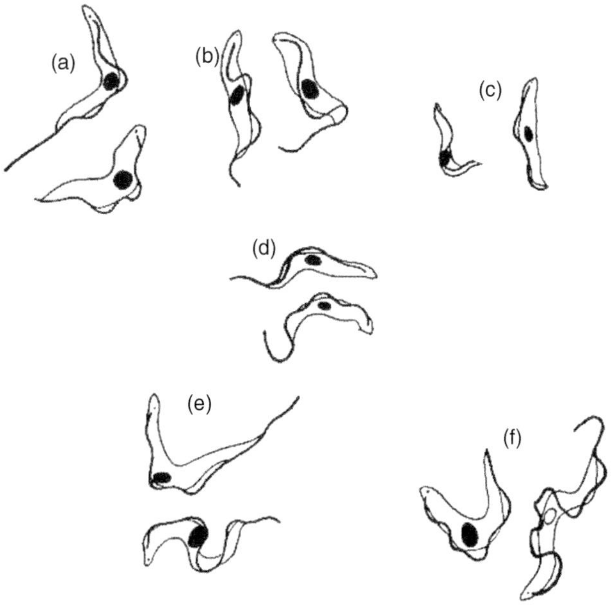

Figure 4.6 Diagrammatic representation of some *Trypanosoma* species. All are drawn to the same magnification. (a) *Trypanosoma brucei*. (b) *Trypanosoma vivax*. (c) *Trypanosoma congolense*. (d) *Trypanosoma equinum*. (e) *Trypanosoma equiperdum*. (f) *Trypanosoma evansi*. Note the small size and lack of a free flagellum in *Trypanosoma congolense*. *Source:* Reproduced from Cameron, (1934) © Copyright A and C Black Ltd.

(a)

(b)

Figure 4.7   Tsetse fly *Glossina* spp. (a) Adult fly. There are currently 23 described species of tsetse flies, and they vary in size from 6 to 16 mm in length. Their coloration consists mostly of the shades of brown and grey. When alive, the eyes of tsetse flies are usually dark brown, unlike those of tabanid flies in which the eyes often brightly coloured. Two characteristic features of tsetse flies are that they fold their wings completely flat when at rest and their mouthparts project forwards from their head. (b) Mouthparts of a tsetse fly. The mouthparts are lowered when the insect feeds. There are two thick palps that serve sensory and protective functions. The robust central labium is equipped with terminal teeth that facilitate skin penetration; the bite of a tsetse fly is extremely painful.

or sexual transmission. However, this was not an evolutionary 'retrograde step' because these modes of transmission broke the link with the African tsetse fly belt and allowed the parasites to spread to other countries. Dyskinetoplastidy arises naturally in wild trypanosome populations by mutation, in response to treatments with certain anti-trypanosome drugs (e.g., triacetylbenzene hydrochloride [TBG]), and through long-term *in vitro* culture.

4.2.2.1 *Trypanosoma brucei*

Trypanosoma brucei is responsible for the disease Human African Trypanosomiasis (HAT), or more colloquially, 'sleeping sickness' and the wasting disease in cattle known as 'nagana'. It was often stated that HAT is invariably fatal in the absence of treatment. However, it is now apparent that some people remain asymptomatic after contracting an infection (Kennedy and Rogers 2019) although it is uncertain how frequently this occurs. Some people become seronegative whilst others remain seropositive, but, in both instances, it is impossible to detect the parasites with the usual diagnostic tests. Those who become seronegative have presumably repelled the parasites and cleared them from their body, but those who remain seropositive are presumably tolerant of their infection. This begs the question of where the infection in the seropositive individuals is residing. It now looks likely that the trypanosomes are located within the skin rather than the vasculature (Capewell et al. 2016). Furthermore, the parasites within the skin can infect tsetse flies when they feed. It is uncertain whether infected but asymptomatic individuals represent a significant reservoir of infection for tsetse flies.

HAT is predominantly a disease afflicting poor people living in rural environments in some of the more politically unstable parts of Africa. Consequently, there is widespread underreporting of the disease. In 2006, the WHO estimated there to be 20,000 cases of HAT every year, although a few years later Brun et al. (2010) suggest that there are between 50,000 and 70,000 cases. Nevertheless, despite the difficulties of working in the affected regions, control programmes are proving effective, and by 2015, there were fewer than 3,000 reported cases (Büscher et al. 2017).

There are three morphologically identical sub-species of *T. brucei*: *Trypanosoma brucei brucei*, *T. brucei gambiense*, and *T. brucei rhodesiense*. The different sub-species vary in their geographical distribution, ability to infect mammalian hosts and their pathology, but they are all transmitted by various tsetse fly species (e.g., *Glossina palpalis*) with different species being of particular importance as vectors in different areas. Tsetse flies are only found in sub-Saharan Africa within the latitudes 14° North and 29° South – an area that encompasses about 10 million km^2. Because of their absolute dependence upon tsetse flies to effect transmission, the parasites are also limited to this region. Tsetse flies have specific environmental requirements to complete their life cycles and therefore do not exploit urban environments. Consequently, people acquire HAT when they work in the fields or visit the countryside/game reserves. An increase in the number of cases of HAT among people living in the Nkhotakota wildlife reserve in Malawi since 2015 was attributed to an increase in the tsetse fly population following the relocation of hundreds of elephants to the reserve. The elephants were moved to both help conserve their numbers and improve the local economy by promoting wildlife tourism.

Trypanosoma brucei brucei is essentially a parasite of wild and domestic animals, and it does not infect humans. Wild game (e.g., kudu [*Tragelaphus strepsiceros*], warthog [*Phacochoerus aethiopicus*]), and some native cattle breeds (e.g., N'Dame, Muturu, Masai Zebu) are 'trypanotolerant' and infections do not always lead to serious disease symptoms. By contrast, *T. brucei brucei* causes serious disease in many introduced varieties of domestic animals and they succumb to a condition called 'nagana' – a word derived from the Zulu language that means 'to be in low or depressed spirits'. It also severely affects horses, sheep, goats, and dogs, and these often suffer an acute disease that culminates in the death of animal within 20 days to a few months of becoming infected. In cattle, *T. brucei brucei* tends to cause chronic disease that lasts several months, and the infected animal may ultimately recover.

Trypanosoma brucei rhodesiense has a close genetic relationship to *T. brucei brucei* and occurs mainly East Africa – principally Tanzania and Uganda. In addition to infecting humans, it also parasitizes many wild game animals. Consequently, *T. brucei rhodesiense* is a zoonotic disease with numerous reservoirs of infection. *Rhodesiense* HAT is usually an acute infection although in some geographical regions a less severe disease occurs.

Trypanosoma brucei gambiense is the principal cause (~90% of cases) of HAT. *Gambiense* HAT usually follows a chronic course over a period of years, and severe nervous system impairment only ensues in the late stages of the disease. It occurs mainly in West and Central Africa with most cases in Democratic Republic of Congo, Angola, and Sudan. Although *T. brucei gambiense* infects various wild animals, the importance of zoonotic transmission in the epidemiology of *gambiense* HAT is uncertain.

Within the mammal host, the trypomastigotes of *T. brucei* exhibit various morphological forms although they all have a prominent undulating membrane (Figures 4.8 and 4.9). The slender forms are 25–35 μm in length, have a pointed posterior end, and a long free flagellum. The mitochondrion of the slender form is poorly developed. It has few cristae and lacks a functional cytochrome chain and tricarboxylic acid cycle (TCA or Krebs' cycle). The slender forms multiply by longitudinal fission until a critical parasite density is reached at which point, they form intermediate forms and then stumpy forms. The stumpy forms are about 15 μm long, they have a broad, blunt posterior end, and lack a free flagellum. The mitochondrion of the stumpy form is more developed than in the slender form and has many cristae, but the cytochrome chain is still absent, and the TCA cycle is incomplete. Intermediate forms averaging 23 μm in length also exist, which have a blunt posterior end and a free flagellum. The stumpy forms do not divide further in the mammalian host. However, their physiological changes prepare them for the very different conditions, they will

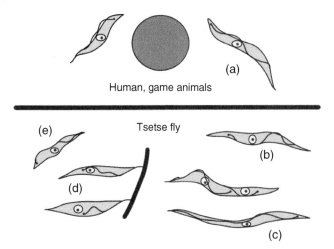

Human, game animals

Tsetse fly

Figure 4.8 Life cycle stages of *Trypanosoma brucei gambiense*. (a) Long and short forms in the blood. (b) Trypomastigote forms in the tsetse fly midgut (~2 days post infection). (c) Slender form in the proventriculus of the tsetse fly (10–15 days post infection). (d) Crithidial stages in the tsetse fly salivary gland (15–25 days post infection). (e) Infectious stage in the salivary glands 20–30 days post infection. *Source:* Redrawn from Chandler and Read (1961), © Wiley-Blackwell.

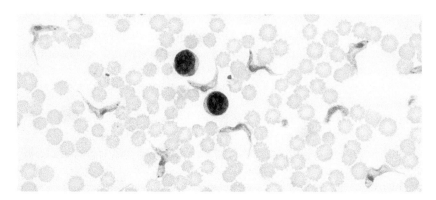

Figure 4.9 Light microscope photograph of trypomastigote stages of *Trypanosoma brucei*. Note the prominent undulating membrane.

experience in the gut of their tsetse fly vector. The density-dependent physiological changes are a consequence of quorum sensing – a phenomenon more usually associated with bacteria. In trypanosomes, quorum sensing operates through peptidase enzymes released by the parasites breaking down host proteins to release oligopeptides (Rojas and Matthews 2019). The identity of the oligopeptides is currently uncertain, but they bring about changes in gene expression within the parasites that result in them changing to the stumpy form. The stimulus is not species specific and in experimental co-infections, the density of *Trypanosoma congolense* modulates the population density at which *T. brucei* differentiates into stumpy forms.

Genomic Regulation in Trypanosomes

In most eukaryotic organisms, the regulatory processes that take place after the conversion of DNA into RNA are of greater complexity and importance than transcription itself. Trypanosomes take this process to its limits and undertake genome regulation almost entirely at post-transcriptional level (Queiroz et al. 2009). The formation of proteins in trypanosomes is not regulated by the rate at which mRNA is synthesised but occurs through factors that control the stability of the mRNA molecules (i.e., alter their half-life and hence concentration) and the rate at which mRNA is translated into protein. RNA binding proteins perform much of this post-transcriptional regulation. Rapid changes in the half-life of mRNA molecules and translational control regulate adaptions to environmental change such as the movement between vector and mammalian host (Schwede et al. 2012).

Once established within their vertebrate host, the trypanosomes rapidly disseminate about the body via the blood and lymphatic system. Unlike *T. cruzi* and *Leishmania*, *T. brucei* remains an extracellular parasite and never invades cells in its vertebrate host. However, it does invade most of the organs of the body by colonising the intercellular spaces (c.f., *T. congolense* that tends to remain within the circulatory system). In humans, *T. brucei gambiense* crosses the blood–brain barrier and colonises the intercellular spaces in the brain. In so doing, it causes the classic symptoms of HAT. By contrast, *T. brucei rhodesiense* does not usually colonise our nervous system although this may be at least partly because the patient dies before this happens.

The development of HAT depends upon the species of trypanosome and its genetic strain, as well as host health and genetic factors (Kazumba et al. 2018). In the case of *gambiense* stage 1 HAT, a red sore develops at the site where the tsetse fly bit and over the subsequent weeks or months the patient develops a fever, their lymph glands swell, and they suffer from aches, pains, and headaches. These symptoms are non-specific, and the disease often remains undiagnosed. In the absence of effective treatment, the disease develops remorselessly to stage 2 HAT in which the symptoms become severe with prolonged fevers, weight loss, anaemia and damage to the central nervous system.

Following their ingestion by a tsetse fly, the parasites differentiate into procyclic trypomastigotes within the midgut region. The main stimulus for the transformation is the drop in temperature (~10 °C) that the parasites experience when they move from the warm mammalian bloodstream to the much cooler insect gut. There are also major changes in the parasite's metabolism in response to the movement from a hot environment in which glucose is plentiful to a cooler one in which glucose is in much lower concentration. In mammals, the trypanosomes have poorly developed mitochondria since they derive their ATP from glycolysis using the abundantly available glucose obtained from their host and most of their glycolytic enzymes are located within their glycosomes. Trypanosome glycolysis within their mammalian hosts is remarkably inefficient (the metabolism of one mole of glucose yields only two moles of ATP) and the pathway ceases at pyruvate, which they excrete. By contrast, in tsetse flies the trypanosome mitochondrion is a much bigger organelle and has well-developed cristae. This is because oxidative catabolism becomes more important as a source of ATP.

Having successfully transformed into procyclic trypomastigotes, the parasites reproduce by longitudinal fission. Interestingly, the procyclic forms can undergo apoptosis – a phenomenon that is normally associated with metazoan animals. Apoptosis also occurs in various other parasitic protozoa, including other *Trypanosoma*, *Leishmania* and *Plasmodium*, but its function is uncertain (Proto et al. 2013). The promastigotes penetrate the fly's peritrophic membrane after which they migrate forward to the proventriculus. At this point, they cease dividing and transform into mesocyclic trypomastigotes, and these penetrate back through the peritrophic membrane and make their way further forward to the salivary glands. Once they reach this point, they transform into epimastigotes. The epimastigotes attach themselves by their flagellum to the epithelial cells lining the tsetse fly salivary gland, multiply by longitudinal fission, and then transform into non-dividing metacyclic trypomastigotes. Some form of mating involving meiosis often takes place during the epimastigote stage, but it is not obligatory and the extent that it occurs varies between populations. It is also affected by the immune status of the tsetse fly and hybrid matings between trypanosome strains is considered most likely if the fly acquires a mixed infection the first time it feeds (Peacock et al. 2016). Hybridization has relevance for the transfer of genetic traits such as drug resistance between trypanosome lineages.

The metacyclic trypomastigote stage is infective to susceptible mammalian hosts and expresses a specific subset of genes coding for variant surface glycoproteins (VSG). When an infected tsetse fly feeds, it injects the metacyclic trypanosomes into the blood stream and the VSG help protect them from the mammalian immune system. The trypanosome life cycle within the tsetse flies therefore involves a complex sequence of migrations and transformations and typically takes about 3–5 weeks. Consequently, effective transmission depends heavily upon the lifespan of the tsetse fly vector. Male tsetse flies usually only live for about 2–3 weeks in the wild and whilst female tsetse flies survive for up to 4 months, most die within 20–40 days.

How Trypanosomes Alter Tsetse Fly Physiology to Facilitate Transmission

Tstetse flies have an extremely effective immune system that usually kills any trypanosomes present within their blood meal soon after it is ingested. Nevertheless, if a tsetse fly is susceptible to infection, the ingestion of even a single trypanosome is sufficient to ensure it becomes a vector (Maudlin and Welburn 1989). Once infected, a tsetse fly remains infected for the rest of its life. Although trypanosomes grow and reproduce in tsetse flies, they do not appear to cause them much, if any, harm. However, the trypanosomes do alter the protein composition and antihaemostatic properties of tsetse fly saliva. This reduces the saliva's capacity to prevent blood coagulation and vasoconstriction. This results in the flies spending more time probing for a blood vessel. This may either provide more time for parasite transmission to take place or encourage the tsetse fly to move more frequently between hosts. The mechanism(s) by which the parasites alter the composition of the tsetse fly saliva are uncertain, but they may trigger a stress response that decreases certain gene expressions (Van Den Abbeele et al. 2010).

At the time of writing, five drugs had approval for the treatment of HAT: pentamidine, suramin, melarsoprol, eflornithine, and nifurtimox. Pentamidine and suramin are used to treat first-stage HAT whilst melarsoprol (Mel B), eflornithine, and nifurtimox are used for second-stage HAT. None of the drugs is ideal, and some have serious side effects. For example, suramin can cause anaphylactic shock and kidney failure, whilst melarsoprol can cause seizures and kills 1 in 20 of the patients who receive it. Although the risks might sound unacceptable, in the absence of treatment, there is an extremely high chance that a patient with HAT will die of the disease. Nifurtimox has the advantages of being easier to administer and less toxic than the other drugs but is prescribed as a combination therapy with eflornithine rather than on its own.

A novel group of chemicals called the benzoxaboroles are highly effective at treating stage 2 HAT but currently they still require registration (Jacobs et al. 2011). They also show potential for the treatment of animal trypanosomiasis (Akama et al. 2018). Not only can benzoxaboroles be taken orally (suramin and melarsoprol must be given as a series of intravenous injections) but they also do not produce the harmful side effects of current drugs. The benzoxaboroles also include chemicals that show promise for the treatment of malaria, cryptosporidiosis, filarial nematodes, and bacteria (Lunde et al. 2019; Sonoiki et al. 2017). Consequently, it may eventually become possible to treat co-infections with a single safe drug.

4.2.2.2 *Trypanosoma congolense*

Measuring only 9–18 μm in length, *T. congolense* is the smallest of the African trypanosomes. It is monomorphic although short and long strains also occur. One of its characteristic features is the absence of a free flagellum although the cell tapers finely at the anterior end, so it is possible to believe mistakenly that one is present. The posterior end is blunt and the kinetoplast marginal. The undulating membrane is not pronounced but this together with the absence of a free flagellum does not restrict its ability to move – although authors differ in their opinion about whether its activity is active or sluggish.

Within the blood and lymphatic system of their mammalian host, the trypomastigotes of *T. congolense* multiply by dividing by longitudinal fission. Several tsetse fly species (e.g., *Glossina morsitans*) are responsible for transmitting *T. congolense* with different species being of particular importance in different areas. Following ingestion by a tsetse fly, the parasites differentiate into procyclic trypomastigotes and undergo a similar migration pattern a sequence of transformations to *T. brucei*. However, *T. congolense* does not invade the tsetse fly salivary glands. Instead, the epimastigotes attach themselves to the walls of the proboscis and after transforming into the metacyclic stage, the parasites migrate to the hypopharynx region. As in *T. brucei*, a form of sexual reproduction occurs in some strains of *T. congolense*.

Trypanosoma congolense occurs throughout southern, East and West Africa and infects many domestic mammals (e.g., cattle, sheep, horses, pigs, dogs) and wild game (e.g., antelopes, warthogs). Except in unusual circumstances, *T. congolense* does not present a risk to humans (Truc et al. 2013). In addition, although the trypanolytic activity of normal human serum normally kills *T. congolense*, some strains are resistant to it (Van Xong et al. 2002).

Trypanosoma congolense is of primary concern for its effect on domestic cattle, and it is a major cause of economic loss to cattle farmers throughout the affected regions. It is less harmful to other domestic animals and wild game. In susceptible cattle, *T. congolense* causes similar symptoms to *T. brucei*. Therefore, when farmers say that their animals are suffering from '*nagana*', it could mean that either parasite is the cause. In addition, co-infections are common. The disease may manifest itself in acute, chronic forms and mild forms. In the acute form, the disease causes anaemia, emaciation and a high parasitaemia in the peripheral circulation. The liver, lymph nodes and spleen enlarge, haemorrhages occur in the heart muscle and kidneys, and the infected animal may die in within 10 weeks of becoming infected. In the chronic form, the symptoms are less severe, and it may be difficult to find the parasites in the blood. There is enlargement of the lymph nodes and liver and signs of degeneration in the kidney, but the infected animal can recover after about a year. In mild infections, it may not be obvious that *T. congolense* is present. The pathology caused by *T. congolense* differs from that of *T. brucei* in that the parasites remain within the circulatory system and the central nervous system is not affected. Anaemia is the most characteristic feature of *T. congolense* infection and results from the destruction of red blood cells in the liver and spleen although other mechanisms (e.g., inflammatory processes) may also be involved (Noyes et al. 2009). Indigenous breeds of cattle, such as N'dama, are not resistant to *T. congolense* but have a genetic ability to limit the development of anaemia (Naessens 2006).

4.2.2.3 *Trypanosoma evansi*

Trypanosoma evansi is monomorphic, 14–33 μm in length and 1.5–2.2 μm in width and morphologically indistinguishable from *Trypanosoma equinum*, *Trypanosoma equiperdum* and the slender forms of *T. brucei*. Molecular evidence suggests that there are two distinct type strains of *T. evansi*, type A and type B, and there are further strains within each type. Types A and B are distinguished by differences in the minicircles in the kinetoplast DNA. Type A is the most common and widespread form whilst type B occurs in camels in Kenya and Ethiopia. Molecular analyses suggest that both strains probably arose independently from West African *T. brucei brucei* (Cuypers et al. 2017). If true, this is unexpected because currently, Type B *T. evansi* only occurs in Eastern Africa.

   Trypanosoma evansi has an extremely wide distribution and occurs in Africa, Asia, and Central and South America. It is particularly pathogenic in horses, but it also causes considerable morbidity and mortality in camels, cattle, pigs, dogs, and cats. It also parasitizes many wild animals such as deer, tapir, and capybara. It is usually mechanically transmitted by biting flies such as tabanids and stable flies. *Trypanosoma evansi* does not reproduce in its insect vectors and they act only as a 'dirty syringe'. In South America, vampire bats can act as both hosts and mechanical vectors of *T. evansi*. Under experimental conditions, Raina et al. (1985) infected both dogs and mice by feeding them meat containing *T. evansi*. The extent to which oral infections occur naturally is, however, uncertain. Reproduction takes place asexually by longitudinal binary fission within the mammalian host.

   The disease caused by *T. evansi* is commonly known as 'surra' – which is the Hindi word for 'rotten' or 'emaciated' although other terms are also used such as *el debab* in many Arabic-speaking countries. It causes the death of many thousands of animals every year and a great deal of morbidity (Aregawi et al. 2019). Horses are particularly susceptible to infections, and there are claims that the inoculation of even a single parasite can prove fatal. The disease is often acute in horses and the animal dies within a few weeks to 2 months. Chronic infections lasting over a year may also occur but also often end with the death of the horse. Surra causes anaemia, emaciation, and oedema that may vary from urticarial plaques on the neck and flanks to widespread swelling of the legs and lower body. The plaques may subsequently become necrotic and bleed whilst encephalitis and demyelination can occur in the brain and spinal cord that results in staggering and paralysis. Affected animals may also express abnormal behaviour such as hyperexcitability, head tilting, and circling.

   Surra is also an important cause of morbidity and mortality in camels in which it tends to be a chronic wasting disease. Characteristic features of infection include fever, anorexia, and the development of oedema. Cattle can be severely affected when exposed to *T. evansi* for the first time, but in endemic areas, the disease in cattle tends to be subclinical and reduce productivity rather than cause major morbidity and mortality. Dogs are susceptible to an acute and rapidly fatal form of the disease that causes nervous signs that are like those of rabies.

4.2.2.4 *Trypanosoma equinum*

Trypanosoma equinum and *T. evansi* are synonymous: *T. equinum* is neither a separate species nor a sub-species of *T. evansi*. However, there are morphological differences between *T. evansi* and *T. equinum*: *T. evansi* expresses both kinetoplastic and dyskinetoplastic forms (i.e., ones retaining only fragments of kinetoplastid DNA) but *T. equinum* lacks a kinetoplast (i.e., it is 'akinetoplastic'). This was one of the reasons for separating them, but molecular studies indicate that genetically they are the same.

   Despite the molecular evidence, we will follow the older parasitology textbooks in dealing with them as though they are two distinct species. This is because the literature often provides different accounts of them in terms of their host range and pathology. *Trypanosoma equinum* only occurs in various South American countries where it causes a disease commonly known as '*mal de caderas*' in Spanish and '*mal de cadeiras*' in Portuguese – which translates as illness of the hips. *Trypanosoma equinum* is principally a parasite of horses. It infects other equids and a variety of domestic and wild animals (e.g., capybara), but these do not tend to suffer severe disease. It is mechanically spread by biting flies, such as tabanids, and in swampy areas where these flies are most numerous, *T. equinum* is a particular problem.

Unlike *surra*, *mal de caderas* normally causes a chronic disease in horses. The condition develops over a period of months but usually has a fatal outcome. Symptoms begin with a fever and loss of weight and then the hindquarters become progressively weaker (hence the name '*mal de caderas*') resulting in staggering and then an inability to walk. Horses can also exhibit conjunctivitis, the eyelids become oedematous (filled with fluid), and transient plaques form on the neck and flanks. In addition, the kidneys, brain, and spinal cord show signs of inflammation and necrosis.

4.2.2.5 *Trypanosoma equiperdum*

Trypanosoma equiperdum is monomorphic and morphologically indistinguishable from *T. equinum*. It exhibits dyskinetoplastidy – that is, it retains only fragments of kinetoplastid DNA. It parasitizes a range of equids although it affects the domestic horse (*Equus cabalus*) more severely than asses, donkeys etc. Some strains infect dogs and one can establish laboratory cultures in rats and mice. It is a sexually transmitted disease and causes a condition called 'dourine': a word deriving from the Arabic for 'unclean'. The use of horses in agriculture and warfare spread *T. equiperdum* around the world. However, following breeding programmes aimed at eliminating infected horses, it currently has a restricted distribution in parts of Africa, Asia, southern Europe, and South America.

Dourine typically manifests itself in three stages. The first stage begins with swelling of the genitalia. There is patchy depigmentation of the penis and vulva, and there is vaginal discharge in mares. The horse may also exhibit a slight fever and loss of appetite. After about a month, the second stage of the disease begins with the development of circular fluid-filled (oedematous) plaques 2.5–10 cm in diameter underneath the skin. They usually form on the flanks although they can develop on any part of the body. The plaques may last a few hours or days after which they disappear and then reappear again later. Because of the rash-like swellings, this is called the 'urticarial stage'. Plaques do not always form but if they do, they are a reliable indicator of the nature of the disease. The onset of paralysis marks the third stage of the disease. Often this begins in the nose and face. It then spreads to affect the rest of the body and can result in complete paralysis affecting all the limbs. The course of the disease may take as long as 2 years and mortality can be as high as 70%.

The relationship between *T. equiperdum*, *T. evansi,* and *T. brucei* remains uncertain. Molecular evidence indicates that *T. equiperdum* and *T. evansi* independently diverged from *T. brucei* on several occasions. According to Cuypers et al. (2017), *T. equiperdum* arose from an East African *T. brucei* ancestor whilst *T. evansi* derives from a West African *T. brucei* ancestor.

Trypanosoma equiperdum is primarily a tissue parasite and not usually present in the circulating blood, but it occurs in smears taken from the genitalia or plaques. Diagnosis has traditionally relied on serological complement fixation tests and in some countries where dourine is endemic, they are mandatory. However, such tests will not reliably distinguish *T. equiperdum* from other trypanosomes such as *T. evansi* and *T. brucei* – which is hardly surprising because they are all closely related.

4.2.2.6 *Trypanosoma cruzi*

Trypanosoma cruzi causes Chagas disease – a potentially fatal infection that afflicts around 11 million people in the New World from Argentina to the southern United States of America. In addition, owing to migration, the disease now occurs in many other parts of the world with important foci in Canada, North America, Europe, and Australia. The name Chagas disease derives from that of Carlos Chagas who first described the parasites in 1910. Chagas initially believed that the parasites underwent schizogony within their mammalian host and hence named them *Schizotrypanum cruzi* – '*cruzi*' derives from the famous Oswaldo Cruz Institute in Brazil. Subsequently, it became clear that the parasite's life cycle does not include schizogony and most workers now refer to it as *Trypanosoma cruzi*.

In addition to humans, *T. cruzi* infects many domestic and wild mammals including dogs, cats, bats, rats, and armadillos. This makes disease control difficult because there are numerous potential reservoirs of infection. *Trypanosoma cruzi* infections are especially common among animals that utilise burrows and hollow trees because these are ideal dwellings for the insect vectors. The parasite is transmitted by blood-feeding reduviid bugs (Order Hemiptera, Family Reduviidae, Sub-Family Triatominae) that because of their appearance are sometimes called cone-nosed bugs. Over 130 species of reduviid bug can act as vectors but only those living near humans are medically important. Although Chagas disease now occurs in many countries, the lack of suitable vectors limits its potential for transmission outside Central and South America. There is therefore concern over the potential for vector species to disperse by air or sea (the bugs can survive for weeks without feeding) to other countries with luggage or farm produce. *Trypanosoma cruzi* can infect several other blood-feeding invertebrates. These include the bed bug (*Cimex lectularius*; Order Hemiptera, Family Cimicidae), the argasid tick *Ornithodoros moubata* and the medicinal leech *Hirudo medicinalis*. However, these are probably not important in the transmission of the parasite to humans. *Trypanosoma cruzi* can also infect vampire bats, but it is uncertain whether they transmit the parasite in their bite. During the acute phase of infection, the parasite occurs within the saliva of dogs, so it is clearly a possibility. However, whether a very ill bat would be capable of flying and feeding is uncertain.

The life cycle of *T. cruzi* differs from that of the other trypanosome species discussed so far in that it involves stercorarial transmission from the invertebrate vector to the vertebrate host (i.e., via the faeces rather than the saliva/bite). Reduviid bugs often defecate during or immediately after feeding, and they void the infective metacyclic stage trypanosomes with their faeces. The bugs feed at night and they often bite their victims around the face. They have therefore gained the common name of kissing bugs. On waking, the natural reaction is to rub/scratch at the bite site, and this rubs the parasite-infected faeces into the wound site. However, most infections result from bug faeces contaminating the victim's fingers and from there transferring to the eyes and mucous membranes where the parasites can penetrate more easily. We can also become infected from organ transplants and the use of contaminated blood during transfusions (Ries et al. 2016). In endemic areas, where the likelihood of acquiring *T. cruzi* infection from a blood transfusion is high, haematologists sometimes treat donated blood with gentian violet. This kills the parasites, but as the blood infuses through the tissues of the recipient, it also has the unfortunate effect of turning the patient purple temporarily. There are isolated reports of oral transmission of *T. cruzi* that date back many years. These arise through consuming food contaminated with infected bug faeces. They were considered rare oddities, but this mode of transmission is apparently becoming more common (Robertson et al. 2016). Furthermore, oral transmission tends to result in acute infections that may prove rapidly fatal. Congenital infection *via* the placenta is possible and can result in spontaneous abortion or serious disease of the infant at birth. Reservoir animals can acquire infections through being bitten, as well as from consuming bugs containing the parasites and from food contaminated with their faeces.

Once *T. cruzi* enters a suitable mammalian host, the trypomastigotes invade both phagocytic and non-phagocytic cells. Indeed, they are apparently capable of invading all nucleated cells. However, most of the pathology arises from their invasion of smooth, cardiac, and skeletal muscle cells, nerve cells, and cells comprising organs such as the liver, spleen, and lymphatics. Within a host cell, the parasites develop within a parasitophorous vacuole. The mechanism by which a trypomastigote enters a host cell varies between cell types and may involve active invasion by the parasite and passive internalization through endocytosis/phagocytosis. Within the host cell, a parasitophorous vacuole forms around the parasites and within this, they transform into the amastigote stage (1.5–4 μm in size). The amastigotes multiply by binary fission and once large numbers are present within a cell, it is sometimes referred to as a pseudocyst. Eventually, the parasites kill their host cell and then they transform back to trypomastigotes (Figure 4.10). The trypomastigotes then invade new cells and repeat the process. This distinguishes *T. cruzi* from *Leishmania* in which the parasites occur mainly within phagocytic cells and remain in the amastigote stage within the vertebrate host. The trypomastigotes of *T. cruzi* are unable to reproduce although large numbers of them may occur in the blood stream, especially during the early stages of infection. Broad and slender morphotypes of the trypomastigotes exist. Both forms are small (16.3–21.8 μm in length including the flagellum) compared with those of some other trypanosomes and they have a characteristic curved 'C' shape. By contrast, their kinetoplast is exceptionally large and the cell membrane sometimes bulges around it.

A reduviid bug vector acquires an infection when it feeds on blood containing the trypomastigote stage. Reduviid bugs can take in large blood meals several times their own body weight, and this increases their chances of ingesting a trypomastigote. In some bug populations, over half of them carry a *T. cruzi* infection. In addition to feeding on blood, reduviid bugs will also attack and feed on one another. Therefore, bugs may acquire an infection by ingesting parasites from one of

Figure 4.10 *Trypanosoma cruzi.* Note the characteristic C shape often seen in the trypomastigote stage.

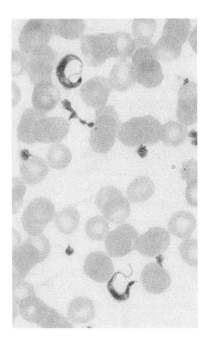

their brethren. However, this is probably not a common means of infection. When the parasites reach the posterior of the insect's midgut, they transform into short epimastigotes. These then divide by longitudinal fission to form long epimastigotes. Ultimately, the epimastigotes move to rectum where they transform into the infective short metacyclic trypomastigote stage. The gut of infected bugs can contain huge numbers of parasites but unlike some of the other trypanosome species, *T. cruzi* does not appear to harm its invertebrate host.

The course of Chagas disease varies considerably, and there are marked disparities between individuals and geographic localities. This suggests that genetic differences on the parts of both the parasite and the host influence the manifestation of disease. There are two phases of the disease: an acute phase and a chronic phase. Initial parasite invasion may cause an acute infection or symptoms so general that it is not obvious that infection has occurred. The acute phase is characterised by high levels of parasitaemia and in a small percentage of cases proves fatal. There is often an initial localised inflammatory response with swelling of the nearest lymph node. If the infection starts at an insect bite wound, a raised red nodule develops called a 'chagoma'. If the infection occurs via the eye, it induces a condition called Romaňa's sign in which the eyelid and preauricular lymph node swell so much that the eye becomes closed. As the acute phase of the disease progresses, the parasites invade all organs of the body. However, the most severe consequences arise from the parasite's tendency to localise within and destroy heart muscle and cardiac ganglion cells. The pathological mechanisms are uncertain. However, damage to cardiac muscle during chronic Chagas disease may have an autoimmune basis. As he became older, Charles Darwin suffered from chronic ill health, the symptoms of which were consistent with him suffering from Chagas disease. Obviously, it is impossible to confirm this, but in his South American journals he recorded being bitten by reduvidid bugs. If the parasites invade the brain, meningoencephalitis can develop with potentially fatal or long-term damage as a result. The patient often develops a fever and their liver and spleen become enlarged; they may also suffer from diarrhoea and exhibit evidence of respiratory infection. The acute phase occurs most commonly in children less than 5-years-old but unless their heart or nervous tissues are severely damaged, most of them recover even without adequate medical treatment.

4.3 Phylum Chlorophyta

Commonly known as the green algae, the Chlorophyta is a paraphyletic phylum – that is, the species within it derive from several different ancestors. Because most of them contain chloroplasts, they are often referred to as plants. Furthermore, these chloroplasts are similar in appearance to those in multicellular plants, such as wheat and roses, have a very similar physiology, and contain both chlorophyll a and chlorophyll b. However, the Chlorophyta are single-celled organisms (although some are colonial) and are usually classed within the Protista rather than Plantae – although this remains a source of debate. Several species have close symbiotic relationships with invertebrates. For example, *Chlorella* spp. lives in association with the cnidarian *Hydra viridis,* and some sea slugs extract the chloroplasts from their algal food and utilise them as photosynthetic organelles within their own cells. Although various algae grow on the pelage of sloths and the skin of certain lizards, there are few reports of them becoming intracellular symbionts of vertebrates. An instance where it does occur is between the alga *Oophila amblystomatis* and the egg masses of certain amphibia. This alga enters the embryos of the spotted salamander *Ambystoma maculatum* and is maternally transmitted (Kerney et al. 2019). The alga utilises nitrogenous waste generated by the host cells and undertakes photosynthesis; the relationship therefore appears to be beneficial to both organisms. The alga

Nannochloris eukaryotum will enter an endosymbiotc relationship with human cells under *in vitro* culture conditions. However, this is mainly of interest for the development of molecular machines (Black et al. 2014) and not something that happens naturally outside the laboratory.

There are isolated case reports of *Chlorella* spp. infecting wounds in humans and other mammals. There are also accounts of fatal disseminated infections in sheep that were presumably acquired via the digestive tract after consuming contaminated drinking water (Ramírez-Romero et al. 2010). These, presumably, represent rare opportunistic infections. Some species of algae lost their chloroplasts during evolution. Amongst these are members of the genus *Prototheca,* which includes species that parasitize mammals and the genus *Helicosporidium* that are parasitic in insects.

4.3.1   Genus *Prototheca*

Members of this genus are closely related to the well-known alga *Chlorella*, but they lack chloroplasts, and most species survive as saprophytes feeding on dead organic matter in a similar manner to free-living fungi. They are found throughout the world and can be isolated from the soil, slime, sludge, gut contents, faeces, marine and freshwater, swimming pools, and virtually anywhere which has high organic matter content (Kano 2020). Some species are facultative parasites that infect various animal species with consequences that range from mild disease to fatalities. *Prototheca wickerhamii* and *Prototheca zopfii* are responsible for most human infections. These are usually associated with patients who are immunocompromised through disease (e.g., HIV infection) or medical treatment (e.g., chemotherapy/ corticosteroid therapy). A new species, *Prototheca cutis*, was described from a patient in Japan (Satoh et al. 2010) and further species will probably be discovered in the future now that the genus is receiving more attention. In 2018, an outbreak occurred in a cancer chemotherapy unit in India that resulted in 12 patients becoming infected with *P. wickerhamii* (Khan et al. 2018).

The algae gain entry to the body via the skin – usually through an existing wound – and cause a localised cutaneous infection. This often manifests as dermatitis with the formation of pustules, ulcers, and erythematous plaques. Occasionally, the infection becomes disseminated throughout the body and causes potentially fatal meningitis (Joerger et al. 2020).

There are isolated but increasing case reports of dogs and cats suffering from illnesses caused by *Prototheca*. These often take the form of gastrointestinal infections that cause diarrhoea, but they can become disseminated elsewhere in the body with often fatal results. In cows, *P. wickerhamii*, *P. zopfii*, and *Prototheca blaschkeae* are responsible for sporadic cases of mastitis in many parts of the world. Prototechosis is not a commonly recognised cause of mastitis, and therefore, it often remains undiagnosed because vets do not think to test for it. This can cause problems because the algae do not respond to normal treatments for mastitis. Indeed, at the time of writing, there was no effective treatment available. Consequently, the course of the disease can be prolonged, and there is a potential for severe economic losses in dairy herds (Jagielski et al. 2019).

4.4   Kingdom Fungi

Some estimates suggest that there may be over a million species of fungi although less than 10% of these have so far been described. Unlike plants, fungi are heterotrophs – that is, they cannot make their own food and must gain their nutrients by breaking down existing organic matter. Most fungi do this by acting as saprophytes, that is, they break down dead organic matter. In addition, many species are in symbiotic relationships with plants and invertebrates, whilst some are parasites of other fungi, plants, and invertebrate and vertebrate animals. Some of these parasitic species are important in human and veterinary medicine, as well as wildlife ecology. For example, *Pneumocystis* (which was once thought to be a protozoan) is a major cause of morbidity and mortality in AIDS patients (Gilroy and Bennett 2011), the skin disease 'ringworm' in cattle is caused not by a helminth but fungi such as *Trichophyton verrucosum* (Pier et al. 1994), and chytridiomycota fungi are responsible for widespread and catastrophic levels of mortality among amphibians in many parts of the world (Fisher and Garner 2020). However, only the Microsporidia will be covered here.

4.4.1 Phylum Microspora

The Microsporidia are a cosmopolitan group of obligate intracellular parasites that infect many invertebrates and vertebrates. There are even accounts of them infecting protozoa but apparently not plants or fungi. Over 1,200 species have been described, but the majority of these are parasites of invertebrates and fish. Several species are of medical, veterinary, and commercial importance. For example, *Nosema apis* and *Nosema ceranae* are major causes of disease in honeybees whilst several species such as *Nosema locustae* (*vs* locusts) and *Nosema algerae* (*vs* mosquitoes) have potential as biological control agents.

Up until the AIDS epidemic, there were few accounts of human infections by microsporidian parasites. However, they subsequently became identified as important causes of morbidity and mortality amongst those suffering from AIDS and other immunosuppressive illnesses. Fourteen species have so far been found to infect humans although *Enterocytozoon bieneusi* is responsible for most clinical cases. It should be noted that there is considerable genetic variation within individual species that influences their host range (Heyworth 2017). Nevertheless, some microsporidia are undoubtedly zoonotic and infect both humans and other mammals and birds. Indeed, some commentators consider them to be extremely important emerging pathogens (Stentiford et al. 2019). This is particularly the case now that global food chains mean that foodstuffs are rapidly transported around the world.

Initially designated as protozoa, subsequent molecular evidence indicated that microsporidia are fungi. Precisely where they fit within the taxonomy of fungi is uncertain although they show some resemblance to the zygomycetes. The zygomycetes also have relevance to parasitologists since they include genera such as *Pilobolus* that helps to spread the infective larvae of the lungworm *Dictyocaulus viviparus* (Doncaster 1981) and *Entomophthora* that have potential as biological control agents of insect vectors. However, the taxonomic status of the microsporidia is far from settled and Ruggiero et al. (2015) consider that the phylum Microspora belongs back within the kingdom Protista.

As with *Entamoeba histolytica*, the microsporidia were once thought to have split off from other organisms at an early stage in their evolution because they did not appear to contain mitochondria. However, they too were subsequently found to contain genes with mitochondrial functions and mitosomes (putative relict mitochondria). They also have some of the smallest genome sizes and the fewest protein coding genes of all the eukaryotic organisms; in *Encephalitozoon intestinalis* the genome is only 2.3 megabases (Mb) in size although in *Glugea atherinae*, a fish parasite, it is almost ten times larger at ~20 Mb.

The spore is the only microsporidian life cycle stage capable of surviving in the external environment. Immediately above the spore's plasma membrane are two protective layers, the first of these is the 'endospore' which contains chitin and is electron luscent when viewed with a transmission electron microscope and above this is the 'exospore' that contains glycoprotein and is electron dense, so it appears dark in transmission electron micrographs (Figure 4.11). The spore walls must provide excellent protection because in some species they can remain infective for over a year.

Figure 4.11 Transmission electron micrograph of a developing spore of the microsporidian *Nosema helminthorum*. The thick spore coat makes sectioning extremely difficult, and they often pull out from the sample. The coiled polar tube is clearly visible as a series of circles either side of the upper portion of the cell. For further details, see text.

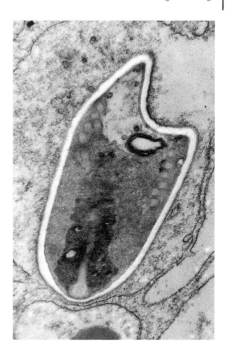

Microsporidia are normally transmitted horizontally when the host encounters the spores; humans usually become infected by ingesting or breathing in the spores. Vertical transmission has not yet been described in humans, but it occurs in some mammals by crossing the placenta or through infecting the eggs while they are still in the ovary in invertebrates. Transovarial transmission is common among endosymbiotic bacteria such as *Wolbachia* but very rare among protozoan parasites (Dunn et al. 2001). Like *Wolbachia*, some of the microsporidia species that are transmitted transovarially affect the sex rations of their hosts. For example, females of the amphipod *Corophium volutator* infected with microsporidia produce predominantly female offspring. They are also more fertile than uninfected females and this will further promote the spread of the parasite through the population (Mautner et al. 2007).

Microsporidian spores, like the other life cycle stages, usually contain either a single nucleus (monokaryon) or two adjacent nuclei (diplokaryon) that function as a single unit. The spores also contain a posterior vacuole, and a structure called the 'polaroplast' that probably derives from modified Golgi apparatus. In addition, there is a coiled polar tubule (polar filament) that attaches to the anterior end of the spore by an 'anchoring disc complex'. The polar tube is hollow and unique to the Microsporidia but in terms of appearance and function it bears more than a passing resemblance to the nematocysts found in jellyfish (Cnidaria). When the spore receives the correct stimulation (presumably a combination of pH and chemical factors), the posterior vacuole and polaroplast absorb water and start to swell. Because the tough spore wall prevents it from expanding, the pressure within the spore starts to rise. Ultimately, the spore wall ruptures at the anterior end where the spore wall is thinnest, and the polar tubule shoots out through the break as if from a harpoon gun. The polar tubule can discharge with sufficient force to pierce the adjacent host cell or alternatively it may be subsequently ingested by receptor-mediated endocytosis. At the same time, the pressure within the spore forces the nucleus and cytoplasm, now referred to as the 'sporoplasm' down the everting polar tubule and thence into the host cell. The spore ensures its proximity to a suitable host cell by binding onto host-cell sulphated glycosaminoglycans (GAGs). In addition, the exospore and endospore both contain an adherence protein called 'endospore protein 1' (EnP1). Alternatively, the host cell may ingest whole spore, but infection still results from the discharge of the sporoplasm into the host cell cytoplasm.

Once within the host cell cytoplasm, the sporoplasm differentiates into a 'meront' and undergoes a series of cycles of asexual reproduction called merogony, which results in the formation of numerous more meronts. In the cases of *Entercytozoon* and *Nosema,* the meronts remain in direct contact with the host cell cytoplasm, whilst in *Encephalitozoon* they are localized within a membrane-bound parasitophorous vacuole of host cell origin. The meronts of both *Nosema* and *Encephalitozoon* divide by simple binary fission, but those of *Enterocytozoon* have a more complex development which produces multinucleate cells. After several cycles of merogony, the parasites start to produce spores by sporogony: the meront transforms into a sporont which produce sporoblasts that then mature into spores. The spores steadily accumulate in the host cell and may eventually fill it. When the host cell membrane eventually ruptures, the spores are released and may infect an adjacent cell or be released into the environment.

In humans, microsporidia usually cause persistent or self-limiting infections of the enterocytes lining the gastrointestinal tract. They can cause extensive damage to the mucosal surface of the intestine and symptoms therefore typically cause diarrhoea, abdominal pain, malabsorption and wasting. There are also records of them invading the eye and causing keratitis in both immuno-compromised and immunocompetent individuals, as well as infecting the skin, kidney, heart, and lungs.

5

Platyhelminth and Acanthocephalan Parasites

5.1   Introduction

In this chapter, we will introduce three of the most important groups of parasitic 'worms'. Namely, the flatworms (flukes), the tapeworms, and the thorny headed worms. The term 'worm' is, in a way, unfortunate because these organisms are only very distantly related to the real worms – the Annelida. Furthermore, most Annelida are free-living although one group, the Hirudinea, includes the leeches, some of which feed on blood and some people consider these to be parasites. The thorny-headed worms or acanthocephalans to give them their scientific name are unusual parasites. They are common parasites, but they rarely infect humans and domestic animals and therefore seldom gain attention in the scientific literature. They do, however, have some fascinating biological traits.

5.2   Phylum Platyhelminthes

Members of the phylum Platyhelminthes are commonly known as the 'flatworms' on the not unreasonable basis that they are dorsoventrally flattened and worm-like in appearance. They are acoelomate soft-bodied animals that are bilaterally symmetrical (i.e., their left side is the same as their right side) with an obvious head-end. Most platyhelminths have a mouth at the anterior end although in some free-living species the mouth is situated at or close to the centre of the body, while in tapeworms the mouth is absent. Tapeworms also lack a gut although this is present in most other platyhelminth species and usually consists of a blind-ending sack or series of branching tubules. The lack of an anus means that waste material is passed back through the mouth.

Table 5.1 Characteristics of the Turbellaria, Monogenea, Digenea, and Cestoda.

	Lifestyle	Number of hosts	Characteristic feature
Turbellaria	Free-living, terrestrial and aquatic species	Not applicable	Predators and scavengers
Monogenea	Ectoparasites	One	Mostly parasites of fish but some infect amphibia
Digenea	Endoparasites	Two or more	First intermediate host is always a mollusc
Cestoda	Endoparasites	Two or more	Lack mouthparts at all stages of development

Platyhelminths do, however, have a ramifying series of tubules that constitute their 'excretory system'. Flame cells (protonephridia) maintain the movement of fluid within the excretory system, and the waste is removed through excretory pores. Because the excretory system removes excess water and ions, it is sometimes referred to as the osmoregulatory system. Most platyhelminths are hermaphrodites, containing both male and female reproductive organs, although in a few species there are separate male and female sexes.

The taxonomy of the phylum Platyhelminthes has undergone substantial changes in recent years. Previously, taxonomists divided the phylum into four classes: Turbellaria, Monogenea, Trematoda (Digenea), and Cestoda (Table 5.1). Subsequent molecular research indicates that the monogeneans, trematodes, and cestodes represent a single monophyletic assemblage that evolved from free-living turbellarians. By contrast, the turbellarians do not form a single coherent monophyletic group and are either paraphyletic or polyphyletic. Currently, according to phylogeneticists, the Monogenea, Trematoda, and Cestoda constitute orders within a superorder Neodermata that sits within the Class Rhabditophora. However, many authors continue to refer to them as four Classes. We shall consider just the Trematoda and Cestoda (Cestoidea) because these contain the bulk of the platyhelminths of medical and veterinary importance.

5.2.1 Trematoda

The Trematoda are commonly known as flukes: those living in the blood are therefore called blood flukes, those living in the gut are intestinal flukes, whilst those associated with the liver and gall bladder are called liver flukes. The word fluke has been in use for hundreds of years to describe the parasite we now know as *Fasciola hepatica*. In this instance, fluke probably derives from the Old English 'flōc' that was a term for flatfish such as the flounder (*Pleuronectes flesus*), to which the liver fluke *F. hepatica* has a superficial resemblance. Before the advent of scientific taxonomy, it was a common belief that animals were related if they shared a similar appearance or mode of locomotion.

The Trematoda divides into two groups: the Aspidogastrea and the Digenea (de León and Hernández-Mena 2019). The Aspidogastrea are also known as the Aspidobothria and the Aspidocotylea. They are of biological interest because several species are apparently making the transition between a free-living and a parasitic lifestyle. Most species are parasitic in fish and turtles, and they include species with fascinating life cycles, but we are unable to discuss them further here. Further details can be found in Kearn (1998).

The majority of the trematodes are members of the Digenea (~18,000 species), and the adults of these species are all obligate parasites of vertebrates. They have complex life cycles that include one or two intermediate hosts. The first intermediate host is invariably a mollusc, in which asexual reproduction takes place (Table 5.2). Intriguingly, in some trematode species the asexual stages developing in the snail intermediate host exhibit social organization (Resetarits et al. 2020). After invading the snail, the parasite reproduces asexually, and some of the progeny develop as reproductive, whilst others form non-reproducing soldiers. The soldiers are smaller than the reproductive, but they are more active and have larger mouthparts. The soldiers congregate in regions where miracidia (the stage that invades the snail host) are likely to penetrate. They attack the asexual stages of other trematodes of both the same species as themselves and those of other species. This indicates that the asexual stages can distinguish 'self' from 'non-self'. That is, they ignore individuals with the same genetic constitution as themselves but attack those of same species developing from a different miracidium. The development of distinct castes is probably widespread among trematodes and they offer excellent models for the study of social organization. After developing and reproducing within the snail intermediate host, digeneans produce cercariae. The cercariae burrow out of their host and, usually, swim off in search of the definitive host. Remarkable numbers of cercariae can be produced: Soldanova et al. (2016) estimate that over 4.5 ton of cercariae per year may be produced in a single pond. Of course, this is presumably a sizeable pond, but it illustrates that parasitic species can represent a considerable nutritional resource that is often overlooked in food web dynamics (McKee et al. 2020).

Table 5.2   Life cycle stages of digenetic trematodes.

Life cycle stage	Description	Reproduction
Adult	Lives in the definitive host. Usually, hermaphrodites but sexes separate in some species. Has mouth and gut though may also absorb nutrients across the body surface. Motile. Produces eggs	Yes. Usually, sexual reproduction but may be parthenogenic
Egg	Contains the miracidium. May hatch in the environment or within gut of the first intermediate host	No
Miracidium	Infective stage. Covered in cilia, motile, invades the first intermediate host	No
Sporocyst	Lacks a mouth and gut; absorbs nutrients across body wall. Reproduces asexually within first intermediate host	Yes. Asexual reproduction to form daughter sporocysts or rediae
Redia	Has a mouth and gut. Motile. Reproduces asexually within first intermediate host. Evidence of caste system in some species	Yes. Asexual reproduction to form daughter rediae or cercariae
Cercaria	Infective stage. Usually motile with a propulsive 'tail'. Often leaves first intermediate host and invades second intermediate host or definitive host	No
Metacercaria	Infective stage. Not motile once encysted and covered with protective wall. Develops in the environment or within the second intermediate host	No

N.B.: Not all species produce rediae and metacercariae.

Olson et al. (2003) divide the Digenea into two clades: the Plagiorchiida and the Diplostomatida. The Plagiorchiida contains the majority of digenean species of which the Families Fasciolidae (e.g., *F. hepatica*), Dicrocoeliidae (e.g., *Dicrocoelium dendriticum*), Paramphistomatidae (e.g., *Calicophoron daubneyi*), and Paragonomidae (e.g., *Paragonimus westermani*) are of particular importance in human and veterinary medicine. The Diplostomatida contains the blood flukes of which the Schistosomatidae (schistosomes) are of importance as the cause of schistosomiasis.

5.2.1.1 Family Fasciolidae

This family includes some of the largest flukes: *Fasciola gigantica* can reach 75 mm in length and 12 mm in breadth whilst *Fascioloides magna* is even bigger and can be up to 100 mm in length and 26 mm in breadth. Adult fasciolids usually live in the bile ducts of ungulates and other herbivorous mammals – although humans are sometimes utilized as definitive hosts as are other omnivores such as pigs and occasionally certain carnivores. *Fasciolopsis buski* is an exception where the adults live in the intestines of both humans and pigs. Adult fasciolids are usually leaf-shaped with the body broadening out immediately behind the anterior end and then tapering gradually towards the posterior (Figure 5.1a). Their tegument is covered with large-scale-like serrated spines and surface folds (Figure 5.1b). The tegument is metabolically active and plays an important role in the absorption and exchange of nutrients and ions. The spines probably facilitate movement by enhancing the parasites grip on the host's tissues. There is an oral sucker surrounding the mouth, and a short distance behind this there is a well-developed ventral sucker. The adult worms lack teeth or cutting mouthparts but have a large muscular pharynx that they use to suck in and remove plugs of host's tissue. Shortly beyond the pharynx, the gut divides up into a series of ever smaller branches throughout the body. These branches ultimately end blindly – there is no anus. They lack a circulatory system, and nutrients have to reach all the cells through diffusion. Consequently, because the worms are so large, if the gut was divided into just two caecae (as is the case with some smaller digenean parasites), the distance between the gut and the most distant body regions would be too great. Like most trematodes, fasciolids are hermaphrodites containing both testes and ovaries. Their vitellaria are well developed and ramify throughout the body in the region beneath the ventral sucker. The eggs have thin shells and a 'lid' (operculum) (Figure 5.1c) through which the miracidium emerges. Fasciolids utilise only one intermediate host, which is an aquatic or semi-aquatic snail. Their transmission to the definitive host is therefore heavily influenced by the availability of their snail intermediate host species and its population dynamics.

5.2.1.1.1 *Fasciola hepatica/Fasciola gigantica*

Fasciola hepatica and *F. gigantica* are economically important as the cause of fascioliasis, commonly known as or 'liver rot', in sheep, cattle, and other ruminants. Human fascioliasis is a common but under-reported disease in many developing countries and the true numbers of people infected could be as high as 35–72 million (Sabourin et al. 2018). *Fasciola hepatica* has a widespread distribution and occurs in Europe, Africa, Asia, Australia, and North and South America. It was once thought that Europeans introduced *F. hepatica* into South America in the fifteenth century. Although they probably did introduce infected animals, the discovery of *F. hepatica* eggs in 2,300-year-old deer coprolites in Patagonia indicates that it was already present in the region (Beltrame et al., 2020). *Fasciola hepatica* also infects numerous domestic and wild mammals, including rabbits, pigs, horses, antelopes, and coypu. This complicates its control because although one can use anthelmintics to rid domestic animals of their infections, snails will continue to be

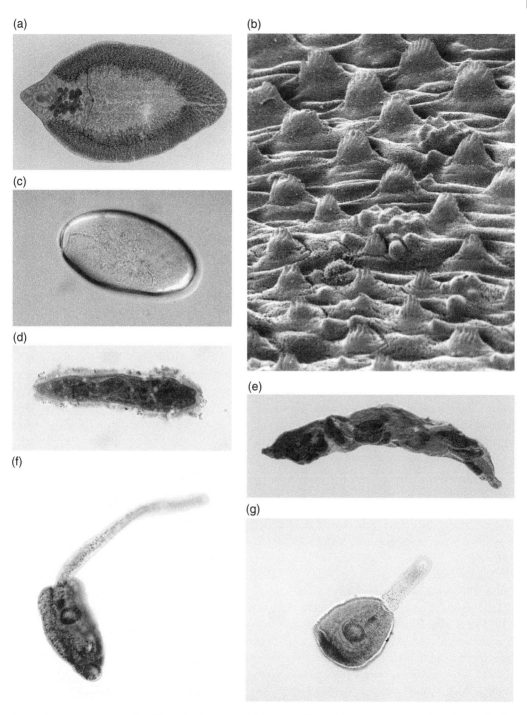

Figure 5.1 *Fasciola hepatica* life cycle stages. (a) Adult worm. The adult worms vary considerably in size but can reach up to 30 mm in length and 15 mm in width (b) Tegument surface of adult worm. (c) Operculate egg. Eggs typically measure 130–150 μm in length and 60–90 μm in width (d) Miracidium. (e) Redia containing developing cercariae. (f) Cercaria. (g) Late cercaria stage in the process of transforming into the metacercaria stage – the tail will be lost.

infected from eggs released by reservoir hosts. *Fasciola gigantica* also has a wide distribution but is most prevalent in the tropics, such as parts of Africa, the Indian subcontinent, and certain islands in the Pacific. In some parts of the world, *F. hepatica* and *F. gigantica* co-exist and hybridise (Ichikawa-Seki et al. 2017). These hybrids are probably incapable of producing sperm and reproduce by parthenogenesis.

The prevalence rates for fascioliasis in sheep and cattle can be as high as 30–90%, and on a global basis it is probably the most important cattle helminth infection. In some countries, fascioliasis is an emerging or re-emerging infection. This is often linked to agricultural practices, such as the building of irrigation systems and dams that aid the spread and proliferation of the snail interme- diate hosts. Lack of clean drinking water and risky crop production practices facilitates human infections. For example, when snail intermediate hosts are present in drinking water and/or ponds in which aquatic crops (e.g., water cress, *Nasturtium officinale*) are grown, and these become con- taminated with the faeces of infected cattle. Furthermore, resistance to the two most widely used flukicides, triclabendazole and oxyclozanide, is increasingly common in many countries.

The adult flukes live in the bile ducts where they feed on blood oozing from wounds that develop on the lining of the bile duct: the worms feed repetitively in the same region until ulcers develop. They do not appear to ingest bile although they absorb nutrients from it across their tegument. Although they can self-fertilize, they usually mate with other worms and there is often a high level of genetic diversity within a single host. This means that if drug-resistant phenotypes arise, these can rapidly spread amongst the local population; at the time of writing, resistance to one of the main drugs, triclabendazole, was an increasing problem. The life cycle (Figure 5.2) begins when adult worms release eggs that pass with the bile into the duodenum and are then voided with the host's faeces. If the eggs fall into freshwater, they continue their development and under ideal circum- stances, they hatch after about 9–10 days to release the free-living but non-feeding miracidium stage. The miracidium is covered in cilia (Figure 5.1d) and actively swims in search of a suitable lymnaeid aquatic or semi-aquatic snail that will act as the intermediate host. At least 30 species of snail world- wide can act as intermediate hosts and this lack of specificity has facilitated the spread of fascioliasis within and between countries. Nevertheless, snail susceptibility varies between populations of the same species (Alba et al. 2019), and therefore, the presence of a particular species does not necessarily mean that it is important in disease transmission. In the United Kingdom, *Galba truncatula* (previously *Lymnaea truncatula*) is the most important intermediate host of *F. hepatica,* whilst in North America *Fossaria modicella* and *Stagnicola bulimoides* are often utilised. *Pseudosuccinea columella* (previously *Lymnaea columella*) is an important intermediate host in both North and South America and has subsequently spread to both Europe and Australia. In many parts of the world, *Radix auricularia* (previously *Lymnaea auricularia*) is the most important intermediate host of *F. gigantica*.

Having located a suitable snail, the miracidium chemically and physically bores its way inside. It then sheds its cilia and transforms into the sack-like mother sporocyst stage. The mother sporo- cyst lacks a mouth and gut and absorbs nutrients across its tegument. Within the mother sporo- cyst, asexual reproduction occurs resulting in the formation of first-generation redia (plural 'rediae'). The rediae are motile and after their release from the mother sporocyst, they move through the snail host's tissues. The rediae have a mouth and gut and ingest snail tissues, although they continue to absorb nutrients across their tegument. The rediae undergo a complex pattern of reproduction in which they form second- and sometimes third-generation rediae. Ultimately, each redia gives rise to several cercariae (Figure 5.1e) and these physically and chemically burrow out of the snail and swim away. A cercaria consists of two body regions: a globular body and a long, stoutly built, 'tail' (Figure 5.1f). The 'body' has a mouth surrounded by an oral sucker and a ventral sucker. There is a pharynx, and the gut divides into two caecae but the cercaria does not feed.

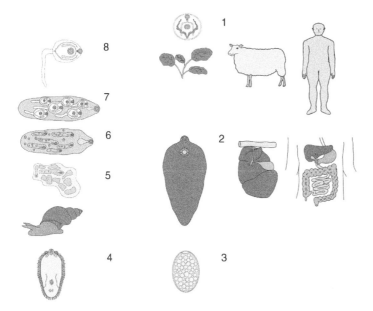

Figure 5.2 Life cycle of *Fasciola hepatica*. 1: The definitive host becomes infected when it consumes vegetation or water contaminated with the metacercaria stage. 2: The young flukes emerge within the small intestine and invade through the gut wall. After migrating through the liver (during which they may cause acute fascioliasis), they reach the bile ducts where they reach adulthood. Adult flukes cause chronic fascioliasis. 3: The flukes release operculate eggs that are passed with their host's faeces. 4: The eggs hatch in water to release the miracidium stage – this actively searches for, and invades, the intermediate host snail. 5: Within the snail, the parasite transforms to the sporocyst stage, and this produces first-generation rediae. 6: The first-generation rediae produce second-generation rediae. 7: The second-generation rediae produce cercariae. 8: The cercariae actively bore their way out of their snail host and swim off. After a short time, the cercariae transform into the inactive metacercaria stage either on semi-aquatic vegetation or among detritus at the base of the pond. Drawings not to scale.

The powerful 'tail' contains striated muscle (unusual in platyhelminths) that enables the cercariae to swim actively. The cercariae of *F. hepatica* usually emerge from their snail host during the night and after swimming for a while they settle on blades of grass, or other plants, just below the water surface. Here they lose their tails and transform into metacercariae (Figure 5.1g). Sometimes they encyst on the surface of the water and, subsequently, sink to the bottom.

The definitive host becomes passively infected when it eats plants contaminated with the metacercariae. Cattle can also become infected when they wade into ponds to drink and, in the process, stir up cysts lying at the bottom and swallow them. After ingestion, the metacercariae hatch in the duodenum in response to specific physical (e.g., temperature) and chemical stimuli (e.g., bile composition). The young flukes then physically and chemically bore their way through the gut and enter the abdominal body cavity. They probably reach the liver indirectly via the body wall, and they then penetrate the liver capsule and begin migrating through the liver parenchyma. The young flukes take about 5–6 weeks to reach the gall bladder and the simultaneous migration of large numbers of flukes causes acute fascioliasis. Precisely how the young flukes move from the liver parenchyma into the bile ducts is uncertain (Moazeni and Ahmadi 2016). Once the flukes reach the gall bladder, they become sexually mature and commence laying eggs. The adult worms live for several years, and during this time they cause chronic fascioliasis. There is no synchrony in the rates at which the young flukes move through the liver and then become mature. Consequently, following an initial infection young flukes will arrive at the bile ducts and then become mature over a period of two or more months.

Environmental DNA (eDNA) and Parasite/Intermediate Host Monitoring

---

The use of environmental DNA (eDNA) is increasingly used in ecology and conservation as a means of monitoring the presence of individual species within an ecosystem and/or, through large-scale metabarcoding, the composition of the biome. Some scientists consider eDNA to be that shed by organisms into the surroundings in their secretions, excretions, and by decomposition of their body after death. Others, consider eDNA can include whole organisms, such as eggs, cysts, and spores present within a sample although, in this case, a preparation step is required to release the DNA from within the organisms.

The benefit of analysing eDNA is that, for example, one need only to take air, water, or soil samples from the sample site and there is no need to physically find and identify the target species. This procedure is now being employed by parasitologists and it is proving especially suitable for those parasites that have aquatic stages. For example, Jones et al. (2018) have shown that it is possible to identify the presence of the trematodes *F. hepatica* and *C. daubneyi* and their snail intermediate host *G. truncatula* from water samples obtained from pastureland. Physically searching for and identifying the snails is difficult and time consuming, and their presence does not necessarily mean that they are acting as intermediate hosts. Similarly, even if the snail population can act as intermediate hosts, only a small minority of snails are usually infected and therefore identifying these would normally involve tedious analysis and the deaths of numerous snails. Consequently, analysing eDNA in water samples offers a potential means of identifying fields that represent disease risks and not undertaking costly control/treatment measures where they are not necessary. Similarly, eDNA can be used to monitor the presence of schistosome species and their intermediate snail hosts within water bodies and thereby the risks they pose and the effectiveness of any control measures (Champion et al. 2021).

5.2.1.1.2 Fasciolopsis buski

Fasciolopsis buski is predominantly found in India, Bangladesh, China, and other Southeast Asian countries. It is a large fluke that reaches up to 75mm in length and 25mm in breadth (Figure 5.3). The definitive hosts are pigs and humans. The life cycle of *Fa. buski* (Figure 5.4) resembles that of *F. hepatica*, except that the adult worms are typically found in the duodenum and upper regions of the small intestine. However, in heavy infections the parasites may also reside in the stomach and as far down the intestinal tract as the colon. *Fasciolopsis buski* also differs from other adult fasciolids in having unbranched digestive caecae. This may be because it lives in its host's digestive tract and can therefore absorb more nutrients across its tegument than flukes that live in the bile ducts. The intermediate hosts are semi-aquatic snails belonging to the genera *Planorbis* and *Segmentina*. Within the snails, sporocysts, rediae, and finally cercariae are formed; the cercariae then bore their way out and encyst as metacercariae on surrounding vegetation. The snails feed on water caltrop (*Trapa natans)* and water chestnut (*Eleocharis tuberosa*) that are widely cultivated throughout Southeast Asia for food and are consumed raw. Furthermore, these plants are often fertilised with human and pig faeces, and therefore, there is a perfect combination of all the elements needed to ensure transmission. Not surprisingly, infection rates can be as high as 60% in both humans and pigs. In addition, conflict and economic problems have resulted in widespread migrations and thereby the spreading of the parasite to regions in which it did not previously occur.

The worms feed on the lining of the gut and the irritation causes excessive mucus secretion. The patient may suffer from non-specific symptoms such as diarrhoea, vomiting, and abdominal pain.

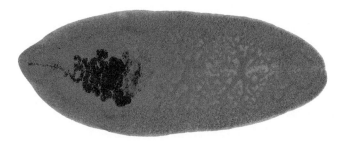

Figure 5.3 *Fasciolopsis buski* adult stage. Adult worms can grow to 75 mm in length and 25 mm in breadth.

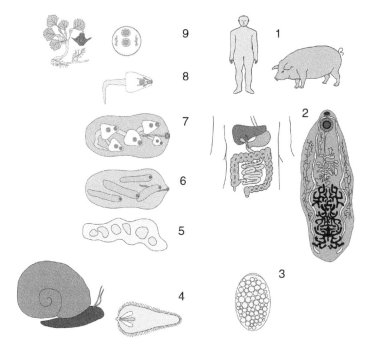

Figure 5.4 Life cycle of *Fasciolopsis buski*. 1: Infection occurs when the definitive host (usually humans and pigs) consumes vegetation containing the metacercaria. 2: Metacercaria excyst in the duodenum and become adult in the duodenum and upper small intestine. 3: Eggs are passed with the faeces. 4: The eggs hatch in water to release the miracidium stage that invades a suitable snail intermediate host, such as *Segmentina hemisphaerula*. 5: Within the snail, the parasite transforms into the sporocyst stage that reproduces asexually and produce the mother redia stage. 6: The mother redia produces daughter rediae. 7: The daughter rediae produce cercariae. 8: The cercariae bore their way out of the snail and swim off. 9: The cercariae encyst on water vegetation, such as water caltrop, to form metacercariae. Drawings not to scale.

In severe cases, there is erosion and ulceration of the gut wall, and this may even become perforated. The pathology coupled with the large size of the flukes can lead to gut obstruction, appendicitis, and in severe cases, the infection can prove fatal. Very severe infections may require surgical intervention (Wu et al. 2020) although the parasites can usually be removed by administering an anthelmintic. In addition, some people develop an allergic reaction to the parasite antigens that result in ascites with oedema developing in the face and legs, and this too can be fatal. However, most people are infected with small numbers of flukes and express few or no signs of disease.

5.2.1.2 Family Dicrocoeliidae

The adult flukes belonging to the Dicrocoeliidae are rather delicate, small- to medium-sized organisms that parasitize amphibians, reptiles, birds, and mammals. They usually reside in the ducts leading from the pancreas and gall bladder, although some species live in the intestine itself. There is a large oral sucker around the mouth and the ventral sucker is situated a short distance behind it. They do not usually have the 'broad shoulders' of the fasciolids and their body tapers to a point at both the anterior and posterior ends. Beyond the pharynx, the gut divides into two simple caecae that end blindly before they reach the end of the body. Their tegument often lacks spines. The first intermediate hosts are usually terrestrial snails, and there is also a second invertebrate intermediate host. Most species are parasites of wild animals and of no medical or veterinary importance. However, *Dicrocoelium dendriticum* is an important pathogen of sheep and there are isolated cases of human infections.

5.2.1.2.1 *Dicrocoelium dendriticum*

This has a widespread distribution including Europe, Asia, and USA and Canada but is currently absent from Australia and Central and South Africa. Its presence in the USA and Canada probably results from the import of infected livestock, and it is considered an emerging infection (van Paridon et al. 2017). The adults (Figure 5.5) grow to about 10 mm in length and 2.5 mm in width and parasitize a wide range of mammals including sheep, cattle, elk, deer, donkeys, and rabbits. It is predominantly a problem in sheep, but the wide range of wild reservoir hosts makes control difficult. The life cycle (Figure 5.6) begins when the adults living within the bile ducts release eggs that are then passed with the definitive host's faeces. Unlike fasciolid eggs, those of *D. dendriticum* do not hatch in the environment and must be consumed by an appropriate terrestrial snail intermediate host. Therefore, unlike *F. hepatica*, infections with *D. dendriticum* are common among animals that graze on well-drained pasture.

In Europe, *Zebrina detrita* is the principal snail intermediate host whilst *Cionella lubrica* is the main intermediate host in North America. However, development may occur in many other snail species. The eggs hatch within the snail's gut to release a miracidium that bores through the gut wall and transforms to the mother sporocyst stage. The mother sporocyst reproduces asexually to produce daughter sporocysts (there is no redia stage), and these in turn produce cercariae that move to the snail's lung. The cercariae are of a type called xiphidiocercaria – that is they have a piercing stylet at the anterior margin of their oral sucker. The cercariae accumulate in groups within the snail's lung where they become coated in mucus that is formed from their own glands and that of the snail host. They are then forcibly ejected from the snail's pneumostome (breathing pore) as 'slime balls' in an event that can be likened to the snail sneezing. The slime balls stick to the vegetation and are attractive to ants belonging to the genus *Formica*. The ants eat the slime ball and the cercariae then penetrate the ant's crop and enter the haemocoel where they transform into metacercariae. Several of them move to the ant's head and one of these penetrates the ant's brain and encysts within the suboesophageal ganglion. The remaining metacercariae travel back to the

Figure 5.5 *Dicrocoelium dendriticum* adult stage. Adult worms grow to 10 mm in length and 2.5 mm in breadth.

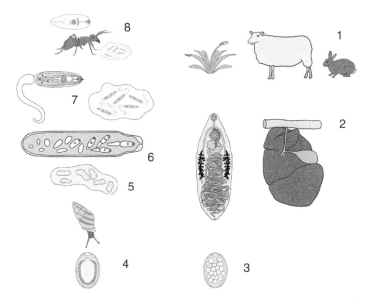

Figure 5.6 Life cycle of *Dicrocoelium dendriticum*. 1: Infection occurs when the definitive host consumes ants containing the metacercaria stage. 2: The young flukes travel up the bile ducts and become adults in its finer branches. Here they reproduce sexually and commence egg laying. 3: The operculate eggs are passed with their host's faeces. 4: The miracidium develops within the egg but is not released until after the egg is consumed by the snail intermediate host. 5: The parasite develops into the sporocyst stage, and this reproduces asexually to produce daughter sporocysts. 6: The daughter sporocysts reproduce asexually to produce cercariae. These sporocysts do not have mouths and the opening at one end represents the 'birth canal' through which the cercariae emerge. 7: The cercariae are ejected in slime balls. Note the well-developed mucus glands in the cercariae and the spike associated with the oral sucker. 8: The cercariae are consumed by ants within which they transform into the metacercaria stage. At the end of the day, infected ants are induced to climb vegetation and then bite onto the plant. They then remain suspended until the temperature rises the following day.

haemocoel of the abdomen and encyst within thick cyst walls. As night falls, ants normally return to their nests, but those infected with metacercariae remain outside. As the temperature drops and the humidity increases, the infected ants climb up vegetation and clamp their mandibles onto the plant they have climbed. They remain firmly attached until the morning arrives and the temperature rises once more. At this point they release their grip and return to their normal tasks. This presumably increases their chances of being consumed by herbivores grazing during dusk or early morning. This begs a question of how the presence of a metacercaria within the suboesophageal ganglion can exert a temperature dependent effect upon its host's behaviour. This might be through the metacercaria changing its position and thereby causing a physical disruption of nerve function and/or by having a temporary effect on the neurochemical transmission (e.g., by secreting certain chemicals). The former hypothesis is supported by the observation using micro-CT imaging that the parasite is in physical contact with the suboesophageal ganglion (Martín-Vega et al. 2018).

After a sheep (or alternative definitive host) consumes an infected ant, the metacercariae excyst within its duodenum. The young flukes then travel up the bile duct until they reach the finer branches where they become adults. Therefore, unlike *F. hepatica*, there is no migratory phase through the liver parenchyma. Several thousand worms can develop within a single sheep and heavy infections cause inflammation, fibrosis, and biliary dysfunction. Long-standing infections can cause cirrhosis and the surface of liver becomes fibrotic. Humans can be infected by

D. dendriticum, but genuine cases are rare (Moure et al. 2017). This is because it is usually obvious that an ant is on one's food, and we seldom consume raw *Formica* spp. ants intentionally. Chocolate-covered ants are an obvious exception. They are considered an expensive delicacy although one can make one's own easily enough. There is little information available on the best species to use. Some websites suggest *Formica* spp. (which would be obviously risky if sourced from some regions) for an acidic sweet-sour taste whilst leaf-cutting ants (*Atta* spp.) are said to impart a lemony flavour.

5.2.1.3 Family Paramphistomatidae

The paramphistomes are intestinal parasites of amphibians, reptiles, birds, and mammals. They are particularly common in countries with warm climates, but some have ranges that extend into northern latitudes. For example, *Paramphistomum leydeni* infects reindeer in Finland. There are several species of veterinary importance, particularly of ruminants. Human infections with paramphistomes are very rare and, despite its name, *Gastrodiscoides hominis* is primarily a parasite of pigs.

5.2.1.3.1 *Genus Calicophoron*: Calicophoron daubneyi

Until recently, paramphistomes were not considered important veterinary parasites in the UK. However, the prevalence of *C. daubneyi* is increasing not only in the United Kingdom but also in Ireland and elsewhere on Western Europe (Jones et al. 2017). The reason(s) for this is/are uncertain although the warmer and wetter conditions associated with climate change may be aiding the reproduction and survival of the snail intermediate hosts. It shares similar environmental requirements and both intermediate and definitive hosts to *F. hepatica*. One would therefore expect that species interaction/competition takes place – although adult *C. daubneyi* live in the reticulum of the rumen rather than the bile ducts.

Adult *C. daubneyi* are rumen parasites and primarily infect cattle although they can also exploit sheep and other ruminants. Their eggs are excreted with the host's faeces and on reaching freshwater they hatch to release a miracidium. The miracidium then searches for a suitable snail host: these include *Planorbis planorbis* and also species such as *G. truncatula* that are also exploited by *F. hepatica*. The life cycle within the snail resembles that of *F. hepatica* and culminates in the release of cercariae that transform into metacercariae on vegetation.

The definitive host becomes infected when it consumes vegetation containing the metacercarial stage. The metacercariae excyst within the duodenum to release the juvenile flukes and these begin feeding on the lining of the intestine. Most of the pathology is associated with this stage of the life cycle although the full extent of the parasite's economic impact on cattle farming is currently uncertain. Cattle infected with large numbers of young flukes suffer diarrhoea, anorexia, bottle jaw, weight loss, and loss of condition. These are non-specific and also typical of *F. hepatica* and many gastrointestinal parasitic infections. After 3–6 weeks, the young flukes migrate forward to the reticulum where they mature – this takes about 3 months after the initial infection. Once they reach the reticulum, they probably feed on the rumen contents rather than the host tissues, and therefore, they do not cause serious pathology.

5.2.1.4 Family Opisthorchiformes

This family contains 33 genera and includes parasites of fish, reptiles, and a range of mammals including humans. The best-known species of medical importance are *Clonorchis sinensis*, *Opisthorchis viverrini*, and *Opisthorchis felineus*. The body of the adult worms is usually very thin, with weak musculature, so they are often translucent. The oral and ventral suckers are poorly developed, and the testes are situated in the posterior half of the worm, usually close to the hind end. The gut divides shortly after the pharynx into two long blindly ending caecae that terminate near to the posterior of the worm.

The Opisthorchiformes have complex life cycles that includes two intermediate hosts (the first of which is a snail) and a definitive vertebrate host. The adults usually live in the bile ducts or intestine and their eggs are already embryonated when passed within the host faeces. Although the first intermediate host is usually an aquatic snail belonging to the family Bithyniidae, Hydrobiidae, or Thiaridae, the eggs do not hatch until after the snail consumes them. However, as the snails are coprophagic (i.e., feed on faeces), it is not usually long before this occurs. After hatching, the miracidium invades the snail's tissues and transforms to a sporocyst. After a few days, the sporocyst produces several rediae through asexual reproduction, and these in turn produce daughter rediae. It is difficult to distinguish between mother and daughter rediae, and therefore it is often uncertain how many generations of rediae arise. Ultimately, each redia forms several cercariae that subsequently burrow out of the snail. The cercariae are of a type termed 'pleurolophorcercous': this means that they have a globular body and a long thick muscular tail, which has dorsal and ventral fin folds that presumably aid swimming. The body has an oral sucker and a ventral sucker and, unusually for cercariae, has two pigmented eyespots. After leaving their snail host, the cercariae hang upside down in the water and allow themselves to sink slowly for a while before swimming back upwards. They continue this sink and swim behaviour but contact with a solid object or experiencing the water currents produced by an animal swimming close by stimulates more activity. This presumably increases the chances of the cercariae making contact with the second intermediate host. Most second intermediate hosts are various species of freshwater fish (usually in the family Cyprinidae) although some species also utilize freshwater crabs or crayfish and even mammals may be infected. In those species that utilize fish as the second intermediate host, the cercariae latch onto the skin of the fish using their suckers. Then, by a combination of releasing the contents of their penetration glands and physical action, they force their way into the fish's body. Afterwards, the cercariae lose their tails and encyst as metacercariae within the musculature although they occasionally enter the gills, the viscera, and the fins. The definitive host becomes infected when it consumes the infected fish. Not surprisingly, fish-eating animals such as certain birds and mammals are common definitive hosts for these parasites. The consumption of raw fish (e.g., sushi, sashimi) is widespread in the Far East and is now popular in western countries too. Consequently, many people are vulnerable to contracting these parasites. *Clonorchis sinensis*, *Opisthorchis viverrini*, and *Opisthorchis felineus* together probably infect about 30 million people with a further 290 million people are at risk of infection.

5.2.1.4.1 *Clonorchis sinensis*

This is an extremely common fluke throughout Southeast Asia in countries such as China, Japan, Thailand, Cambodia, Laos, and Viet Nam. For a review, see Na et al. (2020). In addition to humans, it also parasitizes dogs, cats, pigs, and various other mammals (Figure 5.7). Consequently, there are numerous reservoir hosts, and this makes control problematic. Prevention of infection is also made difficult because the metacercariae are resistant to harsh environmental conditions and can survive in salted and pickled fish. *Clonorchis sinensis* is a medium sized worm that grows up to 25 mm long and 5 mm in breadth. It has a pinkish, almost transparent coloration whilst still alive although older individuals become yellowish through pigment accumulation. Infection of the definitive host (Figure 5.8) commences when metacercariae excyst within the duodenum and then, like *D. dendriticum*, the young flukes make their way up the bile duct to the larger proximal bile ducts and their main branches where they become adults. Heavy infections cause inflammation and fibrosis that disrupts the flow of bile and thereby leads to jaundice. In addition, there is

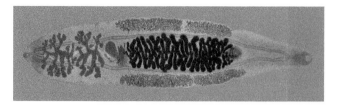

Figure 5.7 *Clonorchis sinensis* adult stage. Adult worms grow to 25 mm in length and 5 mm in breadth.

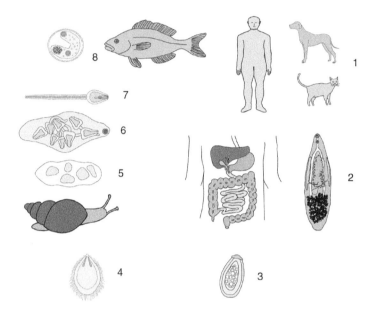

Figure 5.8 Life cycle of *Clonorchis sinensis*. 1: Infection occurs when the definitive host ingests fish containing the metacercariae. In addition to humans, dogs, cats, and several domestic and wild mammals can act as hosts. 2: Metacercariae excyst in the duodenum and the parasites migrate to the bile ducts where they become adults and commence egg laying. 3: Eggs are passed with the host faeces. 4: The eggs hatch in water to release the miracidium stage that invades a suitable snail intermediate host. 5: Within the snail transform into the sporocyst stage that reproduces asexually and produce the redia stage. 6: The redia stage reproduces asexually and gives rise to the cercaria stage. 7: The cercaria stage bores out of the snail host and swims off and invades a suitable fish. 8: Within the fish, the cercariae transform into the metacercaria stage. Drawings not to scale.

pain and the patient can suffer fever, diarrhoea, and malnutrition. Secondary microbial infections can also establish. The liver may become swollen and cirrhotic, thereby disrupting the portal circulation causing the formation of ascites. Sometimes the worms also parasitize the pancreatic ducts and cause pancreatitis (inflammation of the pancreas). Although much of the damage is caused by the physical actions of the parasites, some of the pathology results from the host's immune response to parasite antigens. In common with *O. viverrini* and *O. felineus*, chronic infection with *C. sinensis* is linked to the development of cancer of bile duct (cholangiocarcinoma [CCA]). At present, there is no effective chemotherapy for CCA. Therefore, it is important to reduce the likelihood of its occurrence by preventing infection with parasites associated with its development. The mechanism by which the worms induce cancer is unknown but may be linked to them increasing the susceptibility of DNA to damage by carcinogens (Fried et al. 2011). The identity of

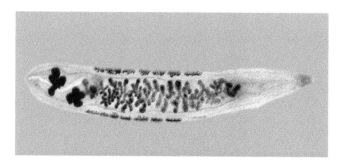

Figure 5.9 *Opisthorchis viverrini* adult stage. Adult worms grow to about 12 mm in length and 2.5 mm in breadth.

these carcinogens is currently uncertain. *Clonorchis sinensis* does not induce CCA in experimental animals (rodents) unless these also receive a chemical hepatocarcinogen. This suggests that the parasite does not act on its own although genetic susceptibility and/or other factors are also probably important in determining whether *C. sinensis* induces cancer.

5.2.1.4.2 *Opisthorchis viverrini*

Opisthorchis viverrini tends to be smaller than *C. sinensis* and grows to only 12 mm in length and 2.5 mm in breadth (Figure 5.9). However, it is very similar in terms of its biology and pathology. Living specimens have a reddish coloration and the testes are lobed rather than branched as is the case in *C. sinensis*. Human infections are particularly common in parts of Thailand and it has been linked to the exceptionally high levels of CCA in the Khon Kaen region (Sriamporn et al. 2004). This is possibly because of the popularity in this region of a fish salad called *koi pla*. An intensive education campaign encouraging people to cook their fish led to a reduction in CCA among women and young people, but the rates remain high among men. This is probably because men continue to consume *koi pla* made from raw fish during communal drinking sessions that take place separately.

5.2.1.4.3 *Opisthorchis felineus*

Opisthorchis felineus is very similar in size and appearance to *O. viverrini* and also to *C. sinensis* in terms of its biology and pathology. However, it has a more widespread distribution than the other two species and occurs in Siberia, Kazakhstan, Europe, and North America, as well as Southeast Asia. Its distribution is spreading through Europe owing to the increased popularity of consuming raw fish and the movement of infected migrant workers from endemic regions (Pozio et al. 2013). It has also been linked with the development of CCA.

5.2.1.5 Family Paragonomidae

The Family Paragonomidae was initially erected in 1939, but many workers continue to place the genus *Paragonimus* in the Family Troglometridae. The genus *Paragonimus* contains several species that infect many mammals and birds and some are zoonotic. Approximately one million people acquire *Paragonimus* spp. infections every year although the numbers in some countries have seen dramatic fluctuations that reflect changes in dietary preferences. In Africa, *Paragonimus africanus* is important, whilst in Mexico and Central America *Paragonimus mexicanus* causes occasional human infections. Similarly, in the Midwest states of USA, occasional human infections with *Paragonimus kellicotti* occur. The taxonomy of the Paragonomidae is proving problematical (Rosa et al. 2020). Consequently, there is a tendency to refer to species complexes. The *Paragonimus*

(a)

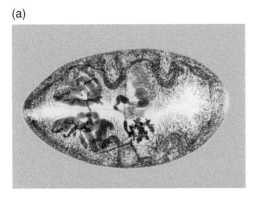

(b)

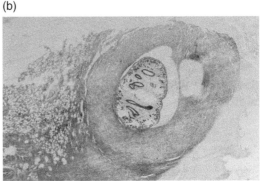

Figure 5.10   *Paragonimus westermani*. (a) Adult stage. Adult worms grow to about 12 mm in length and 6 mm in breadth. Their bodies can be up to 5 mm thick, which is unusually thick for a trematode of this size. (b) Section through a lung sack caused by *Paragonimus westermani*. Note the thickness of the wall surrounding the parasite.

westermani species complex is the best known and the most important as a human pathogen. We shall therefore consider this in some detail and as though it is a single species although there are inevitably some variations. The other *Paragonimus* species and species complexes have a similar appearance and biology.

5.2.1.5.1   *Paragonimus westermani*

This fluke is an important human parasite in Asian countries where freshwater crabs form a significant part of the diet, such as Korea, Japan, China, Taiwan, and the Philippines. For example, the traditional Korean dish, *gejang* consists of crabs marinated in soy sauce. Although marination can kill the parasites, the soy sauce must contain at least 5% salt, and it must last for at least 64 days (Kim et al. 2017). These exacting requirements are often not met and therefore consuming *gejang* is a known risk factor. One cannot physically remove the parasites from the crabs and whilst freezing them for 2 days at −20 °C is effective, many people do not own freezers.

In common with other members of this genus, adult *P. westermani* are oval-shaped and have an unusually stocky body for a trematode that can be up to 5 mm thick (Figure 5.10a). They typically grow up to 12 mm in length and 6 mm in breadth, have a dense covering of spines, and are reddish in colour. *Paragonimus westermani* can be diploid, triploid, or even tetraploid and this influences its ability to undergo sexual reproduction and other biological characteristics. For example, triploid individuals cannot produce spermatozoa, and therefore, they reproduce parthenogenetically. Triploids also tend to develop to maturity faster than diploids and are more pathogenic. This is because each triploid parasite causes the formation of a cyst (see below) and can produce eggs. Triploidy also occurs in several other trematode species including *Fasciola* spp.

Adult *P. westermani* normally live in the lungs although they may also occur in other organs such as the liver, spleen, gut wall, muscles, kidney, and brain (Yoshida et al. 2019). The worms create a space around themselves that becomes surrounded by dead and dying cells (Figure 5.10b), and these (plus the worms) become walled off by a granulomatous immune reaction that is sometimes referred to as a cyst. Because the parasites become enclosed by a host reaction rather than by material produced by the parasites, some workers consider the term 'cyst' in this instance to be inappropriate.

A granuloma usually contains one, two, or three worms: the number varies with the host and other factors – such as whether the parasite is diploid or polyploid. Egg production usually commences before the adults become completely surrounded by fibrous host tissue. However, some eggs are often retained within the lungs, and these also induce a chronic immune response and become enveloped within granulomas. Nevertheless, most eggs reach the bronchial tubes and become transported upwards with sputum. The eggs are then spat out or swallowed and excreted with the faeces.

Once the eggs reach fresh- or brackish water, they hatch after anything from 16 days to several weeks and the miracidia invade aquatic snails such as *Melania* spp., *Semisulcospira* spp., and *Brotia* spp. Within the snail, the miracidium first develops into a sporocyst and then there are two generations of rediae, the last of which gives rise to cercariae. The cercariae are of a type called 'microcercous' because the 'tail' is so reduced that it has become a non-functional appendage. Microcercous cercariae are therefore unable to swim, but they have a large globular body and can crawl across solid surfaces using their well-developed oral and ventral suckers. There is some uncertainty in the literature as to what happens next, with some texts indicating the cercariae bore their way out of the snail and then creep off in search of a suitable crustacean host. Other authors suggest that the cercariae remain within the snail and infect the crustacean second intermediate host when it is consumes the infected snail. A wide range of freshwater crab and crayfish species can act as second intermediate hosts. After penetrating a crustacean, the cercariae transform into thick-walled metacercariae within its muscles and viscera.

Paragonimus westermani and some other *Paragonimus* species exploit paratenic hosts such as poultry and geese. In these cases, the metacercaria excyst in the intestines of the paratenic host and the juvenile worms make their way to the musculature where they remain without developing any further. The life cycle resumes when a definitive host consumes an infected paratenic host. Perhaps unexpectedly, deer sometimes act as paratenic hosts and humans can become infected by eating venison (Yoshida et al. 2019). Although deer are herbivorous, there are occasional reports of them foraging on remains, including those of humans (Meckel et al. 2018), and even anecdotal accounts of them killing and consuming nestling birds. A crab represents a good source of calcium, and it is therefore possible a deer might consume a living or dead one if the opportunity arose.

When a definitive host consumes an infected second intermediate host, the metacercariae excyst in its duodenum and transform into juvenile worms. The juvenile worms then penetrate the gut wall and enter the coelomic cavity: here they often pair up, before making their way through the diaphragm and entering the lungs. If they are unable to pair up, they will often remain in the coelom awaiting the arrival of a suitable mate. The worms become mature about 8–12 weeks after reaching the lungs. However, sometimes the juveniles 'get lost' and ultimately end up in one of the other body organs, which are referred to as ectopic sites.

The consequences of infection with *P. westermani* infection depend upon the number of worms and their location. In humans, cats, and dogs, a small number of flukes in the lungs usually induces few symptoms, whereas numerous flukes cause breathing difficulties, coughing, and the sputum may contain blood. Obviously, the pathology induced by a particular parasite density is influenced by many factors including host health, nutrition, and immune status, as well as both host and parasite genetics. Furthermore, even a single worm locating within the eye, brain, or spine can have serious consequences including blindness, paralysis, and even death.

5.2.1.6 Family Cathaemasiidae: Genus *Ribeiroia*

This family of trematode parasites does not include any species of notable medical or veterinary significance. However, one particular species, *Ribeiroia ondatrae*, is of interest to parasitologists because of its role in causing morphological deformations in frogs. The populations of many amphibian species are declining around the world, and the reasons for this are all too obviously a consequence of habitat destruction, pollution, and the intentional or unintentional introduction

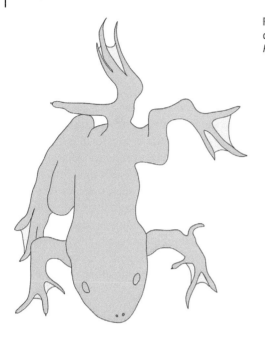

Figure 5.11 Diagrammatic representation of a frog deformed following infection by the metacercariae of *Ribeiroia ondatrae* during its development.

of predators, competitors, and diseases. However, after initial reports in 1995 from schoolchildren who were shocked to find deformed frogs whilst doing pond surveys in Minnesota (USA) large numbers of similarly affected frogs were subsequently found in many other parts of USA and Canada. Some of the frogs had malformed limbs, some had extra limbs, and some had no limbs at all (Figure 5.11). Initially, it was thought that the deformations were caused by environmental pollution, and there were concerns about whether humans might also be at risk. However, it is now almost certain that the majority if not all of the deformities are caused by infection with the metacercariae of trematodes belonging to the genus *Ribeiroia*, and in particular *R. ondatrae*.

Under experimental conditions, the adults of *R. ondatrae* can parasitize various mammals and birds, but the normal definitive hosts are uncertain. The eggs are passed in the faeces and these give rise to miracidia that infect aquatic snails such as *Planorbella* spp. Within the snail, the parasite develops first into a sporocyst, and this gives rise to rediae that in turn give rise to cercariae. The cercariae then leave the snail and infect frog tadpoles, within which they transform into metacercariae. To complete the life cycle, the definitive host must consume an infected frog. Interestingly, the metacercariae tend to cluster around the developing frog's pelvic girdle and hind limbs. Here, they appear to act as physical roadblocks that disrupt the growth of the developing limb buds; similar deformities can be induced experimentally by implanting resin beads (Sessions and Ruth 1990). The metacercariae may also release chemicals that affect the growth of surrounding host cells. Under experimental conditions, exposure of frog tadpoles to *Ribeiroia* spp. induces limb malformations in the majority of frogs that survive to adulthood and the level of malformation relates to parasite density. By contrast, malformations are not induced by metacercariae of the trematode *Alaria* spp. and none occur in the controls (Johnson et al. 1999; Johnson and Sutherland 2003). Obviously, by inducing limb deformities, the parasite compromises the movement of its host. This increases the likelihood of the infected frog being consumed by the predatory bird definitive host – and this has clear evolutionary advantages. However, it does beg the question why so many frogs should suddenly be found with deformities. Scientists and schoolchildren have observed pond-life for generations, but so few people saw deformed frogs in the past that their apparently sudden appearance was considered worrying and newsworthy. The explanations put forward included a sudden increase in parasite abundance, an increase in parasite pathogenicity, and an increase in host susceptibility. However, proving which, if any, of these factors is responsible is not easy. One suggestion is that the increasing use of agrochemicals since the 1950s has resulted in the continual exposure of wildlife to low levels of numerous pollutants. These may not be sufficient to cause mortalities among non-target organisms, but they compromise their immunity and thereby increase their susceptibility to pathogens. Rohr et al. (2008) found good evidence for this through a series of laboratory (mesocosm) and field experiments in which they demonstrated that tadpoles exposed to the herbicide atrazine were more vulnerable to infection with *Ribeiroia* spp. In addition, they showed that atrazine increases the growth of algae (periphyton), and these are an important source of food for the snail intermediate host. Previous workers have shown that amphibians exposed to low concentrations of atrazine are more vulnerable to viral diseases, and it is therefore possible that there may also be interactions with co-infections. Therefore, pollutants may be facilitating an increase in the aquatic snail population and more infected snails will mean that more cercariae will be released into the surrounding waters. This will result in more heavily parasitised developing frogs, and they are therefore more likely to be seriously deformed. Simultaneously, the immune system of the developing frogs may also be compromised by sub-lethal levels of pollutants, and this may facilitate the infection of the frogs by parasites such as *Ribeiroia*.

5.2.1.7 Family Schistosomatidae: *Schistosoma mansoni, Schistosoma japonicum, Schistosoma haematobium*

These parasites are commonly known as blood flukes because the adult worms live in the blood vessels of birds and mammals. The phylogeny of the Family is reviewed by Zarowiecki et al. (2007) and comprehensive discussions of their biology are provided by Basch (1991) and Loker (1983).

Adult schistosomes are unusual among the platyhelminths in having separate sexes (i.e., they are dioecious). The males are shorter but more robustly built than the females (Table 5.3) and have a ventral longitudinal groove called the gynaecophoric canal within which the female worm is normally found (Figure 5.12a). The worms have two suckers, an oral sucker that surrounds the mouth and a ventral sucker situated close to the anterior end and used to grip the lining of the blood vessel they are living in. The suckers of the female worms are less developed than those of the male. The gut divides into two caeca shortly after the mouth, and these then re-join about halfway down the length of the worm and continue as a single tube to the posterior end where it ends

Table 5.3 Morphological characteristics of *Schistosoma mansoni, Schistosoma haematobium*, and *Schistosoma japonicum*.

	Schistosoma mansoni	*Schistosoma haematobium*	*Schistosoma japonicum*
Male size (mm)	$(6–14)\times0.9$	$(10–15)\times0.9$	$(10–22)\times0.3$
Tegument	Coarsely tuberculated	Finely tuberculated	Not tuberculated
Female size (mm)	$(12–16)\times0.16$	20×0.25	$(12–26)\times0.3$
Egg size (μm)	$(140–182)\times(45–73)$	$(112–170)\times(40–73)$	$(74–106)\times(55–80)$
Egg spine	Lateral, long, sharp point	Terminal, delicate, blunt point	Lacks a spine although a very short slightly curved lateral tubercle is present

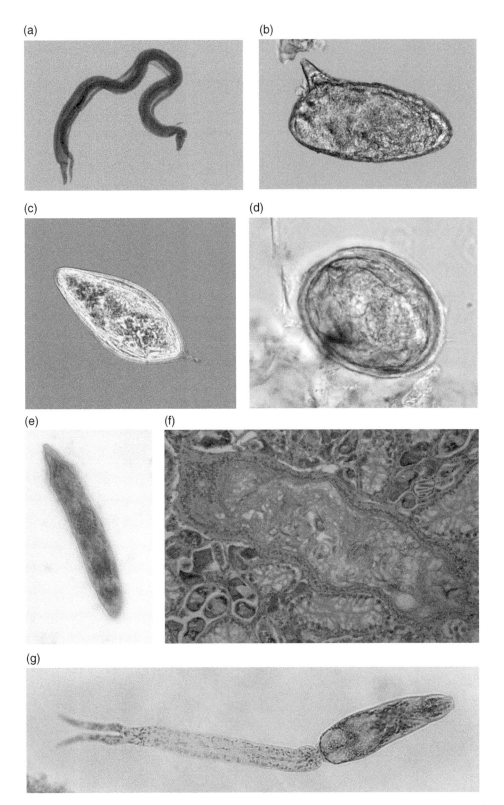

Figure 5.12 Schistosome life cycle stages. (a) Adult *Schistosoma haematobium*. The long slender female is carried in the gynaecophoric canal of the shorter, stouter, male. (b) Egg *Schistosoma mansoni*. (c) Egg *Schistosoma haematobium*. (d) Egg *Schistosoma japonicum*. (e) Miracidium *Schistosoma mansoni*. (f) Sporocyst *Schistosoma mansoni*. (g) Cercaria *Schistosome mansoni*. This morphology is called 'furcocercous' in which the anterior rather globular 'body' is attached to a 'tail' that terminates in two short branches (furca). Unlike the 'body' of most trematode cercariae, schistosome cercaria lacks an oral sucker and the ventral sucker is reduced in size and coated with tiny spines.

blindly – there is no anus. In some, but not all, species, the tegument of the male worms is tuberculated, that is, it is covered with small bumpy projections.

There are at least 100 schistosome species and natural hybrids can occur when a definitive host is invaded by cercariae of two different species. This might be between two animal schistosome species, two human schistosome species, or an animal and a human schistosome species. Some hybrid schistosomes exhibit 'hybrid vigour' such as enhanced growth characteristics. Hybrids between human and animal schistosome species may also have an enhanced host range and therefore greater zoonotic potential (Leger and Webster 2017). In some parts of the world, hybrid schistosomes are responsible for a high proportion of human infections (Sene-Wade et al. 2018). However, not all schistosome species can hybridise and the offspring of hybrid mating are not always viable.

Most of the schistosomes that infect wild animals and birds are not very pathogenic although some of those parasitizing cattle cause chronic disease and loss of condition (De Bont and Vercruysse 1998). By contrast, those infecting humans cause considerable morbidity and mortality. Human schistosomiasis is also known as Bilharzia or bilharziasis, after the German physician Theodor Bilharz who in 1851 identified a schistosome as the causative agent of a disease, which we now know as urinary schistosomiasis. Bilharz named the parasite *Distomum haematobium,* but this was then changed to *Bilharzia haematobium* in honour of its discoverer only for it then to be changed to *S. haematobium.*

Schistosomiasis is one of the most important human parasitic diseases: in 2018, it was estimated that 290.8 million people required preventative treatment. At the time of writing, the mortality statistics were uncertain with figures quoted in the literature varying from 24,000 to 200,000 per annum (WHO 2020). Most cases of human schistosomiasis are caused by just three species of schistosome: *Schistosoma mansoni, Schistosoma japonicum,* and *Schistosoma haematobium.* In addition, *Schistosoma intercalatum, Schistosoma mekongi, Schistosoma malayensis* and *Schistosoma guineensis* can be locally important. *Schistosoma mattheei* will infect humans but is normally a parasite of cattle, sheep, and goats. Kruger and Evans (1990) suggest that the parasites described as *S. mattheei* recovered from humans may actually be hybrids of *S. mattheei* and *S. haematobium.*

In addition to humans, *S. mansoni* also infects several non-human primates including baboons and monkeys, as well as some rodents. The infection rates among non-human primates can be high and, in some instances, there are positive associations between human and non-human primate infections (Richards et al. 2019). However, few humans live in proximity to other primates, and therefore their importance as reservoir hosts from which zoonotic transmission takes place is uncertain. It is also possible that, as is the case for several microbial infections, it is our diseases that pose a risk to non-human primate populations. Similarly, although *S. haematobium* can infect non-human primates, these are unlikely to play an important role in human transmission cycles except in isolated circumstances. By contrast, *S. japonicum* infects over 30 species of domestic and wild mammals including dogs, pigs, cattle, and rats, and therefore zoonotic transmission is probably commonplace.

Adult *S. mansoni, S. intercalatum,* and *S. guineensis* tend to live in the inferior mesenteric veins draining the large intestine, whilst adult *S. japonicum, S. mekongi,* and *S. malayensis* live in the superior and inferior mesenteric veins associated with the small intestine. Adult *S. haematobium* commonly live in the veins of the vesical plexus around the urinary bladder and along the ureters but can also occur in the veins around the rectum.

Schistosome species infecting domestic and wild birds and mammals exist throughout the world, but human schistosomiasis is almost entirely a disease of people living in developing countries with hot or tropical climates. Nevertheless, in 2014, schistosomiasis was diagnosed in two

European families who had never travelled to a country where the disease is endemic. It subsequently transpired that they had all bathed in the River Cavu in Corsica and that the snails in this river were infected with a hybrid of *S. haematobium* and *S. bovis* (Pennisi 2018). There are therefore concerns that the disease could also establish in other European countries because of the natural presence of snails, such as *Bulinus truncatus* that can act as intermediate hosts. The hybrid schistosomes probably reached Corsica when an infected person urinated in the river. Large numbers of legal and illegal migrants from Africa and other regions where schistosomiasis is endemic arrive regularly throughout southern Europe (including Corsica). It is therefore inevitable that schistosome eggs will be entering watercourses and hence potentially initiating local infections. It should be noted that *S. bovis* was endemic in Corsica until at least the 1960s (Calavas and Martin 2014). This is an instance in which molecular genotyping might indicate the source of the Corsican schistosomes.

Climate change will undoubtedly have a major impact upon the epidemiology of schistosomiasis in the near future. Some of the warmer areas of Africa are likely to see a decline in schistosomiasis as a result of drier conditions. By contrast, the incidence of schistosomiasis may increase in those regions that start to experience warmer wetter conditions (Stensgaard et al. 2019).

The life cycle of the three main species infecting humans is similar and illustrated in Figure 5.13. Infection commences when the cercaria stage penetrates the skin. The cercariae of different schistosome species share a similar body plan that is called 'furcocercous' in which there is an anterior

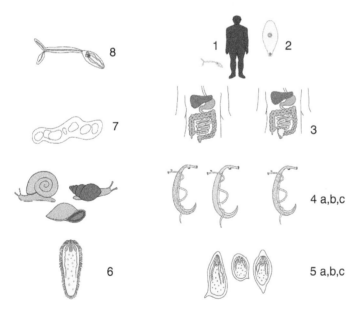

Figure 5.13 Schistosome life cycle. (1) Infection commences when cercariae invade through the skin surface and enter the blood circulation. (2) The parasites then transform to the schistosomulum stage and move through the body to the point at which they pair up and commence egg laying (3). Adult *Schistosoma mansoni* (4a) and *Schistosoma japonicum* (4b) are usually found in the venules around the rectum and mesentery whilst those of *Schistosoma haematobium* (4c) reside in the veins around the bladder. The eggs of *S. mansoni* (5a) and *S. japonicum* (5b) are passed with the faeces, whilst those of *S. haematobium* (5c) are passed with the urine. (6) The eggs hatch when they reach water to release the miracidium stage. The miracidia physically and chemically penetrate certain semi-aquatic snail species and transform to the sporocyst stage (7). The sporocysts reproduce asexually to produce the cercaria stage. (8) The cercariae physically and chemically bore their way out of the host snail and swim off in search of a suitable definitive host. Drawings not to scale.

rather globular 'body' to which is attached a 'tail' that terminates in two short branches (furca) (Figure 5.12g). Unlike the 'body' of most trematode cercariae, the schistosome cercaria lacks an oral sucker and the ventral sucker is reduced in size and coated with tiny spines. The cercariae do not feed, and they die within 1–3 days if unable to find a definitive host.

In those schistosome species that infect us, the cercariae swim to the surface of the water and then become motionless with their forked tail extended. They then passively sink for a short distance before actively swimming back up to the surface. By alternately rising and sinking, the cercariae can be carried in water currents a considerable distance from their snail host. The cercariae have four types of glands: the 'escape glands' are so-called because they are lost after emergence from the snail although their function is not known; preacetabular glands, which contain enzymes and chemicals used in the process of invasion; postacetabular glands, which secrete mucus that may aid purchase whilst invading the definitive host; and the head gland, which produces secretions involved in the post-invasion transformation of the parasite. The host-finding strategies of the different schistosome species varies considerably. Somewhat surprisingly, the attachment of the cercariae of *S. mansoni* is stimulated by L-arginine even though this amino acid is not specific for human skin and is present at very low concentrations (Haas et al. 2002).

The cercariae physically and chemically penetrate through the skin of their definitive host. This normally takes place whilst the skin is fully submerged or from a film of drying water following bathing. The surface tension provided by a drying film of water facilitates invasion. The cercariae can also penetrate the lining of the throat and oesophagus if we consume them in drinking water. The cercariae penetrate beneath the skin surface within 30 seconds, and they then lose their tails and start to transform into schistosomules. This is associated with major changes in the surface covering that enable the parasite to survive the host's immune response and adapt to the new environmental conditions. The schistosomules migrate through the dermis until they locate a venule, which they penetrate to reach the peripheral circulation. Some may also enter the lymphatic system. Ultimately, the schistosomules reach the right side of the heart, and from here, they move to the left side of the heart and thereby gain entry to the systemic (arterial) circulation. They are then swept to the lung capillaries where the many of them die. This is because if they accidentally break through into the alveoli, it is difficult for them to regain access to the circulation. The schistosomules that successfully negotiate the lungs continue their migration until they reach the liver. They are metabolically quiescent whilst migrating through the systemic circulation. However, upon reaching the liver, they feed on blood cells for the first time and their cell division and development recommences. They spend approximately 3 weeks living in the liver sinusoids after which they pair up and move to their species-specific adult breeding site. The migration of the schistosomules and early onset egg deposition is associated with the development of Katayama syndrome. This usually occurs 14–48 days after a first-time infection or a heavy reinfection. It results in night-time fevers, coughing, headaches, and abdominal pain, and because of this nonspecific nature, it is often misdiagnosed.

Adult schistosomes parasitizing large mammals can live for many years, and they probably remain together as monogamous pairs for much if not all this time. However, divorce remains a possibility (Beltran et al. 2009). Monogamy is unusual in the Animal Kingdom and schistosomes are the lowest taxonomic level at which it occurs. Female schistosomes are less muscular than the males, and one suggestion is that they lack the physical strength to move against the flow of blood to reach the sites where they lay their eggs. They therefore must 'pair up' with a strong male before they leave the liver. Furthermore, if a female is separated from her mate, her reproductive system regresses. The reason for this is uncertain, but it is independent of sperm transfer, and it is not species specific – if the female is paired with a male of another species, she can regain her reproductive health. It is therefore possible that by being enveloped in the male's gynaecophoric canal, the female is exposed to the male's muscular pumping action and this has a growth stimulating effect.

In those schistosome species infecting humans, the large oval-shaped eggs (Figure 5.12 b–d) must traverse the tissues between the blood vessel walls and the gut lumen (*S. mansoni* and *S. japonicum*) or the bladder (*S. haematobium*). Adults of *Schistosoma nasalis* (*Schistosoma nasale*) live in the veins of the nasal mucosa of sheep, cattle and horses and the eggs are sneezed out with nasal secretions. Several schistosome species produce eggs that have a prominent spine, and this was once thought to facilitate the egress of the eggs. However, the eggs of some species such as *S. japonicum* lack a spine or it is rudimentary, and these have no difficulty exiting the host's body. It is now known that the eggs move through the tissues by stimulating a dramatic immune reaction, and this is also responsible for much of the pathology associated with schistosomiasis. The function of the egg spine is therefore uncertain but probably facilitates the egg sticking to the side of blood vessels.

By the time they reach the outside world, the eggs are fully embryonated and they hatch when exposed to fresh water. The eggs release the free-living miracidium stage (Figure 5.12e) that swims in search of a suitable snail intermediate host (Table 5.4). Schistosomes are very specific in their choice of intermediate host, and some workers think that they may have originally been parasites of molluscs that subsequently added a vertebrate to their life cycle. Either way, the availability of a suitable mollusc intermediate host is a major factor determining the transmission and distribution of the different schistosome species. The extensive movement of people, and in particular slaves, from Africa to South America during the sixteenth and seventeenth centuries must have resulted in the importation of both *S. mansoni* and *S. haematobium*. However, only *S. mansoni* encountered a suitable local species of snail that could act as intermediate host (*Biomphalaria glabrata*), and therefore it was the only one to establish in the New World. Interestingly, the New World branch of *S. mansoni* is already diverging from its Old-World roots and has evolved an enhanced affinity for the black rat (*Rattus rattus*) as a definitive host (Imbert-Establet and Combes 1986).

The miracidium physically and chemically bores its way into the body of the intermediate snail host, loses its cilia, and develops into the sack-like mother sporocyst stage (Figure 5.12f). The mother sporocyst has no mouth or gut and absorbs nutrients across its epithelium. It grows to about 1mm in length and reproduces asexually to give rise to daughter sporocysts, and these reproduce

Table 5.4 Distribution and intermediate hosts of schistosome species parasitizing humans.

Schistosome species	Snail host	Distribution
Schistosoma mansoni	*Biomphalaria, Tropicorbis*	Africa, Middle East, Caribbean Islands, South America
Schistosoma japonicum	*Oncomelania*	South East Asia
Schistosoma haematobium	*Bulinus, Physopsis*	Africa
Schistosoma intercalatum	*Bulinus*	Africa
Schistosoma mekongi	*Neotricula*	Laos, Cambodia
Schistosoma malayensis	*Robertsiella*	Malaysia
Schistosoma guineensis	*Bulinus*	Central West Africa, São Tomé
Schistosoma mattheei	*Bulinus*	Africa

asexually to produce cercariae. In schistosome species infecting humans, it normally takes about 4–6 weeks from initial infection for the cercariae to be produced. The mother sporocyst continues to produce daughter sporocysts and the snail remains infected for life. This means that a single miracidium can give rise to thousands of cercariae. Once they are mature, the cercariae bore their way out of the snail and swim off in search of the definitive host. Their emergence tends to exhibit daily rhythms related to the time of day the definitive host is most likely to be present. For example, the cercariae of *S. mansoni* and *S. haematobium* tend to emerge mid-morning to shortly after midday when people are often outside collecting water, washing, fishing, paddling, or swimming. By contrast, the cercariae of *Schistosoma rodhaini* emerge at night because this is when its rodent definitive hosts are active. Species that utilise several different definitive hosts may exhibit variable peaks of emergence. For example, *S. japonicum* can exhibit peaks of emergence in the early morning and the late afternoon. The early morning peak presumably coincides with the presence of cattle whilst the later peak ensures that cercaria are present when crepuscular and nocturnal rodent hosts visit the water.

5.3 Class Cestoda

Members of the Cestoda (also referred to as Cestoidea) are commonly known as tapeworms. All cestodes are parasitic during both their adult and larval stages. In most species, the adult lives in the gut of a vertebrate, whilst their larvae exploit one or more intermediate host(s) that is/are another vertebrate or an invertebrate. Some cestodes (e.g., *Diphyllobothrium* spp.) have a succession of larval stages in different intermediate hosts. *Hymenolepis nana* is an exception, which does not have an absolute requirement for an intermediate host. A few species exhibit precocious development of their reproductive organs during the larval stage and produce viable eggs before reaching the definitive host. One of the most extreme examples of this is *Archigetes cryptobothrium* that has dispensed with the need for a vertebrate definitive host entirely and sustains a life cycle within aquatic tubicifid worms (Olson et al. 2008). All the stages in a cestode's life cycle lack a mouth and gut and their body surface consists of a metabolically active tegument across which nutrients are absorbed and waste products disposed. In most species, the adult worm has a scolex at the anterior end that is equipped with attachment organs, such as suckers. Some species are also equipped with hooks that are composed of tough keratin-like proteins. These are mounted on a raised region called a rostellum on the head of the worm. The hooks are strongly curved so that once inserted into the host's gut wall, it becomes difficult to remove the scolex. Beneath the scolex is a short 'neck' from which develop a series of proglottids (segments) that become increasingly mature the further they move away from the scolex. The neck together with the proglottids are sometimes referred to as the strobila. The word strobila is also used to describe the stack of buds that develop asexually during the stationary polyp stage of the life cycle of certain jellyfish (Cnidaria). The plate-like buds become progressively bigger until they eventually drop off and swim away as the ephyra stage and this ultimately develops into the familiar medusiform jellyfish. The word strobila is an example of how scientific terms can have rather contrived origins. It was first used in relation to tapeworms in the nineteenth century as a Latin word *strobilos* deriving from the Ancient Greek (στροβιλοζ) for a pinecone. The reasoning for the association is uncertain but, in longitudinal profile, the sequence of scales in a pinecone increase in size from top to base and thereby has some resemblance to a series of tapeworm segments or the plate-like buds on the jellyfish polyp. The same Greek word was also used for a spinning top, but this association seems unlikely. Ancient spinning tops were ceramic or carved from wood and placing several of these in decreasing size upon one another would produce a structure bearing a very vague resemblance to a series of developing proglottids. However, children whose parents could not afford ceramic spinning tops would use pinecones, so an actual spinning top or an association with spinning is an unlikely explanation for why the word has become associated with tapeworm morphology.

Many tapeworms contain calcium corpuscles. The chemical composition of the corpuscles varies between species and is also influenced by the diet and geographical location of the host. Nevertheless, they are all of the type known as organic-matrix calcium corpuscles because they also contain proteins, lipids, mucopolysaccharides, carbohydrates, DNA, and RNA around which the mineral component develops in concentric rings. In addition to calcium, the corpuscles also contain phosphates, carbonates, and varying amounts of other minerals, such as magnesium and zinc. The corpuscles are typically ovoid and 7–34 μm in diameter although smaller and larger ones occur in some species. In most tapeworms, they develop intracellularly, although in the cysticerci of *Taenia solium* they also develop extracellularly within the lumen of the protonephridial ducts. Their function is uncertain and suggestions vary from them acting as excretory dumps to protecting the parasite from calcification (e.g., through absorbing host calcium and localising it within the corpuscles). The latter role may explain the large number of calcium corpuscles in larval tapeworms that are particularly vulnerable to calcification. In some species, the calcium corpuscles can account for up to 40% of the dry weight of the parasite.

Adult cestodes are usually hermaphrodites and the individual proglottids contain both male and female reproductive organs although these may mature at different times. Usually, the male reproductive organs mature first. Sexual reproduction can occur through mating between two adjacent worms although a worm may also mate with itself when the strobila loops back so that proglottids with mature male and female reproductive organs are touching. In some species, asexual reproduction occurs during the larval stage when protoscolices develop from a germinal membrane.

The classification of the tapeworms is in a state of flux, and it is too complicated to go into the taxonomic disputes here. Fortunately, all species of major medical and veterinary importance belong to just two orders – the Pseudophyllidea and the Cyclophyllidea – that are both within the group known as the Eucestoda. Unfortunately, the validity of the order Pseudophyllidea has been questioned, and some taxonomists suggest splitting it into two new orders: the Bothriocephalidea and the Diphyllobothridea. For our purposes we shall discuss just a few important species that exemplify different aspects of tapeworm biology.

5.3.1 Order Pseudophyllidea/Diphyllobothridea

Most members of this order are parasites of fish although a few species utilise mammals, birds, or reptiles as their definitive host. There are usually two intermediate hosts the first of which is a crustacean and the second is a fish. In many species the adult worms are relatively short but some of them reach remarkable lengths: a specimen of *Polygonoporus giganticus* (also known as *Tetragonoporus giganticus*) recovered from the gut of a sperm whale (*Physeter macrocephalus*) was 30 m in length and had a strobila 5 cm in width at its widest point (Skrjabin 1967). The scolex of an adult pseudophyllidean tapeworm typically has two shallow longitudinal grooves called bothria (Figure 5.14), and in the genus *Diphyllobothrium,* each bothrium clasps one or two of the host's intestinal villi. This causes the villi to degrade and therefore the worm has to regularly change its position. The adult worms therefore cause some direct physical damage and inflammatory reactions.

5.3.1.1 Genus *Diphyllobothrium*

There are several morphologically similar species in the genus *Diphyllobothrium*. It is therefore probable that the literature contains many instances of misidentification. The genus is mostly

Figure 5.14 *Diphyllobothrium latum* adult stage. The scolex of *D. latum* has two sucking grooves (bothria) that are used to grasp the villi lining the small intestine of the definitive host.

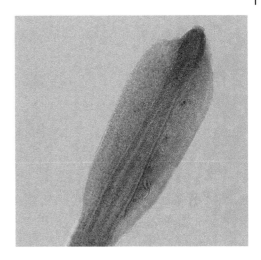

associated with cold, arctic, and sub-arctic regions of the Northern Hemisphere although there are case reports of human infections from tropical regions such as Malaysia. There are also introduced *Diphyllobothrium* populations in the Southern Hemisphere (e.g., Chile, Argentina). There was once a high prevalence of *Diphyllobothrium latum* in Switzerland and this inspired Professor Wegelin, an eminent parasitologist, to write that any girls attending the elite Swiss boarding schools would return home 'better educated than before, but with broad tapeworm into the bargain'. The second intermediate hosts are various species of fish and therefore human infections are common among societies in which there is a tradition of consuming raw or lightly cooked fish. For example, in the Baltic countries many people consume raw slightly salted roe (hard roe = fish eggs contained in the ovarian membrane; soft roe [milt] = fish testes containing the spermatozoa). Some surveys have found that up to 91% of hard roe are infected with plerocercoids (Kearn 1998). The plerocercoid is white and is usually easy to spot when the fish is fresh. However, even light cooking changes the coloration of the fish tissues and identifying the plerocercoids then becomes more difficult.

The best-known species is *Diphyllobothrium latum* and some estimates suggest that it infects up 20 million people. Several other species of *Diphyllobothrium* also parasitize humans, although the number of people they infect is probably only a few thousand (Scholz and Kuchta 2016). These include *Diphyllobothrium dendriticum*, which mainly causes human infections in the Lake Baikal region of Russia and the Arctic regions of North America, and *Diphyllobothrium nihonkaiense* that occurs in Asia, particularly Japan, parts of China, Korea, and the Russian Far East, as well as Canada and North-West USA. The *Diphyllobothrium* species infecting humans are all zoonotic and also parasitize many of the carnivorous mammals found in cold climates that consume fish (e.g., wolves, dogs, and bears). *Diphyllobothrium dentriticum* also infects many fish-eating birds including gulls and corvids (crows).

Although still high, the numbers of human infections in some countries have dropped considerably in recent years, such as in the Baltic states. By contrast, infections are increasing elsewhere owing to the popularity of eating raw or barely cooked fish in the form of sushi etc. This is further aided by the globalization of the fish trade and the migration of infected people to new regions/countries where suitable intermediate hosts occur.

5.3.1.1.1 *Diphyllobothrium latum*

Adult *D. latum* can grow up to 10 m although some texts suggest that they may reach 20 m. The proglottids nearest to the scolex are broader than they are long, but they become square as they approach the posterior end. They are therefore commonly known as broad fish tapeworms.

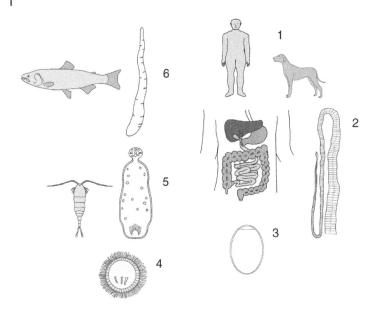

Figure 5.15   Life cycle of *Diphyllobothrium latum*. 1: Infection commences when the definitive host (e.g., humans, dogs) consume fish containing the plerocercoid stage of the parasite. 2: The parasite matures to the adult stage in the small intestine and commences egg laying. 3: Eggs are passed in the definitive host's faeces. 4: The eggs hatch in water to release the free swimming coracidium stage. 5: The coracidium stage is consumed by copepods within which the parasite develops into the procercoid stage. 6: When infected copepods are consumed by fish, the parasites invade their flesh and transform into the plerocercoid stage. Drawings not to scale.

The life cycle (Figure 5.15) begins with the release of eggs into the host's gut via the genital pores on the gravid proglottids, and these are then passed with the host's faeces. The eggs have a superficial resemblance to those of the liver fluke *F. hepatica* since they are ovoid and have an operculum (lid). However, the eggs of *D. latum* are only 67–71 × 44–45 μm, whilst those of *F. hepatica* are about twice this size (130–150 × 63–90 μm). The eggs take several weeks before they are ready to hatch and must be submerged in water. The eggs release a free-living stage called a coracidium that consists of the oncosphere that is covered by a ciliated embryophore. The coracidium does not feed and soon dies unless it is eaten by a suitable copepod crustacean that acts as the first intermediate host. The parasite then penetrates the copepod's gut and develops to the procercoid stage within its haemocoel. The next stage of the life cycle begins when the infected copepod is consumed by a suitable fish that can act as the second intermediate host. These are often members of salmon and pike families (Family Salmonidae and Family Esocidae). The parasite penetrates the gut of the fish and makes its way to the muscles, gonads, and viscera where it develops to the plerocercoid stage. The mature plerocercoid is 1–2 cm in length and has rudimentary bothria. Completion of the life cycle requires the consumption of the infected fish by the definitive host within which the plerocercoid transforms into an adult worm that attaches to the lining of the small intestine.

Although the adult worm can grow to an enormous size, most people infected with *D. latum* suffer few symptoms. They may, however, occasionally pass over a metre of strobila whilst defaecating (Reilly 2020), which can be psychologically rather traumatic. More seriously, some people develop pernicious anaemia, and this is thought to be due to parasite metabolism reducing the availability of vitamin B12 in the gut (Scholz et al. 2009).

5.3.2   Order Cyclophyllidea

Most adult cyclophyllidean tapeworms are parasites of wild birds and mammals although a few of them parasitize amphibians and reptiles. The order also contains some of the most important tapeworms that infect us and our domestic animals. Adult cyclophyllidean tapeworms usually have four circular or oval-shaped suckers (acetabula) that they use to grip the lining of the small intestine. In some species the suckers are supplemented by a rostellum that is armed with hooks (Figure 5.16a). The rostellum is a rounded protuberance at the top of the scolex. It is equipped with muscles that enable it to be retracted and extended, thereby facilitating the use of the hooks. However, some species have a rostellum without any hooks, and others lack a rostellum. A more reliable morphological diagnostic feature for cyclophyllidean tapeworms is the presence of a vitelline gland beneath the ovaries; in other tapeworm species, these glands are scattered throughout the proglottids (Figure 5.16b). The vitelline glands produce components utilised in the manufacture the yolk and shell of the developing embryos.

The typical cyclophyllidean life cycle consists of the adult worm living the gut of the definitive host, producing eggs that pass out with the host's faeces. The eggs are then consumed by the intermediate host, which may be an invertebrate or a vertebrate, depending upon the tapeworm species. The eggs hatch to release an oncosphere that, in common with other Eucestoda, has six hooklets – it is therefore sometimes referred to as a hexacanth embryo or hexacanth larva. Aided by its hooks, the oncosphere penetrates the gut of the intermediate host and moves to the site where it develops into the larval metacestode stage. This can take many forms, and some of them can undertake asexual reproduction (Table 5.5). Completion of the life cycle depends upon the definitive host consuming the intermediate host.

5.3.2.1   Family Taeniidae

The taeniid life cycle typically involves a mammalian carnivore that acts as the definitive host and herbivorous or omnivorous mammalian intermediate hosts. The adult worms live within the small intestine and vary in size from species such as *T. saginata* that grows up to 20 m in length to *Echinococcus multilocularis* that is typically only 1.2–3.7 mm long. The adult tapeworms reproduce sexually, and their eggs are passed with the host's faeces. Subsequently, the eggs are consumed by the intermediate host through faecal–oral contamination. The eggs then hatch within the small intestine of the intermediate host to release an oncosphere that penetrates the gut and is swept in the circulation to a point in the body where it develops into the metacestode (larval) stage. In some taeniid species, the metacestode exhibits asexual reproduction. The extent to which asexual reproduction takes place is related to the size of the adult worm. Those species that produce large adult tapeworms, such as *T. saginata* and *T. solium*, do not reproduce asexually during the metacestode stage. By contrast, medium-sized species, such as *T. multiceps* (0.4–1 m) may exhibit some asexual reproductive capacity during the metacestode stage, whilst species that have very small adults such as *E. granulosus* undergo extensive asexual reproduction (Moore and Brooks 1987). Completion of the life cycle occurs when the definitive host consumes the metacestode stage. In the past, it was often uncertain which larval stage was related to which adult worm. Consequently, larval stages sometimes acquired names that are quite different from those of the adult worm. For example, the larval stage of *Taenia hydatigena* was known as *Cysticercus tenuicollis*, that of *Taenia solium* was known as *Cysticercus cellulosae*, whilst that of *T. multiceps* was called *Coenurus cerebralis*. Unfortunately, some workers continue to use these older names when referring to the larval stages. Consequently, one must be aware of the different names when reading and searching the literature.

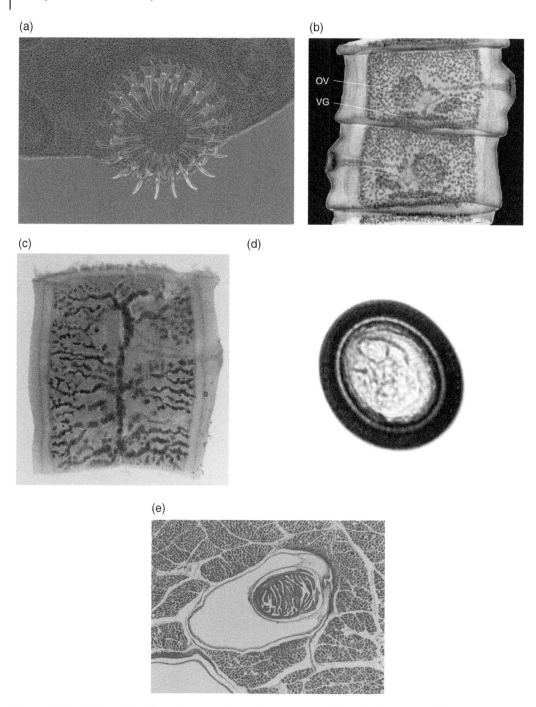

Figure 5.16 *Taenia solium* life cycle stages. (a) Adult stage, scolex. The scolex has four suckers and a rostellum armed with hooks. (b) Adult stage, mature proglottid. In common with other cyclophyllidean tapeworms, the vitelline glands (VG) are situated beneath the ovaries (OV). (c) Adult stage, gravid proglottid. (d) Egg stage. The eggs of *T. solium* and *Taenia saginata* cannot be distinguished by light microscopy (e) Larval stage (cysticercus) formed within skeletal muscle.

Table 5.5 Metacestodes (larval stages) formed by cyclophyllidean tapeworms.

Metacestode	Description	Example
Cysticercoid	Solid bodied with a single invaginated scolex. Asexual reproduction reported in some species, but this is the exception rather than the rule.	*Moniezia expansa, Anoplocephala perfoliata, Hymenolepis nana*
Strobilocercoid	Cysticercoid in which some segmentation occurs behind the scolex	*Schistotaenia*
Cysticercus	A single scolex develops on a germinal membrane and is both introverted and invaginated. The germinal membrane encloses a fluid-filled space (bladder)	*Taenia solium, Taenia saginata, Taenia hydatigena*
Strobilocercus	Like a cysticercus but some segmentation occurs behind the scolex	*Taenia taeniaeformis*
Coenurus	An enlarged version of a cysticercus in which tens or hundreds of protoscolices develop from the germinal membrane through asexual reproduction	*Taenia multiceps*
Unilocular hydatid	Single defined cyst within which brood capsules develop. Thousands or even millions of protoscolices form on germinal membranes of brood capsules	*Echinococcus granulosus*
Multilocular (alveolar) hydatid	Cyst lacks a defined shape and ramifies through the host tissue	*Echinococcus multilocularis*

Taeniids are extremely important in both human and veterinary medicine and several species are zoonotic. We are the definitive hosts for three taeniid species: *T. solium*, *T. saginata*, and *Taenia asiatica*. Molecular evidence suggests that the switch from carnivores to humans as definitive hosts occurred twice during their evolution. *T. saginata* and *T. asiatica* are extremely closely related and can hybridise. They probably originate from a single initial colonization event whilst *T. solium* probably colonized us independently (de Queiroz and Alkire 1998). The adult worms can co-occur in the same individual and inter-specific competition probably influences their transmission dynamics (Conlan et al. 2009). In addition, we are intermediate hosts for several taeniid species including *T. solium*, *E. granulosus*, and *E. multilocularis*.

5.3.2.1.1 *Taenia solium*

Taenia solium (Figure 5.16a–e) is the most important tapeworm species that infects humans, and it is the cause of considerable morbidity and mortality. Owing to stringent meat hygiene regulations, *T. solium* is probably extinct in the United Kingdom and most of northern Europe (Devleesschauwer et al. 2017). Although cases are occasionally reported from these regions, the initial infection can usually be traced back to the patient traveling in a country in which the parasite remains endemic. Infections are common in those countries in which pigs form an important part of the diet and meat hygiene regulations are not enforced. These include South and Central America and parts of Africa, India, Asia, and Melanesia.

Taenia solium is a highly unusual tapeworm in that it exploits humans as both a definitive host and an intermediate host. However, human cannibalism is very rare and therefore those of us who act as intermediate hosts are also 'dead-end' hosts. This is because the larval tapeworms must be consumed by another human to reach adulthood.

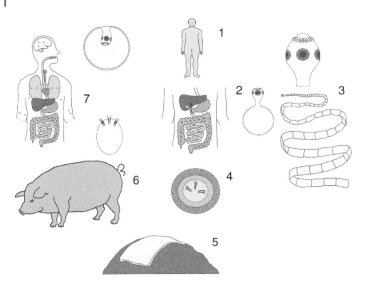

Figure 5.17 Life cycle of *Taenia solium*. 1: Infection commences when a human consumes meat containing the cysticercus stage of the parasite. 2: The cysticercus everts its scolex within the small intestine and attaches to the lining of the gut. 3: The parasite grows to maturity and commences egg laying. The scolex has four suckers and an armed rostellum. 4: Eggs and gravid proglottids are passed in the host's faeces. 5: The proglottids are not motile and remain within the faeces; pigs are the normal intermediate host and are coprophagic. 6: The eggs hatch within small intestine of the intermediate host to release an oncosphere that penetrates the gut wall, enters the circulation, and eventually forms the cysticercus stage within the musculature. 7: Humans may also act as intermediate hosts and the cysticerci develop in skeletal and cardiac musculature, the liver, lungs, brain, and other organs. Cysticerci developing in the brain are particularly pathogenic. Drawings not to scale.

The normal *T. solium* life cycle (Figure 5.17) involves humans acting as definitive host and pigs as the intermediate host – and hence it is commonly known as the pork tapeworm. The adult worm lives in our small intestine and can grow up to 10 m in length although most only reach 2–3 m. The scolex of adult *T. solium* is smaller (~1 mm) and rounder than that of *T. saginata*, and its rostellum is armed with 22–32 curved hooks set out in two rows (Figure 5.16a). The gravid proglottids are shorter than those of *T. saginata* and are typically 10–12 mm in length and 5–6 mm wide (Figure 5.16c). Each gravid proglottid contains about 40,000 eggs; it is not motile and may break up before it is shed with the faeces. Pigs, unlike cattle, are coprophagic. Consequently, it is beneficial for the eggs (Figure 5.16d) to remain within the faeces. By contrast, it is beneficial for *T. saginata* eggs to spread onto vegetation away from the faeces so that they are more likely to be consumed by cattle.

In countries where *T. solium* remains endemic, the levels of sanitation are usually poor and pigs forage freely around villages and towns. Consequently, it is easy for pigs to gain access to human faeces or food scraps contaminated with faeces. Once an egg reaches a pig's small intestine, it hatches to release an oncosphere. This penetrates the gut mucosa and gains access to the blood circulation. It then gets swept around the body until it reaches the site where it will develop into a cysticercus (Figure 5.16e). A cysticercus consists of a fluid-filled, membrane-bound bladder within which is a single invaginated protoscolex. A protoscolex is a 'scolex in waiting' that will eventually evert and attach to the gut wall of the definitive host. Most cysticerci develop within the skeletal and cardiac muscles although the liver, lungs, and brain may also become infected. The muscles of the tongue are frequently infected and meat inspectors often check this as a means of diagnosing its presence. The oval-shaped cysticerci are much bigger than those of *T. saginata* and typically grow to 20 mm by

10 mm in size. The life cycle is completed when we consume raw or undercooked pork that is infested with mature cysticerci. Dogs and cats can also act as intermediate hosts, and where these form part of our diet (e.g., parts of Asia), they can also be a source of human infection (Ito et al. 2002). We usually become infected with cysticerci after consuming the eggs through faecal contamination of our food/drinking water. Because *T. solium* proglottids are not motile and sometimes break up before being passed, the faeces of an infected person may contain a high density of eggs. Contact with such faeces therefore leads to simultaneous infection with numerous cysticerci. Autoinfection may also occur although its frequency is uncertain. This is when reverse peristalsis results in a gravid proglottid being transported back to the stomach. This provides the eggs with the physiological stimuli to hatch without leaving our body. Again, this can result in sudden massive infection. If we ingest *T. solium* eggs, cysticerci can develop in virtually all our body tissues and organs including skeletal muscles, subcutaneous tissues, liver, spleen, eyes, and brain. These cysts take about 3 months to develop to maturity and the consequences depend upon their number and location. Those developing in the brain and spinal cord cause neurocysticercosis that results in seizures, epilepsy and can prove fatal (Fleury et al. 2016). Cysts developing in the brain can grow unusually large and exceed 50 mm in size. This increases their pathogenicity since it results in pressure atrophy of the surrounding tissues and restricts the local blood circulation. There is a suggestion that Julius Caesar (100–44 BCE) may have suffered from neurocysticercosis. He is known to have begun suffering from seizures when he was 54 years-old and he would sometimes lose consciousness whilst in the senate. McLachlan (2010) argues that Julius Caesar probably acquired an infection whilst on his military campaigns and that his obvious infirmity encouraged his enemies in their belief that he was unfit to govern. If this is correct, then it is another example of how a parasite has influenced the course of human history.

5.3.2.1.2 Taenia saginata

Taenia saginata has a world-wide distribution and is one of the commonest tapeworm parasites to infect us. Its prevalence in Western Europe is declining rapidly but it remains common in parts of Africa and the South and Central Americas. In the most affected regions, the prevalence may exceed 10%, particularly amongst children. Some reports indicate that there is an exceptionally high prevalence (~50%) amongst some pre-school-age children in Nigeria. A possible explanation for this may be the practice of weaning children by feeding them undercooked adult food that contains beef (Hendrickx et al. 2019).

Adult *T. saginata* can grow to over 20 m in length although 3–5 m is more common. Despite its large size the scolex is only 1–2 mm, and unlike *T. solium*, the grip provided by the four suckers is not supplemented by the presence of hooks and there is no rostellum (Figure 5.18a). A fully-grown adult worm may consist of over 2,000 proglottids (Figure 5.18b) and the final gravid segments (Figure 5.18c) are 16–20 mm long and 4–7 mm wide. The life cycle (Figure 5.19) begins when a gravid segment detaches and either passes with the host's faeces or may even make its own way out of the anus – the detached proglottids are motile. An infected person often becomes first aware of their condition when they observe a proglottid slowly crawling around in their underwear, across the bedcovers or within the toilet bowl. A single proglottid can contain up to 200,000 eggs, and these disperse into the surroundings as the proglottid crawls around: the eggs are forced through surface cracks that form as the proglottid dries out. The eggs are hardy and can survive exposure to freezing temperatures such as those experienced during a northern European winter (Bucur et al. 2019). Cattle consume the eggs whilst grazing on pasture contaminated with faeces. The eggs hatch within their small intestine to release oncospheres that then penetrate the gut, enter the circulation, and are swept around the body. Ultimately, the parasites penetrate muscle fibres in which they develop into cysticerci (singular cysticercus) (*Cysticercus bovis*) (Figure 5.18d). The

(a)

(b)

(c)

(d)

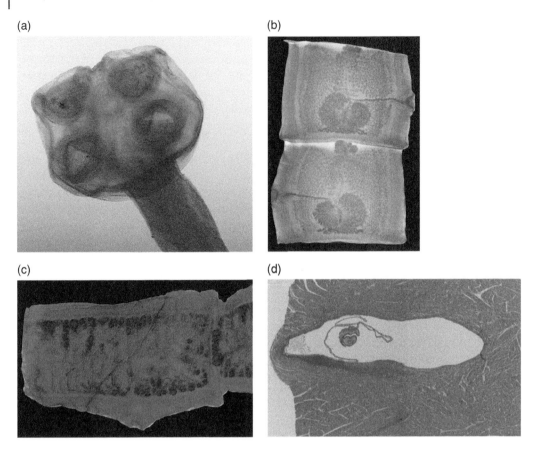

Figure 5.18 *Taenia saginata* life cycle stages. (a) Adult stage, unarmed scolex. (b) Adult stage, mature proglottids. (c) Adult stage, gravid proglottid. (d) Larval stage, cysticercus developing in skeletal muscle.

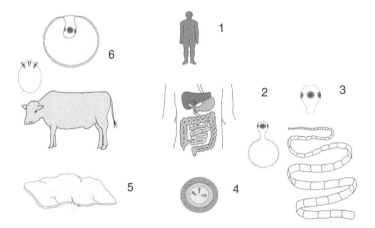

Figure 5.19 Life cycle to *Taenia saginata*. 1: Infection commences when a human consumes beef containing the cysticercus stage of the parasite. 2: The cysticercus everts its scolex within the small intestine and attaches to the lining of the gut. 3: The parasite grows to maturity and commences egg laying. The scolex has four suckers and is unarmed. 4: Eggs and gravid proglottids are passed in the host's faeces. 5: The proglottids are motile and move away from faeces. As they move, eggs are forced out via cracks and the genital pores and these contaminate the vegetation where cows are grazing. 6: The eggs hatch within the cow's small intestine to release an oncosphere that penetrates the gut wall, enters the circulation, and eventually forms the cysticercus stage within the musculature. Drawings not to scale.

heart, tongue, masseter muscles, and intercostal muscles are common sites of infection although non-muscular tissues such as the liver and kidneys may also be parasitized. A cysticercus takes about 2 months to reach the infective stage, by which time it is approximately 2 mm in diameter. The cysticerci have a pearly white coloration and heavily infected meat is sometimes said to be afflicted with 'beef measles'.

Adult *T. saginata* are seldom seriously pathogenic unless the patient is already malnourished or suffers an allergic reaction to the tapeworm antigens. Similarly, cattle with cysticercosis rarely exhibit clinical signs unless they have extremely high parasite burdens. Occasionally, cysticercosis 'storms' occur in which numerous cattle on a farm become infected with large numbers of parasites. This is often a consequence of using untreated human sewage as a fertilizer or the presence of an infected farmer/farmer-worker who is not careful about where he/she defecates.

5.3.2.1.3 *Taenia hydatigena*

Taenia hydatigena has a typical *Taenia* lifecycle involving an adult worm living in the small intestine of dogs and other Canidae, whilst the larva develops as a cysticercus within the omentum, mesentery and sometimes on the liver surface of sheep and goats although deer and other ruminants can be parasitized (Figure 5.20a,b). Pigs can also be infected, especially when they are allowed to roam freely, although the prevalence is usually much lower than that in sheep and goats (Braae et al. 2015). It has a cosmopolitan distribution. It was once common in the United Kingdom but has declined following the advent of effective anthelmintics and greater compliance with the disposal of dead livestock. In addition, there is currently improved control of feral dogs, and dog owners are encouraged to worm their dogs and dispose of their faeces safely.

The adult tapeworm grows to 177 cm but causes relatively little pathology. Whilst still attached to the adult worm, a gravid proglottid contains approximately 31,000 eggs. However, after release, those floating freely in the dog's gut contain only approximately 500 eggs. This indicates that most of the eggs have dispersed throughout the dog's faeces by the time of defecation (Coman and Rickard 1975). This is to be expected since sheep and other ruminants avoid grazing around faeces, so it is important for the eggs to spread onto the surrounding vegetation. The eggs are subsequently ingested by the intermediate host whilst grazing. The eggs hatch in the small intestine to release an oncosphere that burrows through the gut wall and enters the blood circulation. It is then swept to the liver and proceeds to migrate through the liver parenchyma until it reaches the peritoneal cavity. It usually then forms a cysticercus within the omentum although other nearby locations may be utilised. The cysticercus develops as a fluid-filled bladder growing up to 6–8 cm in size although most are much smaller. The cysticercus is not a totally passive entity and the tegument includes muscle tissue that maintains waves of movement across the surface. Presumably, this ensures that the cyst fluid is kept circulating. This becomes obvious if you remove a cysticercus from surrounding host tissue and place it in physiological saline – it then proceeds to bounce around the container.

The life cycle is completed when a dog eats the sheep and, in the process, consumes the cysticerci. Foxes also act as definitive hosts for *T. hydatigena,* but it is uncertain whether badgers can be infected. Therefore, the parasite can be controlled if dogs and foxes cannot consume sheep that die naturally in the fields and dogs are not fed uncooked meat from infected sheep. Worming dogs regularly also prevents pastureland from becoming infected.

The pathology in sheep depends upon the number of parasites developing. Small numbers usually have little consequence for the health of the animal, but their migration through the liver results in bleeding and the wounds heal as fibrotic tracts. Large infections can therefore cause ill health through the damage to the liver and in rare cases even death, whilst the liver would be condemned as unfit for consumption.

(a)

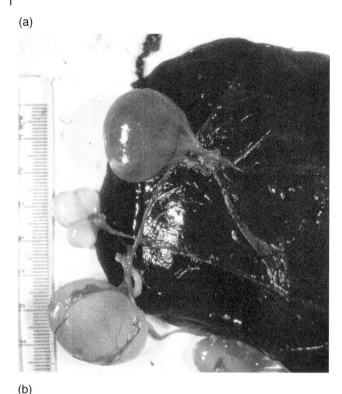

(b)

Figure 5.20 *Taenia hydatigena* larval stage. (a) Cysticerci growing on the surface of the liver of a sheep. (b) Cysticercus dissected from surrounding host tissue.

5.3.2.1.4 *Taenia multiceps*

Taenia multiceps also has a life cycle involving an adult worm that lives in the small intestine of dogs and a larval stage that develops primarily in sheep. A marked difference, however, is that the larval stage develops as a coenurus, and this usually forms within the brain and central nervous system. A coenurus is a large fluid-filled bladder and numerous protoscolices develop asexually on the internal membrane of the coenurus (Figure 5.21a,b). Therefore, one egg gives rise to one coenurus, which can then give rise to numerous adult worms. This coenurus was also once given its own species name, *Coenurus cerebralis*, that has persisted in the literature.

Taenia multiceps is found in most countries in which sheep farming occurs although it is absent from Australia and it tends to have a sporadic distribution. It used to be common in the United Kingdom but is now rare following the introduction of effective anthelmintics, enforcement of regulations on the safe disposal of dead livestock, and more responsible attitudes to dog ownership.

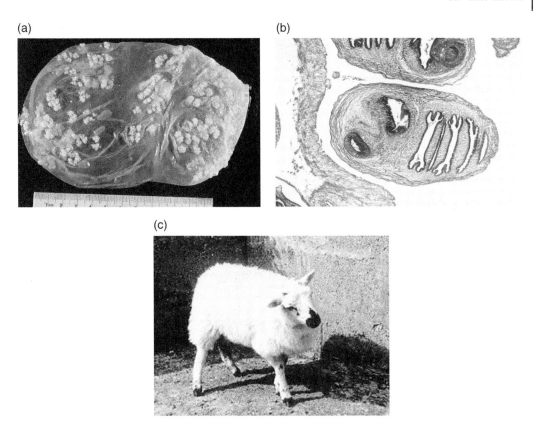

Figure 5.21   *Taenia multiceps*. (a) Coenurus removed from the brain of a sheep. Note the large number of protoscolices budding from the germinal membrane. (b) Section through the coenurus to show developing protoscolices. The suckers are already formed and hooks can be seen in the upper protoscolex. Note the thinness of the germinal membrane. (c) Sheep infected with *Taenia multiceps* coenurus and exhibiting gid.

The adult worm lives in the small intestine of dogs and other canids where it grows to 0.4-1 metre in length. Foxes can also act as effective definitive hosts (Varcasia et al. 2015). Sheep and goats are the primary intermediate hosts although several other farm animals and wild ruminants can also be infected. There are also occasional case reports of human infections. Therefore, unlike *T. hydatigena*, *T. multiceps* is a zoonotic disease, albeit a rare one. The eggs are passed in the faces of the definitive host and ingested by the intermediate host whilst grazing, or in the case of human infections, through faecal contamination of food or drinking water. Within the small intestine, the eggs hatch to release an oncosphere that burrows through the gut wall, gains entry to the blood circulation, and is then transmitted to the central nervous system where it develops into a coenurus. The consequences of infection depend upon the number and size of the coenuri and their location. The growth of a coenurus mimics that of a non-malignant solid tumour in that it restricts the local blood supply and causes pressure atrophy of the surrounding tissue. The latter is particularly marked in the brain because the tissues become crushed against the thick bones of the skull. However, the prolonged pressure causes thinning of the skull. This is probably sufficient to enable foxes to gain access to the brain tissue. It is unlikely that they could do this if presented with the head of an otherwise healthy adult dead sheep. The damage to the brain may cause a condition known as 'gid' (Figure 5.21c). This manifests when the affected animal develops a characteristic high-stepping gait and walks around in circles. A sheep may express this behaviour for a short time and then go back to grazing again. However, as the infection continues, the sheep's ability to control its movement declines and it loses the ability to stand and ultimately dies. In areas of Wales in which the disease was once common, a farmer would sometimes catch an affected sheep, stab its head with a penknife where the skull was thinning, and suck out the coenurus fluid with a straw. This relieved the pressure and thereby enabled the sheep to move normally again. The sheep was then sold as rapidly as possible before the coenurus could grow again. No doubt the practice was/is widespread in other endemic regions. Several genetic variants of *T. multiceps* exist and some of these form coenuri that develop in the musculature, peritoneum, and subcutaneous tissues (Varcasia et al. 2012). The life cycle is completed when a dog eats the sheep and, in the process, consumes the coenurus. As with *T. hydatigena*, control therefore depends upon preventing dogs from consuming dead sheep and on regular worming of dogs.

5.3.2.2 Genus *Echinococcus*

The adults of this genus are amongst the smallest tapeworms: they often have only two to four proglottids and are no more than a few millimetres in length (Figure 5.22a). By contrast, their larval stages, called hydatid cysts, can become extremely large. The best known species are *E. granulosus* and *E. multilocularis*. In addition, there are several other species that are less common and have more restricted distributions. For example, in parts of Africa, *Echinococcus felidis* has lions

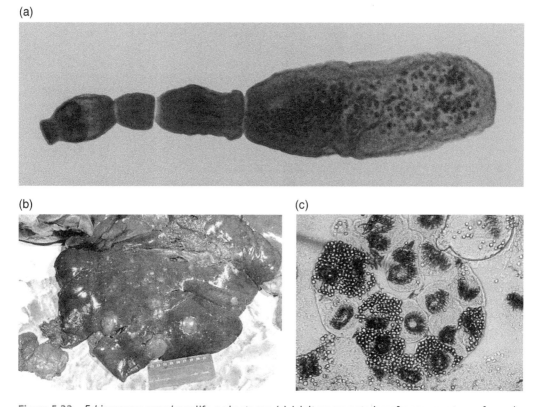

(a)

(b)   (c)

Figure 5.22 *Echinococcus granulosus* life cycle stages. (a) Adult worm, note how few segments are formed. (b) Hydatid cysts growing in the liver of a donkey. (c) Live hydatid sand (developing protoscolices) removed from one of the hydatid cysts.

(*Panthera leo*) and hyenas (e.g., *Crocuta crocuta*) as its definitive hosts and various big game animals, including hippopotamuses, serve as its intermediate hosts.

5.3.2.2.1 *Echinococcus granulosus*

For many years, *E. granulosus* was considered a single species. Subsequently, 10 genotypes were recognized with their own transmission characteristics and some of these were then elevated to distinct species (Romig et al. 2015). This makes the literature rather confusing. Currently, the parasite originally designated *E. granulosus* divides into *E. granulosus* (*sensu stricto*) that contains 3 genotypes, *E. canadensis* (3 genotypes), *E. intermedius*, (2 genotypes), *E. equinus* (1 genotype), and *E. ortleppi* (1 genotype). The term '*sensu stricto*' indicates that an organism fulfils the criteria of the species in its original narrow (restricted) sense. By contrast, if one sees *E. granulosus* (*sensu lato*), this indicates that the organisms are being considered collectively in the widest sense of the definition – in this case all 10 genotypes. Current estimates suggest that over 200,000 people become infected with hydatid cysts every year. *Echinococcus granulosus* (*sensu stricto*) is responsible for about 88% of the cases and *E. canadensis* for about 11%. Mixed species hydatid cyst infections also occur.

Echinococcus granulosus (*sensu stricto*) and *E. canadensis* are found throughout the world, especially in sheep-farming regions, whilst the other species tend to have more localized distributions. The adult worms usually parasitize domestic dogs (*Canis familiaris*) although in some regions it also infects other canids such as wolves (*Canis lupus*) and foxes (e.g., *Vulpes vulpes*). Cats and other felids are usually poor definitive hosts.

The adult tapeworms attach to the mucosa of the definitive host's small intestine using their suckers and the double row of hooks on their rostellum. They are typically 2–7 mm in length and have 3–4 proglottids. The penultimate proglottid is mature whilst the final proglottid is gravid (full of eggs) and usually constitutes about half the length of the worm. After it detaches, the gravid proglottid disintegrates within the intestine. This means that eggs become thoroughly mixed in with the host's faeces. A dog might be infected with hundreds or even thousands of adult worms. If a dog is otherwise healthy and well-fed, the worms seldom cause serious pathology although heavy infestations can result in diarrhoea, weight loss, and poor condition.

The life cycle (Figure 5.23) begins with the passing of the eggs in faeces and the intermediate host becomes infected when it ingests these through faecal–oral contamination. The different *Echinococcus* species tend to be associated with different intermediate hosts, but these are all large herbivorous mammals such as sheep, cattle, camels, and pigs. Humans may also be infected although we are usually dead-end hosts because these days few of us are eaten by dogs etc., after we die. However, earlier in our evolution this would have been more common. The eggs hatch within the intestine of the intermediate host to release the oncosphere stage, and this then penetrates the gut and enters the blood circulation. The oncosphere is then swept around the body to the point at which it will develop into a hydatid cyst. Most cysts develop in the liver (Figure 5.22b) although other organs may be affected including the lungs, spleen, and brain. The cyst is usually unilocular – that is, it is more-or-less spherical and has a thick laminated outer membrane within which is an inner germinal membrane that surrounds a fluid-filled cavity. The cysts grow slowly and usually become 5–10 cm in diameter although there is a record of a 50 cm cyst containing 16 l of fluid being removed from an African patient. Brood capsules, within which protoscolices (singular: 'protoscolex') develop by asexual reproduction, are formed from the germinal membrane (Figure 5.22c). In this way, a single egg gives rise to a single hydatid cyst which, in turn gives rise to hundreds of protoscolices – each of which has the potential of developing into an adult worm.

The consequences of a unilocular hydatid cyst for the health of the intermediate host are dependent upon the number and size of the cysts and the organ in which they develop. Small cysts within

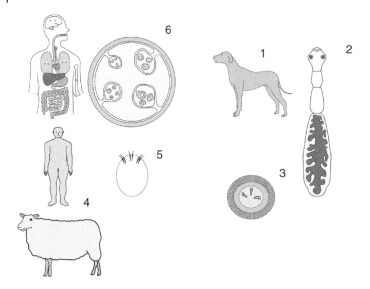

Figure 5.23 Life cycle of *Echinococcus granulosus* (*sensu strictu*). 1: Infection commences when a canine, such as a dog, consumes flesh from an intermediate host containing a hydatid cyst. 2: Within the dog's small intestine, the protoscolices in the hydatid cyst evaginate and attach to the gut wall. There they become adults, reproduce sexually, and commence egg laying. 3: Eggs are passed with the host's faeces and contaminate vegetation and drinking water. 4: The eggs are consumed by the intermediate hosts. These are usually ungulates although humans, may be infected. 5: The eggs hatch in the small intestine of the intermediate host to release an oncosphere that penetrates the small intestine and is transported around the body in the circulation to the place it develops into a hydatid cyst. The hydatid cysts undergo asexual reproduction to produce hundreds of protoscolices. 6: In humans, the liver is the most common place where hydatid cysts develop but the lungs, brain, bones, and other sites may be infected. Drawings not to scale.

the liver may have little impact, whilst the same sized cysts within the brain could prove fatal. In countries in which the parasite is highly prevalent, the infection of livestock results in losses of millions of pounds per annum through poor productivity and condemnation of the liver and other viscera (Wilson et al. 2020). On a global scale, the extent of the losses is only second to those caused by the liver fluke *F. hepatica*.

Hydatid cysts cannot survive for long after the death of their host. Therefore, the life cycle depends upon the definitive host either killing and eating the intermediate host or consuming it shortly after its death (e.g., after it died of natural causes or being slaughtered by humans). Consequently, in most cases, transmission is strongly affected by human activities such as farming practices and our relationships with dogs. However, in Australia, the introduction of *E. granulosus* to the continent has led to the establishment of a sylvatic cycle between kangaroos and dingoes (*Canis lupus dingo*) that is based on predator prey dynamics (Barnes et al. 2007). Human hydatid disease remains a serious problem in those countries where meat hygiene legislation is not enforced and/ or dead farm animals are left where dogs can eat them. For example, Gholami et al. (2018) reported that the seroprevalence of *E. granulosus* in the general Iranian population was 6%.

5.3.2.2.2 *Echinococcus multilocularis*
Echinococcus multilocularis tends to be associated with northern cold climates and has high prevalence in Alaska, Siberia, and Northern China. The disease appears to be spreading in Europe and between 2015 and 2020, new foci were reported in the south-western Italian Alps, Croatia, Denmark, and Sweden. Globally, *E. multilocularis* causes around 18,000 human infections every year of which around 150–200 occur in Central Europe. The adult worms have a similar

Figure 5.24 Multilocular cyst formed by *Echinococcus multilocularis* in the liver of a rodent.

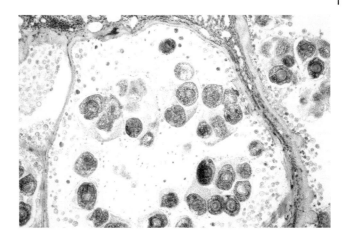

morphology to those of *E. granulosus*, but the larval stage forms a multilocular (alveolar) cyst that consists of a mass of separate vesicles (Figure 5.24). Unlike the unilocular cysts of *E. granulosus,* the alveolar cyst contents are gelatinous rather than fluid. The adult worms usually parasitize the small intestine of foxes such as the red fox (*Vulpes vulpes*) and the arctic fox (*Alopex lagopus*) although they also infect coyotes (*Canis latrans*), domestic dogs (*Canis familiaris*) and wolves (*Canis lupus*). Coinfections of *E. multilocularis* and *E. canadensis* have been reported in coyotes and red foxes. In Germany and other parts of Central Europe, the racoon dog *Nyctereutes procyonoides* is an important definitive host in the transmission cycle. (The raccoon dog was intentionally introduced into Central Europe from its native East Asia on several occasions during the years 1928–1953. Raccoon dogs are hunted for their luxurious fur. Since their initial introductions, they have become abundant and widespread.) The intermediate hosts of *E. multilocularis* are usually wild rodents such as voles (Tribe Microtini) and lemmings (Tribe Lemmini), and parasite transmission is heavily affected by predator-prey (e.g., fox-rodent) relationships.

The multilocular cysts formed by *E. multilocularis* have been likened to neoplasms from the way in which they grow and ramify through the affected tissues. Most cysts develop in the liver, although other organs may be affected. The cysts lack a limiting membrane and consist of a constantly growing sponge-like mass of parasite tissue. Furthermore, like certain cancers, fragments of the germinal membrane can break away and be carried in the blood stream to other regions of the body where they initiate secondary cysts. In humans, the cysts grow slowly over the course of many years but without treatment they will eventually completely destroy the liver and/or any other organ they infect. In humans and other atypical intermediate hosts, the germinal membrane of multilocular cysts usually does not produce protoscolices. There are rare case reports of multilocular cysts developing in dogs (Corsini et al. 2015). This has been ascribed to them consuming a large number of eggs – dogs can be coprophagic, so this is certainly a possible scenario.

Has Successful Anti-Rabies Control Resulted in an Increased Risk of *Echinococcus multilocularis* Infection in Europe?

In parts of Germany, Switzerland, and several other European countries, there has been a dramatic increase in the red fox population since the 1990s and the present day. The reason for the increase is probably a result of a combination of factors, but one of these is almost certainly the success of anti-rabies control programmes. In the past, rabies was an important

(Continued)

> **Has Successful Anti-Rabies Control Resulted in an Increased Risk of *Echinococcus multilocularis* nfection in Europe?** (Continued)
>
> cause of fox mortality, but the risk of the disease being transmitted to humans and domestic animals led to the introduction of widespread control programmes in which baits laced with anti-rabies vaccine were employed. In addition, foxes are changing their behaviour patterns, and they are becoming increasingly urbanised; many large towns and cities in Europe now support thriving fox populations. Unfortunately, the rise in the fox population has often been accompanied by an increasing prevalence of their infection with *E. multilocularis*. For example, a survey of 268 foxes shot in the Starnberg region of Bavaria found that 51% were infected (König et al. 2005). Similarly, in Lower Saxony in Northern Germany, the prevalence of *E. multilocularis* in foxes rose from 12% in 1991 to 20% in 2005 (Berke et al. 2008). There is therefore a concern that foxes are contaminating the environment with tapeworm eggs, and this is the reason for the increasing numbers of *E. multilocularis* infections in humans in parts of Europe (Mueller et al. 2020). Whilst infected foxes are undoubtedly a contributing factor to human disease, we have a closer relationship with dogs, and these can be a significant source of infection (Torgerson et al. 2020). For example, some of highest incidences of human alveolar cyst formation occur in parts of China and Kyrgyzstan in which there are also high levels of infection amongst the dog population. The extremely low winter temperatures in these parts of the world mean that working dogs often share human households rather than remaining outside. This increases human:dog contacts and the opportunity for contaminative transmission of tapeworm eggs in dog faces or contaminating their fur. Furthermore, the lack of firewood means that the meat consumed by both dogs and humans is often eaten raw, dried, or poorly smoked. This increases the chances of dogs consuming metacestodes of both *E. multilocularis* and *E. granulosus*.

5.3.2.3 Family Anoplocephalidae

Adult anoplocephalid tapeworms have a scolex that is equipped with four suckers but lacks a rostellum or hooks. The proglottids are usually wider than they are long, and each proglottid has one or two sets of genital organs. The presence of two sets of genitalia probably increases reproductive output. The eggs have three protective layers: an external 'shell', a middle 'sub-shell layer' containing lipid droplets, and innermost sub-shell membrane. These layers are probably important in protecting the oncosphere when it is consumed by the intermediate host.

The adults of most anoplocephalid species are parasites of herbivorous mammals although *Bertiella studeri* infects a variety of primates including monkeys, baboons, and gibbons, and there are occasional reports of human infections (Lopes et al. 2015b). The larvae develop as cysticercoids within mites (Acari) belonging to the Family Oribatidae. These mites are extremely common on the surface and upper layers of soil and can be found throughout the world. They are detritivores that feed on decaying organic matter and are important in nutrient recycling. Herbivores ingest the mites accidentally whilst grazing and primates presumably ingest them when foraging for roots, tubers, and food lying on the ground. Important anoplocephalid tapeworms include *Anoplocephala perfoliata* that has been implicated in the development of colic in horses, *Cittotaenia pectinata* that can cause fatal infections in rabbits, and *Moniezia expansa* and *Moniezia benedeni* that are common parasites of sheep and cattle.

5.3.2.3.1 Anoplocephala perfoliata

This is an extremely common parasites of horses and other equids. It has a worldwide distribution and often occurs as a co-infection with *Anoplocephala magna* which shares many of its biological characteristics and pathology. The adult worms are not particularly large and grow to only 8–25 cm in length, but heavy infestations may amount to several hundred worms. The adults tend to attach in the region of the ileocecal valve and the gravid proglottids release their eggs into the gut lumen so that they are well mixed in with the faeces by the time they are defaecated. Because horses produce a lot of faeces, the density of tapeworm eggs is often low and infections are therefore difficult to detect by faecal egg analysis. After being passed in the faeces, the eggs are consumed by oribatid mites within which they develop into cysticercoids. Horses become infected when they accidentally consume oribatid mites whilst grazing. Consequently, infection is heavily influenced by the population dynamics of the mites. Horses that live on lush pastureland are reportedly more likely to harbour large infections than those living in arid areas. This is possibly related to the density and survival of the oribatid mites. The access of mites to horse faeces is another key factor. Horse owners are encouraged to remove horse faeces from paddocks every day as a means of preventing the spread of nematode parasites and maintaining pastureland health. This will also reduce the opportunity for mites to become infected. However, this is not possible where horses graze on open land. Artificially controlling the mite population is neither feasible nor desirable since they play an essential role in soil ecology. The lifespan of the adult worms is uncertain, but studies in Central Europe indicate that there is an annual cycle of infection with mostly adult worms present in late Spring and early summer and immature worms predominating in late summer and autumn (Tomczuk et al. 2015).

Once an infected mite is consumed by a horse, it is broken apart and digested within the gut, thereby releasing the cysticercoid(s). The parasite then everts its scolex and attaches to the gut wall. The young worms attach themselves around the caecum but as they become mature, they move to the ileocecal junction where they cluster together in groups. Small numbers of worms probably cause little harm, whilst large numbers induce serious pathology. The level of infection that causes damage depends upon the horse's underlying health and immune status. *Anoplocephala* infection is mostly linked with the development of colic although it also causes general poor growth and loss of condition. Colic in horses may arise from a variety of causes but is characterized by extreme abdominal pain and can be rapidly fatal. *Anoplocephala* induces colic through causing ulceration of caecum, ileocecal intussusception, and ileal impaction. Intussusception occurs when a portion of the gut becomes paralysed, and this results in the neighbouring motile portion telescoping into the paralysed section. The result is that the gut becomes blocked. Ileal impaction is where passage of digesta through the gut becomes obstructed.

5.3.2.3.2 Moniezia expansa *and* Moniezia benedeni

There are at least 12 species of tapeworm belonging to the genus *Moniezia,* and it is possible that some of them are actually composed of several cryptic species. They are mostly parasites of ruminants although some infect warthogs and horses (Beveridge 2014). *Moniezia expansa* and *Moniezia benedeni* are the best-known species. These two species are found throughout the world and are common parasites of sheep, goats, and occasionally cattle; coinfections frequently occur. The adult worms live in the small intestine, and their eggs along with strings of gravid proglottids are shed with the faeces (Figure 5.25a). Wild birds such as starlings aid their dissemination by consuming the gravid segments voided in the animal faeces. The eggs are not affected by passage through the bird's gut and so may be spread over a wide area (Kozlov 1974). The eggs are consumed by oribatid mites (Acari) such as *Galumna*, *Oribatula*, and *Zygoribatula*. Exactly how the mites become infected is uncertain because the anterior section of their gut is probably too narrow to permit them to swallow the eggs whole. It is possible that the mites pierce the eggshell and suck out the embryophore without damaging it and the oncosphere then penetrates the mite's gut (Caley 1975). Within the mites, the parasites develop into cysticercoids over a period of 1–4 months depending on the temperature and can remain viable for at least 15 months. The simultaneous development

(a) (b)

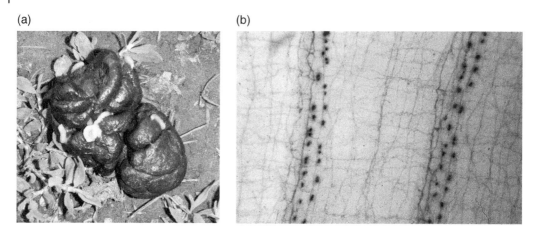

Figure 5.25 *Moniezia expansa.* (a) Proglottids of *Moniezia expansa* in sheep faeces. (b) Mature proglottids stained for acetylcholinesterase activity. The nerve net and interproglottidal glands stain strongly.

of several cysticercoids can prove fatal for the mite. The life cycle is completed when the ruminant definitive host accidentally consumes the infected mites whilst grazing. The adult worms grow rapidly and reach maturity between 30 and 57 days after infection. They typically grow up to about 2 m in length although some texts suggest that they can reach up to 6 m. The scolex is globular in shape and 0.36–0.8 mm at its widest point. It is equipped with four well-developed suckers, but there is no rostellum and there are no hooks. The mature proglottids of *M. expansa* are about 1.6 cm wide, whilst those of *M. benedeni* are about 2.6 cm. A unique feature of the genus *Moniezia* is the presence of interproglottidal glands at the posterior margin of each proglottid that become fully developed in the mature proglottids. In *M. expansa,* the glands run the full length of each segment, whilst in *M. benedeni,* they are concentrated more to the centre. The glands secrete vacuole bound substances and several enzymes including alkaline phosphatase and acetylcholinesterase (Figure 5.25b) (Gunn and Probert 1983). *Moniezia* are the only tapeworm species currently known to secrete acetylcholinesterase although it occurs in many nematode species, such as *Trichostrongylus colubriformis, Nippostrongylus brasiliensis, Necator americanus*) and a few trematodes, such as *Microphallus similis.* Why they should do this remains a mystery although various suggestions have been put forward (Lee 1996). An obvious reason might be to cause a localised interruption in cholinergic nervous transmission, thereby reducing gut motility and facilitating the parasite in holding on to its host's mucosa. However, there is limited evidence that this occurs to any significant extent. A more likely role for secreted acetylcholinesterase is to interfere with the immune response against the worms. Cholinergic signalling plays an important role in the development of both innate and adaptive immunity to helminth infections (Darby et al. 2015). For example, there is a suggestion that the filarial nematode *Wuchereria bancrofti* exerts an immunosuppressive effect by secreting acetylcholinesterase and thereby causing the localised breakdown of acetylcholine, which would otherwise have stimulated phagocytosis and the release of lysosomal enzymes (Bhattacharya et al. 1997). Any interference with innate and adaptive immune responses in the gut will also impact on the gut microbiome, and this emphasises that host: parasite relationships can be extremely complicated.

There is some uncertainty about the lifespan of the adult *Moniezia.* Seddon (1931) stated that it was as short as 65–70 days, whilst Kuznetsov (1968) suggested that it is up to 256 days. The worms probably induce a good protective immune response since heavy infections are usually only found in young sheep. In the past, *Moniezia* infection rates in the United Kingdom were high and surveys often revealed over 80% of sheep were infected. However, the widespread use of modern broad spectrum anthelmintics means that they are now much less common.

There are contradictory reports in the literature regarding the pathogenicity of *M. expansa* and *M. benedeni*. In the United Kingdom, Australia, New Zealand, and USA, *Moniezia* is considered to be of minor importance and removing the worms from lambs has no effect on live weight gain (Dever et al. 2015). Conversely, on the European continent, especially in Russia and the countries comprising the former USSR, *Moniezia* is often considered seriously pathogenic. For example, Averkhin et al. (1974) describe pathological changes in the intestine, pancreas, liver, spleen, kidney, and heart muscle of experimentally infected calves that would suggest immune-mediated pathology in response to parasite antigens. It is possible that the discrepancy may arise from differences in pathogenicity between strains of the parasite and susceptibility of the sheep and cattle definitive hosts. It is also possible that because *Moniezia* are large and obvious, they get the blame for diseases caused by other pathogens such as viruses, bacteria, protozoa, or even nematodes. Whether there are relationships between *Moniezia* and other pathogens/the gut microbiome is currently uncertain.

5.4 Phylum Acanthocephala

The Acanthocephala are commonly known as the 'thorny-headed worms' or 'spiny-headed worms' because they have a proboscis at their anterior end that is armed with an array of hooks (Figure 5.26a,b). The proboscis is eversible and can be retracted into a proboscis sac. Despite its fearsome appearance, the proboscis is used solely for attachment and acanthocephalans lack a

Figure 5.26 Unidentified acanthocephalan. (a) Whole worm (25 mm in length). (b) Proboscis showing spines.

(a)

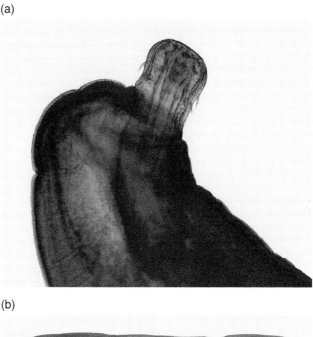

(b)

mouth or gut. They are usually relatively small, sausage-shaped creatures and grow to only a few millimetres or centimetres in length. There are, of course, exceptions and the males of *Macracanathorhynchus hirudinaceus* (a parasite of pigs) can grow to 10 cm, whilst the females may reach 60 cm in length. However, even these are dwarfed by *Oligoacanthorhyncus longissimus* which live in aardvarks (*Orycteropus afer*) and can grow to about 1 m in length.

All acanthocephalans are obligatory endoparasites. Their life cycle usually involves a larval stage that develops in an invertebrate intermediate host, whilst the adult stage lives within the gut of a fish, amphibian, reptile, bird, or mammal. About 1,000 species of acanthocephalan have been described to date, but the biology of most of them is poorly understood. Adult acanthocephalans feed in the same manner as cestodes by absorbing nutrients across their body surface. However, unlike cestodes, acanthocephalans are pseudocoelomates and have separate male and female sexes. Taxonomically, there is morphological and molecular evidence for a close relationship between the Acanthocephala and the Rotifera, although how the different groups are related is uncertain (Ricci 2017). This is a somewhat unexpected association because most rotifers are free-living aquatic organisms and have a fully developed mouth and gastrointestinal tract. Rotifers are generally less than 1mm in length and only a few species achieve 2 mm. However, like acanthocephalans, rotifers are pseudocoelomates.

Several acanthocephalan species are parasites of domestic animals such as *Ma. hirudinaceus* in pigs, *Oncicola canis* in dogs, and *Filicollis anatis* in ducks, geese, and various wild aquatic birds. Human infections are rare and most probably result from accidentally ingesting an infected invertebrate intermediate host (Mathison et al. 2016).

Macroacanathorhynchus hirudinaceus lives in the small intestine of pigs, and its size and pinkish coloration give it a superficial resemblance to the nematode *Ascaris suum*. It is not particularly host-specific and it also infects dogs, other carnivores, and even (rarely) humans. In pigs, *Ma. hirudinaceus* is rarely pathogenic although heavy infestations result in reduced weight gain. If the worms penetrate deep into the intestinal mucosa, they may ultimately perforate the gut giving rise to potentially fatal peritonitis. After mating, the female worms release eggs that then leave in the host's faeces. The eggs are highly resistant to environmental conditions and can survive for several years. The eggs hatch to release the acanthor stage after being consumed by beetle (Coleoptera) larvae such as those of dung beetles and cockchafers. The acanthor bores through the gut of the insect and transforms into the acanthella stage. Subsequently, the acanthella develops into a juvenile and then becomes a cystacanth. At this point further development ceases until the insect is consumed by the definitive host – the cystacanth is then released and attaches to the host's gut wall where it becomes an adult.

Homosexual Interactions in the Acanthocephalan *Moniliformis dubius*

Homosexual interactions are common in many vertebrate and invertebrate groups although their biological significance is a source of constant controversy. Among invertebrates, most same-sex interactions occur between males. These are often dismissed as a consequence of poor sex recognition and followed by the jaundiced observation that whatever their level of evolution, males will attempt to copulate with anything that is about the right size and shape and doesn't move out of the way fast enough! This, however, neglects the fact that the production of sperm, especially in some invertebrates, can represent a considerable metabolic investment and a male mating with another male is potentially wasting both time and energy.

Sexual interactions between adult male *Moniliformis dubius* are common, but in this species, they have a clear, albeit sinister, purpose. The adult worms live in the intestine of rats and other rodents, whilst the larval stages develop in a variety of insects, including the cockroach *Periplaneta americana*: it therefore makes a good laboratory model and has been studied extensively. During adult development, the males and females are spatially segregated with the females developing in the anterior region of the small intestine where there is a higher carbohydrate concentration. When they are mature, the males migrate up the intestine to mate with the females and after mating a male seals the female's vagina with a cement cap using secretions from his cement glands. This prevents the female from mating with another male and thereby reduces the opportunity for sperm competition. All the female worms from a single infection tend to mature at about the same time, which leads to intense competition between males for access to the females. Abele and Gilchrist (1977) observed that male worms would forcibly mate with one another in an attempt to seal off a competitor's genital opening. Although they found that only 2.5% of males were 'capped', they estimated that the true figure could be as high as 50%. They used the term 'homosexual rape', which is a dangerously emotive term but does indicate that the interaction is not accidental and neither is it without purpose.

6

Nematode Parasites

CONTENTS

6.1 Introduction

The nematodes belong to the phylum Nematoda, which is also known as the phylum Nemata. Somewhat paradoxically, the nematodes are simultaneously one of the most abundant, important, and yet neglected groups of metazoan animals. Some estimates suggest that there may be over 100 million species but fewer than 30,000 have been described. Nematodes occupy all terrestrial and aquatic ecosystems whilst one species, *Halicephalobus mephisto*, lives over a kilometre below ground in a gold mine in South Africa. In marine sediments, nematodes may constitute up to 95% of the organisms present and they are responsible for much of the benthic invertebrate biomass. Many nematodes feed on bacteria, fungi, or are detritivores, but there are also examples of predators and many are parasites of plants, invertebrates, and vertebrates.

Most nematodes, especially the free-living species, are microscopic in size, but some of those parasitic in mammals can be several centimetres in length, whilst some giants are over a metre long. For example, the females of *Placentonema gigantissima*, which lives in the placenta of sperm whales, can reach at least 8.4 m in length and 2.5 cm in width. Nematodes are very uniform in their appearance, and their thin, elongate, and cylindrical shape explains why they are commonly known as 'roundworms'. They lack a well-defined head and have a worm-like (vermiform) body that tapers at the anterior and posterior ends. The body is covered by a complex layered cuticle that

Parasitology: An Integrated Approach, Second Edition. Alan Gunn and Sarah J. Pitt.
© 2022 John Wiley & Sons Ltd. Published 2022 by John Wiley & Sons Ltd.
Companion website: www.wiley.com/go/gunn/parasitology2

is secreted by the underlying epidermis and is periodically shed during the juvenile stages to enable growth. The cuticle is proteinaceous but, unlike that of insects, it is not chitinous. Nematodes do, however, contain chitin, and it occurs in their eggshells, their pharynx, and in the sheath of microfilarial nematodes. In common with other higher invertebrates, nematodes are triploblastic and bilaterally symmetrical. That is, they have three primary germ layers during embryonic development, and the left side of their body is arranged the same as the right side. Although some nematodes have annulations (ridges around the body) on the surface of their cuticle there are no body segments and unlike the trematodes and cestodes they do not have suckers. Their body cavity takes the form of a pseudocoelom that contains fluid (haemolymph) at exceptionally high pressure and serves to maintain the hydrostatic skeleton. For example, the pressure within *A. suum* is 6.6–$37.6\,kN\,m^{-2}$, which is considerably higher than that of other invertebrates with hydrostatic skeletons: in earthworms the pressure is only 0.28–$2.8\,kN\,m^{-2}$. The high internal pressure necessitates the cuticle to be extremely strong and affects other aspects of nematode biology. For a more familiar comparison, typical values for human blood pressure would be $16\,kN\,m^{-2}$ systolic and $11\,kN\,m^{-2}$ diastolic.

The mouth of nematodes is usually surrounded by lips/mouthparts that are arranged symmetrically, and there is a muscular pharynx that pumps food into the body. The gut is tube-like and terminates in an anus close to the posterior of the worm. The high internal pressure means that nematodes project their waste with considerable force: *A. suum* can eject a stream of liquid waste several centimetres, and it is not a good idea to peer too close to the hind end of a live worm. All nematodes have unique chemoreceptors at their anterior end called amphids and some also have them at their posterior (caudal) region called phasmids. The amphids are usually well developed in free-living marine nematodes but reduced in size in terrestrial and parasitic species. An amphid consists of an external opening that leads via a duct to an amphidial pouch within which there are sensory neurons. An amphidial gland is also associated with the amphidial pouch. In the free-living nematode *Caenorhabditis elegans,* the amphids work as chemoreceptors involved in taxis (movement towards a stimulus), whilst the phasmids detect repellents and influence movement away from a stimulus. Nematodes have longitudinal muscles but lack circular muscles: they can therefore only bend their body from side to side and are unable to exhibit the creeping style of movement seen in earthworms (Annelida).

A few species of nematode are parthenogenic, but most of them reproduce sexually and have separate male and female sexes (i.e., they are dioecious). The males are usually smaller than the females, and there is often marked sexual dimorphism. Female nematodes typically have a vagina, vulva and a pair of tube-like uteri, oviducts, and ovaries. In parasitic nematodes, the female reproductive system is often extremely productive – a single worm may release hundreds or even thousands of eggs every day. Usually female nematodes release ovoid eggs, but in filarial nematodes, the eggs hatch whilst still inside the female worm and she releases microfilariae. The males have a single testis and the ejaculatory duct empties into the rectum. Nematode spermatozoa are usually large and amoeboid in shape – even if they appear to have a tail, it is not used as a flagellum to provide propulsion. The hind end of male nematodes is usually curved and has several sensory papillae. In some species, these papillae are little more than raised bumps, whilst in others they are much larger or reduced to pit-like depressions. In males belonging to the order Strongylida, the posterior of the worm is modified to form a 'copulatory bursa' that consists of two lateral lobes and a smaller dorsal lobe (Figures 9.1c and 11.1c). The lobes are supported by fleshy rays, and the whole structure envelops the female's genital opening during mating. *Bursa* is the Latin for a purse and the copulatory bursa has some resemblance to an old-fashioned moneybag. Male nematodes are equipped with one or two spicules that are used to dilate the female's vaginal opening during copulation: this is necessary to overcome the high internal body pressure and allows the spermatozoa to be then rapidly and forcibly injected into the female. The spicules are enclosed in a fibrous sheath that opens into the rectum. The spicules vary enormously in size and shape between species, and are therefore very useful in species identification. The term spicule derives from the Latin *spiculum* for a slender sharply pointed object and is used in various contexts in Biology. For example, the sharp calcareous/salicaceous structures in sponges are also known as spicules.

Studying the taxonomy of nematodes is difficult owing to their small size and the limited number of defining morphological characteristics. Furthermore, their morphological similarity belies considerable underlying genetic diversity. Traditionally, the nematodes are divided into two classes: the Secernentea (Phasmida) and the Adenophorea (Aphasmida). The Secernentea exhibit both cephalic amphids and caudal phasmids, whilst the Adenophorea (Aphasmida) have cephalic amphids but lack caudal phasmids. However, molecular analysis indicates that the Adenophorea is paraphyletic. Blaxter et al. (1998) recognised five distinct clades of nematodes all of which include example of parasitic species. They suggested that this indicates that animal parasitism probably evolved independently on at least four separate occasions. Subsequently, Holterman et al. (2006) proposed that the number of clades should be increased to twelve, whilst further revisions to the taxonomy of nematodes were suggested by Kern et al. (2020). Some taxonomists divide the nematodes into two classes: the Enoplea and the Rhabditea. The Enoplea supersedes the Adenophorea and contains those nematodes that, apart from certain parasitic species, have well-developed amphids but lack phasmids. Most nematodes in this class are free living, but it includes the order Trichurida that contains important parasitic genera such as *Trichuris*, *Trichinella*, and *Capillaria*. The Rhabditea supersedes the Secernentea and contains both free-living and many parasitic species. Among the many important and diverse genera within the Rhabditea are *Strongyloides*, *Ancylostoma*, *Necator*, *Ascaris*, *Anisakis*, *Dictyocaulus*, *Dracunculus*, and *Wuchereria*. In view of the large numbers of parasitic species and the uncertainty concerning the higher taxonomy of the nematodes, we will only consider certain genera to illustrate the diversity in the biology of animal parasitic nematodes. Further details on the biology of nematodes are available in Lee (2001).

6.2 Class Enoplea

6.2.1 Genus *Trichuris*

The genus name *Trichuris* derives from the Greek 'τριχηα' (*trica*) for hair and 'ουρα' (*oura*) for a tail. In fact, it is the anterior part of the worm that is thin and slender, whilst the posterior third is thickened (Figure 6.1a–c). This shape is the reason they have the common name of whipworms. The adult worms lack lips, and there are no cutting plates or other forms of mouthparts. The posterior of the males is tightly coiled, and they have a single spicule that is contained within a protrusible membranous sheath covered in a dense layer of short chitinous spines. The females have their genital opening at the junction of thin anterior region and the thicker posterior region. *Trichuris* eggs have a distinctive barrel shape with clear opercular plugs at either end (Figure 6.1d).

There are over 20 species in the genus, and they tend to parasitise groups of closely related hosts and have similar life cycles. For example, *Trichuris trichiura* infects humans and other primates, *Trichuris suis* infects pigs, *Trichuris ovis* infects sheep and other ruminants, whilst *Trichuris muris* infects rodents and is a good laboratory model for studying gastrointestinal nematode infections.

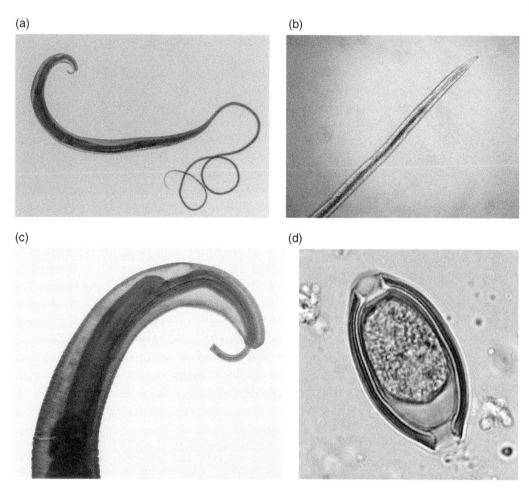

Figure 6.1 *Trichuris trichiura*. (a) Adult male worm. Note the long slender anterior end and much thicker posterior region. (b) Anterior of male worm showing long filariform oesophagus. (c) Posterior of male worm showing detail of the spicule. (d) Egg. Note the clear opercular plugs that are a characteristic feature of *Trichuris* spp eggs. The eggs are typically 50–55 μm in length and 20–25 μm in width and must undergo a period of development after being shed before they reach the infective stage.

The adult worms live in the caecum and their unsegmented eggs are passed with the host's faeces. If they are kept in moist shaded soil, the eggs embryonate to the infective first juvenile stage but do not hatch. The infective eggs are very resistant to environmental conditions and can remain viable for several years. Their robustness means that they have been identified from numerous archaeological sites. These include latrines built by Roman legionaries stationed near Hadrian's Wall in the United Kingdom, the mummified remains of Siberian nomads that date back to the fifth to fourth centuries BCE and coprolites recovered from an Inca settlement in Chile. This indicates that the parasite has a long-standing association with us although, as Mulder (2017) forcefully states, one must be careful when estimating parasite burdens from such specimens.

Transmission occurs through contamination and the eggs hatch after being consumed by a suitable definitive host. After hatching, the larvae invade the crypts of Lieberkühn and penetrate the cells at the base of the crypt. They then tunnel their way back to the surface of the gut lumen. Although juvenile worms may be found in both the small and large intestine, only those in the caecum complete their development to adulthood. Some texts continue to suggest that juveniles developing in the small intestine subsequently migrate to the caecum, but Bundy and Cooper (1989) state that there is no evidence of this. As the worms become adults, the thick posterior region of the worm breaks through the surface mucosa into the lumen, whilst the anterior portion remains buried within the host's tissues.

Some estimates suggest that between 450 and over 1,000 million people are infected with *T. trichiura*. The adult worms have a pinkish-grey coloration; the females are 30–50 mm in length, whilst the males are slightly smaller (30–45 mm). Infections are common in people living in countries with poor sanitation and the incidence is often highest among children. Light worm burdens are not particularly pathogenic but high worm burdens induce *Trichuris* dysentery syndrome. This manifests as persistent diarrhoea that may be bloody and persist for months or even years. It is a particular problem in young children and results in iron deficiency anaemia, stunting and clubbing of fingers, and may even induce rectal prolapse. Co-infections of *T. trichiura* and *Ascaris lumbricoides* are common since their eggs require similar conditions in which to embryonate and they are both transmitted through contamination. Similar pathologies to *T. trichiura* are caused by *T. suis* in pigs, *T. vulpes* in dogs, and *T. ovis* in sheep.

6.2.2 Genus *Trichinella*

For many years, *Trichinella spiralis* was the only recognised species in this genus, but it is now divided into two clades that contains at least 10 species and at least three distinct genotypes (Table 6.1) (Zarlenga et al. 2020). It is therefore sometimes referred to as the *Trichinella* complex. The various species vary in their distribution and normal hosts, but they all share a similar life cycle and have a potentially wide host range – even if it is not always utilised under natural conditions.

The division into two clades is based on whether the host encapsulates the larvae whilst they are developing within muscle cells. The larvae of most species and genotypes of *Trichinella* are encapsulated within nurse cells that are surrounded by a fibrous collagen wall of host origin

Table 6.1 Species and genotypes belonging to the genus *Trichinella*.

Trichinella genotype	Species name	Encapsulation	Remarks
T1	*Trichinella spiralis*	Encapsulated	
T2	*Trichinella nativa*	Encapsulated	Associated with arctic climate
T3	*Trichinella britovi*	Encapsulated	
T4	*Trichinella pseudospiralis*	Non-encapsulated	Infects mammals and birds
T5	*Trichinella murrelli*	Encapsulated	
T6	Not yet named	Encapsulated	
T7	*Trichinella nelsoni*	Encapsulated	Associated with hot climate
T8	Not yet named	Encapsulated	
T9	Not yet named	Encapsulated	
T10	*Trichinella papuae*	Non-encapsulated	
T11	*Trichinella zimbabwensis*	Non-encapsulated	Infects mammals and reptiles
T12	*Trichinella patagoniensis*	Encapsulated	

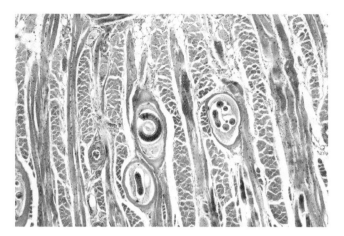

Figure 6.2   Encysted larvae of *Trichinella spiralis* within skeletal muscle.

(Figure 6.2), but in three species (*Trichinella pseudospiralis*, *Trichinella papuae*, and *Trichinella zimbabwensis*) this does not occur. The lack of encapsulation means that the larvae do not survive for long after their host dies. Consequently, non-encapsulated species are more reliant upon the definitive host killing and consuming the intermediate host, or at least discovering it soon after death. It also means that these species are not detected by the trichinoscopy (compression) test that was once widely used in slaughterhouses to detect the presence of *Trichinella*. This test involves placing a muscle sample between two glass slides that are then tightly bound together and examined using a microscope. The encapsulated larvae become calcified with time and their presence is detectable by the gritty feel and observation. Although cheap and simple, trichinoscopy is not particularly sensitive and led to under-reporting of the parasite's presence. One should therefore be careful when comparing old data with that obtained using more recent detection techniques. Trichinoscopy is no longer recommended by the EU for either food safety inspection or surveillance. Currently, enzymic digestion tests are employed that release the larvae from the muscles so that they are observable using a microscope. This approach is more sensitive than trichinoscopy and detects both encapsulated and non-encapsulated species. However, there are sometimes problems using it with frozen meat and if there are low infection levels. Serological and molecular detection techniques are currently employed for research and surveillance, but the digestion method is used in the meat industry for food safety inspection.

Trichinella spiralis has a worldwide distribution and an exceptionally wide host range that includes humans, pigs, dogs, cats, badgers, rats, and numerous other mammals. The spread of the parasite is aided by the global market in live animals, meat, and animal products, as well as an increasing interest in some countries in consuming game animals and wild animals of various descriptions. Most of this trade is legal, but there is also a huge global black-market trade in the meat of domestic and wild animals – and much of this does not undergo meat hygiene inspection.

There are essentially four life cycle types: domestic, sylvatic-temperate, sylvatic-torrid, and sylvatic -frigid, and all of them can include humans – although usually as dead-end hosts. The 'domestic' life cycle principally involves *Tr. spiralis* that is transmitted from pigs to us when we consume raw or poorly cooked infected meat. It is therefore a particular problem in countries in which there remains a tradition of rearing pigs for domestic consumption (Murrell 2016). These pigs are often fed food waste and/or allowed to wander freely and thereby feed on anything that they can find. They therefore have a high probability of becoming infected. We are essentially dead-end hosts because pigs rarely get the opportunity to eat us. Pigs usually become re-infected when farmers feed them scraps of infected pork meat. Pigs are also prone to acts of cannibalism: during fights they will consume one another's tails (and worse). Rats and other rodents can also figure in this cycle through eating discarded infected meat and then falling victim to a hungry pig later. The sylvatic-temperate transmission cycle occurs between wild animals living in temperate climates such as in Europe, parts of Asia, and North America. This cycle typically involves *Tr. spiralis*, *Tr. britovi*, and *Tr. murrelli*, and the parasites cycle between foxes, bears, wild boar, rodents, and other carnivores or scavengers. Infection may take place through consuming either prey or animals that died from natural causes. Humans usually become infected through eating poorly cooked game animals. The parasites may subsequently enter a domestic cycle if flesh from game animals is fed to pigs. Human infections with *Tr. pseudospiralis* occur through similar domestic and sylvatic-temperate transmission cycles. Its prevalence in us is a lot lower than that of *Tr. spiralis* but it has a much wider host range that encompasses birds. However, the records of *Tr. pseudospiralis* infections in birds currently relate to predatory and scavenging species, and these seldom form part of our diet. Therefore, until more extensive surveys are undertaken, it remains uncertain whether birds are a source of human infections. The sylvatic-torrid transmission cycle occurs in tropical Africa and involves *Tr. nelsoni*. Hyenas, lions, warthogs, and bush pigs are thought to be important in the transmission of this parasite, and it is very pathogenic in humans. As one would expect, the sylvatic-frigid transmission cycle occurs in the Arctic and often involves *Tr. nativa*. Normally, this parasite cycles between predators and scavengers such as bears, wild cats, foxes, mustelids, and walruses. Encapsulated *Tr. nativa* larvae can survive freezing at −20 °C, whilst *Tr. britovi* and *Trichinella* genotype 6 also exhibit resistance to very low temperatures. Consequently, one probably could not rely on storing meat in commercial meat freezers to kill off these parasites (Pozio 2016).

6.2.2.1 *Trichinella spiralis*

Trichinella spiralis, like the other *Trichinella* species, is an unusual parasite in that an infected individual serves simultaneously as both a definitive host and an intermediate host. It also has the distinction of being the largest intracellular parasite known. The life cycle begins with the host consuming raw or undercooked meat containing infective first-stage larvae that are encapsulated within the flesh. The larvae are released within the host's duodenum and proceed to invade the glandular crypts of the upper region of the small intestine. The developing worms proceed to thread through adjoining villi by penetrating the cytoplasm of cells (rather than pushing the cells to either side of their body). The juveniles grow rapidly and within 4 days they become sexually mature adults. The worms are slender and there is not the obvious division between the anterior and posterior regions seen in *Trichuris* spp. The females are 3–4 mm in length and 60 μm in diameter, whilst the males are about 1.6 mm long and 40 μm in diameter. The male worm lacks spicules, but he has a pair of lateral flaps at the posterior end that he uses to grasp the female. The female has her genital opening in the middle of the oesophageal region. The male worm has a shorter lifespan than the female, but he has sufficient time to mate with several females. The female lives for 4–16 weeks and during this time the eggs within her uterus mature and hatch *in situ*. Consequently, rather than eggs, she releases hundreds or even thousands of first-stage larvae (80–160 μm in length). The larvae penetrate the intestinal mucosa and then enter the blood circulation, which distributes them around the body. Larvae that reach skeletal muscles proceed to penetrate individual muscle fibre cells. Although all skeletal muscle cells can be infected, the diaphragm, jaws, tongue, extraocular muscles, and larynx are usually the most heavily parasitised.

It takes about 4–8 weeks for *Tr. spiralis* larvae to reach the infective stage, at which point they enter arrested development during which they can remain infectious for many years. After 6–9 months, the nurse cells often start to become calcified and in heavy infections, the calcified nurse cells may be visible as a scattering of grey-white spots within the muscle. The life cycle is completed when the infected muscle is consumed by a suitable definitive host. Therefore, unlike many nematode parasites, none of the developmental stages are exposed to the environment and the whole life cycle takes place within the body of another animal.

Invasion of Muscle Cells by *Trichinella spiralis*

Muscle cells are highly specialised and differentiated cells but upon infection by *Tr. spiralis* larvae, they de-differentiate into nurse cells. The larvae induce this by altering gene expression within parasitised cells and in particular the activity of those genes controlling cell cycle-related factors. In addition, there is down-regulation of genes coding for muscle cell-specific proteins and consequently the muscle cell de-differentiates. That is, it ceases to be a muscle cell and re-enters the cell cycle. De-differentiation is to be expected since this is a normal part of the muscle cell repair process following damage – and the larvae damage the cells in the process of invasion. However, normally, de-differentiation is followed by regeneration, but this does not occur in parasitised cells. Instead, the larvae manipulate their environment to facilitate their own growth. The contents of the cell undergo apoptosis and the sarcomeres and myofilaments break down. This presumably provides nutrients for the worms. The muscle cell does not totally degenerate because there is a balance between the upregulation of apoptosis-genes and anti-apoptosis genes (Babal et al. 2011). In addition, the parasite induces the nurse cell to secrete vascular-endothelial growth factor that stimulates the growth of small blood vessels around the nurse cell, and this presumably increases the supply of oxygen and metabolites. Eventually, the nurse cells become up to 0.4–0.6 mm in length and 0.25 mm in width and contain coiled infective first-stage larvae that are about 1mm in length. Further details are provided by Wu et al. (2013). The ability to induce apoptosis has led to an interest in identifying chemicals produced by *Tr. spiralis* larvae that might be exploited to suppress tumour growth (Luo et al. 2017).

Adult *Tr. spiralis* seldom cause clinical disease and are soon cleared by the host's immune response. However, large infections can cause enteritis with the production of excessive intestinal mucus and diarrhoea. Protein deficiency in the diet reduces the strength of the immune response and thereby prolongs the infection. It also results in in more larvae invading the muscles (Vila et al. 2019).

Small numbers of *Trichinella* larvae encysting within the muscles may cause no problems, but the sudden invasion of numerous larvae can result in serious pathology and even prove fatal. This typically occurs if we eat poorly cooked, heavily infested pork or wild game. The wandering larvae can induce pneumonia and their movement through the heart, eye or brain can have serious consequences. The invasion of skeletal muscles causes myositis (inflammation of the muscles), fever, intense muscular pain, and impairs the ability of the affected muscles to contract. Consequently, invasion of the diaphragm and intercostal muscles compromises breathing, whilst invasion of muscles in the tongue and oesophagus result in swallowing difficulties. There is an ongoing debate concerning the popular belief that the consumption of whisky will protect against contracting *Trichinella*. Campbell and Blair (1974) refer to unpublished studies alleging that pigs are protected from *Trichinella* if given sufficient Irish whisky shortly after exposure. Similarly, the low morbidity and mortality associated with a mass outbreak of trichinellosis in Laos was ascribed to the excessive inebriation of those exposed to infection (Barennes et al. 2008). Sadly, any attempt at discovering the truth behind these claims runs into some serious ethical considerations.

6.3 Class Rhabdita

6.3.1 Genus *Strongyloides*

Members of this genus have an unusual life cycle that alternates between generations that are free-living and those that are parasitic (Figure 6.3). Both free-living and parasitic generations involve four larval stages and the production of egg-laying adult worms. However, in the free-living cycle males and females are formed and sexual reproduction occurs, whilst in the parasitic cycle only females develop, and they reproduce by parthenogenesis.

Over 50 species of *Strongyloides* have been described and their host ranges encompasses amphibians, reptiles, birds, and mammals. Humans are mostly parasitised by *Strongyloides stercoralis,* and some estimates suggest it infects 30–100 million people in over 70 countries (Nutman 2017). *Strongyloides stercoralis* also infects non-human primates as well as dogs and cats and is therefore a zoonotic parasite. Molecular evidence suggests that *S. stercoralis* may have evolved as a parasite

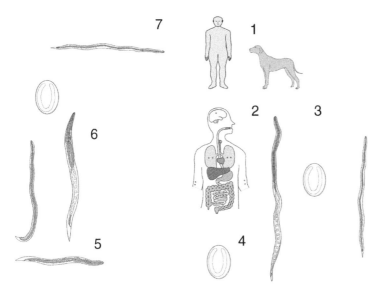

Figure 6.3 Life cycle of *Strongyloides stercoralis*. 1: Infection commences when humans (or other susceptible mammals) ingest the infectious filariform larvae or these invade through the skin. Those invading through the skin migrate through the heart and lungs and ultimately reach the small intestine. 2: Adult filariform female worms develop in the duodenum and small intestine and reproduce parthenogenically. 3: Some of their eggs hatch in the small intestine, invade through the gut wall, and cause disseminated strongyloidiasis. This can result in pathology in various organs of the body including the lungs and brain. These larvae ultimately return to the gut where they mature into adult females. 4: Some eggs and first-stage larvae pass with the faeces. These develop into either free-living rhabditiform male and female worms or infectious third-stage filariform larvae. 5: Free living rhabditiform larvae develop into male and female worms. 6: Eggs resulting from sexual reproduction by free-living female worms. 7: Infectious filariform larvae formed from eggs laid by free-living worms or eggs/larvae passed in the faeces of an infected individual. Drawings not to scale.

of dogs that subsequently spread to humans (Nagayasu et al. 2017). Several other *Strongyloides* species also cause zoonotic infections. For example, *Strongyloides fuelleborni* is usually a parasite of non-human primates but also infects people in parts of Africa and Southeast Asia. A closely related species *Strongyloides fuelleborni kellyi* is currently only known to infect people in certain restricted regions of Papua New Guinea. A previously unknown genotype of *S. fuelleborni* was identified in 2019 in Australia where there are high rates of infections among some Aboriginal communities (Barratt and Sapp 2020). *Strongyloides papillosus* normally parasitises rabbits, sheep, and other ungulates, and although its larvae will invade humans, they probably do not develop to adulthood. Thamsborg et al. (2017) provide further details of *Strongyloides* infections in domestic and companion animals.

6.3.1.1 *Strongyloides stercoralis*

Strongyloides stercoralis has a cosmopolitan distribution and is a common infection wherever sanitation is poor, and people walk barefoot. The free-living larval stages have a rhabditiform pharynx and live in the soil where they feed on bacteria (Figure 6.4a). A rhabditiform pharynx is one in which the muscular pharynx has a prominent posterior bulb that is separated from a smaller anterior swelling called the metacorpus (in appearance this is a bit like two clubs balancing on top of one another). After mating, the female produces eggs that hatch to release the first-stage larvae. The free-living cycle is sometimes referred to as 'indirect' or 'heterogonic' development. Provided the conditions are favourable, some species of *Strongyloides* have repetitive free-living generations in which the larvae develop into free-living adults. In these species, the larvae develop into infective filariform third-stage larvae only when the conditions would no longer support further free-living generations. By contrast, according to Streit (2008), *S. stercoralis* has only a single free-living generation and its larvae invariably develop into infectious third-stage form. These larvae are 490–630 µm in length and have a long cylindrical pharynx (Figure 6.4b). If a suitable host passes by the larvae burrow into its skin. Consequently, we often acquire infections by walking barefoot in areas contaminated with faeces. Infections can also occur through consuming larvae that are contaminating drinking water or food. Larvae that invade through the skin make their way to the veins and enter the venous circulation and are swept to the heart and thence to the lungs. They then penetrate the alveoli and make their way up the bronchioles to the bronchi and thence to the trachea where they are swallowed. When they reach the duodenum the worms burrow into the intestinal villi and may penetrate the crypts of Lieberkühn. Here the females mature and start to produce dozens of eggs every day by parthenogenesis. The adult female worm is about 2.0–2.5 mm in length and 0.04–0.05 mm in width and has a long thin

(a)   (b)

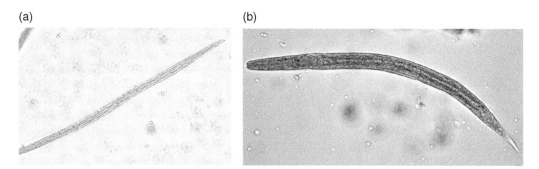

Figure 6.4   *Strongyloides stercoralis* third-stage larvae. (a) Infective form exhibiting filariform pharynx. (b) Free-living form exhibiting rhabditiform pharynx.

cylindrical oesophagus that is about a third the length of the body and lacks a posterior bulb. Most commentators state that none of the various species of *Strongyloides* produce parasitic males, although more work is required to confirm this. The process of host invasion and the subsequent production of parasitic adult worms is sometimes referred to as 'direct' or 'homogonic' development. The eggs released by the female worms are already embryonated and usually hatch within the host's intestine to release first-stage larvae. Some of these larvae pass out with the faeces where they either develop into free-living adults or into infective filariform third-stage larvae. This begs the question of how parthenogenic reproduction can give rise to both male and female offspring, especially as *S. stercoralis* has a sex chromosome: females have six chromosomes (XX) whilst males have only five (XO). Environmental factors such as temperature, food availability, and population density can affect the development of the larvae either through previous impact on the mother or directly upon on the developing eggs/larvae. However, other factors such as the age and genetic constitution of the female worm and the host immune response may also be important. More detail on this topic is provided by Streit (2008). Other first-stage, larvae develop rapidly to the infective third stage and auto-infect the host by penetrating the intestinal mucosa or the peri-anal skin. These larvae then undergo visceral migration to the heart, lungs, and other organs and become adults in the duodenum. Consequently, a host can remain infected for many years through repetitive autoinfection with successive generations of worms. Furthermore, an initial infectious dose of one or a few larvae can ultimately give rise to a large worm burden without the host exposure to another external source of infection.

Strongyloides stercoralis causes pathology at three points during its parasitic life cycle: when it initially penetrates the skin, during migration through the lungs and other body tissues, and as adults in the small intestine. Larvae penetrating through the skin cause haemorrhage and an acute inflammatory response with resultant pain and swelling, whilst migration through the lungs induces symptoms similar to those of asthma – although pulmonary symptoms are uncommon. Low worm burdens of adult *S. stercoralis* in the gut may cause few symptoms but as the numbers of worms increase, they cause pain, nausea, diarrhoea, intestinal bleeding, and obstruct the passage of digesta. Immunocompromised patients can suffer from ongoing cycles of excessive autoinfection (hyperinfection) and the larvae may become disseminated around the body. This can result in the larvae invading other tissues such as the liver, kidneys, and central nervous system with debilitating and potentially fatal consequences. The mortality associated with hyperinfection is around 60% (Chan et al. 2018). Hyperinfection also occurs in people who are not immunocompromised and may be exacerbated by bacterial infections that cause septicaemia and bacterial meningitis (Vadlamudi et al. 2006). The bacteria may invade through the damaged gut epithelium, but it is more likely that they are transported into the body either by adhering to the cuticle of the invading larvae or by passing unharmed through their gut and being voided with their faeces.

Although *Strongyloides* can cause serious pathology, many infected people remain asymptomatic or suffer non-specific or mild symptoms. In addition, because it is difficult to diagnose, many infections go undetected. Therefore, patients can become parasitised early in life and remain unknowingly infected for many years. Hyperinfection may only manifest when the patient suffers a condition that weakens their immune system. These often include factors such as co-infection with the retrovirus Human T-cell Leukaemia Virus Type 1 (HTLV-1), prolonged use of corticosteroids, and alcoholism (de Souza et al. 2020; Rao et al. 2019). If the patient moves to a non-endemic country after infection, the cause of their symptoms may not be expected and therefore tested. Physically detecting the larvae in faecal smears using a microscope is very time-consuming and insensitive, especially when the worm burden is low. Culturing the worms from faecal samples improves sensitivity but adds to the time it takes (Schär et al. 2016). For this reason, *Strongyloides* infections are often not included in surveys of soil-transmitted helminths. Commercial serological tests are available but there are concerns with cross reactivity with other parasites such as filarial nematodes. This is therefore an infection for which molecular diagnostic tests would be ideal (Formenti et al. 2019). That is, if they could be produced cheaply enough to be used on a large scale in the developing countries where the parasite is most a problem.

6.3.2 Genus *Ancylostoma*

The genus *Ancylostoma* is one of several genera of nematodes that are commonly known as the hookworms. It contains several species although the best known of these is *Ancylostoma duodenale* that infects humans. Other important species include *Ancylostoma caninum* that is primarily a parasite of dogs and wild canines, *Ancylostoma tubaeforme* that usually infects cats, and *Ancylostoma braziliense* and *Ancylostoma ceylanicum* that both infect wild and domestic mammalian carnivores. The larvae of all these species can infect us although the worms do not usually develop to adulthood.

Adult *Ancylostoma* have a well-developed buccal capsule that on its ventral margin is armed with 1–4 pairs of large chitinous 'teeth', which terminate in sharp backwardly curved points (Figure 6.5a,b). Most authors state that it is these mouthparts that explain the name '*Ancylostoma*', which derives from the Greek words 'ankylos' (ανκψλοσ = hook) and 'stoma' (στομα = mouth). However, some authors state that the 'hook' refers to the fact that the worms have stout stiff bodies that are bent at the anterior end so that they resemble a crotchet hook. In common with other hookworms, male *Ancylostoma* have a well-developed copulatory bursa that has two broad lateral lobes and a small dorsal lobe, and they also have two long needle-like spicules. Another explanation for the derivation of '*Ancylostoma*' is that the term was originally used in relation to a different genus of hookworm in which the rays of the male's copulatory bursa appear hook-like.

Ancylostoma duodenale has a widespread distribution and occurs in Europe, Africa, Australia, Asia, China, and Japan and both North and South America. The larvae require a combination of moisture and a temperature of over 20 °C to develop, so it is less common in regions with cold or temperate climates. However, hookworm infection is a known risk for Northern European mine-workers because deep mines are hot and humid, thereby facilitating the survival of the eggs and

(a)   (b)

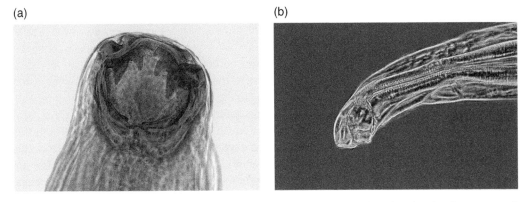

Figure 6.5   Mouth and buccal cavity of *Ancylostoma duodenale*. (a) Dorsal view showing the sharp recurved 'teeth'. (b) Side view to show the muscular pharynx. Note how the anterior of the worm is curved (hooked): this ensures that the body of the worm can be pressed against the surface of the gut and less likely to be swept away by the flow of digesta around it.

larvae. Furthermore, the workers are often not provided with toilet facilities underground (it can be over 1 km from the workface to the surface and mineworkers cannot be expected to make the trek every time that nature calls even if the management are willing to allow it – which they are not). The female worms are 10–15 mm in length and lay up to 30,000 eggs day^{-1}. The males are slightly smaller at 8–11 mm in length. The adults of both sexes have two large ventral teeth and one small ventral tooth on either side of the mouth.

Ancylostoma braziliense is found throughout the tropical regions of the world and is sometimes confused with *A. ceylanicum*. The females are 7–10 mm in length and the males are 6–7.75 mm long. The adults can be distinguished by having one large and one small ventral tooth on either side of the mouth.

Ancylostoma caninum has a cosmopolitan distribution. The females are 14–16 mm in length, whilst the males are 10–12 mm. The adults have a pair of triangular dorsal teeth and a pair of ventro-lateral teeth.

Adult *Ancylostoma* live in the small intestine where they use their mouthparts to grasp onto the mucosa and lacerate the underlying tissue. The worms then feed on the blood that seeps from the wound. The worms do not remain in a fixed position but move around the gut and therefore leave wounds that continue to bleed. In addition, the worms consume more blood than they require and excrete much of it before completely digesting it. In heavy infections, this inefficient process causes significant blood loss, thereby increasing the risk of anaemia.

After mating, the female worms release eggs that are passed with the host's faeces. Most hookworm eggs are very similar in appearance, and it is usually impossible to identify them to species (Figure 6.6a). Provided they are maintained in a warm moist environment, the eggs continue to develop and hatch after about 24–48 hours to release free-living first-stage larvae. These larvae have a rhabditiform oesophagus and feed on bacteria and other material in the faeces and soil. In countries with warm damp climates, a fresh faecal deposit instantly attracts large numbers of coprophagic insects. The insects churn up the faeces and thereby aerate it and incorporate the faecal material into the soil. This improves the conditions for the growth of the hookworm larvae. After 2–3 days, the first-stage larvae moult to the second stage and these then continue to feed before moulting after about 5 days to the infectious third stage (Figure 6.6b,c). The third-stage larvae are referred to as 'filariform' because their oesophagus lacks a terminal bulb or isthmus. They do not feed but have sufficient food particles stored in their intestines to enable them to survive for several weeks. The third-stage larvae usually retain the cuticle shed by the second stage as a protective sheath although this is lost when the worm invades a potential host. The infectious larvae position themselves in the top layer of the soil although they move up or down the profile depending upon soil moisture levels. When they contact a potential host, the larvae penetrate through its skin by a combination of physically boring and releasing enzymes. They are very thin and easily make their way through cracks in the skin or they penetrate via the hair follicles. After passing through the epidermis, they make their way to the blood vessels or lymphatic system and are transported via the circulatory system to the heart and thence to the lungs. The larvae then penetrate the alveoli and enter the air spaces after which they climb up the bronchioles to the bronchi and thence to the trachea after which they are swallowed down the oesophagus. Once the third-stage larvae reach the small intestine, they moult to the fourth stage. This stage has an enlarged buccal cavity that is used to attach the worm to the intestinal mucosa. The fourth-stage larvae moult to the adult stage and once mature these mate and egg-laying commences. It typically takes about five or so weeks from initial invasion through the skin to egg-laying although in some species it takes much longer than this because the larvae enter a period of delayed development. Some texts state that adult *Ancylostoma* can live for several years, but it is more probable that their

(a)

(c)

(b)

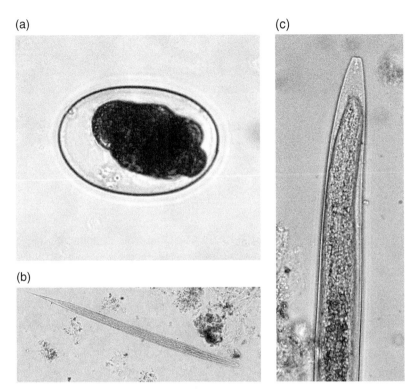

Figure 6.6   Hookworm eggs and larvae. Most hookworm eggs and larvae are similar in size and morphology, and it is therefore impossible to identify them by light microscopy. (a) Egg. (b) Whole infectious third-stage larva. (c) Anterior of third-stage larva to show the retained second-stage cuticle forming a protective sheath.

lifespan is in the order of 6–12 months. Although infections can persist for many years this could be explained by the periodic re-activation of larvae in arrested development.

Arrested Development (Hypobiosis) in Hookworm Larvae

In some *Ancylostoma* species, a variable proportion of the migrating third instar larvae do not proceed to the lungs but disperse from the circulation into the skeletal muscles and enter a period of arrested development (hypobiosis). For example, the larvae of *A. duodenale* may arrest their development for up to 38 weeks so that they do not become adults and commence egg-laying until the external environmental conditions are warm and moist enough for the survival of the thin-shelled eggs and free-living larvae. Similarly, some of the larvae of *A. caninum* undergo developmental arrest whilst migrating through female dogs and then re-activate during pregnancy. These larvae then make their way to the mammary glands and infect the pups through trans-mammary transmission over a period of about 3 weeks after birth. There is a possibility that trans-mammary transmission of *A. duodenale* also occurs in human infections but evidence for trans-placental transmission is currently lacking. Interestingly, in pregnant female dogs, the arrested *A. caninum* larvae are probably not directly reactivated by changes in hormonal titres. According to Arasu (2001), they have receptors that respond to the up-regulation

(Continued)

Arrested Development (Hypobiosis) in Hookworm Larvae   (Continued)

of transforming growth factor-β in the uterus and mammaries brought about by changes in the titres of oestrogen and prolactin. Transforming growth factor-β signalling pathways are highly conserved and exist in both invertebrates and vertebrates. They are involved in many cell processes including cell proliferation and morphogenesis. In the free-living nematode, *Caenorhabditis elegans*, the transforming growth factor-β signalling pathway is involved, amongst other things, in the formation of the resting dauer larvae. The details of the molecular/physiological mechanisms by which *Ancylostoma* larvae resume their development is uncertain but are probably similar to those of the dauer larvae of *C. elegans*.

Arrested larval development is a feature of parasitic nematodes belonging to several genera and is an important factor when considering their treatment and control. If these nematodes share a common basic mechanism, *C. elegans* might provide a simple laboratory model for studying the factors governing the onset into/emergence from arrested development. Prolactin stimulates the reactivation of arrested *Toxocara canis* larvae and thereby facilitates their transmission in the mother's milk, but it is uncertain whether this is a direct effect or occurs indirectly through up-regulating the production of transforming growth factor-β as occurs in *A. caninum* (Jin et al. 2008). In addition, genome analysis indicates that there are major differences between the transforming growth factor-β pathways in *C. elegans* and *T. canis* (Ma et al. 2019).

Hookworms cause pathology at three main points in their life cycle: during the initial invasion through the skin, during their migration through the body, and as adults consuming blood in the small intestine. Larvae that penetrate the skin of their normal definitive host do not usually cause a serious host reaction. However, they may introduce bacteria that set up a localised acute inflammatory response. Penetration usually takes place through our feet although other parts of the body may be affected – for example if we sit, kneel, or lie on the ground. If we are infected by the larvae of species that normally parasitise other animals, such as *A. caninum* and *A. braziliense*, we may suffer from a condition known as *cutaneous larvae migrans*. These larvae are unable to move beyond the basal layer and wander around just underneath the skin surface for a period of weeks or even months until they are eventually killed by our immune system. Whilst they are wandering, they produce long serpentine tracks that are intensely itchy and painful because of the physical damage coupled with the acute inflammatory response. The damaged tissue often becomes secondarily infected by bacteria that are either introduced by the worm or because of scratching. The pathology caused during migration of larvae through the lungs, like that of the adults in the intestine is very dependent upon the number of worms involved. Small numbers may cause so little damage that there are no clinical signs. However, if numerous larvae make their way through the lungs at the same time, they can cause respiratory distress and pneumonitis which can prove fatal. The consequences of infection with adult worms depend upon the numbers involved and the nutrition of the host. An adult *A. duodenale* consumes about 0.26 ml blood day^{-1}, so the presence of a few worms has little impact on health and low worm burdens are even considered by some people to be beneficial to the maintenance of a healthy immune system. However, as the numbers of worms rises then the amount of blood lost can quickly become significant – especially if the host does not have a nutritious diet. The loss of blood results in iron-deficiency anaemia and the faeces can turn red or black depending upon the amount of blood loss. Hookworm infection is also responsible for a range of other symptoms (e.g., fatigue, weakness, pallor) most of which link to the development of iron-deficiency anaemia. One of the more unusual symptoms is 'reverse iris-dilation reflex' in which the pupil of the eye dilates in response to strong light rather than contracting. Hookworm infection has also been suggested to stimulate the consumption of soil – known as 'geophagia' or 'pica'. Chronic infections in children can result in stunted physical growth and mental retardation.

6.3.3 Genus *Necator*

Several species of *Necator* parasitise primates but *Necator americanus* is commonly believed to be the only one that infects humans. It is now apparent that we can also be infected by *Necator gorillae* although this is probably only a risk factor for the small number of people who have close dealings with gorillas. Members of the genus are distinguishable from those of *Ancylostoma* by the structure of their mouthparts. Adult *N. americanus* have a pair of ventral cutting plates rather than teeth, and at the base of the buccal capsule, there is a pair of ventral teeth and a pair of subdorsal teeth (Figure 6.7). In addition, the copulatory bursa in the males is longer and narrower and the rays have a different arrangement; in the females the vulva is situated more towards the anterior of the worm. *Necator americanus* is somewhat smaller and slenderer than *A. duodenale*: the female worms average 10–11 mm in length, whilst the males are typically 7–8 mm.

The life cycle of *N. americanus* (Figure 6.8) closely resembles that of *Ancylostoma* but the larvae do not undergo hypobiosis and the adults can live for five or more years. It is primarily a human parasite although it also infects non-human primates. It consumes less blood than *A. duodenale* – typically 0.05 ml worm^{-1} day^{-1} – but large worm burdens can still cause sufficient blood loss to result in serious disease. Nevertheless, in regions in which the parasite is endemic, many people harbour relatively small worm burdens that cause little harm. In those people who harbour large worm burdens, the pathology of *N. americanus* closely resembles that of *Ancylostoma*.

In Latin, '*necator*' translates as 'murderer' or 'slayer'. *Necator americanus* can therefore be taken to mean 'Slayer of Americans'. This is a bit of a Hollywood overstatement, and the worm is certainly not racially prejudiced. Indeed, it is the commonest hookworm parasite that afflicts humans – some estimates suggest that it infects as much as 10% of the world's population. It has a widespread distribution in parts of North and South America, Africa, Asia, India, Australia, and southwest Pacific

Figure 6.7 *Necator americanus.* Adult stage. Anterior region, note the cutting plates.

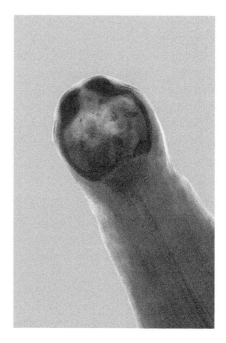

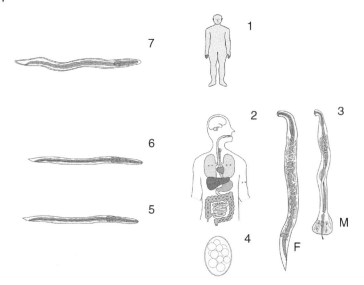

Figure 6.8 Life cycle of *Necator americanus*. 1: Infection commences when the infective third-stage larvae penetrate through someone's skin. 2: The larvae enter the circulation and eventually reach the lungs. Here, they penetrate the alveoli, climb up the trachea, and are then swallowed down the oesophagus. 3: The parasites become adults in the small intestine where they mate and commence laying eggs (F = female, M = male). 4: Unembryonated eggs are passed in the host faeces. 5: The eggs hatch to release the rhabditiform first-stage free-living larvae that feed on bacteria etc. 6: The first stage larvae moult to become second-stage rhabditiform larvae. 7: The second-stage larvae moult to become infectious third-stage filariform larvae. These larvae do not feed and retain the cuticle of the second-stage larvae. Drawings not to scale.

Islands. However, it requires even warmer (25–28 °C) conditions than *A. caninum* to complete the free-living stages of development, and therefore, it is limited to regions with hot/tropical climates. Co-infections of *N. americanus* and *A. duodenale* can occur, but usually one of them is more abundant than the other. One of the theories of how humans came to colonise the Americas is that they migrated across the Behring land bridge. If correct, this would mean leaving behind those parasites whose transmission stages could not survive exposure to cold environmental conditions. *Necator americanus* requires very warm conditions for transmission, and therefore some workers suggest that *N. americanus* was first introduced into the Americas during the period of colonisation by Europeans and particularly through the slave trade. However, native Paraguayan Indians living in comparative isolation have been found suffering from *N. americanus*. This has been used as evidence supporting the theory that human colonisation of South America may also have occurred by trans-Pacific migration of Asiatic populations during pre-Columbian times (Araujo et al. 2008).

6.3.4 Genus *Ascaris*

This genus contains two of the best-known parasites: *A. lumbricoides* that infects humans and *A. suum* that infects pigs. However, a debate about whether they are genuinely two distinct host-specific species has been raging for several decades. For a few years there was general consensus that they are two distinct species that although relatively host-specific and could infect other mammals, they did not reach sexual maturity in them. For example, there are occasional reports of *A. suum* infecting sheep (Gunn 1980). Older literature often does not distinguish between them and consequently one can find references to *A. suum* in humans and *A. lumbricoides* in pigs.

For once, molecular studies appear to be simplifying matters and indicate that *A. lumbricoides* and *A. suum* may be one and the same species. For example, *Ascaris* infections in both humans and pigs share common genotypes and similar miRNA profiles and there are frequent cross transmissions (Alves et al. 2016; Shao et al. 2014). Nevertheless, for the purposes of this book we will stay with the current convention and refer to them as separate species.

Members of the *Ascaris* genus are large robust worms (Figure 6.9a). For example, female *A. suum* can grow up to 41 cm in length and 5 mm in width, whilst the males grow up to 25 cm in length and 3 mm in width (some texts state that they can grow even larger). Their large size means that they have been recognised since antiquity and the name Ascaris is a direct translation of the Ancient Greek '*Askaris*' (*Ασκαριζ*) meaning a worm found in the intestines. Their cuticle is thick and covered in numerous fine striations. The body is rather rigid and has a cream or pale pinkish coloration. Interestingly, they have an unusual haemoglobin in their perienteric fluid that binds oxygen almost 25,000 times more tightly than human haemoglobin. Indeed, it binds so tightly that it is incapable of functioning in oxygen transport. Instead, its function is probably to eliminate oxygen (Imai 1999)! Adult *Ascaris* have three curved lips and on the inner edge of each of them is a single row of tiny teeth (denticles) (Figure 6.9b). The oesophagus is a simple cylindrical muscular tube and does not have a bulb. The posterior of the male is curved and lacks a bursa although there are numerous papillae in the vicinity of the cloaca that probably serve as sensory structures during copulation. The male has two simple spicules that are about 2 mm in length. The vulva of the female opens in the anterior third of the body and leads to two uteri that occupy the bulk of the internal space of the worm.

Both *A. lumbricoides* and *A. suum* have cosmopolitan distributions. *A. suum* is a common parasite of domestic and wild pigs throughout the world. By contrast, human infections with *A. lumbricoides* are rare in developed countries but exceedingly common wherever there is poor sanitation. Some estimates suggest that as many as 800 million people are infected (Brooker and Pullan 2013). Infection usually peaks during childhood and early adolescence. As with many parasites, within a population a small number of people are often much more heavily infected than everyone else. This is to a large part a consequence of genetic susceptibility, and there is a clear link between susceptibility to *Ascaris* infection and loci on chromosome 1 and chromosome 13 (Williams-Blangero et al. 2008). The adult worms live in the small intestine where they consume the digesta; they may consume blood if it leaks from an existing wound, but there is limited evidence that they 'intentionally' damage the gut mucosa to feed on the host tissues or blood. The life cycle (Figure 6.10) begins with the adult worms mating and the females subsequently releasing their eggs that are then passed with the host's faeces. Their reproductive potential is phenomenal: Brown and Cort (1927) estimated that a single female worm may contain up to 27 million eggs at any one time and lay over 200,000 eggs day^{-1}.

Ascaris eggs are roundish to oval (45–75 μm long × 35–50 μm wide) with a thick clear inner shell and a lumpy outer layer that is variously called the 'mammilated', 'uterine', 'albuminous', or 'proteinaceous' layer (Figure 6.9c). This outer layer is often stained yellow or brown due to incorporating bile from within the digesta. When it is passed with the faeces, the egg contains well-defined roundish cells. Unfertilised eggs (Figure 6.9d) are commonly seen in faecal samples. They usually have a more elongated profile, their outer mammilated layer is either pronounced or absent and the contents of the egg are amorphous. These unfertilised eggs are incapable of further development. The fertilised eggs are extremely resistant to environmental conditions. They can survive for at least 5 years under favourable conditions although they are susceptible to desiccation and exposure to UV light. According to Chandler and Read (1961), 'an enterprising German researcher seeded a plot of soil with *Ascaris* eggs; two persons ate unwashed strawberries raised from the plot

(a)

(b)

(c)

(d)

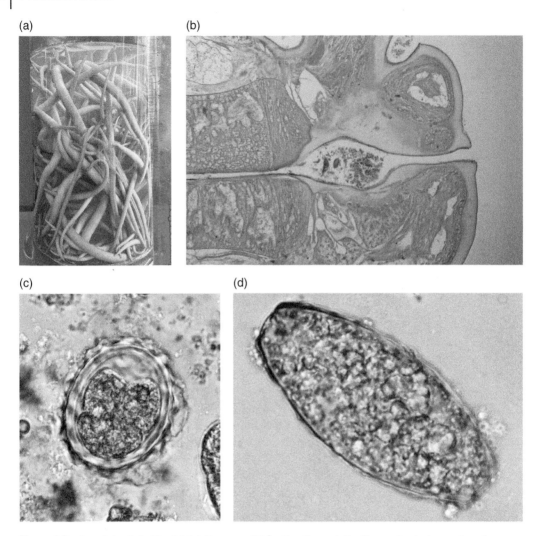

Figure 6.9 *Ascaris lumbricoides.* (a) Adult worms. (b) Section through the lips and anterior region of an adult worm. (c) Fertilized egg. This is an egg from a faecal sample and is therefore unembryonated. Note the 'lumpy' outer layer. (d) Unfertilized egg. Note the more oval shape and the lack of a well-developed outer layer.

each year for 6 years, and each year acquired a few *Ascaris*'. There are more than a few problems with this experimental design, not least on the matter of ethics, but it does suggest that *Ascaris* eggs may remain viable for many years in the environment. The use of untreated human faces as a fertiliser coupled with a lack of safe drinking water to wash vegetables facilitates the transmission of soil-transmitted helminth infections. For example, Amoah et al. (2006) found that *Ascaris* eggs contaminated 65% of spring onion, 60% of lettuce, and 55% of cabbage samples that they surveyed from an urban market in Ghana.

Studies on *A. suum* eggs indicate that embryonation and the early stages of larval development take place outside the host. Within the egg the first-stage larva moults to the second stage and then to the third stage – and it is this which is the infectious stage (Kirchgäßner et al. 2008). However, some earlier workers state that the infectious stage is the second-instar larva and the moult to the

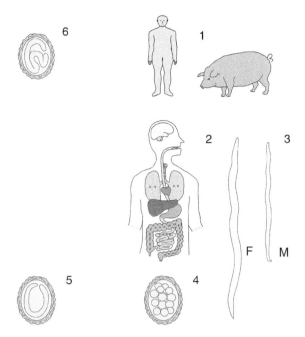

Figure 6.10 Life cycle of *Ascaris lumbricoides/Ascaris suum*. 1: Infection commences with consumption of eggs containing the infective larvae. 2: The eggs hatch in the small intestine and the larvae penetrate the lining of the gut, enter the circulation, and eventually reach the lungs. Here they penetrate the alveoli, climb the trachea until they reach the join with the oesophagus. They are then swallowed and reach the small intestine again. 3: Within the small intestine, the larvae moult to the adult stage and commence sexual reproduction. 4: Unembryonated eggs are passed in the host's faeces. 5: The eggs embryonate in the environment. 6: The larva moults to the infectious stage larva within its egg. There is some debate in the literature whether this is the second-stage or the third-stage. Drawings not to scale.

third instar takes place in the liver during migration through an infected host (Douvres et al. 1969). It is highly likely that *A. lumbricoides* shows a similar sequence of development stages as *A. suum*. Development to the infectious stage takes about 9–13 days under ideal conditions and consequently fresh faeces is not an immediate source of infection. Transmission is through passive faecal–oral contamination. The eggs hatch in the host's intestine and the larvae then burrow through the mucosa and enter the venous or lymphatic circulation. The larvae usually reach the liver via the hepato-portal blood supply about 24 hours after the eggs hatch and after travelling through the hepatic venules they are transported to the heart. Once they reach the heart, the larvae enter the pulmonary circulation. When the larvae reach the lungs, they are about 1.8 mm in length and burrow through the walls of the capillaries and enter the air spaces. The larvae then make their way up the bronchial tree to the trachea and are then swallowed down the oesophagus. Once the larvae reach the small intestine, they moult to the adult stage, mate, and commence egg laying. The time between initial infection and the adult worms commencing egg production is in the region of 8–12 weeks. Most of the work on larval migration has been done on *A. suum* in pigs and there is some variation in the literature about the sequence of developmental stages. Douvres et al. (1969) state that the migrating larvae reach the small intestine still in the third-instar stage and subsequently moult to the fourth stage and then moult again to become adults. By contrast, Roberts (1934) states that the larvae moult to the fourth stage whilst still in the lungs and those that remain at the third stage are killed by the acidic pH of the stomach. The reason(s) why some gastrointestinal nematodes undertake apparently pointless migrations around their host's body only to arrive back at the point where they began remains a mystery. Some workers consider it to be a 'phylogenetic reminiscence' but Read and Sharping (1995) propose that gastrointestinal nematodes whose larvae undergo a period of migration through the host's tissues grow faster than those parasite species that remain solely in the gut lumen. By contrast, Mulcahy et al. (2005) propose that the migration enables the larvae to avoid the host's immune response associated with the gut mucosa. However, this is at the cost of exposing themselves to the host's immune response in the circulation and tissues.

Ascaris lumbricoides and *A. suum* cause pathology at three distinct points during the live cycle: during migration through the liver, migration through the lungs, and as adults in the small intestine. Numerous larvae simultaneously moving through the intralobular veins cause extensive haemorrhage and tissue damage. In pigs, this gives rise to 'white spot' in which functional liver cells tissue are replaced by connective scar tissue. This results in the liver being condemned as unfit for human consumption. More serious pathology occurs in the lungs when the larvae penetrate the air spaces. This may cause Loeffler's pneumonia (pneumonitis) and in severe cases this can be fatal. The consequences of infection with adult worms depend upon the number of worms involved and whether they go wandering. Heavy infections can cause potentially fatal intestinal blockage but more normally the host suffers from malnutrition and digestive upsets. The adult worms release various antimicrobial substances that influence the composition of the surrounding gut microbiome (Midha et al. 2018). It is currently uncertain whether this contributes to the digestive upsets but it would seem probable. In young children, the consequences of malnutrition can lead to stunted growth and cognitive impairment. Wandering worms may induce appendicitis, block the bile duct, or cause physical and psychological trauma when they are coughed up or the worm attempts to exit via the nose. Paparau et al. (2018) describe a horrific case in which a 2-year-old Romanian child died of traumatic asphyxia when several adult worms blocked his oesophagus and trachea.

6.3.5 Genus *Enterobius*: *Enterobius vermicularis*

Enterobius vermicularis is a remarkably common nematode even within developed countries. They are relatively short and thin (~0.5 mm), the females growing to 8–13 mm and the males are even smaller at 2–5 mm (Figure 6.11a,b). Hence, their common names of pinworms and threadworms. The adult worms live in the lumen of the caecal regions but, when they are gravid, the female worms migrate to the rectum to deposit their eggs. Sometimes they return afterwards but very often they continue through the anus and can be found on the perianal regions and within the faeces. The presence of the worms and their eggs in the rectal/perianal region induces an inflammatory reaction and hence intense itching – *pruritis ani*. This in turn results in the infected person scratching the region and getting the eggs underneath their fingernails. The eggs cannot survive for more than a few days in the outside environment. However, they reach the infective stage within 6 hours of being passed and, therefore, an infected person who does not scrub their nails when washing their hands is very likely to reinfect him/herself (and others) through contamination. In addition, underwear, nightwear, and bedsheets can become heavily contaminated with the eggs. From these sources, the eggs can become airborne, for example, when gathering up washing, and thereby contaminate all surfaces within a house/dwelling. Infections can therefore spread rapidly within households, halls of residence, boarding schools etc. The adult worms are short-lived, females living for only 37–93 days, and consequently a persistent infection requires continual re-infection. Rates of infection can be exceptionally high in some situations – even reaching 100%.

(a)

(b)

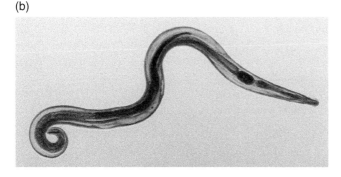

(c)

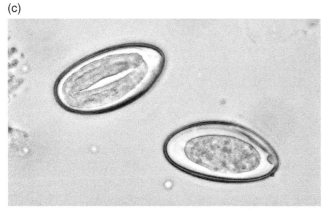

Figure 6.11 *Enterobius vermicularis*. (a) Adult female – these are 8–13 mm in length. (b) Adult male – these are 2–5 mm in length. (c) Eggs – these are typically 55 × 26 μm in size.

As a rule, children aged 5–10 years of age are those most commonly infected, but all ages are susceptible.

The female worms typically migrate to lay their eggs during the night and the resultant intense itching can result in loss of sleep. In addition, the scratching risks damaging the epidermis and facilitating bacterial infections. In girls and women, the worms sometimes move the vulva where they induce vaginitis and have even been recorded reaching the Fallopian tubes.

The eggs of *E. vermicularis* (Figure 6.11c) are not easy to detect using normal faecal egg analytical techniques. However, for many years, the use of Scotch tape has been the method of choice. This involves placing a length of Scotch tape onto the perianal region, flattening it down with a tongue depressor, and then removing the tape and placing it sticky side down onto a microscope slide.

In the 1980s, a new species of *Enterobius* was proposed, *Enterobius gregorii*. It was subsequently identified around the world and co-infections were commonly reported. *Enterobius gergorii* was distinguished from *E. vermicularis* based on the spicule shape in the male worms but doubt was later cast on whether this was a reliable indicator of a genuine difference in species identity.

6.3.6 Genus *Toxocara*: *Toxocara canis*

Adult worms belonging to the genus *Toxocara* have cervical alae, which means that their anterior is shaped a bit like an arrowhead. This, in a roundabout way, explains the derivation of the name *Toxocara*, which is a compound of the Greek τοξο, (*toxo*) meaning a bow or arc and Latin *caro* meaning flesh. It is an extremely common worm with a cosmopolitan distribution in both developed and developing countries. The adult worms (Figure 6.12a,b) live in the small intestines of dogs, foxes, and other caniids where the females grow to about 18 cm whilst the males are up to about 10 cm. The life cycle (Figure 6.13) begins when females release eggs that pass out with their host's faces and embryonate to the infective stage over the subsequent 10–15 days. Transmission is through passive contamination and this why the safe disposal dog faeces is so important for control. If a dog ingests the infective eggs, these hatch in the small intestine and the larvae invade through the mucosa, enter the bloodstream, and make their way to the liver. In young dogs, the larvae then move to the lungs, cross the alveoli, and climb up the bronchi and then the trachea until they reach the join with the oesophagus. They are then swallowed and move down the oesophagus, through the stomach until they finally reach the small intestine. Here, they become adults, mate, and commence reproduction. This may also happen in older female dogs but more commonly larvae enter arrested development within her somatic tissues. When she becomes pregnant, changes in her hormonal balance induce some of the larvae to become mobilised at successive pregnancies. These larvae cross the placenta and invade the developing puppies. This usually happens after day 42 of pregnancy. The larvae make their way to the liver of the puppies and then migrate through its tissues such that they reach the lungs at about the time of birth. From there they move up the trachea and then down the oesophagus, through the stomach, and ultimately reach the small intestine. Here, they reach adulthood about 4 weeks after the puppies' birth. There is some variation in the literature about when different stages of development occur. For example, some state that the larvae moult to the third instar when they reach the liver of a developing puppy, whilst others state that the third instar is the infective stage within the egg. Puppies can also become infected after birth by activated larvae within the female dog entering her mammary glands and contaminating the milk.

Toxocara canis can exploit various mammal and bird species as paratenic hosts. This occurs when they ingest the infective eggs. The eggs hatch in the small intestine and the larvae invade various somatic tissues where they enter arrested development. If a dog eats an infected paratenic

(a)   (b)

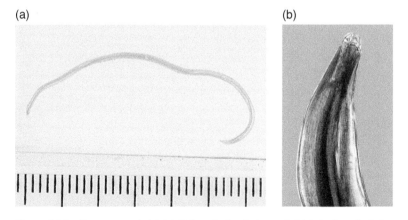

Figure 6.12   *Toxocara canis*. (a) Adult female (scale in mm). (b) Anterior of adult worm.

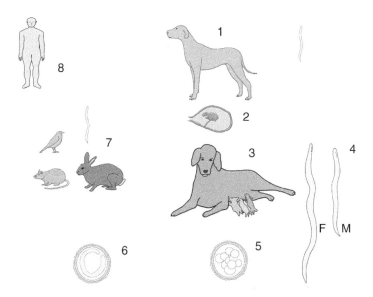

Figure 6.13 *Toxocara canis* life cycle. 1: A dog becomes infected when it consumes either eggs containing the infectious larvae or a paratenic host that is infected with the larvae. The eggs hatch in the small intestine and the larvae penetrate the gut wall, enter the circulation, and migrate through the liver and other somatic tissues. In young dogs, the larvae complete their migration and become adults in the small intestine. This involves reaching the lungs where they penetrate the alveoli and climbing the trachea to the join with the oesophagus where they are swallowed. Once they reach the small intestine they mature to adults. In older dogs, the larvae enter into arrested development within somatic tissues. 2: In pregnant females, arrested larvae are activated and cross the placenta where they can infect the developing pups. 3: In lactating females, activated larvae may also infect the pups via the milk. 4: In both pregnant/lactating females and pups, the larvae complete their migration and become adults in the small intestine where they commence sexual reproduction (F = female; M = male). 5: Unembryonated eggs are passed in the faeces. 6: Eggs embryonate to the infectious larval stage in the soil. 7: Many mammals and birds can act as paratenic hosts. The eggs hatch in their small intestine and the larvae invade their tissues but are unable to complete their development. 8: Eggs can hatch in the human small intestine and the larvae migrate through various somatic tissues. They are unable to develop but their migration can cause serious physical damage. Drawings not to scale.

host, the larvae will resume their development to adulthood. Humans can also act as paratenic hosts (Ma et al. 2018), although we are seldom eaten by dogs these days. Within us, the larvae cause visceral larva migrans, the consequences of which depend upon the number of larvae moving through the body and where they are moving. For example, children are particularly at risk of suffering eye damage if the larvae migrate across the eye. In severe cases, it can result in blindness. Children tend to roll around on the floor and become over-familiar with pets, so they are at high risk of contracting infections.

Toxocara cati, in which the definitive hosts are domestic cats and other felids, has a similar life cycle to *T. canis* although trans-placental and trans-mammary infection is probably not as important for transmission. However, transmission via paratenic hosts, such as mice and other rodents, is probably more important. Zoonotic transmission can also occur.

6.3.7 Genus *Anisakis*

The adults of members of this genus are parasites of marine mammals, whilst the larval stages infect krill and fish. Those species belonging to the *Anisakis simplex* complex (*Anisakis simplex* sensu stricto, *Anisakis pegreffii, Anisakis berlandi* [= *A. simplex* C]) are important in human medicine because ingestion of the third-instar larvae can induce a short-lived but potentially harmful infection of the gastrointestinal tract or a harmful immune reaction.

The adult worms live within the mucosa of the stomach of seals, dolphins, and other marine mammals and large infections can cause serious ulceration. Their eggs are passed with their host's faces and hatch to release free-swimming larvae that are consumed by krill. There is some variation in the literature about whether the eggs hatch to release first-instar larvae or third-instar larvae. After being ingested by the krill, the larvae invade the haemocoel, and there is consensus that it is here where third-instar larvae occur. If a fish or cephalopod, such as a squid, consumes infected krill, it becomes a paratenic host, and the third-instar larvae embed within its mesentery. If a marine mammal consumes an infected fish/squid, the larvae moult twice and become adults in its stomach. If a human consumes a raw or undercooked infected fish, the worms are not immediately killed and survive for about 3 weeks without reaching adulthood.

The industrialisation of the fishing industry has probably contributed to the increase in human infections in recent years. In the past, fish was gutted at sea immediately after being caught, and this removed the majority of the *Anisakis* larvae. However, nowadays fish are often stored whole in refrigerated conditions and gutted when the catch is landed ashore. During the time the fish is in transit, it is thought that many larvae move from the viscera into the surrounding muscle tissue and thereby increasing the chances of human infection.

Anisakiasis is particularly common in Japan because of the popularity of raw fish (and possibly its highly industrialised fishing industry). Infection with live larvae may occur in the stomach or the small intestine and results in abdominal pain and diarrhoea. The inflammatory reaction may even be sufficient to block the intestine. Occasionally, the larvae invade through the gut wall and become lodged in ectopic sites such as the tongue, and pancreas. Anisakiasis has been linked to the development of gastric cancer, but the evidence is not especially strong. The allergic response to the larvae may be triggered by both live and dead larvae and varies from relatively mild to a life-threatening anaphylactic reaction. Similar pathologies are also caused by related nematode species that normally parasitise marine mammals and fish such as the *Anisakis physeteris* complex, *Pseudoterranova decipiens* complex and *Contracecum osculatum* complex.

6.3.8 Family Onchocercidae

The parasites belonging to this group are commonly known as 'filarial nematodes' because of their characteristic larvae. There are about 80 genera within the family and the adult worms are tissue parasites that live in mammals, birds, reptiles, and amphibians although not fish. Most, but not all, species are in a symbiotic relationship with *Wolbachia* bacteria. In those species that are in a symbiotic relationship, the bacteria are essential for the worms to reproduce, and they are also responsible for some of the pathology associated with filariasis (Lefoulon et al. 2016). Targeting the *Wolbachia* bacteria with antibiotics such as doxycycline provides a means of treating these filarial nematode infections. In the absence of their symbiotic bacteria, both adult worms and their microfilariae exhibit increased apoptosis in certain cell types. Exactly how the *Wolbachia* prevents apoptosis is uncertain (Grote et al. 2017). The bacteria are also involved in several biosynthetic pathways necessary for the growth and reproduction of the worms (Cotton et al. 2016).

The adult worms are usually long and slender and have a long cylindrical pharynx. The eggs hatch in the uteri of the females to release 'microfilariae' – these are less developed than the first larval stage of other nematode species, and some workers describe them as 'pre-larval' or 'advanced embryos'. In some species, the microfilariae are enclosed in a 'sheath'. This sheath derives from the egg membrane and is therefore distinct from the 'sheath' of third-stage nematodes such as *A. duodenale* in which it is formed by retaining the cast cuticle of the second juvenile stage. One means of distinguishing between microfilariae is by the presence or absence of a sheath. For example, the microfilariae of *Mansonella* and *Onchocerca* do not have sheaths, whilst those of *Wuchereria* and *Brugia* are sheathed (Figure 6.14). The larvae are transmitted by blood-feeding insects that also act as intermediate hosts within which development to the infectious stage takes place.

Although most species within the family Onchocercidae are parasites of wild animals, it includes the important genera *Onchocerca*, *Wuchereria*, *Brugia*, *Dirofilaria*, *Loa*, and *Mansonella*. Details on the molecular taxonomy of filarial nematodes are covered by Lefoulon et al. (2015) and Morales-Hojas (2009).

6.3.8.1 Genus *Onchocerca*

There are several species within this genus of which *Onchocerca volvulus* is the best known. Humans are the sole definitive host for *O. volvulus,* and it afflicts about 15.5 million people (GBD 2016). It causes the disease onchocerciasis that is more commonly known as 'river blindness'.

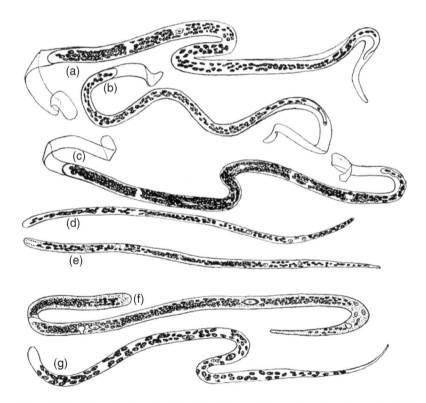

Figure 6.14 Microfilariae of various filarial nematode species drawn to scale. (a) *Wuchereria bancrofti* (sheathed, no nuclei in tip of tail). (b) *Brugia malayi* (sheathed, two nuclei in tail). (c) *Loa loa* (sheathed, nuclei to tip of tail). (d) *Dipetalonema perstans* (no sheath, tail blunt with nuclei to tip). (e) *Mansonella ozzardi* (no sheath, no nuclei in end of tail). (f) *Onchocerca volvulus* (no sheath, no nuclei in end of tail). (g) *Dirofilaria immitis* (no sheath, sharp tail without nuclei in end). *Source:* Reproduced from Chandler and Read (1961), © Wiley- Blackwell.

It is particularly common in parts of Africa (e.g., South Sudan, Ghana, Uganda, Nigeria, Congo, Chad), but it is also found in certain regions of the Middle East (e.g., Yemen) and Central America (e.g., Mexico, Ecuador, Colombia, and Venezuela). The parasite was almost certainly introduced into South America through the slave trade but has since evolved distinctive local characteristics. Adult female *O. volvulus* are up to 50 cm in length and 0.4 mm in width, whilst the male worms are much smaller at 3–5 cm long and 0.2 mm in width. They live in pairs or groups tightly coiled together in the subcutaneous tissues. This can give rise to raised nodules (Figure 6.15a) on the scalp, torso, and limbs that can be anywhere from 0.5 to 10 cm in size. In Africa, most of the nodules develop below the waist, whilst in Central America they tend to form above the waist. This variation relates to the biting behaviour of the local blackfly vectors. The nodules are unsightly but cause little direct pathology. The female worms release microfilariae but unlike many of the strains of *W. bancrofti* and *B. malayi*, these do not exhibit periodicity and they are not sheathed (Figure 6.15b). The microfilariae are 220–360 μm in length and tend to remain within the subcutaneous tissues in which they can live for up to a year. Up until the development of effective anthelmintics, health workers would dissect out the nodules. This reduced the patient's exposure to microfilariae and the chances of them acting as a source of infection to blackflies. However, this had to be done very carefully because if the adult worms were damaged, it would release a sudden influx of parasite antigens into the circulation that might trigger an anaphylactic shock.

In common with other members of the genus, *O. volvulus* is transmitted by various species of blackfly (*Simulium* spp.) (Figure 6.15c). Blackfly larvae are aquatic and live attached to rocks etc. within fast-flowing streams and rivers. Therefore, *Onchocerca* infections are also predominantly found in their locality. However, the adults of some blackfly species can disperse over 500 km from their site of emergence through being borne aloft on seasonal winds. This makes vector control extremely difficult, not least because the breeding sites of the flies may be in a different country to where disease transmission occurs. Like mosquitoes, the females of many blackfly species need to take a blood meal to mature their eggs. There are a few blackfly species that do not need a blood meal, and these, obviously, are not important in disease transmission. Male blackflies do not feed on blood and are therefore not important in disease transmission (apart from fertilising the female flies). Female blackflies have short mouthparts and are pool-feeders: they acquire the microfilariae – which are in the tissues – whilst lacerating skin and imbibing the blood and serum that oozes to the surface. The microfilariae then penetrate their gut and make their way to the thoracic flight muscles where they develop to the infectious third stage. This typically takes about 6–12 days. The infectious larvae then move to the fly's mouthparts and transfer to us the next time the blackfly feeds. Female blackflies typically feed only every 3–5 days and therefore they do not usually transmit an infection until at least their third blood meal. The infectious larvae moult to the fourth stage about 7 days after entering our circulation and then moult again to become adults. The adults mature over a period of 12–18 months and can survive for up to 15 years.

In common with *Wuchereria* and *Brugia*, the various species of *Onchocerca* have an obligatory symbiotic relationship with *Wolbachia* bacteria. If patients infected with *O. volvulus* receive the antibiotic doxycycline, the fertility of the female worms is suppressed for several months. However, because the worms are long-lived, if they regain their fecundity then the patient will continue to suffer from the damage caused by the microfilariae and may also be a source of infection to blackflies. Laboratory studies using the model system of *Brugia pahangi* infecting jirds (*Meriones* spp.) indicate that bacteria living within the worms' ovarian tissues may be protected from antibiotic exposure (Gunderson et al. 2020). The use of antibiotics on their own may therefore have limitations as a long-term treatment. Combining doxycycline with standard ivermectin treatment can be highly effective at both reducing the production of microfilariae and enhancing

Figure 6.15 *Onchocerca volvulus.*
(a) Section through a nodule containing adult worms. (b) Microfilaria (lacks a sheath, no nuclei in tip of tail).
(c) Vector blackfly (*Simulium* spp.)

(a)

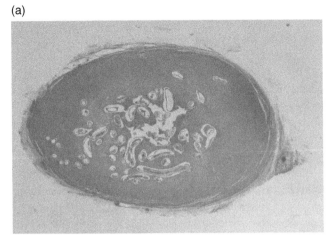

(b)

(c)

the mortality of the worms (Debrah et al. 2015). It does, however, require regular doses of doxycycline, and this limits its use for mass treatment.

The pathology associated with *O. volvulus* is mostly caused by the immune reaction against the microfilariae and their symbiotic *Wolbachia* bacteria. This pathology includes blindness, 'leopard skin', and 'hanging groin' (Chapter 10). Interestingly, in West Africa, two strains of *O. volvulus* are recognised: a severe strain that is associated with serious eye damage, and a mild strain that causes less eye damage and mostly causes dermal pathology. These differences may relate to differences in the strains of *Wolbachia* bacteria harboured by the worms (Higazi et al. 2005) although not to their density in the worms (Armoo et al. 2017). Onchocerciasis may also be a factor in the increase in cases of epilepsy and nodding syndrome reported in the early part of the twenty-first century throughout parts of southern Sudan and northern Uganda (Colebunders et al. 2018). Nodding syndrome is typified by seizures, stunted growth, and cognitive impairment. It gets its name from the way children experiencing seizures temporarily lose muscle tone in their neck so that their head rocks backwards and forwards every 5–8 seconds. The microfilariae do not appear to enter the brain, but it is possible that they could be triggering an autoimmune reaction that results in nerve damage.

Although *O. volvulus* is essentially a tropical disease there are several species of *Onchocerca* that afflict animals other than humans that occur in temperate countries. For example, *Onchocerca gutturosa* is a relatively common parasite of cattle in Europe and elsewhere in the world in which it causes raised nodules on the back ears and neck. It is not, however, considered pathogenic.

6.3.8.2 Genus *Wuchereria*, *Wuchereria bancrofti*

Wuchereria bancrofti occurs in tropical and sub-tropical regions of Africa, India, China, Indonesia, South America, and certain Eastern Pacific Islands. We are its sole definitive host but over 70 species and sub-species of mosquito can act as intermediate host and vector. The adult worms live tightly coiled together so that they form nodule-like masses within the afferent lymphatic ducts – especially in the lower parts of the body. The female worms grow up to 10 cm long but are only 0.3 mm wide, whilst the male worms are 4 cm long and 0.1 mm wide. The female worm releases thousands of microfilariae every day into the lymph, which transports them to the thoracic duct and where they enter the blood circulation. The adults can live for 5–8 years and during this time a female may release millions of microfilariae. The microfilariae are 244–296 μm in length and move actively, but they lack a functional gut and do not feed. The microfilariae exhibit marked periodicity in their appearance in the peripheral circulation that relates to the biting habits of the local mosquito vector(s). The appearance of microfilariae in the circulation is sometimes called 'microfilaremia'. When they are not in the peripheral circulation, the microfilariae retreat the capillaries surrounding the alveoli in the lungs. After ingestion by a mosquito, the microfilariae penetrate its intestine and make their way to the thoracic flight muscles. Here, they develop into the infectious third stage over a period of 7–21 days and during this time they damage the flight muscles. Over part of its range, *W. bancrofti* co-occurs with *Plasmodium falciparum*, and they share the same vector mosquitoes (e.g., *Anopheles gambiae* and *Anopheles funestus*). In Kenya, Muturi et al. (2006) found that a small percentage (~1.1–1.6%) of mosquitoes were naturally infected with both parasites. Although mosquitoes infected with *Wuchereria* had higher *Plasmodium* sporozoite rates, they also had a higher mortality and therefore co-infections would appear to reduce the chances of transmission for both parasites.

The infectious third-stage larvae are 1.4–2.0 mm in length and migrate to the mosquito's proboscis. They then enter our circulation when the mosquito feeds and make their way to the peripheral lymphatic system. Ultimately, they locate themselves in lymph vessels distal to the lymph nodes and continue their development to the adult stage, mate, and commence the production of microfilariae. The microfilariae are usually detected 8–12 months after the initial infection although sometimes disease symptoms manifest in the absence of circulating microfilariae.

Infection with *W. bancrofti* causes the disease lymphatic filariasis. This afflicts approximately 90 million people in 52 countries. Although *Brugia malayi* and *Brugia timori* also cause lymphatic filariasis, *W. bancrofti* is responsible for 80–90% of the cases. Some infections remain asymptomatic, and the individual may or may not express microfilariae in their blood. Those who do not express microfilariae are effectively dead-end hosts and are not a source of infection to others. However, asymptomatic individuals who express microfilariae represent an unsuspecting reservoir of infection. The pathology associated with lymphatic filariasis largely results from our inflammatory reactions to the parasite's symbiotic *Wolbachia* bacteria (Taylor et al. 2010). Lymphatic filariasis manifests as periodic filarial fevers followed by chills and lymphangitis and/or adenolymphangitis – inflammation of the lymphatic channels/lymphatic adenoidal tissue that is situated behind our nasal cavity. The most debilitating pathology associated with lymphatic filariasis results from the blockage of the lymph drainage. For some reason, presumably connected to both our genetic constitution and that of the nematodes, the pathology tends to vary between geographical locations. For example, in Tanzania, hydrocoele develops much more commonly than lymphoedema and elephantiasis. By contrast, in India hydrocoele and lymphoedema are about equally common. Lymphoedema refers to swelling of the body tissues, usually the legs in the case of lymphatic filariasis, owing to restriction of the drainage of lymphatic fluid. Strictly speaking, elephantiasis means to be 'afflicted by elephants' but is really a reference to the enormous swelling that occurs in the affected limbs. There is also some evidence that lymphatic filariasis may increase susceptibility to HIV infection (Kroidl et al. 2016). Debilitating elephantiasis afflicts only a small proportion of those infected by *W. bancrofti* and other filarial nematodes and it is not an inevitability of infection.

6.3.8.3 Genus *Brugia*

There are several species within this genus although *Brugia malayi* is the most important. As the species name intimates, this parasite is found in Malaya, but it also occurs in India, Sri Lanka, Thailand, the Philippines, China, Korea, and Japan. On certain Indonesian islands, *B. malayi* is replaced by the closely related *Brugia timori* (Fischer et al. 2004). Between them, *W. bancrofti* and the various species of *Brugia* are estimated to cause lymphatic filariasis in approximately 38.5 million people and many of these suffer from serious debilitating disease. Adult *B. malayi* are very difficult to distinguish from those of *W. bancrofti* and they also parasitise the lymphatics. However, although they also produce sheathed microfilariae, these can be distinguished from those of *Wuchereria* by being smaller (177–230 µm) and from the positioning of the nuclei in the tail region: in *B. malayi* the nuclei extend to the tip of the tail, whilst in *W. bancrofti* they do not (Figure 6.14). The life cycle resembles that of *W. bancrofti,* but a key difference is that the adults parasitise a range of animals includes monkeys, pangolins (*Manis* spp.), and some wild felids. It is therefore a zoonotic parasite with a range of reservoir hosts, and this can make control more difficult. It does, however, mean that *B. malayi* is the only filarial nematode that one can maintain in laboratory rodents. It develops faster than *W. bancrofti* both within the mosquito vector and the definitive host: microfilariae can appear in the peripheral circulation in as little as 3 months after the initial infection. The pathology resembles that of *W. bancrofti* and is also partly a consequence of the reaction of our immune system to the nematode's symbiotic *Wolbachia* bacteria.

6.3.8.4 Genus *Loa, Loa loa*

Infections with *Loa loa*, loiasis, occur in parts of West and Central Africa. Currently, over 10 million people are thought to be infected. Although the worm can infect monkeys, it is uncertain how important zoonotic transmission is in human infections. The adult worms resemble a long thin silk thread, females can reach 7 cm, whilst the males are typically about half this size and reach up 3.5 cm. The name *Loa loa* translates as 'worm worm', but it is commonly known as the African eyeworm – a moniker it has gained through the frequency with which the adult worm crawls across the cornea (Figure 6.16a). If one is quick, it is possible to pick out the worm at this point but the risk of causing damage to the eye is great. Furthermore, if any of the worm is left behind, then it will decay and cause potentially harmful inflammation. The adult *L. loa* dwell within the subcutaneous tissues but make periodic migrations and at this point they may travel across the eye. Exposing the skin to heat stimulates the worms to move, so it often happens when an infected person sits in front of a fire at night. The female worms release sheathed microfilariae (Figure 6.16b) that make their way to the bloodstream. The microfilariae show diurnal periodicity and move to the capillaries close to the surface during daylight hours and return to deeper layers during the night. This relates to the feeding behaviour of the blood-feeding mango flies that serve as both intermediate hosts and vectors. *Chrysops dimidiata* and *Chrysops silacea* are the main vector species, and they feed predominantly on humans – a factor that further reduces the opportunity for zoonotic transmission. The larvae continue their development within the abdomen of the fly and then move to the fly's mouthparts when they reach the infective stage third stage – this takes about 10–12 days although, as in all things related to invertebrates, this is very much dependent upon temperature. Because the vectors feed during the day, the use of insecticide-treated bednets does not reduce disease transmission. In addition, the fly larvae develop at low densities within the soil in the surrounding jungle. Consequently, it is much harder to target the vector than in, for example, the case of anopheline mosquitoes transmitting malaria.

The pathology associated with loiasis is relatively minor compared to some of the other filarial nematodes although there is evidence of their association with increased mortality (Whittaker et al. 2018). The worms cause itching and skin irritation, whilst their movement across the eye is painful. Unlike *W. bancrofti* and *O. volvulus*, *L. loa* does not have a symbiotic relationship with *Wolbachia* bacteria. Therefore, the pathology it causes results from a combination of physical damage and inflammatory reactions to the worms and their excretory/secretory products. Sometimes, they induce temporary oedematous swellings several centimetres in size called 'Calabar swellings' – Calabar is a city in Southern Nigeria. These swellings are particularly common around the

(a)

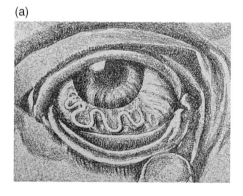

Figure 6.16 *Loa loa.* (a) Drawing of an adult worm traversing the cornea. *Source:* Reproduced from Chandler and Read (1961), © Wiley-Blackwell. (b) Microfilaria. The sheath is obvious and there are nuclei to the tip of the tail.

(b)

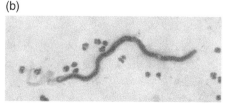

joints but may occur anywhere. Infected people who move from tropical Africa to colder climates are prone to develop them after a few months in the new residence.

Loa loa is particularly important in tropical medicine because patients expressing high levels of microfilariae sometimes react extremely badly, even fatally, to the antiparasitic drug ivermectin. In parts of Africa, the distributions of *Wuchereria*, *Onchocerca*, and *L. loa* overlap and coinfections occur. Consequently, ivermectin cannot be employed as part of a community treatment control strategy to limit *Wuchereria* and *Onchocerca* infections in these regions. Or at least, at risk individuals must be screened out, thereby adding to the costs and logistical difficulties. Similarly, individual patients must be tested for *L. loa* before treatment under medical supervision.

6.3.9 Family Dracunculidae: *Dracunculus medinensis*

This family includes several species that are parasitic on various mammals and reptiles (Cleveland et al. 2018). The most studied species is the only one that infects us, *Dracunculus medinensis*. It is commonly known as the Guinea worm and has a long history but a bleak future. Some people believe that *D. medinensis* was the 'fiery serpent' that afflicted the Jewish people, whilst Moses was leading them through the Sinai wilderness to the Promised Land (Numbers XXI. v. 4–8). However, this is disputed not least because according to the Biblical account "many of the people of Israel died" and whilst *Dracunculus* does cause serious morbidity it has a low mortality. Nevertheless, there are many more convincing accounts from antiquity indicating that *D. medinensis* has been a serious problem for thousands of years (Adamson 1988). It was once a common problem in many parts of Africa, the Middle East, India, and Pakistan with millions of cases recorded every year. However, a highly effective control programme has supposedly reduced it to just a few villages in six sub-Saharan countries. The word supposedly is used because although the WHO has declared India free of Guinea worm, a case was reported in Kerala in 2019 in which a patient was initially thought to be suffering from a diabetic foot ulcer (Pichakacheri 2019). Although the aim to eradicate the disease by 2009 was not met, there were fewer than 1,800 cases in 2010 and in 2020, the Carter Centre (which is orchestrating an international eradication campaign) recorded only 27 human infections. The SARS COVID-19 pandemic that began in 2019 caused serious problems for most disease control programmes, but the numbers of *D. medinensis* infections appear to be continuing to fall. In 2021, there were only four provisional human cases reported between January 1 and April 30 and the numbers of animal infections were also said to be declining considerably. Consequently, it may not be long before *D. medinensis* becomes extinct.

Dracunculus medinensis causes the condition known as 'dracunculiasis'. This can be translated as 'afflicted by little dragons', which is an apt a description of what it feels like to be unfortunate enough to be parasitised by this worm. The adults live in the subcutaneous tissues and particularly in those regions that are most likely to come into contact with the water such as the lower legs, ankles, and feet although other parts of the body can also be afflicted. The females grow to an enormous length (worms of 80 cm have been extracted) although they are very thin – typically about 1.7 mm in width. The males are much smaller and are typically 1–4 cm long although very few have been described. It isn't known at what point in the life cycle that the male and female worms copulate, but the scarcity of males may be because they die soon after mating. If so, the female must have some means of storing sperm or she mates again with males that invade later. The life cycle (Figure 6.17) begins when the eggs hatch within the uterus of the female worm to release first-stage larvae that are 500–750 μm in length. In the older literature, some authors refer to these larvae as microfilariae. The genus *Dracunculus* was once incorporated with the filarial nematodes but has since been moved to a totally separate sub-order – the Camallanina. The vulva is non-functional in gravid females, and therefore, the larvae are unable to exit this way. Instead, the high internal body pressure within the worm causes its cuticle and uterus to rupture near to the anterior end and thereby releases the larvae. The larvae induce a pronounced immune response, but they do not have a symbiotic relationship with *Wolbachia* (Foster et al. 2014). The inflammatory reaction must therefore be in response to excretory/secretory substances released by the parasites. The inflammatory response results in the formation of an ulcer (also sometimes referred to as a 'blister' or a 'bulla') that ultimately bursts and the larvae escape through the wound. If the wound is immersed in water, it stimulates the worm to contract, and this pushes more larvae out of the uterus. While the ulcer is underwater part of the uterus or worm's body often extends out of the ulcer, but this either contracts back in when the affected body part is withdrawn, or the exposed region shrivels and prevents further release of larvae. As the uteri release larvae, the empty regions atrophy, and the posterior sections are pushed forward until the worm has released its full complement of larvae. At this point, the worm dies and the ulcer heals. A traditional means of treating the condition was to grasp the part of the worm extending from the ulcer and wrap it around a thorn or (in the modern era) matchstick. The worm would then be slowly wound out a few centimetres a week until it was all removed. This is a potentially high-risk treatment since if the worm should break it would die *in situ* and this could cause a serious immune reaction as its body decomposed and the wound often became infected with bacteria.

The larvae must be released underwater if they are to survive. They are unable to feed and die within 3 days unless they are consumed by an appropriate species of *Cyclops*. *Cyclops* spp. are common freshwater Crustaceans belonging to the Copepoda. The first-stage larvae penetrate the gut of the *Cyclops* and develop within the body cavity to the third stage: this usually takes about 3 weeks. The life cycle is completed when an infected *Cyclops* is accidentally consumed with drinking water. The larvae are released from the *Cyclops* when it is digested in the host's gut. They then penetrate the gut wall and make their way to the subcutaneous tissues before migrating to the axillary and inguinal regions. During this time, they develop into adults and mating occurs. After about 8–10 months, the fertilised female worms migrate to their final position in the subcutaneous

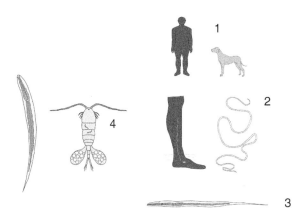

Figure 6.17 Life cycle of *Dracunculus medinensis*. 1: Infection of the definitive host (usually humans but it is a zoonotic disease) commences with the ingestion of infected copepods in the drinking water. 2: The larvae burrow through the gut wall and undergo a migration through the body to the site where they become adults: this is usually in the lower limbs. The anterior of the worm protrudes through an ulcer. 3: When the limb is submerged, the adult female worm releases first-stage larvae into the surroundings. 4: The first-stage larvae are ingested by copepods and these proceed to penetrate their gut and develop into the infectious third stage. Drawings not to scale.

tissues. It takes about 12 months between initial infection and the release of first-stage larvae. Frogs, tadpoles, and fish that ingest infected copepods can act as paratenic hosts although it is uncertain whether this is an important factor in the normal transmission cycle (Cleveland et al. 2017).

Perhaps counter-intuitively for a parasite with an aquatic intermediate host, *D. medinensis* is (or rather was) a particular problem in regions with low rainfall. This is partly because *Cyclops* grows best in still stagnant water. In addition, it is common practice for villagers without piped water to walk into a pond or step-well in order to draw water. This means that the adult worms can release their larvae from ulcers on a person's feet or ankles. Furthermore, *D. medinensis* is a zoonotic parasite and also infects dogs, there are contradictory reports of how many other domestic animals it infects with some mentioning horses, cattle, and a number of other animals. These can act as reservoirs of infection if they are allowed to walk into the local water supply. The parasite has been controlled by a concerted education campaign in which villagers are encouraged to filter their drinking water to remove potentially infected *Cyclops*. In addition, funds have been made available to modify wells and other sources of drinking water to prevent humans and animals from walking into them and thereby infecting the *Cyclops*. In parts of Africa, domestic dogs are emerging as important reservoirs of infection that are limiting the effectiveness of control measures (McDonald et al. 2020). Dog owners can lower the risk of their animal becoming infected with *D. medinensis* by providing it with filtered drinking water. However, feeding fish to the dog increases its risk of infection. This suggests that fish might act as paratenic hosts in this instance. We are more likely to gut and cook fish before eating it whilst raw scraps or whole fish would be fed to the dogs.

The pathology associated with *D. medinensis* infection occurs through three processes: migration of the adult worms to their 'emergence site', failure to migrate to the 'emergence site', and through the formation of an ulcer at the 'emergence site'. The migration of the adult female worms induces an allergic reaction that results in pain, nausea, and diarrhoea. Worms that fail to migrate ultimately die and often become calcified. The consequences of this depend upon the location of the worm. Sometimes there are no harmful effects, but the decomposition of a worm close to joints or important nervous tissue can cause arthritis and even paralysis. The formation of ulcers is extremely painful and debilitating and the ulcers can become secondarily infected by bacteria.

7

Arthropod Parasites

CONTENTS

7.1   Introduction

Arthropods are the most successful metazoan animals in terms of their abundance, numbers of species, total biomass, and exploitation of ecosystems. There are taxonomic descriptions of over a million species of arthropod, and there are probably several times as many awaiting the attentions of biologists willing to study them. Some estimates suggest that there are currently about 1×10^{18} individual arthropods living at any one time, and there is a high possibility that some of them will be living on us. As a group, modern day arthropods are generally small creatures with only a few marine species such as the giant spider crab (*Macrocheira kaempferi*) reaching any appreciable size. However, 460–248 million years ago they included the Eurypterid 'water scorpions', which were among the largest creatures alive at the time and grew up to 2.5 m in length.

Arthropods share many common morphological and developmental features and these are listed in Table 7.1. All arthropods share some of these traits – such as a coelom and a hardened exoskeleton. However, as might be expected from such an abundant and diverse group of creatures, there are always some species that do not possess all the traits. For example, the spiders (Aranea) have well-developed eyes, but they do not have compound eyes. Similarly, some species of parasitic arthropod are so modified that it is difficult to determine their taxonomic position from their external morphology.

Parasitology: An Integrated Approach, Second Edition. Alan Gunn and Sarah J. Pitt.
© 2022 John Wiley & Sons Ltd. Published 2022 by John Wiley & Sons Ltd.
Companion website: www.wiley.com/go/gunn/parasitology2

Table 7.1   Morphological and developmental features exhibited by the majority of arthropods.

Features common to the majority of arthropods
Coelom (haemocoel is the primary body cavity)
Hardened chitinous exoskeleton (cuticle)
Moulting (ecdysis) of exoskeleton to allow growth
Foregut and hindgut lined with cuticle
Jointed appendages
Specialised body segments and appendages that interact with bundles of muscles
Compound eyes
Metamorphosis

The taxonomic relationships within the Arthropoda are constantly being reappraised. To begin with, there is some uncertainty whether the Arthropoda is a monophyletic or polyphyletic clade. That is, whether the current members derive from a single ancestor or from two or more different ancestors. The monophyly hypothesis is based on the observation that arthropods share numerous morphological features, and some DNA evidence suggests a common molecular heritage. By contrast, the polyphyly hypothesis considers there to be too many morphological and developmental differences between the current arthropods for them to have arisen from a common ancestor. Furthermore, it is argued that many of the similarities probably arose through convergent evolution. For example, if an organism has a hard exoskeleton the only way it could move would be with specialised jointed appendages. To further complicate taxonomic considerations, there is now widespread acceptance for the clade Ecdysozoa proposed by Aguinaldo et al. (1997). The Ecdysozoa incorporates all invertebrates that grow by moulting (ecdysis). It therefore comprises of the arthropods, tardigrades, onychophorans, nematodes, nematomorphs, kinorhynchs, and priapulids. Molecular studies are improving our understanding of the phylogenetic relationships between these various groups although some remain controversial (Giribet and Edgecomb 2017).

Within the Arthropoda there is considerable uncertainty as to how the various groups relate to one another. For example, a common means of dividing the living arthropods is to recognise three phyla. These are the Chelicerata that includes the scorpions, spiders, mites, and ticks; the Crustacea that includes the barnacles, lobsters, and crabs; and the Uniramia that includes the centipedes, millipedes (Myriapoda), and insects (Hexapoda). Unfortunately, this nice simple arrangement is no longer tenable. Molecular evidence indicates that the Chelicerata is a distinct entity, but the relationship between the various crustaceans, myriapods, and insects is less certain. The phylum Crustacea is not a monophyletic group, and the insects are actually much more closely related to two obscure groups of Crustacea (the Remipedia and the Cephalocarida) than they are to the millipedes and centipedes. Furthermore, the millipedes, centipedes, and other myriapods are closely related to one another and would probably be best considered as a distinct group within the Arthropoda. These issues will not be resolved for a long time and for the purposes of this chapter, we will consider three taxonomic divisions: Chelicerata, Crustacea, and Hexapoda (Insects).

7.2   Phylum Chelicerata

The phylum Chelicerata consists of three somewhat disparate classes. The class Merostoma contains the horseshoe 'crabs'; the class Arachnida contains the scorpions, spiders, ticks, and mites; and the class Pycnogonida contains the sea spiders. Members of the Chelicerata exhibit the basic arthropod characteristics but share several morphological characteristics that distinguish them from other arthropods (Table 7.2)

From a parasitological perspective, the most important members of the Chelicerata are the Acari that includes the mites, ticks, and chiggers. The Acari are also sometimes referred to as the 'Acarina' and the study of the Acari is known as 'Acarology'. The Acari contains thousands of species that exhibit every possible lifestyle and occupy most ecosystems. There are ongoing disputes about the taxonomy of the Acari although one of the more widespread arrangements is to recognise two divisions (Table 7.3). Although the Acari is sometimes referred to as a single order, it is probably polyphyletic and its various members evolved from various other groups of arachnids. Most Acari are tiny creatures difficult to see with the naked eye and some are even smaller than large protozoa. The largest of the Acari only reach about 2 cm in length and even these often owe their size to ingesting large quantities of their host's blood. Regardless of their lifestyle, most of the Acari feed by ingesting fluids and their mouthparts are adapted for this purpose. Their chelicerae are usually pincer-like or needle-like, whilst their pedipalps take the form of leg-like or pincer-like appendages. However, in some species the pedipalps have regressed to vestigial structures that probably serve little if any function. It is often difficult to make out the two body divisions because the prosoma and opisthosoma have fused and/or the dorsal (upper) surface is covered by a hard chitinous plate.

The Acari tend to have simple life cycles and separate sexes. After mating, the female lays eggs that hatch to release six-legged larvae. Usually, the eggs are laid in an appropriate location, but in some species the eggs are retained in the female's body until they hatch. The larvae moult through a variable number of further developmental stages; these are known sequentially as protonymph, deutonymph, and tritonymph. The nymphal stages and the adults usually have four pairs of legs.

Table 7.2   Distinguishing morphological characteristics of the phylum Chelicerata.

Morphological characteristics of chelicerates
Body divided into two regions: the cephalothorax (prosoma) and abdomen (opisthosoma)
Cephalothorax divided into eight segments
No appendages on the first cephalothorax segment
Chelicerae attached to second cephalothorax segment. Chelicerae are specialised appendages that are used for various functions in different species
Pedipalps attached to third cephalothorax segment. Pedipalps are specialised appendages that perform different functions in different species
Four pairs of walking legs (in adult) attached to cephalothorax segments 4–7. Some species, especially parasitic species have only 2 or 3 pairs of legs as adults
Lack antennae

Table 7.3   Taxonomic arrangement of the Acari.

Super-Order	Parasitic examples
Parasitiformes	Poultry mites (*Dermanyssus gallinae*), Soft ticks, Hard ticks
Acariformes	*Demodex* spp., Scabies mite, Chiggers,

The durations of the different stages of the life cycle are hugely variable and often affected by environmental conditions and, in the case of parasitic species, the behaviour and ecology of their host.

There are numerous species of Acari that are of medical and veterinary importance, especially as vectors of other pathogens and we can only cover a few of these. Those requiring more detail are advised to consult Mullen and Durden (2002).

7.2.1 Family Demodicidae

This is an unusual family of mites that are specialised skin parasites. There are at least 65 species, and it is probable that most mammals have 'their own' species: for example, *Demodex canis* (dogs), *Demodex cati* (cats), *Demodex equi* (horses), *Demodex phylloides* (pigs), *Demodex bovis* (cattle), *Demodex musculi* (mice), *Demodex caviae* (guinea pigs). There is even a marine species, *Demodex zalophi*, which parasitises Atlantic harbour seals (*Phoca vitulina*).

There are two species that parasitise humans: *Demodex folliculorum* and *Demodex brevis*. *Demodex folliculorum* has a long (0.3–0.4 mm in length) carrot-shaped body and lives in the hair follicles of simple hairs (Figure 7.1). By contrast, *D. brevis* has a shorter stumpy body (0.2–0.3 mm in length) and lives in the sebaceous glands of body hairs. They are photosensitive and only come onto the skin surface at night to mate and, potentially, transfer to another host. Despite their small legs and odd body shape, they have been timed at travelling at $16\,\mathrm{mm\,h^{-1}}$. Nevertheless, the female mite lays her eggs within the skin follicles in which she lives, and it is probable that most hosts acquire their infections from their mother. If this is so, then it would make genetic mixing between mite populations difficult. It would therefore be interesting to compare the molecular profile of *Demodex* within and between communities.

Figure 7.1 Adult *Demodex folliculorum.*
Source: Reproduced from Chandler and Read, (1961), © Wiley-Blackwell.

The life cycle of *D. folliculorum* takes about 14–18 days and consists of an egg, larval, protonymph, deutonymph, and adult stage. They are remarkably common and their numbers increase as we age – one estimate suggested that we are all infected by the time we reach 70 (Post and Juhlin 1963). *Demodex* spp. mites feed on host cells and secretions (sebum) within the follicles. An odd feature of their morphology is that they lack an anus. Consequently, their faeces accumulate within their body so that when the mite dies and decays, all the faeces is released at the same time. This may contribute to an inflammatory reaction in those people/animals that are allergic to the faeces.

In dogs, *Demodex* mites can cause demodectic mange, which may become a serious condition. However, most *Demodex* infections in dogs are harmless and development into mange is probably linked to underlying problems with immune health. In humans, *Demodex* mites are often associated with rosacea and blepharitis – chronic inflammation of the eyelid, particularly at the base of the eyelashes (Fromstein et al. 2018). However, some authors consider them beneficial by consuming bacteria within the sebum, and it is only when we mount an excessive immune response against the mites that problems arise.

7.2.2 Family Sarcoptidae

This family contains the well-known scabies mite *Sarcoptes scabiei*. Two other common species are *Knemidocoptes mutans* that causes scaly leg in hens and various other birds and *Notoedres cati* that afflicts domestic and wild cats, as well as dogs and a number of other mammals.

7.2.2.1 Genus Sarcoptes

Unlike *Demodex*, in which there are numerous host-specific species, *S. scabiei* exists as a variety of host-specific sub-species. For example, humans are infected by *S. scabiei* var *hominis*, whilst dogs are hosts to *S. scabiei* var *canis*. We are sometimes infected by other varieties of *S. scabiei* such as those normally found on dogs and horses. These tend to cause us temporary infestations that persist for a few weeks or months. *Sarcoptes scabiei* var *hominis* occurs throughout the world and is an obligate parasite. The mites are protected within their burrows and not harmed by frequent washing in hot soapy water. Scabies is sometimes incorrectly thought to be an indicator of poverty and neglect. In reality, it infests people of all social classes and levels of affluence. Scabies is, however, a risk for people living in close association in cramped unhygienic circumstances such as urban squatter settlements and refugee camps. For a good review, see Arlian and Morgan (2017).

Sarcoptes scabiei is a small almost circular-shaped mite with miniscule legs (Figure 7.2). The front two pairs of legs terminate in structures called pedicels that look a bit like sink plungers. However, the mite's shape belies a remarkable turn of speed and the females have been timed at moving at $2.5\,cm\,min^{-1}$. This might not sound very fast but the female mites are only 350–440 μm in length and the males are even smaller (180–240 μm). Assuming a body length of 350 μm (0.35 mm), this translates as 71.4 body lengths per minute – or the equivalent of $129\,m\,min^{-1}$ ($7.74\,km\,h^{-1}$) for a 1.8-m-tall human.

After mating, the female mite goes in search of a suitable place to make a permanent burrow in which she can raise her family. She digs as far the boundary of the *stratum corneum* and *stratum granulosum* and then begins burrowing horizontally. She also starts laying eggs at a rate of two or three eggs every day up until she dies after about 30 or so days. The eggs hatch about 3 or 4 days after laying. Most of the larvae then crawl up to the skin surface and move a short distance away before digging a shallow 'moulting pocket'. Each larva digs its own 'pocket' and within this it moults to a protonymph and then a tritonymph before becoming an adult. It takes a female

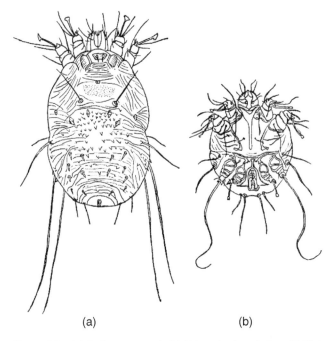

(a)   (b)

Figure 7.2 Adult *Sarcoptes scabei*. (a) Female, dorsal view. (b) Male, ventral view. *Source:* Reproduced from Chandler and Read (1961), © Wiley-Blackwell

S. scabiei var *hominis* mite about 14 days to develop from egg to adulthood whilst the males can take as little as 10 days. The times vary slightly between the different sub-species. After becoming mature, the male mites leave their pocket and wander off in search of female mites. The females, by contrast, remain within their pocket until the males find them and mating occurs. After mating, the females leave their pocket and search for a suitable permanent burrow.

In adults, the mites usually dig where our skin is thin and wrinkled. The majority of infestations (63%) occur on the webs between the fingers and on the wrists, after which the elbows are the second most common site (11%). Infestations also occur on our ankles and buttocks, whilst in women the mites may be found in the skin folds on the breasts and nipples, and in men on the scrotum. Young children have softer skin than adults and the burrows commonly occur on the palms of their hands and soles of their feet, as well as on the neck and head.

The mites feed on the fluid exudates of damaged dermal cells. Although the physical damage is relatively minor, scabies infections cause intense itching. This is a consequence of immune reactions to their cuticular components, as well as their secretions and excretions. In us, this results in the formation of rashes that may develop at sites distant from the actual site of infestation and persist even after the death of all the mites. Although the mites produce highly immunogenic substances, chemicals in their saliva modulate the immune response directed against them, and this probably facilitates their initial establishment (Morgan and Arlian 2010). There are differences between host animals in their response to *S. scabiei* but usually those that recover from an infestation are resistant to re-infection. This is probably a consequence of immunoglobulin-E-mediated responses. There is therefore interest in developing vaccines to treat sarcoptic mange in domestic animals and scabies in humans (Liu et al. 2014).

Scabies infestations are usually transmitted through close physical contact. Although 15–20 minutes of close contact is usually quoted as being required this presumably depends a lot on the numbers of mites involved and the individual circumstances. For example, if the donor is heavily infested, there is a higher chance that sufficient mites will 'make the jump' to establish an infestation sooner rather than later. Transmission can occur via bedding or shared clothing, but this is not a major route of infestation. Scabies can be a particular problem for poor people, especially children, living in tropical countries. This is because they often wear few clothes, and therefore there are more opportunities for skin: skin contact – especially at night if they share a bed, bunk, or hammock.

Sarcoptes scabiei mites are not disease vectors but they often cause secondary bacterial infections. These result from a combination of lesions caused by their tunnelling, damage from the host immune reaction, and the scratching they induce. In humans, hyperinfection with huge numbers of mites gives rise to crusted scabies, which is potentially fatal. In addition, the presence of so many mites mean that people suffering from the condition are highly infectious to others.

In cattle, *S. scabiei* var *bovis* causes sarcoptic mange; this is a notifiable disease in the United Kingdom, and the import of cattle that are suffering from infestation is banned regardless of whether or not they are clinically affected by it. Serious disease is also caused by *S. scabiei* var *ovis* in sheep, *S. scabiei* var *cameli* in camels, and *S. scabei* var *aucheniae* in the alpaca (*Vicugna pacos*). Severe disease is usually associated with animals in poor condition at the end of the winter period. In the UK and several other countries in northern Europe the fox (*Vulpes vulpes*) population suffers from epidemics of sarcoptic mange that has a consequent impact on the population of other wild animals (Scott et al. 2020). For example, Pisano et al. (2019) describe how sarcoptic mange spread from wild foxes in Switzerland onto farm animals and from these onto four farm workers. Cats are seldom infected by *S. scabiei,* but they can suffer badly from *Notoedres cati.*

7.2.2.2 Genus *Notoedres*, *Notoedres cati*

Most members of this genus are parasites of rats and bats although, as its name indicates, *Notoedres cati* is a parasite of domestic cats and many wild felids such as lynx, tigers, and panthers. *Notoedres cati* has a similar globular appearance to *S. scabiei* but is smaller: the females are typically only 200–240 μm in length. It forms deep burrows that can extend beneath the *stratum corneum*. The ears, head, and neck of the cat are the most common sites of infection, but it can spread elsewhere resulting in extensive mange (notoedric mange). The condition is extremely debilitating and the intense itching may lead to alopecia, pruritis, self-mutilation, and even prove fatal.

The life cycle of *N. cati* closely resembles that of *S. scabiei*. The female mite lays 3–4 eggs day^{-1} within her burrow, these hatch after about 4–5 days to release six-legged larvae. The larvae usually climb out of the burrow and wander across the skin surface before digging a shallow burrow of their own. Within this burrow, they moult to the first nymphal stage – this, and the subsequent stages, has eight legs. The first nymphal stage then leaves the burrow and wanders around before digging another shallow burrow within which it moults to the second nymphal stage. After emerging, the second stage nymph wanders around for a time before digging a deeper burrow within which it moults to the adult stage. The adult female mites tend to remain in their burrow whilst the males leave theirs and go in search of females to mate with. The whole life cycle from egg-hatch to becoming an adult can take as little as 12 days.

Transfer of the infection is normally through physical contact. Humans can become infected if they have contact with infected cats. Within humans, the mites normally infect the hands and legs where they can cause intense pruritis. They do not, however, tend to form burrows or establish permanent infections on humans. Epidemic outbreaks among wild felids can cause significant mortalities.

7.2.2.3 Genus *Knemidocoptes*

The name of this genus is also sometimes spelled *Cnemidocoptes*. There are several species that are important pests of domestic and wild birds. *Knemidocoptes gallinae* causes 'depluming itch' of domestic fowl. The mites burrow into the shafts of feathers, thereby inducing an inflammatory response and itching. The feathers break easily and are pulled out by the affected bird resulting in bare patches on the back and wings although the head and neck may also be affected. *Knemidocoptes mutans* causes 'scaly leg' in domestic hens, turkeys and a number of other birds. The mite probably initially infects a bird by crawling onto it from a surface since the lesions usually start in the toes and then progress up the leg. The mites penetrate under the scales and create tunnels in the underlying skin. A combination of the mites' feeding and the induction of inflammatory reactions causes the scales to raise upwards and become loose. The *stratum corneum* becomes hyperkeratotic (i.e., the *stratum corneum* becomes thickened owing to the deposition of excess keratin) and the lesions often become secondarily invaded by opportunistic bacteria and fungi. Consequently, the affected limb becomes deformed, digits become necrotic and fall off, and the bird is rendered lame. Ultimately, the bird may cease feeding and die. The comb and neck may also be affected.

7.2.3 Family Psoroptidae

The family Psoroptidae includes several genera of importance in veterinary medicine. For example, the genus *Psoroptes* includes *Psoroptes ovis* that causes sheep scab, genus *Chorioptes* includes *Chorioptes bovis* that cause chorioptic mange in cattle, horses, and goats, and the genus *Otodectes* that includes *Otodectes cynotis* that afflicts the ears of cats, dogs, and several other carnivores.

7.2.3.1 Genus *Psoroptes*

The genus *Psoroptes* contains several important ectoparasites of domestic animals, but none of them parasitise humans. There are several distinct host-specific species although these are difficult to distinguish from their morphology. Indeed, some of the mites that were previously considered to be separate species such as *P. ovis* in sheep and *Psoroptes cuniculi* in rabbits are actually strains of the same species.

Psoroptes ovis is economically the most important member of the genus as it causes the highly debilitating disease sheep scab. The condition results primarily from the host's immune reaction to the mites' saliva and faeces. The sheep's skin becomes seriously inflamed and the affected animal attempts to relieve the intense itching by rubbing against any hard surface. This damages the skin surface further and leads to secondary infections. The sheep becomes increasingly distracted, ceases to feed and may die of exhaustion and dehydration. The rubbing also transfers the mites onto surfaces from which they can be picked up other sheep.

Psoroptes ovis lives entirely on the skin surface where it feeds on bacteria and skin secretions. The adult mites are small (~0.75 mm) and oval-shaped with relatively long legs and the mouthparts project to form a noticeable cone-shaped point at the front of the body (Figure 7.3). The female mite has pedicels on the first, second, and fourth pair of her legs and long whip-like setae (hairs) on her third pair of legs. By contrast, the somewhat smaller male has pedicels on its first, second, and third pairs of legs and long setae on his fourth pair of legs. The male also has a pair of copulatory suckers towards the rear of its ventral surface – he can spend a long time attached to a female so he needs to maintain a good grip.

The female mite lays her eggs on the surface of the skin. The eggs are approximately 0.25 mm in size – which makes them about a third the size of the adult female. The eggs hatch after about

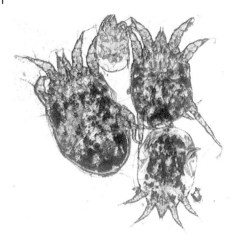

Figure 7.3   *Psoroptes ovis.* Adults and nymphs. The adult mites are small (~0.75 mm) and oval-shaped with relatively long legs and the mouthparts project to form a noticeable cone-shaped point at the front of the body.

1–3 days but may take longer if they are in a cooler microclimate. The eggs release a six-legged larva, which then moults first to a protonymph and then a tritonymph, which then moults to become an adult – the whole process from egg to adulthood taking about 10 days. The adult males will attach themselves to a female, whilst she is still a tritonymph or even a protonymph. After she finally moults to an adult, he will mate with her in a process that typically lasts a day but may take less if there are many more females than males in the population. After mating, the female lives for a further 30–40 days during which she lays about five eggs every day. The rapid rate of reproduction ensures that the acquisition of just a few mites can rapidly lead to clinical disease.

In the United Kingdom, up until 1992, it was a requirement to dip all sheep every year to control the spread of sheep scab. Partly as a response to very vocal concerns raised about the health hazards of organophosphate sheep dips, the procedure was deregulated. Since then, the incidence of sheep scab has increased alarmingly. At present, the main means of preventing sheep from contracting sheep scab are the use of organophosphate dips and injectable macrocyclic lactones, such as moxidectin and milbemycin. Unfortunately, there appears to be increasing resistance to macrocyclic lactones among UK *P. ovis* populations (Sturgess-Osborne et al. 2019). This presents a classic problem of disease control in which a balance must somehow be struck between competing concerns. First, one wishes to protect the welfare of sheep by preventing them from catching a dreadful disease. However, employing organophosphate sheep dips carries serious risks to the farm workers who dip the sheep and also problems with disposing of the used dip afterwards without harming the environment. Employing macrocyclic lactones is currently a safer method but if resistance to them becomes more widespread, some difficult decisions will need to be made.

7.2.4   Suborder Ixodida

The suborder Ixodida contains all the ticks, all of which are parasitic. There are three families within the sub-order: the Argasidae ('soft ticks'), the Ixodida ('hard ticks'), and the Nuttalliellidae. The Nuttalliellidae contains a single obscure species, *Nuttalliella numaqua* that lives in South Africa where it feeds on various hosts including rodents and lizards (Mans et al. 2014). Although ticks can cause serious pathology, especially if present in large numbers, they are of principal importance as vectors of viral, bacterial, and parasitic diseases.

(a)　　　　　　　　　　　　　　　　　(b)

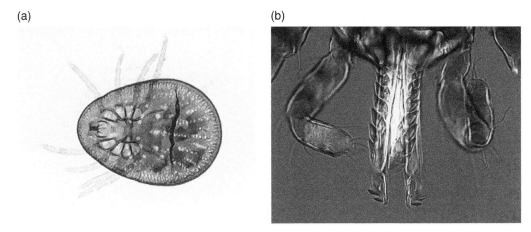

Figure 7.4 *Argas persicus.* (a) Adult stage. This grows up to 10 mm in length and 6 mm in width and is yellow-brown when unfed and slate-blue after feeding. The hypostome cannot be seen because it is hidden by the integument. (b) Hypostome of larval stage. The toothed hypostome is easily visible in the larval stage.

Ticks are generally much bigger than mites and can be seen with the naked eye. Their mouthparts also include a large-toothed structure called a hypostome (Figure 7.4b). The tick uses this to firmly anchor itself to its host. Some ticks also secrete chemicals that act like a glue to further anchor themselves in position. Any attempt at pulling a tick from its host usually results in the ripping the tick apart and leaving the mouthparts embedded in the flesh of its victim. To extract a tick, one needs to twist and pull. However, on pets and domestic animals it is often simpler to leave a tick to drop off of its own accord. The hypostome has evolved from the coxae of the pedipalps that have fused together to form a single structure. A hypostome also occurs in some mites, but it is not toothed and is small and not easily seen. Ticks also differ from mites in possessing a sensory structure called the Haller's organ on the tarsal segments of the first walking legs. The Haller's organ consists of an anterior pit and a proximal capsule. The pit region detects humidity and the capsule region is used for olfaction. Ticks often wave their front pair of legs in front of them, especially when 'questing'. It is therefore probable that the Haller's organs on the tarsi function perform sensory perception in a similar manner to the receptors on the antennae in insects.

7.2.4.1 Family Argasidae

The Argasidae are often known as the 'soft ticks' because they have a leathery appearance and lack a hard scutum (protective shield) on their upper surface. Unlike the 'hard ticks', there are no obvious morphological differences between the sexes apart from the shape of the genitalia. The family contains about 170 species and they are mostly parasites of birds although some species parasitise bats and some exploit various terrestrial mammals. One species, *Argas (Microargas) transversus*, is totally dependent upon the Galapagos giant tortoise (*Chelonoidis nigra*). The argasid life cycle typically comprises of the egg, larval, and anything from 2 to 7 nymphal stages before the ticks become adults. As a rule, the nymphs and adults feed quickly (30 minutes to a few hours) and then drop off their host whilst the larvae may remain attached for several days. Regardless of how long they remain attached, after dropping off the larvae and nymphs usually moult to the next developmental stage, whilst the adults produce another batch of eggs or sperm. This exploitation of numerous hosts makes the ticks potential vectors of several diseases. There are several exceptions to this generalised life cycle. For example, *Argas transversus* is unusual in spending its whole life aboard its tortoise host and even lays its eggs directly onto the tortoise.

7.2.4.1.1 *Argas persicus*

Argas persicus is commonly known as a fowl tick although it will also feed on turkeys, pigeons, ostriches, and a variety of wild birds. Hungry ticks will also bite us although we are not important hosts. *Argas persicus* has a cosmopolitan distribution but is most common in tropical countries. The adult has a flattened oval shape that narrows towards the anterior. It has with well-developed legs and the mouthparts cannot be seen from above because they are hidden by the integument (Figure 7.4a). It is quite large growing up to 10 mm in length and 6 mm in width and is yellow-brown when unfed and slate-blue after feeding.

The female tick lays batches of 20–100 eggs in crevices, and the eggs hatch after about 3 weeks to release the larvae. The larvae have round bodies and six legs. They hunt for a suitable bird host and usually attach underneath the wings where they remain attached for about 5 days. The larvae then drop off and hide for about 7 days during which time they moult to the first nymphal stage. There are two nymphal stages during which the ticks feed and then moult after which they become adults. The nymphs and adult ticks remain hidden in crevices during the day and come out to feed at night. If large tick populations build up in chicken sheds, the repeated disturbance and loss of blood can leave the birds unthrifty and their anaemia may prove fatal. The ticks can also transmit fatal diseases such as fowl spirochaetosis (*Borrelia anserina*). *Argas persicus* are notoriously resistant to starvation and can probably live for years without feeding. This is a common trait among bird ticks. This is because many birds only nest once or twice a year, and therefore the ticks must survive starvation during the long time between nesting seasons.

7.2.4.2 Family Ixodidae

Ticks belonging to the family Ixodidae are known as the 'hard ticks' because they have a hard scutum (chitinous shield) on the upper surface. Adult male Ixodid ticks have a scutum that extends to cover the whole dorsal surface. By contrast, in adult females, nymphs, and larvae, the scutum extends only a short distance behind the head. Unlike soft ticks, in the hard ticks the mouthparts project in front of the body and are clearly visible from above. Their life cycle is generally simpler too and usually consists of the egg, a six-legged larval stage, a single nymphal stage, and then the adult stage. Ixodid ticks tend to be less active than soft ticks and they often adopt a 'sit and wait' strategy that is called 'questing'. Basically, the ticks crawl up vegetation or some other structure that will give them a prominent position and extend their front legs. When a potentially suitable host passes by, the ticks grab onto it and then look for a place to start feeding. Feeding typically takes 4–6 days after which the tick proceeds to the next stage in its life cycle. Whether or not the tick drops off between feeding and the number and species of hosts involved varies between tick species. Ixodid ticks exhibit three different life cycle strategies.

One-Host Ticks: The whole life cycle (apart from egg-laying) takes place on a single individual animal. The larval tick climbs aboard an animal feeds, moults to the nymphal stage that then feeds and moults to the adult stage. The adult ticks then feed and mate after which the female leaves the host animal and lays its eggs on the ground. *Rhipicephalus annulatus,* formerly known as *Boophilus annulatus* (the North American cattle tick) is a one-host tick that feeds on domestic and wild ungulates. It occasionally attacks both humans and dogs although we are not suitable for the maintenance of the whole life cycle. Up until the 1940s, this tick was a major problem in the USA as a consequence of it transmitting 'cattle fever' (*Babesia bigemina*). The tick was subsequently eradicated from the USA but it still exists in Mexico. Consequently, a permanent quarantine zone is maintained at the border between the two countries to prevent the tick from being reintroduced.

Two-Host Ticks: The larval and nymphal stage take place on a single host but the adult stage parasitises a different animal. In this case, the larva climbs aboard an animal, feeds, and moults to the nymphal stage. The nymph then feeds on the same host but having engorged, it drops off and moults to the adult stage on the ground. The adult tick then finds a suitable host on which it will feed and mate after which the female tick drops off and lays its eggs. For example, *Rhipicephalus evertsi* (the red-legged tick) is a two-host tick that feeds on a wide range of mammals including domestic species such horses, cattle, sheep, and goats. Its larvae and nymphs tend to feed in the ears or inguinal region, whilst the adult ticks tend to feed under the tail. *Rhipicephalus evertsi* is common in parts of central and southern Africa and is an important vector of many diseases including *Theileria parva* and *Babesia bigemina*.

Three-Host Ticks: Exploiting a different animal at each stage of development is the most common life cycle strategy among Ixodid ticks. After hatching, the larva climbs aboard an animal, engorges, and then drops off and moults to the nymphal stage on the ground. The nymph then climbs aboard a different animal, engorges, and then drops off and moults to the adult stage, which then climbs aboard another animal, engorges, drops off, and then lays its eggs. The hosts utilised at each stage may belong to the same species or be different species although typically, the tick feeds on a larger host animal after each moult. They achieve this by climbing further up the vegetation at each stage in the life cycle. *Ixodes ricinus,* the 'castor bean tick' or 'sheep tick', is an example of a three-host tick. It is common in Europe and many other countries. The larvae and nymphs parasitise many animals including birds and lizards, whilst the adults commonly latch onto sheep and dogs, as well as numerous other wild and domestic mammals. In common with several other ticks, *I. ricinus* has a long lifespan. The larval period occupies the first year of its life, the nymphal period lasts a further year, and the adult stage lasts for a third year. The adult female tick feeds only once. After fully engorging, she is about 1cm in length and slate-grey in colour and has all the appearance of a castor bean – for which it gets its common name (Figure 7.5). She expands to such an extent that her legs cease to be visible when viewed from above. The male, by contrast, takes several smaller blood meals and given the opportunity, he mates with numerous females. *Ixodes ricinus* is an important vector of many parasitic, bacterial, and viral diseases including *Babesia divergens*, *Babesia bovis*, Lyme disease (*Borrelia burgdorferi*), human granulocytic anaplasmosis

Figure 7.5 Engorged adult female *Ixodes ricinus*.

(also known as human granulocytic ehrlichiosis) (*Anaplasma phagocytophilum*) and tick borne encephalitis virus.

7.3 Phylum Crustacea

The taxonomy of the Phylum Crustacea is extremely complex and frequently revised. Consequently, there is little consensus between textbooks on how the phylum should be arranged. The Crustacea currently contains over 40,000 described species and there undoubtedly many more awaiting the attentions of taxonomists. It is an enormously diverse phylum and representatives can be found in every marine, brackish, and freshwater environment, but there are only a few species that are terrestrial. There is no such thing as a typical crustacean body plan although the majority of them have a distinctive nauplius larva stage. The nauplius larva may remain within the egg or be a free-swimming member of the planktonic community. The nauplius larva has a single median eye on the front of its head and three pairs of appendages: the first and second pairs of antennae and the mandibles.

The Crustacea include many examples of unusual and fascinating parasites, but these tend to afflict other aquatic invertebrates or fish and have little economic or medical importance. We will therefore only mention a few examples of them.

7.3.1 Subclass Copepoda

The copepods are the most abundant metazoan animals in the oceans and huge numbers also occur in freshwater as well. There are in the region of 1,600–1,800 species of parasitic copepod species, and they infest animals ranging from sponges to whales. Most (~75%) parasitic copepods belong to a single order – the Siphonostomatoida. The majority of parasitic copepod species only exploit a single animal during their life but those belonging to the family Pennellidae are unique in having a life cycle that includes both an intermediate host and a definitive host. Although most copepods are small (0.5–10 mm in length), one free-living species approaches a length of 2 cm and some parasitic forms are over 32 cm. An unusual relationship appears to exist between the parasitic copepod *Ommatokoita elongata* and the Greenland shark *Somniosus microcephalus*. The copepods attach to the cornea of the shark and cause lesions that could potentially lead to blindness. However, according to Berland (1961) the 3–5 cm long pinkish white copepods might act as lures that attract char (*Salmo alpinus*) close enough to be caught by the sluggish-natured shark. The largest known copepod is *Pennella balaenopterae* that parasitises whales and grows to at least 32 cm in length (Figure 7.6). In common with several other parasitic copepod species, the body of *P. balaenopterae* bears little resemblance to that of free-living copepods. Overall, the body is long and thin and divided into three regions: the cephalothorax that bears two or three holdfast organs that are deeply embedded in the whale's blubber, a long thin thoracic region (~80% of the total body length) and an abdominal region that is covered in plume-like processes and terminates in two long thin ovisacs that may extend a further 40 cm beyond the tip of the abdomen.

The free-living copepod *Mesocyclops longisetus* is of interest to parasitologists as a biological control agent of mosquito larvae, such as those of *Aedes aegypti*. *Aedes aegypti* is an important vector of dengue fever and yellow fever and occurs in many parts of the world including South and North America, the Middle East, and throughout Africa and Asia. Interestingly, not only is *M. longisetus* extremely effective at eliminating mosquito larvae, but it also releases chemicals that attract gravid female mosquitoes to lay their eggs in water containing the copepods. Torres-Estrada et al. (2001)

Figure 7.6 Parasitic copepod *Pennella balaenopterae*. Key: a: abdomen; c: cephalothorax; n: neck; o: ovisacs; t: trunk. Scale bar = 1 cm. *Source:* Reproduced from Abaunza et al. (2001).

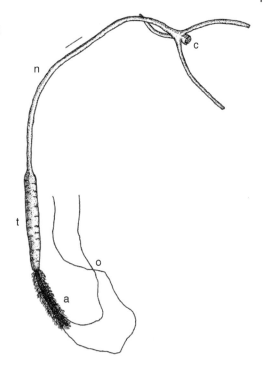

demonstrated that gravid female *A. aegypti* are more attracted to water containing *M. longisetus* or to water that previously held them compared to distilled water. This may increase the numbers of potential prey for the copepod and also increases their usefulness as a biological control agent. *Mesocyclops longisetus* also preys on other species of mosquito, including malaria transmitting anopheline mosquitoes and the technology is available to enable its large-scale cultivation. Several other predatory copepod species are also potential biological control agents against mosquitoes (Baldacchino et al. 2017). This approach avoids the use of harmful insecticides and could potentially provide long-lasting control. However, the history of biological control is littered with well-intentioned introductions that have gone horribly wrong and Coelho and Henry (2017) consider that the intentional distribution of predatory copepods with a catholic diet could pose serious ecological risks.

7.3.2 Infraclass Cirripedia

The Cirripedia includes the barnacles – the only sessile group of crustaceans – and also many parasitic species. The genus *Rhizocephala* are so highly specialised that all traces of arthropod structure have disappeared in the adult. The rhizocephalans mainly parasitise decapod crustaceans (e.g., crabs and lobsters). The female cyprid larval stage is free-living and attaches to the body of a suitable host. It then grows into its host by forming a root-like nutrient absorbing structure known as the '*interna*'. When she is mature, the female parasite forms an external reproductive structure called the '*externa*' that emerges on the host's ventral surface. The male is free-living and after finding a female, attaches himself. He then injects cells, which migrate to a special chamber in the female where they differentiate into a testis. The female is now functioning as a hermaphrodite, and fertilisation of eggs and development to the nauplius stage occurs within the brood chamber. The effects of parasitism on the decapod crustacean host include inhibiting moulting and parasitic castration. Infestation levels can be high in some crab populations and concerns are sometimes expressed that rhizocephalan parasites might affect commercial crab fisheries. By contrast, it has also been suggested that rhizocephalans could be used to control invasive pest crab species (Thresher et al. 2000). However, rhizocephalans produce planktonic larvae and therefore have the potential to spread far from their initial release site and thereby impact on non-target areas. In addition, they are not very host specific and could severely deplete the population of native crab species (Goddard et al. 2005). These are typical of the concerns when considering any parasite for use as a biological control agent.

7.3.3 Subclass Branchiura

The majority of the Branchiura are obligatory ectoparasites of fish in both freshwater and marine ecosystems although a few species parasitise amphibians. They are commonly known as 'fish lice' although this term is also applied to certain parasitic copepods (Pike, 1989), and it should be remembered that the true lice are insects. Even as adults, the branchiurans retain their swimming ability and the female usually leaves her host to lay her eggs on a suitable substrate. After locating a host, the parasites often crawl to a position behind the gill operculum or one of fins where water turbulence is less. They then use their suctorial proboscis to feed on mucus and scales or to pierce the skin and feed on blood. A distinguishing morphological feature of the branchiurans is the presence of two large sucking discs on the underside of the head that are formed by modifications to the first pair of maxillae (maxillules). The discs help the parasite maintain its grip on the host. In addition, immediately behind the sucking discs is a pair of large maxillae that are also used for grip rather than feeding.

The Branchiura consists of four genera, the best known of which is the genus *Argulus* that has representative species throughout the world (Figure 7.7). Many *Argulus* species are not host specific, and some of them (e.g., *Argulus foliaceus* and *Argulus coregoni*) are economically important parasites of farmed fish and high-value ornamental fish such as koi carp. They have simple direct

Figure 7.7 *Argulus foliaceus*. Fish louse. This female specimen was 7 mm in length. The two dark spots towards the anterior are the compound eyes. Slightly behind the compound eyes are two 'suction cups' that are modified mouthparts having evolved from the first pair of maxillae. The proboscis is situated between the two suction cups. Adult fish lice have four pairs of swimming legs. The paddle-like structure at the rear comprises of the abdominal segments (referred to as the urosome in some texts) and the two dark spots are the spermathecae that store sperm received from the male.

life cycles in which the female *Argulus* lays her eggs on a substrate such as stones at the bottom of the water body. The eggs hatch to release metanauplii that actively swim in search of a suitable fish host. Once a metanauplius finds its host, it attaches and undergoes a series of moults until it becomes an adult. The sexes are separate and they usually mate on the host.

When they feed, branchiurans inflict physical damage and also induce inflammation that further contributes to irritation. This causes fish to rub against rocks etc., and thereby induces further physical damage. The wound sites then become secondarily infected by bacteria and fungi. In addition, damage to the fish's skin upsets its ability to regulate its ion balance. The parasites also act as mechanical vectors of viruses such as spring viremia of carp virus (SVCV) (Ahne 1985). This virus is a highly contagious disease of cyprinid fish and is spreading rapidly in Europe, Asia, and the Americas. Farmed fish are maintained at high population densities within confined areas and are therefore vulnerable to suffering from high parasite burdens. This can result in low growth rates and high mortalities. Fish farmers therefore tend to use a lot of pesticides, such as ivermectin, to reduce the abundance of fish lice. Unfortunately, these chemicals can have an adverse effect on the surrounding ecosystem (Davies et al. 1997). Furthermore, if the farmed fish are maintained in sea lochs, the parasites can be transmitted (along with the viruses etc. they carry) to free-living salmon and sea trout with a consequent harmful impact on their population.

7.3.4   Subclass Pentastomida

There are over 130 species of pentastome, and they are all parasitic. Most of them live in the lungs and nasal passageways of vertebrates during their adult stage (Figure 7.8). They grow from 1 to 14 cm in length and are sometimes referred to as 'tongue worms' from the shape of the adult worm.

They are a highly unusual group of invertebrates whose taxonomic position has been a constant source of debate. For many years, they were cited as examples of primitive 'proto-arthropods' along with the tardigrades and the onychophorans. However, their apparently primitive traits derive from their parasitic lifestyle. Molecular evidence has now added sufficient weight to studies on pentastome sperm morphology, embryogenesis, and cuticular fine structure to indicate that they are actually crustaceans.

Figure 7.8 *Armillifer armillatus* (adult stage), Pentastoma. The adult stage is normally found in the lungs of snakes; various species of mammals (including humans) can act as intermediate hosts of the larval stage. This specimen was 6 cm in length.

Although the name 'Pentastoma' translates as '5 mouths', the pentastomes have only a single mouth. In adult worms, the mouth lacks jaws and, in some species, it sits at the top of a snout-like projection. The other four 'mouths' are actually two pairs of much reduced 'legs' that sometimes bear chitinous claws, which the pentastome uses to cling on to its host. Pentastomes feed on blood that they imbibe by sucking. Some species have frontal glands that produce secretions that probably break down host tissue and serve as anticoagulants. Pentastomes often cause haemorrhages and ulcerative lesions where they attach. Their cuticle is non-chitinous, highly porous, and moulted periodically until they attain full adult size.

The pentastome life cycle generally involves the female worm shedding eggs that are then coughed or sneezed up or passed out with the host's faeces. The eggs are then consumed by an intermediate host that, depending upon the pentastome species, may be an insect, fish, amphibian, reptile or mammal. The eggs hatch within the gut of the intermediate host and the pentastome larva penetrates the intermediate host's gut. It then undergoes an apparently random migration until it settles down and moults to the nymphal stage. There can be several nymphal stages, and in some species, the nymph may become sexually mature before it moults to become an adult. The intermediate host must be consumed by the definitive host before the pentastome can moult to the adult stage. After consumption, the infective nymph either penetrates the gut of the definitive host or climbs up its oesophagus and then migrates to the lungs or nasal passageways where it becomes an adult. There are separate sexes although the male worm does not live for long, so most observations are based on females.

The definitive host is usually a snake or other reptile although there are a few species in which it is a mammal, bird, or amphibian. Autoinfection can occur in which a pentastome egg hatches in the gut of the definitive host, but if this happens the pentastome can only develop as far as the nymphal stage. However, as cannibalism is common among some reptiles, it is possible that they act as both intermediate and definitive host. *Linguatula arctica* appears to have dispensed with the need for an intermediate host. This parasite infects the nasal passages and sinuses of reindeer (*Rangifer tarandus*). This is unusual as all other adult pentastomes described to date are found in carnivores. The egg is unable to survive exposure to freezing conditions and therefore cannot survive the long arctic winter. Instead, direct transmission takes place when reindeer calves consume eggs passed by worms infecting older reindeer (Haugerud and Nilssen 1990).

Humans normally develop visceral pentastomiasis through consuming pentastome eggs. The increasing popularity of keeping exotic pets and in particular reptiles means that their owners are at an increased risk of exposure to equally exotic pathogens. There is a large and often illegal trade in animals collected in the wild and then marketed via the internet. These animals are likely to be infested with parasites and some of these, such as the pentastomes, can infect us. The pentastome eggs hatch in our intestine to release the first-stage larva (nauplius larva) that penetrates the lining of the gut aided by a stylet. The larvae then encapsulate within our liver, lungs, and other viscera – although they may reach other organs including the eye and moult through several nymphal stages. Usually, human infections cause few, if any, symptoms and the parasites are only discovered during a medical examination for another complaint or at autopsy. However, heavy infestations cause abdominal pain, coughing, and night sweats and death may occur from secondary septicaemia or pneumonia (Tappe and Büttner 2009).

Dakubo et al. (2008) provide an interesting example of how close association with snakes can increase the risk of acquiring pentastome infections. Many of the clans and tribes of Ghana identify themselves with certain animal or plant totems. This should not be considered as primitivism or superstition and similar practices occur throughout the world. One has only to consider the

British lion, the French cockerel, and the US bald eagle. Some of the Ghanaian clans have the python as their totem and they will not kill or harm pythons. Indeed, pythons are welcomed into the village and they may share a villager's sleeping mat. None of the cases of visceral pentastomiasis described by the authors proved serious, but it would have been interesting to know how the level of infection of the python clans compared to that of clans with different totems.

Humans are most commonly infected by *Linguatula serrata* for which we act as intermediate hosts. *Linguatula serrata* has a widespread distribution and is an unusual pentastome in that the life cycle does not involve reptiles (Figure 7.9). The adult worm develops in the nasal passages and sinuses of dogs, wolves, and other caniids and lives for about 2 years. Heavy infestations cause nasal discharge and constant coughing and sneezing. Where there is a large population of stray and feral dogs, they can act as significant reservoirs of infection. For example, a survey of stray dogs in the city of Shiraz in Iran found that 76.5% of them were infected with *L. serrata* (Oryan et al. 2008). Estimating the numbers of stray animals presents considerable logistical difficulties. However, to provide some examples, Ankara in Turkey may contain in the region of 18,000 strays (Özen et al. 2016), whilst in San Francisco de Campeche in Mexico there may be 65–80,000 (Cortez-Aguirre et al. 2018). Within Europe, Bucharest in Romania is notorious for its numbers of stray dogs, which may number between 50,000 and 100,000 (Voslářvá and Passantino 2012). Stray dogs do not receive veterinary treatment and are therefore potentially a source of many zoonotic infections, some of which are much more important than pentastomes. These numbers are emphasised to give an idea of the scale of the problem in some countries. In addition, tourists and other travellers sometimes adopt stray dogs and bring them home, and this can help spread pentastomes (and other parasites) to countries such as the United Kingdom (Mitchell et al. 2016).

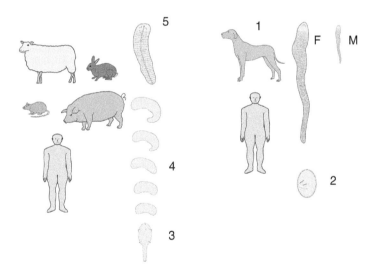

Figure 7.9 Life cycle of the pentastome *Linguatula serrata*. 1: The definitive hosts are usually dogs and other caniids although humans may also be infected. The adult pentastomes live in the nasal passages and sinuses; the females (F) are much larger than the males (M). 2: Eggs are passed out by sneezing or are swallowed and pass with the host's faeces. Many mammals, including humans, can act as intermediate hosts. 3: Infection of the intermediate host occurs through passive contamination. The egg hatches in the intermediate host's gut to release the first-stage larva that penetrates the gut. 4: The first-stage larvae encapsulate in the liver, lungs, or other viscera and then undergo a series of moults. They are often referred to as nymphs at this point in the life cycle. There is some variation in the literature concerning the number of nymphal stages. 5: The definitive host becomes infected when it consumes an intermediate host containing the final nymphal stage.

Infected dogs either cough or sneeze out the eggs or swallow them and pass them out with their faeces. The eggs are then ingested by the intermediate host through faecal-oral contamination of water, food etc. The nymphs can infect virtually any mammal although they are principally recorded from cattle, goats, and sheep. After the eggs hatch in the intermediate host's intestine, the first stage larvae burrow through the gut wall and develop into nymphs that encyst in the viscera. Consequently, we can develop visceral pentastomiasis linguatuliasis if we ingest *L. serrata* eggs. In some instances, this has resulted in a misdiagnosis of a cancerous growth. However, we are 'dead-end' hosts because we are seldom killed and eaten by other animals. By contrast, in Australia a predator-prey transmission chain now occurs between canines and rabbits (Barton et al. 2020). Such transmission chains exist in other countries but the situation in Australia is interesting because both rabbits and *L. serrata* are relatively recent introductions to the country.

Another means by which we can acquire pentastome infections is through the consuming flesh containing their nymphal stage. This gives rise to nasopharyngeal pentastomiasis/linguatuliasis. In Eastern Mediterranean countries it is known locally as 'halzoun' and is associated with the consumption of raw or lightly cooked liver and other visceral organs. Halzoun is a generic term relating to nasopharyngeal parasitic infections acquired from consuming raw liver etc.; it is not specific to pentastome infections but may also result from those caused by the flatworms *Fasciola hepatica* and *Dicrocoelium dendriticum*. There is some debate about the best means of treatment although some physicians suggest gargling with alcohol (Musharrafieh et al. 2018). Similarly, the consumption of snake meat is popular in parts of Africa, Asia, and among survivalists. It is a potential source of infection if the meat is not cooked properly. The nymphs move from the gut to the nasopharyngeal region where they cause pain, inflammatory reactions, and localised swelling, and these result in difficulty of swallowing and breathing. They do not, however, develop into adults.

7.4 Subphylum Hexapoda

The sub-phylum Hexapoda contains two classes: Class Entognatha and the Class Insecta. The Class Entognatha contains several orders of six-legged wingless organisms, which may or may not be true insects. Their common feature is that they do not have sufficient of the traits needed to be included in the Class Insecta. These traits are listed in Table 7.4. The Class Insecta is, thankfully, a single cohesive unit that consists of the winged insects and a few primitive wingless insects. The Hexapoda contains more than 750,000 described species and is the largest group of metazoan animals. In fact, it is three times larger than all the other animal groups combined. Hexapods are essentially terrestrial organisms, and they occur in virtually all terrestrial ecosystems. Some of them live in aquatic habitats but they are, for some reason, virtually absent from the sub-tidal waters of the sea. Much of their success is undoubtedly a consequence of their wings, which has facilitated dispersal, escape from predators, and the location of suitable environmental conditions in which to live and breed.

In most insects, the sexes are separate and following a courtship ritual mating takes place by apposition of the genitals. A few species, principally among the Homoptera (aphids, whitefly, mealybugs) and Hymenoptera (wasps, ants, bees), are parthenogenetic. In these, males are either never produced or are produced in certain generations in response to external stimuli. The female insect usually lays her eggs in a suitable location and then leaves them although in some species there are varying degrees of parental care and sociality. Primitive insect orders undergo hemimetabolous development in which the eggs hatch to give rise to a nymph that looks rather like the

Table 7.4   Morphological features of the Class Insecta.

Body divided into three regions: head, thorax, abdomen

Three pairs of unbranched legs on the thorax, one pair of legs on each thoracic segment.

One pair of antennae

Usually, two pairs of wings carried on the thorax. The wings are absent in some primitive orders and in some of the more advanced orders the wings may be secondarily lost, or one pair has become modified so that only a single pair of wings provides propulsion

Two pairs of maxillae on the head although one pair is often fused to form the labium

Base of mouthparts exposed and mouthparts themselves project from the head capsule

With the exception of one primitive order, the mandibles have two points of articulation

Gut has digestive caecae

Separate male and female sexes

Genital openings located on the terminal or sub-terminal abdominal segments

Usually, two or three median ocelli on the head with compound eyes to either side although these are secondarily lost in some species

Gas exchange occurs via spiracles that open into a branching arrangement of tracheal tubules that deliver air directly to individual cells

adult insect but lacks functional genitalia and wings (if they have them). It may or may not feed on the same food and occupy the same ecological niche as the adult. The nymph goes through a series of moults, and after each moult its resemblance to its parent increases. After the final moult, the insect becomes an adult, and it is only at this stage that its wings and genitalia are functional. Examples of insects that undergo hemimetabolous development include the human louse *Pediculus humanus* and the triatomid bugs that transmit Chagas disease. Advanced insect orders undergo holometabolous development in which an egg hatches to release a larva that looks nothing like its parent and frequently has a different diet and occupies a different ecological niche. The larva is often a simple 'eating machine' with a thin cuticle and limited powers of locomotion and sensory capacity. The larva undergoes a series of moults during which it becomes a bigger version of itself. After the final moult, the larva enters a resting pupal stage during which its body is reorganised, and once this is complete the insect emerges as a fully formed adult. Examples of insects that undergo holometabolous development include the human flea *Pulex irritans* and the flies such as mosquitoes, tsetse flies, and blowflies. With only a few rare exceptions, insects do not moult again once they have become adults and acquired their wings.

The Entognatha does not include any important parasites, and they do not play an important role in the ecology of parasitic diseases that affect us and our domestic animals. We will therefore not consider them any further. By contrast, the Class Insecta includes many examples of insects that are of parasitological importance. This is principally through the blood-feeding insects that act as vectors of diseases such as malaria and leishmaniasis or the coprophilic species that act as mechanical vectors of faecal-orally transmitted parasites such as *Entamoeba histolytica*. Unfortunately, it is beyond the scope of this book to consider the biology of insect vectors and those requiring more details on these should consult Mullen and Durden (2002). There are a number of insect species that are parasitic during part or all of their lives and some of these are of considerable medical and veterinary importance. Parasitic insects are also of interest for their potential to act as biological control agents of pests of agricultural crops and also of other parasitic organisms. It should also be remembered that a great many insect species are beneficial (e.g., for pollinating crops), and they are essential for the normal functioning of many ecosystems.

7.4.1 Order Phthiraptera (Lice)

The Phthiraptera divides into two sub-orders: the Mallophaga, or biting lice, and the Anoplura, or sucking lice. They are all wingless ectoparasites of birds and mammals. They probably evolved from members of the order Psocoptera. The psocopterans include winged species and are harmless free-living individuals that feed on detritus, dry plant matter, algae and fungi. The best known psocopterans are the booklice that are often found consuming the gum in old books and dry starchy matter in store cupboards. Some taxonomists combine the Phthiraptera and the Psocoptera together within a single order – the Psocodea.

The phthirapterans are relatively host specific with most louse species feeding on a single host species or group of closely-related host species. The Mallophaga have triangular-shaped heads that are broader than their thorax and biting mouthparts (Figure 7.10a). They are predominantly ectoparasites of birds although there are a few species that feed on mammals. Most of them use their mandibles to chew on feathers, fur, or dander but some of them will also consume epidermal secretions or gnaw through skin/quills to obtain blood. By contrast, the head capsule of the Anoplura is narrower than their thorax, and they have sucking mouthparts that they use to penetrate the skin and consume blood. The mouthparts of the Anoplura are unlike those of most other blood-sucking insects: the labrum is modified to form a snout-like process called the *haustellum* (Figure 7.11b), and when a louse feeds it everts its *haustellum*. This brings into play a series of chitinous 'teeth' that are now on the outer surface and anchor the *haustellum* in the host's skin. The louse then employs three stylets to form two channels: it secretes saliva down one and sucks blood up the other. The Anoplura have a more restricted host range than the Mallophaga and only parasitise certain placental (eutherian) mammals.

The Phthiraptera are all relatively small insects (0.4–10 mm) that are dorso-ventrally flattened. They are wingless and have typical hemimetabolous life cycles that consist of an egg stage, three nymphal instars, and then the adult stage. The eggs are laid directly upon the host and are usually physically attached to the hairs or feathers. The nymphs resemble their parents but are smaller and

(a) (b)

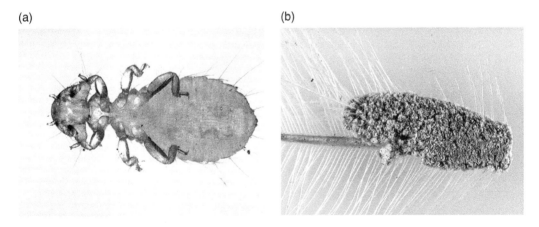

Figure 7.10 *Menacanthus stramineus* Mallophaga, biting louse. (a) Live louse removed from a chicken. Note the triangular-shaped head and its width in relation to the thorax. (b) Masses of eggs at the base of a feather.

(a)

(b)

(c)

(d)

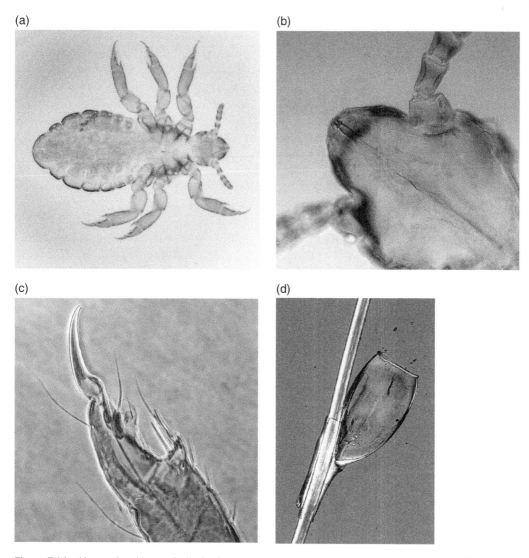

Figure 7.11 Human head louse, *Pediculus humanus capitis*. (a) Adult female louse (female head lice are usually 2.4–3.3 mm in length, the males are very slightly smaller). (b) Head showing haustellum and stylets. (c) Claw on hind leg. The tibia (lower segment in the photograph) terminates in a region called the 'thumb' that bears sensory bristles. Above the tibia is the tarsus that bears a curved, sharply pointed claw. The concave surface of the claw has a rough texture, and the louse grasp onto fine hairs or clothing fibres by pinching them between the claw and thumb. Compare the comparatively delicate arrangement seen here with that of the pubic louse in Figure 7.12. (d) Egg (nit) glued to a hair shaft. This egg has already hatched.

lack functional genitalia. Successive generations of lice develop upon an individual host animal. Transfer between hosts takes place through physical contact although some species of Mallophaga hitchhike attached to the bodies of hippoboscid flies and other insects. As a result, theoretically, there ought to be a close relationship between the lice found on parent hosts and their offspring and closely related species of louse should be found on closely related species of host. However, although there are instances in which this occurs, it does not occur as commonly as might be expected. Even among the Anoplura, which diversified at about the same time as the mammals,

there is considerable conflict between the mammalian and anopluran phylogenies. This is probably because of extinctions and host-switching events over the millennia.

The Phthiraptera have separate sexes. The male genitalia are unusually large and can extend to
half the length of the abdomen. They lack dorsal ocelli, and their compound eyes are reduced in
size or entirely lacking. This is to be expected as lice spend their whole life on the skin of their host
and have no requirement for vision to find their food or to avoid predators. Those Mallophaga species that parasitise mammals have tarsi that are modified to grasp hairs, but those found on birds
usually have unmodified one- or two-segmented tarsi. By contrast, the one-segmented tarsi of the
Anoplura, particularly those of the mid-legs and hind-legs, are modified to form tibio-tarsal claws
adapted for grasping hairs (Figures 7.11c and 7.12b).

The Mallophaga are primarily of veterinary importance, and many species have been transmitted around the world with domestic animals. Important species include the 'cattle biting louse'
Bovicola bovis that is typically found on the top of the head and feeds by chewing on the skin and
scurf at the base of the hairs. The louse population can grow rapidly and causes the affected animal
considerable irritation. This results in rubbing and scratching that in turn leads to hair loss, skin
damage, secondary infections, and loss of productivity. The chicken body louse *Menacanthus*
stramineus is a major pest of poultry around the world. Its yellow coloration also gives it its other
common name 'the yellow body louse'. It is highly pathogenic and can cause mortalities among
young birds. It causes severe inflammation and irritation especially around the vent where it lays
its eggs (Figure 7.10b) but also around the head and neck. The population can increase very quickly,
especially in caged laying hens, and a single bird may be afflicted by tens of thousands of lice.
Menacanthus stramineus can be naturally infected with eastern equine encephalomyelitis virus
but, like most Mallophaga, it is not considered to be an important vector of disease.

The Anoplura includes numerous examples of species of medical and veterinary importance.
The two best-known examples are the human head louse *Pediculus humanus capitis* and the
human body louse *Pediculus humanus humanus*. Both species are common cosmopolitan parasites. The body louse probably evolved from the head louse at about the time primitive humans
started wearing clothing.

The average insect genome is approximately 504 Mb (He et al. 2016) but that of the body louse
only 108 Mb. This small size is predominantly a consequence of losing genes involved in

(a)   (b)

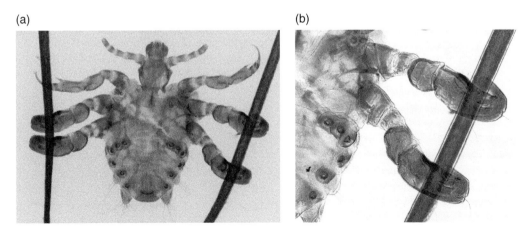

Figure 7.12   Human crab louse *Phthirus pubis*. (a) Adult louse. (b) Claws on mid- and hind legs. Note the
robust structure. This provides a powerful grip on coarse hair but limits movement.

environmental sensing and those coding for detoxification enzymes. This is to be expected in a parasite living permanently upon its host and relying on it to provide food, shelter from the environment and dispersal. Like other animals that feed solely on blood, body lice and head lice rely on symbiotic microorganisms to provide essential nutrients that are low or lacking in blood. In the case of head lice and body lice, they contain an endosymbiotic bacterium *Riesia pediculicola* within a mycetome. This bacterium provides the lice with pantothenic acid (vitamin B5). *Riesia pediculicola* also has an unusually small genome that encodes less than 600 genes (Boyd et al. 2017). These do not include antibiotic resistance genes and it may therefore prove possible to develop new means of controlling lice by targeting their symbiont (Kirkness et al. 2010).

The head louse generally lays its eggs singly upon scalp hairs and these are commonly referred to as 'nits' (Figure 7.11d). The eggs hatch after about 7–10 days and the nymphal stages last a further 7–10 days. The adults live for about 2–4 weeks and during that time the female louse lays between 50 and 150 eggs. Unlike the body louse, the head lice do not usually lay eggs on clothing and the sharing of earphones, hats, scarves, chair head rests, or pillows is probably not a significant means of transmission. The life cycle of the body louse is very similar to that of the head louse but they often lay their eggs on clothing – particularly along the seams. Depending upon the temperature, eggs laid on clothing can remain infective for up to 4 weeks but they cannot survive for more than a month. Therefore, nit-infested clothing that has not been worn for a long time is not infective. The adults of both head lice and body lice are unable to survive off their host's body for more than about 48 hours. Transmission takes place through close physical contact and, in the case of body lice, the sharing of clothes. Poor people living in cold climates often wear multiple layers of clothes that are not changed and washed regularly. Consequently, they can suffer from exceptionally high louse infestations. Louse infestations are also a problem for people living in crowded, squalid conditions such as squatter camps, refugee camps, and prisoners in jails. According to the Old Testament (Exodus 8, v16), the third plague to afflict the Egyptians were lice that were brought forth from the 'dust of the land'. However, these lice were said to afflict 'man and beast', which is unusual since most lice are host-specific and those afflicting domestic animals do not infect humans and *vice versa*. It should be noted that most natural history terms in the Scriptures are controversial and some commentators consider the original Hebrew would be more appropriately translated as gnats, mosquitoes, or sandflies.

Head lice and body lice feed several times every day and thereby cause irritation and inflammation. The constant itching can become debilitating and the affected person literally feels 'lousy'. Heavily infested people may harbour several hundred body lice and their skin becomes thickened and deeply pigmented. This condition is sometimes called 'vagabond's disease'. Head lice have not been implicated in the transmission of disease although the scratching they cause may result in secondary skin infections. Body lice are vectors of *Rickettsia prowazeki* (louse-borne epidemic typhus), *Borrelia recurrentis* (louse-borne epidemic relapsing fever), and *Bartonella quintana* (trench fever). Transmission of louse-borne typhus to us often occurs via louse faeces. The infected faeces may be scratched into the itchy bite sites, rubbed into the eyes or mucous membranes or even breathed in via aerosols. Lice infected with *R. prowazeki* turn red and die and whilst louse-borne typhus is not always fatal in humans, the mortality rate can reach over 60% in malnourished populations. Louse-borne typhus is, not surprisingly, a common disease wherever lice are able to flourish and it often accompanies wars. For example, an examination of a mass grave containing the remains of some of Napoleon's soldiers who died during the retreat from Moscow in 1812 found that many of them were infested with lice and suffered from louse-borne epidemic typhus and trench fever (Raoult et al. 2006). Similarly, much of the literature of WW1 contains vivid descriptions of louse-infested clothing and several million people died of typhus across Europe at the end of the war. Fortunately, the discovery of the insecticidal properties of DDT prevented similar mortalities due to louse-borne typhus during WW2. Nevertheless, it was a major cause of death among concentration camp inmates, such as those of Bergen-Belsen, immediately before and after their liberation in 1945.

Pthirus pubis, also known as the crab louse from its appearance and the pubic louse from its normal dwelling place, is a fascinating creature (Figure 7.12a). Ever romantic, the French refer to it *'papillon d'amour'* although it is doubtful if those who are infected look on it with any affection. It is a small louse that normally only grows to about 2 mm in length. The body is almost as broad as it is long and the middle and hind legs end in large claws that are used for grasping onto coarse hairs (Figure 7.12b). The life cycle takes about 17–25 days from egg-laying to moulting to the adult stage. The female louse glues her eggs to coarse body hairs and several may be attached to a single hair. The eggs take 6–8 days to hatch and there are three nymphal stages that take a total of about 10–17 days to complete.

Crab lice are normally found on pubic hairs but may also attach to other regions of the body where coarse, sparsely growing hairs are found such as the eyebrows, and the beard, moustache and chest hair in men. The current popularity among both women and men for removing hair from around the genitalia is reportedly having serious consequences for crab louse populations in many countries (Dholakia et al. 2014). They are normally transmitted through sexual intercourse although transmission may also occur through other forms of close physical contact such as between parents and children who share beds. The bites of the crab louse cause inflammation and this together with the localised bleeding results in purple or blue-grey coloured spots (macules) forming at the feeding site. This results in irritation and scratching but crab lice are not considered to be vectors of disease. DNA studies conducted by Amanzougaghene et al. (2020) found that some crab lice contained *Bartonella quintana* (causes trench fever), and *Acinetobacter baumannii* (a common cause of hospital acquired pneumonia). However, whether they can transmit a pathogen after ingesting it remains uncertain. Crab louse infestations often correlate with other sexually transmitted diseases, but this is a reflection of sexual behaviour rather than an actual link. In common with other sexually transmitted infections, there is an enormous amount of ignorance about how one acquires crab lice and also what the best means of control are. For example, a survey of US College students found that many of them thought that crab lice could be caught by sharing living space with an infected person or using a toilet seat immediately after them and that crab lice could be effectively treated using antibiotics (Anderson and Chaney 2009). In reality, crab lice are very slow moving and grasp firmly onto their host's hair. Consequently, they are unlikely to become dislodged and need time to transfer between hosts. In common with head lice and body lice, pubic lice cannot survive for long off their host and those found on furniture etc. are probably unwell or dying. Crab lice are best treated using insecticidal creams or lotions.

There are numerous species of Anoplura of veterinary significance. These include the delightfully named but distinctly unpleasant 'little blue cattle louse' *Solenopotes capillatus*. This louse has a world-wide distribution and is one of the commonest lice found on cattle. It normally lives on the face and jaw region. Large numbers congregating around the eyes can give the affected animal the appearance of wearing spectacles. As with most other lice, it causes pathology mainly through irritation that leads to disturbance, scratching and secondary infections.

The 'hog louse', *Haematopinus suis*, is one of the largest of the Anoplura and adults can grow to 5 mm in length (Figure 7.13). It is the only louse to parasitise pigs and is a common and cosmopolitan species. The nymphs attach mainly on the ears and head region, whilst the adults have a more widespread distribution and are also found on the neck, jowls, back, and flanks. *Haematopinus suis* is normally transmitted by direct physical contact. However, they are capable of surviving for

Figure 7.13 *Haematopinus suis.* This is the only louse that parasitises pigs. It is one of the largest of the Anoplura and grows up to 5 mm in length.

up to 3–4 days off their host, so it is possible that transmission may also occur via bedding if pigs are moved into pens that have not been cleaned. *Haematopinus suis* nymphs will also attach to the housefly *Musca domestica* and therefore dispersal by phoresis may also occur.

Low-intensity infections with *H. suis* are seldom a serious problem. However, large numbers of lice cause constant irritation. An affected animal may rub off virtually all the hair from its body and its skin becomes raw and broken. Heavy infestations of young piglets may result in sufficient blood loss to cause anaemia and death. *Haematopinus suis* is also a potential mechanical vector of the bacterium *Mycoplasma suis* (previously known as *Eperythrozoon suis*) that causes porcine infectious anaemia (Acosta et al. 2019)

7.4.2 Order Siphonaptera (Fleas)

The fleas are an unusual group of insects whose relationship with the other insect orders is still uncertain. They are all rather small (1–10 mm) wingless insects that are parasitic on birds and mammals during their adult stage (Figure 7.14a–e). Of the 2,500 or so species described to date, over 70% are parasitic on rodents. They all feed on blood and have mouthparts modified for piercing and sucking. Adult fleas remain permanently on their host but usually move around upon it and feed periodically. However, so-called stick-tight fleas such as the rabbit flea *Spilopsyllus cuniculi* tend to remain attached for long periods of time after firmly anchoring themselves in place with their long mouthparts. Movement of adult fleas between hosts occurs when there is close physical contact. The sexes are separate and male fleas allegedly have the most complex genitalia in the Animal kingdom. Fleas lack wings, and this is usually cited as an example of adaptation to parasitism since these could make it difficult to move about on their host. Wings could also be considered redundant since the fleas' hosts act as their means of dispersal. However, some workers suggest that the Siphonaptera evolved from the Boreidae, which is a family of wingless insects within the order Mecoptera. If this is true, the fleas' wingless condition may be an ancestral condition and not a consequence of their parasitic lifestyle. The extant species of Boreidae are unusual in being mostly found at high altitude habitats where they are active during winter and feed on plant material. By contrast, all other members of the Mecoptera are winged and most of them are either predatory or feed on dead invertebrates. There are, however, considerable morphological differences between the head and mouthparts of the Boreidae and the Siphonaptera. Nevertheless, the suggestion that the two might be sister groups remains controversial. Molecular phylogenetic studies suggest that the fleas have a long history of association with mammals and indicate that they evolved during the Cretaceous

(a)   (b)   (c)

(d)   (e)

(f)

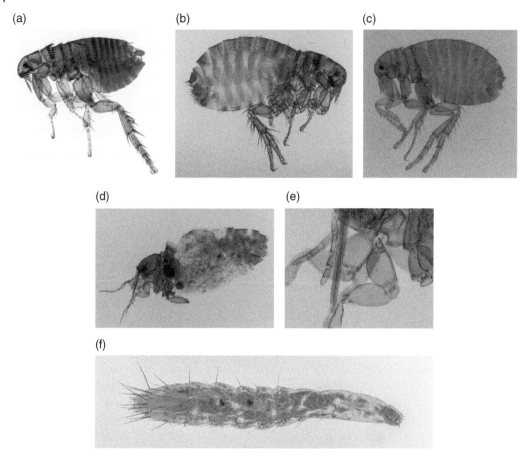

Figure 7.14 Examples of Order Siphonaptera (fleas). (a) *Ctenocephalides felis* (female). Note well developed hind legs and backwardly-pointing spines that interlock with the host's hairs making it difficult to pick them out of the pelage. (b) Dog flea *Ctenocephalides canis*. (c) Human flea *Pulex irritans*. (d) *Tunga penetrans*. The body of the female swells enormously and this contributes to making it difficult to remove the flea. (e) Mouthparts of *T. penetrans*: note how the long, toothed structure helps to anchor the flea firmly in place. (f) Flea larva.

period (145.5–66 million years ago). This would mean that they evolved alongside primitive therian mammals (Zhu et al. 2015).

Fleas lack compound eyes although they usually have clusters of ocelli. They therefore have limited visual capabilities but can detect changes in light intensity. Unlike other groups of insects, their body is laterally compressed, and this adaption enables them to crawl between the hairs and feathers of their host. Many flea species have arrangements of backwardly pointing bristles and rows of hardened (sclerotinised) spines called *ctenidia* in various parts of their body. These enmesh with the hairs/ feathers of their host and thereby make them difficult to extract (Figure 7.14a,b).

Fleas have a sensory structure called a *sensillum* (*pygidium*) on the dorsal surface of their last abdominal segment that detects vibration. This *sensillum* is highly sensitive and enables the flea to detect an approaching 'predator' from changes in air currents and vibrations transmitted down hair shafts. In addition, fleas can run very quickly and their hind pair of legs are enlarged and adapted for jumping – some species can jump 30 cm in a single leap. Fleas are usually light–dark brown in coloration, which makes it difficult to see them within hair/fur/feathers of their host and this, together with their shape and jumping ability make it very difficult to catch a flea. However, a deftly wielded moist bar of soap works is an efficient 'flea catcher'.

Fleas are holometabolous insects whose life cycle consists of an egg, three larval instars (usually), a pupal stage, and then the adult stage. The larval stage is rather worm-like and lacks legs and eyes (Figure 7.14f). The larvae are not parasitic and usually feed on detritus within the host's nest or bedding. The larvae of some flea species do, however, need a blood meal to complete their development. This blood is obtained by consuming the faeces of adult fleas and hungry larvae sometimes prod adult fleas to make them defecate. Having completed the larval period, the larva spins a silken cocoon within which it first enters a resting pre-pupal period and then moults to the pupa. Fleas that parasitise mammals often breed throughout the year but those that parasitise birds only breed during their host's nesting season. After completing their pupal development, the adult fleas do not emerge until they detect mechanical stimuli because this indicates the return of a suitable host to the nest. This explains why houses that have been unoccupied for months, or even years, suddenly become 'alive with fleas' when a new owner moves in.

Most fleas are associated with a particular host species. However, this is seldom a highly specific relationship, and a hungry flea is liable to feed on any warm-blooded animal. For example, in addition to the so-called human flea (*Pulex irritans*) (Figure 7.14c), there are records of about twenty other flea species feeding on us. Nevertheless, a flea's fecundity often declines if it is unable to feed on its preferred host species. This catholic diet can have serious implications for the transmission of zoonotic diseases. For example, bubonic plague (*Yersinia pestis*) is essentially a disease of rodents. However, when the rodents die the fleas leave their host's dead body as it cools and crawl and jump in search of alternative food. The catastrophic outbreaks of bubonic plague in Europe in earlier centuries were associated with the deaths of large numbers of rats and the consequent movement of fleas onto humans.

The human flea, *P. irritans*, is becoming uncommon in industrialised countries. In the United Kingdom and Europe, its place is being taken by the cat flea, *Ctenocephalides felis* (Figure 7.14a). *Pulex irritans* remains common in many developing countries and is a potentially important vector of diseases. For example, in a survey of 12 villages in Tanzania, Laudisoit et al. (2007) found *P. irritans* in 72.4% of houses and, unlike the other flea species that were present, its density was positively correlated with higher frequency of bubonic plague. Although it is called the human flea, *P. irritans* also parasitises sheep, cattle, goats, pigs, and even chickens. According to the sixth century, Byzantine scholar Isodore of Miletus, the word *pulex* derives from *pulvis* meaning dust. This is because fleas were thought to develop from dust, especially after it was sprinkled with urine. The French derive their word for a flea, *puce*, from the Latin *pulex*. In the eighteenth century, the brownish purple clothing colour, that was popular with Marie Antoinette among other aristocracy, gained its name puce because it resembles the colour of fleas. Hygiene at the court of Versailles was notoriously appalling (as it was in many places at the time) and fleas were common. The dog flea, *Ctenocephalides canis* (Figure 7.14b) has also been replaced by *C. felis* in many regions, so the most common flea found on dogs is now usually the cat flea. In towns and cities, feral pigeons (*Columba livia*) often nest on or near to dwellings, and we are sometimes attacked by the pigeon flea (*Ceratophyllus columbae*), which migrates from deserted pigeon nests into our homes.

Although some flea species act as vectors of bubonic plague and murine typhus (*Rickettsia typhi*), the list of other pathogens of medical and veterinary importance for which fleas are the principal vectors is actually rather short. Flea bites can prove intensely irritating, and in sensitive individuals and domestic animals, they induce flea-bite dermatitis. Fleas have a reputation for preferring to bite some people more than others and women are often said to be bitten more frequently than men. However, there is limited experimental evidence of differential attractiveness. Many of the reported differences probably arise because some people react more severely to flea bites. Breathing in aerosolised flea faeces and cuticular components can induce allergic asthmatic reactions. In addition to acting as vectors of disease, fleas are intermediate hosts for the tapeworms *Dipylidium caninum*, *Hymenolepis diminuta*, and *Hymenolepis nana*. For many people, especially in developed countries, the psychological consequences of flea bites can be extremely upsetting. Although we are often willing to 'accept' mosquito bites, flea bites carry a certain stigma. Some people consider them evidence of being dirty and uncivilised, and therefore when they are victims, it results in psychological trauma, anxiety, and depression.

Tungiasis

Tungiasis refers to infection with fleas belonging to the genus *Tunga*. These unusual fleas are commonly known as sand fleas, jiggers, or *chigoes*. There are several species of *Tunga*, but the best known is *Tunga penetrans* (Figure 7.14d,e) that was initially restricted to tropical America. However, in 1872, it found its way to West Africa in ballast sand being shipped from Brazil to Angola. Since then, it has colonised much of sub-Saharan Africa and parts of the Middle East. Although *T. penetrans* was accidentally transported to Bombay and Karachi in 1899, it failed to spread in the way that it did in Africa.

Tunga penetrans is a very small flea, the adults being about 1mm in size. It is typically found in sandy soil that is shaded by vegetation and in the earth floor of huts. After the adults emerge, they latch onto us or a variety of mammals and usually commence feeding on our feet. Much of the pathology is caused by the female flea that, after mating, burrows into the soft skin between our toes and under the toenails. They invade other regions of our body, or that of an animal, if we were lying on the ground. In animals that snuffle along the ground as they walk, such as pigs and dogs, the muzzle may be infected. Unlike many other flea species, the female *Tunga* retains her eggs within her body. Consequently, her abdomen swells enormously. Within 8–10 days, the abdomen of a female *T. penetrans* becomes an almost spherical 6 mm in diameter so that the thorax and head look like a tennis ball sitting on top of a football. The posterior two abdominal segments do not swell and they project out of the host's body – and it is through these that the eggs are ultimately expelled. The eggs subsequently hatch and the larval stages are free-living, but unlike other Siphonaptera, there are only two larval instars.

The flea's feeding causes an inflammatory reaction and its increase in body size puts pressure on the surrounding tissues. Together, this results in intense pain and where large numbers of fleas are present, they effectively 'honeycomb' the affected region and cripple the host. Obviously, this has economic consequences as it reduces the ability of people to work and the productivity of domestic animals. In the past, some military campaigns were brought to a literal standstill when so many soldiers were crippled by tungiasis that operations had to be abandoned. In addition, tungiasis lesions often become secondarily infected and tetanus may occur in individuals who have not been vaccinated. In poor communities in which most people walk barefoot over 50% of the community can be infected. In addition, tungiasis is a zoonotic disease and dogs, cats, pigs, and rats can act as reservoirs of infection. Furthermore, the popularity of exotic holidays also means that tungiasis is sometimes seen in returning travellers in industrialised countries.

7.4.3 Order Diptera (True Flies)

The Diptera is one of the largest insect Orders and contains over 100,000 described species. From a parasitological perspective, it is the most important group of arthropods since the Diptera includes numerous examples of species that are vectors of viral, microbial, and parasitic disease and also those that are either facultative or obligatory parasites of us and our domestic animals.

Adult Diptera are characterised by possessing a single pair of membranous forewings that provide all the propulsion for flight. The hind wings are modified to form tiny club-shaped structures called halteres that function as gyroscopes. Most adult Diptera are active fliers and consequently have a well-developed thorax and large compound eyes. Adult flies usually feed on fluids and have mouthparts adapted for either piercing and sucking or their labium has become expanded to form two fleshy lobes that function a bit like a sponge. The Diptera exhibit holometabolous development and the legless worm-like larvae go through three or more instars depending upon the species. Fly larvae are commonly referred to as 'maggots' although some entomologists dislike seeing this term used in scientific literature. The larvae exhibit a wide variety of feeding strategies and in some species, this includes parasitism on vertebrates or other invertebrates. Currently, most taxonomists divide the Diptera into two suborders: the Nematocera and the Brachycera.

7.4.3.1 Suborder Nematocera

The Nematocera are considered the most primitive of the Diptera. The adult flies are characterised by having antennae that are usually longer than the combined length of the head and thorax and consist of numerous small segments. An exception to this rule is the Simuliidae (blackflies) in which the antennae are more compact but still consist of many segments (Figure 7.16c). Nematoceran larvae have a well-developed head capsule that is equipped with biting mouthparts that operate in a horizontal plane like a pair of pincers. After completing its growth, the last larval instar moults to form an obtect pupa – that is as the adult fly develops its legs and other appendages are fastened down to the rest of the body. Despite this, in some species, such as the mosquitoes, the pupa can remain mobile. The suborder Nematocera includes numerous species that are vectors of parasitic diseases, but there are few that are themselves parasites of medical or veterinary importance (Table 7.5). We will therefore not consider the Nematocera any further here.

Table 7.5 Examples of Diptera that act as vectors of parasitic diseases.

Suborder	Family	Example	Common name	Parasitic disease transmitted
Nematocera	Culicidae	*Culex quinquefasciatus*	Mosquito	*Wuchereria bancrofti* (Lymphatic filariasis)
		Anopheles gambiae	Mosquito	*Plasmodium falciparum* (Malaria)
	Psychodidae	*Phlebotomus martini*	Sandfly	*Leishmania donovani* (Visceral leishmaniasis)
	Simuliidae	*Simulium damnosum*	Blackfly	*Onchocerca volvulus* (River blindness)
Brachycera	Tabanidae	*Tabanus* spp.	Horsefly	*Trypanosoma evansi* (Surra)
	Muscidae	*Musca domestica*	Housefly	Faecal–oral transmitted parasites
	Glossinidae	*Glossina morsitans*	Tsetse fly	*Trypanosoma brucei* (Nagana)
	Calliphoridae	*Calliphora vicina*	Blowfly	Faecal–oral transmitted parasites

7.4.3.2   Suborder Brachycera

The Brachycera consists of the infraorders Tabanomorpha, Asilomorpha, and Cyclorrhapha. The Tabanomorpha contains several families, but it is the Tabanidae that are the most important from a veterinary and medical perspective. The Tabanidae are stoutly built flies with short 'horn-like' antennae and large compound eyes that are often beautifully coloured. They are strong fliers but seldom make much noise – which means that they can approach their victim without drawing attention to themselves. Like mosquitoes, only female tabanids feed on blood and the male lives on nectar, honeydew, and other natural sugary secretions. The male tabanid therefore lacks the rigid hypopharynx, blade-like mandibles, and recurved teeth on the maxillary laciniae that are found on the female flies. Although they might approach silently, the bite of a tabanid is painful, and this means that they are usually disturbed before they have taken a full meal. However, they are also persistent, and after being disturbed they quickly return to the same victim or another one nearby. Consequently, female tabanid flies make good mechanical vectors. A female tabanid fly lays up to several hundred eggs in a single mass or as several smaller masses. The larvae have a discernible head region although it is not as sclerotinised as in the Nematocera, and they are equipped with fang-like mandibles that operate in a vertical plane. Depending upon the species, tabanid larvae may be aquatic or develop in the soil, and they go through up to 13 moults. Some of them are predacious on other invertebrates and cannibalism can be an important factor influencing larval density and dispersal.

The Asilomorpha contains the robber flies that are voracious predators of other insects. These are not important as parasites of us and our domestic animals or as vectors of disease. The infraorder does, however, include several species of bee-fly *Thyridanthrax* (Family Bombyliidae) that are parasitoids of tsetse fly pupae. These can have a significant impact on the tsetse fly population, but it has not proved possible to exploit them as biological control agents (Laird 1977). Flies belonging to the family Acroceridae often have remarkably small heads in relation to their body size, whilst the larvae are endoparasitoids of spiders. The female fly lays eggs in masses at the end of twigs or scattered upon the ground and after hatching the first-instar larvae search for a spider to parasitise. The larvae can crawl along spider webs and jump up to 6 mm by bending and suddenly straightening their body. After locating a spider, the larva climbs up one of its legs and then penetrates its body wall. It then slowly consumes the tissues so that the spider does not die until the larva has completed its development.

The Cyclorrhapha is also known as the Muscomorpha and contains a number of species that are parasitic on us and our domestic animals. It also includes the unusual Nycteribiidae and Streblidae that are highly specialised parasites of bats. Adult cyclorrhaphan flies usually have well-developed compound eyes and their antennae consist of three short segments on the last of which there is a feather-like sensory structure called the 'arista'. Cyclorrhaphan flies typically lay their eggs singly or in batches, and there are three larval instars. The larvae are usually bullet-shaped and lack a head capsule. The larvae lack compound eyes but have lateral ocelli (stemmata) that in some species are arranged in complex arrays. The larvae are equipped with an internal cephalopharyngeal skeleton that includes a pair of mouth hooks. The cephalopharyngeal skeleton attaches to the exoskeleton by a series of powerful muscles that swing it back and forth. In the process of moving the cephalopharyngeal skeleton, the mouth hooks also move back and forth. This drags food into the mouth or enables them to work as 'grappling hooks' to facilitate movement. After completing its third instar, the larva forms a coarctate pupa. This involves the larva moulting but it retains the cast exoskeleton to form puparium. The cast exoskeleton is then sclerotinised, thereby becoming hardened and darkened. The puparium acts as a protective covering within which the pupa develops. The term 'Cyclorrhapha' translates as 'circular seam' because the puparium has a weak point running around it close to the anterior end. When the adult fly is ready to emerge, it exerts pressure at this point, and the puparium breaks open to release the fly. Many cyclorrhaphan flies (i.e., those belonging to the Schizophora) employ an inflatable bag-like structure called the *ptilinum* to facilitate their escape. The *ptilinum* is located at the front of the head and the fly pumps air into it to inflate it. After emerging, the fly deflates its *ptilinum* and retracts it behind the ptilinal suture never to be used again. The size of the adult flies can vary remarkably depending upon the availability and quality of food during the larval stage. There is a certain minimum size below which the larvae fail to pupariate but poorly fed larvae may produce flies that are less than half the size of those that were well fed.

Wound Myiasis

The development of Diptera larvae within a living vertebrate animal is known as myiasis. Wound myiasis refers to larvae that develop within an existing wound or within a wound that they caused to form (Figure 7.15a). By causing wound myiasis, the larvae benefit from the constant warm temperature of their host, which facilitates rapid larval development. They also avoid competition from many other fly species, other insects (e.g., ants and wasps), and vertebrates that would occur if feeding on something that was already dead.

The species causing wound myiasis can be broadly categorised as those that are facultative parasites and those that are obligatory parasites. Facultative species are those whose larvae can develop as detritivores in dead animals or as parasites of living animals. Obligatory species are those that only lay their eggs on living animals. In both cases, there is little host specificity, and they will cause wound myiasis in virtually any warm-blooded animal.

An alternative means of myiasis classification is based on their ability to initiate wound myiasis. Primary species are those that initiate a myiasis through inducing a wound or establishing an infection through invading via one of the body orifices. Secondary species are those that do not initiate an infection but invade a wound once it is already infested with fly larvae. Primary species will also lay their eggs on a wound that is already infested with fly larvae.

A wound does not have to be large before it is suitable for exploitation. Some flies will lay their eggs in a lesion as small as a tick bite. In addition, some fly species are highly sensitive to the odour of blood and can detect a wound from long distances. Farm animals often carry wounds from a variety of causes. For example, surplus males are usually castrated, cattle are often dehorned, and sheep often receive nicks when they are clipped or from catching themselves on barbed wire. Farm animals, like their wild counterparts, are also frequently wounded during conflicts with one another as they seek to establish their place in the flock or herd. For example, head wounds are common in rams. Wounds that are caused by infections such as foot rot in sheep or digital dermatitis in milking cows are also potential sites of infestation. Similarly, leech bites, tabanid fly bites, and lesions caused by rubbing and scratching in response to lice infestations can all be exploited by the flies that cause wound myiasis. The umbilicus of new-born mammals is also often infested.

Wound myiasis in humans is usually associated with personal neglect or incapacity and tends to occur among those who are mentally or physically incapable of looking after themselves. Filth and wounds give off smells that attract flies, and these then lay their eggs on the skin or soiled clothing. One of the more famous historical personages thought to have suffered from myiasis is Herod the Great (73 BCE–4 CE). He is reputed to have died a protracted and miserable death during which his genitals "putrefied and brought forth worms". Anything white and wriggly tends to be called a 'worm', and therefore it has been suggested that, amongst other things, he may have suffered from myiasis of the genitalia subsequent to a bacterial or other infection (Retief and Cilliers 2006). Another noted sufferer of myiasis was St Symeon Stylites (*c*390–459 ce) who voluntarily spent 20 years of his life chained to the top of a stone column not far from Aleppo in Syria. Like many of his contemporaries, he believed that squalor was a sign of holiness and neither washed nor changed his clothes for years. One of the many stories about him are of his gently returning the maggots that fell from his rags and encouraging them to feed because they were 'God's creatures and a manifestation of his will'. Presumably, these fly larvae were actually feeding upon his encrusted filth rather than his flesh since St Symeon lived to a comparatively old age.

Wound myiasis is a potential nosocomial infection in hospital patients if care is not taken to exclude flies and change dressings regularly. Any skin lesion such as a post-operative scar, diabetic ulcer, and even psoriasis can become infected (Dagci et al. 2008). Oral myiasis occurs when fly larvae invade the periodontal pockets and then crawl across the surface of the gums. These cases are relatively rare and usually involve patients with poor oral hygiene, a habit of mouth breathing, and mental impairment (Filho et al. 2018). Tumour lesions such as squamous cell carcinomas can present a particular problem because in the end stage, these are associated with bleeding that may prove fatal. Health care workers therefore tend to minimise changing dressings and wound treatments when their patients reach this point. However, the carcinomas can become infected with fly larvae, and this causes considerable psychological torment to the patient who has to cope with their body being consumed by maggots even as they must come to terms with their own impending death (Sesterhenn et al. 2009). For a detailed account of myiasis in humans, see Bernhardt et al. (2019).

Treatment requires the physical removal of the fly larvae and the cleaning of the affected area. This can prove difficult in large infestations and may require the patient to be anaesthetised. Covering the affected area with liquid paraffin or turpentine oil will block the posterior spiracles of the larvae and thereby encourage them to move onto the surface. This makes them easier to remove. After removal, the wound site needs protecting until healed to prevent further infections, and the provision of antimicrobials may be required to reduce the chances of bacterial infection.

7.4.3.2.1 Family Calliphoridae

The Calliphoridae contains over 1,000 species and includes species with a diverse array of lifestyles. However, those of medical and veterinary importance belong to the groups that are commonly known as the blowflies and the screwworm flies. The origin of the term 'blowfly' is not known. One suggestion is that one of the several meanings for the word 'blow' is to describe a mass of fly eggs. A piece of meat that has fly eggs on it is said to be 'flyblown' and the fly that leaves those eggs is therefore referred to as a 'blowfly'. An alternative explanation is that it was once believed that flies blew their eggs onto meat etc. in a similar manner to someone using a peashooter. The term 'screwworm' derives from the way in which the larvae burrow deeply into the flesh of their host and it appears as if they are screwing down into the tissues. The adults of blowfly and screwworm species are generally harmless as they feed on nectar, plant sap, and fluid decaying organic matter. Some of them will also feed on bodily secretions (e.g., tears) and blood oozing from wounds, but they lack piercing and sucking mouthparts. These species can be a considerable nuisance and cause disturbance. They can also act as mechanical vectors of faecal–orally transmitted pathogens.

The word 'blowfly' is a generic term and is applied to both species that cause myiasis and those that are detritivores or parasites of invertebrates. Among the most important blowfly species causing cutaneous and wound myiasis are *Lucilia sericata*, *Lucilia cuprina*, *Calliphora vicina*, and *Protophormia terrae-novae*. The *Lucilia* species are commonly known as greenbottles because of their shiny green thorax and abdomen (Figure 7.15d). The *Calliphora* species tend to be called bluebottles because of their dark blue coloration, whilst *Protophormia* species are called blackbottles because they are usually extremely dark. However, such terms are not to be recommended as they invariably lead to confusion. The life cycle of those species causing myiasis is relatively similar although the duration of the different stages varies. The female fly lays her eggs in large masses that can contain over 100 eggs upon the host either at the edges of a wound or upon soiled skin. The eggs hatch after a few hours and the larva starts to feed. If there is an existing wound, the larvae will move into this and extend it but if one is not present, they may initiate a wound by inducing an inflammatory reaction that weakens the skin. The larvae do not have strong enough mouthparts to penetrate healthy unbroken skin. There are three larval instars and once a larva reaches optimal size it ceases feeding and drops off. It then voids its gut and crawls across the ground until it reaches a suitable site where it digs into the soil. The larva then contracts and

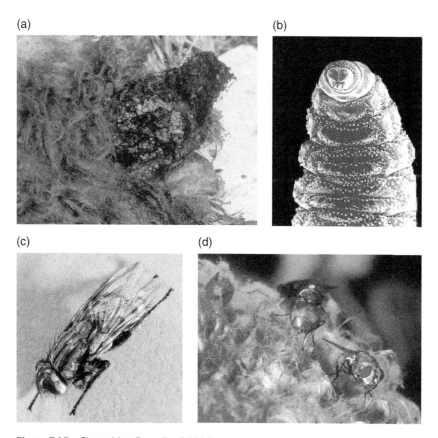

Figure 7.15 Sheep blowfly strike (Irbid, Jordan) caused mostly by the fleshfly *Wohlfahrtia magnifica* larvae. (a) Gross appearance. The sheep is a fatty-tailed Awassi breed, and the larvae have almost destroyed the tail. Note how deeply the larvae have burrowed into the flesh. (b) Third-instar larva of *W. magnifica*. Scanning electron microscope image. Note the bullet-shaped body, prominent body spines and well-developed mouth hooks. (c) Adult *W. magnifica*. (d) Blowflies laying their eggs at the wound site.

undergoes pupariation – although in some species (e.g., *Calliphora vicina*), the larva may undergo a resting diapause stage at this point. Once metamorphosis to the adult stage within the puparium is complete, the adult fly emerges, crawls up to the soil surface, expands its wings, and flies off. The whole life cycle from egg hatch to adult emergence typically takes about 3–4 weeks but is heavily influenced by environmental factors.

Blowfly Strike in Sheep

Although it is called by blowfly strike, this is a generic term and other species of fly may also be responsible such as sarcophagid flies and screwworm flies. Before the introduction of insecticidal sheep dips, blowfly strike was a cause of considerable suffering wherever sheep were raised (Figure 7.15a). Even today, anybody who rears sheep has to keep a careful watch on his/her flock during the 'fly season'. Unless one can halt an infection before it develops too far, a sheep can be literally 'eaten alive' by successive waves of thousands of fly larvae. However, the animal often succumbs to septicaemia before vital organs are affected.

Sheep are usually first 'struck' around their rear end where the wool becomes soiled with faeces and urine. This is known as 'breech strike'. Gastrointestinal parasites that induce diarrhoea increase a sheep's susceptibility to breech strike by increasing the soiling of the wool. Fatty-tailed sheep, such as the Awassi and other breeds that are popular in the Middle East and Central Asia, are particularly vulnerable to blowfly strike because of the soiling that occurs under their tails. Allegedly, some farmers allow the rear end of these sheep to become heavily soiled because it increases the live weight at sale.

Blowfly strike typically begins when a primary fly species such as *Lucilia sericata* or *Lucilia cuprina* is attracted by the smell given off by soiled wool. Bacteria contribute to fleece rot and the odours released by species such as *Pseudomonas aeruginosa* stimulate blowfly oviposition. After landing, the female fly lays her eggs, and once these hatch the first-instar larvae move down to the surface of the skin where they commence feeding on bacteria, detritus, and dead skin tissues. In the process, the larvae release proteolytic secretions that are absorbed across the sheep's skin where they initiate an inflammatory reaction. Consequently, the skin surface becomes raw and inflamed and releases watery exudates. At this point, the skin no longer presents an effective barrier and the larvae start to feed on the underlying tissues. Because there are usually numerous larvae present, an open wound soon forms and is then extended. This attracts both secondary and primary flies, and they arrive and lay their eggs. In addition, the larvae transport bacteria from faeces-contaminated wool onto the wound. This results in the wound becoming septic and toxaemia may develop.

The screwworms include the important species *Chrysomya bezziana* – the Old World Screwworm – and *Cochliomyia hominivorax* – the New World Screwworm. As their common names suggest, *C. bezziana* is predominantly found in Asia and Africa, whilst *C. hominivorax* is found in South America.

7.4.3.2.1.1 Genus *Chrysomya*   There are several species of *Chrysomya* some of which are obligatory parasites during their larval stage, whilst others are either facultative parasites or detritivores. The most important species is *C. bezziana* whose larvae are obligatory parasites of us and other warm-blooded vertebrates. It occurs throughout parts of India, Asia, the Middle East, Africa, and also New Guinea although its distribution and the number of infections it causes are both probably under-reported (Zhou et al. 2019). The adult fly is bright metallic green or blue and 8–10 mm in length and superficially resembles the blowfly *Lucilia*. The adult flies aren't often seen in the wild. This is because they are not attracted to dead animals or the sorts of baits that attract blowflies. The female quickly lays batches of approximately 100–300 eggs at the site of a wound, on skin contaminated with blood, or at a body opening. The eggs hatch after 12–24 hours, and the first-instar larvae commence feeding on blood and discharges emanating from the wound. After 12–18 hours, the larvae moult to the second instar and move into the wound and begin extending it. After a further 12 or so hours, the larvae moult to the third instar and burrow deep into the wound so that only their posterior spiracles remain visible. The third-instar larvae feed communally and voraciously and grow to about 18 mm length over the course of 3–4 days. Once they are mature, the larvae emerge from the wound, drop to the ground, and burrow into the soil where they pupariate. The adult fly emerges after 7–9 days although the development period may last several weeks in cool weather.

7.4.3.2.1.2 Genus *Cochliomyia* The genus *Cochliomyia* contains a single species, *C. hominivorax*. The term '*hominivorax*' translates as 'man-eater' and for very good reasons. The first published report on the fly was written by the French surgeon Dr Charles Coquerel in 1858. He described treating five men who were parasitised by the fly in Cayenne (French Guiana, South America) that was at the time a hellish penal colony. The flies had laid eggs in the nostrils of the men and after these hatched hundreds of maggots proceeded to eat into the men's heads. Despite many attempts at flushing the larvae out, three of the men subsequently died.

Adult *C. hominivorax* are about 15 mm in length and have a bright greenish blue metallic coloration, a yellowish face, and bright orange-red compound eyes. The females mate only once whilst the males mate several times, and this is one of the reasons the fly can be successfully controlled by the release of sterile males (FAO 1992). After mating, the female fly lays her eggs in batches of 10–500 at the edges of wounds so that the eggs overlap the wound. She lays 4–8 batches of eggs over a lifespan that is typically 7–10 days but can extend to over a month. The larvae hatch after 11–24 hours and move into the wound. The larvae feed communally and burrow much deeper into wounds than most fly species that cause wound myiasis and may be found 10 cm or so below the surface. In addition, their body is armed with backwardly pointing spines that make it difficult to remove them. The wounds usually become secondarily infected with bacteria and then give off a characteristic odour that is highly attractive to female *C. hominivorax*. Consequently, the wounds rapidly become deep and extended. The larvae take about 4–8 days to reach the end of the third instar, and at this point they drop off and pupariate in the soil. The adult flies emerge after 7–54 days depending upon the temperature. The whole life cycle from egg laying to adult emergence can therefore take as little as 21 days under ideal conditions.

Cochliomyia hominivorax is predominantly a parasite of wild and domestic animals although humans are occasionally infected. It used to be a major problem in parts of North America and Mexico where it was responsible for annual losses amounting to millions of pounds. For example, in 1976 the cost of surveillance, treatment, and control of *C. hominivorax* was estimated to amount to US$ 300 million per annum in just the state of Texas. Since then, the success of the sterile male release programme to control *C. hominivorax* has eliminated the fly from North America, Mexico, and some South American countries. In the late 1980s, an outbreak of *C. hominivorax* occurred in Libya through the import of infected cattle. The flies can disperse rapidly, and there were fears that the species would establish itself in neighbouring countries with consequent devastating impact on their economies. At the time Libya and USA were politically hostile to one another, but they cooperated to mount a sterile male release control programme that eliminated the fly before it could become a major problem. Interestingly, the advent of CRISPR gene manipulation technology is opening new ways in which it may be possible to control the populations of both *C. hominivorax* and other flies that cause myiasis (Paulo et al. 2019).

7.4.3.2.1.3 Genus *Auchmeromyia* The Congo floor maggot, *Auchmeromyia senegalensis* (also known as *Auchmeromyia luteola*) is sometimes stated as a specific human parasite. However, there are several reports of it attacking warthogs, pigs, and hyenas. The larva feeds on blood but spends the daylight hours hiding in the soil and crevices and emerges at night to feed on us when we are asleep. Presumably, the maggot reverses its activity cycle when parasitising nocturnal animal hosts. In the colonial era, it was a common problem in what was then the French Congo. This region has since become divided up into the Republic of Congo, The Central African Republic, and Gabon. Nowadays, *A. senegalensis* is seldom reported because villagers increasingly sleep on beds rather than mats on the floor. Consequently, the larva's meal is now out of reach. However, social change means that another group of people who were not previously at risk are acting as hosts. The Pygmies who live in the Chailu region of the Congo traditionally led a nomadic existence in the jungles and were not afflicted by *A. senegalensis*. This may have been because they moved too frequently for the flies to establish a population at their camp sites. However, the Pygmies have started to become sedentary and live in poorly constructed camps on the outskirts of Bantu villages. The Pygmies tend to sleep on mats on the floor and are attacked by the maggots, whilst the Bantu villagers are protected as they sleep on beds (Noireau 1992).

7.4.3.2.1.4 Genus *Cordylobia* The Tumbu fly, *Cordylobia anthropophaga*, occurs in Sub-Saharan Africa although there are occasional reports of human infections from Saudi Arabia and Yemen, so it may be more widespread. The adult female is a stout yellow-brown fly and 8–12 mm in length. She lays her eggs on the ground, especially in sandy regions and if the soil is contaminated with urine or faeces. The larvae then hatch and they wait until a suitable host passes by. Despite the fearsome name 'anthropophaga', we are but one of their potential hosts and dogs, and many other mammals are commonly parasitised. The larvae quickly burrow into the skin and each larva forms an individual boil-like lesion within which it develops. The boils typically form on the feet and lower limbs, but other areas of the body may be affected. Obviously, wearing shoes and long clothing minimises the risks of becoming infected. The larva takes about 7–15 days to become a mature third instar by which time it is 12–28 mm in length. The thick-bodied larva (Figure 7.16) feeds head down and has recurved spines that make it difficult to remove. The boil that develops around the

Figure 7.16 *Cordylobia anthropophaga*. Mature third-stage larva. This specimen was 22 mm in length. The larva feeds with its mouthparts (blunt end) buried deep in the flesh of its host. The larva tapers to a narrow cone at the rear and terminates with the posterior spiracles that are exposed to the air. This shape makes it difficult to extract from the skin.

larva has a hole at its centre through which the larva breathes. When it is mature, the larva manoeuvres out of the boil, drops to the ground, and pupariates in the soil. The adult fly then emerges after 3–4 weeks.

The Tumbu fly is notorious for laying its eggs on washing that is hung out to dry. The fly does not lay its eggs on washing that is in full sunlight. It favours clothing left in the shade and still has a smell of urine – such as underwear or nappies. When the clothes are next worn, the larvae hatch in response to the heat of the skin. Many people consider ironing one's underwear to be a waste of time but in parts of Africa it makes good sense because it kills any Tumbu fly larvae or eggs that are attached. There are many possibly apocryphal stories of servants intentionally not ironing the washing of employers who were miserly or ill-treated them.

7.4.3.2.2 Family Sarcophagidae
Flies belonging to the family Sarcophagidae are commonly known as 'Flesh Flies' because some of them are attracted to meat and dead animals. Unlike the blowflies, the sarcophagids are larviparous. That is, the eggs hatch inside the uterus of the female and it deposits first-instar larvae directly onto their food source. The larvae of sarcophagids differ from those of blowflies by having their posterior spiracles recessed within a cavity. Adult sarcophagids are handsome grey-coloured flies with red eyes. Their thorax usually bears black longitudinal stripes, and their abdomen has a chequered or tessellated pattern. Their tarsi have large pluvilii (pads) that gives them the appearance of having big feet. Sarcophagid flies and their larvae are notoriously difficult to identify and therefore care must be taken when reading the literature.

The family Sarcophagidae divides into two subfamilies: the Miltogramminae and the Sarcophaginae, and together they contain over 2,000 described species. Most of the Miltogramminae are parasitoids of other invertebrates although some species are detritivores and a few cause myiasis in vertebrates. The Sarcophaginae contains a number of species whose larvae develop on dead animals and some are either facultative or obligatory parasites of vertebrates.

One of the most encountered sarcophagid flies causing myiasis is the wonderfully named *Wohlfahrtia magnifica* (Figure 7.15b,c). It belongs to the Miltogramminae and is widely distributed in Eastern Europe, the Mediterranean region, and the Middle East (Giangaspero et al. 2011). It is a relatively large fly that grows up to 14 mm in length and its larvae are obligate parasites of mammals and birds. Infections are sometimes referred to as wohlfahrtiosis or Wohlfahrt's wound myiasis. The female fly deposits 80–120 first-instar larvae on wounds or at moist body orifices (e.g., anus and genitalia) and the larvae then quickly burrow into their host. Nosocomial infections of unconscious patients can occur through flies depositing larvae in intubation tubes. The larvae burrow deep into the subcutaneous tissue and cause liquefaction and necrosis of the surrounding flesh. The larvae feed communally and mixed sarcophagid/blowfly infections can occur. There are three larval instars. The mature larvae leave their host after about 5–7 days, drop to the ground, and pupariate in the soil.

Unless treated promptly, *W. magnifica* infestations cause serious pathology. Furthermore, infested wounds soon attract other *Wohlfahrtia*, as well as blowflies. This means that there is rapid extension of an infested wound and septicaemia may develop. In sheep, death may occur within 1–2 weeks of initial infestation. Several other sarcophagid species cause myiasis (e.g., *Sarcophaga argyrostoma*), and there are at least 20 species within the genus *Wohlfahrtia*. Therefore, one should not assume that a sarcophagid causing myiasis must be *W. magnifica*.

7.4.3.2.3 Family Oestridae
This family contains the so-called warble flies and botflies. The term 'warble' as a verb to describe a singing style dates back to the 1300s with roots indicating whirling motion. However, 'warble'

subsequently also became a noun describing a raised swelling. One suggestion is that in this case it derives from an old Swedish word for a boil '*varbulde*' (*var* = pus; *bulde* = swelling). The term 'bot' was already in use in the English language to describe a parasite or maggot in the 1500s but its derivation is unclear. Certainly, the same term was in use elsewhere in Europe at the time – for example a Dutch word for a liver fluke was *leberbot*. The Oestridae is a relatively small family that contains about 150 or so species. They are all obligate endoparasites of mammals during the larval stage and the adults lack fully functional mouthparts – although a few species probably imbibe water or other fluids. Many species are host-specific and infest only a single host species or group of closely related species. Therefore, although the family includes several species of economic importance it also includes some that are in serious danger of extinction. For example, the population of rhinoceroses and many other large mammals has declined so badly in recent years that those animals that are dependent upon them are also being pushed to the brink of extinction. Unfortunately, it is highly unlikely that the public will ever be convinced of the need to preserve parasites. However, all species have a role to play in the ecosystem and their loss represents a further diminution of the diversity of life on Earth. For a detailed account of the oestrid flies, see Colwell et al. (2006).

7.4.3.2.3.1 Subfamily Oestrinae   The larvae of flies belonging to this subfamily usually live in the nasal cavities of their host. They are therefore often referred to as nasal botflies. There are several genera and some of them have fascinating life cycles. For example, the larvae of *Gedoelstia hässleri*, which is a parasite of large African antelopes such as the Blue/Common Wildebeest (*Connochaetes taurinus*), undergo an apparently unnecessarily complicated migratory pathway. The female fly places its larvae into the cornea or conjunctiva of a suitable host, and these then penetrate the tissues and get into the blood system. The first-instar larvae are very small (0.7–0.9 mm in length) and make their way via the bloodstream to the subdural space – this is a thin layer between the *dura mater* and the *arachnoid mater* (two of the three meninges that surround the brain and spinal cord). The larvae then migrate through the cribiform plate and ethmoid bone to reach the nasal cavity where they moult to the second instar and then the final (third) instar. The mature larvae are very large and can reach 31 mm in length. The closely related *Gedoelstia cristata*, which also parasitises various African antelopes, shows a similar migratory pattern. The larvae are deposited in the eye of their host, enter its blood circulation, pass through the heart and are swept to the lungs. The larvae then penetrate the alveoli, climb the bronchi and trachea and make their way to the nasal cavities where they moult to the second instar. Some workers suggest that *G. hässleri* and *G. cristata* may hybridise.

In their normal hosts, *G. hässleri* and *G. cristata* appear to cause little harm but in domestic sheep and cattle (in which they cannot reach maturity), they can cause serious pathology. Within the eye, physical damage and inflammatory reactions results in a condition known as '*uitpeuloog*' (bulging eye disease). This manifests as swelling of the eye, glaucoma, and in severe cases the eyeball may rupture. The migrating larvae also induce inflammatory reactions within veins that result in the formation of thrombi (clots) and the localised softening or loss of brain tissue (encephalomalacia), whilst those in the heart can induce heart failure.

7.4.3.2.3.1.1   Oestrus ovis   The sheep nasal fly, *Oestrus ovis,* is the most important member of the subfamily Oestrinae. It is found throughout the world wherever sheep are reared. The adult fly (Figure 7.17b) is about 12 mm in length and has a stout body with a large stumpy head. It has a dark grey coloration with black spots on the abdomen and thorax and a covering of short brown hairs. The mouthparts are non-functional and they are unable to sting. Nevertheless, the distinctive buzzing noise of an approaching fly causes sheep to panic. They attempt to protect their nostrils by pushing their noses into the ground or huddling together and pushing them into one another's fleece. When she finds an unprotected nostril, the female fly expels her eggs into it through her ovipositor without alighting. The expulsion force causes the eggs to hatch and therefore she deposits first-instar larvae rather than eggs. She typically places up to 25 larvae in the nostrils or sheep and goats.

Humans are occasionally infected by *O. ovis* via the nose and eyes. The larvae are usually unable to complete their development in us although they may cause irritation, inflammation, and pain. Infections of our eyes (opthalmomyiasis) result in conjunctivitis and photophobia, but the larvae normally die as first instars and they do not invade the body of the eye. There are also occasional case reports of infections in dogs, horses, and cattle. In their normal host, the larvae crawl into the nasal passageways where they moult to the second instar, and these larvae then continue to the sinuses where they moult to the final third-instar stage. The larvae feed on nasal secretions and irritate the lining of the nasal mucosa with their sharp mouth hooks. Although they do not feed on blood, they may cause bleeding from the nose (epistaxis) (Dorchies et al. 1998). The larvae become large thick-bodied wedge-shaped maggots up to 3 cm in length (Figure 7.17a). When they are mature the larvae move to the nasal passageways and are then sneezed out after which they pupariate in the soil. The duration of the different life cycle stages depends heavily upon the environmental conditions. The adult flies are active on hot sunny days, and in warm climates, there may be up to three generations a year. By contrast, in colder climates the first-instar larvae cease growing during the autumn/winter and resume their development when the air temperature increases. The ability to suspend their development (hypobiosis) is thought to be an important part of the insect's life cycle and can also be induced by other stressors such as crowding and the host immune response.

The disturbance caused by adult *O. ovis* can lower productivity, but it is the larvae that cause most harm to sheep. Heavy infections can induce a condition known as 'false gid' in which the afflicted animal becomes unthrifty and exhibits a lack of co-ordination and staggers around in circles. More normally, the animal develops a nasal discharge, constantly sneezes and rubs its nose against objects as though attempting to relieve an irritation. Inflammatory reactions to the physical damage caused by the larvae and also to their excretory-secretory products results in rhinitis and sinusitis. The literature often includes suggestions that the fly's name derives from the larvae preventing male sheep and goats from detecting whether the females are in oestrus. This would not be

(a)   (b)

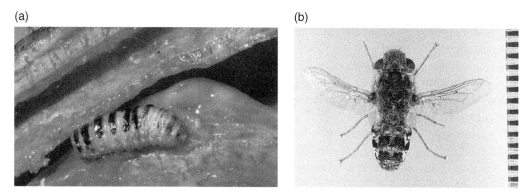

Figure 7.17 *Oestrus ovis*. (a) Third instar larva within the turbinate bones of a sheep. Note how the shape of the larva ensures it is tightly wedged within the cavities. (b) Adult *O. ovis*. The adult fly lacks functional mouthparts and is therefore unable to feed.

surprising, but there appears to be little experimental evidence. Unlike the migrating larvae of *Hypoderma*, the larvae of *O. ovis* appear to promote an inflammatory response so that they can feed on the resultant exudates. In serious *O. ovis* infections, secondary bacterial infections develop that may result in potentially fatal pneumonia. There are suggestions that the larvae may activate viruses such as orf (a zoonotic parapox virus) and the *maedi-visna* virus (Angulo-Valadez et al. 2010). *Maedi-visna* virus is a lentivirus that is usually present as a subclinical infection but when activated, it can induce fatal disease characterised by dyspnoea (*maedi*; breathing difficulty) or neurological (*visna*) signs. In some breeds of sheep, *O. ovis* infection has been linked to the development of tumours (Dorchies et al. 2003). This may also be related to virus activation, but there is currently very limited published information on this and the evidence of an association is not strong.

7.4.3.2.3.2 Subfamily Gasterophilinae   This subfamily contains the well-known horse botflies *Gasterophilus intestinalis*, *Gasterophilus nasalis*, *Gasterophilus pecorum*, and *Gasterophilus haemorrhoidalis* that infect horses, donkeys, and asses. In older literature, the genus name '*Gasterophilus*' is often spelt '*Gastrophilus*'. Other Gasterophilinae infect wild equids such as zebra, whilst some species are specific for elephants and rhinoceroses. In all cases, the final-instar larvae live within the gastrointestinal tract – usually the stomach. Appropriately enough, the rhinoceros stomach botfly *Gyrostigma rhinocerontis* (previously *Gyrostigma pavesii*) is reputedly the largest fly in Africa. Its body can be up to 35 mm in length whilst the larvae can grow to 40 mm.

Adult *Gasterophilus* are usually 10–15 mm in length and covered in dense brown/yellowish hairs so that they bear a superficial resemblance to bumble bees (Figure 7.18d). The adults are strong fliers but they lack functional mouthparts. Therefore, like many other adult oestrid flies, their lifespan is probably short. The female fly has a long stout ovipositor, which she uses to glue eggs to hairs or to the skin on specific regions of the host animal. On horses, *G. intestinalis* usually lays its eggs on the hairs of the fetlocks of the forelegs but may also attach them as high up as the scapular region. *Gasterophilus nasalis* lays eggs underneath the horse's chin whilst *G. haemorrhoidalis* lays eggs on its lips. An exception is *G. pecorum* that lays eggs on vegetation that is subsequently consumed by the host. In their attempts to lay their eggs, the flies disturb horses and they attempt to protect the region the flies are aiming for. In some species, such as *G. nasalis* the eggs hatch spontaneously, whilst those of other species, such as *G. intestinalis*, are stimulated to hatch by being licked by the host.

The first-instar larvae either migrate to the mouth or the eggs hatch in the mouth after being consumed. The larvae then undergo a species-specific series of movements and development within the buccal cavity before re-locating to specific locations in the gastrointestinal tract. Ultimately, third-instar larvae of both *G. pecorum* and *G. intestinalis* attach to the mucosa of the stomach although *G. intestinalis* tends to be restricted to the junction of the glandular and non-glandular regions. The third-instar larvae of *G. nasalis* attach to the lining of the duodenum near the pylorus whilst, as the name suggests, the larvae of *G. haemorrhoidalis* complete their development attached to the lining of the rectum. The larvae take about 10–12 months to develop to maturity after which they detach and pass out with the faeces in the Spring. The larvae pupariate in the soil and the adults emerge after 3–5 weeks.

The third-instar larvae usually cluster together in dense groups. They are barrel-shaped, grow to around 16–20 mm in length, and attach to the mucosa of the gut using large, sharply pointed, sickle-shaped mouth hooks (Figure 7.18a,b). In many species, the second and third-instar larvae are coloured various shades of dark red although those of *G. nasalis* tend be light yellow. This coloration results from the presence of haemoglobin molecules that facilitate the uptake, transport, and storage of oxygen. In *G. intestinalis* larvae, the haemoglobin occurs as an intracellular

(a)
(b)
(c)
(d)

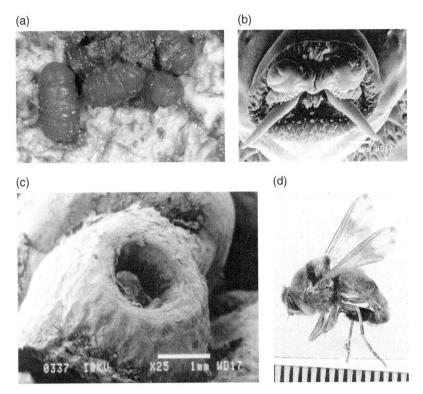

Figure 7.18 *Gasterophilus intestinalis.* (a) Live larvae in the stomach of a donkey. (b) Scanning electron micrograph of anterior of third-instar larva showing curved mouth hooks. (c) Sites of larval attachment in the stomach of a donkey. (d) Adult *G. intestinalis.*

molecule in a variety of cell types during the early stages of larval development but is subsequently restricted to the tracheal cells. A further adaption to their lifestyle is that their posterior spiracles are sunk into a pouch that opens when they are above fluid level.

The migration of the first-instar gasterophilid larvae in the gums/palate/tongue/palate causes bleeding, pain, inflammation, and secondary bacterial infections. There is some debate whether final-instar gasterophilid larvae feed on host tissue/secretions or act as so-called kleptoparasites that obtain their nutrition from ingesting the host's digesta. However, the attachment of the larvae induces inflammatory reactions that results in ulceration, localised fibrosis, and the loss of functional mucosal tissue (Figure 7.18c). *Gasterophilus pecorum* has the most pathogenic larvae because they attach to walls of pharynx and oesophagus for up 10 months, and it is only the older third-instar larvae that migrate to the stomach. Larvae within the oesophagus induce inflammation that result in localised swelling. Consequently, the presence of numerous larvae and/or excessive inflammation can cause constriction of the oesophagus, thereby causing choking and difficulty in swallowing – and this can prove fatal.

7.4.3.2.3.3 Subfamily Hypodermatinae This subfamily includes the well-known 'warble flies' whose name derives from the raised nodule-like swellings caused by the third-instar larvae that are referred to as 'warbles'. There are two species of warble fly of importance in cattle: *Hypoderma bovis* and *Hypoderma lineatum*. They have a widespread distribution in parts of Europe, USA, the Middle East, North Africa, and Asia. Their life cycle begins when the female fly attaches her eggs to a cow. The flies have non-functional mouthparts and are unable to bite but they induce panic in cattle. Frightened animals rush blindly away bucking their hind legs and with their tails in the air. This behaviour is known as 'gadding' and the animals can injure themselves by colliding with fences, gates, and barbed wire. Even if the cattle are not injured, the disturbance reduces feeding and hence affects live-weight gain and milk production. According to Gansser (1956), the flies do not produce a characteristic buzzing while in flight although this does not preclude them from emitting a high-pitched noise that is audible to cattle but not us. Clearly, there must be a distinctive feature of an approaching warble fly that distinguishes it from the many hundreds of other flies that plague cattle during the summer months.

Female *H. bovis* (Figure 7.19c) attach single eggs to hairs on the legs below the hocks, whilst *H. lineatum* attaches rows of six or more eggs to hairs on the legs above the hocks and on the lower parts of the body. Female *H. bovis* therefore have to make more frequent approaches in order to lay their eggs and their repetitive darting flight around the feet of cattle may explain why they are more prone to causing gadding than *H. lineatum*. After hatching, the first-instar larvae of both species crawl down the hairs and penetrate the hair follicles. *Hypoderma bovis* larvae migrate along nerves until they reach the epidural fat that surrounds the spinal cord. By contrast, *H. lineatum* larvae migrate through the muscles and connective tissue until they reach the submucosa of the oesophagus. Both species reach their respective destinations in the late Autumn, and they remain there overwinter during which time they moult to the second instar. In the Spring, the larvae resume their migrations and both species ultimately reach the skin underneath the host animal's back. Here they moult to form barrel-shaped larvae that can grow to about 28 mm in length (Figure 7.19a). The larvae cut a breathing hole through the host's skin through which they breathe and when they are mature, they manoeuvre their way out, drop to the floor and pupariate in the soil. The adults emerge after about 3–5 weeks although those of *H. lineatum* are usually active about a month before those of *H. bovis*.

The host immune reaction against the third-instar larvae results in fluid accumulation and hence the swelling of the warble (Figure 7.19b). It is impossible to pull a larva out of its warble but one can sometimes squeeze them out by exerting pressure from below – rather like squeezing a spot. When the third-instar larvae cut their breathing holes/ emerge the damage they cause cannot be repaired. Consequently, the hide of a heavily infected animal is ruined. In addition, the inflammatory reaction against the larvae means that the surrounding flesh is unfit for human consumption and has to be cut away. The first- and second-instar larvae do not usually induce serious pathology although if the larvae of *H. bovis* die, whilst they are overwintering, their decay induces an inflammatory reaction that may result in nerve damage and paralysis.

Hypoderma bovis and *H. lineatum* are both probably extinct in Great Britain following a successful eradication campaign. The campaign began in 1978 at which time approximately 38% of the cattle were infected and losses to the livestock industry amounted to about £13 million per annum. Legislation was passed (Warble Fly [England and Wales] Order 1982 and the Warble Fly [Scotland] Order 1982) that made the warble fly infections a notifiable disease. Anyone who had an animal that they knew or suspected was infected had to first report it to a Ministry Divisional Veterinary Officer before attempting any treatment. If the infestation was confirmed, then all cattle on the farm over 12 weeks of age had to be treated. Warbles become obvious in the Spring and treatment at this time kills the larvae before they can become adults. Further treatment is also required in the Autumn (before the larvae have time to reach their winter resting sites) to prevent infections developing to the 'warble' stage. In addition, cattle on surrounding farms also had to be treated regardless of whether they had warbles. These measures were so successful that by 1986 less than 0.007% of inspected cattle were infected.

(a)   (b)

(c)

Figure 7.19   *Hypoderma lineatum.* (a) Mature larva removed from the warble (the larva is orientated as it would be in the cow with the posterior uppermost – note the two posterior spiracles). (b) Warbles on the back of a cow. (c) Adult *Hypoderma bovis.*

Hypoderma diana has a widespread distribution in Europe, including the United Kingdom, where it infects various deer species. There are also reports of severe infections in horses and alpaca but it does not appear to infect cattle (Borges et al. 2019). The female flies lay their eggs predominantly on the lower parts of the forelegs. The larvae then penetrate the skin. Over the next 6 months, the first-instar larvae migrate upwards, over the shoulders, and then backwards towards the flanks. They then moult to the second instar in the subcutaneous tissues and remain stationary whilst a warble develops around them. Heavy infestations can be debilitating and are potential sites for wound myiasis. However, there is currently no dedicated control programme. Most deer in the United Kingdom are wild rather than farmed and therefore any control programme would prove difficult. Furthermore, there is a limited market for deer hides and the species poses no risk to cattle.

7.4.3.2.3.4   Subfamily Cuterebrinae   The members of this subfamily are sometimes referred to as the 'New World skin bot flies' because they are restricted to the North and South America. However, some taxonomists feel that the subfamily should include the genera *Neocuterebra* and *Ruttenia* that are parasites of African elephants, and this would make the common name inappropriate. Members of the genus *Cuterebra* are often host-specific and most are parasites of rodents and rabbits. Typically, they lay their eggs near to burrows or other sites of host activity and the eggs hatch in response to the heat of the passing host. The first-instar larvae then quickly attach to the host's hair and make their way to its mouth or nose. The larvae then penetrate the mucosa and spend about 7 days wandering around the pharyngeal region. The larvae then migrate sub-dermally through the host's connective tissue until they reach a species-specific site where they stop, cut an air hole in overlying skin, and develop into the final instar. The larvae of some species of *Cuterebra* can reach up to 30 mm in length, which is exceptionally large in relation to the size of their host and therefore severely debilitating. After reaching maturity, the larva forces its way out through the skin and pupariates in the soil. The various species of *Cuterebra* are of little medical or veterinary importance although cats and dogs are sometimes infected and heavy infestations can prove fatal. The most important member of the subfamily Cuterebrinae is *Dermatobia hominis* although its lifestyle, lack of host specificity, and development distinguish it from other members of the group.

7.4.3.2.3.4.1 Dermatobia hominis *Dermatobia hominis* is found throughout South America and also the islands of Trinidad. Although it is often called the 'human bot fly' it infects many large mammals. It is also the only species among the Oestridae that naturally parasitises carnivores. Nevertheless, *D. hominis* is most important as a pest of cattle, and in 1976, the annual losses to cattle production it caused in Brazil and Central America were estimated to be US$ 200 million per annum.

The fascinating thing about this fly is that the female does not lay its eggs directly onto its host. Instead, she captures flies that are likely to frequent cattle or other mammals. She usually chooses insects that are in the vicinity of the host animal and glues up to 25 eggs to their abdomen or thorax. Day-flying mosquitoes are often exploited, but other flies such as *Musca* spp. and *Stomoxys calcitrans* may also be used. There is even a report of ticks being exploited. She then releases her 'vector' and when this next alights on a mammal, the rise in temperature stimulates the eggs to hatch. The first-instar larvae then drop onto the host and quickly burrow into its skin. If a blood-feeding insect was being employed, the larvae may make use of the puncture wound it caused. The larvae do not migrate and develop slowly at the site of their initial invasion over the next 3 months. A raised nodule forms around the developing larva with a central breathing hole. The structure can become inflamed and resemble a boil. These lesions may be secondarily infected by bacteria and also attract the attentions of screwworms and other flies that cause wound myiasis. Once it is mature, the larva emerges, drops to the ground, and pupariates in the soil.

The larva of *D. hominis* is pear-shaped and its curved spines make it very difficult to pull out of the skin (Figure 7.20). However, covering the breathing hole with petroleum jelly or similar substance can usually stimulate the larva to move out or at least reposition itself so that it can be extracted more easily. Some reports suggest using bacon fat but it is the effectiveness of a substance at blocking the breathing hole that encourages a larva to leave the skin rather than it being tempted out with a tasty morsel. We are usually infected on the head region because this is where the flies

Figure 7.20 *Dermatobia hominis*. Mature larva after removal from the skin. Note the recurved body spines and pear-shaped body that make its removal difficult.

that *D. hominis* exploits are most likely to land. Therefore, wearing a jungle hat can protect against mosquito bites and hence *D. hominis* infections in this region. However, other body parts can be infected and Passos et al. (2004) describe a case in which a larva developed in the penis of a man who routinely wore shorts without any underwear.

7.4.3.2.4 Family Streblidae

The Streblidae are all obligatory ectoparasites of bats (Chiroptera) and are primarily found in the tropics and especially on bats that roost communally in caves. They are sensitive to low temperature and unable to survive on hibernating bats. Bats go into hibernation when the winter temperature drops to +10 °C. During their hibernation, a bat's body temperature falls to slightly above ambient temperature and average temperatures of below +10 °C are too low for streblids. Approximately 240 streblid species have been described, and they are typically small 1.5–2.5 mm in length but some species can grow to 5 mm. Although they are totally reliant upon their host for food, most species are instantly recognisable as Diptera and about 80% of species have retained functional wings. Nevertheless, their compound eyes are poorly developed or absent, and their antennae are extremely small. The wings are folded back like pleats over the abdomen when not in use so that they do not interfere with movement through their host's pelage. All species feed on blood and have piercing and sucking mouthparts. The extent to which Streblid and Nycteribiid flies directly harm their host is not known. However, they can transmit the protozoan parasite *Polychromophilus melanipherus* that causes bat malaria (Obame-Nkoghe et al. 2016).

The eggs hatch within the female's uterus, and the larva is nourished by the secretions of two milk glands in a similar manner to the larvae of tsetse fly. One larva develops at a time and after it reaches maturity, the female fly leaves its host and attaches the larva to a nearby surface. The female fly then returns to its host and the larva pupariates.

Female streblids belonging to the genus *Ascodipteron* are somewhat unusual in that although they are initially winged once they have found a suitable host, they undergo a series of morphological changes. This is very rare among adult insects. First, the female fly locates a suitable site on the host's body and then she uses her well-developed mouthparts to pierce and cut through the skin and into the underlying dermis. During this process, she sheds her wings and legs and ultimately only the posterior of her abdomen protrudes from the bat's skin. The abdomen then swells until the head and thorax are embedded within it. Because female *Ascodipteron* are permanently attached to their host, they have to drop their larvae onto the floor of the bat roost rather than attach them to the surrounding surfaces. Presumably, they also have a limited ability to influence mate choice. The male flies retain their mobility.

7.4.3.2.5 Family Nycteribiidae

Flies belonging to this family are all highly modified blood-feeding, obligatory ectoparasites of bats. It is an intriguing question how and why two families, the Streblidae and the Nycteribiidae, of specialist dipteran parasites of bats should have evolved and the extent to which competition between species exists for a remarkably specialised niche. The Nycteribiidae are relatively small flies and are typically 1.5–5.0 mm in length (Figure 7.21). Bats are well known for their species diversity, and there is a similar level of diversity among the Nycteribiidae: to date 951 species of bats and 274 species of Nyceteribiidae have been described. Species diversity among the Nycteribiidae is promoted by their lack of mobility, and they spend their whole life on a single host. The adult flies lack wings but retain their halteres. Because they lack wings, the thorax muscles have atrophied and the thorax is compressed and reduced in size. Their eyes are either absent or reduced, but they have well-developed antennae and their legs are long and end in grasping claws

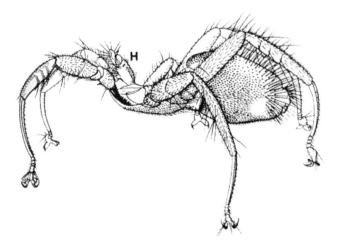

H

Figure 7.21 Adult Nycteribiid. These wingless flies are parasitic on bats. H: Head. Note the long legs that end in large-hooked claws. *Source:* Reproduced from Oldroyd (1964), © Norton Library.

that enable them to maintain a firm grip. The flies can therefore move easily and quickly about their host. When at rest, the head projects upwards and appears to perch on top of the thorax a bit like the gun turret on a tank. Consequently, the insect has to rotate its head forward and downwards so that it can exploit its piercing and sucking mouthparts.

The reproduction of Nycteribiidae is very similar to that of the Streblidae in that the eggs hatch within the female's uterus, and the larva is nourished by the secretions of two milk glands. One larva develops at a time and once it reaches maturity, the female leaves her host and attaches the larva to a nearby surface. She then scuttles back to the host whilst the larva pupariates.

8

Parasite Transmission

CONTENTS

8.1 Introduction

Living as a parasite within or upon another organism offers numerous advantages, but it also poses two fundamental problems. The first is that all life is mortal and therefore either the parasite or its progeny must move between hosts in order to ensure the survival of subsequent generations. Assuming that a means of transfer is available the second major problem is identifying a suitable new host. In particular, endoparasites usually only have one chance after committing to the invasion process. This means that they must identify a suitable host with a high degree of certainty, otherwise the consequences would be fatal to the parasite. Some permanent ectoparasites, such as the human body

Parasitology: An Integrated Approach, Second Edition. Alan Gunn and Sarah J. Pitt.
© 2022 John Wiley & Sons Ltd. Published 2022 by John Wiley & Sons Ltd.
Companion website: www.wiley.com/go/gunn/parasitology2

louse *Pediculus humanus corporis* can move between hosts at several life cycle stages, including the adult stage. However, for the majority of parasites, including many ectoparasites, there is a single infective stage in the life cycle. If the infective stage has to spend some time in the external environment, it usually has adaptations that enable it to survive risk factors such as high/low temperature and desiccation. Unless it is free-living, the infective stage also often has food reserves. It must also express features that enable it to recognise and invade the new host. For example, a contaminatively transmitted parasite egg cannot choose which animal consumes it. Therefore, it must recognise whether it has reached the gut of the correct host before it commences hatching. If it is not the correct host, the egg can continue its journey through the gut and after being excreted it may have another opportunity. Even when transmission relies on organism A being eaten by organism B, and therefore there is no exposure to the outside environment, the transmission stage must be capable of surviving exposure to the acidic pH of organism B's stomach and to digestive enzymes.

The simplest means of moving between hosts is to sit and wait for the host to find you. This can be seen at its most basic in contaminative transmission in which the parasite's involvement consists of surviving until ingested by a suitable host. This is a high-risk strategy, but it is obviously an effective one since it is employed by numerous protozoan and helminth parasites. These parasites usually produce vast numbers of cysts (protozoa)/eggs (helminths) because most will never be ingested by a suitable host. This approach is very energy demanding and wasteful, but this is not a problem for the parasite since the majority of its energy needs are met by the host. Despite this, many parasites have evolved mechanisms, which increase the chances of finding their host. These can include actively searching for a suitable host, modifying host behaviour, and employing the services of an intermediate host, paratenic host or vector. Some parasites can be transmitted in several different ways. For example, although the protozoan *Entamoeba histolytica* is normally transmitted in contaminated food or water, it can also be acquired through the actions of a mechanical vector, during sexual intercourse, or from medical procedures. The transmission of parasites is also affected by numerous variables such as natural environmental factors (e.g., weather), man-made factors (e.g., pollution, cultural and agricultural practices), and biological factors (e.g., behaviour of the host). Many of these factors are interlinked, and therefore, the transmission of parasitic disease is influenced by numerous biotic and abiotic factors.

8.2 Contaminative Transmission

Contaminative transmission is a passive process in which the parasite and host meet through chance. It usually occurs when a host consumes food or drinks water contaminated with the parasite's infective stage. It may also occur through breathing in or accidentally ingesting the infective stage if this becomes airborne in dust. The success of contaminative transmission is therefore heavily influenced by the abundance of the infective stage(s) in the environment. Sometimes contamination can occur in the most unexpected ways. For example, in 2007 there was an unexplained outbreak of Chagas disease among children attending a school in Venezuela. It subsequently transpired that they had become infected after consuming guava juice containing the metacyclic trypanosomes of *Trypanosoma cruzi* (Alarcón de Noya et al. 2010). Chagas disease is normally transmitted by blood-feeding reduviid bugs, so the suggestion of it being transmitted in fruit juice is at first sight nonsensical. However, the bugs feed on various wild and domestic animals, and *T. cruzi* is a zoonotic disease that can cycle both in us and many other mammals. It is now apparent that infected reduviid bugs living among the forests and feeding on wild animals defecated on the guavas. Later, the juice prepared from the guavas was not properly sterilised before being drunk by the children. Further outbreaks of *T. cruzi* associated with the consumption of contaminated fruits and sugar cane juice subsequently occurred elsewhere in Venezuela, as well as in Brazil and Columbia. Contaminative transmission of *T. cruzi* is now recognised as an important source of infection. This is concerning because there is a large market for açai palm collected in the Amazonian forest. Fruit juice which is exported has to be sterilised and there is strict quality control but much of that consumed locally is untreated. It is therefore essential that fruit and fruit products are monitored for *T. cruzi* (de Oliveira et al. 2019). This, however, will depend upon the costs and the presence of infrastructure to ensure regulations are enforced.

Contaminative transmission is a common strategy among gastro-intestinal parasites in which the infective stage is voided in the faeces of one host and subsequently contaminates the food or water of another, which may or may not be the same species. This is referred to as faecal–oral contamination and the means by which this occurs are best remembered by the mantra of the four 'f- words': faeces, food, fingers, flies. For human parasites, contamination often results from faeces (night soil) being used as a fertiliser for salad vegetables. The use of unsuitably composted human faeces as agricultural fertiliser in Europe, Asia, and South America also contributes heavily to the spread of contaminatively transmitted parasitic infections.

Contaminative transmission can also occur when the host cleans itself or grooms another animal and ingests cysts or eggs contaminating the body. We can transmit contaminative parasites by not washing our hands properly after going to the toilet or handling domestic animals, hence the reference to fingers. In developed countries, food hygiene laws are rigorously enforced to reduce the possibility of contaminative transmission, but in developing countries different standards can apply. In a survey of food handlers in Abeokuta in Nigeria, Idowu and Rowland (2006) found that 98.7% of street vendors and 92% of school food vendors were infected with one or more species of faecal–orally transmissible parasites. Furthermore, few of the vendors admitted to washing their hands after defecation and those who did wash seldom used soap as this was looked on as an unnecessary expense. Most of those surveyed considered being infected with parasitic worms to be 'normal' and attempts at getting rid of them to be a wasted effort. Lack of personal hygiene is partly the reason why the prevalence of gastrointestinal parasites is often higher in children. For the same reason, faecal–orally transmitted parasites can also be a problem among those who are senile or otherwise mentally impaired. These groups are also often incontinent, which makes maintaining personal hygiene even more difficult.

Many dung-frequenting flies (e.g., *Musca domestica*) and other insects (e.g., cockroaches) can act as mechanical vectors when they pick up parasite infective stages (particularly on their legs) whilst feeding on faeces and then transport them to the food or body of the parasite's host. Water is also a major means by which faecal–oral transmission takes place. This may be through faeces contaminating drinking water or water used to wash food. Needless to say, faecal–orally transmitted parasites are a common problem among people living in poverty who lack sanitation and access to safe drinking water. Polluted water supplies, such as wells, ditches, and springs are common sources of infection for a variety of gastrointestinal parasites, such as *E. histolytica*. Similarly, poorly maintained plumbing has led to transmission via cracked mains water supplies. This is not restricted developing countries. For example, the tap water supply of the city of St Petersburg in Russia is a notorious source of giardiasis, such that visitors are advised to consume only bottled water, which means not just for drinking, but for cleaning their teeth as well. Also, there are occasional outbreaks of *Cryptosporidium parvum* infection from the mains water in the United Kingdom – usually as a result of a pipeline cracking, poor general maintenance, or inadequate treatment of sewage.

Geohelminths and Geophagy

Helminth parasites whose infective stages are found in the soil are sometimes referred to as 'geohelminths'. This therefore includes species such as *Trichuris trichiura* and *Ascaris lumbricoides* in which the eggs are passively ingested and hookworms such as *Ancylostoma duodenale* in which the eggs hatch in the soil and the larvae actively invade the host by burrowing through the skin or are consumed in water or food. Geophagy is the term used to describe the voluntary consumption of soil or earth. It is remarkably common and occurs in most societies. It is a behavioural trait that goes under the umbrella term 'pica'. *Pica* is the Latin word for a magpie (*Pica pica*) – a bird that has a reputation for collecting bright objects for no obvious reason.

The trait is most common in children and women, especially during the later stages of pregnancy. In addition to earth, other commonly consumed materials include chalk, charcoal, cooking starch, and ice. The common factor is that they are 'crunchy' and serve no obvious metabolic purpose. Geophagy is also common among primates and many other mammals, as well as birds (Krishnamani and Mahaney 2000). It is often stated that people infected with hookworms are prone to geophagy as an adaptation to replace lost iron. This is usually supported by the observation that during the 1800s and early 1900s geophagy was often practised by poor people living in the southern states of USA, and they were commonly afflicted by hookworms. However, there is no clear link between geophagy and iron status. Furthermore, the consumption of excessive amounts of some earths is harmful and interferes with the absorption of essential nutrients, including iron.

One might expect that geophagy would pose an increased risk of contracting geohelminth infections. Although there are reports that this is the case (e.g., Geissler et al. 1998), other studies have failed to confirm it (Young et al. 2007). The lack of a clear link is mainly because those who consume earth do not do so indiscriminately. Usually particular sources of earth are chosen such as from termite mounds or from the walls of huts where there is a lower chance of parasite transmission stages occurring – although in many cases, this is probably not a conscious factor guiding choice. In addition, the earth is often dried or baked in an oven before it is consumed. Finally, of course, the infective stages of geohelminths vary in their ability to withstand environmental stress.

If they can avoid it, most herbivorous animals avoid eating food contaminated with faeces although this may be impossible at high stocking densities or when food is limited (e.g., during a drought). Some parasites have therefore evolved mechanisms that enable them to leave the faeces and contaminate the surrounding 'clean' vegetation. For example, after a proglottid of the beef tapeworm *Taenia saginata* is shed with its host's faeces it crawls away into the undergrowth, dispersing eggs onto the vegetation as it moves; these then await ingestion by a bovine host. The cattle lungworm (nematode) *Dictyocaulus viviparus* has evolved a more dramatic means of escape from the dung pile. In this species, the parasite eggs hatch shortly before or immediately after they are shed with the faeces and then develop into the infective third-stage larvae. At the same time, the fungus *Pilobolus* is also developing on the faecal pat and the nematode reaches its infective stage just as the fungus forms its fruiting bodies called sporangiophores. In order to colonise new dung-pats, the fungus relies on its sporangia being consumed by herbivores (e.g., a cow, sheep or horse): the sporangia pass harmlessly through the animal's gut and then germinate when passed with the dung. The globular sporangiophores of *Pilobolus* develop at the top of long 'stalks' and when they are ripe, they 'explode' thereby propelling the sporangia up to 2 m from the dung pat. The infective larvae of *D. viviparus* climb to the top of the ripe sporangiophores and when these 'explode' they too are catapulted into the surrounding vegetation. Once on clean vegetation, the larvae wait until they are eaten by a suitable host, and they are then able to continue with the next stage of their life cycle. By contrast, there are some animals that feed on faeces (coprophiles), and the infective stages of the parasites of these animals remain within the faeces. For example, the proglottids of the pork tapeworm *Taenia solium*, unlike those of *T. saginata*, remain immobile within the faeces they are shed in. This is because the next stage in the life cycle (usually) depends upon the eggs being consumed by a pig and pigs will eat human faeces.

The 'sit and wait' strategy can mean a lot of waiting, and therefore, the transmission stage is usually capable of withstanding environmental extremes and surviving for months or even years. The eggs and cysts often have thick walls to protect against water loss and environmental extremes, as well as attack from fungi and bacteria. They are also usually provided with abundant metabolic reserves and have a low metabolic rate. In some nematode species, the larvae hatch from the egg and then develop to the infective stage, and this is adapted to survive environmental conditions etc. For example, the infective third-stage larvae of *D. viviparus* do not feed (they rely on stored reserves) and retain the second-stage cuticle as a protective 'coat'. The eggs of the nematodes *A. lumbricoides* and *Ascaris suum* are notorious for their ability to survive in situations that would kill the majority of creatures. For example, they can embryonate in 10% formalin, 7% glacial acetic acid, and 1% hydrochloric acid (Yoshida 1920). Although the eggs of some parasite species, such as *A. suum* and *Trichuris suis*, remain infective for at least 6 years under experimental conditions, it is likely that even robust species such as these survive for much shorter periods in real life (Larsen and Roepstorff 1999).

Faeces is a wonderful resource and after being passed, it rapidly becomes the home of countless bacteria, fungi, and invertebrates. These together with the effects of the weather ensures that it is quickly broken down. In addition, birds that feed on the dung invertebrate fauna tear it apart in search of their prey. This means that even those parasite cysts and eggs that do not have an 'exit strategy' soon become dispersed into the soil and from here they may be washed into streams, rivers, lakes, or ponds. Alternatively, faeces may be deposited directly into the water. Cattle often defecate when they wade into streams or ponds to drink and people living on the banks of rivers etc. will defecate into the water if there is no sanitation. Even many developed countries discharge untreated sewage into the sea or large bodies of water either as a matter of routine or following flooding. In parts of Asia, it is common practice to add human faeces to pools containing farmed fish as a cheap food. Consequently, many sources of water can become contaminated with various faecal–orally transmitted parasites.

Contaminative Transmission of *Cyclospora* spp.

High levels of *Cyclospora* spp. contamination (3.75–11.6%) have been found on coriander and other herbs sold in markets in Viet Nam (Tram et al. 2008). Herbs and salad vegetables are a potentially good source of infection because they are often consumed raw and the opportunity for contamination can occur at many levels. For example, plants may be contaminated when they are growing through the use of human faeces as a fertiliser or faecally contaminated irrigation water. The hands of the person harvesting the plants and/or the containers used to collect and transport them may be contaminated and on market stalls the plants are usually kept fresh by being sprinkled with water, which may also be contaminated. In developing countries with poor levels of sanitation contact with soil has is an important factor in the transmission of *Cyclospora cayetanensis* (Chacín-Bonilla 2008), since flies and other invertebrates can also be expected to pick up and transmit the oocyst to food. The globalisation of the food supply may help to spread the parasite within and between distant regions/countries and the rise in air travel has probably helped spread the parasite to non-endemic regions (Li et al. 2020).

8.3 Transmission Associated with Reproduction

8.3.1 Sexual (Venereal) Transmission

Sexual intercourse is an exceptionally dangerous activity. For example, during intercourse the participants become vulnerable to attack because they focus upon one another rather than their surroundings and their ability to move and defend themselves is compromised. Consequently, in most animals the physical act is usually extremely brief. Sexual intercourse also presents an ideal opportunity for disease transmission because two hosts are in close physical contact. Furthermore, penetrative sex and the exchange of body fluids mean that endoparasitic and intracellular species can be transmitted without exposure to the environment. They therefore do not need special transmission stages in their life cycle. Sexual transmission does, however, pose difficulties of its own. One of these is that it is dependent upon the sexual behaviour of its host. If the host breeding season is short and the animals are not promiscuous, then there are few transmission opportunities for the parasite. For example, the male honeybee *Apis mellifera* only mates once and then he dies. In addition, as already mentioned, the sexual act may be brief and therefore the parasite must be in a position to move quickly when the opportunity arises. Furthermore, because sexual reproduction is vital to the host and the risks of sexually-transmitted infections (STIs) so real, most animals have well-developed immune mechanisms to combat pathogens that attempt to establish themselves via the reproductive organs. For some parasite species, sexual transmission is the principal means of moving between hosts but for many others copulation is just another transmission opportunity.

Pathogen transmission during sexual intercourse usually occurs through one of the following ways: actively moving between the bodies, transmission within the reproductive fluids, and transmission through tissue damage caused during copulation. Although transmission can occur in both directions, the male tends to be the one transmitting the infection, and the consequences of the infection are often worse for the female.

Whenever two or more animals come into physical contact it presents an opportunity for ectoparasites to move between hosts. For social animals, contact may be a regular occurrence, but those living more solitary lives may only come into contact with one another when they need to reproduce. Adaptations for exploiting copulation (as distinct from other forms of contact) are more likely to evolve the more frequently such acts take place. For example, the crab louse *Pthirus pubis* clings onto the thick coarse hairs around our genitalia and in the process of evolution it has gained grasping strength but lost the ability to move quickly. Although crab lice will attach to other regions of the body where there are coarse hairs (e.g., eyebrows, beards), they are primarily transmitted during sexual intercourse when the genital hairs are in close contact. By contrast, the human head louse *Pediculus humanus capitis* is also adapted to grasping onto hair but it has retained sufficient mobility to move between hosts during brief moments of physical contact. Therefore, although the head louse could move between participants at the time of copulation it should not be considered an STI.

Most sexually transmitted parasites are protozoa, the best known of which are *Trichomonas vaginalis* and *Trichomonas foetus*. *Trichomonas vaginalis* cannot form cysts and although it can survive for up to 24 hours outside our body (in urine, semen, and water) contaminative transmission – such as through shared towels or toilet seats – is not considered a common occurrence. It infects both men and women and is typically transmitted within the genital secretions. Because men are often asymptomatic, it can be difficult to persuade them to be tested and/or treated for the infection. However, unless they are tested and treated, they will constantly re-infect their sexual partner(s).

Trichomonas foetus was once of major economic importance in the cattle industry and especially on dairy farms. The infection is relatively benign in bulls but because they can transmit it to cows, they cannot be used for breeding purposes. It is extremely difficult to completely cure an animal of *T. foetus* and therefore because stud bulls can be worth tens or even hundreds of thousands of pounds, their loss can have serious financial implications. The increasing use of artificial insemination has led to a decline in the prevalence of *T. foetus* infections although it remains an important cause of disease in some countries.

Human sexual behaviour encompasses a remarkable variety of activities. Some of these provide the opportunity for the transmission of parasites that might not otherwise be considered STIs. For example, the protozoan *Entamoeba gingivalis* normally inhabits our mouth but it has also been recovered from the uterus – a circumstance the reporting authors conjectured arose from an unfortunate combination of oral sex, an intrauterine device, and an existing bacterial infection (Clark and Diamond 1992). Anal intercourse and anilingus (mouth–anal contact) facilitate the transmission of those gastro-intestinal parasites that produce cysts or eggs that are immediately infective. Protozoan trophozoites may also be transmitted in this way although they are not usually considered a transmission stage. It may enable the parasites to colonise regions of the body where they would not normally be found. This is partly through faecal contamination of the genitalia but also because anal intercourse can damage the lining of the rectum and the resulting bleeding facilitates the transmission of blood and tissue-dwelling parasites. For example, men may acquire cutaneous *E. histolytica* infections of the penis following anal intercourse through trophozoites colonising pre-existing lesions (Parshad et al. 2002). In non-endemic countries, there are increasing numbers of sexually-transmitted *E. histolytica* infections among men-who-have-sex-with-men (Escolà-Vergé et al. 2017) although heterosexual transmission also occurs (Billet et al. 2019). Sometimes a combination of mechanisms is involved during sexual transmission. For example, *Leishmania* spp. are normally transmitted by sandflies but in dogs, venereal transmission can occur through skin lesions on the penis and amastigotes may occur in the semen (Teichmann et al. 2011).

Most mammals acquire the protozoan *Toxoplasma gondii* by consuming flesh containing the tachyzoites/bradyzoites or by ingesting the oocysts passed by cats through contamination. In several male mammals, the parasite is present in their reproductive organs and semen and sexual transmission occurs in sheep, goats, and dogs (Lopes et al. 2013). The parasites are also present in the semen of *Toxoplasma* positive men, and there is a suggestion that oral sex (via fellatio) with such men might present similar risks of infection to eating raw infected meat (Kaňková et al. 2020). At the time of writing, it was uncertain whether the release of *Toxoplasma* in semen is a widespread phenomenon in infected male mammals and whether it represents a common source of infection. Interestingly, it does not occur in cats, which are the definitive hosts for *T. gondii* (Teixeira et al. 2017). People have sexual relations with other animals far more frequently than many of us would like to consider (Munro and Thrushfield 2001). This activity increases the chances for the transmission of zoonotic infections, but, hardly surprisingly, there is limited published information on the topic (e.g., Ergun et al. 2007; Khorvash et al. 2008).

Humans are not the only animals to indulge in risky sexual behaviour. For example, several invertebrates, of which the bed bug *Cimex lectularius* is the best known, reproduce by traumatic insemination. In this case, the male bug stabs the female's abdomen with his *aedeagus* and injects his spermatozoa into her haemolymph. These then migrate to her oviducts where they fertilise the eggs. The female has functional genitalia, but these are only used for laying the eggs. She has an energetic cost associated with wound repair and in a worst-case scenario, the wound can be infected by bacteria with potentially fatal consequences. This form of reproduction therefore offers no obvious benefits to the female. Several other bugs belonging to the family Cimicidae also reproduce by traumatic insemination. Some of these are vectors of various bacterial and protozoan diseases, including *T. cruzi*. It is feasible that an infected male might transfer pathogens to an uninfected female during reproduction, but there is little published information on this.

There are few examples of helminth infections that are sexually transmitted. A rare exception is the nematode *Mehdinema alii*, which infects the 'decorated' or 'Indian house cricket', *Gryllodes sigillatus* (Luong et al. 2000). The adult worms live within the hind gut of the male cricket and form infective *dauer* larvae that migrate to the insect's genital chamber. The larvae are passed to the female during mating and she in turn can then transmit these to other males. Although the nematodes may subsequently reach maturity in the female crickets, this is not thought to happen frequently and the male is the principal host.

8.3.2 Transmission within Gametes

Gametes are unicellular, and therefore usually only intracellular parasites can be transmitted within them. The majority of male gametes, especially in mammals, are very small and contain little cytoplasm, which limits the space available for any parasite. Indeed, the evolution of small male gamete size may be driven by the host's attempt to control the transmission of parasites within them. There are, of course, exceptions. For example, the males of some species of the fruit fly *Drosophila* produce sperm that are over ten times their own body length. Another major factor preventing organisms from parasitising spermatozoa is that sperm competition is intense, so anything that compromises their movement would reduce their chances of fertilising the egg and thus passing on the disease to offspring. It is therefore not surprising that there are few records of spermatozoa being parasitised, and this is usually with viruses or rickettsia-like organisms (Afzelius et al. 1989). Intriguingly, the outer surfaces of the spermatozoa of soft ticks are coated with protists belonging to the genus *Alderocystis*. The relationship is thought to be symbiotic and *Alderocystis* may provide a nutritive function or prevent the sperm being damaged by bacteria (Perotti and Braig 2011). *Alderocystis* is also found in hard ticks but has not yet been observed in direct contact with the sperm. By contrast to spermatozoa, female gametes are often large and a number of pathogenic and non-pathogenic symbionts are transmitted within them. For example, *Wolbachia*, a rickettsia found in many invertebrates, is typically transmitted via the cytoplasm of the egg. Transmission within the gametes is an example of vertical transmission and the reproductive success of the parasite and its host are therefore closely linked.

8.3.3 Congenital Transmission

Congenital transmission is an example of vertical transmission. It refers to transmission that occurs through infection of the developing egg or embryo while it is still within its mother's body or at the time of birth/egg-laying. In invertebrates, it is often referred to as trans-ovarial transmission. In mammals, the infection is either acquired across the placenta (trans-placental transmission) or through passage through the birth canal or contact with the mother around the time of the birth (perinatal transmission). The end result is the same in that the young are born (or hatch) already infected with the parasite. Although vertically transmitted parasites are often portrayed as less pathogenic than those that are horizontally transmitted, this is not always the case and especially so if congenital transmission is not the normal means of transmission. Indeed, because the infection occurs during an exceptionally sensitive time during the host's development, congenital transmission can have fatal consequences. For example, the protozoan *Neospora caninum* can cause abortion epidemics in cattle.

The most common parasite associated with human congenital infections is the protozoan *T. gondii*. Human congenital toxoplasmosis usually results when the mother becomes infected with *T. gondii* for the first time during her pregnancy. It can also result from the reactivation of a dormant *T. gondii* infection although in this case the mother is often immunosuppressed – for example, as a consequence of systemic *lupus erythematosus* and AIDS. Intriguingly, Šárka Kaňková and her co-workers have demonstrated that women who are latently infected with *T. gondii* give birth to more sons than those who are not infected (Flegr and Kaňková, 2020; Kaňková et al. 2007a). Her work was the first to demonstrate that a parasite can influence the sex ratio of human babies. However, this is relatively common occurrence in invertebrate host–parasite relationships. Whether the parasite directly influences the sex ratio of human babies or a third factor – such as changes to the immune system or to the balance of hormones – is not known. *Toxoplasma* certainly influences the hormonal balance in its hosts: for example, in male rats it causes a raised level of testosterone (Tan and Vyas 2016). Kaňková et al. (2007b) found that female mice in the early phase of latent *T. gondii* infection (i.e., gave birth 89–120 days after infection; mice usually give birth ~21 days after conception) also produced more male offspring than uninfected controls. Male mice tend to move further away from their nest than females and also tend to be more active in their search for mates. Consequently, they are more effective at dispersing the parasite and are more likely to be encountered and consumed by cats. However, pregnant mice in the later phase of infection (i.e., gave birth >120 days after infection) produced more females than uninfected controls. They suggested that this was because these pregnant females would be in poorer physical condition because their infection was of longer standing. According to the Trivers-Willard hypothesis (Trivers and Willard 1973) in species in which females are not in serious competition with one another for food and/or mates, it is better for a female in poor condition to produce more female offspring than males. This is because even if the offspring is small and weak, if it is female it is likely to be able to reproduce. By contrast, a male that is small and weak and has to compete with other males is unlikely to reproduce.

Congenital toxoplasmosis can also be a problem in many other wild and domestic mammals including moose, dogs, cats, pigs, sheep, and goats (Bresciani and da Costa 2018). In sheep, as in human infections, it can cause spontaneous abortion, stillbirth, or the lamb is born weak and unable to thrive (dos Santos et al. 2018). It has been suggested that sheep may become infected through consuming concentrate feed that has somehow become contaminated with cat faeces. Sheep are usually fed supplements before mating ('tupping') and also before lambing, and this would fit in with the subsequent congenital infections. However, cats normally bury their faeces and food concentrate is stored in sacks. Although it could be argued that flies etc. might transport oocysts between cat faeces and the sheep concentrate, sheep usually mate in the autumn and give birth in early spring – when insect activity is reduced. As mentioned above, *T. gondii* could also be being spread more commonly than previously thought within the semen of infected rams (Lopes et al. 2013).

Congenital transmissions of malaria caused by *Plasmodium falciparum* and *Plasmodium vivax* can occur (Del Castillo et al. 2017; Tekle et al. 2018). However, their prevalence varies enormously between surveys from 0% to 37% (Uneke 2007). There are occasional case reports of congenital transmission of various trypanosome parasites, including *Leishmania* spp. (Argy et al. 2020), *Trypanosoma brucei gambiense* (De Kyvon et al. 2016), and *Trypanosoma cruzi* (Parra-Henao et al. 2019). However, the frequency of congenital Chagas disease is probably underestimated (Howard et al. 2014). This is because asymptomatic women can transmit *T. cruzi* to their babies and the baby may not display any immediate signs of infection. Unfortunately, it is not possible to treat a pregnant woman for *T. cruzi* because the drugs are potentially teratogenic. Screening pregnant women is therefore important to ensure prompt treatment of the child after birth. These congenital infections can be problematic when the birth occurs in non-endemic countries because the medical staff might not expect these parasites as the cause of a new-borne child's ill health.

Dogs are the normal definitive hosts for the roundworm *Toxocara canis* and transplacental transmission of the larvae to the developing pups is a normal part of the life cycle. Although humans can be infected by the larvae, they are unable to reach adulthood and encyst within our tissues as second-stage larvae. There is an isolated report of them achieving transplacental transmission in humans (Maffrand et al. 2006). *Ascaris lumbricoides* is not associated with congenital transmission but it is possible that larvae might achieve this if they stray from their normal migratory route (Costa-Macedo and Rey 1990). Transplacental infection by nematode *Trichinella spiralis* occurs in several mammals but whether it is possible in humans is uncertain (Saracino et al. 2016). A survey by Gomez-Puerta and Mayor (2017) found that 10 out of 31 Red Brocket Deer (*Mazama americana*) foetuses were infected by the filarial nematode *Setaria bidentate*. Furthermore, many of these worms were already adults. This suggests that congenital transmission is common in this species. By contrast, there are only isolated case reports of congenital transmission of filarial nematodes in humans and these all involve finding microfilariae in skin snips taken from the new borne child. For example, *Wuchereria bancrofti* (Bloomfield et al. 1978) and *Onchocerca volvulus* (Brinkmann et al. 1976).

Transovarial transmission of parasitic infections is relatively common among invertebrates. In some situations, this can have relevance to human and veterinary medicine. This is because several of the diseases that are transmitted by ticks (e.g., *Babesia bovis*) are passed between mother tick and offspring in this way (Antunes et al. 2017). Consequently, the disease can rapidly spread through the tick population and the young ticks may be capable of transmitting an infection when they take their first blood meal.

The protozoan parasite *Histomonas meleagridis* is unusual in exploiting another parasite's reproduction to ensure its own transmission. Infected birds pass *H. meleagridis* trophozoites in their faeces but these cannot survive for more than a few hours in the outside world. Despite this, direct bird-to-bird transmission (presumably via contaminative transmission) occurs in turkeys (McDougald and Fuller 2005). More typically, infection takes place when *H. meleagridis* becomes incorporated into the eggs of the common caecal nematode parasite *Heterakis gallinarum*. This occurs whilst both are still in the infected bird's gut. The protozoa initially parasitise the gut of the nematode but then invade and multiply within its body tissues. Once within the ovaries, the protozoa become assimilated within the developing nematode eggs. Male nematodes are also infected and the protozoa can invade their reproductive system. It is therefore conceivable that males might transmitted them as an STI to the female at copulation. *Histomonas meleagridis* does not invade the spermatozoa but they could be transmitted in the secretions that accompany them. Although the protozoa continue to reproduce within the nematode eggs, they do not kill the developing larva – indeed, the eggs (and the protozoa within them) can remain infective for at least 2 years. Interestingly, if the *Heterakis* eggs are consumed by earthworms, they will hatch and the second-stage larvae invade the earthworm's tissues where they become dormant. When the nematode eggs or an infected earthworm are consumed by a suitable bird host, the eggs hatch or the juvenile nematodes are released in the bird's gut. The nematodes then make their way to the bird's caecum. Once there, the protozoa, which are present in the nematode gut and reproductive tissues, are shed with its faeces and they then invade the bird's tissues.

8.4   Autoinfection

Autoinfection occurs when a parasite produces infective stages that colonise the same individual host as their parent(s). Because the offspring remain within or upon their host, they do not have to overcome all the biological and environmental factors that normally reduce the chances of transmission. However, parasites often induce an immune reaction in their hosts that prevents or limits the ability of other parasites of the same species from establishing themselves. In addition, the parasite must ultimately infect other hosts if it is to spread and thereby ensure its survival as a species.

The intestinal helminth *Enterobius vermicularis* (the 'pin worm') provides a good example. This is an exceptionally common human parasite in both developing and developed countries; prevalence rates can reach 100%. The worm emerges from our body during the night to lay its eggs on the skin around the anal area. Infective larvae develop inside these eggs within a few hours. The eggs cause the skin to itch, so on waking, the infected person scratches their bottom, which transfers the parasite to their fingers. Therefore, if the person puts their fingers straight into their mouth or onto food before washing their hands, they re-infect themselves. Pin worm infections are commonly found in children because they are less likely to consider hand hygiene before sucking their fingers. Furthermore, they are more likely than adults to suffer frequent and repeated autoinfection. Infection may also occur through retroinfection in which the eggs hatch on the anal mucosa and the larvae crawl back up the colon. Contaminative transfer may also occur from soiled bed linen and clothing.

The life cycle of most helminth parasites makes autoinfection unfeasible. For example, many nematode parasites (e.g., *A. lumbricoides*) produce eggs that do not reach their infective stage until after they are shed in the faeces, whilst others require a period of larval development in an intermediate host or vector (e.g., *W. bancrofti*). A notable exception is the nematode *Strongyloides stercoralis*. This worm has an unusually complex life cycle that involves free-living and parasitic cycles with adult reproduction taking place in both of them (Figure 6.4). The adult female worm lives in our small intestine – there are no males in the parasitic part of the life cycle. She produces eggs that hatch soon after they are laid and the larvae develop so rapidly that they reach the infective third stage whilst still in the gut. Some of these larvae are shed with the faeces and some penetrate the gut, enter the circulation, and spread throughout the body. Some of these larvae complete their migration and return to the small intestine where they mature into adult females that give rise to further rounds of autoinfection. Repetitive autoinfection can result in a person remaining infected for many years (Nutman 2017). For example, men who were infected with *S. stercoralis* whilst they were Japanese prisoners of war during WW2 sometimes only became aware of their infection after several decades. In many cases, after the war, these men returned to their home country where there was no risk of infection However, as they became increasingly elderly and their immune system was weakened because of disease or drug treatment, they became susceptible to disseminated strongyloidiasis.

8.5   Nosocomial Transmission

Nosocomial infections are those that are acquired in hospital or as a result of a medical procedure. Hospitals can be unhealthy places as they contain lots of sick people within a very confined space. It is therefore not unusual for patients to acquire an infectious disease whilst being treated for a different condition. The dangers posed by hospital acquired bacterial (e.g., MRSA) and viral (e.g., H1N1) infections are well known, but some parasites can also cause problems. For example, unless hygiene is rigorously enforced there is the potential for the contaminative transmission of many parasitic diseases – particularly gastro-intestinal parasites. This can be a particular concern in countries where a high proportion of the patients are infected with parasites before they arrive at the hospital and they become a source of infection to others. There is also the potential for the transmission of ectoparasites such as lice and mites. For example, even in a developed country such as Japan, nosocomial transmission of scabies (*Sarcoptes scabiei*) has been reported as a problem (Furuya et al. 2016).

The re-use of needles and poor sterilisation of hospital equipment can result in the transmission of several parasitic diseases. A bizarre event occurred in a clinic in Colorado in 1980, when an epidemic of amoebiasis occurred among patients who had undergone colonic irrigation with a contaminated enema machine. Ten of the patients had to have colectomies and seven of them died. Blood transfusions and organ transplants are a potential source of transmission for parasitic infections such as malaria, toxoplasmosis, leishmaniasis, Chagas disease, and strongyloidiasis (Blackburn and Montoya 2019). It is therefore important that blood and tissues from donors are screened for parasitic infections before they are used to treat patients. Nowadays, people travel extensively and therefore one cannot assume that a blood or organ donor would not be infected with an exotic parasite. For example, vector-borne Chagas disease is rare in the USA, but in 2007 a new blood-screening test revealed that a significant number of blood donors were antibody positive for *T. cruzi* and their blood could have been a source of infection if used for a transfusion. Case reports of *T. cruzi* infections via blood transfusions are even reported in European countries including Switzerland (Ries et al. 2016) and Belgium (Blumental et al. 2015). That is, not only are they non-endemic countries but they are on a different continent.

Patients with open wounds and those who are comatose are vulnerable to myiasis – infection by the larvae of blowflies and fleshflies. When the patient is intubated with an endotracheal tube, they are particularly vulnerable to oral myiasis and cases even arise within intensive care units (e.g., Martínez-Rojano et al. 2018). A fly can easily make is way inside a room without being seen by flying through a temporarily open door, by sitting on a trolley as it is wheeled around, or even on someone's clothing.

8.6 Active Parasite Transmission

In some parasite species, there is a free-living infective stage that actively searches for and invades its host. Schistosomes have two free-living infective stages in their life cycle. The schistosome egg hatches to release a free-living stage, the miracidium, which searches for and invades the snail intermediate host. After a series of developmental stages in the snail, another free-living stage, the cercaria, is released. The cercaria searches for and invades the final host within which it becomes an adult worm. These different stages of the life cycle are marked by dramatic changes in gene expression patterns. For example, the cercaria stage is characterised by an upregulation of genes responsible for energy production, movement (e.g., actin), and the protease enzymes that play a part in the invasion process (Hagerty and Jolly 2019).

The success of the active host searching strategy depends upon the presence of a reliably high density of potential hosts. This is because searching is more energy demanding than 'sitting and waiting' and the transmission stage in a parasite's life cycle does not usually feed. Therefore, once the infective stage runs out of energy it loses the capacity to invade and will die. For example, the schistosome miracidium survives for only a few hours after hatching. Because the chances of successful transmission are so low, parasites usually produce huge numbers of eggs. If the host density was low or difficult to predict, packing their eggs with energy reserves sufficient to fuel an active free-living stage would reduce the rate of egg production without necessarily significantly increasing the chances of transmission. An alternative is therefore to have a feeding free-living stage. For example, hookworm eggs hatch to release free-living larvae that feed on bacteria before moulting to the infective stage. In some hookworm species, the infective stage may survive for 1–2 weeks under ideal conditions. However, this transmission strategy also requires a warm, moist, or aquatic environment to avoid the danger of desiccation and exposure to UV light.

The actively invading infective stage must have some means of identifying a suitable host. Schistosome miracidia are attracted by components of the mucus of their snail intermediate host, whilst their cercariae respond to the thermal gradient created by their warm-blooded host. The host identification capabilities of schistosomes are not particularly sophisticated and both the miracidia and cercariae will attempt to invade unsuitable hosts. For example, there is a report of *Schistosoma mansoni* miracidia attacking tadpoles. It has been suggested that it might be possible to control schistosomes using 'decoy snails' that the miracidia invade but within which they are unable to develop. This could be done either by promoting the growth and reproduction in an existing snail species in the relevant area or introducing a new one. However, the decoy may have to be present in very large numbers to be effective.

Both the miracidia and the cercariae of schistosomes have glands containing digestive enzymes and other chemicals and they physically and chemically force their way into their host. However, having gained entry they then require the correct environment to continue their development and without this they die.

Paddling or swimming in open water sources sometimes carries the risk of being invaded by the cercariae of schistosome species that usually parasitise ducks and other aquatic birds. These are often various species of *Trichobilharzia*. These cercariae die shortly after invading our body but the parasites and their secretions trigger an inflammatory response that manifests as a rash, popularly known as 'swimmer's itch' (Tracz et al. 2019). This rash may persist for up to a fortnight and in sensitive individuals can give rise to fever. The surface tension provided by a slowly drying film of water facilitates invasion by cercariae. Consequently, one way of reducing the risk of swimmer's itch is to rapidly dry oneself after bathing in schistosome-infested waters. In Europe, there is currently an increased interest in wild swimming in lakes and rivers and therefore more people are at risk of developing swimmer's itch.

The infective stage larvae of hookworms are attracted by factors such as temperature, carbon dioxide, humidity, and certain chemical components in the skin (Bryant and Hallem 2018). They invade their host using a combination of enzymatic and chemical secretions and physically force their way into its body. In us, invasion with human-adapted hookworm species usually results in no reaction or the formation of a rash at the site of invasion. However, invasion with non-human adapted species (e.g., *Ancylostoma caninum*) causes cutaneous *larva migrans* in which vivid red serpentine tracks underneath the skin indicate the path of the larvae. These larvae ultimately die before they develop any further.

8.7 Hosts and Vectors

8.7.1 Paratenic Hosts

Paratenic hosts are not usually essential for the transmission of a parasite, but they do improve the chances of it taking place. For example, the larvae of certain *Ancylostoma* species (e.g., *Ancylostoma tubaeforme*, *Ancylostoma braziliense*) sometimes invade rodents but they are unable to complete their lifecycle. The invading larvae enter arrested development (hypobiosis) within the tissues of the rodent and can continue developing to adulthood in a dog or cat should it consume the infected rodent (Norris 1971). The transmission cycle can theoretically become even more complicated since the infectious third-stage larvae can also survive in insects and a rodent may become infected if it consumes an infected insect (Little 1961).

In the case of the mermithid nematode *Pheromermis vesparum*, the paratenic host plays a much more important role in the parasite's life cycle (Molloy et al. 1999). Like most mermithid nematodes, the larval stages develop within their insect host, which in this case are social wasps (Hymenoptera: *Vespidae*). Once they have completed their development, they physically cut their way out of their host, usually killing it in the process. This emergence takes place when the adult wasp is next to a stream or pond. Some mermithid nematodes cause a change in the behaviour of their hosts, which results in them being attracted to and falling into bodies of water. The larvae of *P. vesparum* then enter the water, moult to the adult stage, mate, and the female worm starts to release eggs. If the eggs are consumed by the larval stages of aquatic insects such as caddis flies (Trichoptera) and stone flies (Plecoptera), they will remain within them even after the insect has moulted to their adult stage. The nematode's life cycle is completed when the adult aquatic insect is captured by an adult wasp and fed to a wasp larva. The mermithid eggs hatch in the wasp larvae and begin to develop. The mermithid larvae develop slowly within their host such that the wasp can complete its own development to adulthood and be capable of flight. The aquatic insects therefore act as paratenic hosts, enabling the transmission of the mermithid eggs to the wasp host. They also facilitate the dispersal of the parasite when they themselves emerge and disperse from the pond or stream.

8.7.2 Intermediate Hosts

Many parasites have complex life cycles involving two or more species of host within which different developmental stages occur. The intermediate hosts are those within which some (or all) of a parasite's immature (non-sexual) stages occur. The intermediate host is therefore essential, since without it the parasite would be unable to complete its life cycle. Consequently, targeting the intermediate host may prove to be the most effective control strategy. Similarly, the absence of a suitable intermediate host can prevent a parasite from extending its geographical range.

In some cases, such as Schistosomes and the liver fluke *Fasciola hepatica*, the parasite leaves the intermediate host after a period of development. Many parasite species, however, rely upon the intermediate host being consumed by the next host in its life cycle: this may be another intermediate host or the definitive host. In this situation, transmission is determined by predator-prey dynamics. Being parasitised can harm the host's health and thereby increase the chances of it being captured and consumed. For example, the host may become weak and therefore less able to detect or avoid a predator. This is a 'by-product' of being parasitised and cannot be considered a specific adaptation on the part of the pathogen to increasing the chances of transmission. Nevertheless, parasites sometimes induce changes in the morphology, physiology, or behaviour of their hosts that appear to have evolved specifically to enhance the chances of transmission taking place. However, distinguishing evidence of evolution from happen-chance is not always easy. For example, the coenurus (bladderworm) of the tapeworm *Taenia multiceps* causes a condition that is colloquially known as 'gid' in which infected sheep exhibit a characteristic high stepping gait and walk around in circles. Such sheep would be easy for dogs, the tapeworm's final host, to catch. 'Gid' results from the parasite invading the brain and forming a large fluid-filled coenurus that causes pressure atrophy and localised blockage of the blood supply, thereby bringing about the death of brain cells. One could suggest that this is an evolutionary adaptation to enhance transmission, although it could also simply be a fortuitous consequence of the coenurus developing in the brain.

8.7.3 Vectors

Some parasites use another animal, called a vector, to physically transmit them between hosts. Vectors therefore play an important part in their epidemiology and by controlling the vector one can often control the spread of the disease. Excellent comprehensive reviews of the role invertebrates play in vector-assisted transmission of parasites are provided by Lehane (2005) and Mullen and Durden (2002).

In some cases, the parasite-vector relationship is a loose association in which the vector acquires the parasite by accident and carries it to a point at which it can infect the host. For example, when the vector transports the infective stage of the parasite from faeces and deposits it on the host's food. This can therefore be looked on as a form of contaminative transmission and the animal doing the transporting is referred to as a mechanical vector or mechanical transmission host. Importantly, the parasite does not undergo development during its transport and is usually carried upon the vector's outer body surface or passes through its gut. A mechanical vector is therefore distinct from a paratenic host in which the parasite invades the host's tissues.

Flies, such as the housefly *Musca domestica* and the blowfly *Calliphora vicina*, and cockroaches (e.g., *Periplaneta americana*) are important mechanical vectors of various gastrointestinal infections, including parasites such as *E. histolytica*. Their bristly appendages can easily carry protozoan cysts or helminth eggs from faeces to the host's food. Flies may also ingest them, so their habit of vomiting and defecating while feeding is another means of transmission. In many rural areas it is not unusual for dogs, chickens and pigs to wander about farmyards and villages, moving freely in and out of buildings. Chickens will root through faeces looking for invertebrates and undigested plant material, whilst pigs are coprophagic. Dogs will also eat faeces if they are hungry – and stray dogs usually are. Therefore, unless there is good sanitation, chickens, dogs, and pigs will consume human faeces as well as that of other animals. Faeces may also contaminate their outer body surface. Dogs can act as mechanical vectors for many human gastrointestinal parasites, such as *A. lumbricoides*. Although a dog might ingest protozoan cysts such as those of *E. histolytica* through consuming faeces, it is uncertain whether the parasite can survive and remain infective following passage through a canine intestinal tract. Similarly, unembryonated eggs of *A. suum*, *T. suis*, and *Oesophagostomum dentatum* eggs can travel through the guts of pigs and chickens and thereby be further dispersed around the local environment (Olsen et al. 2001).

Blood-dwelling parasites face a particular problem in moving between hosts because they must traverse several layers of tissues both to leave their existing host and gain entry to another one. Consequently, the majority of them employ a vector that is usually a blood-sucking arthropod such as a mosquito, reduviid bug, or leech. In some cases, the vector acts as a 'dirty syringe' and simply moves the parasite directly between hosts. That is, the blood-feeding invertebrate serves as a mechanical vector. For example, the transmission of *Trypanosoma evansi* by tabanid flies. However, in many cases the parasite passes through a series of developmental stages in its vector. In this situation the vector acts as both transmission agent and intermediate host (e.g., *P. falciparum* in the mosquito *Anopheles gambiae*). It is usually a very specific relationship and even strains of the same species of vector may differ in their ability to transmit a particular parasite. For example, numerous factors influence the tsetse fly: trypanosome relationship including the strain of the trypanosome, the species and sex of the tsetse fly (infections establish more readily in males than females) and the presence of and strain of symbionts in the fly.

The Role of Symbionts in the Life of Tsetse Flies and Their Transmission of Trypanosome Parasites

Tsetse flies are important in parts of Africa as vectors of certain trypanosome (protozoan) parasites of us (e.g., *Trypanosoma brucei gambiense*) and domestic cattle (e.g., *Trypanosoma brucei brucei*). Like most other blood-feeding organisms, tsetse flies harbour bacterial symbionts that facilitate the breakdown of the blood meal and provide essential nutrients to the fly (Rio et al. 2016). In the case of tsetse flies, these are principally B group vitamins, vitamin H, folic acid, and pantothenic acid. In the absence of the symbionts, the female tsetse fly cannot reproduce. The symbionts are transmitted from the female fly to her larvae as they develop in her uterus. Tsetse flies have an unusual life cycle in which the larvae develop to maturity within the uterus of the female fly. Therefore, the female fly 'lays' a larva that immediately burrows into the ground and pupariates. This mode of reproduction is called 'adenotrophic viviparity' and it limits the exposure of the developing larvae to external microbes.

Three different symbiont species live within tsetse flies. *Wigglesworthia glossinidia* is an essential primary symbiont whilst the other two, *Sodalis glossinidius* and *Wolbachia* are secondary non-essential symbionts that are nevertheless relatively common. *Wigglesworthia glossinidia* and *S. glossinidius* infect the developing larva via the 'milk' that supports its development in the uterus. A male tsetse fly can also infect a female fly with *S. glossinidius* during copulation. This indicates that there are relatively limited opportunities for the mixing of bacterial genomes. As in many other insects, the *Wolbachia* infect the young through the ovaries and can cause cytoplasmic incompatibility: this means that if a male and female tsetse fly harbour different *Wolbachia* strains, then their offspring will not develop.

The relationship between *W. glossinidia* and the tsetse fly probably began 50–80 million years ago. The bacteria produce several B-vitamins and their presence is essential for the survival of the fly. By contrast, the relationship between *S. glossinidius* and tsetse flies is much more recent and enigmatic. In at least some circumstances, *S. glossinidius* may influence the establishment of trypanosomes in the tsetse fly. Tsetse flies have an effective immune system that protects them from invading microbes. This includes the production of lectins that attach to and kill the invading organisms and toxic reactive oxygen species such as superoxide and hydrogen radicals. However, *S. glossinidius* releases *N*-acetylglucosamine that interferes with the activity of the lectins and scavenges reactive oxygen species, thereby allowing the trypanosomes to establish. Despite this, some surveys of wild tsetse flies find no association between the presence of *S. glossinidius* and trypanosome infection. It is possible that there are differences between strains of *S. glossinidius* in the production of *N*-acetylglucosamine and this may be (to a greater or lesser extent) the reason that there are differences in the susceptibility of tsetse flies to infection with trypanosomes.

The effectiveness of a vector as a transmission agent is heavily determined by its feeding behaviour: that is, what it feeds on, where it feeds, and when it feeds. Vectors that preferentially feed upon humans are said to be 'anthropophilic', whilst those that prefer to feed on other animals are classed as 'zoophilic'. Obviously, anthropophilic species are more likely to be important vectors of human parasitic diseases. If the vector is willing to enter buildings, it is said to be 'endophilic' and this also increases the chances of it transmitting an infection to us. Although some blood-feeding invertebrates are relatively catholic in their choice of victim, most have a restricted host-range. Even within a single species, different strains often show different host preferences. For example, some strains of the bed bug *Cimex lectularius* prefer humans and others prefer rabbits. This is important because the vector-borne parasite relies on the vector to make the 'correct host choice'. Unless the parasite has a wide host range, if it is injected into the 'wrong host' it will die. Similarly, not all hosts of the same species are equally attractive: whether or not we are bitten by mosquitoes depends upon numerous variables including our sex, age, blood group, pregnancy, physical health, and smoking habits (Lehane 2005). Obviously, if the vector does not bite a particular host, then it will not transmit any parasites to it. This can all change if the invertebrate is starving. The flea *Xenopsylla cheopis* normally feeds on rats and in the process can transmit bubonic plague (*Pasteurella pestis*). Plague is fatal to rats and once their population declines the fleas start to feed on us and in the process initiate a human plague outbreak. *Leishmania* spp. promastigotes move to the head and mouth parts of infected sandflies, thus making it physically difficult for the vector to take and digest a complete blood meal. This means that an infected sandfly bites the mammalian host more frequently than an uninfected insect would – thus enhancing the opportunities for transmission of the parasite. Sometimes a parasite exhibits behaviour patterns that increase the likelihood of it being transmitted by a particular vector. For example, over much of its range, the microfilariae of the nematode *W. bancrofti* are only found in the peripheral circulation of its human host late at night – usually between the hours of 10 p.m. and 2 a.m. This probably relates to this being the peak of feeding activity of its principal mosquito vector(s) and outside this time interval the microfilaria retreat to the pulmonary capillaries and other blood vessels deep within the body. However, in some areas, the microfilariae either do not exhibit periodicity or there is a peak during daylight hours, and in these situations, the microfilariae are always present within the peripheral circulation. This probably relates to the biting behaviour of the local mosquito species or strains.

An interesting example of the mechanisms determining how a parasite may become associated with a particular vector is that between *Leishmania* and sandflies. When small numbers of *Leishmania major* promastigotes are artificially inoculated into a suitable vertebrate host they fail to establish an infection. In contrast, the same numbers of promastigotes can establish when injected along with salivary extracts from the sandfly vector. The reason is that the sandfly saliva contains substances that facilitate the establishment of the parasite.

In common with other blood-feeding insects, sandfly saliva includes various vasodilators and anticoagulants. These, along with other chemicals, may contribute to the recruitment of neutrophils and macrophages (potential host cells) to the bite site. Neutrophils are the most numerous of the mononuclear phagocytes and are found within both the circulating blood and the tissues. *Leishmania* can invade neutrophils, and this is probably aided by the presence of endonuclease and hyaluronidase enzymes within the saliva that destroys neutrophil extracellular traps (Chagas et al. 2014; Martin-Martin et al. 2018). Following stimulation, neutrophils extrude long fibrils composed of protein and DNA. Pathogens stick to these fibrils and are then killed without being phagocytosed. After invading a neutrophil, *Leishmania* can develop and either kill it directly, cause it to be ingested by macrophages, or induce apoptosis (programmed cell death). The *Leishmania* may be ingested by macrophages whilst they are still inside neutrophils or after their release (e.g. following apoptosis). In addition, although *Leishmania* infects macrophages, these can kill the parasite by producing superoxide (O_2^-), nitric oxide (NO), and other toxic metabolites. The macrophages are stimulated to produce superoxide etc. by gamma interferon (IFN-γ) produced by activated parasite-specific T-cells. The T-cells are activated by the macrophages presenting them with parasite antigens. Sandfly saliva contains substances that interfere with these processes. For example, the vasodilator Maxadilan causes a reduction in the production of TNF-α and IFN-γ. The inhibition of the host's immune response lasts just long enough to allow the parasite to become established. The facilitation is probably a factor in the co-evolution of *Leishmania* and their sandfly vectors. It is not, however, the same in all species. For example, Maxadilan is not present in the saliva of all sandfly species. Similarly, although neutrophils are probably involved in the early stages of infection their importance in the establishment of the parasite within macrophages is uncertain and may vary between species/ strains of *Leishmania*. Further details are provided by Lestinova et al. (2017).

8.8 Host Factors

8.8.1 Host Identification

Most parasites infect only a single host species or a group of closely related species and therefore they need to identify their host(s) reliably. Vector-transmitted parasites effectively delegate host-identification to their vector, whilst those that are transmitted when their host is consumed depend upon the specificity of a predator–prey relationship. Those parasites that actively search for and invade their host or are transmitted by contamination have a more dynamic involvement in the transmission process since they can reject potential hosts in favour of waiting for a more suitable one to turn up. For contaminatively transmitted gastro-intestinal parasites, the conditions experienced in the gastrointestinal tract determine whether or not excystment (protozoa) or egg hatching (helminths) takes place. Factors include temperature, oxygen reduction potential, carbon dioxide, and composition of the bile. For example, digestive enzymes in the gut of mice digest away the outer layers of the eggs of the tapeworm *Hymenolepis diminuta* and stimulate the hexacanth embryo within to begin secreting its own enzymes (Holmes and Fairweather 1982). In the mouse whipworm *Trichuris muris*, interactions between the parasite eggs and gut bacteria are important in stimulating hatching. However, this is not the case for all *T. muris* strains and in the pig whipworm, *T. suis*, bacteria appear to not influence egg hatching (Vejzagić et al. 2015). The gut microbial flora is incredibly complex and the composition varies between individuals. It is therefore possible that further research will identify a role for bacteria in the hatching of other parasite species, and it may also help to explain why some people are more resistant (or susceptible) to certain parasites than others.

These stimuli are not always totally reliable, and it is not unusual for eggs to hatch in unsuitable hosts in which the parasite is unable to develop to maturity. For example, the eggs of the nematode *Toxocara canis* will hatch in the human gut and the larvae migrate through our body just as they would in dogs, which are the natural host. However, in us the larvae are unable to complete their development by returning to the small intestine and transforming into adults.

Starvation usually induces blood-feeding invertebrates to feed on hosts that they would normally ignore. One would therefore expect that starvation might affect host-choice in parasites that actively invade their host. For example, are recently released cercariae of bird schistosome species less likely to attempt to invade us than those that are several hours old? However, there is limited information on this.

8.8.2 The Influence of Host Behaviour on Parasite Transmission

Unless an infection is vertically transmitted, the only way an animal can become infected with a parasite is by encountering its infective stage. Consequently, anything that increases or decreases this contact affects the chances of parasite transmission. One of the most important factors that determine host: parasite contact is host behaviour. There is evidence that hosts will modify their behaviour to avoid exposure to parasites and some parasites modify their host's behaviour to increase the chances of transmission.

Hosts typically become infected whilst undertaking their normal activities (e.g. eating, drinking) although the manner in which they do these things can increase or decrease their risks of infection. Although there is some evidence of predators avoiding parasitised prey, there is often no selective pressure for them to do so. Indeed, for predators, parasitised prey is often easier to catch and subdue and the energetic costs of becoming parasitised may not be high. For example, rodents infected with *T. gondii* are easier for cats to catch and the parasite usually has little serious impact on the cats. We have the option of cooking our food and provided this is carried out at a high enough temperature and for long enough it will kill parasite infective stages. However, in many cultures certain foods are considered best eaten raw, and this increases the risk of contracting parasitic diseases. For example, the recent worldwide popularity of Japanese sushi and sashimi cuisine – which includes raw fish – poses a risk of becoming infected with several infectious diseases including the nematode *Anisakis* spp. (Bao et al. 2019). Raw meat, and in particular beef and pork, is popular in many continental European countries with the consequent risks of contracting *Trichinella spiralis* and tapeworm infections. For example, the cyclist Lauren Fignon, winner of the Tour de France in 1983 and 1984, put his poor performance in the 1988 race down to a tapeworm he acquired through consuming raw meat. Among the many weird diets being promulgated on the internet is the raw meat diet on the basis that our health would improve if we ate as our ancestors did in the Stone Age. It goes without saying that this also provides an opportunity for acquiring many parasites.

Defecation is something that all animals do and is fundamental to life. Animals vary widely in their defecatory habits, and there are big differences in human societies in their attitudes towards defecation and the subsequent disposal of faeces (Lewin 1999). The study of defecation is therefore a fascinating topic but one that is also sadly neglected in university education. It is, however, important to engage with the subject because it has major implications for the transmission of many parasitic diseases. Some animals defecate indiscriminately (e.g., cows), and this leads to widespread contamination of the environment. For migratory animals or those with a large range, this does not matter since they are unlikely to come into contact with the contaminated areas, but it becomes a big problem when these animals are constrained within the confines of a small field or zoo enclosure. Some animals defecate in a restricted area or construct latrines or dung piles. For example, horses often defecate in one corner of a field whilst both black and white rhinos (*Diceros bicornis* and *Ceratotherium simum*) defecate in a particular spot until they have built a dung pile that can grow to over a metre in height. This behaviour reduces the faecal contamination of the surrounding area but depending upon where and how they are constructed may not reduce the risk of contaminative transmission. Individual deposits of faeces can dry out rapidly and break up, thereby resulting in the death of any parasite eggs, larvae or cysts present, through a combination of desiccation and exposure to UV light. By contrast, large accumulations of faeces present a reduced surface area to volume ratio and therefore remain moist for longer and facilitate the survival of the parasite transmission stages. In North America, racoons (*Procyon lotor*) construct latrines on logs or at the base of trees. The racoons are often infested with the nematode *Baylisascaris procyonis* and its eggs are therefore found in high densities around the racoon latrines. Numerous mammals and birds can act as intermediate hosts for this parasite within which it causes visceral *l. migrans* and potentially fatal damage to the central nervous system. Because the racoon latrines are constructed on the 'runs' of many small mammals, they increase the chances of them becoming infected (Page et al. 1998).

Baby Care and *Strongyloides fuelleborni kellyi* Infection

Strongyloides fuelleborni kellyi has a very restricted distribution, but in parts of New Guinea, the prevalence in children 3–5 years old can reach 100%, whilst in adults the figure is 15–20%. It is especially pathogenic in babies of around 2 months of age and causes a potentially fatal condition known as 'swollen belly syndrome'. The afflicted children can pass enormous numbers of eggs in their faeces – figures as high as 300,000 eggs ml^{-1} have been reported – and they typically present with a distended abdomen and suffer from respiratory distress. This begs the question of how young babies become infected with so many nematodes. One possibility is that they are transmitted to the baby through the mother's milk: trans-mammary infection certainly occurs in other species of *Strongyloides*. Another possible source is through exposure to large numbers of infectious third-stage larvae. The local practice of mothers is to place their babies in string bags that are then kept slung around the body. The bags are lined with banana leaves and other vegetation, but these are not changed very often and therefore become heavily contaminated with faeces. The fact that so many eggs of *S. fuelleborni kellyi* are found in the faeces suggests that few if any of them hatch in the gut and therefore autoinfection via the gut or perianal region is unlikely. However, the eggs hatch soon after they are passed with the faeces and rapidly develop to the infectious third stage, and this coupled with the poor hygiene facilitates the re-infection of the child (Ashford et al. 1992).

Amongst the many fascinating accounts of parasitic diseases related by Desowitz (1987), there is one concerning the transmission of hookworms in a Bengali village. The village had no sanitation and everyone defecated on a nearby patch of land. This land was heavily contaminated with faeces and most villagers walked barefoot. It might therefore have been expected that hookworm infection levels would have been high, with larvae infecting people as they went to defecate. However, this was not the case – although the men had higher worm burdens than women. The low burdens were partly a consequence of the villagers defecating quickly and then washing themselves – as proscribed by their religion. The infective larvae in previously deposited faeces therefore had very little time to locate their host and would also have to invade deeply enough to avoid being washed off the skin surface. The men tended to defecate in the morning whilst the ground was still moist – and thus facilitating the movement of the infective larvae – whilst the women tended to defecate in the afternoon when it was hot and dry and therefore less suitable for the parasite. Furthermore, the men's morning faeces were rapidly dried out, thereby killing many of the parasite eggs and larvae. By contrast, the women's faeces were exposed to the sun for less time and the eggs would hatch and the larvae reach the infective stage during the night – and therefore be in a 'pole position' to infect the men the following morning.

Even if sanitation is provided, it does not mean that it will be used, maintained, or the waste disposed of safely. Even within developed countries, public toilets are frequently few and far between, people behave irresponsibly within them, and they are often poorly maintained. It should not, therefore, be surprising that in developing countries, it can be difficult to persuade people to fund, use, and maintain toilet facilities.

The advent of cheap air travel has meant that millions of people can move rapidly between countries. In the process, they take with them their existing infections and acquire new ones which they then transport across the globe. For example, the movement of people from malaria-endemic to malaria-free countries is a constant source of worry. The fact that there are about 2,000 cases of malaria diagnosed each year in the United Kingdom despite the fact that the disease has not been present in the general population for many years indicates the scale of the problem. In addition, parasite vectors (possibly already infected) can also be transmitted between countries within aircraft either by flying into the cabin whilst it is on the runway or as a 'contaminant' among agricultural produce etc. Once arrived at their destination, these 'stowaways' can transmit infections to people who have never been abroad. For example, there are several confirmed cases of so-called airport malaria in the United Kingdom and Northern Europe where malaria is no longer endemic. There are also legitimate concerns that vectors or intermediate hosts transported between countries might subsequently establish themselves and thereby enable the parasite to set up a transmission cycle as well. During much of 2020–2021, the SARS-COVID-19 pandemic had brought air travel (but not transport of goods) between countries to a virtual standstill. At the time of writing, it was still too early to know when and whether large-scale movements will resume to the same extent as previously. Clearly, this and the heightened concern for disease transmission could have implications for the future transfer of infections.

The popularity of extreme sports and adventure holidays can also put people at risk of contracting parasites that they might not ordinarily come into contact with. For example, jumping or diving into water enhances the risk of infection with the protozoan *Naegleria fowleri*. This is because the activity may damage the mucus membranes lining the nose, which facilitates the entry of the parasite into the body. Cases of infection have been described from all age groups although the majority occur in younger individuals: this is probably a reflection of the fact that they are more likely to indulge in water sports than in any immunological reason. Most cases occur during the warmer months of the year, which is probably because this is when most people will be swimming outdoors and also that *N. fowleri* is a thermophile that grows best at high temperatures. It reproduces rapidly at temperatures above 30 °C and can survive at 45 °C. This has led to the suggestion that global warming may result in an increase in the number of human infections and more cases being reported in countries in which it has not previously been a problem (Maciver et al. 2020).

8.8.3 Religion and Parasite Transmission

There are many faiths in this world and religion plays a major role in the day-to-day lives of millions of people. Religious practices can influence what people eat and drink, their interactions with animals, how they conduct sexual relationships, and where and how they dispose of their dead. Religion can therefore be an important factor in the epidemiology of parasitic infections. Most religions include strict instructions over what foods can be eaten, when it is eaten, and how it should be prepared. For example, the consumption of pork is forbidden in the Jewish religion, Islam, and Ethiopian Orthodox Christianity. The ban on eating pork was initially made in the Old Testament (Leviticus 11, 7–8; Deuteronomy 14, 8) and is also found in the Quran (2: 173; 16: 115). The pig was, and is, considered to be an unclean animal by people practising these religions – a view that is probably not helped by its coprophagic tendencies. By contrast, the Hindu religion forbids the consumption of beef because cows are sacred. Refraining from consuming pork prevents infection by *Trichinella* spp. and *T. solium* whilst not eating beef protects against acquiring infection with the beef tapeworm *T. saginata*. When people feel vulnerable to disease transmission, they show increased aversion to those who are obviously disabled and to animals thought to be associated with disease (Prokop et al. 2010). However, whether the religious lawmakers made an informed decision based on disease prevention is debateable. The consumption of virtually any animal (and unwashed vegetables) carries the risk of disease transmission and according to Leviticus, hares, swans, and owls (amongst many others) are ranked alongside pigs as being 'unclean'. The association of pigs with being unclean is a feature of Near- and Middle East religions. By contrast, many other faiths, and cultures with equal risks of exposure to parasitic disease consider pigs to be highly valuable and therefore have much closer relationships with them. For example, the pig is one of the signs in the Chinese Zodiac and people born under its sign are considered hard working and intelligent. The Romans valued pigs and 'Porcus' was an esteemed name. Among some Melanesian religions, it is believed that humans were initially fashioned in the shape of pigs, but the god Qat subsequently beat the pigs down into their current four-legged state (Codrington 1881).

During the three-day Feast of Sacrifice ('Īd al-Adha), most Muslims who can afford to do so sacrifice a sheep, goat, camel, cow or other bovid to commemorate the prophet Abraham's willingness to sacrifice his son Isma'il. Within many countries that have a sizeable Muslim population, the numbers of animals required vastly exceeds the local supply. Consequently, huge numbers of animals are imported from far and near. This has caused problems in parts of northern Africa and Southwestern Asia where farmers are encouraged to overstock their grazing land. Furthermore, the vagaries of the Islamic lunar calendar mean that the requirement for animals does not always match the cycles of plant growth and animal reproduction. Similarly, large numbers of sheep are shipped thousands of miles from Australia to the Middle East and elsewhere in the Islamic world, and this has raised animal welfare concerns. Although some countries have attempted to ensure that the animals are killed for free by trained slaughtermen or butchers in licensed abattoirs, many animals are killed in backyards or at the side of the road. Far more animals are killed than it is possible to consume and all too frequently blood and offal are simply poured down drains, thrown onto rubbish heaps or, at best, buried under a thin layer of soil. Needless to say, domestic and feral dogs, as well as rats, eat much of that which is thrown away, and there is an explosion in the carrion insect population (Zaidi and Chen 2011). The movement of large numbers of animals between countries over a short period of time, their confinement in small spaces, the absence of meat inspection, the lack of safe disposal of offal, and the abundance of insects capable of acting as mechanical vectors provides ideal opportunities for the spread of animal and zoonotic diseases. However, there is surprisingly little published information on the extent to which the festival influences the spread of parasites that one might expect to be enhanced (e.g., *Chrysomya bezziana* and *Echinococcus granulosus*). Nevertheless, there are increasing concerns about the transmission of microbial pathogens such as anthrax (*Bacillus anthracis*) and viruses such as Rift Valley Fever.

Current religious practices are heavily influenced by decisions made long ago without the benefit of present-day scientific knowledge. For example, the Christian tradition of eating fish on Friday arose because meat was considered to derive from warm animals. Fish are cold and were not thought to be meat and therefore could be eaten on a Fast Day. Unless one cooks (or salts or pickles) fish sufficiently, consuming it carries the risk of contracting parasites such as *Diphyllobothrium* tapeworms and the nematode *Anisakis*. Among European and American Jewish peoples, it is common to prepare gefilte fish on Shabbat, the Passover, and other special occasions. The job of food preparation inevitably fell on the women, and they would, and probably still do, taste the food during its preparation. Consequently, they would often contract *Diphyllobothrium* infections (Desowitz 1987) and in the United States, it was known as Jewish housewife's disease.

Most faiths require that the dead be disposed of by burial or burning. Consequently, although we are the intermediate host for several parasite species, we are normally dead-end hosts. This is because we are seldom preyed on by wild animals and after death our bodies are removed from the disease transmission chain by cremation or deep burial. The Parsees and some Tibetan Buddhist sects arrange for their dead to be consumed by vultures, but these probably have little impact on the transmission of diseases. By contrast, a traditional belief among the Nandi tribe in Kenya is that unless a person's dead body is eaten by a hyena, his *mukuledo* (the personality aspect of the soul) cannot reach the spirit land and is destined to roam forever as a ghost on earth (Hollis 1909). Similarly, the Turkana, another Kenyan tribe, do not have a tradition of burying their dead and corpses are left in the bush where the dogs soon sniff them out. These practices contribute to the transmission of hydatid disease caused by *Echinococcus granulosus* in which humans are one of the intermediate hosts and the adult tapeworm develops in the intestine of dogs and other caniids.

8.8.4   War and Parasite Transmission

Wars almost always result in enhanced transmission disease. The health system comes under strain, and disease control programmes are often aborted. The destruction of infrastructure leads to lack of sanitation and the availability of clean water and food. Consequently, there is an increase in contaminatively transmitted parasitic diseases. There is often a problem with the disposal of dead bodies. This leads to an increase in the abundance of flies etc., and this further increases contaminatively transmitted infections. Furthermore, scavenging of dead bodies by dogs etc. means that we cease to be dead-end hosts. Water that collects in bomb craters and destroyed buildings etc. can provide breeding grounds for the vectors of some parasitic diseases. The mass movement of refugees, who are usually stressed and malnourished, can both transmit infections and expose the refugees to new infections. For example, recent conflicts in Syria, Yemen, Sudan, Afghanistan, and Pakistan have led to both outbreaks in endemic regions and the establishment of leishmaniasis in areas in which it had previously been absent (Sharara and Kanj 2014). In the Khurram Agency in the northwest frontier province of Pakistan, cutaneous leishmaniasis was seldom seen until 2002 when approximately 5,000 new cases were recorded. This was attributed to the influx of large numbers of Afghan refugees. Similarly, within Afghanistan itself, large numbers of new cases of cutaneous leishmaniasis were recorded in Kabul when people returned to the city when it has become (comparatively) more peaceful. With an influx of susceptible people, a lack of effective control measures to reduce the abundance of the sandfly vector, the health infrastructure in a parlous state, and inadequate medical supplies, it is hardly surprising that the disease rapidly became an important emerging infection. Leishmaniasis is also a significant problem for troops serving in Iraq, and they can carry the infection with them when they return home (Stahlman et al. 2017).

War also causes a collapse in veterinary services and therefore negatively affects the health of domestic animals. Furthermore, many parasites are zoonotic, and therefore the lack of control of infections amongst domestic (and wild) animals usually leads to an increase in human infections. There is little published information on whether there has been any change in the incidence of parasitic diseases in domestic livestock in the Middle Eastern countries most affected by past and ongoing military activities (e.g., Iraq, Syria, Libya, Yemen). However, the conflicts have almost certainly had an impact. Furthermore, in the absence of an effective government, animals are moved within and between countries with few, if any, health checks. This will facilitate the transmission of infections.

Wars cause widespread devastation and pollution of the environment, which can have consequences for the transmission parasites. Pollution impacts on parasite transmission in many ways (see later), but there is limited information on pollution caused directly by warfare. For example, after the Gulf War there was controversy over the levels and possible effects of depleted uranium associated with its use in munitions. Sheep and goats are particularly vulnerable to ingesting depleted uranium because they crop vegetation close to the ground and will ingest large amounts of soil in the process. Some studies suggest that depleted uranium is unlikely to enter water and food cycles (Bleise et al. 2003). In support of this, Al-Kinani (2006) found no evidence of meat or milk being contaminated with depleted uranium from farms in the Basra region.

8.8.5   Parasites Influencing Host Behaviour

Although many parasites cause changes in the behaviour of their host, most of these are 'non-adaptive' and confer no advantage to either the parasite or its host (Poulin 2000). These changes result from one or more of the following: impairment of vision, hearing or other sensory systems, physical damage to the central nervous system and/or altering the levels of neurotransmitters, physical damage to the endocrine system and/or altering the levels of hormones, causing starvation through interfering with feeding/drinking or removal of metabolic resources. In some cases, however, the changes in host behaviour appear to be adaptive and promote the chances of parasite transmission. For example, the larval stage (plerocercoid) of the tapeworm *Schistocephalus solidus* lives in the peritoneal cavity of three-spined sticklebacks (*Gasterosteus aculeatus*). These must be consumed by fish-eating birds such as Arctic terns (*Sterna paradisaea*) for the parasite to complete its life cycle and develop into an adult. The plerocercoid grows to an enormous size, which compromises the movement of the fish host and increases its oxygen demand. Consequently, infected fish tend to swim close to the water surface where the oxygen tension is higher. They also show a preference for warmer temperatures, and this means that the parasite grows faster and larger. This is advantageous because the adult parasite's fecundity is linked to the size of the preceding plerocercoid stage (Macnab and Barber 2012). In addition, infected fish exhibit a reduced fright response when attacked and either fail to react to the shadow of a predator approaching from above or quickly return to their former position after being startled. Uninfected fish, by contrast, quickly retreat and hide among underwater vegetation in response to a shadow passing over the surface of the water and do not swim in open water for some time after being startled. Infected fish are therefore much easier for the definitive bird host to catch (Barber et al. 2000).

Another interesting example of how a parasite appears to modify a host's behaviour is that of the protozoan *T. gondii*. This parasite infects most warm-blooded animals, but its natural transmission cycle is probably between rodents and cats. The asexual stages encyst within the nervous tissue of its many intermediate hosts, but the parasitaemia often reaches exceptionally high levels in rats and mice. Infected rodents are said to exhibit changes in behaviour that have been called 'suicidal'. They become more active and lose their normal fear of new objects (neophobia). Consequently, they are more likely to be caught in live traps than those which are uninfected. Even more importantly, some researchers find that the infected rodents lose their aversion to the smell of cats and even become attracted to their odour. Rats and mice infected with *T. gondii* would therefore be much more likely to be caught by cats – which are the only hosts within which the parasite reproduces sexually. According to Boillat et al. (2020), infected rodents lose their fear of predators generally rather than a specific loss of fear of cats. They also found that the *T. gondii* cysts localise predominantly within the cortical regions and the changes in behaviour correlate with the number of cysts and level of neuroinflammation. Unfortunately, the literature on the effect of *T. gondii* on rodents contains numerous contradictory reports. For example, some researchers find no effect on rodent behaviour, and there are different accounts of where the cysts localise within the brain (Johnson and Koshy 2020). Many of these discrepancies arise through differences in experimental design.

Toxoplasma gondii also induces behavioural changes in other animals than rodents although whether they are adaptive is uncertain. For example, southern sea otters (*Enhydra lutris nereis*) suffering from toxoplasmic encephalitis are more likely to be attacked and killed by sharks (Kreuder et al. 2003). This may be because they display aberrant behaviour that attracts the attention of sharks or because they are less able to avoid them when they do attack. There are no records of *T. gondii* infecting sharks or fish. Fish such as anchovies can filter out *T. gondii* oocysts from the surrounding water, but these do not hatch, and they pass through the gut. The fish can, however, act as transport hosts if they are then eaten by marine mammals, such as otters or dolphins, before the oocysts are defecated.

The Effect of *Toxoplasma gondii* on Human Behaviour

Human infections with *T. gondii* have been linked with numerous behavioural changes although one must treat the literature with care. In particular, although we talk of infections, what is actually measured is whether someone is seropositive for *T. gondii*. That is, whether a person has developed antibodies against the parasite, and this says little about their current status of infection. Experimentally manipulating human infections is not ethical. Therefore, one can only establish a statistical correlation between seropositivity and a trait and not that infection causes the trait. Furthermore, it could equally be argued that a behavioural trait made it more likely for a person to become infected by *T. gondii*. Many studies are based on small sample sizes and among populations with low rates of infection a few infected individuals can have a disproportionate influence on the results. For a detailed critique of the methods employed in many of the studies and a discussion of their findings, see Johnson and Koshy (2020).

Some studies suggest *T. gondii* affects men and women differently. For example, in some personality profile tests, seropositive men tend to be jealous and to disregard rules whilst seropositive women are trusting and conscientious (Lafferty 2006). Other studies have indicated that *T. gondii* has different effects on hormone titres in men and women. For example, they suggest that seropositive men tend to have higher testosterone levels than those who are seronegative whilst seropositive women have lower testosterone levels than those who are seronegative. Some studies indicate that people who are seropositive for *T. gondii* are more likely to be involved in serious road accidents and commit suicide (Sutterland et al. 2019). It is tempting to suggest that this may be related to behavioural changes caused by the parasite although whether there is a genuine causal effect remains uncertain.

The development of schizophrenia, bipolar disorder, and various psychological problems has been linked to infection with *T. gondii* although the evidence is controversial (Del Grande et al. 2017). Yolken et al. (2017) found a correlation between exposure to *T. gondii* and recent onset of psychosis. However, they found no relationship between exposure and patients diagnosed with schizophrenia who had not experienced a recent bout of psychosis. That is, there might be a relationship between *T. gondii* exposure and the onset of psychosis. Interestingly, some of the drugs that are used to alleviate the symptoms of schizophrenia (e.g., haloperidol and valproic acid) also reduce the replication of *T. gondii* (Jones-Brando et al. 2003). However, there is a difference between reducing and stopping entirely and the various studies report varying degrees of effectiveness. The treatment of toxoplasmosis is extremely challenging, and there are currently no approved drugs that will eliminate the tissue cysts (Deng et al. 2019). Until such drugs become available, it will be difficult to fully establish a relationship between *T. gondii* infection and certain mental illnesses.

Schizophrenia is a lifelong neuropsychiatric condition that is characterised by emotional and memory problems, difficulties in expressing oneself and controlling one's thoughts and some patients become delusional and hear voices. What causes schizophrenia remains uncertain, but it probably includes a combination of genetic susceptibility and agents that disturb brain chemistry such as infectious agents. In addition, gut dysfunction and changes to the gut microbiome are now thought to play a role in schizophrenia (Severance and Yolken 2020). This begs the question whether gastrointestinal parasites might also have a role to play in either onset or treatment.

Nowadays, humans are dead-end hosts for *T. gondii*, and therefore, if it does influence our behaviour, this cannot be considered adaptive. However, this may not have always been the case. Our primitive ancestors would have been prey for large felines, such as sabre-toothed cats, and it is possible that there was once a *T. gondii* transmission cycle between the two. This could have been enhanced by the parasite, making our ancestors more likely to take risks or place themselves in harm's way – although this is, of course, pure speculation.

8.9 Co-Transmission and Interactions Between Pathogens

Interactions, including transmission events, between hosts and parasites do not happen in isolation of their environment. The susceptibility to infection and subsequent pathology in the host can be affected interactions between different species microorganisms inside it. Individuals harbour various microorganisms, and so it is clear that when passing one particular type of organism to another host, representatives of several species might follow suit. For example, the Bengali villagers in the example mentioned previously could easily be transmitting and acquiring protozoan, bacterial, and viral diseases from the uncovered faecal depositions, even though the account concentrated on helminths. Similarly, a blood-feeding vector may carry more than one pathogen. In some cases, the association between the pathogens is minimal, and there is little obvious effect on the vector or the host. In other situations, the pathogens compete with each other to the detriment of at least one of them. There are also instances where biochemical and immunological interactions between organisms during co-infection cause altered and often serious pathology in the host. There are many potential examples of co-transmission and co-infection involving protozoa and helminths with other parasites or other types of microorganism. A few examples follow to illustrate some of the points, but there are plenty of others.

Some invertebrate species are hosts for more than one species of parasite, which could lead to the mammalian host being infected with more than one infection via the same route and at the same time (co-infection). A documented example of this occurring is when *Anopheles* mosquitoes transmit both malaria parasites and filarial worms. It is theoretically possible for an individual mosquito to carry either or both parasites, and therefore there are a number of ways in which a person infected with malaria and bancroftian filariasis at the same time could have acquired the infections. The most likely scenario is that a person is first bitten by a mosquito carrying *P. falciparum* sporozoites and then later has the misfortune to encounter a different mosquito infected with *W. bancrofti* larvae, or *vice versa*. The mosquito could also be carrying both parasites and pass them on to a human during a blood meal, although it would be virtually impossible to prove either way. There is some evidence that being infected with one of these parasites causes physical or physiological disturbance within the mosquito, making it more susceptible to infection with the other (Manguin et al. 2010). Thus, co-infection of the human host is a possibility; however, this concomitant infection may also affect the fitness of the mosquito, meaning it cannot fly as far or live as long as others. Surveys of insect populations in endemic area tend to find only small proportions of them carrying both parasites. For example, Muturi et al. (2006) studied populations of *A. gambiae* and *Anopheles funestus* in Kenya, where these are the main vectors of *P. falciparum* and *W. bancrofti* and found less than 2% were carrying both parasites. Low rates of dual infection are also reported in humans (Shetty et al. 2018). Although the interactions between species of parasites within us have not been investigated extensively, it has been suggested that the presence of filariae can have an adverse effect on the development of the *Plasmodium*, resulting in lower parasitaemia. This could, in turn, be detrimental to the transmission of the parasites, particularly the protozoan. If there is a relatively low concentration of all life cycle stages in the host's blood, the chances of male and female and gametocytes being present in sufficient concentrations to ensure their uptake by a feeding female *Anopheles* are reduced.

Pigs are commonly infected with *T. gondii* and eating poorly cooked pork is probably a major source of human infections. The parasite seldom causes serious disease in pigs, but there is some concern that co-infections of *T. gondii* with porcine circovirus-2 infections (PCV-2) could result in them developing systemic toxoplasmosis (Klein et al. 2010). PCV-2 afflicts pigs in many parts of the world and causes potentially fatal pneumonia, enteritis, and abortion. It also causes immunosuppression and is commonly associated with co-infections with other viruses and also bacteria.

Dogs seldom suffer from acute toxoplasmosis. When it does occur, it is usually associated with young animals suffering from canine distemper, which is caused by Canine Distemper Virus (CDV) (Calero-Bernal and Gennari 2019). Toxoplasmosis in dogs therefore differs from neosporosis in which serious disease is usually a consequence of a primary infection. Care is needed when interpreting literature that pre-dates the identification of *Neospora caninum* and ascribes a disease condition in dogs to *T. gondii*. CDV is a highly contagious and potentially fatal disease that affects not only dogs but many other carnivorous mammals including racoons, ferrets, and seals. It suppresses the immune system and causes fever, vomiting, diarrhoea, and nervous symptoms such as convulsions and paralysis. CDV is a member of the morbillivirus group that also includes measles virus, rinderpest virus, and phocine distemper virus. CDV is spread via aerosol droplets and contact with secretions and body fluids. Canine distemper is now uncommon among domestic dogs in developed countries owing to the availability of an effective vaccine. However, CDV has not been eradicated, and it is common in developing countries and can have serious consequences for wild animals. For example, massive losses among some seal populations have been attributed to CDV. Seals, like other mammals, are often infected with *T. gondii* (Reiling et al. 2019) and therefore CDV-*T. gondii* co-infections are probably common in some populations. It is uncertain whether co-infections contribute to the severity of the disease in seals. However, in Caspian seals, studies suggest that infection does not necessarily result in disease (Namroodi et al. 2018).

There are three clinical variants of Burkitt's lymphoma: endemic, sporadic, and immunodeficiency related. Endemic Burkitt's lymphoma (eBL) is an unusual facial tumour, which usually occurs in children living in areas, which are holoendemic for malaria. Endemic Burkitt's Lymphoma was first described in Uganda and is more common in East Africa, though high rates are also reported in Malawi and Papua New Guinea. It is almost always associated with infection with the Epstein-Barr virus (EBV). The virus is a gamma herpes virus which is transmitted in saliva and which infects B lymphocytes. In common with other herpes viruses, after the primary infection, EBV persists in the body for life. It is associated with Glandular Fever among teenagers and young adults in Western Europe. However, it is usually acquired much earlier in life in many other parts of the world and infection is often asymptomatic, meaning that most African children over the age of two carry EBV.

The tumour cells in eBL are B lymphocytes, which are transformed to be malignant due to a specific chromosomal translocation between chromosome 8 and usually chromosome 14, 2, or 22, with the first being the most common. This translocation moves the c-myc gene in chromosome 8 to be under the regulatory control of genes coding for Ig heavy (chromosome 14) or light chains and to become an oncogene which is expressed constitutively.

There is an observable association of co-infection with *P. falciparum* and EBV and eBL. The incidence of eBL decreases in areas where malaria control programmes are successful, and it is less likely to occur among children who move away from holoendemic regions (Orem et al. 2007). Also, children with sickle cell trait and other haemoglobinopathies, which afford some protection against malaria infection, are less prone to developing eBL.

Plasmodium falciparum infection stimulates B-cell activity in the germinal centre and therefore more EBV-infected B cells are produced. In addition, *P. falciparum* induces deregulated expression of activation-induced cytidine deaminase (AID) in germinal cells. AID can cut DNA strands and causes mutations by changing bases within a DNA strand (e.g., changing cytosine to uracil). It has an essential role in the germinal centre but once deregulated, it starts to act on new DNA targets, including facilitating c-myc gene translocations (Thorley-Lawson et al. 2016). Therefore, when there is co-infection, more virus-infected B cells are produced, and these are more likely to become malignant, thereby leading to lymphoma. Malaria also suppresses T cells and the immune mechanisms whereby EBV-infected cells would be noticed and eliminated, which again would encourage transformed cells to proliferate.

8.10 Environmental Factors

8.10.1 Natural Environmental Variables

The environment plays a major role in parasite transmission either directly, by affecting the survival of the transmission stage (e.g., cysts, eggs, or free-living larvae), or indirectly, by affecting the distribution and survival of the host (intermediate or definitive) or vector. For example, parasite cysts and eggs usually die quickly if they are exposed to dry conditions, especially if this is coupled with exposure to UV radiation. By contrast, moist conditions favour their survival and facilitate the movement of hookworm larvae. Countries with tropical climates often experience sudden torrential downpours that rapidly overwhelm the capacity of the drainage system, if there is one to cope with the arrival of so much water. This is further exacerbated by the continued use of open sewers and their overflow leads to widespread contamination with raw untreated sewage. This increases the risk of contaminatively transmitted parasites. Although there is a tendency to consider environmental factors individually, in reality they act in concert and can include variables such as soil or water chemistry, as well as atmospheric properties. A good review of how environmental factors affect the transmission of parasites of domestic livestock is provided by Stromberg (1997) although the scenarios he discusses could equally be applied to many parasites of medical importance.

8.10.2 Pollution

Pollution can occur naturally, for example because of volcanic eruptions but it is more commonly associated with human activity. The effects of a pollutant upon an animal are highly case dependent and are affected by numerous abiotic and biotic variables. For example, oestrogenic pollutants enhance the growth of the plerocercoid larval stage of the tapeworm *S. solidus* in male sticklebacks (*Gasterosteus aculeatus*) but not females (Macnab et al. 2016). Similarly, the presence of a pollutant in the environment does not mean that it is biologically available: for example, it might be bound to clay minerals and cannot be absorbed (Table 8.1). Therefore, the effect of a pollutant upon disease transmission will depend upon a complex interplay between the parasite, the host, the pollutant, and other environmental factors. See Morley et al. (2003) for a discussion of how pollution can affect the transmission of digenean trematode larvae. One must also remember that correlation does not necessarily equal causation. For example, Pennino et al. (2020) found that the prevalence of parasites correlates with the abundance of microplastics in anchovies (*Engraulis encrasicolus*) and sardines (*Sardina pilchardus*). As the authors discuss, the factors governing the ingestion and accumulation of microplastics are almost as complex as those governing the transmission of

Table 8.1   Biological and abiotic factors that influence the toxicity of a pollutant to an animal.

Biological factors	Abiotic factors
Species of animal	Type of pollutant
Genetic constitution	Concentration
Age	Distribution
Gender	Environmental variables (e.g. temperature, pH, soil and water characteristics etc.)
Health	Presence and concentration of other pollutants
Nutritional status	
Pre-existing disease	
Reproductive status (e.g. pregnancy/gravid)	

parasites. Microplastics are very variable in size, shape, and constituency, and they can be ingested directly and through the consumption of prey. Their ingestion will also be influenced by where (e.g., depth) and when feeding occurs. Therefore, fish that are most exposed to parasite infection might also be most exposed to microplastic pollution. That is, the correlation is happenchance. Alternatively, the accumulation of microplastics might harm the fish and make them more vulnerable to parasite infection. What is certain is that fish containing both a lot of microplastics and a lot of parasites are less healthy than those with lower burdens.

The mechanisms by which a pollutant might enhance parasite transmission are summarised in Table 8.2. However, this list could equally be reversed to illustrate how the pollutant might reduce the chances of transmission by exerting a deleterious effect on the parasite. The effect of a pollutant on other organisms also needs to be considered. For example, if a pollutant kills the predator, pathogen or competitor of a vector or intermediate host then its population may rise.

Theoretically, parasites with simple life cycles should predominate in disturbed habitats such as those affected by pollution because fewer species are required to ensure the life cycle is completed. There is evidence for this, and some researchers consider parasites to be good indicators of ecosystem health (Sures et al. 2017). Similarly, some parasites accumulate toxins, such as heavy metals, and therefore might be useful as indicators of the risk of bioaccumulation within the local environment. Obviously, monitoring internal parasites is logistically more challenging than monitoring free-living

Table 8.2   How a pollutant can increase the chances of parasite transmission.

Pollutant is poisonous to the host but not the parasite. The pollutant could therefore weaken the host's immune system and makes it more susceptible to the parasite. The increased parasitaemia could increase the chances of transmission (e.g., more infective stages released into the environment/more chance of vector 'picking up' the parasite).

Pollutant is poisonous to both the host and the parasite, and the outcome will depend upon which suffers the most harmful effect.

Pollutant is beneficial to the parasite's vector or intermediate host, thereby increasing the number of opportunities for transmission.

Pollutant is beneficial for the survival of the parasite's infective stage, thereby enhancing the chances of transmission.

Presence of the pollutant increases the likelihood of the parasite and host coming into contact.

organisms. However, to reach an internal parasite, a pollutant must first be absorbed by its host. Internal parasites therefore provide a good indicator of whether a pollutant is bioavailable. By contrast, it would be necessary to dissect internal tissues from a free-living filter-feeding organism to avoid the possibility that the pollutant was not simply stuck to its outer body surface.

Sewage Effluent, *Toxoplasma gondii* and Marine Sentinel Species

Marine mammals are potentially excellent 'sentinel species' as indicators of the health of the oceans and coastal environments. This is because they tend to live for many years, feed at higher trophic levels, and their 'public profile' means that they are more likely to be observed, with and dead or dying animals being reported. In addition, many species have high fat reserves in which lipophilic man-made chemicals, such as DDT, become sequestered.

The decline in the population of several marine mammal species has been attributed to disease. Human activities are thought to contribute to spread and severity of some of these diseases. For example, in the United States, the southern sea otter (*Enhydra lutris nereis*) population remains low despite federal protection. Infectious diseases are estimated to account for an unusually high 38.5% of mortalities. In one seroprevalence study, *T. gondii* was identified in 52% of freshly dead beachcast otters and 38% of live otters sampled along the Californian coastline (Conrad et al. 2005). However, the tests currently used to diagnose the presence *T. gondii* in stranded marine mammals suffer from limitations that can result in either over-estimates or under-estimates (van der Velde et al. 2016). In addition, being seropositive is not the same as suffering clinical disease. Nevertheless, autopsies have indicated that toxoplasmic encephalitis was responsible for 16.2% of the otter deaths (Kreuder et al. 2003). Subsequently, high levels of *T. gondii* infection were reported in northern sea otters (*Enhydra lutris kenyoni*) (Verma et al. 2018). However, in the northern population, *Sarcocystis neurona* is considered a more important cause of mortality.

It is not unusual for cat owners to dispose of the waste from their pet's litter tray down the toilet and flushable litter can be bought that reduces the risk of blocking the pipes. It is therefore possible that *T. gondii* oocysts are surviving passage through the sewage system and being transported into the marine environment where they ultimately cause infections of marine mammals (and possibly birds as well). It is presumed that sea otters become infected from oocysts that are washed into the coastal environment in this way. There are several *T. gondii* genotypes, and these vary in their pathogenicity. It is currently uncertain whether genotype is an important factor in the pathogenicity of *T. gondii* in sea otters. In southern sea otters, 60% carry *T. gondii* with a novel genotype – designated Type X (Conrad et al. 2005). Whether type X is also common among nearby mainland animals is uncertain. By contrast, most northern sea otters are infected with *T. gondii* Type A; this circulates amongst wild terrestrial animals living in their region (Verma et al. 2018).

It is possible that sea otters become infected through ingesting oocysts in the seawater (e.g., whilst grooming) – especially if they are particularly susceptible to the disease. However, it is more likely that they consume prey containing the parasites within their tissues. Birds are natural intermediate hosts for *T. gondii* and infection levels are so high in some seagulls that they have been suggested as sentinel species (Gamble et al. 2019). It is quite likely that sea otters would consume a dead or dying seagull. Although fish are not natural hosts for *T. gondii*, the parasite has been detected by PCR in the tissues of wild caught Mediterranean fish (Marino et al. 2019). The oocysts can also be incorporated into the flesh of filter feeding invertebrates such as oysters (Cong et al. 2017).

8.10.3   Climate Change

There is consensus among most scientists that climate change is a real phenomenon, and many believe it to be mankind's greatest threat. Its manifestations and consequences are too numerous to enumerate here but can be broadly characterised by a prolonged rise in air and water temperatures accompanied by an increasing frequency of extreme weather events, such as floods, cyclones, and droughts. These exacerbate soil degradation, which in turn affects the growth of crops and livestock production. Similarly, rising sea levels threaten coastal communities and a combination of warming temperatures, and acidification is causing changes to the abundance and distribution of the fish stocks that many depend upon. For poor communities, the result is invariably poverty, malnutrition, and economic migration.

Global warming is bringing about dramatic changes in the Arctic regions. Some researchers have attributed this to the northern spread of several pathogens, such as lungworms in caribou and muskoxen (Kafle et al. 2018). Some data, however, are currently ambiguous. For example, although an initial survey in northern Canada found the protozoan parasite *T. gondii* in beluga whales and suggested that this was evidence of a northern spread of the pathogen subsequent to global warming, a follow-up study using a different recording technique failed to find it (Dolgin 2017).

It is impossible to generalise about the consequences of climate change on parasitic diseases because these vary between individual species. This is especially the case for those parasites that have complex life cycles involving two or more species of host. This is because environmental change may affect each of them differently. A rise in temperature usually favours the development rate of the free-living stages of parasites and their vectors, provided they are not simultaneously at increased risk of desiccation. For example, some scientists consider that a combination of mild winters, higher spring and summer temperatures, and changes to rainfall patterns has contributed to an increase in the abundance of ticks in many northern European countries, as well as Russia and parts of the United States. Ticks are themselves important parasites, as well as vectors of many viral, bacterial, and protozoan diseases. However, the data linking tick abundance, disease transmission, and climate change is currently inconsistent (Ostfeld and Brunner 2015). A complicating factor is that warmer weather encourages more people, wearing fewer clothes, to visit the countryside. This means that more people (and their dogs) are likely to be bitten by ticks and are therefore at increased risk of contracting tick-borne diseases such as babesiosis. Warmer temperatures can also bring about changes in agriculture that facilitates the spread of ticks to new areas by providing them with more hosts (e.g., sheep) on which to feed. Similarly, wild animals that are natural tick hosts, such as deer and rodents, may have increased breeding success if a rise in temperature increases the food available to them. An increase in the incidence of tick-borne disease may therefore not be solely a direct consequence of global warming affecting the biology of ticks but of changes in our behaviour and wildlife ecology. By contrast, in parts of Africa, climate change is resulting in increased desertification. This, in turn, is resulting in cattle being moved increased distances in order to find food and water. This is facilitating the spread of cattle ticks and the diseases they transmit, as well as associated cattle schistosomes and other helminth parasites.

Increased rainfall and flooding favours the transmission of faecal–oral transmitted parasites and the reproduction of mosquitoes and other vectors/intermediate hosts that have aquatic stages. For example, a marked rise in the incidence of malaria in Anhui Province in China since 2000 was linked to increased rainfall (Gao et al. 2012). However, one must be careful of mistaking an association with a causation. For example, since 2009, Greece has experienced a resurgence of locally transmitted *P. vivax* malaria. Global warming may have contributed to this, but 2009 was also the year in which European debt crisis began. Greece was badly affected by this, and many people lost their jobs and suffered poverty – and poverty affects both exposure and susceptibility to many diseases.

A Perfect Storm: Did Global Warming Contribute to Disease in African Lions?

In recent years, there have been sudden crashes in the lion (*Panthera leo*) population in some parts of Africa. For example, in 1994 the lion population in the Serengeti National Park declined by about 33%, and in 2001, there was a similar population crash in the Ngorongoro Crater region. The deaths were initially attributed to outbreaks of canine distemper virus that was acquired (probably indirectly) from domestic/feral dogs (Roelke-Parker et al. 1996). However, subsequent analysis of lion serum samples taken before and after the population declines indicated that several outbreaks of CDV had taken place without any noticeable effects on the lion population. This suggested that the lions were acquiring CDV on a regular basis, and the population crashes were therefore unlikely to result from sudden exposure of an immunologically naive population to the disease. It is now thought that the lions were killed by massive exposure to *Babesia* parasites (Munson et al. 2008). This was brought about by changes in the climate, possibly resulting from global warming, that caused unusually prolonged droughts. The lack of rain resulted in the death of vegetation and consequently many of the buffalo and other herbivores on which the lions preyed died of starvation. When the rains eventually restarted the surviving herbivores were in poor health, but the conditions were ideal for the ticks that fed upon them. Furthermore, the ticks were able to feed and reproduce more successfully owing to the weakened state of their host's immune system. Most *Babesia* species are relatively host specific and the lions are unlikely to have been infected with those present in their prey. However, in feeding on buffalo they would be exposed to unusually large numbers of ticks (e.g., *Rhipicephalus appendiculatus*), and these are capable of feeding on lions and transmitting lion babesias such as *Babesia leo* and *Babesia felis*. Babesiosis is seldom a serious condition in lions but they would be in a naturally weak state owing to the lack of prey and this combined with infection with CDV probably reduced the immune system to a point at which a disease that is not normally fatal resulted in mass mortalities. It is not known whether *T. gondii* was also involved, but it is capable of causing fatal disease in them (Ocholi et al. 1989).

Much of the concern about the impact of global warming on parasitic disease has focussed on the potential of malaria to spread to new countries and become a more serious problem within those where it already exists. Increases in temperature and rainfall can make the environment more suitable for the mosquito vectors. Some computer models support this conjecture, but numerous factors other than mosquito abundance influence malaria transmission, and therefore, their validity has been strongly questioned. In addition, it is difficult to disentangle global and local influences on climate change. For example, changes in land use, such as deforestation, can result in local environmental and climate changes that increase the breeding of mosquitoes and the consequent transmission of malaria (Patz and Olson 2006). Societal changes also influence the transmission of malaria. For example, malaria was common in parts of the United Kingdom and Northern Europe up until the early 1900s but has since disappeared. The decline in malaria was not related to changes in the climate or the disappearance of the mosquito vector, and these countries remain theoretically vulnerable to the establishment of the disease (Kuhn et al. 2003). However, despite the arrival of thousands of people infected with malaria into the United Kingdom, every year the disease shows no signs of re-establishing itself here. This is, in part, a result of the improved standards of living and the quality of the health service, which mean that people with symptomatic malaria are quickly identified and treated and therefore do not become a source of infection for the mosquitoes.

More detailed accounts of how climate change is likely to impact on the transmission of parasitic diseases are provided by Booth (2018); Short et al. (2017); Hotez (2016); Lafferty and Mordecai (2016).

9

Immune Reactions to Parasitic Infections

9.1   Introduction

Our immune system enables us to distinguish 'self' from 'non-self' and mount defensive responses against perceived threats. In common with other metazoan animals, in addition to our own cells and fluids, we have an associated microbiome that is essential to our health. Consequently, our immune system must also distinguish between those microbes that are beneficial and those that are harmful. The immune system is therefore analogous to a police force that must identify and neutralise threats to society (i.e., our survival). However, a crude immune system, like a crude police force, would do a lot of damage. Therefore, the immune system needs to be sophisticated and avoid harming the body's own tissues and those microbes that are 'useful members of society'. Immune systems are therefore complicated and include many checks and balances.

Parasitology: An Integrated Approach, Second Edition. Alan Gunn and Sarah J. Pitt.
© 2022 John Wiley & Sons Ltd. Published 2022 by John Wiley & Sons Ltd.
Companion website: www.wiley.com/go/gunn/parasitology2

The immune system identifies and removes viruses, bacteria, parasites, inanimate materials, the body's own dead or dying cells, and any other potentially harmful substance. Our immune mechanisms therefore determine whether or not a parasite establishes itself, whether it grows and reproduces, the progression of an infection, our subsequent response to other pathogens, and whether one can design an effective vaccine against the parasite. The immune reaction to an infection is also important in determining the amount of damage (pathology) it inflicts. For example, a strong immune reaction may remove the parasite and therefore reduce pathology or it may seriously harm us and even increase our vulnerability to other pathogens. Within this chapter, we will consider our own immune mechanisms and those in other animals and in Chapters 10 and 11, there will be a more in-depth consideration of how the immune response both limits and contributes to the pathology associated with parasitic diseases.

The most basic level of immunity is 'innate' or 'non-specific' immunity that consists of structural and physiological features that prevent an invading organism entering the body. As its name suggests, the innate response does not distinguish between potential pathogens and all invading organisms experience the same level of challenge. This is the only form of immunity found in invertebrates. It provides an immediate response to an invading pathogen, but it does not confer any long-lasting protection. 'Adaptive immunity', also referred to as 'acquired' or 'specific' immunity, is restricted to vertebrates and triggered by a response to specific antigens. Aspects of the innate immune response are immediately active or fully mounted within a short time after the immune challenge but in adaptive immunity, there is a longer time gap between recognition of the antigens and production of an immune response. Another difference is that an effective adaptive immune response can provide prolonged protection against a particular microorganism. Immunology is a fast-moving and complicated subject, but we have attempted to keep things simple whilst still including examples of some of the most recent advances. Inevitably, some of the explanations of how the immune system operates contained in this chapter are not complete and will be superseded or modified within a short time of publication: the science of immunology should be considered as 'work-in-progress' rather than a series of absolute truths. Those requiring a more in-depth coverage of vertebrate immunity, and in particular that of humans are advised to consult Delves et al. (2017), whilst the immunology of animals of veterinary importance is covered by Tizard (2017). Lamb (2012) provides an excellent coverage of immunity to parasitic infections.

9.2 Invertebrate Immunity

There is a great diversity of invertebrates, and therefore, it should not come as a surprise that there are also considerable differences in their immune response to invading pathogens. The information given below is therefore not applicable to all invertebrates. Further details on invertebrate immunity are available in Kojour et al. (2020), Ligoxygakis (2017), and Söderhäll (2011).

The outer body surface of invertebrates provides protection from both biotic and abiotic factors. For example, nemerteans and molluscs constantly secrete mucus that prevents pathogens and predators from contacting the body wall, whilst nematodes and arthropods have a thick cuticle to protect them. In the case of arthropods, the cuticle can extend to line the foregut and hindgut, whilst in insects a peritrophic membrane protects the mid-gut epithelial cells. Invertebrates do not have a closed circulatory system – that is, they lack the network of veins, arteries etc. that are found in vertebrates. In addition, many of them have hydrostatic exoskeletons – i.e., they owe their body shape to the pressure of fluid within their coelom (fluid-filled body cavity). Consequently, puncturing their body wall results in them shrivelling like a punctured balloon as fluid leaks through the

Table 9.1   Principal invertebrate immune mechanisms.

Haemolymph coagulation entraps the invader
Invader is phagocytosed or encapsulated by blood cells
Activation of the pro-phenoloxidase cascade results in melanisation that isolates and/or kills the invader
Production of reactive oxygen radicals kills the invader
Lectin-complement system kills the invader
Agglutinin-lectin system agglutinates the invader
Activation of Toll-like receptor pathways and/or Imd-Relish pathways results in the production of antimicrobial peptides

Source: After Iwanaga and Lee (2005).

wound – and this is fatal. Many arthropods therefore have coagulation systems that immediately seal off wounds, and this can trap pathogens attempting physically to penetrate into the body.

A pathogen that breaches an invertebrate's outer defences encounters several innate immune mechanisms that can kill or isolate it (Table 9.1). These mechanisms depend upon pattern recognition receptors (PRRs) such as Toll-like receptors (see later) detecting the presence of common cell surface antigens (Wang et al. 2019). One can divide PRRs into those secreted into the haemolymph and those attached to cell membranes. The PRRs found in the haemolymph are mostly involved with detecting invading pathogens, whilst those attached to cell membranes stimulate the cellular immune response and transduce signals between the extracellular and intracellular environment.

In many invertebrates, phagocytic cells that float freely within the haemolymph detect invading pathogens and will engulf and digest them if they are small enough. If the pathogen is too large for this, numerous phagocytes converge on the invader until they surround it in a process called encapsulation. This is accompanied by activation of the pro-phenoloxidase cascade at the site of invasion and/or on the surface of the pathogen. This results in the formation of melanin (melanisation) and reactive oxygen species such as superoxide (O_2^-) and hydroxyl radicals ($^\cdot$OH). The reactive oxygen species are toxic, whilst melanisation strengthens the capsule/nodule around the pathogen and probably also restricts oxygen and nutrient availability.

Invertebrates produce a range of antimicrobial peptides (AMPs) in response to the activation of Toll-like receptor pathways and/or Imd-Relish pathways. However, the first Toll-like receptor to be identified, Toll-1, was from the fruit fly *Drosophila* in which it is involved in embryonic development. Subsequently, eight further Toll-like receptors were discovered in *Drosophila* with roles in initiating innate immune reactions against fungi and bacteria. Toll-like receptors are present in numerous organisms, including mammals. They are proteins that are located within membranes and have both an intracellular domain and an extracellular domain. Organisms vary in their number of Toll-like receptor genes, but they are absent from platyhelminths. They probably evolved in the Cnidaria (jellyfish, corals, Hydra) but for some reason the Platyhelminthes lost them. Toll-like receptors probably initially served non-immune functions and it was only later, following the development of the Coelomate animals that they became involved with mediating innate immune responses (Brennan and Gilmore 2018).

Gómez et al. (2017) list several hundred invertebrate AMPs in their InverPep database, and therefore, we will only consider a few of them. Defensins are small (18–45 amino acids) AMPs produced by most animal phyla, as well as fungi and plants. Those found in arthropods are structurally distinct from those in vertebrates and are primarily active against Gram-positive bacteria. Cecropins are also small peptides consisting of 29–42 amino acid residues but they have a more restricted distribution. An initial report stated that pigs produce cecropins but it subsequently transpired that these actually originated from *Ascaris suum* nematodes living in their intestines (Pillai et al. 2005). Some AMPs kill bacteria by interacting with and destroying their cell walls, but it is uncertain whether they all act in this way. In addition to acting on bacteria, some AMPs also have antiparasitic activity, but it is unclear how they exert this effect. The presence and levels of these AMPs may determine (at least partly) the effectiveness of a strain or species of invertebrate to act as a vector or intermediate host. For example, the stable fly *Stomoxys calcitrans* coexists with tsetse flies over parts of Africa but despite sharing the same hosts and having similar physiologies, the stable flies only act as mechanical vectors for trypanosomes such as *Trypanosoma evansi.* Usually, when *S. calcitrans* feeds on animals heavily infected with *Trypanosoma brucei*, the parasites fail to establish an infection in the flies and in those in which it is successful all the parasites die within 24 hours (Taylor 1930). One possible reason for this is that stable flies produce an antimicrobial protein, stomoxyn, which is lethal for trypanosomes. Stomoxyn does not occur in tsetse flies, and this may contribute to their susceptibility (Boulanger et al. 2006).

Some workers suggest that vector-borne pathogens might become controllable by releasing genetically engineering vectors that express or over-express particular AMPs and are therefore refractory to infection. For example, the bird malaria *Plasmodium gallinaceum* is unable to develop in *Aedes aegypti* mosquitoes genetically engineered to over-express defensin-A and cecropin-A (Kokoza et al. 2010). There are safety and ethical issues surrounding the mass-release of genetically engineered vectors of pathogenic diseases (Resnik 2018). In 2020, there was considerable controversy in the USA when the EPA (Environmental Protection Agency) approved the biotech company Oxitec to conduct trials involving the release of genetically modified *A. aegypti* in Florida and Texas. At the time of writing, there had been only a few similar field trials in other countries and inevitably the proponents claimed encouraging results, whilst organisations such as Genewatch (www.genewatch.org) state that the results are not as good as claimed and pose serious risks.

There is limited information on the immune responses of nematodes to pathogens. However, many parasitic species live within the microbe-rich gastrointestinal tract and must possess immune mechanisms that protect them from infections. The coelomocytes of nematodes exhibit a variety of morphologies, but there are usually very few of them (typically 2, 4, or 6) within the pseudocoelom and their filopodia (fine cytoplasmic processes) anchor them to the inner surface of the body wall. The coelomocytes in the nematodes are therefore incapable of encapsulation although they will take up bacteria. Their role within the nematode immune response is therefore uncertain. The free-living nematode *Caenorhabditis elegans* expresses a range of AMPs such as defensins, as well as producing reactive oxygen species in response to pathogen challenge. The AMP cecropin P1 is present in both *Ascaris lumbricoides* and *A. suum*. It is primarily active against Gram-negative bacteria such as *Pseudomonas aeruginosa* and to lesser extent Gram positives such as *Staphylococcus aureus* (Baek et al. 2016). *Ascaris suum* also produces defensins, lysozyme, and a variety of other antimicrobial substances. The worms release some of these into their surroundings, and this may affect the composition of the host's gut microbiome (Midha et al. 2018).

Information on the immune responses of parasitic trematodes and cestodes is even more limited than that on nematodes. However, it would be surprising if they did not exhibit an ability to distinguish self from non-self and the means to prevent pathogens from invading their tissues. Free-living planarians possess phenoloxidase activity, AMPs, C-type lectins, and rapidly phagocytose, and remove bacteria present in their food or injected into their body (Abnave et al. 2014; Gao et al. 2017).

Although they do not produce antibodies, there is increasing evidence of 'immune memory' (also referred to as 'immune priming') among invertebrates (Milutinović and Kurz 2016). In addition, some invertebrates pass immunity to their offspring (Tetreau et al. 2019). For example, exposure of *Anopheles gambiae* mosquito larvae to the microsporidian *Vavraia culicis* results in them producing offspring that are more resistant to infection by the microsporidian and less likely to transmit malaria (Lorenz and Koella 2011). Invertebrates therefore exhibit traits similar to those of adaptive immunity in vertebrates although the physiological mechanisms underlying it are currently uncertain. In some invertebrates, 'memory' results from a sustained response to the initial infection although whether this really 'memory' or an 'ongoing response' is difficult to establish. In others, there is better evidence of 'memory' because once the level of the initial immune response declines, a secondary challenge results in a faster and stronger immune reaction. In both the above, the physiological nature of the immune response to initial and secondary challenge are the same. However, in some invertebrates, the two responses are different. For example, when *Schistosoma mansoni* miracidia initially infect the snail *Biomphalaria glabrata*, they trigger a cellular immune response that encapsulates the invading parasites. However, a subsequent challenge stimulates a humoral response characterised by the production of fibrinogen-related proteins (FREPs) and lipopolysaccharide binding protein/ bactericidal/permeability increasing protein (LBP/BPI – lipopolysaccharide binding protein/bactericidal/permeability-increasing protein), and biomphalysin (Pinaud et al. 2019). In most experiments on invertebrate immune memory, the nature of the initial and secondary challenge is the same, and it is therefore uncertain whether the 'memory' involves any specificity. In real life, an invertebrate is likely to experience frequent challenges from various pathogens.

9.3 Vertebrate Immunity

A common practice is to consider the vertebrate immune system as two distinct branches – that is the innate and adaptive components – although in reality they work together and influence one another (Table 9.2). For example, natural killer cells (NK cells) are a subset of the lymphocytes that form part of the initial innate immune response, but they also exhibit adaptive processes and are important in shaping the adaptive immune response.

9.3.1 Innate Immunity

One can divide innate immunity into physical, microbial and physiological aspects. Physical barriers, such as the skin and the mucus secreted over the epithelium of the gut and respiratory tract, prevent many infectious agents from gaining access to the body. The natural microbial flora that live externally upon the body surfaces and within the intestinal and genital tracts also plays a major role in both preventing and facilitating infections. The resident microbial flora help prevent invasion through a combination of competition and producing antimicrobial substances. There is also evidence that the human gut microbiome modulates susceptibility to soil transmitted helminths (Huwe et al. 2019).

Physiological defence mechanisms include the secretion of chemicals onto the outer body surface and those employed when an invading organism breaches the outermost physical barriers and gains entry into the body. They include AMPs such as defensins in the lungs and gastrointestinal tract, leukocytes, the complement system, acute phase proteins, and the acute inflammatory response. Leukocytes are located within tissues (e.g., tissue macrophages, dendritic cells) and circulate within the blood (e.g., eosinophils, neutrophils, natural killer cells and lymphokine-activated killer cells). These cells recognise invading organisms and kill them by either phagocytosing and then digesting them or releasing noxious chemicals such as hydroxyl radicals (˙OH) and antibiotic peptides that kill them.

Table 9.2 The sequence of major immune responses that occur in a typical mammal following parasite invasion.

Time after invasion	Type of immune response	Nature of immune response
Minutes to hours	Innate immune response	Recognition by leukocytes
		Non-specific phagocytosis (neutrophils, macrophages)
		Cytokine release (e.g. alpha tumour necrosis factor [TNFα], interleukin-1, interleukin-6, interleukin-8, and others)
		Release of toxic chemicals (e.g. hydroxyl radicals, nitric oxide)
		Complement activation (alternative pathway)
Minutes to hours	Early induced response, inflammation	Recruitment of neutrophils, monocytes, natural killer cells etc.
		Differentiation to effector cells
		Cytokine release
		Release of toxic chemicals (e.g. hydroxyl radicals, nitric oxide)
		Complement activation
Days to weeks	Adaptive immune response	Transport of antigen to lymph nodes
		Antigen presentation
		Recognition by B- and T-lymphocytes
		Production and release of antibodies
		Lymphocyte maturation
		Clonal expansion
		Differentiation to effector cells (cytotoxic T lymphocytes etc.)
		Activation of natural killer cells
		Major histocompatibility complex-1 expression (MHC-1)
Months to years	Protective immunity	B-cells differentiate into memory cells

Source: Modified from Schmid-Hempel (2008).
Note: The actual times taken for individual reactions to take place after pathogen invasion are very case-dependent.

Pathogen recognition is possible because leukocytes have pattern recognition receptors that recognise unique 'pathogen-associated molecular patterns' (PAMPS). There are several categories of pattern-recognition receptors although one can divide them broadly into those that function as 'signallers' and those that promote endocytosis. Among the 'signalling' pattern recognition receptors are the membrane bound Toll-like receptors and the cytoplasmic NOD-like receptors. For example, stimulation of the Toll-like receptors on a macrophage induces it to secrete inflammatory cytokines. There are various Toll-like receptors, and each type is primed to recognise a panel of internal and external stimuli. To date, 10 Toll-like receptors have been characterised from humans and 13 from mice. Toll-like receptor 4 (TLR4) responds to lipopolysaccharide and when stimulated it initiates a signalling cascade resulting in the production of proinflammatory cytokines and the differentiation of T helper 1 cells (Th1 – see later). Other Toll-like receptors initiate similar cascades and, in addition, there is expression of co-stimulatory molecules such as CD14, CD40, CD80, CD86, and major histocompatibility complex II. CD14 is expressed on the membranes of macrophages, neutrophils, and dendritic cells and also exists in a soluble form. CD14 is similar to

Toll-like receptor 4 in that it detects bacterial lipopolysaccharide and is therefore involved in the immune response to *Wolbachia* bacteria that is important in the pathology of filarial nematode infections. Among the endocytic pattern recognition receptors are the mannose receptors on the surface of macrophages and dendritic cells that initiate phagocytosis via the complement system. Although pattern recognition receptors are clearly important for the detection and removal of pathogens, some parasites subvert or exploit them for their own purposes. For example, *Leishmania donovani* exploits the mannose receptor to gain entry to macrophages (Horta et al. 2020).

Dendritic cells gain their name from their wispy processes that gives them a tree-like appearance. There are four types of dendritic cell: Langerhans cells (in the skin), interstitial dendritic cells, myeloid cells, and lymphoid dendritic cells. Immature dendritic cells recognise invading pathogens and phagocytose them after which they transform into mature dendritic cells, process their captured material and move to the secondary lymphoid tissues (spleen and lymph nodes). Once there, they present antigens and non-specific stimulatory signals to naive T helper cells, T killer cells, and B lymphocytes, thereby activating them. For example, within the skin, the Langerhans cells constantly monitor antigens passing through the *stratum corneum*. Interestingly, both *Leishmania* and HIV-1 compete for the same membrane receptor (DC-specific ICAM-3-grabbing nonintegrin [DC-SIGN]) in order to gain entry into dendritic cells (Andreani et al. 2012).

The complement system is a triggered multi-component enzyme cascade found in the plasma. Once activated, the complement system initiates numerous events. For example, components of the cascade attract phagocytes to invading microorganisms, increase the permeability of capillaries, and mediate the acute inflammatory response. Tissue injury, an infection, or anything stimulating an inflammatory response causes macrophages and T cells to secrete interleukin-6, and this induces an increase in the levels of acute phase proteins such as C-reactive protein within the circulation. C-reactive protein binds to lysophosphatidylcholines present on invading microorganisms and dead cells. The act of binding activates the complement cascade and the target is opsonised – and this stimulates phagocytes to engulf it. *Leishmania donovani* exploits C-reactive protein to gain entrance to macrophages and it stimulates the transformation from the promastigote to amastigote stage. Lipophosphoglycan on the promastigote cell membrane binds to C-reactive protein, and this increases their uptake into macrophages in a process mediated by their mannose receptors but without activating the cells.

Innate Immunity to Trypanosome Infection

There are numerous species of animal trypanosomes and they occur in many parts of the world (including the UK) but very few species infect humans. This is because humans naturally express apolipoprotein L-1 (APOL1) in their serum that has trypanolytic activity (Pays and Vanhollebeke 2009). The structural similarity of APOL1 to certain other regulatory molecules indicates that it was probably originally involved in normal apoptosis mechanisms (programmed cell death) and subsequently evolved a role in the immune response against pathogens. In susceptible trypanosome species, APOL1 creates anionic pores in the membrane of the parasite's lysosomes. There is a higher ion concentration inside the lysosomes compared to the surrounding cytoplasm, and therefore osmosis occurs and water flows through these pores into the lysosomes. This causes the lysosomes to swell uncontrollably and burst. The subsequent release of lysosomal enzymes causes lysis and death of the trypanosome. *Trypanosoma brucei rhodesiense* is probably able to avoid this fate and therefore infect humans because it expresses a VSG-like protein (variant surface glycoprotein) that neutralises the effects of

APOL1 (Gibson 2005). In *T. brucei rhodesiense* the serum-resistance associated gene (SRA) codes for VSG-like protein but this gene does not occur in *Trypanosoma brucei gambiense*. It is therefore uncertain how this species, which also infects humans, avoids the effects of APOL1.

Trypanolytic activity probably evolved during the course of human evolution (Thomson et al. 2014). In Africa, ground-dwelling early hominids would have been continually exposed to tsetse flies and other biting flies and therefore challenged by trypanosome parasites. Indeed, there is a suggestion that susceptibility to trypanosome infections may have influenced the evolution of early hominids. Several African primates express trypanolytic activity although, surprisingly, it is absent among our nearest living relatives the chimpanzees (*Pan troglodytes*) and bonobos (*Pan paniscus*). One suggestion for this is that chimpanzees are less exposed to the bites of tsetse flies because they spend much of their time living in trees rather than on the ground. In support of this hypothesis is the fact that Gorillas (*Gorilla gorilla*), which spend most of their lives on the ground, do express APOL1. However, the theory starts to fall apart when one considers that chimpanzees actually spend a lot of time on the ground – sometimes over 60% – and usually travel between resting and feeding sites on the ground rather than by swinging through the trees.

Factors that confer resistance to pathogens are sometimes associated with increased risks to other diseases. For example, the 'faulty' genes that result in sickle cell anaemia persist because they confer increased resistance to malaria. Similarly, although APOL1 confers resistance to trypanosome infections, genetic variants in the APOL1 gene are associated with kidney failure. Indeed, the reason that African Americans are 4–5 times more likely to develop kidney failure than those of European descent appears linked to genetic variants in the APOL1 gene (Friedman and Pollak 2016).

9.3.2 Adaptive Immunity

There are two types of adaptive immunity: passive and active. Passive adaptive immunity occurs when an animal receives antibodies (immunoglobulins) from another organism. In mammals, this typically occurs when the developing foetus receives antibodies from its mother across the placenta and within colostrum. The levels of these antibodies subsequently decline over a period of weeks or months and the conferred immunity is lost. The process allows the neonate to survive challenge from infections in the first few months of life while it is very vulnerable to disease and before its own immune system is fully functional. Active adaptive immunity occurs when the animal mounts its own immune response to challenge by live pathogens – as in an infection – or to dead or disabled pathogens – as in a vaccine.

Whether it is passive or active, adaptive immunity represents a specific response to specific antigens and is primarily mediated by lymphocytes. In humans, there are three types of lymphocytes (Table 9.3) and these have a variety of functions. All of the lymphocytes produce cytokines although the composition varies between cell types. There is a complex cross-regulation between the cytokines and the cytokines themselves affect the activity of both immune and non-immune cells, as well as, in some cases, acting directly upon pathogens.

One can distinguish between lymphocytes at a molecular level by their membrane receptors, each lymphocyte carrying the receptor for a particular ligand (antigen). Binding of a ligand to the receptor stimulates the lymphocyte to divide and hence the numbers of lymphocytes possessing this particular receptor increases in a process called 'clonal expansion'. Most of the cells are 'effector cells' that perform the job they are designed to do and then die after a few days. Some of them,

Table 9.3 Classification of lymphocytes.

Lymphocytes	Response to stimulation
B lymphocytes	Develop into B cells that secrete antibodies (immunoglobulins)
T lymphocytes	Regulate other immune cells or kill virus-infected cells
Natural killer (NK) lymphocytes	Kill virus-infected cells and tumour cells.

however, become 'memory cells': these are long-lived cells and their subsequent stimulation enables rapid clonal expansion in which effector cells and memory cells form more quickly than in the initial (primary) immune response. These memory cells enable the mounting of a protective secondary immune response against subsequent infections by a particular pathogen. The aim of vaccination is to induce this without exposing the person or animal to the wild type infection.

B-lymphocytes are produced in the bone marrow and when activated they form plasma cells (effector cells) that secrete antibodies into the surrounding plasma. This gives rise to 'humoral immunity'. These antibodies are glycoproteins that are also referred to as immunoglobulins or, collectively, gamma globulins (γ globulins). In mammals, there are five classes of immunoglobulins (Table 9.4), and these in turn divide into subclasses. Teleost (bony) fish lack immunoglobulin A (IgA) but produce their own immunoglobulin known as IgT (trout or teleost) or IgZ (zebra fish). Immunoglobulin T, like IgA, is important in mucosal immunity and its secretion into the intestine influences the composition of the gut microbiome. Immunoglobulins that are secreted lack a trans-membrane region and are known as soluble immunoglobulins. Immunoglobulins that are bound to the membrane surface of lymphocytes are known as membrane immunoglobulins or surface immunoglobulins. The basic immunoglobulin monomer is a Y-shaped molecule, the stem of which is the 'Fc region' (Fragment, crystallisable region) and the two arms are the 'Fab regions' (Fragment, antigen-binding region). The structure of the Fc region varies between the different classes of immunoglobulins although within a class it is constant. The Fc region binds to specific receptors (Fc receptors) and other immune molecules, such as those in the complement cascade,

Table 9.4 Classification, distribution, and function of mammalian immunoglobulins.

Immunoglobulin	Distribution and functions
IgG	Major (~75%) immunoglobulin in plasma and extravascular spaces in adult humans; associated with secondary immune response; transmitted to foetus via the placenta and colostrum, activates complement, binds to Fc receptors
IgA	Second highest level in serum; present in external secretions such as intestinal mucus, saliva, tears, and colostrum; does not activate complement; binds to Fc receptors on some cells
IgE	Least common immunoglobulin. Involved in allergic reactions: induce mast cells to release histamine and other chemical mediators; does not activate complement; binds to Fc receptors on eosinophils
IgM	Third highest level in serum; activate complement; first immunoglobulin made by foetus; first class of antibody raised in primary infection; react to blood group antigens; binds to Fc receptors
IgD	Fourth highest level in serum; present on surface of B lymphocytes; does not activate complement; role(s) uncertain although can activate basophils and mast cells

Table 9.5 Mechanisms by which immunoglobulins bring about the removal of pathogens and toxins.

Soluble immunoglobulins attach to antigens present on objects and thereby identify them to phagocytic cells as things they should phagocytose. That is, the objects become 'opsonised'

Soluble immunoglobulins bind to antigens and cause them to 'clump': this effectively 'increases the size of the target' and thereby increases the chances of phagocytosis

Immunoglobulins bind to certain toxins and thereby render them non-toxic

Following the binding of an antigen to an immunoglobulin, the Fc end of the immunoglobulin activates the complement cascade

IgE is bound to the surface of mast cells and the attachment of an antigen to the IgE induces the mast cell to release histamine and other chemicals that mediate the inflammatory response

The binding of an antigen to an immunoglobulin brings about the activation of antibody-dependent immune cells via Fc binding. For example, this can result in the activation of B cells and the production of antibodies or the initiation of phagocytosis by macrophages

---

thereby initiating a variety of physiological reactions. For example, if the Fc receptor resides on a phagocyte, then the binding of an immunoglobulin with the appropriate Fc characteristics will initiate phagocytosis. By contrast, if the Fc receptor is on a cytotoxic cell, then the binding of the immunoglobulin to the Fc receptor will cause destruction of the cell.

Antibodies are not themselves harmful to invading pathogens, but they initiate a variety of physiological processes that result in their removal from the body (Table 9.5).

9.3.3 Cell-Mediated Immunity

Antibodies are only effective against extracellular pathogens. This is because once a pathogen enters a cell its antigens are no longer accessible. Therefore, another branch of the immune system, the 'cell-mediated immune response', in which T lymphocytes play a central role, deals with intracellular parasites. T lymphocytes have receptors on their outer cell membrane called 'T-cell receptors' that bind to cells displaying antigens as part of a major histocompatibility complex (MHC) on their cell membrane.

Major histocompatibility complexes are a family of membrane proteins present in most vertebrates. The MHC complexes act as signals to T lymphocytes by taking fragments of an infected cell's own proteins, as well as those of any invading pathogen and displaying these on the cell surface. In humans, there are two types MHC complex: Class I and Class II. Class I MHC proteins occur in all nucleated cells, and they interact with cytotoxic T cells and natural killer cells. Antigens derived from intracellular pathogens presented by MHC class I proteins are detected by cytotoxic T cells and natural killer cells and these then kill the infected cells. They accomplish this by secreting the cytolytic protein 'perforin' and the serine protease 'granzyme B'. Perforin forms a pore in the target cell membrane through which granzyme B enters. Once inside the target cell, granzyme B induces it to undergo apoptosis. Cytotoxic T cells carry a surface glycoprotein called CD8 and hence another name for them is CD8+ T cells. Class II MHC proteins are found predominantly on B-lymphocytes, macrophages, and dendritic cells and they interact with T helper lymphocytes (Th cells). Mature T helper cells express a surface glycoprotein called CD4: hence, the moniker CD4+ T helper cells (CD4+ Th cells). T helper cells do not directly kill antigen-presenting cells, but instead produce cytokines that influence the behaviour of other immune cells.

There are three types of T helper cell: T helper 1 (Th1), T helper 2 (Th2), and T helper 17 (Th17). These cells have different roles and they cross regulate one another (Table 9.6). The Th17 cells are a

Table 9.6   T helper lymphocytes and some of their functions.

	T helper cell 1 (Th1)	T helper cell 2 (Th2)	T helper cell 17 (Th17)
Main cell type interaction with	Macrophages	B lymphocytes	Interleukin-17 receptors and interleukin-22 receptors are widely distributed. They include T cells, B cells, vascular endothelial cells, fibroblasts, and several other cell types
Cytokines released	Interferon gamma (IFN γ) Tumour necrosis factor (TNF), interleukin-3 Interleukin-10	Interleukin-3 Interleukin-4 Interleukin-5 Interleukin-6 Interleukin-9 Interleukin-10 Interleukin-13 Interleukin-25 Interleukin-33	Interleukin-17 Interleukin-17F Interleukin-6 Interleukin-21 Interleukin-22 Tumour necrosis factor (TNF)
Main effect	Stimulation of cellular immunity through promoting phagocytosis by macrophages and the production of cytotoxic T cells and natural killer (NK) cells	Stimulation of humoral immunity through promoting the proliferation of B lymphocytes and antibody production	Induce tissue inflammation (e.g., through recruitment and activation of neutrophils
Other effects	IFN-γ stimulates dendritic cells and macrophages to release interleukin-12. Interleukin-12 stimulates Th1 cells to release IFN-γ in a positive feedback loop. IFN-γ inhibits the production of interleukin-4 and thereby reduces the Th2 response	Interleukin-4 stimulates the production of itself and other Th2 cytokines. Interleukin 10 inhibits macrophages from presenting antigens, producing cytokines, and killing pathogens using nitric oxide (NO). Interleukin-13 is important in the expulsion of gut helminths and the response to schistosome eggs	Stimulate target cells to produce antimicrobial substances (e.g., β defensin-2 and β defensin-3). Mediates protection against specific pathogens such as fungus *Candida albicans*

distinct group of CD4+ Th cells and important in initiating and regulating inflammation. They inter-act with cells bearing interleukin 17 receptors and interleukin 22 receptors: these occur on numerous cell types. Therefore, activation of Th17 cells results in a powerful tissue response. CD4+ CD25+ FoxP3+ regulatory T cells (Tregs) provide a further level of control. Tregs suppress the activation of other immune cells and are therefore important in preventing excessive immune responses and autoimmune conditions. They should not, however, be considered a simple brake whose level of activation directly correlates with that of the pro-inflammatory process to stop it getting out of control. Tregs are involved in the immune response to many parasites and parasites may in turn influence the activity of Treg cells to improve their chances of survival (Maizels and McSorley 2016). In patients infected with *Plasmodium vivax* and *Leishmania donovani*, the number of activated Treg cells increases and their number is associated with the parasite load (Bhattacharya et al. 2016; Bueno et al. 2010).

9.4 Innate Immunity to Parasites

9.4.1 Physical Factors

The physical integrity of the outer body surfaces and the lining of the gastrointestinal tract, lungs, etc. prevents many potential pathogens from establishing infections. For example, first-instar larvae of the New World screwworm fly, *Cochliomyia hominivorax*, cannot penetrate normal healthy skin. Consequently, the female flies lay their eggs on pre-existing wounds. After hatching, the larvae invade the wound, and extend it, and this attracts further flies to lay their eggs on the wound. All animals therefore have mechanisms for rapidly sealing and repairing wounds before these serve as entry portals. Similarly, within the gastrointestinal tract, the cytokine interleukin-10 is important for the maintenance of the integrity of the gut epithelium. If the levels of interleukin-10 fall too low then the epithelium becomes more permeable, and this facilitates invasion by parasites such *Entamoeba histolytica* (Redpath et al. 2014).

9.4.2 Chemical and Microbial Factors

A host can protect itself from pathogens by releasing secretions over its outer and inner body surfaces. These may protect the host through being either toxic and/or preventing pathogens from making contact with the body. For example, the constant secretion of mucus can make it difficult for invading pathogens to make contact with the underlying epithelial cells. The highly acidic nature of the stomach contents is sufficient to kill many organisms and most gastrointestinal parasites pass through this region rather than dwell here on a permanent basis. Sometimes, specific stimuli induce the secretions. For example, mice normally produce the mucin 'Muc5ac' in their lungs but not in the intestine. However, mice infected with the gut nematode *Trichuris muris* produce a lot of Muc5ac in their caecum. Genetically engineered mice that are unable to express Muc5ac cannot expel *Tr. muris* and remain vulnerable to infection despite mounting a strong Th2 type immune response (Hasnain et al. 2011).

Bodily secretions often contain microbes, and these have varying influences on host : pathogen relationships ranging from preventing infection to facilitating invasion. Within the vagina, the pH of the vaginal fluid influences colonisation by the protozoan parasite *Trichomonas vaginalis*. The normal vaginal pH is relatively acidic (pH 4.5), whilst *Tri. vaginalis* grows better in a more alkaline environment. Maintenance of the acidic conditions depends upon a combination of oestrogen levels that promote the secretion of lactic acid and the resident bacterial flora which secrete lactic acid as a waste product. The bacterium *Lactobacillus acidophilus* is an important member of the vaginal flora (and yoghurts) and contributes to the lactic acid present in vaginal secretions. Several other species and phylotypes of *Lactobacillus* also commonly co-occur and they vary in their contribution to vaginal health. There are five principal types of vaginal microbiome: these differ in their microbial diversity and the relative abundance of *Lactobacillus* species. Infection with *Tri. vaginalis* is frequently associated with the type IV microbiome that has a very diverse microbial flora including many anaerobic species, fewer lactobacilli, and an elevated pH. The type IV microbiome is also associated with bacterial vaginosis and hence damage to the vaginal epithelium. It therefore becomes a chicken and egg scenario of whether *Tri. vaginalis* infection disturbs the microbiome and predisposes a woman to bacterial vaginosis or whether altering the composition of the microbiome changes the pH, damages the vaginal epithelium, and thereby predisposes her to *Tri. vaginalis* infection (Mercer and Johnson 2018). Interestingly, many *Tri. vaginalis* harbour a double stranded RNA virus – *Tri. vaginalis* virus – and a mycoplasma *Mycoplasma hominis* and host reaction to these results in an enhanced inflammatory response. The immune response to *Tri. vaginalis* infection therefore represents a complex interplay between the responses to the parasite, the presence or absence of its virus and mycoplasma partners, and the microbes that comprise the vaginal microbiome.

9.4.3   Acute Inflammatory Response

To exploit its host a parasite must first attach to and then physically penetrate its host's body – even if this is only to insert its feeding apparatus. In the process, the parasite advertises its presence through the production of specific chemicals and causing cell damage. It is at this point that the parasite engenders an initial acute inflammatory response – this can be divided into a series of stages:

1) Leukocytes detect the invader or cell damage and phagocytose the debris and foreign material. They also release cytokines and other inflammatory mediators such as tumour necrosis factor (TNF), interleukin-1, and interleukin-6.
2) The surrounding capillaries dilate and become more permeable to proteins and fluid. This facilitates the movement of immune cells, proteins etc. into the damaged area where they mediate the defensive response. The damaged area therefore becomes semi-liquid and oedema can develop.
3) Leukocytes migrate to the site of the infection. Neutrophils are the main mediators of the acute immune response.
4) Mast cells release immune mediators that amplify the response.
5) If effective, this localises the foreign material and stops it from spreading further into the body. The leukocytes clear the infection through a combination of phagocytosis and secreting inflammatory cytokines.
6) If the amounts of cytokines released becomes excessive, they enter the general circulation and affect organs distant from the site of invasion. The rise in cytokine titres in the circulation results in a systemic inflammatory response. For example, interleukin-1 is involved in the development of fever and anorexia, whilst interleukin-6 stimulates hepatocytes in the liver to release acute phase proteins.
7) Clearance of the infection results in the inflammatory response subsiding and tissue repair commencing. However, if the object or organism inducing the response remains, then the acute phase progresses to chronic inflammation that is characterised by the formation of granulation tissue and fibrous scar formation.

Many parasitic protozoa express adhesin molecules on their cell surface that enable them to attach to host cells. For example, in *E. histolytica* these include lipophosphoglycan (LPG) and lipopeptidophosphoglycans (LPPG) bound to the parasite cell membrane by glycophosphatidylinositol (GPI) anchors. Secretion of these molecules together with their GPI anchor into the surrounding area influences both the parasite's virulence and the host's ability to mount an immune response. For example, Toll-like receptor-2 and Toll-like receptor-4 recognise the GPI anchor and membrane molecules such as glycoinositol phospholipid, lipophosphoglycans, and lipopeptidophosphoglycans, and therefore initiate an inflammatory response. In the case of *E. histolytica* this results in the recruitment of numerous neutrophils to the affected region. Although inflammation can be protective, it may also result in localised tissue damage and during the initial stages of invasion by *E. histolytica*, this manifests as abdominal cramping and/or diarrhoea that is symptomatic of amoebic dysentery. Neutrophils kill invading organisms by a combination of phagocytosis and releasing toxic substances such as reactive oxygen species.

Although neutrophils can kill *E. histolytica*, the toxic chemicals they release damage the gut epithelium – and this may facilitate parasite invasion (Nakada-Tsukui and Nozaki 2016). The end result is therefore a balancing act between the killing of invading *E. histolytica* and the parasite's abundance and ability to invade (virulence). As we have seen, numerous other factors can influence this balance. For example, although neutrophils can kill *E. histolytica* trophozoites, the trophozoites can kill the neutrophils to an extent that varies between parasite strains. One calculation suggests that a single trophozoite of a highly virulent strain can destroy several thousand neutrophils (Guerrant et al. 1981).

Activation of Toll-like receptors is an important first step in the immune response to many parasitic protozoa infections (Ashour 2015). For example, glycophosphatidylinositol anchors and glycoinositolphospholipids derived from *Trypanosoma cruzi* activate Toll-like receptor-2 pathways to initiate changes in macrophage function. Similarly, one of the main causes of the pathology associated with malaria results from the excessive immune response triggered by glycosylphosphatidylinositol anchors acting via Toll-like receptor-2 pathways.

Another important adhesin molecule found on the cell surface of the trophozoites of *E. histolytica* is a lectin known as Gal/GalNAc-lectin because it recognises galactose (Gal) and *N*-acetyl galactosamine (GalNAc) – which are common components of the mucosa and epithelial cells. Gal/GalNAc stimulates both cell-mediated and humoral immune responses. The lectin stimulates Toll-like receptor 2 pathways to produce interleukins 1β, 6, 8, and 12, as well as gamma interferon and tumour necrosis factor. Gal/GalNAc-lectin also induces the secretion of IgA from plasma cells in the lamina propria that can prevent parasite adherence. Indeed, experimental trials indicate that vaccination with *E. histolytica* Gal/GalNAc lectin generates a protective immune response against the parasite (Houpt et al. 2004). The secretory IgA, however, also down-regulates the inflammatory response and influences host-gut microbial flora interactions. In addition, the parasite secretes cysteine proteases that break up the IgA and render it inactive.

Apart from membrane components, a variety of other substances can also stimulate the acute inflammatory response. For example, the DNA of *E. histolytica* stimulates an acute inflammatory response by activating Toll-like receptor-9 on macrophages, thereby inducing them to secrete tumour necrosis factor (Ivory et al. 2008). Chitin fragments can stimulate an inflammatory response through activating Toll-like receptor-2 pathways that then induce macrophages to produce the proinflammatory cytokine interleukin-17 (Uribe-Querol and Rosales 2020). Chitin is a major component of arthropod cuticle, the cuticle of nematodes, and the cell walls of fungi, and therefore this may be an important feature of the immune response to these organisms.

9.4.4 Cell-Mediated Immune Reactions

The nature of the immune response to parasitic protozoa is somewhat different to that raised against helminths. This is because protozoa are small enough to phagocytose and some of them are intracellular. By contrast, helminths are too large to phagocytose and most are extracellular parasites. The larvae of *Trichinella spiralis* living inside striated muscle cells are an obvious exception although these cells are unusually large and the larvae are still too big to phagocytose.

A Th1 response is protective against many extracellular (e.g., *E. histolytica*) and intracellular protozoan parasites (e.g., *Leishmania* spp., *Toxoplasma gondii*, *Plasmodium falciparum*). By contrast, a Th2 response tends to favour their survival and multiplication. The Th1 response promotes cell-mediated immunity by activating neutrophils and macrophages and stimulating them to produce nitric oxide. By contrast, a Th2 response induces cytokines that suppress the production of gamma interferon (IFN-γ). Because IFN-γ has an important role in stimulating neutrophils and macrophages etc., its suppression effectively down-regulates the cellular immune response – which is protective against these parasites.

Helminth infections in mammals induce a strong CD4+Th2 response in which there are elevated levels of interleukin-4, interleukin-5, and interleukin-13. These cytokines bring about the mobilisation and expansion of effector cells such as mast cells, eosinophils, and basophils. There are also raised levels of total and parasite-specific immunoglobulin E (IgE). Although interleukin-4 is usually thought crucial for initiating the Th2 response, the situation is probably more complicated and other Th2 inducing cytokines might be equally or more important – at least in some helminth infections. An effective Th2 response is essential for clearing many helminth infections but if the response is excessive or prolonged it causes pathology. The simultaneous induction of regulatory T cells provides some level of control although this may also prevent a strong enough Th2 response to clear the infection (Sorobetea et al. 2018).

9.5 Adaptive Immune Reactions to Parasites

Although many parasites induce antibody responses, whether these are protective depends upon numerous parasite and host factors such as genetics, age, health and so on. Host adaptive immune responses therefore vary considerably between parasites and between individuals to a particular species of parasite. For example, antibodies against *Tri. vaginalis* are detectable in the serum and vaginal fluids of infected women but even after repeated infections, these do not provide protective immunity (Nemati et al. 2018). In other cases, concomitant immunity develops in which the host is unable to remove the resident parasite population but becomes resistant to newly invading parasites (Brown and Grenfell 2001). For example, immune reactions against filarial nematode infections usually target the infective third-stage larvae rather than the adult worms. In this situation, the resident parasite reduces or prevents competition but remains able to reproduce and therefore act as a source of infection for other hosts. By contrast, the invasion of infective larvae of the nematode *Haemonchus contortus* (Figure 9.1a–c) precipitates the expulsion of the adult worms in a phenomenon known as 'self-cure'. Adult *H. contortus* live in the abomasum sheep; they are relatively large worms, growing up to 3 cm in length, and they move about the surface of the mucosa, which they pierce with their lancet-like mouthparts to feed on blood. Their eggs pass with the host faeces and after hatching on pasture-land, they develop to infective third-stage larvae. The sheep then consume the larvae whilst grazing. Heavy rain provides ideal conditions for egg hatching and larval survival, and it is associated with the sudden mass-expulsion of the adult worms. This is probably because large numbers of invading third-stage larvae trigger an increased immunoglobulin E (IgE) titre that results in the expulsion of the adult worms. Nevertheless, the invading larvae sometimes continue to develop to adulthood, and there are suggestions that the 'self-cure' reaction may actually be a result of a non-specific immune reaction to a constituent of freshly growing grass rather than *H. contortus* larvae (Taylor et al. 2007).

In the case of gastrointestinal helminth infections, protective immunity is usually a consequence of Th2-mediated responses although there are considerable differences between species in their effectiveness. By contrast, chronic infections with tissue helminths tend to generate a mixed Th1–Th2 immune response (Gazzinelli-Guimaraes and Nutman 2018). A characteristic feature of both gastrointestinal helminth infections and tissue helminth infections is the stimulation of elevated levels of parasite-specific immunoglobulin E (IgE). Elevated IgE titres are not normally a feature of other infectious diseases although they are of allergic reactions. For example, a rise in IgE titre is associated with an immediate hypersensitivity reaction and is preventable by the administration of antihistamine drugs. Unfortunately, IgE, with a few exceptions, seldom provides effective

(a)

(b)

(c)

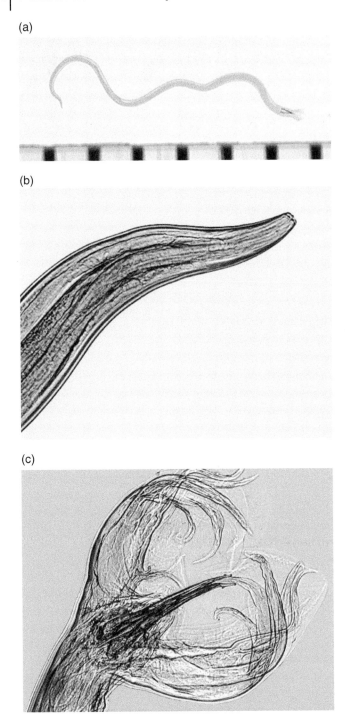

Figure 9.1 *Haemonchus contortus*. (a) Whole worm (male), scale in mm. (b) Anterior of worm. (c) Posterior of worm showing copulatory bursa and spicules.

protection against helminth infections. One of these exceptions is the expulsion of *T. spiralis* from the intestinal lumen of rats, following the secretion of IgE from the plasma into the gut lumen via transporting epithelial cells. There is also some evidence for IgE having a protective effect against tissue-dwelling *T. spiralis* in rats but in humans, the role of IgE in immunity to *T. spiralis* is less certain (Watanabe et al. 2005).

Gastrointestinal helminth infections also cause elevated titres of other immunoglobulins and in particularly IgG1 and IgG4, which, together with IgE, are under the control of Th2 cytokines although the importance of these varies between species. For example, immunity to the nematode *Nematodirus battus* is associated with increased titres of IgM, IgG1, and IgA. This nematode lives in the duodenum of sheep and induces a strong protective immunity. The immunoglobulins act in concert with other immune changes and in particular the mass shedding of the villi lining the duodenum. The nematodes coil around the villi and thereby use it as a 'holdfast' – with the loss of their anchor the worms are quickly shed (Winter et al. 1997). The strength and longevity of the immune response means that *N. battus* is only pathogenic in previously uninfected lambs although adult sheep may be infected and remain a source of infection for others. Interestingly, there is not an increase in the levels of IgA associated with infection with adult hookworms. This is possibly because proteases released by the nematode destroy the IgA once it is secreted. IgA is important in protecting against harmful gut bacteria. Therefore, a decline in IgA levels may influence the development of these infections. For example, a hookworm (*Uncinaria* spp.) enteritis plus bacteraemia complex found in the California sea lion (*Zalophus californianus*) population in US waters has killed many seal pups (Spraker et al. 2007). However, although enteritis is a feature of severe hookworm infestations in humans and a range of other mammals, there is limited information on the extent to which bacteria contribute to the pathology of the helminth infection.

9.6 Microbiomes and Host Immune Reactions to Parasites

A normal 'healthy' microbiome is essential for many aspects of our physiology. Most regions of our body have a resident microbiome, the composition of which is dynamic rather than fixed. Parasites also have their own microbiomes in the form of intracellular symbionts and, in metazoan parasites, microbes that live within or upon their bodies. Therefore, a host : parasite relationship has potentially four players: the host, the parasite, the host's microbiome, and the parasite's microbiome. The microbiome, of both host and parasite, can influence all aspects of host : parasite interactions. This includes invasion, survival, virulence, and the development of a protective host immune response in which a 'change' by one of the players results in a change by the other three, which then results in a further change by all four. In addition, the parasites may also feed on the host's microbiome, as is the case for many gastrointestinal parasites, and release antimicrobial substances. The situation is even more complex than this because an organism commonly harbours more than one species of parasite. Therefore, the parasites may be interacting both directly with one another and indirectly via their host and their respective microbiomes.

Disentangling the relationship between host, parasite(s), and microbiomes is extremely difficult. In some instances, the host microbiome is protective and prevents a pathogen from establishing. For example, in bumblebees, the composition of the gut microbiome influences the ability of the trypanosomatid parasite *Crithidia bombi* to establish an infection (Mockler et al. 2018). Alternatively, the presence of a microbiome may be essential for the parasite. For example, it is impossible to establish experimental *E. histolytica* infections in germ-free guinea pigs (Phillips et al. 1955). The microbiome in the guinea pig gut produces many glycosidases that break down complex carbohydrates into smaller molecules that the amoebae absorbs and utilises. Although the amoebae produce their own glycosidases, presumably the activity of the host microbiome enhances their metabolism and growth. In addition, *Bacteroides fragilis*, which is a common constituent of the human gut microbiota, expresses polysaccharide A that binds to Toll2-like receptors on CD4+T cells and thereby induces the expression of regulatory T cells (Treg). These Treg cells depress the inflammatory response, and this may help the establishment of *E. histolytica*.

In a similar manner, *in vitro* experiments show that defensins released by intestinal Paneth cells can kill *Giardia duodenalis*. However, under *in vivo* conditions these antimicrobial substances also affect the intestinal microbial flora and the flora have a feedback effect on the release of defensins and the establishment of *Giardia*. Similarly, although immunoglobulin A (IgA) contributes to the immunity against *Giardia*, animal models show that mice that are unable to release IgA into their intestine can still eradicate the parasite (Solaymani-Mohammadi and Singer 2010). In addition, the gut often contains a variety of helminth parasites and there can be feedback effects between the immune responses to the protozoa and the helminths. For example, mast cells are important for the development of immunity against *Giardia*. However, although *T. spiralis* causes an increase in the numbers of intestinal mast cells, *Giardia* establishes more effectively in nematode-infected hosts. This is probably because helminth infections engender a predominantly Th2 immune response, and this either alters the gut environment or reduces the immune response against *Giardia* (Von Allmen et al. 2006).

Sometimes the parasite does not induce the inflammatory response directly. For example, although ivermectin is effective at treating human filariasis, it can result in harmful side effects. This is because when dead and dying worms release their symbiotic bacteria, *Wolbachia*, these trigger a systemic inflammatory response. This is because the bacteria induce an inflammatory reaction through stimulating Toll-like receptor-4 and CD14 pattern recognition receptors. Live worms also release the bacteria and these stimulate pattern recognition receptors, thereby resulting in the recruitment of neutrophils (Taylor et al. 2001). An excellent review of the immunological background to parasite coinfections is provided by Mabbot (2018).

9.7 Avoiding the Host Immune Response

Many parasites survive within their host for a long time – sometimes extending to five or more years. For example, in humans, adult schistosomes typically live for about 3–6 years although there are cases of people still passing eggs over 20 years after their last possible exposure to infection. Such long lifespans are notable in themselves, since the majority of invertebrates have relatively short lives. However, they are even more remarkable for organisms living in an ecosystem 'designed' to kill them. The reason that some parasites survive for so long is the same as the reason they survive as parasites in the first place – they have evolved mechanisms enabling them to avoid the harmful consequences of the host immune response (Maizels and McSorley 2016). This may be because the host does not generate an immune response or that the immune response is not protective. Immunosuppressive mechanisms often result in increased susceptibility to other pathogens and can compromise the effectiveness of vaccines against unrelated organisms (Wait et al. 2020). For example, chronic helminth infections that generate a predominantly Th2 response can compromise the effectiveness of vaccines that require a Th1 response (e.g. hepatitis B, BCG vaccine against *Mycobacterium tuberculosis*).

Table 9.7 Summary of the main mechanisms parasites employ to avoiding the host immune system.

Means of reducing risk	Parasite example
Hide in a 'safe house'	Intracellular parasites develop within vacuoles and therefore they are 'hidden' from some (if not all) aspects of the immune system. For example, *Leishmania* spp. amastigotes reside within parasitophorous vacuoles in macrophages
Stay quiet and do not draw attention to yourself	The parasite encysts within tissues and either do not grow/ reproduce or do so very slowly. For example, tissue cysts of *Toxoplasma gondii*
Camouflage	The parasite acquires host antigens or manufactures the same/similar substances and therefore no longer 'appears foreign'. For example, adult schistosomes
Keep changing your disguise	The parasite constantly alters its surface coat. For example, *T. brucei* in the vertebrate host
Put out false information	The parasite manipulates the host immune system. For example, *E. histolytica* affects Treg activity
Neutralise any threats	The parasite produces chemicals that harm specific immune cells and/or neutralise the chemicals they produce. For example, *E. histolytica* trophozoites degrade IgG
Exhaust the enemy	The parasite overwhelms the host immune system

One can crudely divide the mechanisms parasites employ to avoid the host immune response into the same categories that a spy might use when in enemy territory (Table 9.7) and many parasites use a combination of these mechanisms.

Hide in a 'safe house'

Many of the most successful protozoan parasites are intracellular parasites, and this protects them from some aspects of the immune system. For example, *Leishmania* amastigotes reside within macrophages, *Plasmodium* trophozoites live within red blood cells, and *Toxoplasma gondii* tachyzoites invade a wide variety of cell types. Many intracellular parasites reside within parasitophorous vacuoles rather than being free within the cytoplasm of their host cell. These parasites are either able to survive the fusion of lysosomes with the vacuole (e.g., *Leishmania*) or they prevent fusion taking place (e.g., *Toxoplasma gondii*). Whilst within their vacuoles/host cell cytoplasm the parasites avoid circulating antibodies but remain vulnerable to cell-mediated immune responses. Nevertheless, these intracellular parasites require mechanisms of leaving their host cell and rapidly re-infecting a new cell to avoid exposure to the wider immune system.

Stay quiet and do not draw attention to yourself

The onset of an effective immune response stimulates some parasites to encyst within the host. Within the cysts, the parasite is to some extent protected against the immune system and either ceases to grow or reproduces very slowly. In this way, the parasite may survive for many years until either the host's immune system starts to fail or another animal in the parasite's life cycle consumes the host. In the case of *Toxoplasma gondii*, the onset of a protective immune response induces the parasite to forms tissue cysts (zoitocysts) containing bradyzoites. The tissue cysts probably persist for life in some hosts although their presence may be sustained by periodical re-activation, transformation into tachyzoites, followed by the formation of new tissue cysts.

Camouflage

The immune response detects invading organisms by identifying their 'foreign molecules'. Therefore, a parasite avoids the host immune response if it avoids recognition as 'foreign'. Parasites do this is by hiding 'vulnerable' antigens beneath a protective coat, capturing host antigens, or manufacturing 'look-alike' molecules. Schistosomes best exemplify this. Despite their full exposure to the forces of the immune system within blood vessels, adult schistosomes have long lifespans. This is a consequence of the larval and adult parasites expressing a range of evasion strategies – including 'camouflage'. By contrast, schistosome eggs engender a pronounced immune response that is essential if they are to traverse the tissues between the blood vessels and the lumen of the gut/bladder.

In schistosomes, immune evasion strategies commence immediately after the cercariae invade through the host's skin. The cercariae shed their outer surface, known as the glycocalyx (which is highly immunoreactive) and replace it with an inner plasma membrane and outer secreted bilayer called the membranocalyx. One of the schistosome immune evasion mechanisms is to camouflage their outer layer with host antigens. They capture some of these antigens from the host and incorporate them into the outer membrane. In the case of *S. mansoni*, these include red blood cell antigens (e.g., major histocompatibility complex class I antigens), serum glycoprotein ligands, and IgG. In addition, they manufacture 'host-like' molecules such as some resembling α2-macroglobulin. The schistosomula become progressively less susceptible to the host immune system as they mature during their migration first from the skin to the lungs and then to their final destination. This may be partly a consequence of the progressive acquisition and expression of host molecules although there are also ongoing developmental changes in the parasite's tegument and in its antigenic composition.

Interestingly, IgM antibodies formed against the eggs of *S. mansoni* cross-react with surface antigens on the schistosomulum. However, these do not appear to contribute to concomitant immunity – i.e., prevent further infections taking place whilst there is an existing infection. Instead, the antibodies appear to compete for binding sites on the schistosomula and thereby reduce the effectiveness of antibody-dependant, eosinophil-mediated killing. IgM binding to the surface tegument also occurs in juvenile *Fasciola hepatica* in which it prevents eosinophil adhesion. The strong Th2 response engendered by the eggs of *S. mansoni* and consequent down-regulation of the Th1 response may also favour the establishment of the schistosomula because a Th1 response appears to be protective against this stage. Nevertheless, concomitant immunity is a well-known phenomenon in human schistosomiasis although it is usually only apparent in adults and the underlying mechanism(s) remain uncertain (Buck et al. 2020)

Keep changing your disguise

Some parasites express a variety of surface proteins, and therefore the host is presented with a constantly moving target in which having mounted an antibody response to one common parasite antigen, the parasite then expresses a new surface protein to which a new antibody response is required. This requires three features:

1) Numerous homologous genes that code for surface antigens that stimulate strong immune responses. That is, genes coding for molecules that are similar in structure but stimulate different antibody responses.
2) A mechanism that ensures that the parasite expresses only one of these antigens at a time.
3) A means of switching from one surface antigen to another.

This approach to evading the immune response is particularly suited to protozoa because they can rapidly proliferate. Consequently, parasites bearing new antigenic variants quickly replace those killed by the immune response. For instance, *G. duodenalis* confuses the antibody response against it by expressing variant-specific surface proteins (VSPs) (Gargantini et al. 2016). However, *T. brucei* is the best example of a parasite that avoids the host immune response by expressing novel antigens. Within its mammalian host, *T. brucei* adopts the trypomastigote form, and a layer of invariant antigens covers its cell surface. However, a dense coat of a single variant surface glyco-protein (VSG) shields it from the external environment. The host recognises the VSG coat as 'for-eign' and mounts an immune response against it, but the VSG coat is more akin to a fluid coating than an array of molecules fixed in position. The trypomastigotes constantly endocytose their VSG coat, and if antibodies attach to the VSG molecules, then they remove and degrade them before sending the VSG molecules back to the surface. Nevertheless, high antibody titres kill the parasites because they overwhelm the ability of the parasites to remove antibodies from their VSG coat. However, the parasites have an archive of over 1,000 VSG-coding genes, so they periodically alter the composition of their coat. Consequently, the parasites express a constantly changing, immuno-logically distinct challenge to their mammalian host (Mugnier et al. 2016). This almost limitless ability to change the VSG coat also makes the development of an anti-*Trypanosoma brucei* vaccine a huge challenge.

Trypanosomes express VSG genes at the telomeres of their chromosomes. Telomeres are regions of repetitive DNA found at the end of the chromosomes in all eukaryotes, and in *T. brucei*, these consist of tandem 5′-TTAGGG-3′ (T_2AG_3) repeats. The expression of a new VSG takes place when a previously silent VSG gene is copied into one of the bloodstream form expression sites (BES), replacing the 'old' VSG gene in the process. There are about 15 bloodstream expression sites situated on the telomeres although only one of these is active at a time. (The metacyclic trypanosomes found in the tsetse fly have a separate class of expression sites called metacyclic expression sites.) The initial stage in the process of inserting a new VSG gene into the active expression site is the formation of a DNA double-strand break in one of the 70 base pair repeats that are adjacent to each VSG gene. Double-strand breaks are one mechanism of inserting new genetic material into a length of DNA and occurs in various organisms. In addition to the VSG gene, several other expression-site-associated genes co-transcribe into the bloodstream expression site. The function of these genes is uncertain but may include coding for specific cell surface receptors/transporters.

Once the parasite enters the tsetse fly vector, the transformation to the procyclic form is essential for the trypanosome's survival. In the majority of cases, the parasite fails to establish itself in the fly. When the parasite adopts the procyclic form, it also changes its surface coat. It sheds its VSG coat and replaces it with an invariant procyclin coat. There are two major types of procyclins: EP and GPEET. The EP procyclins consist of extensive tandem repeats of glutamic acid (E) and pro-line (P). The GPEET procyclins consist of internal pentapeptide repeats (GPEET = glycine : pro-line : glutamic acid : glutamic acid : threonine). Different glycophosphatidylinositol (GPI) structures anchor the procyclins to the cell membrane than those that anchor the VSGs. There are only six or seven genes coding for EP procyclins (the number depends on the strain) and two genes that code for the GPEET procyclins compared to the hundreds of VSG genes. However, the procyclin coat also changes during development, and it serves to protect the parasites from the tsetse fly gut hydrolase enzymes although they probably have other functions too.

This 'changeability' is not limited to protozoa and similar mechanisms occur in some helminths. For example, adult *S. mansoni* continuously replace their outer membranocalyx and it has a half-life of about 5 days under *in vivo* conditions. This probably clears potentially harmful antibodies from the outer surface but may not be rapid enough to prevent the parasite suffering damage from the immune system. However, the parasite can repair itself (Saunders et al. 1987). The 'new' membranocalyx is not antigenically identical to the one it replaces and the *S. mansoni* genome contains numerous genes encoding membrane proteins and enzymes that produce the glycans found in the membranocalyx.

Despite the immune avoidance mechanisms deployed by schistosomes, the host is still able to mount an antibody response against them. However, these antibodies are not protective against the adult worms – probably because they are unable to bind with sufficient antigens for long enough to have an effect. The continual shedding of the membranocalyx helps to prevent potentially harmful antibodies accreting on the outer surface. Evidence for this comes from the observation that the drug praziquantel only cures schistosome infections if the host produces antibodies against the parasite. This is because praziquantel physically disrupts the parasite's outer surface membranes, and this exposes the previously hidden parasite antigens. The antibodies then initiate immune processes that cause the death of the parasite.

Put out false information

A classic means of avoiding detection is to convince the authorities that there is not a problem worth investigating. Parasites achieve a similar effect by manipulating the activities of the Tregs, which suppress the activities of other immune cells (Maizels and McSorley 2016). For example, the filarial nematode *Litomosoides sigmodontis* maintains long-lasting infections in its mouse host by stimulating the activity of Treg cells. If the numbers of these cells decline, the immune system promptly removes the parasites. Similarly, Treg numbers increase in mice following their infection with the nematode *Strongyloides ratti*. The increase in Treg activity suppresses the immune system at a critical time in the parasite's life cycle and facilitates establishment in its host, reducing the numbers of Treg cells at later stages in the infection does not improve resistance. Similarly, some filarial nematodes, such as *Brugia malayi*, as well as the gastrointestinal nematode *Nippostrongylus brasiliensis*, produce cysteine protease inhibitors (cystatins) that suppress the ability of dendritic cells to present antigens to T cells and thereby reduce the T-cell response.

Adult *F. hepatica* shed tegument coat antigens every 2–3 hours. These antigens include numerous oligomannose *N*-glycans. The antigens induce the expression of the SOCS3 gene (suppressor of cytokine signalling gene 3). This forms part of a negative feedback system that regulates the cytokine signal transduction via Toll-like receptor pathways and STAT3 (signal transducer and activator of transcription 3) pathways. Consequently, there is inhibition of the phagocytic capabilities of dendritic cells, as well as their capacity to prime T cells (Vukman et al. 2013). The tegument coat antigens bind to the mannose receptor – a transmembrane glycoprotein present on cell membrane of macrophages, dendritic cells, and a variety of other cell types. However, tegument coat antigens induce SOCS3 expression and supress cytokine production in mannose receptor knockout mice (Ravida et al. 2016). Therefore, the tegument coat antigens are acting through other receptors.

Leishmania spp. parasites exploit and manipulate the immune system in a wide variety of ways – starting with living within some of the most important immune cell types such as the macrophages and dendritic cells. Dendritic cells infected with *Leishmania major* produce the immune regulatory enzyme indoleamine 2,3 dioxygenase (IDO). This enzyme activates Treg cells and suppresses the ability of the dendritic cells to communicate with T cells and T-cell activity towards parasite antigens. The parasites therefore down-regulate both the innate and adaptive immune responses against them (Makala et al. 2011).

Neutralise any threats

Many parasites produce chemicals that either kill specific immune cells or neutralise the chemicals that they produce. For example, when the trophozoites of *E. histolytica* breach the gut epithelium and invade the body, they are exposed to new immune cells and humoral factors, including immunoglobulin G (IgG) and the complement cascade. Upon infection with *E. histolytica*, over 95% of people mount a specific IgG response but it is not protective because the parasites secrete cysteine proteases that degrade the immunoglobulins. Similarly, although the parasites activate the complement system and this can kill them, some strains of *E. histolytica* express cell surface characteristics that help them avoid this immune mechanism. For example, lipophosphoglycans and lipopeptidophosphoglycans in the parasite cell membrane form a physical barrier to the complement system. In addition, complement-resistant parasites express Gal/GalNAc lectin on their surface that prevents the binding of complement-components that would initiate lysis to the parasite cell membrane. A further means by which *E. histolytica* avoids the immune response is by 'receptor capping'. This involves the parasite moving membrane receptors to which immune components have bound to the posterior pole of cell and then releasing them into the surrounding medium in membranous vesicles. For further details, see Nakada-Tsukui and Nozaki (2016).

Larval and adult *S. mansoni* avoid attack by the complement system through expressing inhibitors such as paramyosin and other proteins that prevent the activation of the complement terminal pathway. In addition, they produce serine protease that prevents complement opsonisation and thereby reduce the risk of attack by neutrophils, eosinophils, and macrophages. They also produce serine protease inhibitors that possibly inhibit protease enzymes produced by neutrophils (Skelly and Wilson 2006). Similarly, adult *F. hepatica* secrete a protease enzyme, cathepsin L1 that breaks apart host immunoglobulins. This prevents antibody-dependent cell-mediated cytotoxic killing of the flukes by eosinophils. The importance of this makes cathepsin L1, and other excretory/secretory enzymes released by *F. hepatica* involved in immune evasion, potential targets for vaccine design (Ortega-Vargas et al. 2019).

Adult hookworms change their feeding site every few hours. This may be in response to localised inflammatory responses although these often involve an unusually low number of neutrophils. In the case of the hookworm *Ancylostoma caninum*, this is may be because they secrete a glycoprotein called 'neutrophil inhibitory factor' that prevents activated neutrophils from adhering to vascular endothelial cells (Magalhães et al. 2020). It also prevents the neutrophils from secreting hydrogen peroxide (H_2O_2). Neutrophil inhibitory factor there has the potential to reduce the inflammatory response in the area in which the hookworm is feeding. This has led to it being investigated for use in treating inflammatory conditions (Barnard et al. 1995). Nevertheless, infections with the closely related *Ancylostoma ceylanicum* (and some reports of *Ancylostoma caninum*) result in heavy infiltrations of neutrophils at feeding sites. Intriguingly, the larvae of *A. caninum*, and those of several other hookworms, release *Ancylostoma* secreted protein (ASP2) that serves to attract neutrophils and other leucocytes (Wiedemann and Voehringer 2020).

The eosinophil chemokine eotaxin-1 attracts eosinophils to immune targets and is important in the clearance of filarial nematodes and many other helminth infections (Gentil et al. 2014). Several helminths produce enzymes that digest eotaxin-1, including the hookworm *Necator americanus* and the metacestodes of *Echinococcus multilocularis,* whilst the nematode *Heligmosomoides polygyrus* down-regulates eotaxin-1 although the mechanism(s) is uncertain (Cadman and Lawrence 2010).

Exhaust the enemy

Malnourishment through lack of food, micronutrients, disease, or a combination result in depression of both the innate and specific immune responses. For example, malnutrition results in a reduction in T-cell function and complement activity (Keusch 2003). A genetic component also influences the effects of malnutrition on immune function (Clough et al. 2016). Consequently, pre-existing infections become worse and the host becomes vulnerable to opportunistic infections. Many parasitic diseases cause malnutrition although this is an indirect consequence of infection rather than an adaptation. In some parasites, however, overwhelming the host's immune system forms part of their survival strategy. For example, *T. brucei* and *Trypanosoma congolense* depress the immune system of their hosts partly a consequence of the variant surface glycoproteins (VSG) overstimulating the host macrophages and thereby inducing them to produce excessive amounts of TNF and other inflammatory substances. Macrophages receive constant stimulation from both the soluble VSGs shed by the parasites all the time and the membrane-fixed VSGs released *en masse* whenever large numbers of parasites are killed (Antoine-Moussiaux et al. 2009; Onyilagha and Uzonna 2019). The latter occurs when the host mounts a successful immune response against the dominant VSG form expressed by the trypanosomes at the time. The disease tends to take a cyclical course in which the parasites multiply until the host develops sufficient antibodies to clear the majority of parasites that express the dominant VSG form. The rise in the TNF titre to harmful levels results in suppression of the immune system, anaemia, cachexia, and localised cell death.

9.8 Immunity to Malaria

We cover immunity to malaria separately and in some detail because it exemplifies many aspects of the immune response to parasites. The complex life cycle of *Plasmodium* parasites involves the invertebrate mosquito vector and, within the vertebrate host, two distinct cell types (hepatocytes in the liver and red blood cells) and both intracellular and extracellular stages. This means that at various stages in their development the parasites experiences several very different immune mechanisms.

9.8.1 *Plasmodium: Anopheles* Interactions

Transmission of human malaria relies upon specific species of *Anopheles* mosquitoes. Furthermore, vector competence varies between different strains of individual species of mosquito. This indicates that genetic factors are important in determining the ability of mosquitoes to transmit malaria. Indeed, both parasite genotype and mosquito genotype influence whether *Plasmodium falciparum* infects a mosquito (Lambrechts et al. 2005). In addition, non-genetic factors, such as temperature, diet, and gut microbial flora all impact upon the ability of *Plasmodium* species to establish infections in their mosquito vector (Lefèvre et al. 2013). Interestingly, the composition of the mosquito midgut microbiota depends upon the environment where the larvae develop (Boissière et al. 2012).

The sequence of *Plasmodium* developmental stages within a mosquito begins in her midgut. Following the ingestion of a blood meal, the red blood cells break apart and this releases the male and female gametes present inside them. After their release, these gametes are the only *Plasmodium* life cycle stage that spends a prolonged period outside a host cell – within the vertebrate the gametocytes live within their host's red blood cells. The gametocytes transform into gametes and fuse to form a zygote. The zygote then transforms into the motile ookinete stage that invades the mosquito gut epithelial cells. This complex series of events is a slow process, and some estimates suggests that during this time there is up to a 300-fold loss in abundance. Consequently, it is a potential target for a transmission-blocking vaccine. Within the vertebrate host, the immune system has relatively little impact on the gametocytes because they reside within red blood cells. Nevertheless, cytokines such as TNF can kill them and within the mosquito gut host-derived complement-mediated and antibody-mediated events can kill gametocytes after their release from red blood cells (Bennink et al. 2016).

After an ookinete invades an epithelial cell, there follows a series of developments and multiplications that culminate in the parasite migrating to the insect's salivary glands. Therefore, within the insect vector, the parasite has both intracellular and extracellular stages and must survive exposure to two different immune challenges: the midgut lumen and the tissue/ haemolymph. However, the major limiting step that determines vector competence is the ability of the ookinete to invade the midgut epithelial cells and then establish itself as an oocyst. Some estimates suggest that only about 2% of ookinetes successfully form oocysts.

Within the gut lumen of a mosquito, *Plasmodium* parasites face challenges directly from the constituents on the gut microbiome and indirectly from the mosquito immune response to the gut bacteria. Direct challenge comes from bacteria such as *Serratia marcesans* and *Enterobacter* spp. that in the midgut of *Anopheles albimanus* prevent *P. vivax* from completing its development (Gonzalez-Ceron et al. 2003). Indirect challenge comes from the arrival of a blood meal within the gut triggering a rapid increase in the numbers of microbes and this triggers the mosquito Imd (immune deficiency) signalling pathway. The result is the production of AMPs that harm the ookinetes, as well as the microbes (Zakovic and Levashina 2017). It is also possible that the ookinetes may be recognised by the Imd pathway and targeted directly.

Once it penetrates the mosquito's midgut epithelium, the parasite triggers the c-Jun N-terminal kinase (JNK) pathway (Garver et al. 2013). The JNK pathway kills the parasites by two means: (1) it identifies parasitised cells to other components of the pathway and that then induce the infected cells to undergo apoptosis; (2) it marks the ookinetes for destruction by the mosquito's complement system. The latter is the most important process and reviewed in detail by Povelones et al. (2016). The mosquito complement system is, in some ways, analogous to the vertebrate complement system and consists of a variety of proteins that circulate with the mosquito haemolymph. Of particular importance is thioester-containing protein 1 (TEP1) – this has close similarity to complement factor 3 (C3) that is important in activating the complement system in the vertebrate innate immune response. TEP-1 is a pattern recognition receptor (PRR) and binds to the cell membrane of ookinetes. Once bound, it initiates reactions that culminate in the parasite being melanised or lysed. Precisely how these processes are accomplished is currently uncertain. Because the *Anopheles* complement system is so effective at destroying ookinetes, it is not surprising that the parasites have evolved immune avoidance mechanisms. For example, *P. falciparum* strains expressing the Pfs47 surface protein tend to show increased resistance to the TEP1-mediated immune response. In addition to TEP1, several other PRR present in the mosquito haemolymph bind to *Plasmodium* parasites and initiate immune reactions against them. For example, they possess numerous fibrinogen-related proteins (FREPs) and one of these, FBN9, is associated with anti-*Plasmodium* activity in the mosquito *Anopheles gambiae* (Dong and Dimopoulos 2009). There are FREPs that act as pattern recognition receptors within the innate immune response in both invertebrates and vertebrates. Interestingly, although vertebrates utilise FREPs in coagulation, they do not fulfil this role in invertebrates (Hanington and Zhang 2011). It is worth remembering that the immune mechanisms present in those mosquitoes that transmit human malaria may not be the same in other mosquito species and certainly not in all insects.

9.8.2 *Plasmodium*: Human Interactions

Within humans, *Plasmodium* parasites face challenges from a combination of our innate, humoral, and cellular immune mechanisms.

9.8.2.1 Innate Immune Mechanisms Against Malaria

Our innate immune mechanisms provide variable levels of protection and occur predominantly among populations in which malaria is endemic, and there is constant exposure to infection. They are inherited traits and are not up-regulated in response to infection. For example, the merozoites of *P. vivax* and *Plasmodium knowlsei* only penetrate red blood cells carrying the Duffy buffer blood group antigens Fy^a and Fy^b. Most West Africans do not express these antigens and they are therefore resistant to these species of *Plasmodium*.

Plasmodium falciparum has difficulty invading red blood cells that express the sickle-cell trait (genotype Hb AS) genotype. In addition, even if the parasites establish within Hb AS genotype red blood cells, the infected cells become deformed (sickle) and therefore the spleen destroys them. In cases of mild malaria, the expression of the Hb AS genotype also contributes to the development of adaptive immunity. The mechanism(s) by which this occurs are uncertain, but the slower development of the parasites may facilitate the development of antibodies to antigens expressed on the surface of infected red blood cells. The sickle cell trait arises from a point mutation on the gene coding for β-globin. Because of the alteration to the β-globin, it forms haemoglobin that polymerises at low oxygen saturations and gives rise to deformed red blood cells. In people who are heterozygous for Hb AS, only about 30% of the haemoglobin is affected, and they do not usually express any clinical abnormalities and they gain protection from *P. falciparum*. However, the homozygous condition can result in alterations to over 80% of the haemoglobin. The consequence of this is an increased likelihood of developing the potentially life-threatening condition 'sickle-cell anaemia'. Furthermore, should a person expressing a high proportion of haemoglobin mutations become infected with *P. falciparum*, they are at an increased risk of suffering severe malaria and dying (McAuley et al. 2010).

The blood disorder thalassemia also results from defective synthesis of globin. There are two basic types of thalassemia although within each there are subtypes, depending upon the number of genes affected and the nature of the alteration. α-Thalassemia mostly results from deletions rather than mutations on the genes coding for the α globin chains, whilst β-thalassemia can result from any of over 90 different mutations affecting the genes coding for the β globin chains. Both α- and β- forms of thalassemia influence malaria infection. The level of protection, if any, they provide varies between the subdivisions and the species of *Plasmodium* (Weatherall 2018). In the case of α-thalassemia, the red blood cells lack complement factor-1 receptors (CR1) and consequently are less likely to form rosettes (Cockburn et al. 2004). Infection of a red blood cell by *Plasmodium* results in it binding to CR1 on uninfected red blood cells. This results in the cells clumping to form 'rosettes' (rosetting) and the blockage of capillaries that is a feature of cerebral malaria. Expression of β-thalassemia is associated with reduced infection and growth of *P. falciparum* in the red blood cells and a similar reduced tendency to exhibit resetting. α-Thalassemia is common throughout sub-Saharan Africa and although it does not prevent a person becoming infected or developing high parasitaemia, those with α-thalassemia appear to be less likely to suffer from severe anaemia (Wambua et al. 2006).

Intriguingly, people who exhibit a mutation in the gene coding for the Fc-gamma receptor-IIb that results in threonine being substituted for isoleucine at position 232 (FcγRIIb^{T232}) have both an increased susceptibility to the autoimmune condition *systemic lupus erythematosus* (SLE) and enhanced protection against infection by *P. falciparum*. Reportedly, macrophages derived from individuals homozygous for this mutation 'phagocytose malarial parasites more avidly' (Willcocks et al. 2010). The Fc-gamma receptor-IIb is an immunoglobulin-G receptor that occurs on a range of immune cells such as macrophages and B cells and is involved in a range of regulatory processes such as the control of phagocytosis and the production of inflammatory cytokines. If the receptor does not function properly, it can result in uncontrolled inflammation – and this is a feature of SLE. This condition is particularly common among women in Southeast Asia and parts of sub-Saharan Africa. It may be that conservation of this trait, like sickle cell anaemia, and thalassemia, results from the enhanced protection provided against malaria.

9.8.2.2 Antibodies Against Malaria

Although we develop immunity to malaria, this is never completely protective. There are two main theories of how this immunity develops. The first of these, the 'strain specific' hypothesis, is widely accepted but the so-called cross-reactive or strain transcending hypothesis also has its adherents (Doolan et al. 2009). The basis of the strain specific hypothesis is the observation that children growing up in malaria endemic regions gradually develop immunity to the local strains of malaria. This probably arises from repeated exposure to infection. Within a geographical area, different species of malaria exist as a variety of strains and immunity therefore develops slowly (~10–15 years) because the immune system needs repeated exposure to the various strains in order to develop and maintain sufficient memory cells to mobilise a response against future infections. This therefore explains the observation that a person resistant to malaria within their own region is often susceptible to malaria elsewhere. Similarly, a person who moves away from a malaria region may be susceptible to malaria when they return. By contrast, the 'cross-reactive' hypothesis has more limited supporting evidence and is currently based mainly upon observations on transit migrants (people passing through the country) in Indonesian West Papua. According to this hypothesis, sudden heavy exposure to infection results in protective adaptive immunity developing rapidly in 1–2 years in previously non-immune individuals. Furthermore, this immunity is not limited to the local strains (i.e., it is 'strain transcending'). The distinction between the mechanisms by which adaptive immunity develops is important because it affects the design and use of vaccines.

As mentioned above, infection with malaria induces the formation of antibodies against the parasite. Therefore, in a previously exposed individual, antibodies will target the sporozoites injected by a mosquito. Under natural circumstances, this is probably not protective. However, under experimental conditions vaccines containing irradiated sporozoites or pre-erythrocytic stage antigens generate a substantial degree of immunity to subsequent challenge infections. Vaccine studies indicate a role for CD8+ T cells in the destruction of sporozoites mediated through the production of gamma interferon (IFN-γ). The next stage in the *Plasmodium* life cycle involves penetration and replication within hepatocytes in the liver. CD11c+ monocyte cells detect the infected liver cells and present parasite antigens to parasite-specific CD8+ T cells – these then mount a protective immune response against the infected liver cells (Kurup et al. 2019).

Having left the hepatocytes, the malaria parasites invade red blood cells, and this is associated with the development of a humoral response based on polyclonal B-cell expansion. This is associated with the production of large amounts of immunoglobulins – especially IgG, IgA, and IgM. However, only a small proportion of these immunoglobulins are *Plasmodium*-specific. These antibodies bind to parasite antigens present on the surface of red blood cells and merozoites released when infected red blood cells disintegrate. The antibodies exert their protective effects in several ways although their relative importance is not yet clear. For example, they can block merozoites from invading fresh red blood cells, facilitate antibody-dependent cell destruction through cytophilic antibodies binding to in the infected cells or merozoites, thereby encouraging them to be phagocytosed by macrophages and other immune cells (especially in the spleen). The binding of antibodies to the antigens can also activate the complement cascade and thereby destroy the parasite or infected cell. In addition, the antibodies also reduce the pathology associated with malaria by blocking the binding of infected red blood cells to surface of blood vessels, and they inhibit induction of TNF by malaria toxins.

Mammalian red blood cells lack nuclei, and therefore their cell membranes lack major histocompatibility complex class I and class II presenting molecules. One might therefore expect that cellular immunity would play little part in the immune response against the schizont and merozoite stages of the malaria life cycle. However, CD4+ T cells are essential for the expression of protective immunity against the red blood cell stages of the malaria life cycle both as 'helper cells' for B-lymphocytes and as producers of inflammatory cytokines and activators of macrophages. Mouse models indicate that activation of CD8+ T cells follows previous exposure to *Plasmodium* antigens, and these then activate macrophages through the release of gamma interferon and/or induce perforin-mediated destruction of the infected cell (Chan et al. 2014). Although many of the precise mechanisms of the immune response to human malaria remain uncertain, what can be certain is that the immune response involves a combination of humoral and cellular immune mechanisms and their relative importance in conferring protective immunity probably varies with the circumstances.

9.8.2.3 Why Humans Do Not Develop Protective Immunity Against Malaria

Although we generate immune responses against all stages of the *Plasmodium* life cycle that develop within us, these are never fully protective. One reason for this is that the parasite expresses variant immunodominant antigens (i.e. antigens that generate the greatest antibody response) on the surface of the infected red blood cells. For example, the genome of *P. falciparum* includes a large number of '*var* genes' that encode for a protein that becomes incorporated into the cell membrane of infected red blood cells. This protein, '*P. falciparum* erythrocyte membrane protein 1' (PfEMP1), is an important target in the development of adaptive immunity and both activates and suppresses dendritic cells, monocytes, and T cells (Hviid and Jensen 2015). PfEMP1 has an important role in facilitating the attachment of infected cells to the walls of blood vessels and thereby, presumably, reducing the likelihood of the parasite reaching the spleen where it would be killed. Antibodies formed against PfEMP1 can prevent infected red blood cells sticking to the walls of blood vessels. The *var* gene is present at numerous loci on different chromosomes and the parasite switches between different antigenic variants of PfEMP1 during the blood stages. In addition to the problems caused by antigenic variation, antigenic diversity also compromises the development of a protective immunity. For example, another important antigen, 'merozoite surface protein-1' has numerous alleles and antibodies formed against one 'version' do not necessarily recognise another (Hisaeda et al. 2005). Malaria also induces immunosuppression that compromises the immune response to vaccines and some other infectious agents. Immunosuppression results from a combination of processes including, stimulating the production of the anti-inflammatory cytokines interleukin-10 and transforming growth factor-β and inhibiting the maturation of antigen-presenting cells. There is also evidence that malaria causes immunosuppression through inducing the activation of regulatory T cells (Hviid and Jensen 2015).

9.9 *Schistosoma* spp. and Hepatitis C Virus Interactions

In the past, there were few effective drugs available to treat schistosomiasis and some of those used, such as niridazole and tartar emetic, were expensive and had serious side effects. These drugs were usually given by injection and the re-use of needles facilitated the spread of Hepatitis C virus (HCV). This became a particular problem in Egypt between the 1950s and the 1980s (Elgharably et al. 2016). This no longer happens because of the availability of a much cheaper and effective drug, praziquantel. Nevertheless, the legacy, coupled with the transmission of HCV from other medical procedures (e.g., blood transfusions), means that Egypt has the highest HCV prevalence in the world. There is, however, no evidence that infection with either schistosomiasis or HCV predisposes a patient to the other infection (Abruzzi et al. 2016). Schistosomiasis: hepatitis B virus (HBV) co-infections are also common in Egypt and elsewhere in the world although the extent to which the two pathogens interact is less certain (Gasim et al. 2015). HBV also spreads through contaminated needles and transfusions but is usually not pathogenic in adults. However, in children it often causes long-lasting infections and serious liver pathology. There is currently an effective vaccine for HBV but not HCV.

Both schistosomiasis and HCV affect the liver and people who are chronic carriers of both tend to develop serious liver damage. Patients co-infected with schistosomes and HCV express higher viral loads and suffer more severe liver complications than those only infected with the virus (Strickland 2006). This probably relates to the impact of schistosomiasis on the patient's immune system. Although the majority of patients infected with HCV develop chronic infections, some of them recover. The ability to resolve HCV infection is associated with a strong Th-1 immune response involving CD4+ and CD8+ cells. Unfortunately, the Th-1 immune response is downregulated in chronic schistosomiasis, which could explain why patients co-infected with both *S. mansoni* and HCV are more likely to develop liver fibrosis and cirrhosis. This immunosuppressive effect does not depend on the presence of an active helminth infection: blood donors with detectable levels of both anti-schistosome and anti-HCV antibodies (indicating past infection) have a reduced CD8+ T-cell response compared to donors who test positive for only one of these antibodies (Strickland 2006). A separate study of apparently healthy blood donors found a high rate of low positive anti-HCV results in people who did not have any indicators of active HCV but did have anti-schistosome antibodies (Agha et al. 2006). The authors suggested that since schistosomes induce polyclonal B-cell activation, these blood donors might have been producing antibodies from past infections, at a relatively low level. However, they also found autoimmune markers in some of these patients, again probably induced by the schistosome infection, which they speculated might cause false positive results in laboratory tests.

9.10 HIV-AIDS and Parasitic Infections

HIV stands for 'Human Immunodeficiency Virus' and it gives rise to the condition 'acquired immunodeficiency syndrome' – commonly known as 'AIDS'. Up until recently, most people infected with HIV ultimately developed AIDS and died – often from opportunistic pathogens. Although there is still no cure for HIV, the development of various anti-retroviral drugs culminated in a combination treatment regimen 'highly active antiretroviral therapy' (HAART), which enables many infected people to live more-or-less normal lives without succumbing to either the virus or associated infections. Nevertheless, AIDS remains a major cause of mortality in many developing countries. In 2018, there were approximately 38 million people living with HIV around the world and an estimated 1.7 million people became newly infected by the virus. Parasites contribute to the acquisition of the virus, the progression of the disease, and both morbidity and mortality associated with AIDS (Clark and Serpa 2019).

HIV is an RNA virus belonging to the Family *Retroviridae*, Genus *Lentivirus*. There are two types – HIV-1 is the commonest and has a worldwide distribution, whilst HIV-2 is less infective and occurs predominantly in West Africa. In common with other lentiviruses, HIV can remain quiescent (latent) as a provirus within the host DNA for months or even years but when stimulated undergoes extremely rapid multiplication. Activation of infected CD4+ cells promotes HIV replication and co-infecting pathogens contribute to this in a variety of ways (Mabbot 2018):

1) Co-infecting pathogens activate CD4+ T cells as part of the adaptive immune response
2) Co-infecting pathogens activate the CD4+ cell through inducing pro-inflammatory cytokines
3) Co-infecting pathogens activate the CD4+ cell through Toll-like receptor signalling

HIV only infects cells that express the CD4 molecule on their cell membrane. Consequently, it is restricted to the CD4+ T cells, dendritic cells, and monocytes. In addition, HIV-1 and HIV-2 exist as variants that differ in their ability to infect different CD4+ cell types and the cell types vary in their sensitivity to the lytic effects of the virus. When activated, the virus replicates rapidly and destroys its host cell, thereby releasing progeny that repeat the process on other CD4+ cells. There is some uncertainty whether the virus directly or indirectly contributes the decline in CD4+ cells. Initial thoughts were that the destruction of infected cells by the virus led to increasing rates of proliferation within the thymus in order to replace the lost cells. As the disease progressed to AIDS during the chronic phase of infection, the rates of destruction would ultimately exceed the rate of production, and this manifested as the decline in the numbers of circulating CD4+ cells and consequent immunodeficiency. However, an alternative hypothesis is that HIV indirectly induces the activation of large numbers of CD4+ cells that then become targets for infection. According to this model, HIV rapidly destroys CD4+ T cells in the intestinal mucosa and consequently the gut bacteria and/or their products (e.g., membrane components) translocate into the circulation. This stimulates the immune system, which results in the activation and proliferation of CD4+ and CD8+ T cells – and the activated CD4+ T cells become 'targets' for HIV. Furthermore, the infection and destruction of the CD4+ cells would result in lymph node fibrosis and thereby reduce the capacity of the body to replenish the CD4+ cells in the intestinal mucosa. The suggestion that HIV impairs the gut mucosal barrier and compromises the local immune response would explain why those who are HIV positive are so susceptible to gut pathogens.

In both the above scenarios, the loss of CD4+ T cells is associated with immunosuppression – and hence the host becomes vulnerable to various other pathogens. However, the situation is more complex than this because massive losses in CD4+ T cells occur long before untreated HIV infection develops into acute AIDS. Consequently, Okoye and Picker (2013) suggest that the losses of CD4+ T cells are, for a while, compensated for by CD4+ memory T-cell regeneration. In this scenario, immunodeficiency results from the subsequent failure of CD4+ memory T-cell homeostasis and ultimately the collapse of immune processes.

9.10.1 Parasites and the Transmission of HIV

The transmission of HIV usually occurs during sexual contact because of the opportunity it presents for the exchange of infected blood and body fluids. Transmission also occurs through needles contaminated with infected blood (e.g., poor sterilisation of medical equipment, intravenous drug users who share needles, and needlestick injuries), transfusions using infected blood, and vertically from mother to child. There is no evidence that HIV is transmissible by mosquitoes or any other blood-feeding invertebrate. Although the virus occurs in the genital secretions of both men and women, the presence of lesions on the genitalia of one or both partners further facilitates sexual transmission. For example, in women, infection with *Trichomonas vaginalis* causes inflammation of genital tract that damages the surface epithelium and thereby predisposes them to HIV infection, as well as increasing their chances of transmitting HIV to their sexual partner(s). It can also facilitate the acquisition of other sexually transmitted diseases such as herpes and gonorrhoea, as well as the development of pelvic inflammatory disease (PID), in women. In turn, these infections further increase the risks of contracting and transmitting HIV. In a similar manner, the urogenital lesions associated with *Schistosoma haematobium* infection also facilitate the acquisition of HIV (Furch et al. 2020). For example, a study in Tanzania found a 17% prevalence of HIV among women afflicted by urogenital schistosomiasis, but only 5.9% amongst those not infected by the parasite (Downs et al. 2011). However, it is currently uncertain whether HIV infection increases susceptibility to schistosomiasis.

Persons who contract HIV-1 in sub-Saharan Africa tend to progress more rapidly to AIDS than those living elsewhere in the world. The reason(s) for this are uncertain but are unlikely to be a simple consequence of different HIV-1 variants circulating within the local populations. One suggestion is that people living in sub-Saharan Africa are more likely to harbour parasites and especially helminth parasites (Bentwich et al. 2000). These parasites cause long-lasting chronic infections that result in the activation of CD4+ cells – and HIV replicates at a higher frequency in such cells. Furthermore, helminth infections modulate the immune system to favour their own survival. In the process, this can result in reductions in the Th1 and Th17 responses that are protective against HIV. In addition, they can cause the formation of Treg cells that supress the immune response. Taken together, these events can facilitate the establishment of an HIV infection (Salgame et al. 2013). Both parasitic helminth and HIV infections frequently induce malnutrition that manifests as reductions in the levels of protein, energy, micronutrients, and vitamins. This can compromise the patient's immune system (Bourke et al. 2016) and thereby enable the further progression of their underlying disease and predispose them to other diseases.

9.10.2 Parasite-HIV Co-Infections

As HIV infection progresses towards AIDS, there is a major decline in the cellular immune system and consequently many parasites that were previously unable to establish themselves become serious pathogens. These so-called opportunistic parasites are a significant feature of AIDS and a major cause of both morbidity and mortality. In addition, previously latent infections can become reactivated and cause serious disease. However, the consequences of HIV infection vary considerably between parasite species. For example, the prevalence of *Isospora belli* infections is usually in the region of 1–2% among those who are HIV+ve but may be as high as 27% (Stark et al. 2009). It can be particularly pathogenic should the HIV status develop into AIDS. Similarly, *Cryptosporidium* and *Giardia* have 'emerged' as common causes of severe debilitating diarrhoea in those with HIV-AIDS. *Acanthamoeba* spp. infections are usually only highly pathogenic and cause encephalitis in HIV+ patients, those otherwise immunosuppressed, and those already suffering a debilitating medical condition (Kalra et al. 2020). By contrast, many cases of *Balamuthia mandrillaris* encephalitis occur in both otherwise healthy individuals and those who are immunosuppressed (Lorenzo-Morales et al. 2013). There is often a high prevalence of *E. histolytica* among AIDS patients (Samie et al. 2010). It is uncertain whether this is because AIDS-induced depression of the immune system increases susceptibility or because *E. histolytica* can be sexually transmitted and is a risk factor for men who have sex with men (Fernández-Huerta et al. 2019). By contrast, there are only occasional case reports of serious infections with *Sarcocystis* among immunocompromised individuals (Anderson et al. 2018). This is just as well because there is currently no recommended prophylactic or therapeutic treatment for *Sarcocystis* infections.

9.10.2.1 *Leishmania*-HIV Co-Infections

There is an 'unholy alliance' between HIV and the protozoan parasite *Leishmania* in which co-infections result in mutual activation and enhanced pathology. This is perhaps not surprising since both HIV and Leishmania infect CD4+ T cells and both can remain as latent infections until they receive stimulation. However, the effect occurs more frequently in visceral leishmaniasis-HIV co-infections than cutaneous leishmaniasis-HIV co-infections.

Visceral leishmaniasis-HIV co-infections may arise through a primary infection with either pathogen – this causes a decline in immune function and thereby facilitates the establishment of the other infection. Alternatively, the virus may activate a latent *Leishmania* infection or the *Leishmania* may activate a latent HIV infection. The two pathogens influence one another through a wide variety of immunopathological mechanisms. For example, HIV infection disrupts normal monocyte and macrophage functions by reducing chemotaxis and intracellular killing activity that therefore facilitates the invasion and multiplication of *Leishmania*. Similarly, when *Leishmania* interacts with specific cell-surface receptors on HIV-infected CD4+ lymphocytes and macrophages it results in the activation and multiplication of the virus. The pathogens can also influence one another by affecting the Th1:Th2 balance. In leishmaniasis, a dominant Th1 profile with the production of large amounts of interleukin-2 and interferon-gamma is protective. Conversely, those who are otherwise immunocompetent, but mount a dominant Th2 response in which the cytokines interleukin-4, interleukin-5, and interleukin-10 predominate, are more susceptible to the parasite. A dominant Th1 profile is also protective against HIV and the progression of HIV infection to AIDS is associated with a move towards aTh2 dominance and a decline in interleukin-2 and interferon gamma production. Consequently, HIV infection can promote the development of leishmaniasis and leishmaniasis can promote the development of HIV. The two pathogens also influence one another through a range of other immune mechanisms discussed by Andreani et al. (2012) and Okwor and Uzonna (2013).

9.10.2.1.1 *The Increasing Problem of HIV-Leishmania Co-Infections*

There is an increasing overlap in the geographical distributions of leishmaniasis and HIV/AIDS, and in consequence, there is an increasing incidence of *Leishmania*/HIV co-infection. Indeed, there are reports of *Leishmania*/HIV co-infections from over 35 countries. This is partly because of leishmaniasis moving from its predominantly rural focus to urban areas, whilst HIV/AIDS, which has been a predominantly urban disease, is affecting rural populations more severely. For example, visceral leishmaniasis is a serious problem in the suburbs of large cities in Brazil, Bangladesh and India. In parts of South America, the urbanisation of visceral leishmaniasis is attributed to:

1) Felling large areas of rainforest. This has led to a change in the feeding habits of the sand-fly vectors of leishmaniasis. For the example, the sand-fly vectors of *Leishmania braziliensis* have become more endophagic. That is, whilst they previously fed principally on wild animals they now feed more frequently inside houses, barns, and other buildings. Consequently, they bite humans more frequently and (peri) domestic animals that act as disease reservoirs are increasingly infected.

2) Movement of poor people from rural regions to the outskirts of cities. Some of these people bring existing leishmaniasis infections that sand-flies then transmit to others. In addition, they often live in badly constructed housing on the city: jungle boundary where the natural hosts and sand-fly vectors occur.

3) Many of the people moving to the cities are poor, badly educated males who are the societal group most likely to engage in risk-taking behaviour. They are therefore the group most likely to contract HIV, and HIV infection is associated with an increased likelihood of contracting leishmaniasis. In addition, they are also likely to experience inadequate health care and advice.

By contrast, in countries such as Ethiopia, Kenya and India, the HIV/AIDS pandemic is becoming progressively more common in rural areas. This is partly because migrant workers bring back HIV infections when they return to their villages. They then transmit the infection to the surrounding community. In addition, in Ethiopia there is large-scale movement of itinerant labourers between highland urban and semi-urban areas to lowland agricultural regions. Visceral leishmaniasis is endemic in the farming regions and the labourers become infected. Furthermore, there is a high prevalence rate of HIV infection among the migrant workers, and this increases susceptibility to leishmaniasis and the likelihood of co-infection (Diro et al. 2014).

Within southern Europe, there was an increase in *Leishmania*-HIV co-infections during the 1990s although this has since declined because of the availability of HAART chemotherapy for HIV infection. In this case, the rise was predominantly a consequence of parenteral transmission of *Leishmania* via blood transfusions and, especially, needle sharing amongst intravenous, HIV and *Leishmania* co-infected drug users. At one stage in southern Europe, up to 70% of all adult cases of visceral leishmaniasis related to HIV/AIDS, and up to 9% of all AIDS cases suffered from newly acquired or reactivated visceral leishmaniasis (Monge-Maillo et al. 2014).

9.10.2.2 Malaria-HIV Co-Infections

Sub-Saharan Africa suffers the highest number of cases of malaria in the world and the incidence of HIV infection in many countries in the region is devastatingly high. Both diseases disproportionately affect poor people with limited education who have limited access to medical advice and treatment. Cases of malaria-HIV co-infection are therefore common (Kwenti 2018). There are reports of HIV infection resulting in more severe malaria and of increased HIV viral titres during episodes of malaria. However, the relationship between the two diseases and their influence on one another is uncertain. This is partly because in developing countries the diagnosis of malaria is often based on clinical signs rather than specific tests. HIV-AIDS can itself result in fevers and it increases susceptibility to a wide range of opportunistic infections – some of which also cause pyrexia (fever), which could be confused with malaria.

Because AIDS compromises the immune system, one might expect it to increase susceptibility to infection with and recrudescence of malaria. Women who grow up in malaria-endemic regions develop immunity to the local species and strain of *Plasmodium,* but this immunity wanes during pregnancy. Consequently, they can suffer from pregnancy-associated malaria. HIV-1 co-infection with malaria can result in marked changes in the levels of antibodies against *P. falciparum* (Jaworowski et al. 2009). This may explain why the placenta of women who are HIV positive is more likely to be invaded by the malaria parasite than those who are HIV negative. This in turn can increase their chances of transmitting HIV to their child (Brahmbhatt et al. 2008).

Malaria infection of people who have HIV results in a short-lived increase in the plasma viral load and this rises with the severity of the attack. The rise in viremia could conceivably increase the chances of HIV transmission although a person in the grips of a malaria paroxysm is unlikely to be sexually active. Indeed, the progression of HIV among populations with malaria is similar to those where it is not such a problem. Consequently, it is still a matter of debate whether malaria contributes to the spread HIV. In addition, many variables, such as HIV-subtype, strain of parasite, and host nutrition and genotype will influence malaria: HIV interactions. It is therefore difficult to make direct comparisons between many of the studies.

9.10.2.3 *Toxoplasma*-HIV Co-Infections

Toxoplasma gondii is an exceptionally common parasite. Some estimates suggest that approximately 30% of the world's population have antibodies against it. Consequently, *T. gondii*: HIV co-infections are undoubtedly common too (Wang et al. 2017). *Toxoplasma gondii* does not affect susceptibility to HIV infection because in most infected people it exists as dormant cyst containing bradyzoites. In addition, immune reactions against *T. gondii* involve the Th1 immune response and the formation of inflammatory cytokines such as gamma interferon and interleukin-12 (Sasai et al. 2018). These immune responses are also protective against HIV. Although there is some evidence of an association of *Toxoplasma* infection and risk-taking behaviours – and therefore, potentially including those likely to lead to HIV infection – there are also studies that find no associations (e.g., Sugden et al. 2016).

By contrast, the onset of immunosuppression associated with AIDS facilitates the activation of *T. gondii* bradyzoites with serious and potentially fatal consequences. Once their CD4+ T-lymphocyte count falls below $50\,\text{cells}\,\mu l^{-1}$ there is a heightened risk that a patient's existing *T. gondii* infection will be reactivated (although some reports suggests that the risk is already increased at $200\,\text{CD4+ cells}\,\mu l^{-1}$). The replication of the parasites within brain cells results in encephalitis and abscess formation. The consequences depend upon the size of lesions and the part of the brain affected. Symptoms can include headaches, speech defects, confusion, seizures, and, ultimately, the condition can prove fatal. Up until the widespread use of HAART therapy, toxoplasmic encephalitis affected up to 40% of AIDS patients and 10–30% died of the condition (Tenter et al. 2000). It is now possible to prevent reactivation of dormant *T. gondii* infections by prophylactic treatment with trimethoprim-sulfamethoxazole (TMX-Sulfa). As a result, the rate of toxoplasmic encephalitis has declined considerably among AIDS patients although it remains a risk. In developing countries where access to HAART is limited, toxoplasmic encephalitis remains a serious problem.

9.10.2.4 Microsporidia-HIV Co-Infections

The initial descriptions of microsporidian infections in AIDS patients occurred in 1980s. Subsequently, the number of reported cases and clinical spectrum of diseases increased remarkably (Stentiford et al. 2019). For example, although *Enterocytozoon bieneusi* was not discovered until 1985 (Desportes et al. 1985), it rapidly became one of the most frequently identified pathogens in the intestinal tracts of patients with AIDS. This suggests that some, if not all, of those species causing problems for immunocompromised individuals may be zoonotic infections (Prasertbun et al. 2017). In the developed world, the introduction of HAART therapy resulted in a reduction in the number of HIV-positive individuals developing AIDS, and there has been a corresponding reduction in the number of cases of microsporidia infections. However, there are increasing reports of these parasites in immunocompetent people, and they are an important cause of diarrhoeal diseases in developing countries.

10

Pathology Part A: Factors Influencing Pathogenesis, How Parasites Cause Pathology, Types of Pathology

CONTENTS

Parasitology: An Integrated Approach, Second Edition. Alan Gunn and Sarah J. Pitt.
© 2022 John Wiley & Sons Ltd. Published 2022 by John Wiley & Sons Ltd.
Companion website: www.wiley.com/go/gunn/parasitology2

10.1 Introduction

Pathology is the study of the mechanisms that cause morbidity and mortality. The term 'pathogenesis' describes the cellular processes and reactions that bring about the diseased state, whilst 'pathogenicity' refers to the ability of an organism or substance to cause disease. We often talk about organisms being weakly or highly pathogenic depending upon the amount of damage they cause to their host. The term 'virulence' is also used as an indicator of a parasite's potential to cause disease. This is usually a reflection of the parasite's ability to reproduce within its host. For example, *Trypanosoma evansi* is highly virulent in horses and reportedly the inoculation of even a single parasite can prove fatal. Pathology, like immunology, requires one to learn some basic vocabulary before one can make sense of the literature. We therefore begin by introducing some of the commonest types of pathology and then, in Chapter 11, discuss how parasites affect individual organs of the body. Throughout both chapters, we emphasise how pathology results from a complex interaction of host factors, parasite factors, co-infections, and the environment. Those requiring more detail on specific aspects of pathology should consult specialist textbooks such as Kumar et al. (2020) for human pathology, Coleman and Tsongalis (2017) for molecular pathology, and Zachary (2016) for veterinary pathology.

10.2 Factors Influencing Pathogenesis

The outcome of any close relationship depends upon how the two people/organisms involved interact with one another. It may be harmonious or it can turn toxic for one or both parties. A host: parasite relationship is no different and depends upon both host factors and parasite factors. The host factors determine susceptibility or resistance to a pathogen and parasite factors influence its virulence and ability to avoid the host immune system. In addition, as in human relationships, the two parties do not live in isolation and the outcome of their relationship will also depend upon their simultaneous interactions with other organisms around (and within) them, the availability of food, the environment, and past experiences. Consequently, the extent to which a parasite is pathogenic depends upon many interacting variables (Table 10.1).

10.2.1 Host Factors

Numerous host factors influence whether an organism becomes infected with a parasitic disease (Chapter 8) and also their subsequent reaction to it (Chapter 9). It is therefore usually impossible

Table 10.1 Host factors and parasite factors that determine the pathology associated with parasitic infections.

Host factors	Parasite factors
Genetic constitution	Genetic constitution
Age	Growth rate
Reproductive status	Reproduction
Gender	Production of harmful substances
Underlying health	
Immune status	
Presence of co-infections	

to identify any single factor as being of fundamental importance since most of them interact to varying extents. For example, the gender of the host can affect its behaviour and therefore the likelihood of it encountering the infectious stage of a parasite. However, immune status, and therefore the ability to prevent a parasite from establishing itself, is also influenced by gender.

Host genetic constitution plays a major role in determining susceptibility to many parasitic diseases. For example, scabies (*Psoroptes scabiei*) occurs throughout the world and for those of us who are immunocompetent becoming infected with the mite is an unpleasant experience but seldom has serious consequences. However, for some reason, Australian Aboriginals are particularly susceptible to developing a highly debilitating and potentially fatal condition called 'crusted scabies' (Roberts et al. 2005). Similarly, although many people in Mexico contract amoebic dysentery, hepatic amoebiasis is primarily found in young adult males and in particular those who express the human leucocyte antigen class II alleles HLA-DR3 (Santi-Rocca et al. 2009). In Bangladesh, susceptibility to *Entamoeba histolytica* is linked to mutations in the genes coding for the leptin-receptor. A single substitution of an arginine for a glutamine at position 223 is sufficient to increase susceptibility to amoebic dysentery in children and hepatic amoebiasis in adults (Duggal et al. 2011). The hormone leptin is best known for its role in regulating body weight, but it has numerous other physiological functions. These include acting as a pro-inflammatory cytokine that promotes a Th1-type response and inhibiting apoptosis. Apolipoprotein E (ApoE) is the main form of apolipoprotein found in our brains where it functions in cholesterol and triglyceride metabolism and intracellular signalling. It exists in three main isoforms: ApoE2, ApoE3, and ApoE4. Expression of ApoE4 occurs in all human populations, but its frequency is highest in Sub-Saharan Africa and some other parts of the world where there are high rates of malaria transmission. It appears to provide a level of protection against *Plasmodium falciparum*, but it is also associated with an increased risk of developing Alzheimer's disease, cardiovascular disease, and atherosclerosis later in life (Fujioka et al. 2013). This has led to suggestions that the expression of ApoE4 has been retained in our populations as a trade-off between improving the chances of survival during early life against an increased risk of ill health as we become older. The ApoE4 isoform has other potential benefits and is also associated with higher levels of vitamin D. In experimental studies with genetically engineered knock out mice, the absence of ApoE in the brain is also protective against developing cerebral malaria (Kassa et al. 2016). This suggests that both a predominance of ApoE4 or a total absence of ApoE are protective against developing cerebral malaria. Because ApoE performs various important metabolic functions in the brain (and elsewhere in the body), total absence has not arisen through natural selection.

Sometimes it is difficult to determine the relative importance of immune competence and behavioural characteristics influencing disease susceptibility. Children under the age of fifteen account for about half of all reported cases of *Balamuthia* encephalitis, and whilst this may relate to their immune status, it could also reflect their greater likelihood than adults to come into contact with contaminated soil or water while playing and poorer personal hygiene. In the USA, about half of all cases of *Balamuthia* encephalitis occur in people of Hispanic descent although Hispanics comprise only about 13% of the population (Cope et al. 2019). This may be a consequence of a genetic susceptibility but could also be related to the fact that as a group, Hispanics in the USA tend to be poor and engage in manual work such as farm labour and gardening that might increase their exposure to *Balamuthia*.

10.2.2 Parasite Factors

Genetic constitution influences all aspects of an organism's biology, and different strains of a parasite can vary markedly in their infectiveness and pathogenicity. For example, strains of *E. histolytica* vary in their ability to induce amoebic dysentery (Escueta-de Cadiz et al. 2010). In mice, Type I strains of *Toxoplasma gondii* are highly virulent and infection with just a few tachyzoites can prove fatal. By contrast, the Type II and Type III strains are non-lethal and cause chronic latent infections. *Toxoplasma gondii* infects macrophages, among other cell types, but can be killed by them if the parasitophorous vacuole in which they reside is disrupted. This can occur through physiological processes mediated by immunity-related GTPase proteins (IRG). The virulence of the Type I genotype appears to be due, at least in part, to its ability resist the IRG-mediated disruption of their parasitophorous vacuole. This allows the parasites to survive within even activated macrophages and thereby compromise the cells' role in the immune response and, ultimately, kill them. In addition, the parasites can 'abandon ship' before the macrophages exert their innate 'killing mechanisms'. Under *in vitro* conditions, the tachyzoites inhabiting macrophages undergo up to seven division cycles over 2–3 days before the host cell is destroyed. However, in acutely infected mice, the macrophage can be destroyed after only 0–2 division cycles – this takes about 6 hours and results in rapid macrophage destruction. The parasites are stimulated to leave their host macrophages by the arrival of other, non-infected macrophages that could potentially 'switch on' the parasitised cells. More details of the invasion process are provided by Portes et al. (2020).

It is uncertain whether the various strains of *T. gondii* differ in their pathogenicity in humans and other animals. In most mammal and bird intermediate hosts, *T. gondii* causes subclinical infections, whilst clinical disease is more frequent in some species than others. For example, pigeons and canaries are more likely to suffer clinical disease than other bird species (Dubey 2002). In these species, symptoms include loss of condition, anorexia, conjunctivitis, and encephalitis. Most surveys of human populations indicate that the majority of people are parasitised by the Type II genotype of *T. gondii*. In addition, it is the most prevalent genotype (>60%) in AIDS patients and cases of congenital toxoplasmosis. By contrast, it is less prevalent (41.7%) in cases of ocular toxoplasmosis (Hosseini et al. 2019). However, most of us become infected through consuming infected meat, and therefore this could simply reflect that this is the commonest genotype that infects the animals we consume.

10.3 Mechanisms By Which Parasites Cause Pathology

Typically, a parasite begins causing damage when it invades its host or starts feeding upon it. However, some parasites cause their most serious pathology when they begin reproducing. Parasite antigens and physical damage to host tissues both initiate an immune response that suppresses or expels the parasite or induces further pathology – or both. A parasite may therefore cause harm directly (e.g., through removal of metabolic resources) and indirectly (e.g., through inducing a harmful immune reaction) (Table 10.2). In addition, owing to the interconnectedness of physiological processes, damage to one organ or region of the body can have severe consequences for the functioning of other organs and physiological systems. In order to develop appropriate treatments for parasitic diseases, it is important to understand both the mechanism(s) by which a parasite inflicts damage on its host and the host's response.

10.3.1 Direct Damage

Parasites cause direct physical damage by feeding on the host's tissues, boring through tissues during their migrations, and by the physical act of growing, thereby causing pressure atrophy and restricting of the circulation of blood/lymph to a region of the body. Physical damage represents a loss of nutrients, and there is also a nutritional cost associated with the repair processes. The

Table 10.2   Mechanisms by which parasites cause pathology.

Direct	Indirect
Physical presence	Overstimulation of immune system
Consumption of host tissues	Stimulation of autoimmunity
Competition for nutrients	Compromise of the immune system
Production of harmful secretions/excretions	Alteration of local environment (e.g., pH)
	Alteration of homeostasis
	Increase susceptibility to other diseases
	Disruption of microbiome

seriousness usually depends upon the number of parasites involved, the time interval over which the damage occurs, and the region of the body affected. Of course, if the balance is too severe, then there is a point where the relationship is no longer be advantageous to the parasite. For example, the females of some isopod parasites belonging to the genus *Cymothoa* (e.g., *Cymothoa exigua* and *Cymothoa borbonica*) locate within the buccal cavity of their teleost fish host and attach to the basihyal – a bone that serves as the tongue. As the isopod grows, it totally destroys the basihyal and could thereby bring about its own demise should its host starve to death. However, the parasite then proceeds to act as a replacement fully functional basihyal, so the fish does not starve – although its growth is compromised (Parker and Booth 2013).

Acute parasitic disease usually results from large numbers of parasites arriving at a sensitive location over a short time interval. This may prove fatal for the host. By contrast, chronic parasitic disease is associated with fewer parasites and long-standing infections that may or may not resolve and may or may not prove fatal. For example, acute fascioliasis in sheep results from the simultaneous migration of large numbers of immature *Fasciola hepatica* through the liver. This usually occurs 2–6 weeks after the ingestion of 2,000 or more infective metacercariae and death may occur suddenly without any previous symptoms. If a sheep ingests fewer metacercaria, say 500–1,500, over a longer time interval, then it is more likely to suffer sub-acute fascioliasis. In this case, the disease manifests 6–10 weeks after infection. Although it may ultimately prove fatal, the infected animal does not die so quickly. Finally, chronic fascioliasis in sheep is associated with the growth of adult flukes in the bile ducts. Although chronic fascioliasis can be fatal, it is more commonly a cause of progressively poor health and lack of condition. Chronic fascioliasis typically results from the ingestion of 200–500 metacercariae and clinical disease becomes apparent 4–5 months after infection. Chronic fascioliasis is not as dramatic as the acute disease but, as with other chronic parasitic infections, it afflicts more animals and is economically more important.

Parasites release various secretions and excretions into and onto their hosts. For example, they release them to facilitate invasion, migration through the body and feeding, to overcome host immune defences and in the process of excreting waste products. These substances can cause pathology either directly through damaging cells and metabolic processes or through the less direct means of stimulating a damaging inflammatory response. For example, the infective larvae of hookworms and the cercariae of schistosomes release protease enzymes that help them penetrate their host's skin. In the process, they kill host cells and engender an inflammatory response. Similarly, the protozoa *E. histolytica*, *Trypanosoma brucei*, *Trypanosoma congolense*, and *Trypanosoma cruzi*, all release cysteine peptidases into the surrounding medium. In addition to acting on their substrates, these enzymes also stimulate an immune response. This results in a localised increase in the concentration of activated host-derived pro-inflammatory cytokines. Trypanotolerant cattle have a better immune response against cysteine peptidases than susceptible breeds and consequently the enzymes are potential targets for the development of vaccines.

10.3.2 Indirect Damage

Mounting an immune response, even an effective one, is energetically expensive. Indeed, it has often been observed that sheep selectively bred to be resistant to gastrointestinal nematode infections (as determined by expressing low faecal egg counts) tend to have lower growth rates than unselected control sheep (Williams 2011). In addition, immune-induced pathology is a major cause of the pathology associated with some parasitic infections. For example, we tend to express the chronic phase of Chagas Disease (caused by *Trypanosoma cruzi*) when we are adults and this manifests as an 'indeterminate', a 'cardiac' or a 'digestive' form. Most adults have the indeterminate form and they may be unaware of their infection. Indeed, it is sometimes impossible to detect any abnormalities using conventional electrocardiograph procedures and radiology. Many of these people survive in this state for years and eventually die of some other cause. In some people, however, the disease progresses slowly over several years to the cardiac or digestive forms. In these two forms, the pathology results primarily from the destruction of peripheral and central nervous tissues, although there can also be significant damage to the skeletal, cardiac, and smooth muscle. Much of this damage is immune-mediated. However, it is uncertain how much results from inflammatory responses induced directly by the parasite and how much from them engendering an auto-immune response. In the cardiac form, the patient may die suddenly without expressing any previous outward signs of disease or they might suffer from a steady decline in heart function over a prolonged period, which may also end in death. Damage to the nerves supplying the heart results in a loss of muscle tone so it becomes 'flabby'. In addition, the size of the heart can increase (cardiomegaly) because of hypertrophy of the muscle fibres, inflammation and oedema. A characteristic feature is the formation of a 'vortical lesion' at the apex of the left ventricle. Histologically, one can see the deposition of fibrous tissue within the heart and the focal accumulation of inflammatory cells – a condition known as fibrous myocarditis.

10.4 Types of Pathology

10.4.1 Abortion and Obstetric Pathology

Spontaneous abortion is remarkably common in humans. Some estimates indicate that around 40% of conceptions fail to establish as recognisable pregnancies and of those that do establish around 15% abort at some stage during the pregnancy. There are numerous reasons for spontaneous abortion, among which are certain parasites. Parasites may induce abortion by crossing the placenta and infecting the developing foetus, causing damage to the placenta, and by reducing general health or inducing metabolic disorders in the mother.

Malaria is particularly dangerous for both an expectant mother and her unborn child. This is because it can induce numerous clinical complications (Schantz-Dunn and Nour 2009). One of the complications is the development of placental malaria that is especially a problem for women in their first pregnancy and is mainly associated with *P. falciparum* infections. The red blood cells parasitised by *P. falciparum* express ligands on their cell surface that bind with receptors lining vascular endothelial cells. In pregnant women, this results in the infected cells becoming sequestered within the intervillous spaces of the placenta. Consequently, their density rises to levels much higher than those in the peripheral blood stream and the placenta may turn black from the deposition of hemozoin – a waste product of *Plasmodium* digestion. In response, large numbers of immune cells infiltrate the intervillous spaces where they cause inflammation and pathological changes to the placenta (Sharma and Shukla 2017). For example, membrane glycosylphosphati-dylinositol lipids derived from *P. falciparum* stimulate a subset of natural killer cells called Vα14NKT cells that are connected to the induction of abortion (Ito et al. 2000). Vα14 NKT cells are associated with the immune responses to tumours and a variety of pathogens, allergic reactions, and autoimmune diseases and upon activation produce tumour necrosis factor (TNF-α). TNF-α (and certain other inflammatory cytokines) induce placental coagulopathy (i.e., bleeding that results from defects in the clotting mechanism), and this interferes with the supply of nutrients to the developing foetus. This can result in low birth weight, probably through reducing intrauterine growth rate. There is also an increase in the frequency of spontaneous abortion and premature labour during malaria epidemics (Menendez et al. 2000).

Leishmania and *Trypanosoma* also form membrane glycosylphosphatidylinositol lipids that induce Vα14 NKT cells. Leishmaniasis is implicated in causing abortion in dogs, whilst *Trypanosoma evansi* causes abortion in a variety of domestic animals including camels, horses, and buffalos (Löhr et al. 1986).

In humans, *Toxoplasma gondii* is the parasite most frequently identified with causing abortion (Ambroise-Thomas and Petersen 2013). It does this by passing across the placenta and infecting the developing foetus. Infection of the foetus is most likely (60–90%) during the third trimester of pregnancy. Fortunately, this is also the period when the chances of them causing a deleterious outcome are less. By contrast, the risk of infection during the first trimester is only 10–15% but if it does occur there is a much greater chance that the outcome could be fatal for the developing baby. Congenital infections can result in spontaneous abortion or still birth, and if the baby is born alive it may suffer from a range of morphological and neurological conditions. At its worst, the baby is born with potentially fatal birth defects such as hydrocephalus and microcephalus. Congenital infection can also result in other disorders of the nervous system, including mental impairment. Some of these may not become apparent until several years after birth.

Neospora caninum is a major cause of abortion in cattle in many parts of the world (Reichel et al. 2013). Two forms of abortion occur: epidemic abortion and endemic abortion. Epidemic abortion is where a large proportion of the herd suffer abortions within a short period of time. This is sometimes called an 'abortion storm'. It probably arises from the cattle experiencing a sudden massive challenge infection: for example, if their field or food/water supply is suddenly contaminated with large numbers of infective oocysts. Endemic abortion is where the abortion rate remains above 5% $year^{-1}$ over a period of several years. The maintenance of disease within a herd over many years probably requires periodic horizontal input. Therefore, it is likely that both epidemic and endemic abortion occur in localities where the *N. caninum* is common. Furthermore, *N. caninum* exhibits considerable genetic diversity and therefore the various strains might vary in their pathogenicity.

10.4.2 Abscesses and Ulcers

An abscess develops when a pathogen initiates an infection within an enclosed space that results in an inflammatory reaction and the death of cells (i.e., necrosis). The pathogen and the dead cells become surrounded by inflammatory cells (especially neutrophils) and acute inflammatory exudates (pus). At this stage, it is called an acute abscess, and it is said to be purulent because of the fluid exudates. The purulent discharge varies in colour and composition and may be thick or thin and may be grey, yellowish, or even brown in coloration. If the abscess does not grow significantly, over time it becomes a chronic abscess in which scar tissue gradually replaces the inflammatory exudate. However, the pathogen may remain alive at the centre of the abscess and therefore complete healing cannot occur. An ulcer, by contrast, occurs when a pathogen sets up an infection on the outer surface of the epithelium. For example, on the surface of the skin or the lining of the gastrointestinal tract. The infection then proceeds to invade inwards and may penetrate deeply. This also results in the death of cells and an inflammatory reaction so that the wound becomes purulent. The protozoan parasite *E. histolytica* causes both ulcers and abscesses: initially, it forms flask-shaped ulcers on the surface of the colon and, subsequently, after spreading through the body, it can cause abscesses in the liver, lungs, and brain.

10.4.3 Anaemia

Anaemia becomes outwardly obvious with the development of pallor in the mucus membranes owing to the reduction in the numbers of circulating red blood cells. The affected person or animal becomes lethargic and the pulse rate increases because the heart pumps the blood around the body faster to compensate for its lower oxygen carrying capacity. Depending upon the amount of blood lost and the time interval over which it occurs, anaemia can vary from being a mild chronic condition to rapidly fatal.

Parasites engender anaemia in various ways (Table 10.3). Helminth and arthropod parasites often cause anaemia by feeding on blood and/or damaging blood vessels during their migration through the host's body. For example, hookworms lacerate blood vessels lining the gut and consume more blood than they require. Furthermore, they secrete anticoagulants into the wounds and therefore these continue to bleed after they finish feeding. Consequently, a host parasitised by large numbers of hookworms can lose a significant amount of blood. Similarly, when young *Fasciola hepatica* flukes migrate through the parenchyma of the liver, they damage blood vessels and cause widespread bleeding.

In humans, infection with the tapeworm *Diphyllobothrium latum* presents a risk of developing megaloblastic (pernicious) anaemia. However, although approximately 40% of infected people exhibit low levels of vitamin B12, less than 2% of them actually develop anaemia. Megaloblastic anaemia results from the tapeworm secreting substance(s) that cause the disassociation of vitamin B12 from intrinsic factor within the gut. Intrinsic factor is a glycoprotein secreted by the parietal cells in our stomach. Its role is to bind with vitamin B12 within the ileum and thereby enable it to

Table 10.3 Mechanisms by which parasites cause anaemia.

Cause of Anaemia	Example
Haemorrhage of red blood cells	*Ancylostoma duodenale*
Destruction of parasitised red blood cells	*Plasmodium falciparum*
Autoimmune destruction of uninfected red blood cells	*Trypanosoma brucei*
Deficiency of metabolites needed to produce red blood cells (e.g., Lack of iron and vitamin B12)	*Diphyllobothrium latum*
Aplasia (failure to produce sufficient red blood cells)	*Plasmodium falciparum*
Hypersplenism (red blood cells are sequestered in the spleen and destroyed)	*Plasmodium falciparum*

be absorbed across the gut wall. Because the ability of vitamin B12 to bind with intrinsic factor declines, less vitamin B12 is absorbed. The decline in vitamin B12 levels results in the formation of large red blood cell precursor cells (megaloblasts) that give rise to unusually large red blood cells – and these are then destroyed because they are defective. Megaloblastic anaemia is also associated with the development of both neurological and heart diseases. The worms also selectively absorb vitamin B12 but why they have such an affinity for the vitamin is not known. Megaloblastic anaemia is seldom induced by other tapeworms of the genus *Diphyllobothrium* (Scholz et al. 2009).

Parasitic protozoa induce anaemia through various mechanisms. In the case of *E. histolytica*, extensive ulceration of the colon and large intestine can lead to sufficient blood loss to cause anaemia. However, anaemia is more often associated with those protozoa that live within the bloodstream, sometimes within the blood cells. For example, malaria causes anaemia through the destruction of both parasitised and non-parasitised erythrocytes, the inability of the body to recycle the iron bound in hemozoin (also spelled haemozoin), and an inadequate erythropoietic response (formation of new red blood cells) by the bone marrow. Malaria also causes coagulopathy (i.e., it interferes with the clotting process) and thereby causes susceptibility to uncontrolled bleeding. The reason(s) why so many non-parasitised red blood cells are destroyed in response to malaria infection is unclear, but some evidence suggests that it is linked to an autoimmune response. Destruction of erythrocytes leads to an increase in blood bilirubin, a breakdown product of haemoglobin. When excretion cannot keep up with formation of bilirubin, the yellow bile colour is evident under the skin and whites of the eyes (jaundice). Hemozoin is taken up by circulating leucocytes and deposited in the reticuloendothelial system. In severe cases, the viscera, especially liver, spleen and brain, become blackish as a result of pigment deposition. After ingesting hemozoin, macrophages suffer impairment in phagocytic activity and this compromises the cellular immune response. Identifying the presence of hemozoin does, however, offer the potential for new diagnostic tests for malaria because it is a characteristic feature of *Plasmodium* infection. Kumar et al. (2020) describe field trials of a novel device that can detect hemozoin and thereby identify a malaria infection in as little as one minute.

In cattle, and other animals susceptible to *Trypanosoma brucei brucei*, much of the pathology caused by the parasites is associated with the development of anaemia. Indeed, a key feature of trypanotolerant breeds and species is their ability to prevent the level of parasitaemia rising to a point at which it causes severe anaemia. Trypanotolerant breeds of cattle may, however, exhibit loss in productivity if subject to serious challenge by trypanosome parasites (Taylor et al. 2007). There is some debate over the principal cause of the anaemia and the chances are that it is multifactorial, including variables such as host species, host and parasite genetic constitution, and individual circumstances such as host age, health, and nutrition. One possibility is that the trypanosomes release substances that kill red blood cells. Another is that parasite-derived VSGs become attached to red blood cells, thereby causing them to be identified as 'foreign' and therefore attacked by the host immune system. In laboratory studies, the development of anaemia in mice infected with *T. brucei* has been linked to the release of excessive amounts of inflammatory cytokines such as tumour necrosis factor. For more details on the pathology associated with trypanosomiasis, see Magez and Radwanska (2014).

10.4.4 Anorexia

Anorexia is the voluntary reduction in food intake (i.e., loss of appetite) although some workers prefer to use the term 'inappetence' so as to avoid confusion with the human psychological condition '*anorexia nervosa*'. Anorexia is a feature of many infectious diseases, including those caused by parasites. At first sight, this appears to be counterintuitive because infections impose a considerable metabolic cost on their host. However, anorexia is a common response by both invertebrates and vertebrates to many viral, bacterial, protozoan, and helminth infections. It is not, however, an invariable response and there is no obvious reason why some infections induce anorexia but others do not. For example, exposure of the fruit fly *Drosophila melanogaster* to both live and dead *Listeria monocytogenes* induces anorexia, but their exposure to another species of bacterium *Enterococcus faecalis* does not (Ayres and Schneider 2009). The induction of anorexia therefore probably depends upon both host and parasite characteristics at the time of the infection. Because anorexia is such a common reaction to so many infections, it is probably an ancestral response that has been conserved because it benefits the host. However, the nature of the benefits has proved difficult to ascertain and anorexia may place a considerable cost upon the host. For example, in the case of gastrointestinal helminth infections of domestic animals, the losses caused by anorexia may account for 40–90% of the decline in productivity (Greer 2008). Nevertheless, the consequences of anorexia for both the host and the pathogen depend upon numerous interacting factors and it should not be considered invariably detrimental to the host. Indeed, there is evidence that attempting to circumvent the onset/length of anorexia might even help select for more virulent pathogens (Hite and Cressler 2019). Wang et al. (2018) demonstrated how anorexia might actually reduce the pathology associated with malaria. They found that mice infected with *Plasmodium berghei* were protected from developing cerebral malaria if their ability to undertake glycolysis was artificially reduced. The reduction in glycolysis did not affect the parasitaemia or anaemia and they suggested that the protection probably arose from there being less 'fuel' to power the inflammatory processes that lead to infected red blood cells becoming sequestered within capillaries in the brain. If this is correct then it has implications for the clinical management of patients at risk of suffering cerebral malaria.

Viruses and bacteria tend to induce anorexia shortly after infection but it seldom lasts long. By contrast, in protozoan and helminth infections anorexia often does not commence until several weeks after infection and is of longer duration. For example, in rainbow trout (*Oncorhynchus mykiss*), food intake relates to the density of the protozoan parasite *Cryptobia salmositica* in their blood. Chin et al. (2004) found that 3 weeks after initial infection the food intake of the fish reduced significantly from 1.33 to 0.94% initial body weight and by week 4 had dropped to 0.8% body weight – this corresponded to peak parasitaemia. The physiological mechanisms underlying anorexia are poorly understood and probably differ between hosts. Several workers have found evidence that the immune system is involved in inducing anorexia. For example, mice infected with the nematode *Trichinella spiralis* normally exhibit anorexia, but this does not occur in mice lacking CD4+ T lymphocytes or the receptor for interleukin-4 (McDermott et al. 2006). Anorexia can be induced in non-infected humans and other mammals by treatment with tumour necrosis factor-α (TNFα). Both TNF-α and interleukin-1 have been implicated in the development of anorexia associated with influenza. The circulating levels of TNF-α and interleukin-1 increase during malaria, and it is therefore possible that these cytokines may be responsible for the onset of anorexia (Clark et al. (2008).

10.4.5 Apoptosis

Apoptosis is a form of programmed cell death found in all metazoan animals and some protozoa. It fulfils various functions in processes as diverse as embryogenesis and the response to pathogenic infections. Apoptosis is controlled by various extrinsic and intrinsic pathways. The extrinsic pathways are triggered by the interaction of specific cell surface receptors with ligands. For example, the best understood pathway involves the binding of the ligand CD95 (Fas) with the CD95

(Fas) Receptor to bring about the formation of the 'death-inducing signalling complex' (DISC). The death-inducing signalling complex then brings about the activation of caspase enzymes that cause cell cytolysis. For this reason, caspases are sometimes referred to as executioner enzymes. The intrinsic pathways are associated with the release of cytochrome c into the cytosol following the breakdown of the outer membrane of the mitochondria. The cytochrome c then binds to 'apoptosis-activating factor' that is present in the cytosol. This then sets in train a series of reactions that result in the activation of caspase enzymes. Apoptosis is also initiated by cytotoxic T cells via other pathways such as the granzyme-B pathway and the perforin pathway. Cells that undergo apoptosis are recognised and engulfed by phagocytic cells, so the breakdown products are not released into the surrounding media.

Some parasites owe at least part of their pathogenicity to their ability to induce apoptosis. For example, when the trophozoites of *E. histolytica* make contact with intestinal epithelial cells, they induce them to undergo immediate apoptosis. This almost certainly contributes to the virulence of the parasites and facilitates invasion. In the case of cerebral malaria, hemozoin, the waste product of *Plasmodium* metabolism, becomes internalised by neurons and astrocytes within which it induces apoptosis (Eugenin et al. 2019). Apoptosis is also a feature of pathology induced by many other parasitic protozoa including *Trypanosoma brucei*, *Trypanosoma cruzi*, and *Toxoplasma gondii* (Happi et al. 2012; Pidone and dos Santos 2020; Wang et al. 2014).

Helminth parasites can also induce apoptosis in their host's cells. For example, the cysticerci of the tapeworm *Taenia crassiceps* release a substance that induces apoptosis of germ cells within the seminiferous tubules of male mice (Zepeda et al., 2011). Apoptosis is a normal physiological process within the testis as a means of controlling germ cell numbers but infection with *T. crassiceps* vastly increases its occurrence. So many cells die that numerous macrophages infiltrate the seminiferous tubules to dispose of the dead cells. Although sufficient germ cells die to compromise male fertility, this would not be considered an example of parasitic castration because it only occurs when a large number of cysticercoids are present.

Apart from being induced by substances released by parasites, apoptosis can also be a normal part of the host immune response to infection. For example, if a cell infected with an intracellular parasite undergoes apoptosis, it prevents the parasite from reproducing and may kill the parasite. Intracellular parasites therefore often have physiological mechanisms that control or prevent or apoptosis of the cells they are living in. For example, *Leishmania donovani* inhibits macrophage apoptosis (Abhishek et al. 2018) and *Theileria parva*-infected lymphocytes lose the ability to undergo apoptosis (Tretina et al. 2015). At the opposite extreme, some parasites compromise their host's immune system by inducing apoptosis in certain immune cells (Zakeri 2017).

10.4.6 Autoimmunity

Autoimmunity is when our immune system identifies our own cells or tissues as foreign and attempts to remove them by mounting an immune response against them. The reasons for this self-destructive act are often uncertain but sometimes it may be linked to the absence of parasites rather than their presence. Parasites and other pathogens have probably played an important role in the evolution of immune systems (Zaccone et al. 2008). Furthermore, their presence may even be necessary for immune systems to continue to function properly. For example, there is increasing evidence that we have a tendency to develop allergies and autoimmune diseases if our immune system does not receive regular stimulation from parasites and other pathogens. Indeed, there is a significant correlation between the presence of risk alleles for both inflammatory bowel disease and Crohn's disease and what Fumagalli et al. (2009) refer to as 'pathogen richness'. There is therefore considerable interest in the potential of employing parasites (or chemicals derived from them) in the treatment of allergies and autoimmune diseases – and this is discussed in Chapter 12.

A few parasites, however, rather than protecting against autoimmunity appear to induce it. For example, some patients suffering from *Cyclospora* infection subsequently develop neurological complications such as Guillain-Barré syndrome (Richardson et al. 1998) and Reiter's syndrome (Connor et al. 2001). Guillain-Barré syndrome is an autoimmune disease that causes demyelination of axons in the peripheral nervous system. There are six subtypes, but the most common is acute inflammatory demyelinating polyneuropathy (AIDP) that results in progressive paralysis and although those affected may recover with treatment it can be fatal. It is normally associated with viral (e.g., influenza) and bacterial (e.g., *Campylobacter jejuni*) infections. There are a few isolated case reports linking Guillain-Barré syndrome other parasitic diseases, notably *Toxoplasma gondii* infection (Bossi et al. 1998), visceral leishmaniasis (Attarian et al. 2003), and acute *P. falciparum* malaria (O'Brien and Jagathesan 2016). Reiter's syndrome is another autoimmune condition that follows infection – it is often associated with the bacterium *Chlamydia trachomatis* although gastrointestinal infections may also initiate the condition. It usually manifests as inflammatory reactions in the eyes (e.g., conjunctivitis), the urinary tract (e.g., dysuria), and arthritis, particularly of the large joints. Doctors therefore often remember the condition using the mnemonic 'can't see, can't pee, can't climb a tree'. There is currently very little evidence of an association of Reiter's syndrome with parasitic protozoa or helminths, but whether this is due to lack of suitable research is uncertain. Autoimmune responses are also thought to be contributory factors in the pathology associated with malaria (Rivera-Correa and Rodriguez 2017) and Chagas Disease (De Bona et al. 2018).

10.4.7 Calcification

Calcification refers to the extracellular deposition of calcium salts in tissues other than bone. This may occur as a consequence of metabolic disorders such as those affecting the levels of parathyroid hormone or following chronic infections when calcium salts are deposited in injured tissues. In the latter situation, it is sometimes referred to as dystrophic calcification and it is stimulated by the presence of dead and decaying (necrotic) tissue. The consequences of calcification depend upon its extent and where it occurs. Within lymph nodes, calcification may have no functional consequences but if it occurs within the valves of arteries, it impedes blood flow. The tissue-dwelling cysts of larval tapeworms found in humans and other mammals often have calcium salts deposited around them. The calcification facilitates the identification of hydatid cysts (*Echinococcus granulosus*) and the cysticerci of *Taenia solium* in human patients using imaging techniques such as ultrasound, CT (computerised tomography), and MRI scans (magnetic resonance imaging). Similarly, by cutting through specific sites and feeling for gritty deposits of calcium, meat hygiene inspectors can detect the presence of *Taenia saginata* cysts in the heart muscle of cattle or those of *T. solium* in the masticatory muscles of pigs. Other tissue dwelling parasites such as *Trichinella spiralis* larvae and *Anisakis* spp. larvae can become calcified, whilst calcification can occur in the bile ducts and liver as a consequence of infection with *Fasciola hepatica* and *Fasciola gigantica*.

10.4.8 Cancer

Usually, cell division ceases after withdrawing the stimulus that causes it. Sometimes, however, the stimulus causes permanent genetic change in the affected cells and they cease responding to the normal processes that control cell growth. These 'uncontrolled' cells are referred to as having become 'neoplastic' and they form a lump or ramifying mass of cells called a neoplasm. In common language, the cells are said to have become 'cancerous'. The term cancer derives from the Latin *cancer*, which means 'a crab' because the cancerous region was once thought to spread through the body in a series of 'pincer-like' movements. There are numerous reasons why some cells become neoplastic and genetic constitution is often important in determining susceptibility and disease outcome. Similarly, there are numerous neoplastic diseases which vary from the relatively benign (that have a good prognosis) to the highly malignant that can be rapidly fatal. The International Agency for Research on Cancer (IARC) has devised a scheme in which stimuli (agents) are grouped according to their potential to induce neoplasia in humans. Many parasites affect cell division and differentiation through constant mechanical irritation, inducing immune responses and/or secreting noxious chemicals but relatively few of them induce neoplastic diseases (Table 10.4).

One of the most studied examples of parasite-induced neoplasia is the transformation of *T. parva*-infected lymphocytes in cattle into rapidly multiplying cells. This results in them behaving like lymphoblastomas (neoplastic cells) and causing the development of leukaemia-like pathology (Tretina et al. 2015). Indeed, *T. parva*-induced lymphocyte transformation is a potentially useful model for studying the development of naturally arising lymphoblastomas in humans. Tumours may form in the brain, kidneys, and other organs, which can then invade surrounding tissues and/or metastasise to other regions. The exact mechanism by which transformation is brought about is uncertain but *T. parva* induces its host lymphocyte to over-express several regulatory kinases such as jun-NH_2-terminal kinase, Src family kinases, and casein kinase II. Casein kinase II is involved a wide variety of cell functions and is implicated in cell transformation, development of tumours, and protection of transformed cells from apoptosis.

The transformation of a cell, whether by oncogenes or viruses, brings about the activation of the cell's apoptotic mechanisms. Consequently, the transformed cell can only survive if it is able to avoid these mechanisms. In the case of *Theileria*-infected lymphocytes, the presence of the living parasite is necessary to maintain the transformed state and should the parasite die then apoptosis occurs. For example, following drug treatment, the lymphocyte reverts to its normal resting phenotype and then dies through apoptosis. One mechanism by which the transformed state is maintained is because *Theileria* induces the activation of c-Jun NH_2-terminal kinases 1 and 2 (JNK1, JNK2), which bring about the phosphorylation, and thereby activation, of c-Jun. Both activation of

Table 10.4   IARC classification of the ability of agents to induce cancer in humans (as of 2020).

Group	Ability to cause neoplasia in humans	Number of agents listed in 2020	Parasite example
1	Definitely	120	*Schistosoma haematobium* *Clonorchis sinensis* *Opisthorchis viverrini*
2a	Probably	88	*Plasmodium falciparum*
2b	Possibly	313	*Schistosoma japonicum*
3	Not classifiable	449	*Schistosoma mansoni* *Opisthorchis felineus*

https://monographs.iarc.fr/agents-classified-by-the-iarc/.

the JNK enzyme pathway and induction of the c-Jun (that forms part of AP-1 early response transcription factor) help prevent apoptosis in the transformed lymphocytes. In addition, *Theileria* brings about the activation of the transcription factor NF-κB within transformed lymphocytes, and this, among other things, brings about the formation of inhibitors of caspase family proteases that are involved in apoptosis. Consequently, the death of the parasite results in a decline in the activity of the JNK enzyme pathway, the inhibition of c-Jun and NF-κB, and these events together with a decline in other anti-apoptotic proteins, make the cell susceptible to apoptosis. For more details, see Tretina et al. (2015).

The relationship between helminth parasites and cancer is poorly understood (Brindley and Loukas 2017), and there are only a few of them for which there is a clear link between infection and neoplasia. This is perhaps surprising because helminth infections tend to stimulate a predominantly Th-2 immune response and this is associated with a poor prognosis for several common cancers, such as breast cancer and lung cancer (Lippitz 2013).

The trematode parasites *Clonorchis sinensis*, *Opisthorchis viverrini*, and *Opisthorchis felineus*, all infect our bile ducts and are recognised as important in the development of cholangicarcinoma (Pakharukova and Mordinov 2016). Although rare, this is a highly malignant neoplasia of the intrahepatic bile duct and most patients die within 6–12 months of diagnosis. As yet, the mechanism(s) by which the parasites induce neoplasia is uncertain but, as in most cancers, it probably results from a combination of host, parasite, and environmental (e.g., diet) factors (Salao et al. 2020). The adult worms can live for many years and during this time their feeding inflicts repetitive lesions to the lining of the bile ducts. Therefore, one might expect that the constant physical irritation combined with the secretion of digestive enzymes would predispose the host's cells to becoming cancerous. Somewhat paradoxically, *O. viverrini* also secretes a growth factor that promotes wound healing and angiogenesis. Unfortunately, it also appears to stimulate cancer pathways (Haugen et al. 2018; Smout et al. 2015). *Opisthorchis* also produces genotoxins such as catechol oestrogens and oxysterols (Brindley and Loukas 2017). In sheep, the flukes *Fasciola hepatica*, *Fasciola gigantica*, and *Dicrocoelium dendriticum* also cause serious damage to bile ducts and an inflammatory condition called hyperplastic cholangitis, which is also a feature of human infections with *C. sinensis, O. viverrini*, and *O. felinus*. The sheep parasites are not usually associated with causing cancer although most sheep (and other farm animals) are slaughtered whilst still comparatively young. That is, they may die before the parasite infection induces obvious cancerous lesions.

The granulomas caused by the nematode *Anisakis* spp. are sometimes misdiagnosed as gastric (and rarely gynaecological) cancer (Nogami et al. 2016), and there is some debate about whether the worms are themselves carcinogenic. Under *in vitro* conditions, worm extracts can cause proliferation of epithelial cells, and in animal models, they induce changes in the expression of cancer-related miRNAs (Corcuera et al. 2018). Nevertheless, at the time of writing, it remained uncertain whether anisakiasis represents a cancer risk.

The schistosome species that infect humans vary in their tendency to induce neoplasms although the reasons for this are uncertain. *Schistosoma haematobium* infection has a proven association bladder cancer. It is the major cause of cancer in Egyptian males and second most common neoplasm in Egyptian women after breast cancer (Mostafa et al. 1999). The mechanism(s) by which *S. haematobium* induces neoplasia is uncertain but almost certainly involves a combination of host, parasite, and environmental factors (Ishida and Hsieh 2018). Much of the pathology associated with schistosomiasis results from the host immune reaction to the eggs. Consequently, host genetic factors that result in an excessive immune reaction may contribute to the likelihood of cancer developing. Bladder pathology is also associated with the expression of a protein kinase gene in the parasite's genome (Huyse et al. 2018). This enzyme is involved in egg production, so a deleterious outcome from an infection might be at least partly associated with strains of *S. haematobium* that produce lots of eggs (a female schistosome typically produces around 300 eggs day^{-1}).

As with *Opisthorchis*, *S. haematobium* eggs cause constant irritation and fibrosis and they form genotoxic catechol oestrogens (Brindley and Loukas 2017). Similarly, the inflammatory response to *S. haematobium* results in the formation of N-nitrosamines, reactive oxygen radicals, and a variety of other toxic chemicals that can damage DNA. However, other *Schistosoma* species and various other parasites that are not considered carcinogenic also induce the formation of *N*-nitrosamines etc. In addition, there is a possibility that the ingestion/inhalation of *N*-nitrosamines and other mutagenic chemicals (e.g., through smoking) may increase the chances *S. haematobium* infection leading to bladder cancer. Diabetes is also a risk factor for developing bladder cancer, and the prevalence of type 2 diabetes in Egypt is remarkably high (Khalil et al. 2018). However, whether there is an association between schistosomiasis, diabetes, and bladder cancer is currently unknown. Unsurprisingly, *S. haematobium* infections result in changes to the composition of the bladder microbiome and patients infected with *S. haematobium* tend to pass higher levels of bacteria such as *Klebsiella* spp., *Staphylococcus aureus*, and *Escherichia coli* in their urine than those who are uninfected. It is possible that bacteria facilitate the induction of neoplasia through the formation of carcinogenic compounds but, as yet, there is little conclusive evidence of their association with bladder cancer (Ishida and Hsieh 2018).

10.4.9 Castration

Castration refers to the prevention of an organism's ability to reproduce and may be permanent (e.g., through the removal of the sexual organs) or temporary (e.g., transient loss of function of the sexual organs). Several parasite species damage the genitalia of their host sufficient to prevent reproduction (e.g., *Wuchereria bancrofti*) but these typically involve many individual parasites. Parasitic castration is restricted to specific host: parasite relationships in which a single individual parasite limits or prevents host reproduction. Most instances of parasitic castration involve parasites of invertebrates although there are a few cases of it occurring in fish and amphibians. Parasitic castrators tend to be very host specific and are much larger in proportion to the size of their host than other parasites. Usually a single individual parasite weighs considerably less than 1% of its host's body mass but parasitic castrators typically weigh over 1% and may reach up to 39%. Most cases of parasitic castration appear to arise through the removal of nutrients by the parasite, although it is possible that this is supplemented to a greater or lesser extent by the secretion of substances that interfere with the development of the host's gonads. By comparison there are far fewer recorded instances in which the parasites physically consume the gonads or mechanically prevented them from functioning (Lafferty and Kuris 2009).

In humans and many other mammals, castration inflicted whilst growth is still taking place can lead to an increase in body size. This is because metabolic reserves that would have been devoted to reproduction are used to enhance growth. In some parasite: host relationships, parasitic castration leads to increased growth, and this is termed 'gigantism'. Presumably, this is facilitated if the host manages to acquire and metabolise more resources than the parasite requires for its own purposes. It is most likely to occur during the early stages of infection when the host is actively growing but the parasite is still relatively small. Gigantism is often associated with parasitic infections of invertebrates, and it is particularly well-known in snail: trematode parasite associations (Chapuis 2009). There are far fewer recorded instances of parasitic castration-associated gigantism in vertebrates, although three-spined sticklebacks (*Gasterosteus aculeatus*) infected with the cestode *Schistocephalus solidus* grow faster and have better body condition than those that are uninfected (Arnott et al. 2000). The host does not benefit from gigantism since it is unable to reproduce. By contrast, the parasite benefits because its host is better able to compete for food and has a longer life span – and this would enhance the parasite's reproductive output. It is highly likely that parasites that induce gigantism are able to avoid or subvert the host immune system because one would expect a large and otherwise healthy individual to be capable of mounting an effective immune response against infections.

Whether or not parasitic castration leads to gigantism, castration benefits the parasite because it means that the host does not divert energy to its own reproductive efforts. However, this has a serious implication for the population dynamics of the host species, since infected individuals continue to feed and occupy space but do not leave any offspring. Consequently, where there is a high population of parasitic castrators there is selection for early reproduction in the host species (Ebert et al., 2004). This ensures that the host leaves some offspring before the castrator completely inhibits its reproductive potential. In some instances, should the castrator die before its host then it is possible that the host might regain some reproductive potential provided its reproductive organs were not completely destroyed by the parasite. As might be expected, destruction or inactivity of the reproductive organs is often accompanied by physiological and behavioural changes such as feminisation in males.

An interesting example of parasitic castration of vertebrates is the relationship between the crustacean isopod parasite *Anilocra apogonae* and the five-lined cardinalfish *Cheilodipterus quinquelineatus* (Fogelman et al. 2009). *Anilocra apogonae* does not live in the genitalia and is actually an ectoparasite that attaches itself at a specific site immediately behind the fish host's head. Surprisingly, it does not appear to induce a localised inflammatory response or make the infected fish susceptible to secondary infections. Nevertheless, infected fish are smaller in size than those that are uninfected. This suggests that the parasitism represents a considerable metabolic burden. Infected female fish develop small gonads and produce few eggs, whilst infected males fail to mouth-brood – which indicates that they are not successful in either mating or attracting a mate. It is uncertain whether *A. apogone* acts as a castrator through extracting so many metabolites that the fish are unable to develop their gonads or it secretes chemicals that interfere with gonad development.

10.4.10 Delusional Parasitosis

The mistaken belief that one is suffering from a parasitic infection has been described in many societies around the world. The affected person often refers to mites, lice, or insects crawling around under their skin. It is a recognised psychological disorder and is referred to as 'delusional parasitosis' or 'Ekbom Syndrome'. It occurs in men and women of all ages but is more common in middle-aged or elderly women. It may be the primary psychological disturbance or secondary to another psychological condition (e.g., schizophrenia, drug abuse) or illness (e.g., dementia, brain tumours). Very often the patient persistently approaches a dermatologist before being referred to a psychiatrist. However, many patients refuse such referrals. Although it is sometimes said to be rare, a survey of 144 UK dermatologists found that 56% of them had seen at least one case of delusional parasitosis in the previous 5 years (Driscoll et al. 1993). The patient is often frustrated and exhausted because 'nobody believes them' and they cannot 'get rid of the parasites'. Many patients suffering from delusional parasitosis present the doctor treating them with the so-called matchbox-sign. This is where the patient turns up with a small container, piece of sticky tape or other trapping device that is alleged to house the parasites. They are never convinced if they are told that 'there's nothing there' or 'it's just a piece of dust, hair etc'. There is always a risk that a patient could harm themselves through excessive use of insecticides and other chemicals to kill the 'parasites' or through physical attempts to dig them out. The condition is very difficult to treat and has considerable adverse impact on the patients' work, family, and social life (Waykar et al. 2020).

10.4.11 Diarrhoea

Large volumes of fluid enter the digestive tract every day via the consumption of food and water and via the body's own secretions (e.g., saliva, pancreatic fluid). In adult humans, about 9 l of fluid enter the small intestine every day but about 90% of this absorbed and much of the remainder is absorbed in the large intestine. Consequently, only about 100–200 ml of water is lost with formed faeces. However, in the case of diarrhoea there is production of excessive watery faeces. This is a major cause of morbidity and mortality in us and all animals and can be caused by a large number of infectious and non-infectious agents. Diarrhoea caused by infections such as parasites is often accompanied by fever and the consequent sweating leads to the loss of more water and electrolytes. Numerous parasites cause diarrhoea including not only the obvious gastrointestinal parasites such as *E. histolytica*, *Giardia duodenalis*, and *Strongyloides stercoralis* but also blood parasites such as *L. donovani* and *Trypanosoma brucei gambiense*. The loss of too much fluid over too short a period of time results in a drop in the blood volume (hypovolaemia), and this compromises the heart's ability to keep the brain supplied with oxygen and nutrients. In addition, the losses in electrolytes (particularly sodium, potassium, chloride, and bicarbonate ions) disturbs metabolic processes. Babies and small animals are particularly vulnerable to diarrhoea because the fluid and electrolyte losses are greater as a proportion of their size compared to adults and larger animals.

There are two principal causes of diarrhoea: secretory diarrhoea and osmotic diarrhoea. Both types may occur at the same time. Secretory diarrhoea results from a continued or excessive secretion of water and electrolytes (especially chloride ions) into the small intestine that is coupled with a reduction in the absorption of sodium ions. This results in a net loss of water and electrolytes. Osmotic diarrhoea results from the presence of an osmotically active substance in the gut that causes the movement of water and electrolytes from the extracellular fluid into the gut.

The consequences of diarrhoea vary depending upon the amounts of fluids and electrolytes that are lost (Table 10.5).

Isotonic dehydration is the most common form of diarrhoea-associated dehydration, and it occurs when water and sodium ions are lost in the same proportion as they occur in the extracellular fluid. Consequently, there is no change in the serum sodium ion concentration, but hypovolaemia occurs owing to the loss of fluid volume. In humans, once our fluid deficit approaches 5% of body weight the physical effects of dehydration become apparent: there is thirst, the mucous membranes start to dry out, and the heartbeat increases. As the fluid deficit approaches 10%, we may start to lose consciousness and the signs of potentially fatal hypovolaemic shock appear. Hypertonic dehydration, sometimes referred to as hypernatraemic dehydration, occurs when the losses of water and sodium ions are not in the same proportion to their occurrence in extracellular fluid: more water is lost than sodium ions. Consequently, there is a rise in serum sodium ion concentration. This results in an extreme thirst that is greater than would be expected from the level of apparent dehydration. The reverse situation occurs in hypotonic dehydration in which more sodium ions are lost than water. In this situation, there is a fall in the serum sodium ion concentration, which is associated with a risk of seizures. Metabolic acidosis usually occurs when the diarrhoea is accompanied with damage to the kidneys. In this situation the kidneys lose their ability to replace the bicarbonate ions lost. This may occur due to the pathogen damaging renal function or through hypovolaemia restricting the blood supply to the kidneys. Because the serum bicarbonate ion concentration starts to fall, there is

Table 10.5   Metabolic consequences of diarrhoea.

Consequence of diarrhoea	Cause	Features
Isotonic dehydration	Loss of water and sodium ions are in the same proportion to their occurrence in extracellular fluid	Hypovolaemia
Hypertonic (hypernatraemic) dehydration	Loss of water exceeds the loss of sodium ions	Increase in serum sodium ion concentration. Seizures may develop
Hypotonic (hyponatraemic) dehydration	Loss of sodium ions exceeds the loss of water	Decrease in sodium ion concentration. Seizures may develop
Metabolic acidosis (base-deficient acidosis)	Excessive loss of bicarbonate ions	Extracellular fluid becomes more acidic. Breathing rate increases, vomiting
Potassium depletion	Excessive loss of potassium ions	Decrease in serum potassium ion concentration. Weakness, paralysis, erratic heartbeat

a reduction in the blood pH. Potassium depletion, or hypokalaemia, is a common feature of diarrhoea and is particularly a problem in malnourished children who are often already potassium deficient. Provided the loss of potassium ions is accompanied by the loss of sufficient bicarbonate ions, it is possible that the symptoms of hypokalaemia will not develop. This is because the loss of bicarbonate ions brings about a reduction in the pH in the extracellular fluid because of the rise in hydrogen ion concentrations. The body attempts to buffer this by exchanging extracellular hydrogen ions with intracellular potassium ions and therefore the serum potassium ion concentration may not change despite the loss of potassium ions in the diarrhoea.

If the serum potassium concentration declines, it can have serious consequences. Many cells, but especially the excitable cells (nerves and muscles), rely on the balance of ions between the inside and outside of their cell membranes to function properly. For example, disturbing the ion balance interferes with the ability of nerve cells to repolarise after transmitting action potentials. Hypokalaemia can therefore result in muscle weakness, cramping, problems with breathing, and, in extreme cases the heartbeat becomes erratic, paralysis occurs, and the patient may die. Hypokalaemia is a known risk factor associated with several gastrointestinal parasite infections (e.g., *Isospora* spp., *Strongyloides stercoralis*) in which there are non-specific symptoms such as profuse watery diarrhoea (not usually bloody), fever, vomiting, and weight loss. When disturbances in electrolyte balance such as hypokalaemia are accompanied by hypovolaemia the situation is potentially lethal – and this is typically what happens in AIDS patients afflicted with cryptosporidiosis.

10.4.12   Elephantiasis

Elephantiasis refers to the gross swelling of the extremities or genitalia. Typically, it is associated with lymphoedema that accompanies infections with certain filarial nematode infections, such as *W. bancrofti*. The nematodes block the drainage of lymph, which therefore accumulates in the blocked vessels and dependent region. Lymph contains proteins and interstitial cells that are being transported back to the general circulation. Consequently, lymphatic oedema has much higher protein content than oedema resulting from the accumulation of serum (e.g., 'bottlejaw'), and this may provide a stimulus for fibrogenesis. In elephantiasis induced by *W. bancrofti*, the overlying skin becomes thickened and warty owing to the formation of fibrotic tissue.

The term elephantiasis is also used in conjunction with other medical conditions that result in gross swelling. For example, the bacterium *Chlamydia trachomatis* serovars L1, L2, and L3 cause *lymphogranuloma venereum* that results in lymphatic obstruction that then manifests as genital elephantiasis. Podoconiosis can cause similar swelling to the feet and lower limbs as lymphatic filariasis and is known as 'endemic non-filarial elephantiasis'. In this case, however, the causative agent is not an infectious agent but the result of walking barefoot on irritant alkaline clay soils (Deribe et al. 2018). Mineral particles within the soil enter the circulation and are phagocytosed by macrophages within the lymphatic system. Subsequently, there is an intense proliferation of macrophages, blockage of the lymphatic system and fibrosis. Unlike filarial elephantiasis, podoconiosis is usually limited to the feet and lower leg rather than the whole limb and genital involvement is rare. The congenital condition neurofibromatosis type 1 results in extensive hypertrophy of the skin, soft tissues, and underlying skeleton. This results in the gross enlargement of a whole extremity and is known as 'elephantiasis neuromatosa'.

10.4.13 Fever

Fever is a common non-specific reaction of the body to infectious organisms. Its function, at least in part, is to increase the rate of metabolic reactions important in host defences. However, the extent to which fever is beneficial or harmful if very case dependent. As we become older, our ability to develop a fever declines and severe infections may result in only a modest rise in temperature. The cytokine 'tumour necrosis factor-alpha' (TNF-α) has an important role in the development of fever associated with several parasitic diseases. TNF-α is produced mainly by macrophages and, amongst a variety of effects, causes the destruction of tumours (hence the name), the migration of neutrophils and macrophages towards sites of inflammation and stimulates the killing of microbes, and acts as a pyrogen inducing fever. TNF-α crosses the blood–brain barrier where it stimulates the production of prostaglandin E2, which then acts on the hypothalamus to increase the body's set point temperature. TNF-α is not the only cytokine associated with the development of fever. For example, interleukin 1b and interleukin 6 act as potent pyrogens.

Patients naturally infected with *Plasmodium vivax* exhibit a rise and fall in levels of circulating TNF-α that correlate with their changes in body temperature. There is a time gap between initial infection and the onset of fevers because the first few cycles of erythrocytic schizogony do not produce enough toxins to generate a TNF-α response sufficient to raise the body temperature. The threshold that induces a response varies between individuals. Someone who is immunologically naive for malaria usually exhibits fever at a very low parasitaemia (e.g., 0.001%) – by contrast someone living in malaria endemic regions may have a parasitaemia as high as 15% without developing a fever.

Patients suffering from malaria typically exhibit periodic fevers (paroxysms) every 48 or 72 hours depending on the species of *Plasmodium* responsible. Somewhat confusingly, these time intervals are given the terms 'tertian' and 'quartan', respectively. The reason is historical. The Ancient Greeks recognised the periodic nature of the fevers associated with malaria but did not know the cause and their medical writings subsequently influenced the Romans. The Romans referred to the first day of an event as day 1 and therefore 48 hours after day 1 would be day 3 and hence the term 'tertian', whilst 72 hours later would be day 4 and be termed 'quartan'. Intermittent fevers are also a feature of other parasitic diseases such as visceral leishmaniasis and the early stages of Human African Trypanosomiasis. It is therefore essential to obtain a definitive diagnosis in order to provide effective treatment. It should also be remembered that *Leishmania*-malaria and *Trypanosoma*-malaria co-infections occur.

Plasmodium falciparum, *P. vivax*, and *Plasmodium ovale* cause tertian malaria (i.e., the fever occurs on day 1 and day 3), whilst *Plasmodium malariae* causes quartan malaria (i.e., the fever occurs on days 1 and 4). *Plasmodium knowlesi* produces fevers every 24 hours and therefore they are said to be quotidian. The time gaps correlate with the maturation of a generation of merozoites and the rupture of the red blood cells containing them (Kwiatkowski 1990). In *P. vivax*, the cyclic periodicity of erythrocytic merozoite production is very pronounced, but in *P. falciparum* it can vary between strains. Fever is probably stimulated by the waste products of the parasites that are released when the erythrocytes break up. Their sudden presence in the bloodstream triggers a burst of TNF-α from activated macrophages. Fever is one of several consequences of overproduction of TNF-α. TNF-α toxicity can account for many of the other symptoms of malaria, such as nausea, muscle pain, head-ache, and loss of appetite.

10.4.14 Fibrosis

Fibrosis refers to the formation of excessive connective tissue and is typically associated with the repair of wounds and scar formation (Beschin et al. 2013). The immune system is involved in the development of fibrosis: interleukin-13 has a key role through activating fibrogenic pathways and up-regulating collagen production. Fibrogenesis is a common feature of many parasitic infections (e.g., schistosomiasis, fascioliasis, amoebiasis), and the loss of functional tissue is permanent. Consequently, where there is extensive fibrosis, normal organ function becomes compromised with potentially fatal consequences. For example, the healing of amoebic ulcers caused by *E. histolytica* in the gut involves fibrosis. If the gut lesions are extensive, the replacement of functional tissues with connective tissue may be sufficient to impair peristalsis. In very severe cases, the impairment results in potentially fatal gut blockage. Similarly, fibrosis in the liver resulting from schistosome infections interferes with blood circulation and leads to cirrhosis.

10.4.15 Granulation

Granulation refers to the process by which a foreign object that is proving difficult to eliminate becomes walled off by a surrounding mass of immune cells. The object and the cells that come to surround it are then called a granuloma. The granulation process is associated with a Th2 immune response and is marked by the arrival of large numbers of alternatively activated macrophages (Beschin et al. 2013). Alternatively-activated macrophages are those that are activated by interleukin-4 and interleukin-13. After arriving, the macrophages transform into non-motile epitheloid cells, and these surround the target. The granuloma may also include other cell types, such as Langerhans giant cells, lymphocytes, eosinophils, and fibroblasts. Inflammatory granu-lomatous responses are a common reaction to many parasitic infections. Examples include the granulation of liver abscesses caused by *E. histolytica* and the granulomas that form around *Schistosoma* eggs and the tissue dwelling stages of nematodes and tapeworms. Although the 'inten-tion' of the granulomatous response is to isolate the parasite, it may not stop harmful antigens 'leaking out' and it also may be detrimental to the host. For example, the filarial nematode *Setaria tundra* induces granulomatous peritonitis in reindeer (*Rangifer tarandus tarandus*). This has led to the deaths of thousands of reindeer in many arctic regions in recent years. This parasite is cur-rently extending its range, possibly because of global warming, and is now considered a major threat to reindeer herding (Haider et al. 2018). Within the liver abscess granuloma caused by

E. histolytica trophozoites, the parasites produce secretions that prevent macrophages from functioning normally. For example, the suppressed macrophages do not respond to IFN-γ stimulation, they do not produce interleukin-12 and other cytokines, and they do not express nitric oxide synthase – and hence form nitric oxide. Consequently, the suppressed macrophages are unable to kill the trophozoites.

The schistosome species infecting us vary in their locations and the seriousness of the disease they cause. In addition, much of our knowledge concerning the immune response to schistosomes derives from mouse models. Consequently, it cannot be assumed that our immune response to all schistosome species is the same. Nevertheless, serious disease is, in most cases, associated with our immune reaction to the parasite eggs. The eggs need to pass through blood vessel walls and then through various tissues into the urine (*Schistosoma haematobium*) or faeces (*Schistosoma mansoni* and *Schistosoma japonicum*) before being passed into the environment. However, some eggs become trapped in the walls of the bladder and the intestine or are swept to the liver and other body tissues. For example, in women, granulomas that develop within the genital tract can disrupt normal menstruation and induce preterm labour, and infertility (Nour 2010).

The newly laid egg of a schistosome is not immunogenic. It is only when the miracidium inside the egg starts to develop and releases lytic and antigenic secretions that it engenders an immune reaction. This typically begins 5–6 days after being laid by the female worm. Within a previously uninfected host, the onset of egg laying is characterised by an acute inflammatory response that reaches a maximum after about 8 weeks and is then down-regulated as the disease progresses to the chronic stage of infection. The egg antigens engender a Th2-type immune response in which there is enhanced production of interleukin-4 (IL4), interleukin-5 (IL5), interleukin-10 (IL10), and interleukin-13 (IL13). IL4 is the most important cytokine driving the formation of the granulomas, whilst IL10 modulates the immune response. In the absence of IL10, there is an excessive Th2 response that results in more severe pathology. The levels of IL4 determine the size of the granulomas, and this, together with their number and location, has implications for the pathology. This is because the granulomas are many times the size of the eggs (Figure 10.1). Therefore, the presence of numerous granulomas causes swelling of the affected organ and obstruction. *Schistosoma japonicum* lays its eggs in batches and the presence of several eggs close together results in the formation of large granulomas and hence serious pathology. By contrast, *S. mansoni* tends to cause less serious pathology than *S. japonicum* because it lays its eggs singly. *Schistosoma mansoni*,

Figure 10.1 Granulomatous reaction (arrow) surrounding schistosome eggs.

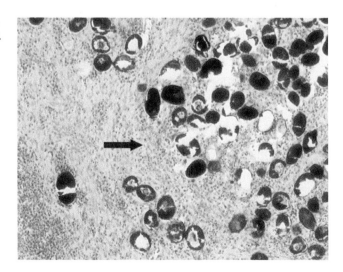

therefore, tends to engender smaller more dispersed granulomas. Once a schistosome egg becomes surrounded by a granuloma, the miracidium within it can survive for at least 7 days. However, schistosome eggs cannot survive unhatched for long and the miracidium ultimately dies of 'old age' even if the immune response is not successful in killing it. The importance of the Th2 response, and in particular the levels of IL4 in determining the pathology have led to suggestions that tipping the Th1 : Th2 balance in favour of the Th1 response might lessen pathology even if the worms remain present and continue producing eggs. In hosts chronically infected with schistosomes, the immune reaction eventually down-regulates and this limits the granulomatous response – and hence the pathology. At this point fibrosis is initiated. For more details of the host immune reaction and pathogenesis associated with schistosome eggs, see Costain et al. (2018).

10.4.16   Hyperplasia

Cell growth is under the control of various chemical growth factors that interact with specific cell surface receptors. Stimulation of these receptors triggers various internal cell signalling pathways and influences the way in which the cell develops and divides. Anything that interferes with this process can result in altered growth patterns such as hyperplasia, hypertrophy, and metaplasia. Hyperplasia refers to an increase in the number of cells owing to an increase in the rate of cell division. This can arise from various causes such as chronic irritation that gives rise of callouses on the skin and as a response to toxins (e.g., cholera toxin A1 subunit causes thyroid hyperplasia). Hyperplasia may not occur equally throughout a tissue and localised hot spots can result in nodules forming between areas of normal tissue. Many infections, including parasitic diseases, cause hyperplasia. For example, in sheep chronic fascioliasis (*Fasciola hepatica*) causes hyperplastic cholangitis. The epithelial cells lining the bile ducts undergo hyperplasia and are thrown into a series of folds, whilst the term cholangitis indicates that the bile ducts are inflamed. Hyperplastic cholangitis restricts the flow of bile through the bile ducts, and in serious cases, the flow may stop entirely. In addition, the pathological consequences of infection are exacerbated by the loss of plasma proteins that leak across the deformed mucosa. In humans, hyperplastic cholangitis is often associated with bacterial infections – the bacteria having travelled up the bile duct from the lumen of the gut (Carpenter, 1998). Because adult *F. hepatica* interfere with the normal flow of bile and damage the lining of the bile ducts, they may facilitate the establishment of bacterial infections. The bacteria would then contribute to or exacerbate the development of hyperplastic cholangitis. Obviously, hyperplasia can only occur in tissues that have the capacity to divide. Consequently, it is not a feature of tissues comprised of non-dividing cells such as nerve cells and cardiac muscle.

10.4.17   Hypertrophy

Hypertrophy refers to an increase in cell size and functional capacity. The increase in size is brought about by an increase in protein synthesis that results in an increase in the cell's structural components. Cells that are unable to divide often respond to an increased functional demand by undergoing hypertrophy. For example, the increase in muscle bulk exhibited by body-builders largely results from in an increase in the size of the skeletal muscle cells. Parasitic infections can also cause an increase in cell size although it is seldom beneficial to the host. For example, *Trypanosoma cruzi* induces hypertrophy of cardiac muscle cells (cardiomyocytes) in humans, whilst *Toxoplasma gondii* induces hypertrophy of the muscularis externa in the colon of rats. Both hypertrophy and hyperplasia may occur independently or together depending upon the cell type and stimulus.

10.4.18 Hypoplasia and Hypotrophy

Hypoplasia arises when an organ or tissue contains fewer than normal cells, whilst hypotrophy indicates that the cells comprising an organ or tissue are smaller than normal. However, there is some variability in the literature and some authors use hypotrophy to indicate that an organ or tissue is reduced in size because of disease. For example, both *P. falciparum* and *P. vivax* malaria are described as causing hypotrophy of the placenta and of the developing foetus. This can lead to stillbirth, premature birth, low birth weight, and high mortality shortly after birth. *Trypanosoma cruzi* can cause hypoplasia of bone marrow cells, and this contributes to the development of anaemia and thrombocytopaenia (low levels of platelets [thrombocytes]).

10.4.19 Inflammation

The earliest record of the four cardinal signs of inflammation is contained within the treatise *De re Medicina* (literally 'Concerning Medical Matters') compiled by Aulus Cornelius Celsus sometime around the start of the first century CE. Within Book 3 of the eight-book treatise, he discusses fever in some detail and his sentence '*Notae vero inflammationis sunt quatuor, rubor & tumor, cum colore & dolore*' continues to be repeated in medical textbooks – although usually in translation – 'Indeed the signs of inflammation are four, redness and swelling, with heat and pain' (Major 1954). The redness and heat of the acute inflammatory response result from the increased blood supply to the affected region and the dilation of the surrounding blood vessels. The swelling is a consequence of the accumulation of fluid exudates. Pain results from a combination of chemicals released during the inflammatory process and also pressure from swelling affecting sensitive nerve endings. TNFα, IL1, and several other cytokines influence pain perception. In terms of pathology, there are two broad categories of inflammation: acute inflammation in response to the initial cell damage and chronic inflammation in response to a damaging stimulus that cannot be resolved. Acute inflammation is induced in response to tissue damage and the death of cells, and the physiological basis is discussed in detail in Chapter 6. An acute inflammatory response is therefore commonly associated with parasites when they physically penetrate their host and when they start to feed or release harmful chemicals. For example, when the larvae of the hookworm *Ancylostoma braziliense* penetrate our skin they cause a condition known as *cutaneous larva migrans*. This manifests as serpentine (wavy, snake-like) tracks at the dermal-epidermal boundary that extend at up to 2 cm a day and may reach 20 cm in length and 2–4 mm in width. The burrowing larvae cause physical damage and release protease enzymes that initiate an acute inflammatory response in which eosinophils figure heavily. The region around the tracks becomes inflamed, red, and swollen. The tracks also cause intense itching (pruritis) and can be secondarily infected with the bacteria such as *Streptococcus pyogenes* that leads to cellulitis. Cellulitis is a serious inflammation of the dermis and subcutaneous layers of the skin and should never be confused with the development of cellulite (skin dimpling), which is an obsession within the fashion media. *Cutaneous larva migrans* is also caused by a number of other parasites although the form and development of the tracks varies between species.

Although the inflammatory response is meant to prevent pathogens from establishing, it can also be the means by which they gain access or leave their host. For example, when the skin around rear end of sheep becomes fouled with faces and urine, it stimulates the growth of bacteria, and consequently, there is an inflammatory reaction in the underlying skin. This weakens the skin surface and makes it easier for the larvae of blowflies and fleshflies to penetrate and initiate a myiasis. Similarly, schistosome eggs provoke a massive immune response that results in tissue damage, and this facilitates the eggs traversing from inside blood vessels to the lumen of the gut or bladder.

The progression from an acute inflammatory response to chronic inflammation is exemplified in the manner in which trophozoites of *E. histolytica* invade the liver. On reaching the liver, the parasites induce an acute inflammatory response that results in the influx of large number of neutrophils. As in the intestine, the neutrophils are both protective in that they kill the parasites but also a source of pathology because the inflammatory response also kills many liver cells. The parasites also experience another immune response in which invariant natural killer T cells (iNKT cells) are vitally important. Indeed, mice that lack the ability to mount an effective iNKT response develop larger liver abscesses following *E. histolytica* infection (Lotter et al. 2009).The iNKT cells are stimulated by lipopeptidophosphoglycan (LPPG) in the parasite cell membrane and in response they rapidly release a range of cytokines such as IFN-γ, which promotes inflammation. For example, IFN-γ activates macrophages and stimulates them to produce inducible nitric oxide synthase (iNOS) and hence the secretion of nitric oxide – which is toxic to the amoebae. Unlike many other activated iNKT cells, those activated by *E. histolytica* LPPG do not release IL4. IL4 is anti-inflammatory and promotes a Th2 response. If the acute inflammatory response does not kill all the trophozoites, they feed and reproduce and thereby cause the formation of purulent abscesses. Eventually, a Th2 immune response predominates and a chronic inflammatory response is established. This represents a balance between ongoing tissue damage and the processes of healing and scar formation. During this process, the abscesses become granulated and fibrous scar formation takes place.

10.4.20 Jaundice

Jaundice results from disruptions to the normal metabolism and excretion of bilirubin. Bilirubin is a breakdown product of the blood pigment haemoglobin and is produced in the spleen, liver, and bone marrow. If there is excessive breakdown of haemoglobin or the breakdown products cannot be excreted fast enough, then the levels of free or conjugated bilirubin in the circulation start to rise. This manifests as yellowing of the skin and the whites of the eyes (sclera). An increase in the level of free bilirubin (unconjugated hyperbilirubinemia) results from the generation of excessive amounts of bilirubin. This can occur following the destruction of large numbers of red blood cells such as happens during malaria. By contrast, conjugated bilirubin is produced by the liver cells (hepatocytes) and a rise in the levels of conjugated bilirubin in the circulation results from disruption of bile secretion. It can, however, also result from a destruction of liver cells. For example, chronic fascioliasis leads to blockage of the bile ducts and thereby obstructs the flow of bile into the duodenum. Within the gut, commensal bacteria convert bilirubin into urobilinogen, and this is what imparts the brown colour to faeces. Conjugated hyperbilirubinemia is therefore characterised by the excretion of pale or even white faeces. It also results in problems with fat metabolism because bile is important for their digestion and the urine becomes dark because conjugated bilirubin (unlike free bilirubin) is water soluble and can therefore be excreted in the urine. It can also cause pruritis since the bile salts can be deposited in the skin where they cause intense itching. Malaria sometimes causes conjugated hyperbilirubinemia although it is usually observed less frequently than the unconjugated form. There are several case reports of conjugated hyperbilirubinemia in neonates suffering from congenital malaria.

10.4.21 Metaplasia

Metaplasia refers to the transformation of one fully differentiated tissue type into another that may be equally or more differentiated. For example, in response to chronic irritation, mucous-secreting epithelium can be replaced by squamous epithelium (squamous metaplasia) or fibrous tissue is replaced by bone (osseous metaplasia). Squamous epithelial cells are simple scale-like cells found on the outer surface of many tissues. Squamous metaplasia is often accompanied by hyperplasia, so the squamous cells exhibit hyperplasia and the outer surface of the tissue is therefore thrown into a series of folds. For example, *S. haematobium* induces squamous metaplasia within the urinary bladder, and this is considered to be a precancerous lesion. Similarly, the liver fluke *O. viverrini* induces goblet cell metaplasia and epithelial adenomatous hyperplasia (a tumour-like inflammatory lesion).

10.4.22 Pressure Atrophy

Atrophy is where there is a decrease in the size of cells and consequently a reduction in the size of the affected organ or tissue. In relation to parasitic diseases, pressure atrophy often occurs when the size of the growing parasite physically restricts the flow of blood/ nerve impulses reaching a tissue or organ. Pressure atrophy is therefore a common pathological feature of parasites such as the larval stages of those tapeworms that form large cysts. For example, the pressure exerted by the coenurus of the ovine parasite *Taenia multiceps* kills surrounding brain cells by a combination of restricting their blood supply and physically pressing them to death. Some idea of the pressure being exerted by the growing coenurus can be obtained by considering that it is sufficient to cause thinning of the overlying skull. And sheep skulls are notoriously thick! Similarly, the hydatid cysts of *E. granulosus* can grow to a large size and cause pressure atrophy on the surrounding tissues. The consequences depend upon the size of the cyst and its location. For example, in us, a large cyst growing within our liver or spleen can induce pressure atrophy on the diaphragm. Ultimately, the diaphragm may rupture and the cyst then prolapses into the pleural cavity (von Sinner and Stridbeck 1992).

10.4.23 Psychological Disturbance

We often consider parasitic diseases solely from the physical harm they inflict. However, many of them also affect our mental well-being. This may be because they directly or indirectly affect our brain or the drugs used to treat them may have harmful effects. For example, *Toxoplasma gondii* has been linked with the development of schizophrenia. Similarly, the drug Lariam (mefloquine) used to treat malaria is notorious for its side effect of inducing adverse neuropsychiatric disturbances such as anxiety and paranoia in some people (Croft 2007). It's use amongst military personnel has been particularly controversial (Bathie 2020). In the US armed services, the prescription of Lariam ceased in 2009 except to those who have contraindications for doxycycline. Lariam continues to be prescribed to UK Armed Forces but typically accounts for less than 0.5% of all antimalarials. Much less appreciated are the ways in which being infected by parasites makes us feel about ourselves and the ways in which people interact with us. People who pass worms or lengths of tapeworm proglottids in their faeces often state that it left them feeling distressed and 'unclean'. This is even more pronounced in those who suffer obvious outward manifestations of their infection, and this soon translates into a sense of unworthiness. This is because in the popular imagination, a lack of cleanliness is associated with poverty and poverty is associated with a lack of intelligence. For example, in England it has long been the custom to refer to someone who has done something stupid as a 'nit'. A 'nit' is also the term for the egg cases of head lice and having hair 'infested' with nits was once considered a sign of poverty – and hence stupidity. We now know that head lice breed very well on the heads of the cleanest children who live in the most affluent of households. However, this does not reduce the sense of anguish engendered in parents when they discover head louse infections in their children. Interestingly, there is increasing evidence that parasitic diseases retard the mental development of schoolchildren. For example, the intense itching caused by scabies and louse infections makes it difficult to pay attention in school. There are several reports linking gut helminth infections to cognitive development (Bethony et al. (2006). Initially, it was thought that this was through affecting the child's nutritional balance. This will undoubtedly be an important factor in some instances, but there is now increasing evidence that the composition of the gut microbiome influences brain development through the gut-brain axis (Carlson et al. 2018). It is therefore likely that gut helminths may influence their host's cognitive development by affecting the composition of the gut microbiome (Guernier et al. 2017). Furthermore, for people of all ages, parasitic infections can prove painful, and this makes it difficult to relax and disturbs sleep. Sleep deprivation soon leads to exhaustion, an inability to control emotions, irrationality, and confusion. Parasitic infections also cause the production of inflammatory cytokines and there is increasing evidence that the levels of some these, such as TNF-α and interleukin 6, correlate with major depressive disorders (Ma et al. 2016).

Human African Trypanosomiasis (HAT) is commonly known as sleeping sickness because it disturbs the normal patterns of sleeping and waking. The infection alters the circadian secretion of prolactin, renin, growth hormone, and cortisol – chemicals that are important in regulating our sleep-wake cycle – but the precise mechanism is uncertain. Work by Rijo-Ferreira et al. (2018) on mice infected with *T. brucei* indicates that the parasite accelerates the host's circadian clock in a manner similar to that seen in circadian clock gene mutations. Whether the parasites are interacting directly with host cells or secreting one or more chemicals that induce the effect currently remains to be determined. What is certain is that a similar effect has not been described in any other parasitic infection. In addition, HAT also disturbs the sleep process. Normally, sleep consists of a series of cycles, each of which consist of five stages. During stages 1–4, our sleep becomes progressively deeper until we enter the fifth stage that is known as 'rapid eye movement' (REM). During the REM stage our brain activity increases and sleep is lighter. This is the period during which we experience dreams. After the REM stage we return to 'stage 1 and the cycle is repeated. Interestingly, HAT alters the sleep pattern and REM periods tend to occur at the start of the sleep cycle rather than the end.

Stage 2 HAT causes various psychiatric and mental disturbances. Sometimes the changes in behaviour are so subtle that they are only recognisable to friends and close family members. However, some patients become violent, have uncontrolled sexual impulses, suffer from hallucinations, and attempt suicide. Psychotic behaviour is said to be particularly common in patients of European heritage which suggests that genetic factors are important in the way the disease progresses. Incorrect diagnosis of the condition among immigrants to Europe has resulted in patients being admitted to psychiatric clinics. Even if the patient is correctly diagnosed and successfully treated, he/she may suffer long-term mental impairment and physical disability, such as paralysis, because of brain damage. Another unusual feature of HAT is disruption of sensory perception resulting in a condition known as 'hyperaesthesia' in which even a mild touch is experienced as extremely unpleasant. This may be the reason many patients complain of intolerable 'itching'.

Parasites that cause outwardly obvious deformities such as elephantiasis, cutaneous leishmaniasis, and crusted scabies frequently lead to feelings of low self-worth, depression; affected people tend to become withdrawn and avoid contact with other members of the community. These feelings are reinforced by our ancestral tendency to avoid people who are obviously ill or different from us. Even within supposedly educated and affluent westernised societies people who suffer physical deformities frequently experience public comment and ridicule when they venture outside their homes. The physical scars can remain long after a person has been cured of an active infection and lead to long-term mental ill health. The costs associated with this is seldom considered in calculations of disability-adjusted life years (DALYs) but as Bailey et al. (2019) have shown in their study on the consequences of cutaneous leishmaniasis for major depressive disorder, they can be remarkably high.

11

Pathology Part B: Damage to Specific Organs; Co-Infections and Pathogenesis

11.1 Introduction

Many parasites reside within specific tissues or organs and thereby caused localised pathology. However, over time this damage may have more widespread implications elsewhere in the body. We do not provide a comprehensive coverage of the pathogenesis in all the body's organs because this would make the book much too long. Indeed, a fascinating textbook has been written covering solely the parasitic infections that afflict the human eye (Kean et al. 1991). Instead, we will summarise how parasites cause pathology in some of the principal organs and explain how this results in morbidity and mortality.

11.2 Damage to Specific Organs

11.2.1 Bladder

The most serious parasite-induced pathology that afflicts our bladder results from infection with *Schistosoma haematobium.* This parasite lives in the veins that serve the bladder, pelvic organs, and rectum. As is the case with other schistosome species, the adult worms cause relatively little damage. However, the eggs induce a dramatic immune reaction that results in the formation of granulomas and fibrosis. In the case of *S. haematobium,* this gives rise to a condition known as 'urogenital

Parasitology: An Integrated Approach, Second Edition. Alan Gunn and Sarah J. Pitt.
© 2022 John Wiley & Sons Ltd. Published 2022 by John Wiley & Sons Ltd.
Companion website: www.wiley.com/go/gunn/parasitology2

schistosomiasis' although the eggs can be transported elsewhere such as the liver and lungs and cause pathology there. Within the bladder, the granulomatous reactions and deposition of fibrotic tissue restricts the flow of urine. This can give rise to secondary inflammatory reactions in bladder, ureters, and kidneys, which makes urination painful (dysuria) and blood is often passed in the urine (haematuria). Indeed, urogenital schistosomiasis was once so common in Egypt that the passing of blood by young boys was considered a natural event and equivalent to the onset of menstruation in girls. Urogenital schistosomiasis is also associated with an increased risk of developing bladder cancer.

Few parasites live within the bladder. One that does, the nematode *Trichosomoides crassicauda*, is a common parasite of rats but often overlooked because it is difficult to detect and causes little pathology. The adult female worms are very thin (~200 μm) and grow to about 10 mm in length. They live within the epithelium and submucosa of the bladder where they induce hyperplasia within the epithelium, but there is very little inflammation. They release eggs intermittently and these pass out with the urine. Transmission is through passive contamination and the eggs hatch in the rat's stomach to release larvae that burrow through the gut wall and reach the circulation. They then move to the lungs and from there to kidneys and then down the ureters to the bladder. The larvae can cause focal haemorrhages during their migration, the severity of which depends upon the number migrating at the same time. In a reverse of the situation in schistosomes, the male is much smaller than the female (1.5–3.55 mm) and lives within the female's uterus.

Of more importance in veterinary medicine, the nematode *Capillaria plica* infects the bladder and sometimes the pelvis of the kidney of dogs and other canines. Cats are occasionally infected by *C. plica,* but they also have 'their own' species *Capillaria feliscati* (although this may be a synonym for *C. plica*). Some taxonomists consider *C. plica* and *C. feliscati* should be moved from the genus *Capillaria* to form their own genus '*Pearsonema*'. Unfortunately, *C. plica* is sometimes referred to as 'the bladder worm' with the consequent opportunity for causing confusion with the more general term 'bladderworm', which is used to describe the larval stage of certain tapeworms. Adult *C. plica* live in the submucosa of the bladder and ureters where they induce a mild inflammatory reaction and oedema. *Capillaria plica* is not usually considered very pathogenic, but there are case reports linking infection to the development of glomerular amyloidosis in dogs. As its name indicates, this condition is associated with the deposition of amyloid protein within the glomeruli in the kidneys. Amyloid is an abnormal extracellular fibrillar protein that can be composed of various peptides. It is not a substance with a defined peptide sequence. The glomeruli are where the blood is filtered, and the accumulation of amyloid interferes with the filtration process. More normally, *C. plica* infections cause nothing more serious than frequent urination and urinary incontinence. However, they are linked to the development of the inflammatory condition 'cystitis'. Most instances of *C. plica*–associated cystitis arise in conjunction with infections of the bladder with gut bacteria such as *Klebsiella* spp., *Enterobacter* spp., and *Escherichia coli*. The damage that the worms cause to the lining of the bladder, either directly or through inducing inflammation probably makes it easier for the bacteria to invade. However, in heavy worm infections, cystitis can occur in the apparent absence of bacteria.

11.2.2 Brain and Nervous System

The brain is protected from most pathogens by the blood–brain barrier although some can breach it. Owing to the importance of the brain in controlling the rest of the body, these infections can prove rapidly fatal. For example, cerebral malaria is a common complication of infection with *P. falciparum* and may account for 10% of *P. falciparum* malaria cases admitted to hospital and 80%

of such deaths. Similarly, before the introduction of HAART therapy, up to 30% of *Toxoplasma gondii* seropositive HIV-infected patients subsequently developed toxoplasmic encephalitis – usually with fatal consequences. Another common parasitic disease afflicting the central nervous system is neurocysticercosis caused by larvae of the tapeworm *Taenia solium*. Neurocysticercosis causes seizures and depending upon the location and number of the cysts can prove fatal. Other parasites that occasionally invade our brains include the zoonotic nematodes *Angiostrongylus cantonensis*, *Baylisascaris procyonis*, and *Gnathostoma spinigerum*. These parasites all cause eosinophilic meningitis – that is the number of eosinophils within the cerebrospinal fluid increases and the meninges (the membranes that protect the brain and spinal cord) become inflamed. The signs and symptoms are like those of other forms of meningitis including headaches, fever, mental confusion, an inability to flex the neck, photophobia, nausea, and vomiting. In serious cases, the disease can prove fatal.

Cerebral malaria sometimes has a gradual in onset, but it is more commonly sudden. A progressive headache may be followed by an uncontrollable rise in temperature to >42 °C, and psychotic symptoms or convulsions, especially in children. Coma and death may ensue within a matter of hours. Initial stages of cerebral malaria are easily mistaken for a variety of other conditions, including meningitis and acute alcoholism, usually with disastrous consequences. Cerebral malaria results from the sequestration of parasitised erythrocytes (i.e., they are effectively withdrawn from the circulation) in small blood vessels within the brain (cerebral microvasculature), where they cause obstruction, local induction of inflammatory cytokines, or both. The infected cells become 'sequestered' because the parasites produce proteins that attach to the host red blood cell membrane where they form protrusions called knobs. Knob formation begins during the early trophozoite stage of parasite development and results in the infected red blood cell becoming 'sticky'. This is because proteins in the knobs undergo receptor-ligand binding to uninfected red blood cells and to glycoproteins such as CD36 and thrombospondin that are found on the endothelial lining of blood vessels. (The endothelium is a layer of flat epithelial cells that lines the inner surface of blood vessels.) Obstructing the flow of blood to specific areas of the brain results in localised stagnant hypoxia (lack of oxygen owing to the blood flow stopping) coupled with a reduction in the supply of glucose and the rate at which waste products, such as CO_2, are removed. Because brain cells are highly metabolically active, even transient reductions in the supply of oxygen can rapidly prove fatal to them. The consequences of this depend upon the part of the brain affected. Because anaemia is one of the consequences of malaria, there are fewer fully functional blood cells in the circulation, thereby reducing its oxygen-carrying capacity – this could exacerbate the effects of sequestration of red blood cells.

TNF-α is probably involved in the pathology of cerebral malaria because it upregulates endothelial molecules such as intercellular-adhesion molecule-1 that, along with other receptors, increases the binding of infected erythrocytes to the walls of blood vessels. Clinically, there is sometimes rapid recovery from profound coma without significant resultant neurological problems, implying that a reversible process contributes to the coma. This would be an unlikely occurrence if the coma resulted from stagnant hypoxia. However, many patients suffering from cerebral malaria experience serious neurological complications upon recovery. For example, of the children who survive, 5% have permanent neurological defects, such as mental blindness. For a more detailed account of malaria pathogenesis, see Ghazanfari et al. (2018) and Bernabeu and Smith (2017).

Plasmodium vivax malaria is usually considered to be more benign than *P. falciparum* malaria although there are reports of new forms emerging in different parts of the world that cause serious, potentially fatal disease (Mukhtar et al. 2019). Like *P. falciparum* malaria, these apparently new forms of *P. vivax* malaria can cause cerebral malaria and serious anaemia. Not only is this of concern from a medical point of view but it also poses a question about how it causes this pathology. This is because *P. vivax* has not previously been considered to induce sequestering of infected blood cells within blood vessels in the brain. Co-infections of *P. vivax* and *P. falciparum* frequently occur, and the disease course tends to be more severe than in single species infections.

When *Trypanosoma brucei* invades the brain, it causes meningoencephalitis and disrupts the normal functioning of the blood–brain barrier. This is accompanied by demyelination of nerve axons and a marked rise in the number of the astrocytes (astrocytosis) (Rodgers 2010). Astrocytes are glial cells that play an important role in nervous transmission but do not themselves generate or conduct action potentials. Astrocytosis is a feature of many brain pathologies and is probably connected to the repair process. The parasites release chemicals that induce apoptosis in the vascular epithelial cells and microglial cells within the brain – that is, the affected cells 'commit suicide'. This is particularly marked in the regions of the cerebellum and brain stem. Apoptosis, or programmed cell death, is a normal part of cell ageing and organ development when it is a controlled process. Apoptosis induced by chemicals released by pathogens such as *T. brucei* is uncontrolled and can lead to serious pathologies. In the case of trypanosomes, procyclin and procyclin derivatives induce apoptosis. Patients with *T. brucei gambesiense* stage 2 Human African Trypanosomiasis (HAT) often exhibit changes in behaviour. They suffer from seizures, their sleep pattern is disturbed, and they become confused, disorientated, and ultimately lose their ability to move; they then become comatose and die. The cerebellum is involved in the coordination (but not initiation) of movement, attention, mental imagery, some aspects of learning, as well as emotions and use of language. Damage to the vascular epithelium interferes with the blood supply to the brain cells in the cerebellum (and elsewhere in the brain). This, coupled with damage to the nerve cells themselves, could explain at least some of the symptoms observed in stage 2 HAT. Similarly, the brain stem is responsible for the control of vital life functions, such as heartbeat, blood pressure, and breathing, so damage to this region can have fatal consequences. Microglial cells engulf and destroy invading micro-organisms within the central nervous system and remove cell debris. The loss of microglial cells could therefore reduce the immune response against trypanosomes. *Trypanosoma brucei rhodesiense* HAT begins in a similar manner but the course of the disease is usually much more rapid. As a rule, the patient develops severe fevers and swelling of the lymph glands, and they lose a lot of weight. They usually die before expressing the behavioural changes and disrupted sleeping patterns that are characteristic of late-stage *T. brucei gambiense* HAT.

The encysting of *Toxoplasma gondii* in the brain of a healthy person seldom causes significant pathology. However, if their immune system becomes compromised (e.g., due to AIDS), then the parasites start to replicate, and this gives rise to focal lesions (abscesses). The consequences depend upon the location of the lesions, their size, and number. The lesions destroy brain cells and interfere with the blood supply to other regions of the brain. The patient starts to complain of headaches and then develops fever, suffers from seizures, and may die. Neurological conditions can also develop in people who are not immunocompromised and retinochoroiditis (inflammation of the retina and the underlying choroid layer) is a relatively common feature of *T. gondii* infections acquired after birth (Kean et al. 1991). There appears to be a relationship between *T. gondii* infection (testing seropositive) and suffering from epilepsy (Sadeghi et al. 2019). However, there is currently a need for more good quality data and the mechanism is uncertain.

The cysticerci of *T. solium* can develop in various regions of the brain and spinal cord and the pathology varies somewhat dependent upon where they occur. Whilst they are still immature, the cysticerci cause little host reaction. However, once they become large, they can cause pressure atrophy on the surrounding tissues. The cysticerci can withstand the host immune response by employing immune avoidance mechanisms that subvert complement activation and cytokine production.

They are even able to absorb the antibodies formed against them and metabolise these as a source of amino acids. However, with time (sometimes after several years) the host immune response breaks through the parasite's defences and the cysticerci start to degenerate. At this point there is a marked inflammatory reaction against the cysticerci, and mononuclear cells invade the cyst wall and penetrate into the cyst fluid. It is this host inflammatory reaction, which is predominantly responsible for the seizures and other pathology associated with neurocysticercosis. Although many people with calcified cysticerci within their brain suffer no ill effects, these represent a future risk of generating an inflammatory immune response and hence seizures (Nash et al. 2017).

Neurocysticercosis should be considered if an adult with no previous history of the condition suffers an attack of epilepsy and had travelled in a country where *T. solium* is endemic. This is because neurocysticercosis can account for 30% of cases of acquired epilepsy in these regions (Ndimubanzi et al. 2010). Cysticerci developing within the ventricles (ventricular neurocysticercosis) induce inflammatory reactions that result in the development of ependymitis granularis and fibrillary astrocytosis. Ependymitis granularis refers to an inflammation of the ependymal cells that causes a patchy loss of the ependymal cells lining the ventricles and maintaining the flow of cerebrospinal fluid. Fibrillary astrocytes are a type of glial cell located in the white matter and they proliferate in response to injury in a similar way to fibroblasts elsewhere in the body. Their multiplication is sometimes called astrogliosis or simply gliosis. These inflammatory reactions cause the degenerating cysticerci to stick to the walls of the ventricles and this together with the damage to the ependymal cells interferes with the flow of cerebrospinal fluid. There follows a build-up of cerebrospinal fluid in the intracranial cavity that results in a condition known as hydrocephalus. Hydrocephalus results in a rise in intracranial pressure that in prolonged cases may be sufficient to cause localised thinning of the bones of the skull. The rise in pressure also interferes with nervous transmission and causes the death of nerve cells – which in turn results in gliosis. The consequences depend upon the region of the brain affected and the speed and severity of the increase in intracranial pressure. Typically, it results in seizures, difficulty in walking, and incontinence.

Degenerating *T. solium* cysticerci in the basal cistern can induce a condition known as arachnoiditis. The basal cistern is a fluid-filled cavity between the *arachnoid mater* and *pia mater* situated at the back of the mid-brain. It is also known as the interpeduncular cistern. The *arachnoid mater* is one of the three meninges (protective tissue membranes) that surround the central nervous system (the other two meninges are the *pia mater* and *dura mater*). Arachnoiditis therefore refers to an inflammation of one of the meninges and it can be extremely painful and debilitating. The consequences depend upon the region of the arachnoid mater affected and the extent of the inflammatory damage. For example, movement, bladder, and bowel control can all be affected, and it may even lead to impotence. The inflammatory reactions can also cause obstruction of the flow of cerebrospinal fluid and the development of hydrocephalus. Inflammatory damage to surrounding blood vessels can also occur – this is known as vasculitis – and if the veins rupture it can result in a stroke.

Many parasites cause damage to the various branches of the nervous system. This might be direct physical damage or indirect damage through stimulating a harmful immune response (Halliez and Buret 2015). In some cases, the damage is through the production of neurotoxins, and this is exemplified in the case of tick paralysis. Tick paralysis is caused by several ixodid tick species and affects many animals including humans, cats, dogs, sheep, birds, and even snakes. It occurs in many countries and has been linked to over 46 species of ticks. Australia is blessed with an unusual abundance of poisonous creatures including *Ixodes holocyclus* that has a particular propensity to cause tick paralysis (Hall-Mendelin et al. 2011). *Ixodes holocyclus* does not induce harmful reactions in koalas, kangaroos, and bandicoots, which are its natural hosts. However, its attachment to us and our domestic animals can prove fatal. Indeed, some estimates suggest that up to 100,000 domestic animals die of tick paralysis every year in Australia. Ticks gain no benefit from killing their host and paralysis often does not develop until after the tick has engorged. Furthermore, even for *I. holocyclus*, tick paralysis is an unusual rather than invariable feature of attachment. It is therefore possible that the development of tick paralysis is heavily dependent upon a combination of individual tick and host features. All ticks inject various chemicals with their saliva when they feed. These facilitate the uptake of blood and reduce the host immune response. In hosts that have not co-evolved with the tick, some of the components of tick saliva induce a particularly harmful reaction.

In humans, tick paralysis tends to affect children more than adults although this may reflect their tendency to play and roll around in fields where they are likely to be bitten by ticks. It is a potentially fatal condition and tends to evolve slowly. It therefore differs from the typical toxicosis that follows being stung by a wasp or bitten by a snake in which the reaction is almost immediate. Tick paralysis caused by *Dermacentor andersoni* and *Dermacentor variabilis* usually improves after removing the tick. However, in the case of *I. holocyclus*, full weakness may not occur until two or more days after the tick drops off naturally or is removed. The affected person becomes increasingly unsteady and flaccid paralysis, then extends symmetrically to affect the other muscles of the body and death may occur through respiratory paralysis. Tick paralysis can also be complicated by the development of inflammation of the skeletal muscles (myositis) and heart muscles (myocarditis). Tick paralysis can be mistaken for Guillain-Barré syndrome although a distinguishing feature is that tick paralysis usually causes paralysis of the extra-ocular muscles that bring about eye movements (ophthalmoplegia). The presence of a tick still attached to the body is, of course, another good indicator.

11.2.3 Gastrointestinal Tract

The gastrointestinal tract extends from the lips to the anus and encompasses many distinct morphological and physiological regions each of which is associated with a particular guild of parasites. However, much of the pathology caused by parasites results from them physically damaging the walls of the tract, blocking the secretion of substances, blocking the absorption of metabolites, and/or blocking the movement of food through the gastrointestinal tract. In addition, the gut contains a complex microbiome, and the pathology caused by gastrointestinal parasites is often facilitated or exacerbated by interactions with its constituents.

Physical damage to the walls of the tract usually results from parasites invading or feeding on the cells lining the gut. For example, in parts of the Middle East, some people develop a condition in known as 'halzoun' (parasitic pharyngitis) in which parasites attach to the pharynx (Musharrafieh et al. 2018). The feeding of the parasites causes physical damage and the acute inflammatory response that manifests as pain and bleeding. Some reports ascribe the condition to adult *Fasciola hepatica* acquired through eating raw or poorly cooked liver containing the parasites, whilst there also cases in which the developmental stages of the pentastome (Crustacea) parasite *Linguatula serrata* were responsible. Similarly, within the small intestine, adult hookworms cause lesions and bleeding, whilst the invasive trophozoites of *Entamoeba histolytica* cause flask-shaped ulcers in the colon.

The pathology associated with the nematode parasite of cattle *Ostertagia ostertagi* provides an interesting example of how damage to one region of the intestinal tract can have implications for the physiology elsewhere in the body. *Ostertagia ostertagi* is an important parasite of cattle and to a lesser extent sheep in many parts of the world. The adult worms live in the abomasum and are commonly known as brown stomach worms from their coloration when fresh. The female worms are 8.3–9.2 mm in length and their eggs are passed with the host's faeces. The eggs hatch in the faeces to release free-living first-instar larvae that feed and grow before moulting to the second stage. These also feed and grow before moulting the infective third stage. The third-stage larvae retain the cuticle of the second stage as a protective sheath. They may remain in the faecal pat for months but when the weather is wet and moist, they migrate onto the surrounding vegetation where they are ingested by grazing cattle. After being ingested, the infective third-stage larvae exsheath in the rumen and continue their development upon passing into the abomasum. The abomasum is the ruminant equivalent of the human stomach. Here, the larvae develop within the lumen of abomasal glands – these perform a similar function to the gastric glands – that is they secrete mucus from mucus cells, pepsinogen from chief cells, and hydrochloric acid from parietal cells. The developing larvae damage the parietal cells and induce the mucus cells to undergo hyperplasia. Once they reach adulthood, the worms leave the abomasal glands and move the mucosal surface of the abomasum. The emergence of the adult worms is associated with a reduction in the secretion of hydrochloric acid and a massive rise in the secretion of the hormone gastrin into the blood. The reduced secretion of hydrochloric acid means that the pH of the abomasum rises from about pH 2 to 7, and therefore, proteins are no longer denatured and thereby rendered easier for intestinal proteases to metabolise. In addition, pepsinogen is not converted into active pepsin. Furthermore, viruses, bacteria, and other potential pathogens that would normally have been killed by the low pH are able to survive and pass on to the small intestine. The damage to the epithelium of the abomasum caused by a combination of the parasite and the acute inflammatory response also renders it more permeable, so pepsinogen leaks into the circulation and plasma proteins leak into the lumen of the gut. If the damage is extensive there can be substantial loss of plasma proteins, and this is reflected in low serum albumen levels (hypoalbuminaemia). There is some uncertainty whether the physiological changes following *O. ostertagi* infection represent an adaptation by the parasite to exploit its host or an attempt by the host to limit the damage caused by the parasite. For example, by reducing the acidity of the abomasum, the parasite could be creating a less hostile environment in which to live. However, raising the levels of gastrin in the circulation not only stimulates gastric secretions it also stimulates the formation of new parietal cells and a new mucosa. Nevertheless, in serious infections the affected animal loses considerable amounts of protein from a combination of the losses of serum proteins and an inability to fully exploit the bacterial protein from the rumen (most microbes passing from the rumen to the abomasum would normally be killed by the acidic pH and then digested). This is further compounded by anorexia and as a result muscle wastage occurs because the animal must metabolise muscle protein to manufacture essential serum proteins.

Food is propelled from the mouth to the anus by waves of muscular contractions called peristalsis. The smooth muscles lining the gut have their own inherent rate of contraction, and this is modulated by input from the autonomic nervous system. Parasites can interfere with these muscular contractions by increasing or decreasing their force and frequency and/or by stopping them entirely. For example, diarrhoea results in an increase in the force and frequency of contractions. If the gut lining is seriously irritated, it may induce the formation of an intussusception in which the gut telescopes back on itself. Once the leading edge has become trapped, peristalsis drives this further forward (in the wrong direction). The formation of an intussusception is extremely painful and can result in a potentially fatal blockage of the gut. It can also compress the blood supply and give rise to an infarction, which is also potentially fatal. An 'infarction' is the term used to describe where there is blockage of the blood supply (usually the arterial supply) to a region, and this leads to coagulative necrosis in the affected tissues. In domestic horses, adults of the tapeworm *Anoplocephala perfoliata* tend to attach at the ileocecal junction where they induce the formation of ulcers. These ulcers are then thought to precipitate intussuception – and this is manifested as 'colic'. Theoretically, any agent that causes localised enteritis or disruption of peristalsis could initiate intussusception and other parasites, such as *Parascaris equorum*, have also been implicated, while the importance of *A. perfoliata* is far from proven. In humans, intussusceptions also usually occur at the ileocecal junction, and they are much more common in children than in adults. Parasites are not a major risk factor for the development of intussusception in human medicine although there are several case studies of them being caused by the nematode *Anisakis* spp. (Yorimitsu et al. 2013). Excessively forceful peristaltic contractions in the rectal region can result in rectal prolapse in which the posterior region of the rectum everts from the anus. In humans, rectal prolapse is a known risk factor associated with heavy *Trichuris trichiura* infections whilst *Gasterophilus* nasalis infections have been blamed for high levels of rectal prolapse in donkeys in Ethiopia (Getachew et al. 2012).

High densities of large parasites can physically slow down and block the movement of digesta through the gut. This usually occurs in the small intestine because this is where the biggest parasites tend to live. For example, large numbers of *Ascaris lumbricoides* or adult *Taenia* tapeworms can block our small intestine. Similarly, *Ascaris suum* can block the intestine of pigs and *Moniezia expansa* can block the intestine of sheep. Blockage can also result from extensive scarring taking place following previous damage to the gut wall in which smooth muscle is replaced by non-contractile connective and fibrous tissues. For example, amoebic dysentery causes ulceration of the colon and when these ulcers heal the extensive scarring can interfere with peristalsis.

Peristalsis can also be compromised through damage to the autonomic nerves that modulate its activity. This is a particular feature of the digestive form of Chagas disease (*T. cruzi*). It principally manifests itself as a disturbance to the normal functioning of the oesophagus and colon – although other regions of the gut may also be affected. In its more extreme forms, it results in the afflicted regions swelling enormously and irreversibly to form conditions such as megaoesophagus and megacolon. The pathology in the oesophagus typically results from damage to parasympathetic ganglia in the lower oesophagus. Consequently, the muscles that the nerves innervate receive a reduced and/or abnormal nervous input. This results in the muscles losing their tone and becoming flabby and thereby causes problems with peristalsis and failure of the lower oesophageal sphincter. In addition, there is an increase in muscle thickness within the oesophagus and inflammatory foci form along with the deposition of fibrous tissue. Over time, the oesophagus dilates, and the patient becomes unable to swallow (dysphagia). Mega-oesophagus-associated Chagas Disease can afflict people of all ages and may predispose patients to oesophageal cancer. Megacolon develops in a similar manner and results from the destruction of the myenteric plexus of the colon. Consequently, the muscles the nerves innervate become flabby, fibrous tissue is laid down, and inflammatory foci develop. The colon swells enormously, peristalsis becomes impaired, and the afflicted region ceases to function effectively. The condition initially manifests itself as constipation but because it is principally found in middle-aged adults (30- to 60-years-old), it is often not diagnosed until well developed. This is because constipation is a non-specific symptom that is common among this age group and can result from numerous causes. As the condition worsens the patient often suffers from cachexia (malnutrition resulting in weight loss) that may prove fatal.

Some helminth infections could be risk factors for us developing acute appendicitis. Adult *A. lumbricoides* are notorious for their tendency to roam. This can result in the anterior of the worm becoming stuck within the entrance to the appendix. The worm then irritates the appendix and induces an inflammatory reaction and/or it dies *in situ*. The decomposing worm then causes inflammation that develops into appendicitis. However, there is some uncertainty over the extent to which helminths initiate potentially fatal appendicitis. For example, Aydin (2007) found *Enterobius vermicularis* in 3.15% of appendectomies but in none of these had acute inflammation occurred. By contrast, da Silva et al. (2007) recorded *E. vermicularis* in 95.8% of appendectomies and stated that there was evidence of acute inflammation in 50% of the worm-infected appendices.

11.2.4 Gall Bladder and Bile Ducts

Adult *F. hepatica* live in the gall bladder and bile ducts where they cause ulceration and inflammation (cholangitis). In extreme cases, the bile ducts become obstructed, thereby causing bile to accumulate in the liver and an increase in the bilirubin concentration in the blood and hence jaundice. A reduction in the secretion of bile into the intestine results in a reduction in the ability to digest fats. *Fasciola hepatica* infections in humans and experimental rats has been linked to the development of gallstones. The formation of gallstones can lead to the blockage of the bile ducts, inflammation (cholecystitis), fibrosis, and there may be secondary bacterial infection. Not surprisingly, the condition is very painful.

Gallstones take a variety of shapes, sizes, and compositions. They typically comprise varying proportions of cholesterol, calcium carbonates, calcium phosphates, and the pigment calcium bilirubinate. In humans, approximately 80% of gall stones are cholesterol stones and ~20% are pigment stones. Pigment stones usually result where there is liver cirrhosis and an excessive secretion of bilirubin by the liver. Black pigment gallstones are common in sheep, particularly lambs and adult ewes but, perhaps surprisingly, there appears to be no link between them and fascioliasis. *Clonorchis sinensis* infects our liver, gall bladder, and bile ducts and is associated with the formation of gall stones. For some reason, these tend to be composed of calcium carbonate – which is not a common form of gallstone (Qiao et al. 2014). On rare occasions, the nematode *A. lumbricoides* migrates from the small intestines and enters the biliary tract. When this happens, the bile duct becomes inflamed and produces pus (pyogenic cholangitis) and gall stones develop. Treatment with some of the commonly used anthelmintics can prove difficult because they do not reach the biliary tract in sufficient amounts to expel the parasite.

11.2.5 Genitalia

Parasitic diseases that afflict human genitalia are not only physically debilitating and painful, but they often impose extra psychological burdens. For example, genital malformations can lead to the afflicted person being openly ridiculed or excluded from society. Furthermore, if they are not already married then their chances of doing so are greatly reduced. Marriage is tremendously important in many traditional societies as it confirms one's status as part of the community and is often the only opportunity for sexual relationships. For those within a sexual relationship, genital malformation may make intercourse painful or even physically impossible, and this in turn creates psychological problems. For example, people who are severely affected by genital elephantiasis can be become withdrawn from society, depressed, incapable of finding work, and may commit suicide (Dreyer et al. 1997). If the pathology induces open wounds to the genitalia, it also increases the chances of both contracting and spreading sexually transmitted diseases.

The filarial nematodes *Wuchereria bancrofti* and *Onchocerca volvulus* can both induce elephantiasis of the genitalia although the condition occurs much more frequently in men than women. *Wuchereria bancrofti* is responsible for most cases of filarial nematode-induced genital pathology, whilst *O. volvulus* is particularly associated with causing a condition known as 'hanging groin' in which folds of skin containing lymph nodes droop down from the body. In men, *W. bancrofti* is normally described as causing elephantiasis but in fact it causes several distinct genital pathologies including hydrocoele, chylocoele, and, of course, elephantiasis of the scrotum (Richens 2004). Hydrocoele arises from an accumulation of pale-yellow serous fluid within the scrotal sack, and this causes the scrotum to swell. The development of hydrocoele is very common in areas in which *W. bancrofti* is endemic. A chylocoele is somewhat like a hydrocoele except that the swelling results from an accumulation of milky lymphatic fluid.

Scrotal elephantiasis develops through the accumulation of lymph within the scrotum: this process is known as lymphoedema. Because the flow of lymph is compromised, it reduces the ability of the body to respond to infections in the affected area and allows the establishment of bacterial infections. These bacteria contribute to the pathology of elephantiasis by causing acute inflammation. There are also ongoing inflammatory reactions against the adult filarial worms and their *Wolbachia* symbionts although the microfilariae may be absent during the later stages of the disease. All this inflammatory activity causes the deposition of fibrous connective tissue and granulomatous tissue within the scrotum. This causes the skin to become thickened and susceptible to surface cracking. The cracks are then invaded by bacteria and fungi that induce further acute inflammatory attacks. Treatment to prevent microbial infections is therefore highly beneficial in reducing the pathology associated with elephantiasis. The progression of the disease is probably influenced by prenatal exposure to *W. bancrofti* antigens and to exposure during childhood (Dreyer et al. 2000).

Prolonged *Schistosoma haematobium* infections can give rise to elephantiasis of the penis in men, but genital pathology is usually much worse in women (Hegertun et al. 2013). All parts of the female reproductive system can be affected by the inflammatory reactions against the trapped schistosome eggs. For example, damage to the uterus can result in disturbance of menstruation whilst damage to the placenta may result in abortion – usually in the second trimester. If the fallopian tubes are affected, there is an increased risk of ectopic pregnancy whilst ulceration of the vagina and vulva is not only painful, but the open lesions increase the risk of contracting and transmitting sexually transmitted diseases such as HIV (Richens 2004). The damage caused by schistosomiasis may not become apparent until many years after the initial exposure. Consequently, although it is not recommended by the WHO, there is an argument to be made for travellers to be screened for the parasite if they have a history of paddling or swimming in lakes where there is a high incidence of infection among the local people. Bailey et al. (2011) describe two cases of serious genital pathology owing to schistosomiasis in British women that was acquired at least 8 years previously – probably through swimming in Lake Malawi whilst they were on holiday.

Relatively few parasites specifically reside within the genitalia – although there are exceptions such as *Trichomonas vaginalis*. In women, *T. vaginalis* causes a profuse yellow discharge, irritation, and inflammation of the vulva and vagina. Petechial lesions on the cervix can be so serious that the condition is known as 'strawberry cervix'. Infection rates in men can be high but it usually causes little pathology in them. The lack of pathology in men is probably owing to the higher zinc content in the male genital tract, as well as unidentified components of the secretions from the prostate gland and seminal glands, that are toxic to *T. vaginalis* (Langley et al. 1987). Several parasite species that normally reside elsewhere in the body can be sexually transmitted (e.g., *E. histolytica*). These species are typically associated with inflammatory reactions and are almost invariably more pathogenic in women than men. Within humans, owing to the proximity of the vagina and anus, it is not unusual for some gastrointestinal parasites to cause gynaecological problems. For example, the pinworm *Enterobius vermicularis* can cause vulvovaginitis and urinary tract infections in young girls. This is probably through initiating inflammatory reactions within the genital tract and transporting bacteria such as *E. coli* from the anus that then cause genital infections (Cook 1994).

Parasites can also affect the genitalia through their effects on the endocrine system. For example, Human African Trypanosomiasis (HAT) can result in temporary or permanent impotence in men. It can also cause a reduction in the size of their testes and cause fat to become redistributed in a similar pattern to that of women. Consequently, men can develop gynaecomastia in which they form prominent breast tissue. This may be related to the parasites accumulating in the testes where they stimulate a strong, and potentially damaging, immune response (Carvalho et al. 2018). Women suffering from HAT may experience shrinkage of the sexual organs, menstrual problems, and become infertile (Ikede et al. 1988).

11.2.6 Kidney

Not many parasites specifically reside within the kidney. Of those that do, the best known are the nematodes *Dioctophyma renale* that infects dogs, many carnivores, and occasionally humans and *Stephanurus dentatus* that infects pigs. *Dioctophyme renale* is a remarkably large worm: the females can grow to over 1 m in length and 1.2 cm in width. The infective larvae penetrate the gut wall and mature within the peritoneal cavity after which they invade the kidney. Despite their large size, the worms do not always induce clinical signs since they usually only infect one of the kidneys – mostly the right one. The feeding of the worms can result in the destruction of the parenchyma of the infected organ, but most animals can survive healthily with only one functional kidney. The damage to the tissues can result in red blood cells being passed in the urine (haematuria). The loss of kidney function also interferes with the formation and release of urine to the bladder. The consequent retention of nitrogenous waste products such as urea within the blood ('uraemia') can result in a wide variety of problems elsewhere in the body. For example, uraemia is associated with calcium deposition in the skin that causes itchiness and discoloration, inflammation of the cardiac muscles, sleepiness, nausea, anorexia, nervous disorders, seizures, coma, and may be fatal. It also interferes with cellular immunity and therefore affects the response to many infectious agents.

Stephanurus dentatus is usually restricted to pigs. The adult worms are fairly large – although much smaller than *D. renale*: the female grows to about 4.5 cm in length and is only about 2 mm in width. The larvae of *S. dentatus* cause serious pathology to the liver and other body organs during their migration, but the adult worms are not considered particularly pathogenic. The worms usually live within the peri-renal fat rather than the kidney itself and they are usually found as pairs within cysts that may reach up to 4 cm in diameter and are filled with green pus. They can, however, cause chronic inflammatory reactions in the ureters that result in the deposition of so much fibrotic tissue that it almost completely blocks them.

Red blood cells infected with *P. falciparum* are 'sticky' and therefore become attached to the walls of capillaries within various organs of the body. If large numbers of infected red blood cells are sequestered within the capillaries of the kidneys, they block the blood flow, and this results in the death of cells from oxygen and nutrient deprivation. The epithelial cells of the proximal and distal tubules are particularly sensitive to a lack of oxygen and their death causes acute tubular necrosis and this manifests as acute renal failure which can be fatal. Acute renal failure in malaria patients is often seen in adults and rarely in children. It is often associated with a reduction in production of urine – this is known as 'oliguria' – and a patient may pass less than 400 ml urine in 24 hours. In extreme cases, urine production may cease – this is known as 'anuria'. However, oliguria is not always a feature of malaria-induced acute renal failure, and the infection is then referred to as 'non-oliguric'. In some parts of the world there have been increases in the incidence of acute renal failure associated with malaria and it often has a poor prognosis. Acute renal failure can also be caused by other species of malaria such as *P. vivax* and it may be presented as a feature of multiple organ failure or on its own.

Malaria can also induce acute renal failure through various other mechanisms, such as a decrease in the volume of blood plasma (hypovolaemia) that arises through loss of body fluids owing to sweating and decreased fluid intake. In severe cases of malaria this gives rise to a condition known as hypovolaemic shock. This results in a decreased supply of blood to the kidneys and consequent acute tubular necrosis. Severe cases of malaria are also associated with the destruction of large numbers of infected red blood cells within the spleen and general circulation. The rise in levels of haemoglobin and its breakdown products in the circulation can exceed the ability of the liver to metabolise them, leading to large amounts being excreted via the kidneys. Consequently, the urine becomes dark and in conjunction with severe fever, jaundice, and vomiting, the effect is called 'Blackwater Fever'. The condition is extremely serious, and it is commonly accompanied by oliguria and acute renal failure. The latter explains why it has a mortality rate of around 50%. Blackwater Fever was once common but is rarely seen these days. The decline was once attributed to it being linked to the use of quinine to treat malaria. Therefore, as quinine was replaced by other drugs, so the number of cases of Blackwater Fever also declined. However, cases of Blackwater Fever continue to occur in patients who do not receive quinine. One suggestion is that it results from some form of autoimmune response, possibly involving IgG1 Malaria antibodies, that causes the rapid haemolysis of large numbers of red blood cells. However, Olupot-Olupot et al. (2017) reported a high frequency of Blackwater Fever among children in Eastern Uganda and raised the worrying possibility that it might be linked to the use of artemisinin-based combination therapies. It is therefore likely that Blackwater Fever manifests from a complex interaction of host, parasite, and treatment interactions (Shanks 2017).

Glomerulonephritis is a general term to describe various medical conditions and despite the suffix '*itis*' they do not all involve direct inflammatory damage to the glomeruli. Glomerulonephritis compromises the normal filtration of the blood although there are differences between glomeruli and not all glomeruli may be affected equally. Depending upon the type of damage, glomerulonephritis may lead to nephrotic syndrome, nephritic syndrome, or acute renal failure. Nephrotic syndrome is characterised by the loss of protein in the urine, and this results in hypoalbuminaemia (low blood protein concentration) and oedema. Nephritic syndrome results in a rise in blood pressure and blood urea, whilst the patient may pass blood in their urine. There many case reports of glomerulonephritis in association with various parasitic infections. These include filariasis, schistosomiasis, amoebic liver abscesses, Chagas disease, *Plasmodium falciparum* malaria, visceral leishmaniasis, and *Strongyloides stercoralis* infection. In many cases, the pathogenic mechanism is uncertain but probably results from immune complex deposition within the glomeruli. Alternatively, or additionally, circulating parasite antigens might become trapped in the glomeruli and damage occurs following the subsequent binding of circulating antibodies.

11.2.7 Liver

The liver is the largest internal organ in humans and many other mammals. It is vitally important in regulating the chemical composition of the blood, detoxifying metabolites, metabolising carbohydrates, lipids, and proteins and synthesising bile. The liver also contains fixed phagocytes called Kupffer cells that are embedded within the wall of the sinusoids and remove pathogens, cell debris and toxins from the circulation. Damage to the liver therefore manifests as jaundice, a failure to synthesise essential molecules, and a failure to detoxify potentially harmful molecules. To diagnose liver damage, liver function tests analyse the levels of liver-specific enzymes and metabolites in circulating blood. For example, a rise in alkaline phosphatase indicates damage to the biliary (bile) system, whilst a rise in gamma-glutamyl transpeptidase is a non-specific indicator of liver cell (hepatocyte) damage.

Although outwardly the liver appears homogeneous, the different regions have different metabolic properties and therefore the consequences of liver damage depend upon which group of hepatocytes are affected. The products of digestion, and anything else that can pass through the gut wall, enter the capillaries and these drain into the hepatic portal vein, which in turn gives rise to capillaries that flow through the liver. Consequently, any pathogens that succeed in penetrating the gut and entering the circulation are immediately swept to the liver. Once the blood has passed through the capillaries within the liver, it enters the hepatic vein and thence enters the general circulation. This flow arrangement of mesenteric capillaries, hepatic portal vein, hepatic capillaries, hepatic vein is known as the 'hepatic portal system'. Numerous parasites cause pathology in the liver when they migrate through its tissues (e.g., *F. hepatica*), parasitise its cells (e.g., *Plasmodium vivax*), cause pressure atrophy (e.g., hydatid cysts of *Echinococcus granulosus*), and induce harmful immune reactions (e.g., eggs of *Schistosoma mansoni*).

Clinically, there are four basic manifestations of liver pathology:

1) Acute hepatitis in which there is sudden onset of a massive inflammatory response. People or animals suffering from acute hepatitis are often tired, there is abdominal pain and there are signs of jaundice. Acute hepatitis typically results from pathogens or toxins that cause the death of numerous liver cells over a short period of time.

2) Chronic hepatitis results from prolonged inflammatory responses that cannot be resolved. There is abdominal pain but, in humans, jaundice may not occur until a late stage in the condition and is then considered a poor prognostic indicator.

3) Fibrosis results from ongoing chronic inflammation. It involves functional hepatocytes being replaced by non-functional fibrotic tissues. Where fibrosis occurs within the bile ducts, it leads to the blockage of the flow of bile from the liver to the duodenum. This results in the faeces becoming pale or even white because the haemoglobin breakdown product bilirubin no longer enters the digestive tract (conjugated hyperbilirubinaemia).

4) Cirrhosis is the end point of liver cell destruction in which there is extensive scarring, and the liver vascular architecture is disrupted. This disrupts the flow of blood through the liver and leads to an increase in the blood pressure in the portal system. This condition is known as portal hypertension and can result in the development of splenomegaly and ascites.

Acute hepatitis is a common feature of the simultaneous migration of large numbers of parasites through the liver. For example, the young flukes of *F. hepatica* in sheep or the movement of *A. suum* larvae in pigs. The movement of numerous parasites causes widespread cell death, bleeding, necrosis, and an acute inflammatory response. This can lead to acute liver failure, which is often rapidly fatal. Chronic hepatitis arises from long-standing infections and/or the accumulation of damage over time. Extensive liver damage severely compromises liver function and may lead to potentially fatal cirrhosis. However, because the liver can self-repair to some extent, clinical disease may not manifest until after severe damage has occurred. For example, dogs naturally infected with *Leishmania infantum* may exhibit evidence of damage from liver-function tests and post-mortem histology but show no clinical signs of liver pathology (Rallis et al. 2005). This emphasises the old legal adage that 'absence of evidence is not evidence of absence'.

The liver is a major site for the synthesis of biomolecules that are then released into the circulatory system to be utilised elsewhere in the body. For example, plasma albumin and most of the plasma globulins (apart from the immunoglobulins) are manufactured in the liver. Consequently, damage to the hepatocytes compromises the production of these chemicals. For example, in sheep, sub-acute fascioliasis causes a decline in the serum albumin concentration – this is known as hypoalbuminaemia. This reduction in serum protein concentration means that there is a decline in the oncotic pressure (a type of osmotic pressure engendered by proteins), and therefore, there is a reduction in the amount of water being dragged into the blood from the surrounding tissues. Consequently, the amount of fluid within tissues starts to rise, and this manifests as the development of oedema (fluid accumulation). In sheep and cattle, a common site of fluid accumulation is in the submandibular region, and this condition is commonly known as 'bottle-jaw'. Fluid accumulation within the peritoneal cavity is known as ascites.

Adult *S. mansoni* and *S. japonicum* live in the mesenteric venules. Therefore, those eggs which they release that do not find their way to the lumen of gut are transported via the circulation to the liver. Within the liver, the eggs become trapped in the intrahepatic venules. They then generate a granulomatous reaction, and the surrounding tissue becomes fibrotic. This is sometimes called 'pipe stem fibrosis' or 'portal tract fibrosis', and it results in localised blockage of the flow of blood. Consequently, the supply of venous blood from the spleen to the liver is reduced, and there is a rise in the blood pressure within the blocked vessels. This results in portal hypertension, which in turn results in enlargement of the spleen (splenomegaly) and ascites.

The Kupffer cells found in the liver sinusoids are important in the pathology of several parasitic diseases. They are probably the portals via which *Plasmodium* sporozoites gain access to the hepatocytes. The sporozoites invade and traverse the Kupffer cells and then ultimately invade hepatocytes within which they undergo the exoerythrocytic stages of development. The Kupffer cells are rendered insensitive to pro-inflammatory signals and the parasites induce them to undergo apoptosis. During malaria, the Kupffer cells exhibit hyperplasia – probably in response to *Plasmodium* spp. toxins and the breakdown products from damaged red blood cells. They also accumulate haemozoin granules that reduces their ability to phagocytose things. The Kupffer cells are important in combating pathogens and clearing cell debris and toxins from the circulation. Consequently, the loss of functional Kupffer cells reduces the body's ability to deal with the malaria infection, and there is also increased vulnerability to other infectious agents. In visceral leishmaniasis, the Kupffer cells, along with other mononuclear phagocytes, are invaded by the parasites. The liver becomes enlarged (hepatomegaly) owing to the accumulation of rapidly dividing parasites within the phagocytes. In addition, the enlargement and distortion of the infected cells disrupts the architecture of the liver and thereby compromises its function. In addition, visceral leishmaniasis can also induce liver fibrosis and portal hypertension.

11.2.8 Lungs

Few parasites are transmitted as air-borne infections and therefore most of them arrive in the lungs via the host's blood circulation. For some of them, the lungs are merely another organ to be passed through during their migration. For example, for the gut helminths *A. suum*, *A. lumbricoides*, *P. equorum*, *S. stercoralis*, and *Ancylostoma caninum*, the migrating larvae must break through the alveoli into the lumen of the bronchioles, climb up the trachea until they reach the join with the oesophagus, and then move down the gastrointestinal tract to the small intestine where they will become adults. The larvae are small in relation to the size of the lungs, so a few of them cause little noticeable pathology. However, the movement of lots of larvae over a short time results in considerable damage. This manifests as focal haemorrhages and associated acute inflammation. The destruction of the alveoli leads to the development of emphysema. Emphysema can arise from many causes in addition to parasitic infections. It is characterised by a permanent enlargement of the air spaces distal to the terminal bronchioles and destruction of the alveoli. So-called generalised emphysema does not involve the formation of scarring. However, the term 'emphysema' is also used to describe other forms of pathology in which scarring, and fibrosis accompany the dilated air space. Obviously, scarring can be a feature of parasite-induced emphysema as the lungs attempt to repair the physical damage caused by the parasites. Whatever the mechanism that induces emphysema, the result is breathing difficulties because of the loss of elastic recoil and a reduced area available for gaseous exchange. Consequently, emphysema is characterised by an increase in the breathing rate to increase the amount of carbon dioxide expelled (the rate and depth of ventilation are mainly controlled by the partial pressure of carbon dioxide in the blood). However, owing to the reduced surface area for gaseous exchange, this is not successful and affected people and animals soon become breathless and hypoxic (i.e., the oxygen supply is insufficient to meet cellular needs) when attempting to undertake exercise. In chronic cases, in which worms migrate through the lungs over a prolonged period, there may be the formation of eosinophilic granulomas. These also affect breathing through disrupting the alveoli: capillary interface and reducing the area available for gaseous exchange.

For some parasites, the lungs are the site where they become adult and reproduce. The best known of these are the nematode parasites known as the lungworms; these include species such as *Dictyocaulus viviparus* (cattle), *Dictyocaulus filaria* (sheep), *Aelurostrongylus abstrusus* (cats), and *Metastrongylus apri* (pigs). Humans do not generally suffer from nematode lungworm infections. Several trematode parasites known as the lung flukes also become adults in the lungs. For example, we can suffer infections with *Paragonimus westermani*, *Paragonimus skrjabni*, and *Paragonimus mexicanus*. These cause signs and symptoms like those of tuberculosis, including chronic coughing, chest pain, and haemoptysis – the coughing up of blood (Calvopina et al. 2017). Diagnosis can be further complicated by the fact that TB-*Paragonimus* co-infections are relatively common.

The most important lungworm is *Dictyocaulus viviparus* (Figure 11.1), which infects cattle, and we will use this to illustrate how pathology within the lung can develop. Dictyocauliasis is primarily

(a)

Figure 11.1 *Dictyocaulus viviparus.*
(a) Adult male worm, scale in mm.
(b) Anterior of adult male worm.
(c) Posterior of adult male worm showing bursa and spicules.

(b)   (c)

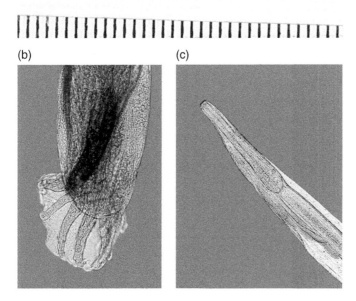

a disease of young cattle on their first season on grass. This is because the initial infection generates a strong protective immunity. The severity of the disease depends upon the rate of intake of infectious larvae. There are three distinct phases: pre-patent, patent, and post-patent. The pre-patent phase begins 8–25 days after the initial infection when the larvae start to leave the circulation and penetrate the alveoli. This induces bleeding and an acute inflammatory response in which inflammatory cells penetrate the lung epithelium, and there is accumulation of fluid (pulmonary oedema) within the lungs and the development of interstitial emphysema. Between the blood capillaries and the alveoli is a fluid filled 'interstitial space' or 'interstitium'. This facilitates close proximity between the two tissues with the minimal involvement of connective tissues or inflammatory cells. However, physical damage to this region caused by the parasites, coupled with the consequences of the inflammatory response, results in this delicate balance being disrupted. In addition, during the repair process interstitial tissue replaces the normal alveoli and capillaries. The damage to alveoli and capillaries coupled with the deposition of interstitial tissue reduces lung elasticity and the area available for gas exchange – and hence emphysema develops. The development of pulmonary oedema is equally serious since it too reduces the area available for gaseous exchange and can lead to hypoxia and fatal respiratory failure. Assuming that the infection is not fatal, the migratory path of the worms is marked by the changing focus of the inflammatory response. Initially, there is alveolitis as the migrating larvae break through into the air spaces, and this is then followed by bronchiolitis and bronchitis as the worms move to bronchi where they moult to become adults.

The arrival of the adult *D. viviparus* at the bronchi 26–60 days after infection marks the start of the patent period. The worms cause the epithelial cells lining the bronchi to undergo hyperplasia and produce excessive amounts of mucus. The mucus then becomes mixed with air so that the bronchi become filled with a white froth. This induces intense coughing as the infected animal attempts to clear its blocked-up lungs. British farmers know the disease as 'husk' because of the characteristic harsh coughing it causes. The coughed-up mucus results in discharges from the mouth and nose, and the animal suffers from anorexia, malnutrition and develops a rough coat. When the worms commence reproduction, they produce eggs that hatch almost immediately to release first stage larvae. The larvae do not feed and are coughed-up and then swallowed and pass out with the faeces. However, some of the eggs and larvae are swept down into the alveoli where they induce aspiration pneumonia. This is characterised by the development of an intense inflammatory response and the alveoli become filled with inflammatory exudates.

Although much of the pathology associated with dictyocauliasis is associated with the host immune response, this usually results in the adult worms being expelled and a strong protective immunity to re-infection. This marks the onset of the post-patent period (around days 61–90) and is associated with the repair of the damage of tissue damage. Lesions to the bronchi and bronchioles become fibrotic and therefore coughing may continue owing to impaired breathing. However, the tissue damage is usually eventually resolved and normal breathing returns. Nevertheless, in about 25% of animals suffering serious infections, there is a return of clinical symptoms of respiratory distress and these may prove fatal. The main cause of this appears to be hyperplasia of type 2 pneumocytes (alveolar cells). There are two types of pneumocyte found in association with walls of the alveoli: type 1 and type 2. They are both epithelial cells, the type 1 pneumocytes are squamous epithelial cells that line the alveolar wall and are the principal sites of gaseous exchange. The type 2 pneumocytes (also called septal cells) are rounded or cuboidal in shape and located between type 1 pneumocytes. The type 2 pneumocytes secrete surfactants that lower alveolar surface tension and thereby reduce the likelihood of the alveoli collapsing. Hyperplasia of the type 2 pneumocytes produces a thickened layer of epithelial cells that is less permeable for gaseous exchange, and it is accompanied by the development of interstitial emphysema and pulmonary oedema. This is sometimes referred to as 'post patent parasitic bronchitis'. It is uncertain what causes the type 2 pneumocytes to proliferate, but it may be a response to chemicals released by dead or dying lung-worms. In serious cases, squeaks and crackles can be heard in the posterior lobes of the lungs when the animal breathes, and death often occurs within 24–96 hours of the onset of illness. This is a result of acute heart failure brought on by respiratory distress. Serious disease and death may also occur during the post patent period because secondary bacterial infections (e.g., *Pasteurella multocida*) establish in the lesions caused by the lungworms and initiate bacterial pneumonia.

Some parasites establish infections in our lungs through being swept there by the circulation and/or through growing into them from surrounding tissues. For example, both *E. histolytica* and *E. granulosus* can form pulmonary infections either because the trophozoites or oncospheres, respectively, reach the lungs via the blood, or through extensions of abscess or cyst growth in the liver. *Entamoeba histolytica* establishes ulcers in our lungs in a similar manner to the liver and this is known as pulmonary amoebiasis. It causes a dry cough, purulent, chocolate-coloured sputum, and breathing difficulties. Most hydatid cysts develop in the liver, but the lungs are the second most common site in both humans and other intermediate hosts. Pulmonary hydatid cysts principally cause problems when they become large and start to compress the surrounding lung tissue. This leads to coughing, chest pain, and haemoptysis (coughing up of blood from the respiratory tract). A growing hydatid cyst can erode the thoracic aorta, and where blood loss is severe, a patient can drown in their own blood. If the cyst ruptures into a bronchus or the pleural cavity, it is referred to as a 'complicated cyst' and the leakage of the cyst fluid can set up an allergic reaction. 'Complicated' pulmonary hydatid cysts can lead to the expectoration of cyst fluid, repetitive haemoptysis, and anaphylactic shock. Pulmonary hydatid cysts also facilitate the establishment of fungal infections such as *Aspergillus* spp. that typically colonise pre-existing lung cavities caused by tuberculosis and other diseases (Manzoor et al. 2008). Pulmonary *Aspergillus* infections are normally associated with immunocompromised people, and the development of fungal hyphae inside the patient leads to necrosis, pneumonia, and infarction through colonising the blood vessel walls.

11.2.9 Skin

The skin is the largest organ in the body and provides an effective barrier to most pathogens. It is also a highly complex ecosystem that contains a huge variety of microorganisms the composition of which varies between regions of the body and between individuals. This 'skin microbiome' has an important role to play in preventing infections and in the development of the host's immune system (Grice and Segre 2011). Much of the parasite-induced pathology associated with the skin is caused by ectoparasites such as leeches, mites, and ticks and fly larvae causing cutaneous myiasis. Skin pathology can also result from actively invading parasite larvae (e.g., cutaneous larva migrans), skin infections such as dermal leishmaniasis, the presence of larvae of filarial nematodes such as *O. volvulus* as they wait to be picked up by their vector, and even immune reactions to parasites that dwell far from the skin. Skin pathologies may also result from the metabolic consequences of damage inflicted to other organs. For example, the fluke *Dicrocoelium dendriticum* is thought responsible for photosensitisation in some sheep populations (Sargison et al. 2012). The flukes damage the liver/bile ducts, and this probably causes an increase in the concentration of certain photolabile metabolites in the circulation. The nature of the chemical(s) is uncertain and might relate to those derived from plants that are no longer effectively detoxified/excreted fast enough (most cases of photosensitisation in sheep relate to the consumption of poisonous plants). Within the outer layers of the skin, the chemicals react with light to produce toxic substances that cause inflammation and necrosis.

It Started with a Leech Bite

Any break in the skin surface represents an opportunity for secondary invasion by bacteria, fungi, and opportunistic parasites. Slesak et al. (2011) describe a fascinating case in which precisely this occurred to an unfortunate farmer living in village in northern Laos. Like many poor farmers, he often walked barefoot and one day he was bitten on his left foot by a leech. Leeches are common in tropical countries such as Laos and being bitten by a leech is an everyday experience for anybody living and working in the countryside. Unfortunately for him, however, the bite site became infected by a fungus belonging to the genus *Fonsecaea*; species belonging to this genus are common opportunistic pathogens that can survive as saprophytes in the soil or in humans as a parasite. *Fonsecaea* and related fungi cause a condition known as 'chromoblastomycosis' in which slow-growing 'cauliflower-like growths', nodular lesions, and plaques develop. Over the course of several years, these growths steadily expanded until they reached as far as his knee. Chromoblastomycosis is not painful or itchy, but it is debilitating because it interferes with movement and lesions on the skin surface can be colonised by other infectious agents. In the farmer's case, they were infected by the bacterium *Escherichia coli* and by larvae of the fly *Chrysomya bezzinana* (Figure 11.2). The farmer therefore developed cutaneous myiasis to add to his woes and even more unfortunately *C. bezziana* is a particularly aggressive deep-burrowing maggot that consumes healthy tissues, as well as those that are necrotic.

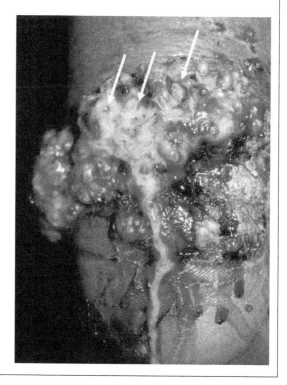

Figure 11.2   Heel of a man suffering from a co-infection of chromoblastomycosis and cutaneous myiasis. The rear ends of some of the larvae of *Chrysomya bezziana* are arrowed. *Source:* Slesak et al. (2011), Figure 11: 14/ Springer Nature/CC BY 2.0.

Ectoparasites damage the skin and underlying tissues when they feed, as well as inject their saliva and this induces an inflammatory reaction. However, the skin exhibits numerous different inflammatory patterns, and there is no typical generic response to ectoparasites. For example, the skin follicle mites *Demodex folliculorum* and *Demodex brevis* can cause a range of different skin conditions. *Demodex folliculorum* lives communally within our hair follicles, whilst *D. brevis* lives a solitary life within our sebaceous glands. Both mite species can occur at the same time and as many studies have not distinguished between them, we will simply refer to them as '*Demodex*'. *Demodex* mites are often found in the hair follicles of people suffering from rosacea, and it is possible that the mites contribute to its development. Rosacea is an acne-like condition in which there is reddening (erythema) of the face – especially the cheeks, chin, forehead, and nose – within which papules and pustules develop. The facial dermatitis condition *pityriasis folliculorum* that gives the skin a 'frosted appearance' has also been linked to infection with *Demodex* infection. *Pityriasis folliculorum* is usually described as occurring in elderly women who use heavy make-up but not much soap. However, Dominey et al. (1989) describe six cases in women aged between 31 and 46 who washed daily in soap and water and used little make-up. Other forms of dermatitis linked to *Demodex* infections include scalp folliculitis, seborrheic dermatitis, perioral dermatitis, and micropapular-pruritic dermatitis. The mechanisms by which the mites induce pathology are uncertain and may relate to blocking the hair follicles and sebaceous glands. This might be through blocking the follicles and glands by their physical presence and/or inducing hyperplasia or immune reactions that block them. Mite faeces and cast exoskeletons could also induce a more widespread immune reaction. However, because many of us who are infected show no evidence of pathology, there is still some doubt as to the role of *Demodex* in causing various forms of dermatitis. A lot probably depends upon the relationship between *Demodex* and the skin microbiome. The mites can act as vectors for bacteria such as staphylococci that are involved in the development of dermatitis. In addition, a symbiotic bacterium (*Bacillus oleronius*) found within the mite stimulates an inflammatory response. High incidences of *Demodex* infection occur in patients suffering from various forms of cancer, but there is currently no suggestion that the mites initiate cancers. The high incidences probably reflect a compromised immune system and/or skin changes that favour the mite's growth and reproduction.

Dogs have their own Demodex mite – *Demodex canis*. It usually does not cause serious problems but sometimes it gives rise to 'generalised demodectic mange'. This is a serious inflammatory condition in which serum, puss, and blood ooze from the skin surface. It is therefore sometimes referred to as the 'red mange'. It is probably linked to an inborn immunodeficiency that makes certain dogs susceptible to developing the condition. However, there is also evidence that *D. canis* induces immunosuppression (Kumari et al. 2017). In addition, as with other examples of *Demodex*-associated dermatitis, generalised demodectic mange usually involves secondary invasion by bacteria, especially staphylococci.

In the case of scabies, the mites (*Sarcoptes scabiei*) produce short (2–15 mm long) wavy-line shaped burrows immediately underneath the surface of the skin. In humans, these typically occur on the fingers, wrists, penis, and scrotum. Chemicals present in the mites' saliva, faeces, and eggs induce a pronounced inflammatory response in which eosinophils infiltrate both the epidermis and dermis. These eosinophils are responsible for the intense itching (pruritis) associated with scabies infections. Papules and vesicles form at the site of infection, and there is localised eczema. The itching results in scratching, and this facilitates secondary bacterial infections – particularly with streptococci, such as *Streptococcus pyogenes*, and staphylococci such as *Staphylococcus aureus*. This can result in the development of impetigo (highly infectious, yellow, crusted blisters on the skin epidermis), abscesses, and cellulitis. Consequently, scabies is a risk factor for facilitating acute post streptococcal glomerulonephritis and, more rarely, *Staphylococcus* glomerulonephritis.

An important difference between the two is that *Staphylococcus* glomerulonephritis only persists whilst there is an active *Staphylococcus* infection. In the past, scabies-associated glomerulonephritis was a serious problem of sometimes epidemic proportions (Svartman et al. 1972). In elderly and immunocompromised individuals, there is a risk that scabies infection may manifest itself as 'crusted scabies'. This condition is also known as 'Norwegian scabies' because the condition was first described in a group of Norwegian leprosy patients in 1848. In crusted scabies, the mites multiply rapidly and induce hyperkeratotic lesions in which the surface of the skin becomes covered in dry scales – hence the term 'crusted scabies'. The explosive increase in the mite population is thought to occur because the immune system is compromised. However, it can also develop in people who are immunocompetent but are presumably genetically susceptible to this condition. The crusted lesions can spread to cover the whole body and are associated with swelling of the lymph nodes (lymphadenopathy). Crusted scabies induces a strong eosinophilic response and there are elevated levels of immunoglobulin E and IL4. Interestingly, IL4 stimulates the production of keratinocytes, and this may contribute to the hyperkeratosis. There is a high risk of secondary infections, and the condition can be fatal.

The mite *Psoroptes ovis* infects sheep and several other mammals, including rabbits and camels. However, it is of most importance in sheep as the cause of the highly pathogenic condition 'sheep scab'. Unlike *S. scabiei*, *P. ovis* does not burrow beneath the skin or feed on host tissues. Instead, *P. ovis* lives on the surface of its host's skin and feeds on the exudates and bacteria found on the skin surface. Nevertheless, it causes equally serious pathology. This is because the mite's faeces induce a hypersensitivity reaction in sheep. This manifests as a dramatic acute inflammatory response in which copious fluid exudates are released onto the skin surface where they form yellow crusty deposits. The inflammation results in intense itching and an infected sheep can become so distracted that it develops anorexia and incessantly rubs itself against any available scratching surface. The combination of skin inflammation and scratching leads to wool loss over much of the body and there is a risk of secondary infections. Ultimately, the infected animal loses weight and body condition and in serious cases the sheep dies.

Adults of the filarial nematode *O. volvulus* live in the subcutaneous tissues of our bodies where they become encapsulated within nodules composed mainly of collagen fibres. These nodules are sometimes called 'onchocercomas' and an infected person may carry up to a 100 of them. The worms live tightly coiled within the nodules and are surrounded by either a 'fibrin lake' or solid granulation tissue. Although they are encapsulated, the worms are not safe from the body's immune system and immunoglobulins are often bound to their outer surface. For some reason, in Africa the nodules usually develop in the pelvic region although they may also be found in spine, chest, and knees. By contrast, in Venezuela the nodules develop predominantly on the head and neck. Regardless of where they develop, apart from causing disfigurement, the adult worms are often said to cause little harm. However, the adult worms also release huge numbers of microfilariae, and these are responsible for much of the pathology associated with onchocerciasis. These larvae migrate to superficial layers of the skin where they can be picked up by the blackfly vector, which is a surface-pool feeder. Some microfilariae, however, also find their way to the eyes, lymph nodes, and other organs of the body. Whilst they are alive, the microfilariae do not usually induce serious pathology but after dying, they stimulate a dramatic inflammatory response.

Onchocerciasis results in lymphadenitis – inflammation of the lymph nodes – and eventually the nodes become enlarged and fibrotic. This disrupts the flow of lymph and can result in the development of 'hanging groin'. Many of the lymph nodes in African patients contain microfilariae and lymphadenitis probably results from the inflammatory response to them. The inflammatory response against microfilariae within the skin results in dermatitis of which there are two types: hyper-reactive onchodermatitis (also called 'sowda') and hypo-reactive onchodermatitis (Murdoch 2018). The inflammatory reaction is caused by the body's immune response to the symbiotic *Wolbachia* bacteria that are released from the microfilariae after their death. Some people mount a strong Th2-biased immune response against *O. volvulus* antigens, which results in the death of numerous microfilariae. However, killing the microfilariae results in the release of large quantities of *Wolbachia* antigens, which then stimulate a pronounced inflammatory reaction. By contrast, people who tolerate the presence of numerous microfilariae express a less-pronounced Th2 response, and they also exhibit a suppressive Th1/Th3 response. Consequently, they develop less-severe hypo-reactive dermatitis because fewer microfilariae are killed and therefore fewer *Wolbachia* antigens are released. However, people cannot be characterised as being either hyper-reactive or hypo-reactive. Many infected individuals exhibit a response that is at some point between these two extremes and people who are currently hyper-reactive may subsequently become hyporeactive. The inflammatory reaction causes dermatitis (onchodermatitis) and intense pruritis. This results in severe scratching and consequent the risk of secondary bacterial infections. With time, the skin becomes thickened, loses its pigmentation and elasticity, and starts to crack. This is sometimes called lichenification although it is colloquially referred to as 'lizard skin' (Figure 11.3). It is not fatal but people suffering onchodermatitis often become isolated and develop feelings of shame and low self-esteem (Okoye and Onwuliri 2007). Consequently, where onchodermatitis is common, it can have considerable social and economic consequences.

Microfilariae damage cells whilst migrating through the eye but much of the pathology probably results from inflammatory reactions against the dead and dying parasites. Depending upon where the damage occurs, several pathologies develop including keratitis (inflammation of the cornea), chorioretinitis (inflammation of choroid and retina), iridocyclitis (inflammation of iris and ciliary body), and optic atrophy (loss of optic nerve fibres). In severe cases, the damage to the eyes results in blindness. Because the blackfly vectors of onchocerciasis breed in fast-flowing rivers, the disease commonly afflicts villagers who live close to these rivers. Until effective control of the blackfly vector was established, many villagers near to blackfly-infested rivers became blind at an early age and the condition was therefore known as 'river-blindness'.

Urticaria, or hives, refers to the sudden formation of a raised itchy rash on the skin surface. The rash forms in response to the release of histamine and other inflammatory mediators from

Figure 11.3 'Lizard skin' resulting from hyper-onchodermatitis in an 11-year-old boy. Note the areas of de-pigmentation on his lower legs and thickening on his thighs and penis. The child is starting to suffer from 'hanging groin' and his scrotum already hangs 20 cm from the pubis. His penis is showing evidence of elephantiasis and is 15 cm long. *Source:* Conor, D.H. et al., (1985), 7, 809–819/with permission of Oxford University Press.

cutaneous mast cells that cause blood vessels to dilate and plasma to leak out. Urticaria can develop for various reasons but is often associated with a response to allergens and infections. It can develop anywhere on the body and varies in size from small discrete sites to wide patches encompassing several centimetres. Urticaria is associated with a rise in circulatory immunoglobulin E (IgE), and this is also a characteristic of helminth infections. One might therefore expect that there would be good links between them. However, the evidence is often poor or contradictory (Kolkhir et al. 2016). Nevertheless, urticaria is mentioned in association with several human parasitic infections including *D. dendriticum* (Blatner et al. 2017) and *S. stercoralis* (Forrer et al. 2017).

11.2.10 Spleen

The spleen is part of the lymphatic system and represents the largest mass of lymphatic tissue in our body: in adults the spleen usually weighs 100–200 g. Underneath an outer layer of connective tissue, there are two types of tissue called 'white pulp' and 'red pulp', respectively. The white pulp consists mostly of lymphocytes and macrophages, which are involved in the trapping and presenting of antigens. The red pulp consists of venous cords and splenic cords within which defective and/or worn out circulating red blood cells, platelets, and macrophages are destroyed. In addition, the red pulp is a major storage site for blood platelets, and during foetal development, red blood cells are formed here. The spleen is therefore important for the normal functioning of the immune system and for the removal of damaged and potentially diseased cells from the circulation. Splenectomy is associated with a decreased ability to produce gamma-interferon and TNFα during infections. Consequently, if we have a splenectomy, we have an increased the risk of contracting opportunistic parasitic infections such as *Babesia divergens* and bacterial infections.

Enlargement of the spleen, splenomegaly, is a feature of many parasitic infections including malaria, leishmaniasis, toxoplasmosis, and schistosomiasis. Splenomegaly is often accompanied with hypersplenism in which large numbers of circulating red blood cells, white blood cells, and platelets are sequestered within the spleen. In addition, many of the red blood cells are prematurely destroyed within the red pulp. This results in a condition known as pancytopenia in which there is a decline in the number of circulating red blood cells, white blood cells, and platelets. Consequently, pancytopenia results in an increase in the plasma volume (because there are fewer circulating cells), anaemia, and there is an increased risk of bleeding. There is also an increased risk from infections with encapsulated bacteria such as *Haemophilus influenzae* and *Streptococcus pneumoniae*. Pancytopenia also causes hyperplasia in the bone marrow as the body attempts to replace the lost cells.

During malaria, our body attempts to remove infected red blood cells by destroying them in the spleen. Unfortunately, the process is not particularly selective and probably ten times as many uninfected cells are simultaneously destroyed. This self-destruction of healthy red blood cells contributes to the development of anaemia. The sequestering and destruction of red blood cells within the spleen is also thought to bring about the development of splenomegaly. However, there are also suggestions splenomegaly in malaria may also occur through a concomitant increase in haematopoiesis (red blood cell formation) within the spleen. In malaria endemic regions, splenomegaly is usually restricted to children. Visceral leishmaniasis (*Leishmania donovani*) causes the destruction of phagocytic cells, and this induces spleen hyperplasia because the organ attempts to rapidly replace the lost cells. This is undertaken to the detriment of the manufacture of red blood cells and leads to splenomegaly (enlarged spleen size) and hypersplenism (overactive spleen function). The resultant disruption of spleen function and development of pancytopenia (reduced numbers of red blood cells, white blood cells, and platelets in the circulation) explain why patients suffering from visceral leishmaniasis often die of secondary infections. Splenomegaly can occur in toxoplasmosis, although it is not common. It mostly occurs in children under 16 years old and probably results from a proliferation of lymphoid tissues. In the case of schistosomiasis, splenomegaly usually results from portal hypertension. However, any schistosome eggs that get swept to the spleen will induce granulation and fibrosis that contributes to the development of splenomegaly. The splenic vein takes blood from the spleen, the stomach, and pancreas and then joins the superior mesenteric vein to form the hepatic portal vein. Schistosome eggs in the liver cause fibrosis, which obstructs the movement of blood into the liver, and therefore there is a rise in blood pressure within the hepatic portal vein (portal hypertension). Although the blood pressure rises, the blood flow reduces or ceases (stasis). Portal hypertension induces hyperplasia within the spleen that manifests itself as splenomegaly and hypersplenism. Schistosomiasis is also associated with the development of thrombocytopenia – that is a reduction in the number of blood platelets and hence susceptibility to poor clotting and abnormal bleeding. This could be related to the development of hypersplenism associated with splenomegaly although there is also evidence that it arises because of an autoimmune reaction in which parasite antigens cross-react with platelet antigens.

11.3   Co-Infections and Pathogenesis

A recurring theme within this book is that parasites seldom act independently. In addition to interacting with other parasites, they also interact with microbes and viruses present within their environment. Interactions between microbes and parasites can influence the course of disease at many levels. It may involve interactions between the parasite and one or more pathogenic microbes or between the parasite and the host microbiome. In addition, horizontal gene transfer probably occurs regularly in both directions between microbes and parasites (Wijayawardena et al. 2013). The pathology caused by a parasite can increase susceptibility to other infectious agents and these in turn will affect the subsequent pathology caused by the parasite. Similarly, bacteria or viruses can facilitate the establishment of parasites and enable a parasite to cause more damage than it would do if acting on its own. For example, unless they receive effective treatment ~90% of people suffering from kala azar (*L. donovani*) will die – often from concurrent bacterial and other opportunistic infections that cause septicaemia and pneumonia. Over half of patients with visceral leishmaniasis acquire secondary microbial infections such as *Staphylococcus aureus*, *Pseudomonas aeruginosa,* and *Haemophilus influenzae* – often these are the result of nosocomial infections acquired whilst they are in hospital.

In sheep, infection with gastrointestinal nematodes is associated with an increased prevalence of mastitis (Kordalis et al. 2019). Similarly, acute fascioliasis increases their susceptibility to the gram-positive bacterium *Clostridium novyi* that causes a necrotic hepatitis – often known as 'black disease'. *Clostridium novyi* is a common soil bacterium and there are various types, some of which produce toxins. Black disease is caused by *C. novyi* type B that produces a toxin called α toxin. Sheep ingest the spores with soil-contaminated food, and these can subsequently pass through their gut epithelium. They are then picked up by macrophages and transported to the liver and other organs. *Clostridium novyi* requires anaerobic conditions to grow and these are provided by the necrotic lesions caused by the migrating flukes. It is possible that the metacercaria may also become contaminated with the bacteria. Alternatively, or additionally, bacteria within the sheep gut might attach to the tegument of the young flukes and then be transported by them to the liver. In cattle, fascioliasis is associated with an increased susceptibility to bacillary haemoglobinuria due to *Clostridium haemolyticum*. *Clostridium haemolyticum* is sometimes referred to as *C. novyi* type D but it is usually only recorded from sheep in experimental conditions. In a similar manner, during their migration *Ascaris* larvae can transport bacteria such as *E. coli* from the gut to the lungs where the bacteria contribute to the pathology (Adedeji et al. 1989).

There is some debate in the literature about the interaction between *F. hepatica* infections in cattle and bovine TB (bTB, *Mycobacterium bovis*). Some studies indicate that infection with *F. hepatica* reduces the sensitivity of the standard bTB test although a review by Howell et al. (2019) concluded that in most cases this is slight or there is no effect. There is stronger evidence of parasites affecting the way in which humans respond to *Mycobacterium tuberculosis* (Mtb). For example, our immune response against schistosome eggs seriously compromises that against Mtb, thereby facilitating its move from latency to disease (DiNardo et al. 2016). Similarly, the levels of Mtb-specific immunoglobulins M and G, along with several other Mtb immune markers, are reduced when the nematode *S. stercoralis* co-infects with Mtb. This could potentially compromise TB surveys and vaccine trials (Anuradha et al. 2017).

There are many examples of microbe-parasite interactions amongst the parasitic amoebae. For example, if our gut wall is damaged by pathogenic bacteria, then it becomes more vulnerable to invasion by *E. histolytica*. In addition, amoebae that ingest bacteria such as *Shigella dysenteriae* (a common cause of severe diarrhoea) are more likely to bind to and ingest gut epithelial cells. This is because the ingested bacteria stimulate the trophozoites to increase their production of a lectin (galactose/*N*-acetyl D-galactosamine-inhibitable adherence lectin – usually abbreviated to Gal/GalNAc) on the surface of their cell wall. This facilitates the amoebae in binding to gut epithelial cells (Verdon et al. 1992). In addition, the amoebae are also stimulated to produce more cysteine peptidase enzymes. These enzymes are released into the surrounding medium and bring about a localised increase in the concentration of pro-inflammatory cytokines. This host inflammatory response is probably important in facilitating the invasion of the parasite. Pathogenic enteric bacteria and *E. histolytica* are commonly found in poor communities that lack adequate sanitation and safe drinking water. Therefore, their interaction could be an important factor in whether a person develops amoebic dysentery and if so, the severity of the infection.

In humans, the gum disease periodontitis is cyclical in nature, with periodical flare-ups followed by remission. The main causative agent is the bacterium *Bacteroides gingivalis* although other microbes can also be involved. It has been suggested that *Entamoeba gingivalis* facilitates disease progression by transporting bacteria between sites of infection within the mouth. Although amoebae in general feed on bacteria, some of them, such as *Shigella dysenteriae* can survive within vacuoles in *E. histolytica* for prolonged periods. Possibly, a similar phenomenon occurs in *E. gingivalis*. Özüm et al. (2008) isolated *E. gingivalis* from a patient suffering from a brain abscess, along with the common oral bacteria *Eikenella corrodens* and *Prevotella* spp. Whether all three organisms arrived in the brain at the same time, or one was the initial cause of the abscess and the others arrived subsequently is difficult to say. Nevertheless, it highlights the close association of amoebae with bacteria in causing pathology. Necrotic periodontitis is a serious and painful disease that is often associated with HIV-1 infection. HIV-1 infection does not necessarily lead to a change in the oral bacterial flora but Lucht et al. (1998) found that 77% of HIV-infected patients with periodontal disease were infected with *E. gingivalis*. By contrast, they found that none of the HIV-positive individuals without periodontal disease were infected with *E. gingivalis*.

It is uncertain whether microbe associations contribute to the pathology of *Naegleria fowleri*, but bacterial symbiont populations live inside other *Naegleria* species. It is probable that *N. fowleri* (and other amoebae) facilitate the survival of the bacterium *Legionella pneumophila* (the causative agent of Legionnaires' Disease) (Dobrowsky et al. 2016). However, whether they influence one another's pathogenicity is currently unknown. Both pathogenic and non-pathogenic *Acanthamoeba* form associations with various bacterial endosymbionts and they also support several pathogenic species such as *Legionella* spp. and *Chlamydia* spp. Non-pathogenic species of *Acanthamoeba* therefore have the potential to act as transport hosts and contribute to the survival of pathogenic microbes. In addition, *Legionella* bacteria that have been in association with *Acanthamoeba* exhibit increased antibiotic resistance. Methicillin-resistant *Staphylococcus aureus* (MRSA) can be internalised by *Acanthamoeba*, replicate within them, and exhibit unchanged or increased antibiotic resistance afterwards (de Souza et al. 2017). Whether *Acanthamoeba* species are important in the spread of pathogenic microbes and the development of antimicrobial resistance is uncertain.

12

The Useful Parasite

CONTENTS

12.1   Introduction: The Goodness of Parasites?

Parasites are normally considered unwelcome guests whose presence is invariably harmful. Although, as we have seen in Chapter 1, defining and measuring harm is problematic, it can still appear perverse to consider parasites as beneficial. The term 'the goodness of parasites' was initially coined by David Lincicome whilst he was Professor of Zoology at Howard University, Washington, DC, USA. Lincicome and his co-workers performed numerous experiments using various protozoan and nematode parasites of rats and mice that are summarised in Lincicome (1971). He observed that the parasitised animals often performed better in terms of body weight gain, longevity, and certain physiological parameters than those that were parasite-free. He therefore felt that parasite–host relationships should not be considered one-way and that parasites could (sometimes) provide paybacks to their host. However, some of his conclusions were criticised for being based on animals fed very specific diets and low levels of parasitaemia. Nevertheless, it is increasingly apparent that parasites sometimes prove useful to their hosts and that low parasite burdens may help maintain a healthy immune system. In some cases, a parasite may even contribute to its host's survival during difficult environmental conditions. For example, the rickettsia *Anaplasma phagocytophilum* enhances the survival of its tick vector, *Ixodes scapularis*, by inducing the expression of a gene that codes to produce antifreeze glycoproteins (Neelakanta et al. 2010).

Consequently, infected ticks survive the harsh winters of the Northeast and Upper Mid-West of the United States of America better than those that remain uninfected. *Anaplasma phagocytophilum* is an intracellular parasite that infects humans, dogs, and various other mammals. It causes lameness and flu-like symptoms, and infections are sometimes mistaken for Lyme disease. Another instance of parasites enhancing the survival of their host is that of worker caste ants of *Temnothorax nylanderi* infected with the larval stage of the tapeworm *Anomotaenia brevis* (Beros et al. 2021). In this case, infected ants often live considerably longer than those that are uninfected. This is probably because the larvae induce changes in gene activity. Longevity is probably further facilitated by changes in the behaviour such that infected ants remain within their nest rather than venturing outside. These changes probably benefit the tapeworm because the definitive hosts are woodpeckers that drill into ant nests to feed on the ants within. Therefore, the longer an infected ant remains within its nest, the better the chances are that it will be eaten by a woodpecker.

Within this chapter, we will consider some of the ways in which parasites are useful to us. For example, some helminths have potential for the treatment of certain autoimmune diseases whilst blowfly larvae and leeches are commonly employed in biosurgery. Parasites are also potentially useful sources of novel pharmacologically active substances. Within agriculture, parasites are used in the biological control of pest invertebrates and plants. Parasites have even been implicated in criminal investigations and some workers have seen fit to coin the term 'forensic parasitology'. Thankfully, parasites have not proved particularly useful as biological warfare agents although that is not for want of people trying.

12.2   The Importance of Parasites for the Maintenance of a Healthy Immune System

In industrialised countries, there has been a rapid rise in the number of people suffering from allergic and autoimmune conditions (e.g., Type 1 diabetes) since the 1960s and 1970s. For example, about 20% of children living in these regions currently suffer from asthma, allergic rhinitis (e.g., hay fever), or eczema. There is every indication that the prevalence of these conditions will continue to increase into the future. The reasons for these increases remain unclear but almost certainly involve many interacting intrinsic and extrinsic factors. For example, numerous genetic factors influence our susceptibility to developing allergies and autoimmune diseases and the way these are expressed (Ceccarelli et al. 2017). However, these genetic factors do not act in isolation and environmental factors influence their expression. For example, people who migrate from Africa and Asia to live in developed countries have an increased susceptibility to autoimmune diseases. Therefore, something about the environment of developed countries appears to make people susceptible to allergies and autoimmune diseases. For many years, the hygiene hypothesis was thought to provide a sensible explanation for this. The hygiene hypothesis was initially invoked to account for why children in the United Kingdom brought up in large families have a lower tendency to suffer from eczema and hay fever. It was suggested that in small family units there is less opportunity for transmission of infectious organisms. Furthermore, in developed countries, improvements in sanitation, better standards of living, and the widespread use of antibiotics have dramatically reduced the transmission of infectious diseases. Therefore, children living in small family units in clean homes do not receive the levels of immunological challenge from infectious organisms that used to be commonplace. The hygiene hypothesis proposes that when the body's immune system is no longer challenged by infectious organisms on a regular basis, it is more likely to become activated by inappropriate stimuli, and therefore there is an increased susceptibility to allergies and autoimmune inflammatory diseases. Support for this proposal comes from the observation that autoimmune conditions such as multiple sclerosis (MS) and type 1 diabetes are rare among poor people living in Africa and Asia but common in industrialised countries. However, there are epidemiological discrepancies, and many scientists now question the validity of the hygiene hypothesis in its original form. Instead, a refinement called the 'old friends' hypothesis is considered to provide a better explanation. This proposal contends that it is not reduced exposure to infectious pathogenic organisms that is responsible for the increasing prevalence of autoimmune diseases. Instead, they result from a failure to acquire and maintain healthy microbiomes, particularly within our gut (Scudellari 2017). Precisely, how our microbiomes reduce our likelihood of developing autoimmune diseases and allergies is uncertain but will undoubtedly involve many interacting components of the immune system.

Autoimmune diseases and allergies can be debilitating, and in extreme cases, they cause lifelong disability and may prove fatal. They are very difficult to manage, and their treatment often involves drugs that have harmful side-effects. There is therefore an urgent need for new approaches to treating these conditions. Intentionally infecting people with parasitic helminths appears to have potential for alleviating some autoimmune diseases and allergic reactions. The concept arose from the observation that there is an inverse relationship between the prevalence of gastrointestinal helminth infections and autoimmune diseases such as Crohn's disease and ulcerative colitis (see later).

We naturally produce self-reactive immune cells such as self-reactive T cells and B cells that make autoantibodies. Most of these self-reactive cells are usually either eliminated or suppressed by physiological control mechanisms. For example, the majority of developing self-reactive T cells are destroyed in the thymus. However, some survive and differentiate into specialised regulatory cells that are important in normal immune function. The production of self-reactive immune cells should therefore not be considered as a pathological event. Autoimmune diseases arise when the control mechanisms governing the production of self-reactive immune cells start to weaken and they attack the body's organs/tissues. Some autoimmune conditions are associated with an increase in the production of T-helper 1 (Th1) cell cytokines (e.g., gamma interferon [INF-γ]) and a reduction in the levels of cytokines produced by T-helper 2 (Th2) cells (e.g., interleukins 4, 5, and 13 [IL-4, IL-5, IL-13]). Helminth infections appear to be particularly effective at reducing the pathology associated with these autoimmune diseases.

Allergic reactions are those in which the body mounts an excessive immune response to a harmless antigen (allergen), such as pollen, shellfish, or peanuts. There are two classes of allergic reaction: IgE-mediated and non-IgE-mediated. The IgE-mediated allergic response is the most common and is characterised as a Type 1 hypersensitivity reaction. In people who are susceptible to developing an allergic reaction, their first exposure to an allergen does not cause a harmful response. However, it does result in CD4+ T-helper 2 cells stimulating B-cells to produce IgE antibodies against the allergen. These antibodies bind to mast cells and basophils and thereby sensitise them. Consequently, the next time the allergen enters the body, it binds to the IgE, and this stimulates the mast cells and basophils to release histamine and other pharmacologically active chemicals that cause vasodilation, oedema, itching etc. There are two types of non-IgE-mediated response: Type II (cytotoxic-mediated response) and Type III (immunocomplex response). Both responses are mediated by IgM and IgG, and they take longer to be expressed than the Type 1 hypersensitivity response.

Helminths interact with their host's immune system in many ways. For example, their feeding and movement causes physical damage and thereby induces an inflammatory reaction. Similarly, many helminth secretions and excretions are immunogenic. Helminths stimulate a predominantly Th2 type cell-mediated immune response in which there is increased release of the cytokines IL-4, IL-5, IL-10, and IL-13 into the circulation. IL-4, -5, and -13 are involved in the activation, growth, and differentiation of eosinophils and also stimulate the production of immunoglobulin E (IgE) by plasma cells. By contrast, IL-10 tends to have an anti-inflammatory effect although it also stimulates the production and survival of B cells. There is a negative feedback relationship between the Th2 and Th1 responses and IL-4 formed during the Th2 response brings about a reduction in the Th1 response. IL-10 also reduces the Th1 response by indirectly inhibiting the production of IFN-γ and several other cytokines. There is, however, an antagonistic relationship between IL-10 and IL-4, so the two cytokines are not acting in concert.

In addition to stimulating an immune response, gastrointestinal helminth parasites directly or indirectly influence the gut microbial flora through their feeding and secretions/excretions. This may in turn influence the host's immunity. For example, disruptions to the gut microbiome (gut dysbiosis) have been linked to the development of autoimmune conditions such as Type-1 diabetes and inflammatory bowel disease, as well as allergic reactions (Lazar et al. 2018). Substances with antimicrobial activity are present in the excretory/secretory products of many gut helminths, including *Trichuris suis* (Abner et al. 2001) and *Ascaris suum* (Midha et al. 2018). Similarly, gastrointestinal helminths can influence the release of antimicrobial peptides by their host into the gut (Zaiss and Harris 2016). Furthermore, some parasitic nematodes can sense and selectively manipulate the composition of the surrounding gut microbiome (Rausch et al. 2018). One might therefore expect that the removal of gastrointestinal helminths using anthelmintics would influence the composition of the gut microbiome. The evidence, however, is variable. A study of rural Kenyan villagers found that removal of the hookworm *Necator americanus* resulted in an increase in the relative abundance of *Clostridiales* and a decrease in the abundance of *Enterobacteriales* but a similar change was not associated with the removal of *Ascaris lumbricoides* (Easton et al. 2019). Whether this relates to the release of blood into the gastrointestinal tract by the hookworms and/or other factors is uncertain. Other studies using different helminth species have found them to have varying or no effect on the composition of the gut microbiome (Loke and Lim 2015; Schneeberger et al. 2018). The beneficial effect that is sometimes observed between helminth infection and certain disease conditions is therefore helminth/case-dependent and not one of simply promoting a Th2 response and thereby reducing the Th1 response.

Allergies that initiate the Type 1 hypersensitivity reaction are typified by an overactive Th2 response in which eosinophils contribute to the inflammation by releasing toxic chemicals that include eosinophil-derived neurotoxin (EDN). Eosinophil-derived neurotoxin has ribonuclease and antiviral activity and can be used as a physiological indicator of the severity of asthma. Helminth infections cause eosinophilia in which the proportion of eosinophils in the general circulation can increase from 2 to 5% of the white blood cell population to as high as 40% (Huang and Appleton 2016). However, despite this, some helminth infections appear to protect against the development of allergies. For example, mice infected with *Schistosoma mansoni* gain protection against allergic airway inflammation (Obieglo et al. 2018). This is perhaps surprising because *S. mansoni* eggs induce an exceptionally strong Th-2 response. An alternative explanation for the negative association between helminth infection and predisposition to allergies could be that those who are atopic are more resistant to helminths than the general population. Atopy is an IgE-mediated hypersensitivity reaction which may occur in parts of the body not in contact with the allergen. The allergens induce reactions in the skin (atopic eczema/atopic dermatitis), eyes (allergic conjunctivitis), nose (allergic rhinitis), and lungs (asthma). Indeed, some genes appear to be involved in both the control of atopy and susceptibility to helminth infection. However, further work needs doing using more genes and various helminth infections.

12.2.1 Type 1 Diabetes Mellitus

Sometimes known as 'insulin-dependent diabetes mellitus (IDDM)' and 'juvenile diabetes', type 1 diabetes mellitus results from the autoimmune destruction of the insulin-producing *beta* (β) cells in the Islets of Langerhans in the pancreas. What precipitates the autoimmune response is uncertain, but several genes and environmental factors are implicated. Without insulin-replacement therapy, type 1 diabetes is fatal. Initially, it was thought that the Th1 inflammatory response caused most of the pathology and the Th2 response was protective. However, although an excessive Th1 response is clearly harmful, it is the balance between the Th1 and Th2 responses that determines the onset and progression of the disease.

The incidence of type 1 diabetes has increased in recent years in many industrialised societies, and some workers suggest that this may in some way be associated with a corresponding decline in helminth infections. Experiments with non-obese diabetic (NOD) mice indicate that various helminths or their products can protect them from developing spontaneous type 1 diabetes (Berbudi et al. 2016). The reason for this protective effect is not solely because helminths cause a predominantly Th2 response and a reduced Th1 response. Neither the pathophysiology of type 1 diabetes nor the immune responses to helminth infections are quite that simple, and many other factors are also involved. For example, helminth infections also affect the population and function of various other components of the immune system such as IL-10, regulatory T cells (Tregs), invariant natural killer cells, and alternatively activated macrophages.

Although lymphocytes produce IL-10 during the Th2 response, it is predominantly formed by monocytes and certain other immune cells including mast cells and a subset of the Tregs. IL-10 acts as an immunomodulator – that is, it modulates the immune response by influencing the activity of other components of the immune system. Its effects are predominantly suppressive, and it prevents inflammatory responses from 'getting out of hand'. However, it can also have a stimulatory role in the immune response against some infectious agents. In mouse models of type 1 diabetes, IL-10 has a protective effect and suppresses the development of the disease. Some helminth infections induce IL-10, and this may help prevent the development of autoimmune conditions including type 1 diabetes. For example, children infected with *Schistosoma haematobium* produce elevated levels of IL-10, and this may be a reason why they are less likely to develop atopy (van den Biggelaar et al. 2000). Up-regulation of IL-10 also occurs in filarial nematode infections, but there is less consistent evidence for it occurring in response to gastrointestinal nematodes.

Like IL-10, Tregs act as a 'brake' preventing the immune system from becoming over-activated and initiating autoimmune reactions. The population and function of Tregs is therefore important in the development of conditions such as type 1 diabetes and asthma. In mouse model systems, the filarial nematode *Litomosoides sigmodontis* induces Tregs, and these may contribute to the parasite preventing the development of type 1 diabetes and allergen sensitivity.

Natural killer T cells (NKT) are distinct from natural killer cells and co-express both a specific (αβ) T-cell receptor and some of the molecular markers normally found on natural killer cells. Invariant natural killer T cells (iNKT) are a subset of NKT cells. When stimulated they produce a variety of immunologically active substances, such as IL-4 and IFN-γ, and reductions in their numbers link to the development of several autoimmune conditions – again including type 1 diabetes. In mice, *Schistosoma mansoni* infection induces iNKT cell activity as does the injection of *S. mansoni* antigens into non-obese diabetic mice. The *S. mansoni* antigens protect these mice from developing diabetes, and this may, at least in part, derive from increased iNKT cell activity.

Macrophages are classically activated by gamma interferon (INFγ) and other Th1-type signals although some are activated by Th2 signals such as IL-4 and IL-13. These latter macrophages are referred to as 'alternatively activated macrophages' (AAMΦs), and they play an important role in the development of allergies and autoimmune disorders. Parasitic helminths induce a Th2 response and the production of IL-4 and IL-13. Consequently, AAMΦs are also produced as part of the immune reaction against parasitic helminths. An elevated Th1 response is associated with the development of type 1 diabetes and the raised levels of IFNγ brings about an increase in the numbers of classically activated macrophages, which then contribute to the pathology. Helminth infections bring about elevated levels of AAMΦs, and these then suppress the Th1 response and can thereby help protect against the development of type 1 diabetes mellitus. For example, Espinoza-Jiménez et al. (2017) found that mice infected with *Taenia crassiceps* or treated with antigens isolated from them are protected from developing type 1 diabetes. They ascribed this as, at least in part, to the induction of raised levels of AAMΦs. Increasing the numbers of AAMΦs is potentially helpful in ameliorating other pathologies. For example, Terrazas et al. (2017) considered the increased numbers of AAMΦs engendered by *T. crassiceps* antigens as responsible for protecting mice from experimental autoimmune encephalomyelitis.

12.2.2 Type 2 Diabetes

Type 2 diabetes is characterised by hyperglycaemia. Its incidence has increased remarkably since the 1960s in both developed and developing countries, and it is currently a major cause of morbidity and mortality. Type 2 diabetes is also increasingly common among pet domesticated animals (e.g., cats, dogs, rabbits) and those kept in zoos (e.g., primates, elephants). The extent to which diabetes occurs within wild animals is uncertain because a diabetic animal is unlikely to survive for long under natural conditions.

After we eat a meal, glucose enters our bloodstream and the rise in glucose concentration stimulates the β-cells in our pancreas to secrete insulin. This reduces the endogenous secretion of glucose by our liver, increases the uptake of glucose by cells, and increases the conversion of glucose to glycogen. Consequently, the glucose concentration within our blood remains stable, and this is important for many biological processes. If the normal level circulating insulin (for a healthy individual) fails to exert its expected physiological effect, then we are said to be expressing insulin resistance. This occurs when our muscle, liver, and fat cells exhibit a reduced sensitivity to insulin. Hyperglycaemia commences once the β-cells are no longer able to secrete sufficient insulin to prevent the blood glucose concentration rising to abnormally high levels. Hyperglycaemia may also occur through dysfunctional β-cells exhibiting reduced sensitivity to circulating glucose levels and therefore producing insufficient insulin.

Numerous interacting genetic and societal factors contribute to the risk of suffering from type 2 diabetes. These include obesity, lack of exercise, excessive consumption of high-energy diets (junk food), genetic risk factors, and smoking. The underlying pathology is associated with inflammatory responses. For example, a high fat diet causes changes to the composition of the gut microbial flora, and this can lead to dysbiosis in which the body mounts an inflammatory reaction against the gut microbes. It is possible that this may in turn lead to metabolic deregulation and contribute to insulin resistance and the development of type 2 diabetes (Sharma and Tripathi 2019). Organisms that induce changes to the composition of the gut microbiome may therefore be important in the development and treatment of type 2 diabetes.

Various pathogenic bacteria and viruses, including SARS-COVID-19, have been implicated the development of type 2 diabetes, but these are probably of importance in a minority of cases. By contrast, people suffering from type 2 diabetes have a vastly increased susceptibility to microbial and fungal pathogens. The interactions between protozoan, helminth, and arthropod parasites and type 2 diabetes are poorly understood (Berbudi et al. 2016). Some reports suggest that helminth infections are less common among those suffering from type 2 diabetes. For example, in India, Aravindhan et al. (2010) found a lower prevalence of lymphatic filariasis among people suffering from type 2 diabetes. Undoubtedly, the immunological and physiological changes associated with type 2 diabetes will influence the ability of parasites to establish an infection. However, one must also consider other factors too. Type 2 diabetes is commonly associated with obesity and obese individuals are usually less active. They are therefore less likely to venture far from home or engage in employments that would expose them to many parasite infections. Nevertheless, experimental studies using rodent models indicate that the type 2 immune response and IL-10 induced by helminth infections improve insulin sensitivity and glucose tolerance and reduce the formation of fat cells (De Ruiter et al. 2017). An initial pilot study by Rajamanickam et al. (2020) involving humans infected with *Strongyloides stercoralis* also indicates that helminth infections may protect against type 2 diabetes through altering the balance of pro-inflammatory cytokines. Preventing people from becoming or remaining obese would certainly help reduce the incidence of type 2 diabetes, and helminths might be helpful in this situation too. For example, in mouse-models, infection with the nematode *Heligmosomoides polygyrus* protects mice against the developing obesity when they are fed a high-fat diet (wen Su et al. 2018). This is thought to be, at least partially, through the worms inducing increased numbers of alternatively activated macrophages.

12.2.3 Irritable Bowel Syndrome

Irritable bowel syndrome (IBS) is a chronic condition involving altered bowel habits in which the patient experiences abdominal pain and constipation or diarrhoea. The prevalence varies considerably between countries. However, variations in diagnostic criteria and methodological approaches make it difficult to compare studies. In addition, many people with IBS do not seek medical help or do not receive an accurate diagnosis if they do. Not surprisingly, even within a country, there can be big differences between estimates of the lowest and highest prevalence: for example, France 1.1-4.7; UK 6.2-21.7; USA 3-20.4 (Canavan et al. 2014). There is limited accurate information on the prevalence of IBS in developing countries, but the general impression is that it is probably increasing, as people adopt the 'western lifestyle'.

Some doctors consider IBS to be a psychosomatic disorder. However, there is strong evidence that it results from changes in the composition of the gut microbiome and associated inflammatory responses (Menees and Chey 2018). Some cases of IBS arise from taking antibiotics and it is possible that gastrointestinal infections that change the composition of the gut flora could induce dysbiosis. In the latter situation, the condition is known as post-infectious IBS. The strongest evidence for post-infectious IBS relates to bacterial and viral infections (Shariati et al. 2019), but protozoan and helminth parasites may also be involved. In Norway, people infected with *Giardia duodenalis* are reportedly more likely to subsequently develop diarrhoea-associated IBS (Nakao et al. 2017). However, in Asia, where *Giardia* is much more common among the general population, evidence of a link between it and IBS is less clear (Ghoshal et al. 2010). The support for a relationship between *Entamoeba histolytica* infection and the development of IBS is also uncertain. This is partly because many studies have not distinguished pathogenic *E. histolytica* from *Entamoeba dispar* and *Entamoeba moshkovskii*.

The gut protozoan *Blastocystis* spp. has been linked to post-infectious IBS but much of the evidence is contradictory (Cifre et al. 2018). This probably relates, at least in part, to problems with identification and as a result the taxonomy of *Blastocystis* spp. is uncertain. Although initially identified as *Blastocystis hominis*, it is now recognised that there are numerous genotypes and many of these are zoonotic. The organism is currently classified as *Blastocystis* spp. and divided into various subtypes (Tan 2008). Even within a subtype, it can express numerous morphological forms. It has a complicated life cycle and its cysts are transmitted via faecal contamination. It has a worldwide distribution although it is more prevalent in developing countries where there is poor sanitation. Accounts of its pathogenicity vary enormously with various claims of it being pathogenic, commensal, or an opportunist that becomes parasitic if the circumstances are right. This is almost certainly linked to differences between the *Blastocystis* subtypes although some variation probably arises from the investigators using different criteria when analysing their data. Interestingly, infection with *Blastocystis* spp. tends to be associated with a healthy gut microbiome (Audebert et al. 2016). This suggests that if it is associated with IBS, it is not acting through inducing changes to the gut microbial flora.

The role of helminth parasites in the development of IBS is uncertain. As a rule, the prevalence of IBS tends to be low in countries in which gut helminth infections are common. This is even though these countries also tend to have high levels of infections with *Giardia* and the many bacteria and viruses that cause gastrointestinal upsets. It is therefore, at first sight, surprising that people living in developing countries do not have high prevalence of post-infection IBS. This has led to suggestions that gut helminth infections might protect against the development of IBS – provided the numbers of worms remain low. This protection might result from the helminths inducing a predominantly Th2 type immune response within the intestinal mucosa, which prevents the development of a chronic inflammatory response and/or through their interactions with the gut microbiome.

12.2.4 Inflammatory Bowel Disease

Because their names are so similar irritable bowel syndrome (IBS) and inflammatory bowel disease (IBD) are often conflated, but they are actually two different conditions. IBD is a generic term for a group of chronic inflammatory diseases of which Crohn's disease and ulcerative colitis are the principal forms. Crohn's disease is an autoimmune condition and can affect all parts of the gut from the mouth to the anus, although it is usually found in the region between the end of the small intestine and the beginning of the large intestine. The pathogenesis of ulcerative colitis is less certain, and the disease is restricted to the colon and rectum. Both diseases probably arise from inflammatory responses to the gut microbial flora rather than a specific infectious agent.

Crohn's disease is typified by an elevated Th1 type inflammatory response in which excessive amounts of IFNγ and tumour necrosis factor alpha (TNF-α) are produced. By contrast, elevated levels of interleukin 8 (IL-8) in the colon mucosa are implicated in the development of ulcerative colitis. IL-8 is an inflammatory cytokine produced by circulating macrophages and several other cell types but not by T lymphocytes. It has several functions including the attraction of migratory immune cells and the activation of neutrophil granulocytes. IL-10 probably also has a key role to play in the development of ulcerative colitis although exactly how is uncertain. For example, in patients with ulcerative colitis, the levels of IL-10 may be either reduced or considerably increased depending upon the stage of the disease (Tatiya-aphiradee et al. 2018).

Both Crohn's disease and ulcerative colitis are increasingly common in developed countries, but they are relatively rare in developing nations. Whether this relates to the higher prevalence of helminth diseases in poorer countries is uncertain. In rodent models of inflammatory bowel disease, rats or mice given an intra-rectal injection of trinitrobenzene sulphonic acid (TNBS) develop an inflammation in their colon (colitis) that results from a Th1 inflammatory response. However, if the rodents are infected with the nematodes *Trichinella spiralis* or *Heligmosomoides polygyrus*, the tapeworm *Hymenolepis diminuta*, or the trematode *Schistosoma mansoni*, then they are protected from developing TNBS-induced colitis (Pomajbíková et al. 2018; Weinstock and Elliott 2009). This may relate to the Th2 response against the helminths causing a down-regulation of the Th1 response. However, the pathogenesis of IBD is linked to changes in the composition of the gut microbiome (Zhang et al. 2017). Therefore, helminth infections may alternatively/additionally influence the course of IBD by affecting the composition of the gut microbiome.

12.3 The Use of Parasites to Treat Medical Conditions

The fact that one disease can influence the outcome of another has been known for millennia. Although the onset of a second affliction is seldom good news for the patient, there are instances in which it can prove beneficial. Rufus of Ephesus writing in either the first century BC or the first or early second century AD (the dates aren't certain) stated that periodic fevers (by which he was probably referring to quartan malaria) provided an excellent cure for epilepsy and melancholia. This has sometimes been cited as an early observation of the potential of malaria therapy for the treatment of mental illnesses (see later). However, he also felt quartan fever was beneficial for asthma, tetany (muscular spasms rather than the disease tetanus), and certain skin diseases. The linking factor is that they were all generally considered to be 'autumn diseases' and caused by an imbalance in the levels of 'black bile' (Major 1954).

Perhaps the most widely known story of parasites being employed for medical purposes is that of unscrupulous 'quack doctors' selling diet pills that contain tapeworm eggs. Indeed, the web is still awash with rumours, horror stories, and concerned advice about the so-called tapeworm diet. The reality is that consuming tapeworm eggs would, except for *Hymenolepis* (*Vampirolepis*) *nana*, not give rise to an adult tapeworm in humans. This is because after hatching the egg of a cestode releases a hexacanth embryo that penetrates the gut of the intermediate host and encysts elsewhere in the body; the larval cestode does not develop into an adult worm until it is consumed by the definitive host (Chapter 5). Cestode eggs do not therefore as a rule hatch in the gut of their definitive host. The only exceptions (of medical importance) to this are *Taenia solium* and *H. nana*. In the case of *T. solium*, we can act as both an intermediate host and the definitive host. However, if we consume *T. solium* eggs, these will cause us to develop cysticercosis and not an infection with adult tapeworms. Indeed, cysticerci may develop in a dangerous place (such as the brain), but this would not necessarily help us to lose weight! *H. nana* infects rodents and humans and is unusual in not requiring an intermediate host to complete its lifecycle – although tenebrionid beetles can act as intermediate hosts. After an egg of *H. nana* hatches in our gut, or that of a rodent, its hexacanth embryo burrows into one of the villi lining the small intestine and transforms into a cysticercoid. After a few days, the scolex everts and the parasite matures into an adult tapeworm. *Hymenolepis nana* is, however, a small tapeworm averaging only 4 cm in length and 1 mm in width. Although it is common in children in many parts of the world, it does not usually cause any problems unless the worm burden is exceptionally high – so it would not be much use as a diet aid. It has, however, been proposed as potentially useful in helminth therapy for the treatment of autoimmune conditions.

12.3.1 Helminth Therapy

Autoimmune diseases are notoriously difficult to treat using conventional medicine, and some of them condemn the patient to a lifetime of pain and disability. These conditions predominantly afflict people living in the developed world, and therefore, there is a huge market for an effective treatment. Epidemiological data and experimental studies using rodent models indicate that parasitic helminth infections may protect against the development of certain autoimmune conditions (Sobotková et al. 2019). Consequently, it may prove beneficial to intentionally infect a patient with

Table 12.1 Potential advantages and problems associated with helminth therapy.

Potential advantages	Problems
Effective against conditions that are difficult to treat with conventional medicines	Ethical considerations
Broad spectrum of disease targets	Lack of large double-blind clinical trials to prove effectiveness
Long lasting control	Could become pathogenic
Worms can be easily and cheaply removed using conventional anthelmintics should concerns arise	Potential to infect others
Parasites might be genetically engineered to improve effectiveness/render them infertile	Acceptability to patients/governments
Simple to administer	Logistical difficulties of sourcing sufficient eggs/larvae for large-scale studies
Cost effective	Cost effective?

helminth parasites – so called, helminth therapy, or to treat them with chemicals derived from them (Table 12.1). Even if this does not cure their condition, it might at least alleviate suffering.

Helminth therapy is usually discussed in relation to IBS and IBD, and to a lesser extent with multiple sclerosis (MS), and allergies (Maizels 2016). However, its potential is much more widespread. It is increasingly apparent that the gut microbiome plays an important role in the immunopathology of many diseases. For example, disturbances to the composition of the gut microbiome are a key factor in triggering the autoimmune demyelination associated with MS (Calvo-Barreiro et al. 2018). Helminths almost certainly influence the immune system, at least in part, through their interactions with the gut microbiome. Consequently, helminth therapy may prove effective for the treatment of many other diseases associated with immune dysfunction. For example, neuropsychiatric diseases such as schizophrenia and autism spectrum disorders that result from neuroinflammation (Abdoli and Ardakani 2020). In support of this, a proof-of-concept study by Hollander et al., (2020) indicated that infecting adults suffering from autism spectrum disorder with *Trichuris suis* led to improvements in some behavioural traits (e.g., repetitive behaviour) but not all (e.g., social-communication).

Intentionally infecting a patient with parasites poses ethical considerations. A fundamental feature of the Hippocratic Oath is that doctors should do no harm and it could be argued that infecting someone with parasites is precisely doing that. Furthermore, one could also say that medical professionals should be in the business of curing diseases and not spreading them. However, a discussion of medical ethics is beyond the scope of this book. On a practical level, once an infection is established, there is a risk of it becoming pathogenic or transmitting to other people. Patients who are immunocompromised through disease (e.g., AIDS) or drug treatment (e.g., following organ transplant or cancer therapy) could be at greater risk of adverse reactions if they became infected either directly or indirectly (i.e., they contracted the parasites from someone undergoing helminth therapy). Indeed, the supportive evidence from human clinical trials using live parasites is limited (Harnett and Harnett 2017), and there is a risk of them exacerbating autoimmune conditions (Briggs et al. 2016).

Because of the concerns listed above, some workers feel that it would be safer to identify the components of the parasites that confer protection against autoimmune diseases or the precise mechanisms by which they provide protection. Patients could then be treated with specific drugs whose safety and pharmaceutical properties are known and their dosages controlled. Indeed, there is experimental evidence from mouse models that treatment with helminth excretory/secretory molecules is effective. For example, the glycoprotein ES-62 derived from the filarial nematode *Acanthocheilonema viteae* restores the normal gut microbiome and reduces the pathology associated with various autoimmune diseases (e.g., inflammatory arthritis) and allergic reactions (e.g., contact dermatitis) in mouse models (Doonan et al. 2019). Nevertheless, the counterargument is that helminth parasites have complex immunological relationships with their hosts. Therefore, it is unlikely that any protective effect will be conferred by just one or a few helminth-derived substances. In addition, auto-immune diseases are, by their nature, long-lasting conditions that one can control but not cure. Parasites are effectively living drug pumps that deliver a regular dose of chemicals that is more convenient for the patient than taking pills or receiving injections. Furthermore, if the helminths exert their effects through influencing the gut microbiome, they will probably need to be physically present for a prolonged time. Although many people dislike the idea of worms living inside them, those suffering from painful conditions that are refractory to traditional pharmaceutical drugs are often willing to consider any treatment that offers them some relief. Nevertheless, Muslims and Jews are unlikely to accept the use of parasites such as *T. suis* that have been sourced from pigs. In addition, some people object to the use of animals to develop pharmaceutical drugs and would therefore be unwilling to accept helminth therapy. An additional problem limiting the acceptance of helminth therapy is that so far, most studies have been small scale and their results are variable. Consequently, some scientists remain sceptical of the usefulness of infecting people with living parasites (Maizels, 2020).

To date, most work on helminth therapy has been done using the whipworm *Trichuris suis* and the hookworm *N. americanus*. Both nematodes can cause serious pathology but low worm burdens in otherwise healthy, well-nourished individuals probably cause little damage. There are also a few studies that have employed the human whipworm, *Trichuris trichiura,* and the tapeworm *Hymenolepis diminuta*. As Sobotková et al. (2019) state, these species appear to have been chosen by happenchance and there may be more effective parasites that could be used.

Trichuris suis is a common parasite of pigs and occurs throughout the world. Although it can infect humans, there are contradictory reports on whether it achieves sexual maturity and reproduces in us. Nevertheless, molecular studies on *Trichuris* eggs recovered from human faecal samples in Thailand suggests that natural infections with *T. suis* occur, and it can complete its life cycle in us (Phosuk et al. 2018). Pigs, and humans, become infected by ingesting *T. suis* eggs that contain the infective larvae. At 34 °C, it takes 19 days for an egg passed in pig faeces to develop to the infective stage, and at 20 °C, this increases to 102 days. Consequently, even if someone harboured sexually mature *T. suis*, the chances of them re-infecting themselves or somebody else (or even a pig) are limited provided basic hygiene is practised. In the closely related *Trichuris muris*, hatching is stimulated by an interaction between specific receptors on the eggshell surface and *Escherichia coli* and certain other bacteria in the small intestine (Hayes et al. 2010) and it is likely that a similar interaction occurs in other *Trichuris* species.

After hatching in their normal pig host, *T. suis* larvae penetrate the intestinal mucosa where they remain for several days. Subsequently, the larvae move down the intestine to the caecum where they mature into adults. Some of the literature states that *T. suis* is non-invasive, but this is not strictly true. The larvae of *T. suis* invade the intestinal mucosa and in pigs the adult worms burrow their anterior region into the mucosa of the large intestine. Van Kruiningen and West (2005) raised concerns that the larvae might migrate from the intestinal mucosa and cause pathology elsewhere in the body, but there are no reports of this occurring.

There have been several clinical trials of *T. suis* as a treatment for IBD, including some that used double-blind, randomised, placebo-controlled studies, and the results are promising (Sobotková et al. 2019). However, trials for its use to treat allergic rhinitis have proved disappointing (Croft et al. 2012). There is limited information on the use of *T. suis* to treat MS. This is partly because the nature of the commonest form of the disease, relapsing remitting multiple sclerosis (RR-MS), makes it difficult to assess the effectiveness of treatment regimes. As its name indicates, RR-MS patients naturally experience periods of illness followed by periods of remission. It can therefore be difficult to ascertain whether a treatment is efficacious, or the patient is naturally entering a period of remission. Nevertheless, preliminary studies indicate that it is safe and may prove useful (Fleming et al. 2019).

Necator americanus is a common hookworm parasite that afflicts many people in temperate and tropical regions of the world. It is particularly common in North and South America. It also infects dogs and pigs, but we are its principal host. The adult worms attach themselves to the mucosa of the small intestine using their cutting plates and feed on blood by slicing through the underlying blood vessels. Heavy infestations can result in anaemia, but small numbers cause little damage. Its life cycle resembles that of other hookworms (Chapter 6). Because we are a natural host for *N. americanus*, once an infection establishes in us it can last for several years and during that time, we regularly pass viable eggs into the environment. However, the eggs are not the infective stage and provided we undertake basic hygiene there should not be a risk of self-re-infection or transmission to others.

Because of logistical problems, to date, there have only been a few small-scale trials using *N. americanus*. There have been positive reports for the treatment of IBD (Croese et al. 2006), and it has proved protective against asthma in animal models although the results from human trials have been less impressive (Feary et al. 2010).

12.3.1.1 Helminth Therapy in Practice

In Europe and the USA, several hundred people with IBD have now been treated with *T. suis* eggs and the outcomes have been mostly positive (Reddy and Fried 2009, 2007; Summers et al. 2005). The eggs are sourced from pathogen-free pigs to reduce the likelihood of co-transmitting other diseases. Because *T. suis* does not usually survive in our intestine for more than a few weeks, patients are typically given doses of 2,500 eggs every 3 weeks for several months (e.g., 24 weeks) during which their condition is monitored. If there are concerns about the parasites establishing a permanent infection or pathogenicity, they can be easily removed using conventional anthelmintic drugs, such as albendazole. There is still a need for further double-blind studies, the determination of the optimal treatment regime, and the patient factors that affect the likely success of the treatment. In addition, because *T. suis* is a cosmopolitan species, it is probable that genetic differences exist between populations. It would therefore be interesting to know whether there are strain differences that vary in the level of protection they confer. The treatment requires the rearing of pigs that are free of other pathogens, the eggs then need to be isolated and rendered sterile (whilst retaining their viability), and the patient needs to re-establish their infection at regular intervals. Consequently, the method is not particularly cheap although the same is true for many of the drugs prescribed to treat autoimmune diseases. For example, in the USA, most of the drugs currently used to treat MS cost over US$ 70,000 patient^{-1}year^{-1}.

There are still relatively few clinical trials of *N. americanus* to treat people with autoimmune conditions. To date, these studies have used eggs derived from people who are already infected with the parasite. Some commentators suggest that there is therefore a theoretical risk of transmission of viral infections, although exactly which viruses has not been stated and there is no evidence that it has occurred. A greater problem is securing sufficient eggs to undertake large-scale studies. The eggs are allowed to hatch and once the larvae have developed to the infective stage they are applied in a patch to the patient's skin and permitted to invade or inoculated under the skin surface. There is often a localised reaction at the site of invasion and some patients report abdominal pain when the worms initially establish themselves in the gut. Successful colonisation is indicated by finding hookworm eggs in the patient's faeces and the number of eggs gives a crude indication of the worm burden. Mortimer et al. (2006) found that a single 'dose' of 10 larvae was sufficient to have a positive effect on patients' asthma. Croese et al. (2006) mostly used doses of 25 larvae and re-inoculated some patients after 27–30 weeks. They found that the condition of most of the patients with Crohn's disease improved following infection with the worms.

Although proponents of helminth therapy consider it to pose little risk to the wider community, there will always be lingering concerns. These could be alleviated somewhat if a way was found to dose patients with single-sex infections or render the worms sterile. Non-destructive sex determination of nematode eggs and larvae is probably unfeasible with current technology. There is, however, considerable expertise in the production of sterile insects in pest management (e.g., for the control of the screwworm *Cochliomyia hominivorax*), and this technology might be applied to parasitic nematodes destined for helminth therapy. An increasing number of parasite genomes have now been sequenced, and it may become possible to genetically engineer parasites to improve their effectiveness for use in therapy and/or render them infertile. However, this presents further ethical considerations because of the potential for genetically engineered organisms to enter the environment.

The advantages of helminth therapy remain 'potential' because there is insufficient evidence to demonstrate that it works better than currently available drugs. Cost effective is placed as both a potential advantage and a problem because the costs of production are uncertain. It is unlikely to be cheap but many of the drugs currently used to treat autoimmune conditions are extremely expensive.

12.3.2 Maggot Therapy

Maggot therapy – also called 'larval therapy' and 'biosurgery' – refers to the application of fly (Diptera) larvae to wounds to speed up the rate at which they heal. It is currently mainly employed in human medicine, but there are case studies describing its use in veterinary medicine (Lepage et al. 2012; Vigani et al. 2011).

There is a long, albeit sporadic, history of using fly maggots (larvae) to encourage wound healing though, perhaps surprisingly in this era of high-tech medicine, maggot therapy is currently undergoing something of a Renaissance. Australian aborigines and other native peoples around the world have used fly larvae to treat wounds for centuries. In Western medicine, the first recorded observation of the potential of fly larvae to promote wound healing was made by the French surgeon Ambroise Paré (1510–1590) who noted how the wounds of soldiers injured in the Battle Saint Quentin (1557) healed faster if they were infested with maggots (Fleischmann et al. 2004). The army surgeon John Forney Zacharias used fly larvae to treat wounded soldiers in the American Civil War (1861–1865), and there were several subsequent reports in the medical literature on the beneficial effects of fly larvae on wound healing (Wollina et al. 2000). Maggot therapy was used extensively in Europe and North America in the 1930s, but it virtually ceased following the discovery of penicillin and other antibiotics. However, antibiotic resistance is spreading rapidly among many pathogenic bacteria, and some strains are now resistant to all currently available antibiotics. This is causing considerable concern in hospitals throughout the world.

The use of fly larvae to treat wounds is counter-intuitive since wound myiasis is a serious and potentially fatal condition for us and other animals. To intentionally apply fly larvae to an existing wound, and thereby establish a wound myiasis, could therefore be considered irresponsible. The important difference between a naturally occurring myiasis and one that is established as part of a medical intervention is that in the latter situation, the fly larvae are sterile, their numbers are carefully controlled, and they are removed before they can damage healthy tissue.

Maggot therapy usually employs the larvae of *Lucilia sericata* (*Phaenicia sericata* in American literature) that are reared under sterile conditions. Many other fly species also cause wound myiasis, but there are few comparative studies to determine whether any of them are equally useful or better than *L. sericata*. The larvae of *Lucilia cuprina*, are often stated to be 'too aggressive' and therefore likely to quickly move on and damage healthy surrounding tissues. However, Tantawi et al. (2010) (unintentionally) used *L. cuprina* successfully to treat diabetic ulcers when their *L. sericata* stock cultures became contaminated with them. The calliphorid fly *Sarconesiopsis magellanica* has been shown to be highly effective in laboratory trials (Díaz-Roa et al. 2016). Similarly, there is a case report of *Cochliomyia macellaria* larvae being used to treat necrotic wounds in a dog (Masiero et al. 2020). However, because there is now so much knowledge and expertise available on the employment of *L. sericata*, it is unlikely to be replaced on anything more than a local scale unless the alternative species offers appreciably greater advantages.

Superficial and sometimes serious skin wounds were a feature of everyday life for our ancestors. Consequently, our bodies, like those of other animals, are capable of remarkable levels of self-repair. This is especially true when we are young and healthy, but the ability declines with age and is seriously compromised by medical conditions such as diabetes, AIDS, and other immunosuppressive illnesses. The formation of any wound results in the death of cells and is followed by an inflammatory reaction in which cell debris and, if present, any fragments of the object causing the wound are removed. During this stage, immune cells (e.g., lymphocytes, macrophages, and neutrophils) produce inflammatory cytokines such as interleukin-1α, interleukin-1β and TNF-α. The inflammatory substances cause the blood capillaries to dilate and become more permeable and thereby allow white blood cells (leucocytes), cytokines, and the chemicals responsible for the blood clotting process to reach the injured region. Whilst the inflammatory process is ongoing, tissue repair commences with the stage known as 'organisation' or 'proliferation' in which the blood supply is restored to the damaged region. In this stage, any blood clots are replaced with granulation tissue that contains delicate capillaries and proliferating fibroblasts (these cells secrete collagen, elastin, and other proteins that provide structural stability). Finally, there is the stage of regeneration and scar formation although the capacity for this varies between tissue types and the nature of the wound. If the stimulus causing the wound persists (e.g., microbial infection), then the events described above occur concurrently rather than sequentially, and this results in chronic inflammation. For example, persons suffering from diabetes are prone to develop ulcers on their feet and legs because of their susceptibility to vascular disease and microbial infections. Similarly, those who are bedridden often develop pressure sores owing to the blood supply to affected regions of the body being restricted. The sores then become infected by bacteria and can ulcerate. In both cases, the ulcers often fail to heal naturally and persist for months or even years; they may even become worse or be a source of potentially fatal septicaemia. For healing to occur, the dead and dying tissue must be removed in a process called 'debridement' and any underlying infection controlled. Surgical debridement can be painful for the patient and is not always successful. Furthermore, antibiotic resistance means that many drugs have lost their efficacy. The larvae of *L. sericata* feed on bacteria and dead tissue and therefore effectively debride the wound. In the process they also eliminate the awful smell that is associated with such wounds. This is a not insignificant consideration since suffering from foul-smelling wounds is extremely unpleasant for both the patients and those who must care for them. It can also lead to depression and self-isolation that further impacts upon the patient's health and recovery.

12.3.2.1   Maggot Therapy in Practice

Maintaining laboratory cultures of *L. sericata* is relatively easy if the only purpose is to breed the flies. However, maggot therapy relies on the production of first-instar larvae and ensuring that there are sufficient of these when required presents a much more complicated challenge. The maggots remain at this stage for a very short time, and this means that the rearing process must be carefully regulated, and there needs to be a fast and efficient delivery chain from supplier to the hospital. The eggs and maggots cannot be stockpiled or stored at low temperature until required. The logistical difficulties of large-scale production and distribution means that there are few major commercial suppliers – for example, BioMonde Ltd in the UK and Monarch Labs in the USA (Stadler 2020). In many countries, hospital laboratories produce their own maggots or there are small-scale operations supplying the local hospitals. This limits the scope for using maggot therapy. Tatham et al. (2017) have suggested that in countries like Australia, pilotless drone aircraft would be the ideal solution for rapidly distributing maggots for use in debridement therapy from a central rearing laboratory.

At its simplest, a hole is cut in a hydrocolloid dressing the size and shape of the wound and this is applied to the wound site. This confines the larvae to the wound site and protects the surrounding healthy skin; 4–8 sterile first-instar larvae (~2 mm in length) per cm^2 of wound are placed on a sheet of semipermeable nylon mesh (e.g., LarvE™ NET, Tegapore™: 3M), and this is then laid on the surface of the wound. The hydrocolloid dressing is then covered with waterproof adhesive tape so as to confine the larvae at the wound site. After 3–4 days, the dressings and the larvae, which have usually grown to approximately 9 mm by this time, are gently washed out of the wound. The procedure is then repeated until the wound is clean, and all the necrotic tissue is removed.

A less messy approach is to confine pre-set numbers of maggots within a loose enclosed mesh, a bit like a tea bag, and this is applied to the wound site. The maggots are usually purchased pre-bagged from the supplier, which speeds up the delivery of the treatment. The maggots can feed through the mesh, but they are unable to escape. After 3–4 days, the mesh is removed and disposed of as clinical waste (Wilson et al. 2019). According to Steenvoorde (2005), free-range maggots perform debridement more effectively but the speed and simplicity of using pre-bagged maggots is a major consideration when deciding on treatment options.

Whichever approach is adopted, most wounds only require one to three treatments. It works extremely well for surface wounds but is not suitable for deep-seated abscesses. Interestingly, laboratory studies suggest that maggot therapy can be effective at treating skin lesions associated with dermal leishmaniasis (Cruz-Saavedra et al. 2016).

Most patients are positive about the benefits of this treatment although a few find it psychologically repugnant. During treatment, most patients experience either a decrease or at least no increase in their level pain beyond a 'tickling sensation' although some do find the treatment painful.

12.3.2.2   How Maggot Therapy Works

Maggots contribute to wound healing through physically consuming bacteria and dead tissues and by producing secretions and excretions that promote wound healing and have antibiotic properties (Nigam and Morgan 2016). For example, *L. sericata* larvae secrete/excrete various amino acids, such as L-histidine, and amino-acid-like compounds, such as L-valinol, that potentially contribute to wound healing by stimulating angiogenesis. They also secrete a novel serine protease and other chemicals that enhances fibrinolysis and thereby contribute to the wound healing process (van der Plas et al. 2014). In addition to their naturally produced pharmaceutically active chemicals, blowfly larvae can be genetically engineered to express specific peptides and proteins. For example, transgenic *L. sericata* larvae can be produced that secrete human platelet derived growth factor-BB (Linger et al. 2016). This peptide promotes wound healing. Whilst this approach holds great potential for improving the effectiveness of maggot therapy, its applicability will depend upon the public acceptance of genetically modified organisms and the risk of them escaping into the environment. Not all blowfly larval secretions and excretions are beneficial, and Elkington et al. (2009) identified a secretory protein produced by *L. cuprina* that had an immunosuppressive effect through inhibiting the proliferation of T lymphocytes.

In addition to consuming bacteria, blowfly larvae secrete/excrete numerous antimicrobial peptides and proteins (Cytryńska et al. 2020). For example, lucifensin has activity against various Gram-positive bacteria, although not Gram-negative bacteria. *Lucilia sericata*, like other blowfly larvae, excretes ammonia but the consequences of this in the controlled circumstances of larval therapy are uncertain. The application of gauze packs soaked in 1–2% ammonium bicarbonate solution to purulent wounds reportedly promote healing (Robinson 1940). However, in sheep suffering from blowfly strike (where there are hundreds of larvae present), the wound site becomes alkaline and unionised ammonia is absorbed into the bloodstream where it has an immunosuppressive effect (Guerrini 1997). The role of pH in the wound healing process is poorly understood (Wallace et al. 2019). Some of the protease enzymes involved in wound healing have alkaline pH optima (e.g. matrix metalloproteinases) and keratinocytes and fibroblasts grow and migrate more effectively at slightly alkaline conditions. However, if the proteinases are too active, they can delay healing. Mild acidification of the wound site can reduce colonisation by bacteria and is usually considered favourable for wound healing (Power et al. 2017).

12.3.3 Leech Therapy

Leeches have been used in both western and eastern medicine for over two thousand years. Indeed, the enthusiasm of western doctors for leeches was once so great that they themselves were often referred to as 'leeches'. The horse leech, *Hirudo medicinalis* (Figure 12.1) became extinct in some regions owing to over-collection for use in human medicine. This popularity was driven by the widespread belief that many ailments could be cured by removing either an 'excess of blood' or blood containing poisons. Leeches were applied, sometimes in large numbers, to specific parts of the body to remove the offending blood. As modern western medicine developed, it was realised that bleeding patients whether by the application of leeches or cutting a vein usually had no therapeutic benefit. Indeed, the loss of blood could be harmful. Consequently, the use of leeches in western medicine largely ceased. However, in recent years, it has been shown that leeches can prove useful in the treatment of certain medical conditions (Sig et al. 2017) and in veterinary medicine (Sobczak and Kantyka 2014). As usual, this has led scientists to coin another ugly and unnecessary word, 'hirudotherapy', to describe their use.

During surgery, particularly that involving the reattachment of body parts, one can join arteries that supply blood but seldom join the veins that take it away. This is because the veins are usually too small and fragile to be stitched together and consequently blood accumulates within the affected region. This is known as venous congestion and can lead to the patient experiencing pain and a delay in healing. In some cases, the region becomes necrotic, and the limb has to be amputated. By applying leeches, one can remove the excess blood from the reattached body part until the

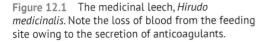

Figure 12.1 The medicinal leech, *Hirudo medicinalis*. Note the loss of blood from the feeding site owing to the secretion of anticoagulants.

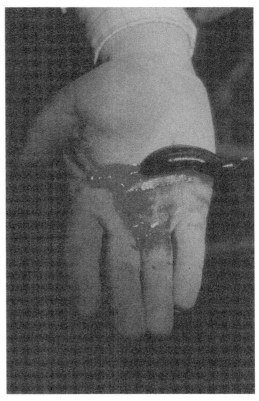

venous supply naturally re-establishes. Leeches inject a local anaesthetic when they bite and therefore the procedure is relatively painless. In addition, the presence of anticoagulants in their saliva ensures that the wound continues to bleed long after removing the leech. An individual *H. medicinalis* can extract up to 15 ml of blood during a single feeding session, but it is the continued slow loss of blood afterwards that is most beneficial.

Leech therapy is particularly useful for the surgical re-attachment of the body's extremities and projections (e.g., fingers, ears, nose, penis) in which problems arise because blood tends to pool within the re-attached part. Leeches have even been used after surgery to reattach a man's tongue following it being bitten off during fight (Kim et al. 2007) (Figure 12.2). Leech therapy can also reduce the pain and symptoms of some forms of joint osteoarthritis (Hohmann et al. 2018). In Asian traditional medicine, leeches are employed to facilitate the healing of ulcers (e.g., diabetic ulcers) and wounds, whilst in Iran there are reports of their use in the treatment of the lesions caused by cutaneous leishmaniasis (Koeppen et al. 2020). Many of these accounts are isolated case reports or the results of small-scale studies but they suggest that leeches have an even greater potential in western medicine than reconstructive surgery.

12.3.3.1 Leech Therapy in Practice

In Europe and North America, the most frequently used species is the horse leech *Hirudo medicinalis* but elsewhere in the world other species are often employed. For example, *Hirudinaria manillensis* is commonly used in Asia and *Aliolimnatis michaelensis* in South Africa. *Hirudo medicinalis* is a protected species in the UK and its numbers have also declined in other parts of the world because of over-collection, habitat loss, and changes in agricultural practices: medicinal

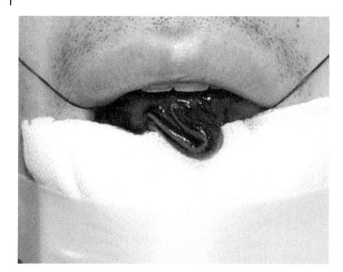

Figure 12.2 After this man's tongue was reattached, leeches were applied to relieve congestion until the venous blood supply was re-established. *Source:* Kim et al., (2007), 60: 1152–1155/with permission of Elsevier.

H. medicinalis are therefore usually specially bred by commercial operatives. However, molecular studies have indicated that many of the leeches supplied as *H. medicinalis* are *Hirudo verbena*.

The procedure is best undertaken in a warm, quiet environment with low-light conditions. Leeches are sensitive to light and are likely to try and escape rather than feed if disturbed. The leech is gently grasped in a gloved hand and encouraged to attach to the appropriate site. Leeches have soft smooth bodies that are easily damaged, so it is better to use a dry gauze pad to manipulate them rather than forceps. Alternatively, one can place a leech in a small container such as the barrel of a syringe or a cupping glass and up-end this at the attachment site. The container is then and held in place with tape. Usually, the leech will attach itself but making a small incision or adding a drop of sugar solution can act as an extra stimulant. Surrounding the bite site with dry gauze or dressings helps prevent the leech wandering off. *Hirudo medicinalis* typically feeds for anything from 20 minutes to 2 hours after which it drops off. Attempting to pull a leech off whilst it is feeding is not recommended since, like ticks, it would leave its mouthparts embedded in the body. If the leech must be removed, it is best to stroke its head region with a cotton bud soaked in saline, vinegar, or alcohol, but this needs to be done carefully otherwise the leech could vomit its ingested blood into and around the wound site. After a leech drops off, it is killed and disposed. Re-using leeches would carry too much of a risk for the transmission of blood-borne diseases. New leeches are applied on successive days until a functional venous blood supply has become established: this typically takes 3–7 days.

12.3.4 Malaria Therapy (Malariotherapy)

Malaria therapy was initially developed to treat the disease syphilis. Syphilis is a serious disease, but it is currently treatable with antibiotics. These days, many people consider syphilis to be just one sexually transmitted disease amongst many. It is therefore difficult to appreciate just how common syphilis once was among the general population and how infection often progressed remorselessly to insanity and an early death. Syphilis is caused by the spirochete bacterium *Treponema pallidum* and is usually transmitted through sexual intercourse, although it can also cross the placenta in pregnant women and cause congenital syphilis. Syphilis manifests as a bewildering variety of symptoms and as a result in the medical profession it was known as 'the great pretender'. In the later stages of the disease, it causes neurosyphilis – also referred to as 'general paresis', 'general paralysis of the insane', and '*dementia paralytica*' – in which the patient suffers from progressive dementia and loses control of their voluntary movements. Until the widespread availability of effective antibiotics in the 1940s, patients who had reached this stage of the disease were often consigned to an asylum where they usually died within 4 years of arrival.

Julius Wagner Ritter von Jauregg (1857–1940) (usually referred to as Wagner-Jauregg) was awarded the Nobel Prize for Medicine in 1927 for his work in developing malaria therapy for the treatment of neurosyphilis. Wagner-Jauregg worked as a psychiatrist at a clinic in Vienna and noted how the mental health of some of his patients improved following illnesses that induced high fever. He decided to carry out a more in-depth series of observations and recorded that patients with neurosyphilis showed considerable improvement in their condition following bouts of fever associated with erysipelas and tuberculosis. He tried various means of inducing fever in his patients, but these had little effect on their mental well-being. He ascribed this to the fevers not raising the body temperature high enough. He thought that malaria would be the ideal means by which this could be achieved but lacked a source of the infection. Eventually, in 1917 a soldier was admitted to his ward suffering from shellshock who had caught benign tertian malaria (*Plasmodium vivax*) whilst serving in the Crimea. He took blood from the soldier and injected it into nine patients suffering from neurosyphilis. After they had experienced at least seven episodes of fever he treated them with quinine. Six of the patients subsequently improved so much that they were able to leave the clinic and return to work. They were not cured of syphilis and four of them suffered relapses in later years, but it should be remembered that their prognosis at the time of their treatment was exceedingly bleak. Wagner-Jauregg was encouraged to continue with his studies and publish his findings, which were met with considerable enthusiasm. Malaria therapy (sometimes referred to as 'malariotherapy') was quickly adopted in Europe and around the world and became the principal means of treating neurosyphilis. Indeed, malaria therapy continued to be used in specific cases in the USA until the 1960s and in the United Kingdom in the 1970s.

Malaria therapy entered the news briefly in the 1990s when Dr Henry Heimlich suggested deploying it to cure Lyme disease, cancer, and AIDS (e.g., Heimlich et al. 1997). Dr Heimlich was famous for developing the 'Heimlich manoeuvre' for treating choking and drowning – although the procedure is no longer recommended. He was a master of self-publicity but held some unorthodox views that made him unpopular with many in the medical profession. Denied the opportunity to undertake work on his theories in the USA he, together with his co-worker Edward Patrick, set up operations in China, Ethiopia, and Gabon. However, the World Health Organisation was less than impressed and a report published by the Commission on Macroeconomics and Health stated that their work was an example of 'unscrupulous and opportune research which exploits the want of a given population' (Bhutta 2002). Interestingly, whole-body hyperthermia, raising the body temperature to 41.8 °C for up to 150 minutes, has been proposed for the treatment of various medical conditions, including depression (Woesner 2019). However, there is no suggestion that this should be done by infecting the patient with malaria.

Plasmodium ovale came to be the most frequently used species of *Plasmodium* for malaria therapy because it induces a high fever but seldom causes serious pathology. However, in the early years of development all the other species were employed, including *Plasmodium falciparum*. The procedure was not without its risks, and some people treated this way suffered harmful effects from the malaria and a few died. Unless they received the right antimalarials, those treated with *P. vivax* or *P. ovale* would have experienced recurrent episodes of malaria due to reactivation of liver hypnozoites. Even so-called benign forms of malaria can induce serious pathology and those receiving malaria therapy were already weakened by syphilis. However, the risks posed by malaria had to be counter-balanced by the almost certainty of imminent, unpleasant death from syphilis. Nevertheless, deliberately infecting someone with a known pathogen does raise ethical issues and malaria therapy has always been controversial.

Initially, transmission was undertaken by withdrawing 2–5 ml blood from the vein of an infected donor and injecting this subcutaneously into the recipient. Subsequently, other approaches were adopted including intravenous or intramuscular injections, and exposure to infected mosquitoes. The recipient was allowed to undergo a series of fevers after which they were treated with quinine. The results were rather variable and influenced by the level of fever induced and the severity of the patient's syphilitic symptoms. As a rule, about 30% of patient's experienced full remission, 20% partial remission and the remaining 50% showed little or no improvement (Whitrow 1990). Patients suffering advanced neurosyphilis with irreparable brain damage might be given several extra years of otherwise good physical health but would not recover their mental faculties.

The physiological and immunological basis of malaria therapy remains uncertain. Patients who experienced the most frequent fevers and the highest temperatures tended to have the best outcomes, but temperature alone is unlikely to have been the sole factor. *Treponema pallidum* lacks a heat-shock response and in culture all the spirochetes die if maintained at 41 °C for 24 hours. However, this temperature is considerably more than a patient would experience during their episodes of malaria. In addition, other fevers that induce lower temperatures than malaria can also improve the condition of patients with neurosyphilis. Possibly, the immune reaction against malaria also directly or indirectly harmed the spirochetes.

There is clearly some immunological link between the two organisms, possibly related to pathological mechanisms. Patients with *T. pallidum* infection produce high titres of antibody to a protein used in laboratory tests called cardiolipin. A similar protein is probably released by cells in the body in response to the detrimental effects of the bacterium and it may also be contained within the treponeme itself. This causes the production of anti-cardiolipin antibodies, which although not specific, is taken as a marker of active syphilis infection and used as the basis of the VDRL (Venereal Disease Research Laboratory) and the RPR (Rapid Plasmin Reagin) tests (also sometimes called the 'Wasserman reaction'). Interestingly this anti-cardiolipin antibody has also been noted in patients with other infections, including viral diseases, as well as trypanosomiasis and malaria. In these cases, the antibody is produced at a lower level than during syphilis and only lasts for about 6 months after the acute infection. This means that if the VDRL or RPR tests are used to determine whether a person might have syphilis, then the laboratory needs to be given information about any recent history of malaria to help them identify possible false-positive results. In all cases, a specific anti-*Treponema pallidum* antibody test would also be carried out, so a correct diagnosis should be assured.

12.4 Parasites as Sources of Novel Pharmaceutically Active Substances

Parasites produce many secretions and excretions, and some of these include compounds with useful pharmacological properties. In the past, it was difficult to isolate and characterise these molecules. For example, between the 1930s and 1950s there was considerable interest in both Russia and America in the potential of extracts derived from *Trypanosoma cruzi* to treat certain forms of cancer (Krementsov 2009). The results of these studies were ambiguous, with some workers claiming remarkable effects and others finding none or even harmful toxicity. This is hardly surprising considering the political problems that limited co-operation between workers in the two countries, the logistical problems of preparing the *T. cruzi* extracts, and the genetic diversity of *T. cruzi*.

Although products based on *T. cruzi* extracts were eventually marketed, the active ingredient(s) were never identified and interest in this line of research ultimately ceased.

Nowadays, we can manufacture recombinant molecules in bacteria/yeast and thereby obtain increased yields of better characterised chemicals for use in clinical trials. This is leading to the identification of many novel compounds produced by parasites that appear to have potentially useful properties. However, at the time of writing, these were all at the experimental stage of development. For example, *T. cruzi* produces a chemical, *T. cruzi* calreticulin (TcCalr), that appears to protect against tumour formation (Ramírez-Toloza et al. 2020). Whether this is the chemical that was responsible for the effects observed by earlier workers is impossible to say. There is also a suggestion of using attenuated *T. cruzi* clones as vectors carrying antigens that induce T cell immunity against cancer cells (Junqueira et al. 2011). This makes for the intriguing prospect of a parasite being both a source of an anticancer drug and a means of delivering anticancer drugs.

The protoscolices of the tapeworm *Echinococcus granulosus* are also a source of chemicals with anticancer properties. For example, they produce a protease inhibitor, EgKI-1, that inhibits the growth and migration of various cancer cell types both in *in vitro* assays and in *in vivo* mouse models (Ranasinghe and McManus 2018). According to Lund et al. (2016), the peptide, FhHDM-1, secreted by adult *Fasciola hepatica* improves the disease outcomes of mice models of type 1 diabetes and relapsing-remitting multiple sclerosis. The authors suggest that this occurs through modulating the activity of macrophages and their cytokines and thereby reducing the proinflammatory response rather than by reducing the production of self-reactive T cells or their cytokines. There is/are one or more chemicals present in the excretions/secretions of adult *F. hepatica* that protect mice from developing allergic asthma (Finlay et al. 2017). However, whether FhHDM-1 is involved is currently uncertain. The ability of parasites to modulate the host immune response by chemicals such as FhHDM-1 has also led to the suggestion that they might produce substances that would be useful for the treatment of SARS-COVID-19 (Siles-Lucas et al. 2020). The lethal effects of this virus are principally a consequence of engendering a 'cytokine storm' that inflicts massive inflammatory response in the lungs. If chemicals derived from parasites could reduce the production of cytokines such as interleukin-6, then they might prove extremely helpful. However, at the time of writing, the work was still ongoing. The flatworm *Opisthorchis viverrini* is notorious for causing cancer. However, it also produces a granulin-like growth factor that induces angiogenesis and thereby promotes wound healing (Haugen et al. 2018).

There has been much more success in developing pharmacologically useful products from blood-feeding parasites. Any organism feeding on mammalian blood must overcome the problem of blood clotting, which could lead to them becoming stuck to their victim or their mouthparts becoming blocked. When a blood vessel is injured, whether by a needle a hookworm's mouthparts, it initiates a sequence of physiological processes which, if able to continue to completion, would stop the loss of blood. Together, these processes are known as haemostasis and can be divided into three interlinked events: vasoconstriction, the formation of a plug of platelets, and the production of a web of fibrin threads that coats and stabilises the platelet plug.

The formation of a platelet plug commences when circulating blood platelets are exposed to von Willebrand factor, collagen and other platelet activating factors contained in the connective tissue that underlies the surface epithelium of blood vessels. Once activated, the platelets release the contents of their secretory granules (these include adenosine diphosphate, serotonin, and the prostaglandin thromboxane A2) into the surrounding plasma and they activate more platelets. Activated platelets change their shape from spherical to stellate (star-shaped) and become 'sticky', so they become stuck to the collagen and to one another. The activated platelets also bind to the soluble plasma protein fibrinogen, and this then cross-links with glycoproteins to form insoluble fibrin that binds the platelets together and stabilise them, so they form a secure plug.

Fibrin formation (i.e., coagulation) involves a complex sequence of reactions that are divided into the contact activation pathway (intrinsic pathway) and the tissue factor pathway (extrinsic pathway). Both pathways involve enzymes and glycoprotein cofactors which, once activated, catalyse the subsequent reaction until finally fibrin monomers become cross-linked to form insoluble fibrin polymers and a clot starts to form. Each of the steps in the pathways is under the control of a specific enzyme or glycoprotein that is known as a coagulation factor. Most of the enzymes are serine proteases, but factor XIII is a transglutaminase and factor V and factor VIII are glycoproteins.

Most blood-feeding (hematophagous) organisms produce chemicals that inhibit platelet aggregation and one or more of the serine protease enzymes (coagulation factors) involved in the tissue factor pathway. For example, the leech *H. medicinalis* produces the protein Calin that interferes with platelet aggregation and the polypeptide hirudin that interferes with fibrin formation. Calin prevents the formation of a platelet plug by interfering with the interaction between platelets and collagen. Hirudin is produced in a variety of isoforms and complexes with α-thrombin (coagulation factor IIa) and thereby prevents it from catalysing the conversion of fibrinogen to fibrin (Hildebrandt and Lemke 2011).

Chemicals that interfere with blood clotting are of enormous interest for their potential for treating thrombotic diseases – that is, those in which a thrombosis (blood clot) forms within a blood vessel and thereby reduces or stops blood circulation in the affected region – such as stroke, myocardial infarction, and deep vein thrombosis. Hirudin is currently employed to treat conditions as varied as topical bruising and heart disease. It has the advantages of low toxicity, it does not interfere with any coagulation factors apart from α-thrombin or any other physiological processes, it does not appear to stimulate an immune response, and is eliminated completely via the kidneys. Recombinant derivates of hirudin, such as desirudin and bivalirudin, are now available. In addition, recently discovered 'hirudin-like factors' in leeches are currently under investigation for potentially useful therapeutic properties (Müller et al. 2020).

Blood-feeding nematodes produce various anticoagulant proteins/peptides (NAPs) although to date only a few species have been investigated and their molecules characterised. The hookworm *Ancylostoma caninum* produces a protein called HPI-1 (Del Valle et al. 2003) that prevents platelets from aggregating together via fibrinogen and various specific inhibitors of coagulation factors VIIa and Xa (Ledizet et al. 2005; Syed and Mehta 2018). Similarly, *Ancylostoma duodenale* produces a peptide, AduNAP4 that inhibits coagulation factors Xa and XIa. *Ancylostoma ceylanicum* produces a peptide, Ace-HPI that inhibits platelet aggregation but has little effect on platelet adhesion (Huang et al. 2020).

Macaque monkeys injected with a potentially lethal dose of Ebola virus do not develop the disease or express only mild symptoms if they are subsequently treated with rNAPc2 derived from *A. caninum* (Geisbert et al. 2003). Viral haemorrhagic fevers such as those caused by Ebola virus and Marburg virus cause disseminated intravascular coagulation (DIC). This develops through over-activation of the blood coagulation cascade, which leads to the formation of an excess of thrombin and the production of small blood clots in blood vessels throughout the body. These small blood clots disrupt the blood flow to organs such as the kidneys and consume so many platelets and coagulation factors that the normal clotting process is disrupted, and the patient suffers from abnormal bleeding in the gastrointestinal tract, lungs, and skin. NAPc2 and its effect by inhibiting the factor VIIa/tissue complex through a unique mechanism. NAPc2 initially interacts with coagulation factor Xa and the resultant binary complex then inhibits factor VIIa of the factor VIIa/tissue factor complex. rNAPc2 also inhibits angiogenesis and both primary and secondary metastatic tumour growth in mice. Interestingly, rNAP5, also derived from *A. caninum*, specifically inhibits coagulation factor Xa but has no effect on tumour growth (Hembrough et al. 2003).

12.5   Parasites as Biological Control Agents

Biological control is the use of an organism to control the population of another organism. The targets are usually pests, which may be plants, invertebrates, or vertebrates, and the control agents are predators, pathogens, parasitoids, and parasites. Biological control can be highly effective, particularly in closed environments such as greenhouses, and may be integrated into other control practices. However, there are many instances in which attempts at biological control have caused serious ecological damage without affecting the target pest. Consequently, before biological control is adopted, it must first undergo stringent safety checking (Hajek et al. 2016). Bale et al. (2008) provide a good review of the practice of biological control whilst Leung et al. (2020) discuss how genomic techniques could potentially improve its effectiveness.

The characteristics of a good control agent, whether it is chemical or biological, are that it should be specific to the pest, not harm non-target organisms, rapidly eliminate the pest from the place it is causing a problem, have a long-lasting effect on pest reduction, and not induce resistance. Chemical control agents often suffer from a lack of specificity, harm other organisms, and pests often develop resistance to them. Parasites are often specific to their hosts, there is seldom a problem with resistance developing, and they may leave eggs or cysts that ensure control of subsequent generations of the target pest. However, as a rule, parasites induce chronic disease and do not cause epidemics. Consequently, although a parasite may cause illness, an infected pest could still be growing where it isn't wanted, consuming a crop, or causing or transmitting a disease. Parasites can also be difficult to rear in large enough numbers to use on a commercial scale, and there can be problems in spreading them in the environment so that they are able to infect the pest. Therefore, although there are several examples of parasites providing exceptionally good control of the target pest species, they are not used as extensively as other pathogens such as viruses, bacteria, fungi, or parasitoids.

There are basically three ways of practising biological control: importation (classical), conservation, and augmentation. When organisms are accidentally or intentionally imported into a new country, they often leave their natural enemies behind, and if there are no local control agents, they reproduce rapidly and become pests. Classical biological control aims to exert control on the introduced species by exposing the pest to its original control agent or a suitable alternative. Conservation biological control is a long-term strategy that involves manipulating the environment to encourage the growth and survival of the natural enemies. As its name suggests, augmentation biological control involves increasing the abundance of the control agent. This is often done by artificially rearing large numbers of the control agent and then releasing them at an appropriate time and place. The population of the control agent is therefore increased to a level considerably above that, which would occur naturally to ensure a rapid decline in the pest population. This method, therefore, requires constant human intervention and is essentially using the control agent in a similar manner to pesticides.

An organism may leave its predators behind when it arrives in a new country, but its parasites often travel within or upon it. If the parasite finds a suitable vector or intermediate host (or does not require one), then it may continue to be transmitted or even exploit new host species. Consequently, there are few examples of parasites being used in classical biological control, due to the accompanying risks to other animals or plants. There are similarly few examples of specific conservation practices being adopted to improve the survival of parasites in the environment. Indeed, any attempt to improve the survival and transmission of the parasite of a pest runs the risk of also enhancing the transmission of the parasites of other organisms. Most examples in which parasites have been used as biological control agents therefore fall into the category of augmentation.

Many of the parasites used as biological control agents are nematodes although there has been some success using microsporidia (now classified as fungi) for the control of mosquitoes, locusts, fire ants, and several other pest insects (Bjørnson and Oi 2014). Nematodes are effective for the control of specific insect pests and of certain slugs and snails. Those species infecting insects, such as *Heterorhabditis* and *Steinernema*, are often referred to as 'entomopathogenic nematodes' and they are used to control various insect pests such as vine weevils (*Otiorhynchus* spp.) and 'leather-jackets' (Tipulidae) (Gaugler 2018).

12.5.1 Life Cycle of the Entomopathogenic Nematodes *Heterorhabditis* and *Steinernema*

When the free-living infective third-stage larvae of *Heterorhabditis* and *Steinernema* locate a suitable insect host, they invade it via its mouth, anus, or spiracles; *Heterorhabditis* larvae can also penetrate via the intersegmental membranes. The infective stage is sometimes referred to as a '*dauer*' larva, and it retains the second-stage cuticle as a protective sheath. *Dauer* is a German word that is usually translated as 'duration' or 'period'. The term '*dauer*' larva is applied to third-stage nematode larvae that have entered a period of arrested development. This state is associated with dispersal and an increased ability to withstand harmful environmental conditions. *Dauer* larvae are also formed by non-parasitic nematode species such as *Caenorhabditis elegans* in response to crowding, starvation, and other adverse circumstances. These *dauer* larvae disperse by attaching to other animals. The formation of *dauer* larvae might therefore represent one of the first stages in the evolution of parasitism in nematodes (Ogawa et al. 2009).

In common with several other entomopathogenic nematode species belonging to the order Rhabdita, *Heterorhabditis*, and *Steinernema* have unusual symbiotic relationships with bacteria that contribute to their pathogenicity. Once within the insect host, the *dauer* larvae penetrate its haemocoel, and there they regurgitate bacteria stored within the anterior region of their gut. These bacteria rapidly multiply within the insect and this, together with the toxins they produce was once thought to be responsible for killing the parasitised insect within 24–48 hours of infection. However, it is now apparent that *Steinernema carpocapsae*, and probably other species too, produce their own toxins that can kill their insect host (Lu et al. 2017). The nematodes are therefore probably not reliant upon symbiotic bacteria to kill their host. The symbiotic bacteria also produce antibiotic substances that prevent the corpse being colonised by other bacteria species and feeding deterrents that make it unpalatable to ants (Zhou et al. 2002). The nematodes are not harmed by the bacteria toxins, and they feed upon the bacteria that multiply within the corpse.

Heterorhabditis nematodes develop into self-fertilising hermaphrodites, whilst in *Steinernema* the sexes are separate. Depending upon the size of the host and the number of nematodes, they may go through 2–3 generations within the dead insect. Rhabditid nematodes, such as *Heterorhabditis* and *Steinernema*, have an unusual style of reproduction called *endotokia matricida* in which the eggs hatch within the mother's uterus and the young then eat her alive. The onset of *endotakia matricida* is delayed if the food supply is plentiful, but once initiated the food supply does not affect its duration (Johnigk and Ehlers 1999). After hatching within their mother's uterus, the first-stage larvae commence feeding on her tissues killing her in the process and eventually leave her body when they reach the third stage. If food is abundant, the larvae will then continue their development to adulthood and reproduce within the body of the insect to form a second generation. If, however, the food supply is limited, the nematodes emerge from their mother's body as *dauer* larvae, which leave the insect corpse *en-masse* and enter the soil. The *dauer* larvae remain within the soil and are incapable of developing further until they have infected a suitable host. In some nematode species, the *dauer* larvae employ a 'sit and wait' strategy, whilst in others the *dauer* larvae actively search for suitable hosts using chemical cues.

Rhabditid Nematode–Bacteria Relationships

Different species of symbiotic bacteria are associated with different Rhabditid nematode species. This may be a single-species relationship (i.e., monoxenic) or an association with two or more bacteria species. For example, *Heterorhabditis bacteriophora* has an association with the bacteria *Photorhabdus luminescens* and *Photorhabdus temperata*. Although it was once thought that many nematodes were in a single species relationship, it is increasingly apparent that a 'pathobiome' comprising of several species of bacteria exacerbates the pathogenicity of the nematode (Ogier et al. 2020).

The bacteria associated with entomopathogenic nematodes are probably essential for the worm's survival and contribute to varying extents to the insect host's death. By contrast, the slug parasite *Phasmarhabditis hermaphrodita* (Figure 12.3a,b) forms associations with assemblages of bacteria without affecting its pathogenicity (Rae et al. 2010). This may (at least partly) explain why *P. hermaphrodita* also differs from the entomopathogenic nematodes in being a facultative parasite that can survive as a saprophyte feeding on slug faeces and as a parasite of living slugs.

The bacteria associating with entomopathogenic nematodes cannot survive on their own in the soil and must form a symbiotic relationship with the nematodes. Initially the bacteria attach themselves by their fimbriae to the posterior region of the intestine of a developing female worm. Fimbriae are projections from the cell wall of bacteria and have a variety of functions but are particularly important in attachment and invasion. Within the nematode gut, the bacteria form a biofilm and then invade the cells of the rectal gland. During the development of the nematode, the rectal glands lyse, and the bacteria are released into the mother nematode's body cavity where they are ingested by the larvae growing within her. The bacteria initially invade the cells of the pharyngeal-intestinal valve of the larvae and then move to the lumen of the intestine. Although the bacteria are essential for the survival of the nematode, the association is not without its costs and *dauer* larvae that are bacteria-free survive longer than those which are infected (Chapuis et al. 2012).

(a)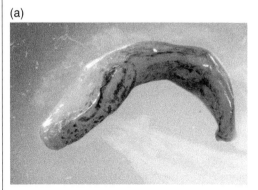

(b)

Figure 12.3   Slug parasitic nematode *Phasmarhabditis hermaphrodita*. (a) Slug (*Arion vulgaris*) that has died following parasitisation by *P. hermaphrodita*. The slug's body was full of nematodes. (b) *Phasmarhabditis hermaphrodita* isolated from the dead slug.

12.6 Parasites as Forensic Indicators

Forensic science is the use of scientific techniques to answer legal questions. Unfortunately, the term 'forensic' has become hijacked to describe virtually any analysis of past events regardless of whether they involve legal or criminal activity. One can therefore find terms such as forensic musicology, forensic astronomy, and numerous other variations on this theme. Forensic biology has more *bona fide* credentials and is a branch of forensic science that utilises biology to investigate legal cases. There is a tendency to focus on crimes such as murder, but the legal cases in which biology is important range from drug trafficking to wound analysis and from the illegal trade in protected species to sexual assault (Gunn 2019). Given that humans and all other organisms are infected by parasites, one would expect them to provide useful forensic indicators. In fact, there are few instances of this although this is mostly because their full potential remains underexploited.

When someone dies suddenly and unexpectedly, there should be a thorough investigation to determine the cause. Although some parasitic diseases have fatal outcomes, they are seldom of rapid onset. One notable exception to this is cerebral malaria; there are several cases in the forensic literature in which a person was found dead for no obvious reason in which malaria was subsequently found to be the cause. For example, Menezes et al. (2010) describe a case in which the body of a dead man was discovered by a railway line in India. The cause of death was not immediately obvious but was later shown to be from malaria. There are few reports of parasites being used intentionally to harm or kill people. One of the most commonly cited cases is that of Eric Kranz who in 1970 deliberately contaminated the food of his student roommates with *Ascaris* eggs. Kranz had been in dispute with his roommates, and during one altercation, he said that he would infect them with parasites. This was dismissed as an idle threat, but when his roommates were hospitalised with respiratory distress, he fled the area. The doctors treating them had no reason to suspect that the cause was ascariasis and therefore gave them antibiotics, which of course had no effect. It was only when larvae were detected in sputum that they coughed up that the cause was established, and an effective treatment regime instigated. There was circumstantial evidence that the *Ascaris* infections were not acquired by chance and Kranz was subsequently arrested and charged with attempted murder. Kranz was a post-graduate student at MacDonald College, Quebec, Canada studying parasitology and allegedly stole the *Ascaris* eggs from a laboratory at the college (Breeze et al. 2005). The fact that this is such an isolated case suggests that it is only when there is the rare coincidence of a parasitologist with a homicidal grudge that parasites are likely to be used for nefarious purposes. However, it could also indicate the difficulty of distinguishing malicious from accidental parasite transmission. If Kranz had not made his initial threats and then disappeared when his victims fell ill, he may have escaped prosecution. For example, in South Africa, it is alleged that women sometimes gain revenge on unfaithful menfolk by lacing their drink with the eggs of the tapeworm *Taenia solium* (Kriel 2003). *Taenia solium* is common in South Africa and unless the intended target found a whole proglottid at the bottom of his beer glass it would be difficult to prove the source of the infection. The extent to which this is a common crime, or an urban myth is uncertain – as is the ability of *T. solium* eggs to survive in alcoholic beverages.

Parasites may, in some circumstances, link victim and assailant together. For example, crab lice (*Phthirus pubis*) are normally only transmitted through sexual contact, and it is possible to isolate human DNA from within their gut. Consequently, the finding of crab lice on the body of 'person X' containing the DNA of 'person Y' indicates that X and Y must have had sexual contact. This could be useful where there is a claim of sexual assault but no semen or other evidence of an association between the two people. However, there do not appear to be any documented cases where this has been undertaken. The fact that parasites are hard to rear artificially in the laboratory and do not tend to cause epidemic disease means that they are not considered suitable for biological warfare or bioterrorism. However, during WW2, the Japanese developed 'bombs' containing fleas infected with plague bacilli.

In addition to determining the cause of death, one of the most important questions to be answered in any case of 'suspicious death' is when the victim died. This is usually determined from physiological changes, the state of decay, and/or the development of blowfly larvae and other invertebrates upon the body (Gunn, 2019). Various ectoparasites could also be added to this list of forensic indicators. For example, as the corpse starts to cool down, lice (e.g., *Pediculus humanus corporis*) start to leave the body. The presence of these lice on the clothing or moving off the body is therefore an immediate indicator that the person or animal died only a short time previously. Those ectoparasites that are unable to leave the dead body, such as the skin follicle mites *Demodex folliculorum* and *Demodex brevis*, eventually die of starvation or succumb to the chemicals produced during the decay process. Whether or not these mites are still alive has therefore been suggested as an indicator of the time since host death. However, Özdemir et al. (2003) were unable to find an obvious relationship between time of death and mite survival. Similarly, the factors that affect mite survival after host death still need to be elucidated.

The Use of Parasitoid Wasps to Determine the Minimum Time Since Death

The parasitoid wasp *Nasonia vitripennis* (Figure 12.4) lays its eggs in the pupae of a range of blowfly species using her ovipositor to bore a hole through the puparium and then lays her eggs on top of the developing pupa. This occurs 24–30 hours after the pupa has formed, i.e., the point at which the blowfly third-instar larval cuticle separates from the pupal cuticle and forms the puparium. After hatching, the wasp larvae feed on the fly pupa, killing it in the process. The wasps pupate within their host's puparium and the adults chew their way out after 10–50 days, depending on the temperature. Because *N. vitripennis* only lays its eggs during a very restricted time, recording when the adult wasps emerge can be used in forensic science to determine the post-mortem interval (Grassberger and Frank 2003). This is because if one knows how long the wasps take to develop, it can be calculated when they laid their eggs and on that date the pupa would be 24–30 hours old from which point it would be possible to determine when the blowfly eggs were laid and therefore approximate the date on which the person died.

Figure 12.4 Scanning electron micrograph of the parasitoid wasp *Nasonia vitripennis*.

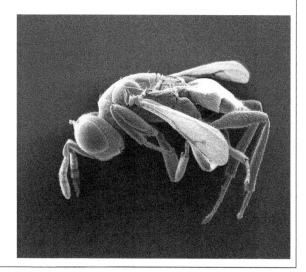

Many forensic investigations require the determination of the provenance of a person, animal, or plant, who/which may be living or dead. In this scenario, any associated organism(s) that have specific environmental requirements or have a restricted geographical distribution can provide an indication of past movements. Similarly, their developmental stage can indicate the passage of time. Many parasites fit these criteria and therefore have forensic potential. For example, the presence of the tick *Rhipicephalus australis* upon a horse enabled Kwak and Schubert (2019) to demonstrate that it was moved illegally across state boundaries in Australia. When the horse arrived in Victoria from Rockhampton, it was discovered to be infested with both adult ticks and nymphs. This tick species is found in several parts of Australia, but it does not occur in Victoria, and it is illegal to import an infested animal into the state. However, the horse owner claimed that it was not infested when it left his property and must have become infested whilst in transit or upon arrival. This claim could be disproved because *R. australis* is a one host tick and after climbing on board its host remains there until it drops off to lay its eggs. Because it takes at least 11 days for the ticks to reach adulthood and adults were present on the horse, the horse must have become infested at least 11 days previously. The transport documentation indicated that the horse was back in Rockhampton at this time, and it is therefore here that it became infested. Consequently, the horse owner was guilty of moving an infested illegally across state boundaries.

The illegal trade in wildlife and exotic flora is worth millions of pounds per year and often involves organised crime. One problem is where disputes arise whether an animal or plant was collected from the wild (and therefore its trade is illegal) or was captive bred (and could therefore be traded subject to the local legislation). Organisms collected from the wild usually harbour parasites which are often absent from captive bred species owing to medical treatment or the absence of a suitable vector or intermediate host. For example, wild Komodo dragons (*Varanus komodoensis*) usually harbour haemogregarine parasites, but these are absent in captive animals (Gillespie et al. 2000). In some instances, an animal may be protected in one country or region but can be hunted and collected in another. Parasites might prove useful indicators in such cases. For example, Criscione et al. (2006) found that the genotype of parasites infecting trout (*Oncorhynchus mykiss*) provided a better indication of the source population than the host's own genotype. Many of us are infected with helminths, and therefore their DNA can potentially prove useful in attempting to determine a person's geographical origin or movements. Burgos et al. (2017) have devised multiplex PCR systems for microsatellite loci that can distinguish between *Ascaris lumbricoides* populations. One can detect helminth DNA in the faeces of an infected person, so it would not be necessary to treat them with an anthelmintic to obtain the worms.

In human forensic cases, there is increasing interest in the potential of viruses as geographical indicators either of birthplace or recent travel history (Inoue et al. 2014). This can be useful in situations in which unidentified bodies are found, and when a person claims a certain nationality, but there is no documentation. The presence of parasites and their genotype might similarly provide evidence of an association with a particular country or region, but at the time of writing, there were no published studies in which this has been done in a forensic context. There are, however, examples from the archaeological literature. For instance, the presence of the eggs of the nematode *Trichuris trichiura* in the mummified remains of Iron Age nomads in Southern Siberia has been presented as evidence that they must have been in contact with agricultural communities many hundreds of miles away in China or Central Asia (Slavinsky et al. 2018). This is because the nematode is not endemic to Southern Siberia due to its dry climate and the infection is acquired through contaminative transmission. Consequently, it is suggested that the nomads must have acquired their infection by either travelling to the nearest endemic region or consuming food originating from there.

13

The Identification of Protozoan and Helminth Parasites

13.1 Laboratory Testing for Parasitic Infections: Introduction

It is important to identify a parasite that might be causing symptoms in a patient, or being carried by a vector, as fully as is feasible. In some situations, establishing the genus of the parasite on morphological grounds is enough to start appropriate treatment or implement suitable control measures. In other circumstances, it is necessary to determine not only the species but the strain of organism. This can guide treatment, as that strain might be resistant to some anti-parasitic treatments. It can also be incredibly useful for epidemiological purposes in tracking outbreaks of a particular infection.

The first problem to address in a human or animal patient is do they have an infection? This sometimes seems obvious if they have symptoms such as pyrexia or diarrhoea which started suddenly, but of course non-infectious agents such as chemical toxins can have similar effects. In contrast, chronic infections are often diagnosed by accident (when a person goes for a general health screen) or when the more serious consequences of long-term infection start to appear. A routine full blood count, FBC (also known as complete blood count, CBC), and tests for urea and electrolytes (U&Es) can help to establish whether the person has an infectious disease.

The next thing to decide is the nature of the infectious agent. Could it be a protozoan or helminth? The patient's history is useful here. Have they travelled from a country where there is no transmission of certain infection to an endemic area? Is there any sign of an insect or tick bite? What have

Parasitology: An Integrated Approach, Second Edition. Alan Gunn and Sarah J. Pitt.
© 2022 John Wiley & Sons Ltd. Published 2022 by John Wiley & Sons Ltd.
Companion website: www.wiley.com/go/gunn/parasitology2

they been eating? The routine blood tests might help. For example, eosinophilia is indicative of a helminth infection, while abnormal liver function tests point to hepatic damage. Symptoms can be very specific, but often they point to a range of possibilities. A person presenting sudden onset of pyrexia, rigors, and vomiting might have malaria, influenza, typhoid fever, or another infection.

Thus, the scientists engaged in the laboratory identification of parasites, particularly those of medical and veterinary importance are making a vital contribution to human healthcare and animal husbandry. Firstly, we will look at techniques that are carried out within a laboratory setting and then discuss assays which have been adapted to allow testing by the patient's bedside or 'in the field'. We will not provide detailed practical instructions of how to identify specific parasites, rather to outline them and discuss their applications, with selected examples.

Those requiring more detail on the isolation and identification of parasites of medical importance should consult Cheesbrough (2009), Leventhal and Cheadle (2019), Garcia (2021), or the more comprehensive Garcia (2006). The DPDx-CDC Parasitology Diagnostic Web Site (http://dpd.cdc.gov/dpdx) is also a very good source of information, and there is specific guidance about laboratory investigation of some parasite diseases on the WHO website (https://www.who.int). For investigation of infections in animals, readers are directed to Hendrix and Robinson (2016), Mehlhorn (2016), or Ponnudurai and Rani (2017).

13.2   Importance of Correct Identification

Many parasitic diseases cause non-specific clinical signs and symptoms such as fever, diarrhoea, weight loss, and lassitude. Consequently, a doctor or vet would seldom be able to give an immediate diagnosis from an examination of a patient/animal without requesting further tests. Contrary to popular perceptions, doctors, and vets seldom perform diagnostic tests themselves. Instead, they send samples to laboratories where professionals undertake the identification, and they then inform the doctor/vet of the results. Once the cause of the disease is identified, it is possible to begin appropriate treatment and thereby prevent the condition becoming worse. Drugs are often expensive and always carry the risk of inducing harmful side-effects, so it is important to ensure that they are only used where appropriate and at the correct dose. This also avoids the likelihood of drug resistance developing in the parasite and of drug residues being excreted and entering the environment. Therefore, the World Health Organisation strongly recommends that even people living in areas where malaria is endemic should have a diagnostic test to confirm *Plasmodium* infection before treatment is prescribed.

For a patient with symptoms of diarrhoea, it would be necessary to determine whether the causative agent was a bacterium, a virus, or a parasite through examination of a faecal specimen. In a case where cysts or trophozoites of the protozoan *Entamoeba* spp. were found in the sample, then it would be necessary to establish the species of the isolate. This is because *Entamoeba coli* is a harmless commensal, whilst *Entamoeba histolytica* is a potentially dangerous parasite (although strains differ in their pathogenicity). Once treatment has commenced, diagnostic techniques are useful for monitoring its success and can confirm when a cure is achieved.

Correct identification is important in epidemiological studies and in monitoring the effectiveness of control programmes (Table 13.1). Laboratory tests on suitable samples can identify animals and people who are infected with parasites but not expressing clinical symptoms. These people/animals can be sources of infection for others in the population and may also be vulnerable to developing disease later themselves. Similarly, correct identification can determine potential sources of infection with a particular parasite, whether a control programme for a certain disease is required and

Table 13.1   Instances in which correct identification is important.

Patient management: determination of most appropriate therapy, drug dosage, and duration of treatment

Prevention of long-term complications by identifying an infection before it becomes serious

Identification of asymptomatic individuals

Identification of human/animal reservoirs of infection

Identification of contamination of food/the environment with transmission stages

Identification of vectors/intermediate hosts

Surveillance and epidemiological studies

Monitoring the need for/success of control measures by assessing parasite prevalence in host and vector/intermediate hosts

Identification of genetic markers for drug resistance/pathogenicity in the parasite population

whether that programme is delivering the required results. It can also be useful to identify parasites within invertebrates that are acting as vectors or intermediate hosts. For example, the effectiveness of a programme to control the filarial nematode *Brugia malayi* through targeting its *Anopheles* vector would be enhanced by examining mosquitoes collected in the area for the presence of the infective filarial worms. Some diseases are classed as 'notifiable'. This means that there it is a legal requirement to inform the relevant authorities of their presence, so that control measures can be instigated. This is clearly a very serious matter, and it is therefore essential that any diagnosis is accurate and timely. For example, in countries where malaria is not usually transmitted, it is usual to have a legal requirement for cases to be notified to the health authorities; similarly in the United Kingdom, warble fly infections in cattle must be sent to the government agency dealing with animal health. The system of notification of infectious diseases is a very important part of surveillance. If cases of the disease are reported to a central authority, it becomes clear at an early stage whether it is an isolated, sporadic case or whether there is an outbreak with a particular pattern of dispersal. There are two main reasons why certain diseases are classed as 'notifiable'. One is when the disease is being controlled through a vaccination programme. If there are suddenly a lot of cases of that infection, then it would indicate a problem with the vaccine or that the appropriate people/animals had not been receiving the vaccine. The other reason is when it is an unusual and serious infection that could potentially spread/establish within the country. Malaria is a notifiable disease in the United Kingdom because mosquitoes capable of spreading the disease are present in the country, and malaria used to exist there until after the turn of the twentieth century. By contrast, Human African Trypanosomiasis is an equally serious disease, but it is not classed as 'notifiable'. This is because tsetse flies are not found here and therefore Human African Trypanosomiasis could not spread among the population. With the advent of molecular techniques, it is now possible to identify parasites not only to species level but also to determine their genetic characteristics. Consequently, one can now monitor parasite populations for their susceptibility/resistance to drugs, the presence of genes that are linked to pathogenicity, and track outbreaks linked to particular strains of organism.

13.3   Properties of an Ideal Diagnostic Test

The properties of an ideal parasite diagnostic test are similar to those of any other test for a medical or veterinary condition and are listed in Table 13.2. Obviously, the test must be accurate, and the results must be unambiguous, otherwise inappropriate treatment may be given or a control

Table 13.2 Properties of an ideal parasite diagnostic test.

Tests for a clinically useful marker

Gives reliable results – i.e., scientifically accurate and clinically credible

Gives reproducible results – i.e., if the sample is re-tested the results are the same each time

Has high sensitivity and high specificity

Easily Interpreted

Quick and simple to perform

Safe to both patient and operator

Does not cause patient discomfort

Does not require unstable chemicals

Does not require delicate instruments

Cost effective

Reagents are readily available

programme may be compromised. This is particularly important where pathogenic and non-pathogenic species live in the same environment and choice of test to fit the situation is key. As already mentioned, it is important to distinguish between *E. coli* and *E. histolytica*. But there are three *Entamoeba* spp. with morphologically identical cysts: *E. histolytica*, *E. dispar*, and *E. moshkovskii* and standard light microscopy cannot distinguish between them, so another method is needed. Similarly, if the test cannot distinguish between a parasite of wild animals and a very similar parasite of humans it can result in a costly and misdirected control programme. There are many *Plasmodium* species that infect wild birds and mammals, and these are transmitted by a range of mosquitoes. If a test was unable to distinguish between mosquitoes transmitting bird malaria and mosquitoes transmitting human malaria a great deal of energy would be spent combating the wrong mosquitoes. Sensitivity and specificity must be considered, as there is usually a balance between them in any given assay. Where parasites may be present in low numbers within the sample being analysed, then sensitivity is more important. However, low parasite abundance in a blood, tissue, or faecal sample does not always mean that the parasite is present in low numbers throughout the body. Similarly, even small numbers of the infectious stages of some parasites can cause potentially fatal disease.

The ideal test must be simple to perform, because whenever anything becomes complicated there is a greater opportunity for errors to occur. More straightforward methods also require less training before someone can conduct the test correctly, and they may also be suitable for automation. The diagnostic test should also be safe for the person performing the assay and not use potentially hazardous chemicals. It should also be harmless for the patient and not cause discomfort. For example, tests that necessitate the collection of spinal fluid (e.g., some tests for Human African Trypanosomiasis) require medical supervision are unpleasant for the patient and carry risks of causing nerve damage. If the test is painful, there is a chance that the patient might not return for follow-up investigations. Tests that require unstable chemicals and delicate instruments are seldom suitable for use in developing countries outside major cities. Cost is a major factor in any diagnostic test (just as it is in drug design and parasite control). The definition of affordability will depend upon the target population and the number of tests that are required: an 'expensive' test that is only undertaken a few times every month may be considered affordable, whilst a modestly priced test that must be undertaken many times every day can quickly become a serious burden on resources. A technically simple test, which uses relatively cheap equipment and reagents, such as blood smear analysis, can become expensive because it takes a skilled person 10–20 minutes to examine each smear. However, when considering cost, the overall expense to the health or veterinary care service should be considered. For example, if a sophisticated and high-priced test provides an answer quickly enough to reduce the number of nights the patient must stay in hospital or allows them to be treated with a cheaper drug, then it can be 'cost effective'.

The test must produce a result in time to be clinically relevant to patient care. This is a particularly important consideration in situations in which the patient could die within hours unless the correct treatment is given, such as where cerebral malaria or amoebic encephalitis is suspected. Most parasitic diseases are, however, chronic and the patient would not die if treatment was delayed. Nevertheless, a long gap between taking samples and diagnosing the infection can mean that the infected person/animal suffers more pathology and is a source of infection for others/vectors/intermediate hosts. It is also important to provide a timely diagnosis to ensure patient compliance. Whenever people are requested to 'come back in a fortnight', it is a certainty that a proportion will not. The longer the time interval between the sample being collected and the result being communicated, the more chance there is that some patients will not receive appropriate advice and may possibly miss vital treatment.

It is impossible to develop a diagnostic test that fulfils all the above criteria. For example, a test may be accurate, sensitive, and quick very expensive. Alternatively, it may be accurate, sensitive, and uses cheap reagents but so complicated that it can only be undertaken by highly trained individuals and takes a long time to perform. As in all aspects of life, the test used in a particular situation is often a compromise between competing requirements. It will also have to be chosen from the list approved by the regulatory authorities. Laboratories are not 'free-agents' and must abide by the regulations of the country within which they operate. Many countries stipulate which tests are approved for the diagnosis of particular diseases in approved or accredited settings. Any test used to make clinical decisions should have been fully evaluated. Use of a poorly performing test could affect treatment and management of an individual with a parasite infection, but also a diagnostic assay of sub-standard quality could have a 'knock-on effect' on many aspects of parasite control.

There is a bewildering array of laboratory techniques that can be applied to parasite identification, but it is helpful to think of them in two main categories. The first is the traditional approach of looking at the whole organism at a specific life cycle stage and classifying it according to morphological characteristics. The second is to assay for a defined part of the organism such as cell surface antigen, enzyme, or piece of its genetic sequence. There are also some diagnostic tests, which detect the immune response to the parasite infection. However, these tests tend to be of limited value, because although they are technically straightforward, their results are often difficult to interpret.

13.4 Isolation of Parasites

As described in the earlier chapters, parasitic protozoa and helminths often have remarkably complicated life cycles. Even the simpler protozoa with a single host, usually exist in two morphological forms, while helminths can go through many different stages as they grow and develop into adult worms. Laboratory identification of a particular parasite requires an appreciation of its life cycle, so that the most appropriate specimen can be collected from a potential host or from the environment at the most suitable time. Host and vector invertebrates also have multi-staged life cycles and may interact with protozoa, helminths, or mammalian hosts during some or all these stages. Consequently, diagnostic tests may be necessary to identify parasites from their vectors/intermediate hosts.

Wherever possible, it is always preferable to begin the identification process by observing the whole organism. The 'traditional techniques' involved in macroscopic and microscopic examination of samples and the identification of any parasites present, using key morphological features, have changed very little since they were originally developed in the late nineteenth and early twentieth centuries. Methods for sample preparation and protocols for preparing stains are thus similar in many parts of the world and standardised.

The first step in identification of a parasite can be macroscopic examination by eye or using a low-resolution (dissecting) microscope or hand lens. If the parasite is large, this can provide enough information to identify it. For example, observation of whole adult tapeworms or their proglottids, fly larvae from a case of myiasis, and scabies mites is often sufficient to identify the organism to species level. However, if it is important to further classify by strain or subspecies, then further work is required.

Many pathogenic parasites, however, are too small to be seen without the aid of a microscope in some or all their life cycle stages. This means that samples of fluid or tissue must be taken from the infected host and examined. An understanding of the life cycle and the effect of the pathogen on the host helps to decide which specimen to collect for laboratory investigation and the expected morphological appearance of any parasites which might be present. Chemical stains are often used to show up the parasite and highlight the internal structure of the cells, which helps to identify them.

Many protozoa and helminths can be detected in the peripheral blood of infected hosts by light microscopy. Blood films are prepared from whole blood – either from a small incision (e.g., finger prick in humans) or venous blood collected into a bottle containing anti-coagulant. The films are made on glass slides and then stained before examination using the microscope. In diagnostic parasitology, one usually prepares 'thick' and 'thin' films from a pin-prick blood sample. A larger volume of blood is used to make a thick film and the cells are not fixed, which causes red blood cells to lyse and release any intracellular organisms present. This is useful for making the initial decision about whether a parasite is present in the host. However, this lack of fixation means that the morphology of the parasite can be distorted, which is why the thin film is more useful for the full identification of species.

Fixed and stained slide preparations can also be made, in a similar way, from skin and tissue biopsies, bone marrow aspirates, lymphatic samples, and other body fluids, depending on the pathology and life cycle of the parasite in question within the host. Considerable time and skill are often required to detect and identify parasites in stained films. Nevertheless, direct observation of the organism is often still the 'gold standard' test method against which other assays are evaluated. It also has the benefit of establishing whether two or more parasite species are present without requiring a separate test.

Protozoan cysts, and helminth eggs and larvae from intestinal parasites, can often be detected in faeces (or urine for some species) of infected hosts. In some situations, other forms can be found as well. For example, trophozoites of gut protozoa can be excreted, particularly during acute stages of the illness, and after anthelmintic treatment adult worms are passed out in faeces. The simplest way of preparing a stool or urine sample for microscopic examination is to make a direct 'wet preparation'. If there is a high concentration of parasites in the sample, this can be sufficient to provide a qualitative result. To detect trophozoites of protozoa such as *Entamoeba* and *Giardia*, a 'warm' stool is the most useful specimen. This is, as the name implies, taken straight to the laboratory within minutes of the patient passing it, as it easier to see the motile, but transparent trophozoites as they move around in the wet preparation. Similarly, to diagnose the sexually transmitted protozoan *Trichomonas vaginalis*, wet saline preparations are made from fresh (preferably less than 1 hour old) genital swabs. The live flagellate trophozoites can easily be seen moving around in the sample. If the slide needs to be preserved for examination, then it can be fixed in alcohol and treated with a suitable stain such as Giemsa.

Most intestinal parasites can be identified from non-motile stages such as protozoan cysts or helminth eggs. Consequently, it is generally not necessary to have a totally fresh specimen and a sample up to a day old can be suitable for examination. It is usual to employ a method to concentrate the sample before making a saline preparation to maximise the opportunity to detect parasites, which might be present in relatively low numbers. Several methods have been developed, but in clinical diagnosis, the formal-ether concentration technique is the most used. Specific preparation methods are also used in surveys (usually for helminths) when attempting to quantify the concentration of eggs, which relates to worm burden, such as the Kato-Katz method (Garcia 2021). For field work on parasites of primates, Pouillevet et al. (2017) recommended the zinc sulphate McMaster floatation method.

In some species of parasite, the shedding of eggs or cysts is intermittent. The abundance of eggs and cysts within a faecal sample collected at a random time might therefore be so low that they cannot be found – even though the helminth or protozoan is present and causing obvious symptoms in the host. For example, in the case of human schistosomiasis, asking the patient to carry out some vigorous exercise, such as running up and down stairs, helps to dislodge the ova resulting in a higher egg concentration in the subsequent faecal or urine sample. In patients with clinically symptomatic and severe giardiasis, it may be impossible to find *Giardia* cysts by conventional microscopy even when samples are taken on consecutive days at different times. Direct fluorescent antibody testing (DFA) offers increased sensitivity but this is more expensive and a simpler approach is the 'string test'.

Giardia String Test (Entero-Test)

Although it is commonly known as the '*Giardia* string test', this technique can also be used to detect the nematode *Strongyloides stercoralis* and obtain samples of duodenal fluid for microbial or physiological analysis. The Entero-Test consists of a gelatine capsule within which is a weight attached by a slipping mechanism to a long length ('string') of nylon line. One end of the nylon line protrudes from the capsule, and this is taped to the patient's cheek. The patient then swallows the capsule, and as it passes down the gut, the nylon line unravels behind it. Within the stomach the gelatine capsule is dissolved, and the normal gut contractions carry the weighted line to the duodenum. After about 4–6 hours, the weight is released, and the nylon line gently pulled back up the oesophagus. The weight continues down the gut and is passed out with the faeces. The presence of yellow-green bile on the end of the line indicates that it reached the duodenum, whilst a low pH indicates that it never left the stomach. Mucus and fluid attached to the end of the line should be examined straight away for the presence of moving *Giardia* trophozoites/*Strongyloides stercoralis* although stained permanent mounts can also be made. The attached material can also be analysed for bacteria such as *Salmonella* using standard procedures. The test requires the patient to remain in the surgery for several hours and he/she can have difficulty swallowing the capsule and may experience retching or vomiting as it is removed. It is therefore not used for the routine testing for *Giardia* unless the disease is suspected but cysts and trophozoites cannot be detected in the faeces. The Entero-Test can also be used with dogs and other domestic animals although it tends to be employed as a research tool rather than for routine diagnosis.

Another interesting method of collecting parasites to take to the laboratory is used when diagnosing human pinworm, *Enterobius vermicularis*, infection. The adult nematodes lay their eggs onto the skin surface in the perianal area and the peak time for egg laying is during the night. Since the area is itchy, people commonly re-infect themselves by scratching their bottom and then putting their fingers in their mouth before washing them – a habit, which is common in, but not restricted to, small children! The best way to obtain a sample from someone with suspected pinworm infection is to put a piece of clear adhesive tape on the skin in the perianal area, in the morning, before that part of the body has been washed. The tape is peeled off carefully and then put it face down onto a microscope slide; this can then be examined at ×10 and ×40 for *E. vermicularis* eggs. Other more specialised techniques applied to faecal parasites include the agar plate method for isolation and subsequent detailed examination of *Stronyloides* and related helminth species and the *Schistosoma* hatching test to determine viability of an egg (Garcia 2021).

13.5 Identification from Gross Morphology

Blood and tissue smears need to be stained before they can be examined. Parasitic protozoa can usually be observed at ×40 although it usually requires ×100 oil immersion to be able to see their diagnostic features.

Helminth eggs tend to be large enough to be detected in faecal samples at ×10 magnification under a light microscope and examined for key identifying features at ×40. For example, even small helminth eggs such as those of the nematode *Capillaria phillipinensis* are $45 \times 21\,\mu m$, whilst most eggs are much larger than this and those of the trematode *Fasciola gigantica* are $190 \times 90\,\mu m$. In contrast, protozoan cysts found in faeces are smaller and mostly transparent, so they can be harder to find. In wet preparations, the addition of a drop of iodine is commonly used to highlight internal structures within the cysts. However, the addition of iodine does not enhance the recognition of helminth eggs, since the stain simply covers the thick walls, obscuring internal structures. Also, the ova of some helminth species, such as *Trichuris* and *Taenia*, already have natural brown colouration.

Wet saline preparations dry out within about 15 minutes, but it is sometimes necessary to keep the slide for longer and/or use additional stains, in which case it can be fixed in alcohol for a few minutes before further processing. Slides that need to be kept for further examination and future reference are stained in a permanent stain such as Trichrome stain.

Cryptosporidium cysts are small ($5-7\,\mu m$ in length) and often present in low concentration in faeces, so to enhance detection, the modified Ziehl-Neelson stain can be applied to a fixed smear (Garcia 2021). After staining, the cysts appear as distinctive pink oval-shaped structures. This draws the microscopist's eye when there only are few per microscopic field and makes them easier to see. Care, however, is needed because several other structures that might be present, such as bacteria spores, fungal spores, and even fat globules, can also take up the stain. Therefore, experienced operators are necessary to correctly interpret microscope slides. Bacterial spores are smaller than *Cryptosporidium* oocysts, whilst fungal spores are usually larger ($6-10\,\mu m$). Yeasts will sometimes take up the stain if the staining was too intense. Another widely available method is to treat the fixed faecal smear with fluorescent-labelled antibodies raised against *Cryptosporidium* cyst surface antigens; the reagent is often used in combination with anti-*Giardia* antibodies. Similarly, microsporidia cysts shed in faeces are extremely small ($1-4\,\mu m$, depending on species) and detecting them by light microscopy is difficult. Therefore, modifications to established staining protocols such as Field's or Trichrome stain have been developed. More reliable diagnosis and identification

Table 13.3 Limitations of traditional light microscope techniques for parasite identification.

Limitation	Example
Sensitivity	Where parasite abundance in the sample is low, they might not be detected. For example, in *Plasmodium malariae*, erythrocytic blood stages are often present in very small numbers in peripheral blood and could be missed on a blood slide. Similarly, when a patient is infected with *E. histolytica* but not suffering from acute dysentery, cysts might be excreted at low levels and the sensitivity of testing may be as low as around 60% (Stark et al., 2008).
Resolution	When parasites are very small (e.g., *Cryptosporidium*), it can be difficult to locate them and identify diagnostic features. This problem is exacerbated by low concentration in the sample.
Specificity	Closely related parasites share many common morphological traits and are difficult to distinguish from one another. There are no observable morphological differences between some species. The eggs of *Taenia saginata* and *Taenia solium* are the same size, shape, colour, and structure, and it is impossible to determine species from microscopic examination of the eggs alone. Similarly, cysts of the pathogen *E. histolytica* and its non-pathogenic relatives *Entamoeba dispar* and *Entamoeba moshkovskii* are indistinguishable by microscopy of the cysts.
Operative skill	To be effective, the technique depends upon the skilled and highly trained operatives.
Quality microscopes	To identify parasitic protozoa, a well-maintained high-quality microscope is required that is equipped with an oil immersion lens.

of species where that is necessary can be achieved using transmission electron microscopy (Garcia 2021). However, transmission electron microscopes are expensive, they require skilled operators, sample preparation is slow and uses toxic chemicals, and the process of examination is time consuming. Consequently, transmission electron microscopy is not suitable for routine diagnosis.

The techniques used in preparation of samples and examination for parasites by microscopy have been in use for around 150 years (albeit with some refinements and variations to suit local conditions and resources). This means that their reliability, reproducibility, and limitations are well documented. Most of the light microscope procedures outlined above are relatively inexpensive to carry out, since they use standard laboratory equipment and reagents. To identify parasites from their morphological characteristics does require skill and training, but with over a century of collective experience among parasitologists, it is not usually difficult to find someone with the expertise willing to teach others. Light microscopy also suffers from problems of sensitivity, resolution, and specificity (Table 13.3). In the case of *Leishmania* spp. in blood and tissue samples, all three problems apply. The parasites are very small (2–4 µm) and can be hard to distinguish from platelets and artefacts in stained slides. They tend to be present in low numbers, and it is not possible to identify *Leishmania* to species level solely based on morphology.

13.5.1 Morphological Identification of *Entamoeba* spp.

Trophozoites of *Entamoeba* spp. may be visible in a 'warm stool' during the acute stage of an infection. Determination of species can be difficult since all amoebae naturally vary in shape quite considerably. Active *E. histolytica* trophozoites ingest red blood cells, which the non-pathogenic species do not, so if these are observed, it can aid in identification. Viable parasites can be grown in culture, which would allow further investigation, but this is not used in routine laboratory diagnosis. Cysts may be found in a saline smear prepared from the faeces of an infected person. Concentration of the sample, for example using the formal-ether method, before preparing the smear improves the sensitivity of this test (Saidin et al. 2019). Cysts are spherical and transparent, so iodine stain is used to enhance their detection and highlight the internal structures, which aid identification. Cysts of *E. coli* are 15–25 μm in diameter and contain up to eight nuclei. In contrast, *E. histolytica*, *E. dispar*, and *E. moshkovskii* cysts are smaller – 10–15 μm – and contain one to four nuclei. Cysts also contain chromatin bodies comprising of ribosomal material, but these are more noticeable in immature cysts and less prominent in *E. coli* than in the other species (Garcia 2021). Permanent stains such as Trichrome stain highlight chromatin bodies and a skilled microscopist can usually distinguish whether the cyst is from *E. coli* or one of the other three species. However, when a smaller cyst is detected, it is very important to know the exact species to apply the appropriate treatment and management of the patient. This requires more specific testing, such as molecular techniques.

13.5.2 Morphological Identification of *Plasmodium* spp. and *Babesia* spp.

The genera *Plasmodium* and *Babesia* contain some of the most important parasites in human and veterinary medicine. However, the different species vary in their pathogenicity and require very different treatments and control measures. The 'gold standard' method in identification of these parasites is microscopic examination of thick and thin blood films. On staining with Giemsa, Field's or a similar stain, the parasite's nuclear material appears dark pink/red, and the cytoplasm shows up as blue. If parasites are present in the peripheral blood sample, they should be visible at ×100 magnification, but accurate detection and identification often require considerable skill. In malaria diagnosis in humans, both thick and thin films are usually prepared. The thick film is useful for determining the presence or absence of *Plasmodium* spp. parasites in the patient's blood and for assessing the level of parasitaemia. However, careful examination of the thin film is more help when attempting to identify the species accurately. The most common life cycle stage seen in the early stages of acute infections is the ring form trophozoite. When the blood sample is taken while the patient's parasitaemia is quite high, it is usually relatively easy to spot infected red cells, as there will be several of them all looking similar per microscopic field. For example, 2% parasitaemia translates into an average of 2 infected cells per 100 red blood cells examined. However, where the concentration of parasites in the blood is lower, for example at a very early stage in the schizogony cycle or in response to effective treatment, it is sometimes hard to discriminate between cells infected with *Plasmodium* spp. parasites and other features, such as unfortunately positioned platelets. Related apicomplexan parasites also infect red blood cells and can have a similar appearance to malaria parasites.

Differentiation of *Plasmodium* species in blood films depends on observation of a range of key features. For example, high parasitaemia at the early trophozoite stage including multiple infections of red blood cells is characteristic of *Plasmodium falciparum*. In contrast, during *Plasmodium malariae* infection, low parasitaemia is typical. *Plasmodium ovale* changes the shape of infected red cells, so that they tend to oval, with uneven edges. *Plasmodium vivax* trophozoites become enlarged and amoeboid in appearance, increasing the size and shape of infected red cells, which can develop to be as large as white blood cells. As the infection progresses, stains reveal dots ('stippling') in infected red cells, which also given indication of species. *Plasmodium falciparum* infection is associated with blue Maurer's dots (or clefts), while in *P. vivax* and *P. ovale*, pink Schuffner's dots are seen. The zoonotic species *Plasmodium knowlesi,* which is transmitted in Southeast Asia, is morphologically very similar to *P. malariae.*

Humans occasionally acquire *Babesia* infections, notably *Babesia microti* (also called *Theileria microti*), *Babesia divergens*, and *Babesia bovis*, and these can be confused with *Plasmodium* spp., particularly in the early stages of infection. However, *Babesia* trophozoites have thicker cytoplasm and in a more ovoid formation rather than a ring shape. Infected red blood cells do not become enlarged, and there is no stippling. As it matures, the trophozoite divides, resulting in up to 8 merozoites per infected red cell. These merozoites do not always separate after division, which can lead to some interesting formations, including the four-ringed 'Maltese cross'. It is possible to identify species of *Babesia* accurately from blood slides, given time and skill. However, it can be difficult as there are fewer distinguishing features than between the various species of *Plasmodium* and parasitaemia in peripheral blood can be rather low. The species of *Babesia* are divided into 'small' and 'large' piroplasms according to their morphology, which can give an indication of the species. For example, in dogs, *Babesia canis* is described as a 'large' species, while *Babesia gibsoni* is a 'small' one. Inside the red cells, *B. canis* parasites take on a pear-shaped appearance and often appear in pairs; they are about twice the size of *B. gibsoni* merozoites, which tend towards the four-pronged, Maltese cross formation.

13.5.3 Morphological Identification of *Taenia* spp. Tapeworms

The Taeniidae family of tapeworms contains the genera *Taenia* and *Echinococcus,* and there are important human and animal pathogens within both. Humans are the definitive host for some species in this family and accidental intermediate (sometimes called 'aberrant') hosts for others. Despite morphological similarities between the species, the pathology that arises within the host because of infection can be different, so it is important to identify them to species level.

A faecal sample from a person or animal infected with adult tapeworms can sometimes contain whole worms, which have been eliminated from the host, or portions of the strobila, which have broken off naturally. The species of the host can give some indication of the possible identity of the cestode. For example, we are the definitive host for *Taenia saginata* and *Taenia solium*, whereas dogs are the definitive hosts of *Taenia pisiformis*. Therefore, you might find an adult *T. pisiformis* in the faeces of a dog, but not in the faeces of the dog's owner. The length and shape of an intact worm is a useful taxonomic indicator and if a scolex (head) is found within the host's faeces, then this can be examined microscopically for anatomical structures, which indicate species. It is important to search for the scolex in the patient's faeces following treatment with an anthelmintic to confirm that he/she is cured. If the scolex remains attached to the gut lumen, then it will grow a new strobila and continue producing eggs.

Where sections of the adult strobila are found but no scolex, there is another way of distinguishing between some species, which is to look for gravid proglottids (i.e., segments containing fertilised eggs). The proglottid contains a uterus, which has branches containing the eggs. By carefully injecting the proglottid with a dye such as Indian Ink, it is possible to count the number of main uterine branches, which is characteristic of the species. For example, *T. solium* has fewer than 13 branches, while *T. saginata* has more than 13.

In the absence of the scolex or gravid proglottids, one can search for tapeworm eggs in the faeces. First, the eggs are concentrated using a suitable technique and a saline wet preparation should be prepared and examined for the presence of helminth eggs at ×10 and ×40 magnification. In terms of making a diagnosis of 'tapeworm infection', this can be helpful. The eggs are orangey brown in colour, a characteristic round, bordering on oval shape, about 40 μm in diameter and with a thick striated wall. By focussing up and down, it is possible to see the hooks inside the embryonic worm. Although only *T. solium* has hooks as an adult, all species have this feature at the embryonic stage.

13.5.4   Morphological Identification of Filarial Nematode Infections

Observations of the patient can provide useful information in filarial infections. In lymphatic filariasis, adult helminths can be observed in the lymphatic vessels in the pelvis and legs on an ultrasound scan. *Loa loa* worms are sometimes visible in the eye. Microfilariae can be observed and identified in blood films, but an understanding of the timing of the parasitaemia is important when collecting samples. The concentration of some microfilariae in peripheral blood exhibits 'periodicity', which means that it varies throughout a 24-hour period, and this variation depends on the nematode species. Some species are usually present in higher numbers during the night – 'nocturnal periodicity' (e.g., *Wuchereria bancrofti*, *Brugia malayi*, and *Brugia timori* in humans). This means that to detect and identify these species, the person collecting the blood sample must wake the patient up at midnight! (Interestingly, this periodicity seems to be a response by the parasite to physiological conditions within the host over a 24-hour period of being active, eating and sleeping, such that the peak parasitaemia timing is found to be reversed in people who regularly work night shifts). Other microfilariarae are at peak concentration in the middle of the day (e.g., *Loa loa* in humans), which means that the patient's lunch might be interrupted for blood collection) or mid-afternoon (e.g., *Dirofilaria* spp. in canids). However, some species do not show marked periodicity such as *Mansonella* spp. in humans.

Microfilariae can be identified in thick blood films made using blood collected from finger pricks. However, whenever possible, it is better to take an anti-coagulated whole blood sample, which should be concentrated by centrifugation or filtration before processing, to enhance sensitivity. The microfilariae of all species look like 'small worms' in blood films, but it is possible to distinguish between species by the arrangement of their nuclei, which is why staining is so useful (Mathison et al. 2019).

The best place to find *Onchocerca* and *Mansonella* microfilariae is usually skin snip biopsies around nodules, which might contain adult worms. If the adult worms are in the pelvic region, then the microfilariae may be found in the urine. Again, a slide is prepared and stained with Giemsa or similar stain, allowing the morphology of the worm, sheath, and organisation of the nuclei to be observed.

13.6   Biochemical Techniques for Identification

The biochemical variation in enzymes in different strains or closely related species of organisms can be exploited to aid accurate laboratory identification. Different species produce isoenzymes, which have distinct separation patterns in electrophoresis. Research into isoenzymes of isolates, which were assumed to be different strains of *E. histolytica* spp. in humans, led to the conclusion that there were two distinct species of *Entamoeba* – the pathogenic *E. histolytica* and the non-pathogenic *E. dispar*. This formed the basis of a diagnostic test, but this has now been superseded by molecular methods such as PCR (Saidin et al. 2019).

Differences in isoenzymes have also been exploited in malaria detection and species identification. The observation that *Plasmodium* has a specific parasite lactate dehydrogenase enzyme, which is different to that of the host cell and that *P. falciparum* produces a version which is biochemically distinct from other *Plasmodium* spp. led to the development of the OptiMAL test for human malaria (Moody et al. 2000). This is an 'antigen capture' rapid test that is an excellent means of providing a rapid preliminary positive/negative indication of *P. falciparum* infection (Mathison and Pritt 2017). Confirmatory tests are usually still required, and it is somewhat less reliable as an indicator of infections with other species of malaria.

The technique of Matrix-Assisted Laser Desorption/Ionisation Time-of-Flight – Mass Spectrometry (MALDI-TOF MS) to analyse the protein profile of microorganisms was developed in the late 1980s. A pure sample of the organism under investigation is mixed into a matrix. This is a solution of very specific molecules in crystallised form and solvents. The sample of interested is added and proteins are ionised using laser beams. The time each fraction takes to move through a mass spectrometer ('time of flight') under known conditions is recorded. From this, the mass to charge ratio (m/z) is calculated for each protein and a profile of the organism is built up. It needs a sample of the organism to test, and this is usually obtained by *in vitro* culture. A small amount of the isolate is placed in the analyser, and the resulting mass spectroscopy reading is compared with the information in the MALDI-TOF database (Singhal et al. 2016). It is now widely used in diagnostic microbiology laboratories for species identification of bacteria and fungi (Patel 2015). It has also proved useful in distinguishing between species of the rare pathogenic alga, *Prototheca* spp. (Ahrholdt et al. 2012).

The MALDI-TOF technique is quick and easy to perform and, once the system is up and running, the cost per test is relatively low. The protein profile of an organism can also indicate strain variation and whether that isolate might be resistant to antimicrobial drugs. One of the disadvantages is that there must be something in the database to compare the isolate with – but the information is constantly updated as new strains are found. The main problem for its application in diagnostic parasitology is the requirement to culture the organism, which is often tricky and may involve growing it in laboratory animals. Since parasites are much more complicated biologically than bacteria and fungi, there is also the issue of making sure that proteins produced at each stage of the life cycle are catered for in the database profile. Therefore, in parasitology, it is likely to be more useful in reference laboratories and as a research tool. Nevertheless, some promising findings have been reported using MALDI-TOF for species identification in several protozoa including *Leishmania* spp. (Mouri et al. 2014), *Cryptosporidium* spp. and *Giardia* spp. (Singhal et al. 2016). It has also been used to help identify the species of mosquitoes, ticks (Yssouf et al. 2016), and fleas (Singhal et al. 2016) from isolated body parts (e.g., legs). Progress in adapting the technique to identification of helminths has been more limited (Feucherolles et al. 2019). This is due to a combination of factors such as the difficulty of culturing helminths *in vitro*, the fact that many research projects would rely on stored samples, which have been preserved in a fixative and the limited information in the database.

13.7 Immunological Techniques for Identification

The use of immunologically based assays to detect both antibodies against specific pathogens and the antigens themselves in clinical specimens has proved very successful in diagnostic microbiology. Whatever the format, the assay is always based on providing the right conditions for a specific antigen–antibody reaction and a suitable system to visualise and measure the reaction. Tests to detect parasite antigens are especially useful when direct observation of the organism in a specimen by microscopy is difficult. One of the most used methods is enzyme-linked immunosorbent assay (ELISA), also known as immunoassay (EIA), to detect parasite cell surface antigens. Antibodies are raised against specific parasite antigens, and these may be linked (conjugated) to an indicator molecule such as a dye or an enzyme that initiates a chromogenic (colour forming) reaction. This enables the presence of the antibody to be determined. Alternatively, the antibody is not conjugated, and it is detected through a secondary (conjugated) antibody that is raised against it. Either the sample containing the presumed antigens, or the antibodies, can be immobilised onto a solid support (i.e., solid phase) such as the wells of a microtitre plate. Then, depending upon which component is the solid phase, either antibodies or the sample antigens are added. If the parasite antigens are present, then the antibodies will bind to them. The wells are then washed to remove unbound reagents, and the presence of the antigen–antibody complex is demonstrated through the indicator molecule or through adding a secondary conjugated antibody. If the parasite antigen is not present, there will be no reaction. Immunofluorescence antigen testing (IFAT) uses the same principal with fluorescent dyes acting as the indicator molecules and is widely used to test for the presence of parasites in faeces, tissues, and body fluid samples.

ELISA is a commonly used laboratory technique to detect the presence of *E. histolytica* in faecal samples (Saidin et al. 2019). The solid phase is a monoclonal antibody raised against an *E. histolytica* – specific cell surface antigen, such as the Gal/GalNAc lectin. This has been widely evaluated in studies in areas of low and high prevalence throughout the world and is reported to have generally good sensitivity and specificity. However, the formalin used in the concentration of faecal samples (and sometimes to preserve samples during transport to the laboratory) can have a detrimental effect on the lectin, and if this occurs, it necessitates a new aliquot of the faeces. Although the more severe and life-threatening clinical conditions such as dysentery and amoebic liver abscess are associated with *E. histolytica*, abdominal symptoms sometimes occur because of *E. moshkovskii* and *E. dispar*. There is some evidence for cross reactivity between these three *Entamoeba* species in some commercially available ELISA tests, so some samples might give a false-positive result for *E. histolytica* (Saidin et al. 2019). Also, the epidemiology and relative distribution of the three species remain uncertain. Thus, while the ELISA tests allow the identification of *E. histolytica*, it might still be important to determine whether *E. dispar* or *E. moshkovskii* is present in a sample – and if so, which one.

Although the detection of helminth eggs in faeces is a relatively quick and simple way to diagnose tapeworm infection, it is not always very sensitive, as an infected person or animal does not excrete eggs continuously. Consequently, an immunoassay technique to detect *Taenia* antigens in human faeces and *Echinococcus* antigens in canine faeces was developed during the 1990s (Allan and Craig 2006). This 'coproantigen' ELISA is a capture assay, designed to detect antigens from eggs or proglottids in a faecal sample. The solid phase is an antibody raised in rabbits, against antigens extracted from adult worms. The assay has better sensitivity than microscopy, but it cannot distinguish between the species due to antibody cross reactivity to genus-specific antigens contained in the worm extract (Praet et al. 2013). Guezala et al. (2009) designed a *T. solium*-specific assay using a different antigen preparation to raise antibodies. This is clinically useful because it enables identification of patients at risk of developing cysticercosis.

ELISA and IFAT methods can also be used to detect serum antibodies, which are widely used in diagnostic microbiology as markers of infection. Protozoa and helminths are more complex organisms than viruses and bacteria, and they present a variety of antigenic components to the host immune system during their life cycles. This means that unfortunately, the immune response to most parasites is poorly understood, which has made the development of assays for specific antibodies difficult. While in assays to detect antibodies to virus or bacterial infections, the solid phase is purified, usually monoclonal antibody, in the case of parasitic infections it is often not possible to identify the key antigen, which the host responds to. Also, as the parasite changes morphologically during its life cycle stages, it also changes biochemically, and the host response to a particular antigen might be transient. For these reasons, there are a limited number of assays to detect antibody response to protozoa and helminths and those which are available usually detect total antibody or IgG, rather than IgM, which would indicate current or recent infection. Nevertheless, the detection of parasite-specific antibodies is particularly useful in the diagnosis of neurocysticercosis.

Neurocysticercosis is normally acquired by ingesting *T. solium* eggs, so the patient may not be harbouring an adult worm themselves meaning faecal analysis would not be helpful. For human *T. solium* cysticercosis, the enzyme-linked immunoelectrotransfer blot (EITB) is considered the best method of antibody detection. It incorporates purified preparations from seven different cysticercus proteins and the antibody–antigen reaction is detected by chromatography. There are also some sandwich ELISA assays available, which are simpler and cheaper to run, but they are often less sensitive than EITB (Garcia et al. 2003). All serological assays measure total or IgG antibody, which is the main class of IgG produced in response to *Taenia* infections. This makes serology limited for screening or epidemiology because it cannot distinguish between active and resolved, past infection. However, serological assays can be very useful for the initial diagnosis of cysticercosis in conjunction with clinical symptoms and a CT or MRI scan. Quantitative formats of these assays are available, which can be used to monitor the progress of patients on treatment. Similarly, assays have been developed to detect cysticercosis caused by *T. saginata* in cattle and *T. solium* in pigs (Bustos et al. 2019). However, their performance can be variable (Garcia et al. 2018), so test kits should be carefully evaluated before use.

The application of serology to detect markers of schistosomiasis in humans has proved even more complicated. Researchers have developed tests to detect serum antibodies against all the stages of the *Schistosoma* spp. associated with human infection – cercariae, schistosomulae, adult worms, and eggs, as well as antigens produced during the infection (including circulating anodic antigen [CAA] and circulating cathodic antigen [CCA]) (Hinz et al. 2017). With this array of options, it is theoretically possible to distinguish the acute IgM response from the later IgG antibody in the patient and to use detection of antibodies to the various life cycle stages of the worm to indicate whether the person has a new or chronic infection or has recovered from treatment. Unfortunately, the performance of the tests themselves, in terms of laboratory quality assurance parameters, is extremely variable and often not up to acceptable standards for making clinical diagnoses. For example, evaluation of tests to detect IgG against cercarial antigens against *Schistosoma mansoni* found that they were quite sensitive, but there was cross reaction against antigens from bird and animal schistosomes (causing cercarial dermatitis), as well as, strangely, hepatitis B markers; conversely, assays for IgG against *Schistosoma haematobium* cercarial antigens showed good specificity, but low sensitivity (Hinz et al. 2017). Performance also seems to be affected by whether the samples are taken from people in an endemic area or travellers returning to non-endemic regions presenting at travel clinics or hospitals for tropical diseases. This makes it hard to make decisions about whether and when a particular test is clinically helpful.

13.8 Molecular Techniques for Identification

An obvious means of identifying organisms is through their unique DNA profile. However, sequencing the whole genome would be far too costly and time consuming to undertake on a routine basis. Consequently, scientists look for regions of the genome that exhibit variability. There is no single locus that is suitable for all organisms, but amongst the eukaryotes the mitochondrial genes for cytochrome oxidase I (COI) and cytochrome oxidase II (COII) have proved effective for differentiating between many animal species. Some parasitic protozoa lack functional mitochondria, and these tend to be differentiated based on various markers although the 18S ribosomal RNA (18S rRNA) gene has proved particularly useful. There are standard protocols for the extraction and analysis of DNA, and this facilitates standardisation between laboratories. The amount of parasite DNA within a sample is extremely small, but the region of the genome of interest can be amplified using nucleic acid amplification techniques (NAATs), such as the polymerase chain reaction (PCR). The primers used to identify the beginning, and end of the sequence of interest can be labelled with differently coloured fluorescent dyes. Consequently, after separation (e.g., electrophoresis), the PCR products can be detected by exposure to a laser beam that induces fluorescence at specific emission wavelengths that are then detected in a recording CCD camera.

Quantitative (real time) PCR is a PCR-based technique that can be used to both amplify and quantify the targeted DNA. Two of the most common means of quantification are the inclusion of a fluorescent dye (e.g., SYBR® Green) in the PCR reaction that intercalates with the DNA as it is produced and the TaqMan® assay. The TaqMan® assay uses custom-designed sequence-specific primers and is therefore better at detecting individual species or variants within a species than the fluorescent dye intercalation method. It is also possible to include several different primers within an assay to detect multiple targets – this is known as a 'multiplex assay'. An example are kits intended to simultaneously detect *E. histolytica, Giardia duodenalis,* and *Cryptosporidium parvum/ hominis, and Dientamoeba fragilis* in one faecal sample (Autier et al. 2018). In both techniques, with each cycle of the PCR process more DNA is produced, and this is measured as an increase in fluorescence. The DNA product is therefore 'quantified' as it increases in 'real time', and hence the terms 'real time' and 'quantitative' PCR. Once sufficient DNA is produced, it can be subject to further analysis such as sequencing or Southern Blotting. A good quality check is to subject the PCR products to DNA dissociation (melting) curve analysis to confirm that the correct product was formed: artefacts would yield a differently shaped dissociation curve. A new type of PCR analysis called digital PCR could replace quantitative PCR in the future (Huggett et al. 2015). This procedure amplifies DNA exclusively from a single template and converts the signals to a digital output (*c.f.* the linear output generated by conventional PCR). Consequently, digital PCR is more effective than real-time PCR at detecting and amplifying low copy number templates, which would be the situation if there was a low parasite density. In addition, it is easier for the operator to quantify the amount of DNA present and undertake statistical analysis of the PCR product (Pomari et al. 2019). Further details on molecular diagnostic techniques can be found in Buckingham (2019).

In addition to identifying suitable regions of the parasite genome, it is important to be able to extract the DNA for analysis. Faecal samples present difficulties because they contain bile salts and plant polysaccharides that interfere with standard DNA extraction and PCR techniques. The parasite DNA also must be differentiated from the enormous amount of bacterial and viral DNA that is naturally present in faeces. However, there are now commercially available kits for the extraction of DNA from faeces such as the QIAGEN QIAmp® DNA Stool Mini Kit. This will also extract DNA from other samples containing high levels of PCR inhibitors.

Molecular techniques are available for the identification of an increasing number of parasites. These tests usually have better sensitivity and specificity than conventional light microscopy and antigen detection immunological methods. They can therefore be particularly valuable when examining samples for the presence of parasites, which might be present in low concentrations in the sample (e.g., *Giardia* spp. cysts), are small and difficult to spot on a microscope slide (e.g., *Leishmania* amastigotes spp.), and where it is difficult to distinguish between species morphologically or antigenically (e.g., *Entamoeba* spp., *Taenia* spp.). Nevertheless, molecular methods are not invariably an improvement on simpler morphological techniques. For example, a survey of laboratories across Europe found that microscopy was more reliable than molecular methods for the detection of taeniasis in faecal samples (Gómez-Morales et al. 2021). For a detailed consideration of the issues involved, see Papaiakovou et al. (2019).

Tests for the detection and characterisation of parasite DNA have allowed diagnostic laboratories to made important contributions to patient care and epidemiology. These include detection and then determination of the species of *Leishmania* parasites in humans, animals, and in sandfly vectors (Weirather et al. 2011), the distinction between *Plasmodium malariae* and *Plasmodium knowlesi* infections in humans in Southeast Asia (Singh and Daneshvar 2013) and the identification of anthelmintic resistance nematode strains in sheep and cattle (Ramos et al. 2020).

Commercial manufacturers now produce user friendly kits and affordable analysers for the detection of a growing repertoire of parasite infections using NAATs. Although assays are expensive relative to other diagnostic methods, this is often outweighed by the advantages of rapid accurate results. Furthermore, the standardised protocols facilitate mechanisation, thereby reducing labour costs. As with many things in life, as more and more laboratories adopt molecular techniques, more kits are produced, the market increases, and the cost decreases. A common criticism of molecular techniques is that they can only detect the parasite being tested for, but this can be overcome to some extent through multiplex assays. Examples of this include multiplex PCR assays designed to detect various soil transmitted helminths in humans (Stracke et al. 2019) and *Plasmodium* spp. in both humans and *Anopheles* mosquitoes (Del Puerto et al. 2020). This approach can also be helpful in surveys – a sample from one individual can be tested for several possible parasite infections. Tu et al. (2021) used this approach to analyse blood samples taken from goats in Thailand to assess prevalence of *Plasmodium caprae*, *Babesia* spp., and *Theileria luwenshuni* simultaneously. There are also multiplex 'panels', such as the gastrointestinal panel, which can detect various bacteria, viruses, and protozoa associated with diarrhoea (Park et al. 2017).

How Diagnostic Techniques Can Influence Epidemiological Studies

Molecular diagnostic techniques sometimes make us re-assess our understanding of parasite epidemiology. For example, in *Giardia duodenalis*, phylogenetic analysis of sequences of loci within the small subunit 18S ribosomal RNA, particularly the regions designated *gdh* and *tpi*, suggests that this parasite exists in several distinct 'strains', which fall into one of eight genetic groups (also called 'Assemblages' i.e., A–H), based on genotype and host mammal(s). The genotypes associated with human infections fall into two broad groupings called variously in the literature 'Polish group' and 'Belgian group', 'Group 1/2' and 'Group 3', 'Genotype A' and 'Genotype B' or 'Assemblage A' and 'Assemblage B', with most authors tending to favour the latter. Isolates with sequences which place them in Assemblage A have been recovered from humans, livestock, dogs, cats, beavers, guinea pigs, slow loris and for Assemblage B humans, dogs, beavers, slow loris, chinchillas, and siamang (Thompson 2004). All human isolates fall into Assemblages A or B. Within Assemblage A, there are two distinct 'subgroups' I and II. Group AI contains isolates from humans and some other mammals, but AII comprises genotypes found in humans only. Assemblage B is a more diverse grouping. The other groups have more restricted host ranges – for example, Assemblage C and D genotypes have only been found in canids – and there is no evidence of human infection. Assemblage E is associated with hoofed animals, F with cats, G with rodents, and H with seals and sea lions (Cacciò et al. 2018). In fact, there is now evidence of sufficient difference genetically that some authors have proposed separating the Assemblages into individual species, compromising a *Giardia duodenalis* species complex!

The ability to identify strains of *Giardia duodenalis* more precisely provides the opportunity for clearer insights into the organism's epidemiology. For example, it is now possible to trace sources of outbreaks accurately and to ascertain whether zoonotic transmission has occurred (Thompson et al. 2008). The evidence from the limited number of such studies that have so far occurred is intriguing. Cases have been found of humans being infected with 'animal' assemblage types of *Giardia*. For example, parasites identified as assemblage E have been found in faecal samples taken from humans with symptoms of diarrhoea and in contaminated drinking water. There have also been outbreaks of giardiasis involving humans and farm animals where the same strain of an assemblage E *Giardia* was isolated from all patients. Transmission among animal hosts also seems to occur; assemblage F (cat) types have been detected in humans, cattle, and pigs in various studies (Cacciò et al. 2018). Strains considered to be human-specific have been noted in dogs (Heyworth 2016). It is worth noting that most of these reports are sporadic cases, concerning small numbers of subjects and more extensive systematic surveys might help to bring more clarity.

One does, also, need to be careful with studies based on recovering cysts from faeces because there is always the possibility that they might have got there by accident. For example, dogs are sometimes coprophagic and some breeds are notorious for their propensity for rolling in farmyard manure etc. – and would therefore consume protozoan cysts and helminth eggs when grooming afterwards. Therefore, they could act as transport hosts – the *Giardia* cysts consumed with faeces travelling through the gut and not hatching. Similarly, dogs, like cats and other carnivores, will consume the guts of prey and may therefore pass the parasites of their prey animal with their own faeces.

Although as mentioned above it is not feasible or necessary to probe samples for the whole genome of an organism in routine diagnostics, the technique of whole genome sequencing (WGS) or next generation sequencing (NGS) have been introduced into clinical microbiology (Balloux et al., 2018) These methods use high-throughput analysers to sequence all the genetic material in each sample. This generates a large volume of sequences, mostly in fragments that are pieced together and then compared by a powerful computer to a sequence database to find any potential organisms of interest. This is a 'catch all' method, which is therefore a solution to the problem posed by standard NAATs of needing to choose sets of probes in advance and therefore possibly missing significant pathogens present in the sample. New organisms can also be found using WGS/NGS – for example, that is how the SARS-CoV-2 virus, which causes COVID-19 was first identified (Zhu et al. 2020).

In parasitology, these techniques have been used in research, including investigation of specific host-pathogen interactions (Kooyman et al. 2019), strain variation (which is helpful in epidemiology) (Figueiredo de Sá et al. 2019) and development of anti-parasitic drug resistance (Doyle and Cotton 2019). One interesting area that genomics is contributing to is the assessment of the gut microbiome. The bacterial composition of the gut is now recognised to be vitally important to many physiological processes. Undoubtedly, there are also interactions occurring between these bacteria and gut protozoa and helminths. Furthermore, there is also a complex gut virome consisting of viruses that infect bacteria and other prokaryotes, as well as those that infect commensal and parasitic protozoa and helminths. The gut should therefore be considered as a highly complex ecosystem (Marzano et al. 2017). Once these interactions are more completely understood, it could lead to new treatments for bowel conditions or changes in dietary advice for people/animals with particular pathological conditions.

It could be incredibly useful to use WGS to track strains of a parasite such as *Cryptosporidium* during an outbreak (Morris et al. 2019), but it is more difficult to apply to parasites than to bacteria or viruses. One reason is that hosts usually excrete oocysts intermittently, and so they are found in relatively small numbers in individual faecal samples. This often does not yield enough DNA to do the extraction and sequencing successfully. The other issue is with the genome itself. Parasite genomes are much larger than for bacteria, so there are more chances for error in the sequence reassembly process. Also, this means it is harder to sequence the whole organism, and so far, there are fewer published whole genetic codes for protozoa and helminths. However, WGS is a very powerful tool, with the potential to enhance diagnostic parasitology services.

13.9 Diagnostic Testing of Parasitic Infections Outside the Laboratory: Introduction

It is important to identify accurately parasitic infections to help treat individual patients and as part of control programmes. The breadth and depth of tests, which can be done in a main laboratory, offer the best way to achieve this. There will be specialised equipment and staff with suitable expertise to choose the right tests and interpret the results. However, there are occasions when the patient, clinician, and laboratory scientist are a long way from a fully equipped laboratory. Light microscopy to detect parasites in patient samples can be done in small clinics and primary care settings. The repertoire would be a bit limited, and the drawbacks of microscopy include relatively low sensitivity and lack of discrimination of species for some organisms. However, the techniques involved are standardised, and because natural light can be used, a reliable source of electricity is not essential. Thanks to imagination and innovation by scientists, it is now also possible to do versions of some of the more complex assays 'out in the field'. Available options include immunochromatographic (also called lateral flow), tests and portable versions of molecular detection methods (Momčilović et al. 2019). It is worth noting the terminology used in the literature when you are researching this topic. Laboratory diagnostic scientists tend to use point-of-care testing (POCT) and other healthcare professionals sometimes use near patient testing (NPT). Parasitologists often prefer the term rapid diagnostic test (RDT). All these names are ways of describing testing outside of the main laboratory.

To be helpful for patient care, any test designed for use outside of the laboratory must be straightforward to use and produce a result that can easily be interpreted. Ideally, it should be user-friendly enough that you do not need specialised scientific knowledge or laboratory experience to carry it out. Someone who has been fully trained should be able to do the procedure and obtain a clear result. As in all aspects of life, there is a trade-off between speed, ease of use, and accuracy, so RDTs are not the most appropriate method to use in all situations. There is a certain amount of compromise (particularly with sensitivity). Since they are a miniature version of standard assays, the cost per test is often greater than for the main laboratory equivalent. However, technical advances in recent years have vastly improved the test performance, and as their use becomes more widespread, they are becoming more affordable. The repertoire of assays is increasingly steadily, so RDTs are already available for many pathogens. This chapter will consider examples of RDTs for selected parasites to highlight the types of tests available, their advantages and limitations, explore how they are evaluated, and how they can be used.

13.9.1 Immunochromatographic (Lateral Flow) Tests

Immunochromatographic tests (ICTs) are essentially immunoassays, with all the components condensed down onto a cellulose strip (the home pregnancy test is a good example of this format). They are also often called lateral flow devices (LFDs) and that is the term which will be used in this chapter.

The LFD test strip typically consists of a long rectangular strip of nitrocellulose, and this is placed in the sample, which may be blood, serum, urine, faeces, saliva etc. The contents of the sample then migrate up the test strip by capillary action as in chromatography. A common LFD design is to have the sample molecules first meet a 'sample pad' within which are adsorbed detection molecules such as a colloidal metal or dye bound to the antigen/antibody designed to detect the sample molecule of interest – this is known as the 'detection conjugate'. The antibody/antigen in the sample interact with the antigen/antibody in the sample pad and the resultant complex continues up the LFD strip. The complex is then captured by immobilised antibodies/antigens at the test and control reagent lines. The test line captures the antibody–antigen–conjugate complex, whilst the control line captures the 'detection conjugate' and thereby confirms that it is present and has migrated up the strip. If the antibody/antigen of interest is present in the sample, there will be two lines, but if only the control line is present, then the antibody/antigen is absent. If there is no control line, it indicates that the test has not worked and must be repeated. Most LFD strips can be read with the naked eye within 30 minutes.

In the same way as standard immunoassays, LFDs can be set up to test for antibody or antigen in a patient's sample. However, as noted in the previous chapter, antibody detection is not always the optimal way to diagnose a parasite infection and most devices test for parasite antigens. A variety of specimen types can be used. They all need some sort of preparation before being added to the device. Blood and urine are the easiest to work with, as they are already liquid, but other samples such as faeces and throat swabs can be pre-treated to make them suitable for testing. In this section, we will consider LFDs designed to detect antigens for selected blood and faecal protozoa and examples of devices to test for antibodies against helminths. The strips can be (and often are) set up to test for more than one substance at a time. Examples of this include a test to detect both IgM and IgG antibodies raised against *Toxoplasma gondii* in blood (Mahinc et al. 2017) and tests strips which are designed to pick up antigens from both *Giardia* spp. and *Cryptosporidum* spp. in faeces (Bitilinyu-Bangoh et al. 2019).

13.9.1.1 Detection of Parasites in Blood with LFDs
13.9.1.1.1 *Detection of* Plasmodium *spp. with LFDs*
The discovery that *Plasmodium*-specific proteins could be detected in peripheral blood of patients with malaria infection led to the development of LFDs to detect this parasite. These were introduced into routine diagnostic parasitology in the late 1990s (Cunningham et al. 2019). The first kits which came onto the market detected a protein produced uniquely by *P. falciparum* called histidine-rich protein-2 (HRP-2). This was excellent as a rapid test for the most serious species of *Plasmodium* but had the major disadvantage of being unable to detect any other malaria species. Therefore, the blood sample from a patient with non-*falciparum* malaria would produce a negative result. The test set up was subsequently refined to include a second marker, an aldolase produced by all *Plasmodium* species – usually referred to as 'pan-malarial antigen' (PMA). Where the test trip showed a reactive line for the PMA but not the HRP-2 that would indicate a diagnosis of malaria; a follow up test such as microscopy of a thin blood film could then be carried out to determine the species. This was an acceptable format. It is important to diagnose *P. falciparum* malaria as soon as possible due to the possibility of the development of cerebral malaria. It is also necessary to determine the exact species in the case of infections with other *Plasmodium* species to ensure that the patient receives the correct treatment (e.g., against hypnozoites in the case of *P. vivax* or *P. ovale*). Formats of the LFD based on HRP-2 detection are in widespread use (Cunningham et al. 2019). Unfortunately, some strains of *P. falciparum* have now developed a gene deletion, which changes the histidine-rich protein such that most LFDs cannot detect it (Verma et al. 2018).

The second type of assay relies on the fact that there is a specific *Plasmodium* isoform of lactate dehydrogenase, usually called parasite LDH (p-LDH). The test is designed to detect both *P. falciparum* specific p-LDH and a p-LDH produced by all species of *Plasmodium*, which means that it should be able to pick up any species, but also specifically identify *P. falciparum*.

Both the HRP-2-based and the p-LDH-based assays use the antigen capture principle with monoclonal antibodies raised against the target antigens. A small amount of the patient's blood is lysed and added to the test strip allowing the liquid to move along by capillary action. When antigen in the sample meets antibody on the strip, it becomes bound. An inbuilt negative control shows whether the test has been performed correctly and all reagents are reacting as expected (Figure 13.1).

These LFDs to diagnose malarial infection are used in two ways. In areas where malaria is an uncommon infection, most patients will have acquired the disease while visiting a country where it is endemic. Expertise in identifying *Plasmodium* spp. by microscopy among laboratory staff in these areas may be limited, but also it is very important to make a quick diagnosis. Therefore, LFDs are very useful and can be followed up with blood slide microscopy or PCR for more detailed information, such as confirmation of species and level of parasitaemia. The other situation where LFDs are invaluable is in places where there is no laboratory on hand. This is particularly effective in malaria endemic areas, where healthcare and community workers can be trained to carry out the test and make assessments regarding treatment and management of patients (Boyce and O'Meara 2017).

In 2010, the World Health Organisation produced a policy recommending that no one should be given anti-malarial treatment unless the *Plasmodium* infection had been confirmed by a parasite test (WHO 2015). This led to more companies developing and marketing LFDs for malaria testing

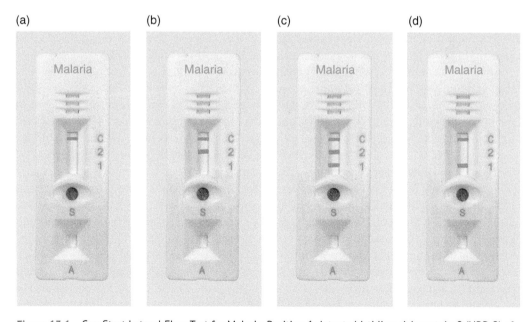

Figure 13.1 CareStart Lateral Flow Test for Malaria. Position 1 detects histidine-rich protein-2 (HRP-2) of *P. falciparum*; Position 2 detects *Plasmodium* parasite lactate dehydrogenase (p-LDH) for all *Plasmodium* species. (a) Uninfected – there is only a positive reaction at the control line. If no line is detected here, it indicates that the test has not worked and any lines occurring elsewhere are invalid. (b) Lines at the control and position 2 indicates non-*falciparum* malaria. (c) Lines at the control and both positions 1 and 2 indicate either *P. falciparum* or a mixed infection of *P. falciparum* and another species of malaria. (d) Lines at the control and position 1 indicates *P. falciparum* infection.

and thus came a requirement for careful and systematic evaluation. Performance of the kits is assessed in comparison with microscopy by a skilled experienced scientist and a polymerase chain reaction. This led to a significant increase in the quality of the tests available world-wide. The Foundation for Innovative New Diagnostics (FIND) also introduced a system for sampling batches (called 'lot testing') of each kit to make sure they perform as expected (Cunningham et al. 2019). The acceptable limit of detection for malaria LFDs is 200 parasites per microliter of blood. Jimenez et al. (2017) have pointed out that there is now a need for quality assurance standard reference reagents to be available more widely, particularly to monitor whether LFDs in use can pick up variant strains. Improvements in the test kits mean that they are better at detecting *P. falciparum* and *P. vivax*, which is good news for parasitologists, healthcare professional, and of course the patients affected by these organisms. However, the ability of LFDs to pick up *P. ovale*, *P. malariae*, and *P. knowlesi* infections is currently still less than optimal.

13.9.1.1.2   Detection of Leishmania *spp. with LFDs*

As discussed previously, identification of *Leishmania* spp. parasites in patient samples can be quite tricky. However, precise identification is important for individual patient care and for epidemiological purposes. Microscopy of blood or tissue biopsy samples requires skilled laboratory scientists has relatively low sensitivity and cannot determine exact species (Reimão et al. 2020). Molecular methods such as polymerase chain reaction are much more accurate and lead to full species identification. The genomes of many *Leishmania* species have been sequenced and published (Cantacessi et al. 2015). Consequently, whole genome sequencing can be performed on samples taken from patients suspected of suffering from *Leishmania*, but this is only available in specialist and reference laboratories.

Visceral leishmaniasis (VL) is a systemic infection, and it does not have any obvious clinical features that distinguish it from many other illnesses (e.g., malaria). Therefore, diagnosis relies on laboratory tests to confirm any suspected infection. This is often achieved through detection of antibodies raised against *Leishmania* antigens. Enzyme immunoassays which detect antibody raised against a parasite antigen called rK39 have proved helpful in the diagnosis of visceral leishmaniasis (Maia et al. 2012). Other similar antigens such as rKE16 have also been employed in immunoassays but the scientific principle behind the tests is the same. They are particularly useful in endemic areas where access to 'state of the art' molecular tests might be limited. This has been adapted into a lateral flow format, which can be used outside of the laboratory and several kits are now available. Their performance has been evaluated in comparison to main laboratory serology assays and PCR of blood and biopsy samples. Their accuracy is rather variable with relatively low sensitivity but also poor specificity in some cases (Kassa et al. 2020). Cunningham et al. (2012) evaluated a range of kits and highlighted differences in sensitivity (i.e., their ability to detect antibodies to *Leishmania* if they were present in the patient sample). The samples they tested came from three areas where VL is endemic (and where LFDs could be useful), namely India, Brazil, and East Africa. Their results showed that the tests performed best at detecting *Leishmania* in samples taken from patients in India. This seemed to be in part due to variations in species and strains of *Leishmania* being transmitted in the three areas. This example highlights the importance of considering, and keeping up with, differences in the organism itself when developing any diagnostic test in microbiology. Since these tests are looking for antibodies in blood (or serum), it is perhaps not surprising that samples from patients who are immunocompromised (such as people living with HIV) often return a false negative result. This is a concern from a diagnostic point of view, since *Leishmania* and HIV co-infection is a considerable public health issue in many countries where VL is endemic. Thus, a negative LFD test result should not be taken as proof that the person does not have VL. Also, for any positive LFD result, more samples must be collected for follow-up laboratory tests to determine the exact species of *Leishmania* infecting the patient and plan their treatment and management. Rapid LFD kits to detect canine VL are also available (Laurenti et al. 2014). These are proving useful in diagnosing leishmaniasis in sick dogs and screening stray dogs in endemic areas. It is important for both animal welfare and to reduce the transmission of zoonotic *Leishmania* infections between dogs and humans.

While the clinical picture for cutaneous leishmaniasis (CL) is clearer than for VL, laboratory tests are also important. As noted previously, the established method of detection of *Leishmania* in blood or skin biopsy samples is an expert job; examination of a stained slide to look for the *Leishmania* parasites takes skill and time. Analysing samples by PCR would be the best option where it is available, but that is not always feasible. Therefore, a rapid test for CL could clearly lend itself to some situations. Rapid LFD kits are available, which are designed to the detect specific parasite antigen peroxidocin that is produced by amastigotes. Scrapings are collected from suspected CL lesions, lysed in a special buffer, and the resulting suspension is applied to the test strip. Evaluations of the LFDs in comparison with PCR have repeatedly shown that it shows good specificity, and therefore, a positive test result can be taken as reliable. However, their sensitivity tends to be poor, which means that samples showing a negative result must be tested again using a different method in case it is a false negative. This is a key limitation of LFD tests for CL (Bennis et al. 2018). Some studies have compared LFD tests with testing using with a rapid and relatively straightforward molecular test method called loop-mediated isothermal amplification (LAMP). Schallig et al. (2019) recommended the LAMP assay, while Vink et al. (2018) devised an algorithm in which the two tests were used in conjunction with one another. Samples which gave a negative result using LFD were then tested with the LAMP, and this combination reportedly picked up over 90% of CL cases. At the time of writing, this approach is likely to be more expensive, but it is more accessible and so could lead to more people receiving and early diagnosis and suitable treatment.

In conclusion, while LFDs have a role in diagnosing leishmaniasis, they should be used in conjunction with other clinical and laboratory tests. Nevertheless, they will pick up some cases of VL and CL that would otherwise have been missed.

13.9.1.2   Detection of Parasites in Faeces with LFDs
13.9.1.2.1   Detection of Faecal Protozoa with LFDs
Microscopy to detect faecal protozoa has several limitations. Faeces is a difficult sample to work with and protozoan cysts are often present in low numbers, so a concentration method is often recommended. The second problem is that cysts are small and hard to spot. Some solutions to this have been found, including immunoassays such as IFAT and ELISA to detect faecal antigens, as well as molecular methods for detection of parasite genomes. To date, most of the focus has been on the diagnosis of giardiasis, cryptosporidiosis, and amoebiasis. These all cause acute and chronic infections, which although potentially life threatening, can usually be treated effectively if diagnosed in good time. Most of the acute and serious gastrointestinal infections worldwide are caused by bacteria and viruses, so diagnostic scientists tend to have less experience with the laboratory identification of the cysts and trophozoites of protozoans. They are also less common in developed countries, so assays which do not involve microscopy are useful and cost effective (the kit is expensive but so is staff time and microscopy takes longer).

Laboratory-based immunoassays of faecal protozoan parasites have been available for over 20 years and have been applied to investigations of infections in travellers and outbreaks (e.g., those associated with contaminated water sources or food). Initially, there were problems with preparing the faecal samples to ensure that the protozoan cysts could be detected from among the faecal debris. However, these have largely been resolved (through improvements to components such as the lysis buffer), and therefore, it is now possible to adapt them to LFD technology. Most kits are currently designed as 'duplex' assays and are set up to detect *Giardia* spp. and *Cryptosporidium* spp. on the same test strip. There are also some LFDs that can test for *Entamoeba* spp. as well (Van den Bossche et al. 2015). When using that format, follow-up PCR is required for any samples testing positive to determine whether the patient is infected with a pathogenic amoeba (see Chapter 3). Kits are usually evaluated in comparison to PCR and the results tend to show good specificity, but sensitivity is sometimes reported to be 70% or less. This means that positive results can be trusted, but negative results should be followed up with a different test. However, when assessed against standard microscopic methods, the rapid tests for these faecal protozoa come out quite well. Goudal et al. (2019) found that an LFD showed sensitivities of over 85% for both *Giardia* and *Cryptosporidium*, which is acceptable, particularly in situations where skilled laboratory staff or facilities are not readily available. Nevertheless, negative results should be investigated further. However, LFDs now perform well enough that they are being used in the diagnosis of acute diarrhoea among children, in places where neither microscopy nor PCR investigations are readily available (Bitilinyu-Bangoh et al. 2019).

13.9.1.2.2   Detection of Helminths in Faeces with LFDs

Helminths are more complex organisms than other infectious agents and usually larger in size. This can make laboratory detection and identification using standard methods easier than for small protozoa. In contrast, the relatively sophisticated biology of helminths and the gaps in our understanding of the host immune response have made it harder to devise rapid tests for this parasite group.

Tests for antibodies against helminths have limited use, since most assay systems cannot distinguish between past infections, new infections, and existing chronic ones. Despite this some LFDs for anti-helminth antibodies have been developed. Examples include a test for antibodies in blood raised against *T. solium*. Van Damme et al. (2021) suggested that this LFD antibody test has the potential to be useful as part of a suite of tests to find patients with cysticercosis in endemic areas. Another interesting approach is to test for antibodies against microfilariae instead of collecting skin snips (see Chapter 6). Using LFDs would be a good way of doing this since it would allow testing to be done in clinics rather than central hospitals. Thus, treatment and management of patients could be started more quickly. However, their reliability still needs improvement. Studies suggest that the kits are not reliable enough yet – sensitivity can be low compared with other detection methods (Hotterbeekx et al. 2020), and cross reactions between antibodies produced by patients against related species leads to poor specificity (Hertz et al. 2018).

A rapid test method to detect serum antibodies against *Schistosoma mansoni* cercarial transformation fluid has been evaluated in areas where bilharzia is endemic. Interestingly, since the LFD reacts to both *S. mansoni and S. haematobium* infection, it has limited use on its own for diagnosis of individuals (Coulibaly et al. 2013), but it has the potential to be useful in field epidemiological surveys (Nausch et al. 2014). Researchers have come to similar conclusions about rapid tests that detect various antigens found in the urine of patients with active schistosomiasis infection. These markers include circulating cathodic antigen (CCA) (Ochodo et al. 2015) and circulating anodic antigen (CAA). Most studies conclude that these LFDs can be useful when incorporated into a suite of diagnostic tests but are not suitable to use on their own to make clear decisions about individual patients (de Dood et al. 2018). However, the manufacturers of diagnostic test kits do take note of data from research studies and are always working to improve their assays.

Point-of-Care testing can involve simple tests for general markers of infection or illness, rather than specific tests for any infectious agents. These are usually easy to do and low cost, but obviously do not give the full clinical picture. Nevertheless, there is the possibility that they can be helpful to decide which patients need further investigations. For example, Ochodo et al. (2015) considered 90 studies of rapid testing for *Schistosoma* infection and concluded that using a urine dip stick to detect haematuria was the best place to start in identifying patients with *S. haematobium* infection. Patel et al. (2021) tried a similar approach. They evaluated the use of LFDs for faecal calprotectin (FC) and faecal occult blood (FOB) as proxy markers for infection with soil transmitted helminths (e.g., *Trichuris trichiura*, *Ascaris lumbricoides*, and hookworm). They tested faecal samples from patients already diagnosed with *T. trichiura* infection, and they estimated the worm burden using the Kato-Katz technique. They found that the FC and FOB results did not help to indicate the worm burden, nor did they help to predict whether an individual patient was co-infected with both *T. trichiura* and *A. lumbricoides*.

13.9.2 Point-of-Care Nucleic Acid Amplification Tests (NAATs)

As discussed previously, methods to detect the genetic material of parasites are usually more sensitive and specific than other laboratory methods. They can be set up to find the organism's genome in a sample from a patient or vector or the environment even when it is present at low concentration. An assay based on assessing the genome is sometimes the only reliable method to distinguish between similar parasite species (e.g., *Entamoeba histolytica/dispar/moshkovskii*). However, the polymerase chain reaction is expensive in terms of equipment, reagents, and staff time. There is a requirement for 'clean areas' to avoid cross contamination with DNA/RNA from the laboratory environment – including from the staff operating the analysers! This means well-trained specialist staff, good facilities, and plenty of space are needed. In other areas of microbiology, particularly virology, NAATs such as polymerase chain reaction have become the standard testing method in routine diagnostic laboratories. The technology has gradually been adapted for use in 'benchtop' analysers, which can be used as point-of-care tests (POCTs) either within the laboratory (for processing urgent samples) or outside on hospital wards and in clinics. The whole of a molecular assay including DNA/RNA extraction, all the amplification steps, then reading and processing the results, all take place inside a small machine that can stand on a table or bench. Like LFDs, NAAT POCTs can be performed by someone trained in the required steps and who can follow the protocol, without their need to understand the scientific principles of the assay. These tests are relatively expensive in comparison to other types of tests. If the cost per test is calculated, then the main laboratory PCR will still usually be cheaper. However, once a greater range of pathogens can be included in POCT kits and they become more widely used, the costs will decrease. They offer a clear advantage for the patient or disease control programme by achieving a reliable result in a faster time than the main laboratory test would take. Some systems can provide a result in an hour. Parasitology diagnostics have benefitted from the advances made in tests for other types of microorganism. While LFDs can be set up to detect two or three pathogens in one test strip, POCT molecular methods can include the primers for 10 or more different organisms. For example, it was mentioned above that there are LFDs, which can simultaneously detect *Giardia* spp., *Cryptosporidium* spp., and *Entamoeba* spp. in faecal samples; multiplex PCR assays for investigation of diarrhoea include these three in a single assay, which can also detect a range of viruses and bacteria (Buss et al. 2015).

Loop-mediated isothermal amplification (LAMP) is a NAAT, which uses a slightly different method to make copies of the target DNA. In PCR, a target section of the DNA of interest is selected and primers are designed to attach to each end of that section. During the amplification of this sequence, there is a requirement to change the temperature for the de-annealing, DNA replication, and re-annealing stages. The LAMP system uses primers at more sites (usually between 4 and 6). This means that a larger piece of target DNA is being copied each time. The primer sites are selected so that there are inner and outer sequences which produce alternating complementary strands. The amplification process uses strand displacement activity to produce two complementary single-stranded DNA chains, which continually loop back on themselves. Thus, they provide the template for the next round of replication (so continual de-annealing is not needed), and the outcome is a double-stranded copy of the target sequence. Therefore, there is no requirement to keep changing the reaction temperature. The optimal temperature for the operation of the chosen DNA polymerase enzyme is used. In most cases, this is the enzyme of *Bacillus stearothermophilus* (Bst), which operates at around 65 °C. This means that the analyser can be electronically simpler and physically more compact. LAMP machines are small and portable and can run on battery power, which means that can be used in situations where mains electricity is not available, or the supply is unreliable. The reactions happen more quickly than for standard PCR, so LAMP is also a relatively rapid test and can give a result in an hour (Notomi et al. 2015). However, the DNA must be extracted from the sample before being loaded into the LAMP machine. The output of amplified DNA can simply be measured by the turbidity in the sample tube (on the basis that only the DNA sequence defined by the primers will increase in concentration during the reaction). Alternatively, the contents of the reaction tube can be tested using more complex methods such as gel electrophoresis – either at the testing site or in laboratory that has the facilities to do this. Therefore, unlike for benchtop PCR, LAMP requires access to basic laboratory facilities. Nevertheless, all the components required to carry out the basic LAMP procedure can easily be transported between sites.

Whole genome sequencing using nanopore technology is becoming increasingly popular in microbiology diagnostics. Portable versions of the MinION equipment are available, and these have been used in remote areas of the world for sequencing of plants and animals, as well as pathogens (e.g., Mongan et al. 2020). This has the potential to be adapted to point-of-care parasitology diagnostics in future but, at the time of writing, its use was still at the experimental stage.

Selected examples to illustrate the impact of POCT NAATs on parasite diagnosis will be discussed here. It is an exciting area with new advances being documented regularly, so readers should check the scientific literature for developments. For example, progress is being made on developing a LAMP test for schistosomiasis (García-Bernalt Diego et al. 2019), *Trypanosoma cruzi*, (Besuschio et al. 2017), and human African trypanosomiasis (Hayashida et al. 2015) although at the time of writing these were still at the prototype stage of development.

13.9.2.1 Detection of *Trichomonas vaginalis* Using POCT NAATs

Trichomonas vaginalis is a flagellate protozoan of humans, which is usually a sexually transmitted infection (Chapters 3 and 8). Many people carry the parasite asymptomatically, but it can cause clinical problems. Accurate diagnosis is important because effective treatment is available and if both symptomatic and asymptomatic individuals can be identified, they can be cured and therefore prevented from passing it to their partners. It is also useful to know the prevalence of *T. vaginalis* for epidemiological purposes. This will highlight the extent of the problem and make the case to target resources towards diagnosis and treatment, thus saving suffering and money later. Many people infected with *T. vaginalis* are asymptomatic and it can be difficult to diagnose the infection through microscopy (and sometimes culture). Consequently, it was not until a PCR test became available that a true picture of the prevalence was revealed in sexually active populations (Muzny et al. 2014). It has been estimated that global prevalence of *T. vaginalis* infections among people between 15 and 49 years of age is 5% in females and at least 0.6% of males; the risk is higher in people who are HIV positive (Masha et al. 2019). Therefore, point-of-care testing in genitourinary medicine (GUM) clinics and screening programmes could be very useful to identify people with *T. vaginalis* infection. There are LFD formats to detect *T. vaginalis* antigen in swabs and urine samples. These tests can yield more accurate results than microscopy (Gaydos et al. 2017), especially in asymptomatic patients. The advantage of POCT tests for GUM clinics is that patients are often just passing through and may be embarrassed about their infection. Therefore, it is beneficial if they can receive their result, advice, and treatment on the same day. A person in that situation is more likely to wait around for a few hours than return a week later. The benchtop PCR format that was developed for rapid screening for infections such as Methicillin Resistant *Staphylococcus aureus* (MRSA) and influenza has been adapted to test for *T. vaginalis* and is performing extremely well (Badman et al. 2016). It is effective in detecting the parasite in both genital swabs and urine samples and from patients with or without symptoms. It can provide a result in under 2 hours, which lends itself to a system where patients attend a clinic for sample collection and then test results in a single day. There are also POCT tests for *T. vaginalis*, which use isothermal helicase-dependent amplification. The analysers are compact and portable, and the test can provide a result in under an hour. These POCT methods would be incredibly useful in resource poor areas with high prevalence of *T. vaginalis*. Once the kits come into more general use around the world, it is to be hoped that this will reduce their cost, so that they can be more widely distributed. An interesting recent development is a multiplex POCT PCR intended to detect T. vaginalis along with two bacterial sexually transmitted infections, namely *Neisseria gonorrhoeae* and *Chlamydia trachomatis* (Morris et al. 2021).

13.9.2.2 Detection of *Plasmodium* spp. Using POCT NAATs

Accurate diagnosis of malaria infection is extremely important for patient management, disease monitoring, and control. POCTs are making a valuable contribution to the World Health Organisation's strategy to reduce malaria transmission and mortality by 90% by 2030 (WHO 2021). The currently available LFD formats for malaria can be helpful in initial diagnosis, but they cannot distinguish between all *Plasmodium* spp. Molecular methods are ideal for species identification, but main laboratory PCR facilities are often unavailable in the places where malaria is most prevalent. Thus, there is a clear role for POCT NAATs in the global strategy to combat malaria. The LAMP method has been applied to malaria testing since the mid 2000s. At the time of writing, there were over 40 commercially available test kits being used in both endemic and non-endemic areas (Morris and Aydin-Schmidt 2021). Picot et al. (2020) reviewed evaluation studies of LAMP kits compared to microscopy, LFDs, and PCR. They concluded that LAMP generally performed well compared to the other test methods. Sensitivity was as good as or better than microscopy, LFDs, or main laboratory PCR. However, most of the currently available LAMP methods are set up to detect *Plasmodium* genus or *P. falciparum*. Consequently, they present the same shortcomings with distinguishing between other species which the LFDs have. As malaria transmission reduces, the epidemiological surveys for prevalence will be important and LAMP would be a good candidate test for this type of work. Species identification will become more significant when rates of infection are lower, so further refinements to the test parameters to address this would be useful. There are some protocols in which the initial test is for *Plasmodium* genus and then positive samples are followed up with species analysis. However, as for the LFDs, these concentrate on *P. falciparum* or *P. vivax* and not the other three species. This is a major limitation of LAMP for malaria, especially in areas where *P. malariae* and *P. knowlesi* are in co-transmission. In that situation, the main laboratory PCR is still the better test method.

13.9.2.3 Invertebrate Vector and Intermediate Host Monitoring Using POCT NAATs

Laboratory investigations of parasite vector and intermediate host species are an essential tool in monitoring the spread of infections and implementing control programmes. When diagnosing infections in human or animal patients, one wants to obtain a result as quickly as possible, which is why POCTs are useful. While there is less urgency to identify whether an individual invertebrate is carrying a particular organism, there are logistical issues with collecting samples in the field and transporting them to a test facility. So POCT methods could be useful here too. Zaky et al. (2018) report the use of a PCR kit that can be carried around in a rucksack. They developed a simple DNA extraction method, and the amplification stage uses a battery-powered thermocycler, which is controlled via a mobile phone. The amplified product is tagged with biotin (during the PCR process). It is then visualised using an LFD format. The PCR product is placed on a cellulose strip, which includes a coloured marker which is linked to streptavidin. The product is then detected, thanks to the biotin-streptavidin binding reaction, which is commonly exploited for immunological-based laboratory tests. In this case, the authors used the kit to detect *Brugia malayi* in *Aedes aegypti* mosquitoes. Although it is relatively more expensive per test than the main laboratory PCR assay, this portable version is an attractive option. Other portable versions of PCR equipment are being tested for suitability in the field to assess prevalence of infections in invertebrate vectors, particularly in the investigations of epidemic viruses such as Zika and Dengue (Rutkowski et al. 2020). There has been some progress towards developing a LAMP test for snails infested with schistosome species although these remain hampered by problems with standardisation (Kamel et al., 2021). At the time of writing, eDNA (Chapter 5) appeared to offer greater promise for monitoring the presence of schistosome species in the environment and in snails. Other POCTs for parasite infections have benefitted from efforts to improve them for bacteria and viruses first, so portable PCR could become more affordable and accessible for parasitologists studying vectors in years to come.

14

Parasite Treatment

CONTENTS

14.1 Introduction

There are numerous species of parasites and a huge variety of substances used in their treatment. We are therefore not attempting to list and compare individual classes of drugs and their modes of action. Good reviews of these topics are provided by Pink et al. (2005), Renslo and McKerrow (2006), and Abonwa et al. (2017). Garcia (2006) provides detailed information about the treatment and prevention of human parasitic infections, and Mehlhorn (2016) does the same for parasites of domestic animals. We will focus upon the general concepts and introduce some of the most recent advances that may prove useful in the future. In many cases, the chemicals under investigation exemplify how developments in fields such as cancer research and the economy influence parasitology.

14.2 The Ideal Antiparasitic Drug

The properties required in an antiparasitic drug are in many respects no different from those looked for in most other pharmaceutical products (Table 14.1).

An Antiparasitic Drug Should Kill All the Target Organism(s): If we, or our domestic animals, suffer from a parasitic infection, the first desire is to kill all the parasites as quickly as possible to stop them doing further harm. Unfortunately, this is not always feasible. It is common for a host to harbour two or more life cycle stages of the same species of parasite. For example,

Parasitology: An Integrated Approach, Second Edition. Alan Gunn and Sarah J. Pitt.
© 2022 John Wiley & Sons Ltd. Published 2022 by John Wiley & Sons Ltd.
Companion website: www.wiley.com/go/gunn/parasitology2

Table 14.1 Properties ssof an ideal antiparasitic drug or treatment regime.

Kills 100% of the parasites (including all life cycle stages) in the host

Broad spectrum of activity

Kills quickly

Provides long-lasting protection

Simple to administer (e.g., does not require invasive procedures or medical supervision)

Requires only one or two treatments to achieve a cure

Safe (does not cause harmful side-effects)

Does not have contraindications (i.e., does not interfere with other medical/veterinary treatments or cause problems if the host is pregnant or is suffering from an underlying medical/veterinary condition)

Affordable to the individual/population

Acceptable to the individual/population

Chemically stable with a long shelf life

Does not enter the food chain

Does not harm the environment

person may harbour the hypnozoite, merozoite, schizont, and gametocyte stages of *Plasmodium vivax*. Drugs often vary in their activity against these stages, either because they have differences in their physiology and/or because they reside in different tissues and are therefore exposed to different concentrations of the drug. For example, praziquantel is effective against adult schistosomes but not very good at killing the developing schistosomulae. If any parasites from the initial infection remain alive after drug treatment, they might multiply and cause disease again or become a source of infection for other animals or humans. In addition, the exposure of parasites to sub-lethal levels of drugs increases the risk of resistance developing.

An Antiparasitic Drug Should Exhibit a Broad Spectrum of Activity: Having a broad spectrum means that a chemical fulfils a range of tasks. Broad spectrum drugs are beneficial because they can be used to treat a variety of parasites. For example, avermectins, such as ivermectin and doramectin, are excellent in this regard, as they are active against various helminths, as well as ectoparasites, such as lice, fleas, and ticks. Multiple parasite infections are common, so a broad-spectrum drug provides a considerable saving because it reduces the need for several drugs given on multiple occasions. However, if a drug with a broad spectrum of activity enters the environment, then there is a possibility that it will harm non-target free-living organisms.

An Antiparasitic Drug Should Kill the Target Organism(s) Quickly: Obviously, an antiparasitic drug needs to be effective against the intended pathogen. However, it also needs to act fast. When the host is gravely ill, it is important to remove the parasites before they cause host death. Even in less serious situations, if parasites are left alive, they will continue to cause pathology. Furthermore, if capable of growth and reproduction, they will cause potential problems for the host and act as a source of infection for others. If it kills its target quickly, a chemical also reduces the chances of resistance developing against it and, ideally, a parasite 'interacts' very briefly with a drug before dying. This is because exposure to sub-lethal levels facilitates the evolution of resistance so if a parasite dies quickly, it is unlikely to evolve a physiological means of counteracting the drug. However, the sudden death of large numbers of parasites within the body's tissues may itself cause problems. This is because the release massive amounts of parasite antigens can trigger harmful immune reactions or even a potentially fatal anaphylactic shock. For example, diethylcarbamazine is no longer recommended for the treatment of onchocerciasis because it can have harmful consequences. These arise from it causing the sudden death of all the microfilariae in the host. In heavy infections, this means that thousands of microfilariae are dying at the same time. These result in an acute inflammatory reaction to symbiotic *Wolbachia* bacteria released by the dead and dying microfilariae. The consequences for the host include unbearable itching, fever, tachycardia, and hypotension.

Drugs are seldom administered as pure compounds. Instead, they are 'formulated' with a cocktail of chemicals, the composition of which varies with the intended means of delivery (e.g., liquid, tablet, or injection), and alters the effectiveness of the drug. The formulation influences a drug's stability, toxicity to both host and target parasite, rates of absorption and excretion, bioavailability, and pharmacokinetics. For example, formulations of insect growth regulators, natural pyrethrin, and synthetic pyrethroid insecticides often include piperonyl butoxide (PBO) because it acts as a synergist and enhances their activity. A relatively new area of research is to extend the usefulness of drugs to which resistance has evolved by employing 'resistance reversers' within drug formulations. Resistance reversers interfere with the physiological processes that confer a parasite's resistance to a particular drug or group of related compounds. For example, a recently evolved resistance phenotype in *Plasmodium falciparum* can be overcome by exposing the parasites to a histone methyltransferase inhibitor (Chan et al. 2020).

An Antiparasitic Drug Should Provide Long-Lasting Protection: The longer the protection, the less risk there is of the host becoming re-infected. This is important for both the host's health and that of others for whom it could be a source of infection. It also reduces the time and costs associated with providing treatment. Ivermectin and some of the other anthelmintics used to treat ruminants can be given in a specially designed canister called a 'slow-release bolus' that is placed into the rumen using a special device. The bolus releases a set amount of drug every day over a prolonged period. This provides long-lasting protection where there is a constant risk of reinfection. However, it is unsuitable for lactating cows because the drug would appear in their milk.

A Drug Should Be Simple to Administer: Ease of administration is important to ensure compliance and so that people and animals can be treated quickly and with the minimum of fuss. For us, drugs in tablet or liquid form are preferable because we can take these without supervision. Furthermore, there is a good chance of us taking the recommended dosage and completing the course of treatment. By contrast, injections, especially intravenous or intra-peritoneal, usually must be given by trained medical personnel – and this limits the situations in which the drugs can be dispensed. Also, many people are more willing to take a pill than receive an injection. By contrast, oral dosing animals is seldom easy – as anyone who has ever tried to get a pill inside a recalcitrant cat will attest. Sometimes one can deliver a drug via an animal's food, but it is then difficult to control the dose it consumes. For domestic livestock such as sheep and cattle, special 'dosing guns' deliver a known quantity of 'drench' (i.e., drug) to the back of the animal's throat so that it has to swallow it. In this way, the farmer can dose large numbers of animals relatively quickly. Injections, especially intramuscular injections, are often preferred where large numbers of animals need treating. A trained person can rapidly inject many animals with minimal handling – and this reduces the stress to all concerned. An even simpler means of drug delivery is through 'pour on' or 'spot on' formulations in which the drug is poured or spotted onto the back or neck of the animal, and it is then absorbed across the skin. For example, eprinomectin (an avermectin type drug) can be poured onto the back of sheep to control both gastrointestinal nematode infections and the sheep nostril fly *Oestrus ovis*.

A Drug Should Require only One or Two Treatments to Effect a Cure: The fewer the number of treatments required, the better the chance of patient compliance. This is especially the case if the treatment must be delivered at a medical centre or veterinary surgery. Once someone starts feeling better or their pet appears to be improving, there is a high chance that they will not bother, or forget, to return to complete the treatment regime. This not only increases the possibility that the parasite will persist but also increases the chances of resistance developing. For farm animals, such as beef cattle and sheep that often graze far from the home farm, gathering the animals is a time consuming and complicated exercise that is stressful for the animals. Therefore, it is something that is minimised wherever possible.

A Drug Should be Safe and Not Cause Harmful Side Effects: Antimicrobial and antiparasitic drugs are intended to kill living organisms and therefore there is always a risk that they will also harm the host. Selectively toxicity is harder to achieve for antiparasitic drugs than for those targeting other microorganisms because both the host and parasite are eukaryotes. Antibacterial agents exploit the differences in cell biology and metabolism between the eukaryotic host and the prokaryotic bacteria. Therefore, it has been easier to develop antibiotics than anthelmintics.

A drug is less likely to harm the host if it selectively acts on a physiological process that only occurs in the target parasite. For example, cyromazine (Vetrazin®) interferes with the deposition of chitin into the cuticle whilst diflubenzuron inhibits chitin synthesis. These drugs are therefore effective against various ectoparasitic arthropods in which they interfere with the moulting process whilst they are safe to vertebrates because they lack comparable metabolic pathways. However, all chemicals are toxic if taken in sufficient concentration, and it is very rare for a drug to be so specific that it only interacts with a single physiological process. Patients are unlikely to complete a treatment regime if the drug induces unpleasant side effects such as nausea and vomiting. If the drug is so toxic that it must be given under medical supervision – such as antimonials for the treatment of leishmaniasis – then it further reduces the situations in which the drug can be employed and adds to the cost of treatment.

A Drug Should Not Have Contraindications: Ideally, a drug should enter our body, do its job, and leave with the minimum of fuss. However, to reach its target, a drug must interact with the chemicals and processes that constitute our physiological environment. This will be influenced by factors such as our age, gender, genetic constitution, reproductive state, diet, health, other infections we have, and other drugs we are taking. Our bodies are complicated, they are unique to us, and the actions and effectiveness of a drug can be influenced in many ways. A contraindication indicates that a drug should not be used in a particular situation. This might be because the drug would cause harm or because the drug would not work effectively.

One of the first contraindications would be if the patient reacted badly to the drug. For virtually any drug, there are usually a few people who demonstrate a hypersensitivity reaction towards it. For example, allergic reactions to penicillin and other antibiotics are well known. Obviously, persons exhibiting hypersensitivity reactions should be treated with an alternative drug that the patient can tolerate.

Ideally, a drug should not influence the activity of another drug. However, drug interactions are relatively common, and in extreme cases, a therapeutic dose becomes harmful or fails to work. For example, patients may be prescribed for the blood thinning drug warfarin to treat or prevent venous thrombosis but if they also take ivermectin it can result in excessive thinning (Gilbert and Slechta 2018). Similarly, praziquantel is contraindicated in patients receiving the antibiotic rifampin (commonly used in the treatment of tuberculosis). This is because rifampin induces a strong cytochrome P450 response that would prevent praziquantel reaching a therapeutic dose in the blood stream.

The activity of drugs can be affected by what we eat and drink. For example, grapefruit juice is well known for lowering the levels of liver enzymes that metabolise drugs. Consequently, if grapefruit juice is drunk shortly before or after taking a drug, its levels in the circulation might rise. Sometimes this might be harmful and sometimes it may be beneficial. For example, drinking grapefruit juice when taking praziquantel results in higher levels of praziquantel in the blood. This may be beneficial, but more studies are required to verify this.

The presence of other infections must also be considered. For example, praziquantel is contraindicated if the patient has a helminth infection within their eyes, such as ocular cysticercosis. This is because praziquantel is a broad spectrum anthelmintic and could potentially kill any helminth present in the body. Therefore, its use to treat schistosomiasis might also result in the death of *Taenia solium* cysticercoids within the eye. The sudden release of antigens from dead and dying cysticercoids might lead to irreversible eye damage.

The host's genetic constitution can influence the way it responds to a drug. For example, glucose-6-phosphate dehydrogenase deficiency is extremely common, especially in tropical countries. This is possibly because it provides protection against malaria. However, its expression means that a person is susceptible to developing haemolytic anaemia if prescribed the antimalarial drug primaquine. It is therefore essential that patients are screened before receiving this drug. Unfortunately, screening facilities are often lacking in developing countries, and this limits the situations in which it can be used (Recht et al. 2018). This is unfortunate because primaquine is the only antimalarial that kills the hypnozoite stages of *P. vivax*, as well as sterilising the mature gametocytes of *P. falciparum*. It therefore both prevents relapses of *P. vivax* and the transmission of *P. falciparum*. Similarly, ivermectin is safe for use in most mammals, but in certain breeds of domestic dogs, very low concentrations (sub-therapeutic dose) induce neurological symptoms that include hypersalivation, ataxia, blindness, respiratory distress, and can prove fatal. The reaction is most pronounced in sub-populations of Border Collies: not all Border Collies are susceptible, and the condition also occurs in other breeds such as Collies, Australian Shepherds, white German Shepherd dogs, and Shetland Sheepdogs. Veterinarians used to adopt the adage 'white feet – don't treat' although this is now recognised to be too sweeping a generalisation and the adage is now 'white feet – test before you treat'. The condition results from a deletion mutation in the MDR1 gene that encodes a protein called *P*-glycoprotein. *P*-glycoprotein is a transporter protein present on the apical border of a variety of cells including intestinal cells, biliary canalicular cells, renal tubular cells, the placenta, and testes. However, in the case of ivermectin sensitivity, it is the role of p-glycoprotein in the blood–brain barrier that is most important. *P*-glycoprotein is a key constituent of brain capillary endothelial cells and transports numerous structurally unrelated compounds that include ivermectin. *P*-glycoprotein therefore regulates the movement of many drugs across the blood–brain barrier. Dogs that are homozygous for the MDR1 deletion are unable to control the movement of ivermectin into and out of the brain and the levels can rise to a point at which it exerts toxic effects (Merola and Eubig 2012). These breeds of dog can also suffer from adverse sensitivity to other drugs that are substrates for *p*-glycoprotein. Dogs that are homozygous for the deletion mutation can also have problems excreting ivermectin and other *p*-glycoprotein substrate drugs via the bile or urine, and this could result in the concentration of the drug within the circulation remaining higher than in other breeds of dog.

Teratogenicity – the ability to damage a developing foetus – is a major concern for any chemical used in human or veterinary medicine. Some antiparasitic drugs are safe for us to take when we are adults but are not suitable for treating women who are pregnant or are likely to become pregnant. This is because of the potential risk they pose to the developing foetus. For example, metronidazole (amoebic dysentery) and miltefosine (leishmaniasis) are potentially teratogenic and therefore unsuitable for treating infections during pregnancy.

A Drug Should Be Affordable: Cost is a major consideration for any treatment regime but especially so for poor people. The definition of 'affordable' varies between individuals, countries, and situations. Even if a drug is ideal in every way, if it is too expensive, then it becomes irrelevant to all but those who are rich. Drugs that are still in patent are usually expensive because the company making them must make sufficient profit to recoup the costs of manufacture and development, as well as fund the development of new drugs. In addition, pharmaceutical companies are not charities and must make a profit for their investors. Nevertheless, sometimes a company makes specific drugs available at reduced price as part of a control programme. Once a drug is out of patent, it can be manufactured by any commercial concern and its cost usually declines. In these situations, it can radically alter the approach to the treatment and control of parasitic diseases. This is exemplified in the cases of albendazole and praziquantel. Albendazole is a broad-spectrum benzimidazole anthelmintic that is used to treat gastrointestinal helminth infections and has the added advantage of being highly effective against *Giardia*. Praziquantel also a broad-spectrum anthelmintic but is most used to treat schistosome infections. In the early 2000s, tablets of albendazole typically sold for about US $0.20 tablet^{-1}, whilst those of praziquantel cost about US $3.00 tablet^{-1}. By the end of the decade, the drugs were out of patent and costs had fallen to about US $0.02 tablet^{-1} for albendazole and US $0.07 tablet^{-1} for praziquantel. Consequently, it became more expensive to identify infected individuals through mass diagnostic screening than it was to simply treat a whole population. The emphasis therefore changed from how to identify the individual in need of treatment to treating everyone regardless of whether they needed it. This approach is possible with drugs such as albendazole and praziquantel because they have been used for many years and are known to be safe with relatively few instances of harmful side effects. However, one must be conscious of the fact that by indiscriminately treating more people there is an increased risk chance of unnecessarily dosing one of the very few individuals in whom the drug has a harmful effect, and there will be more drug residues entering the environment.

A Drug Should be Acceptable to the Individual/Local Population: Even if a drug is cheap and highly effective, it must still be acceptable to those receiving it. The slightest rumour of a medicine being prepared using pigs, pig cells, or pig proteins can make it unacceptable to some Jewish or Muslim communities. For some vegans (and others), anything prepared using animals or animal tissues would be unacceptable. Similarly, some people will avoid anything prepared using genetic engineering, genetically engineered cells, or cell lines derived from aborted foetuses (even if those cell lines were initially obtained many years previously). A consideration of the target population/market is essential at the earliest stage of drug design.

A Drug Should Be Chemically Stable with a Long Shelf Life: The cost of a drug is partly linked to its chemical stability. This is because a drug that is stable, has a long shelf life, and does not need refrigeration is usually going to be cheaper than one that lacks some or all these attributes. Therefore, it can be bought in bulk, stored, and transported at low cost, and made instantly available when needed. This is a major consideration in developing countries where the temperatures are high, the distances long, and the electricity supply unreliable.

A Drug and/or Its Metabolites Should Not Enter the Food Chain: Drugs should be absent or below permitted levels in the food we purchase for consumption. This can present problems with drugs given to farm animals that remain in their circulation for prolonged periods – and this is often a pre-requisite for long-lasting protection. In the case of lactating animals, drug residues may be found in the milk, whilst in those reared for slaughter they may be present in the meat. In food producing animals, there is therefore a compulsory withdrawal period after any drug treatment. The withdrawal period varies between drugs and can vary from several weeks to zero in the case of milking cows treated with eprinomectin. Obviously, a long withdrawal period may compromise profitability because it dictates when an animal or its products can be sold.

A Drug Should Not Harm the Environment: The tendency is always to concentrate on the immediate need. However, all the drugs that go into us and our animals are ultimately passed out in one way or another and they then enter the environment. When drugs are used on a large scale, for example in mass medication campaigns or in treating whole flocks of sheep or herds of cattle, this results in significant amounts of drugs entering the environment. Sometimes a drug is metabolised completely but very often its breakdown products or the unmetabolised drug is passed in the urine or faeces.

The environmental impact of avermectin drugs is a controversial topic, with some workers considering them to be serious pollutants, whilst others believe them to be a minor problem that affects only certain ecosystems at certain times of year. Although avermectins such as ivermectin are very safe in human and veterinary medicine (except in animals homozygous for the MDR1 deletion), most of the drug is rapidly excreted unmetabolised with the animal's faeces. In addition, residues can persist in the environment for weeks under suitable conditions and therefore affect susceptible invertebrates both in the soil and in surrounding water systems (Fernández et al. 2011). This may be considered beneficial since they can affect the development of larvae of parasitic nematodes and of pest insects such as stable flies (*Stomoxys calcitrans*) and horn flies (*Haematobia irritans*). However, the residues can also harm susceptible non-target invertebrates such as dung beetles that are particularly sensitive, and other insects that breed in dung. This may lead to dung decaying slower than it would normally and a reduction in the insect population means that there is less food for insectivores such as certain bats and birds. If dung decays slowly, it also results in excessive fouling of the pastureland. Sheep and cattle avoid plants that are contaminated with faeces, and therefore, the pasture becomes covered in rank grass and the quality of the pasture declines. This problem is particularly acute in dry climates such as parts of Australia in which dung beetles are vitally important for the breakdown of dung.

14.3 Pharmaceutical Drugs

Most pharmaceutical antiparasitic drugs are manufactured synthetically although some are derived from microbes. For example, the avermectins and milbemycins are derived from fermentation products of the actinobacteria *Streptomyces avermitilis* and *Streptomyces cyanogriseus,* respectively. Most of the current antiparasitic drugs used in human medicine were initially developed for the treatment of other medical and veterinary conditions or for the control of crop pests: it was only subsequently that their antiparasitic properties were identified. Although this may appear to place an undue reliance upon serendipity, it must be borne in mind that the market for new antiparasitic drugs is relatively small. In 2018, the global market for antiparasitic drugs and chemicals for veterinary applications was US $7 billion annum^{-1} (Selzer and Epe 2020) – this is many times more than the market for human antiparasitic drugs. Pet owners will spend considerable sums of money on their animals, whilst there are quantifiable economic returns for the treatment of farm animals. Therefore, it makes more commercial sense to develop drugs for these markets than it does for humans: in 2008 the market for antiparasitic drugs for dogs and cats was put at US $3.4 billion (Woods and Knauer 2010). These may sound like large sums of money, but they must be set against the much bigger markets for other pharmaceutical products. For example, the market for heart disease drugs in the USA was estimated to be US $273 billion year^{-1} in 2010 with a prediction that it could rise to US $818 billion year^{-1} by 2030. Furthermore, the enormous cost of developing a drug from initial synthesis to final marketing means that there is little commercial incentive for introducing new antiparasitic drugs. The cost of bringing a new antimalarial drug to the market has been put at US $300 million, and the chances of a candidate drug failing during phase II clinical trials are as high as 50%.

Despite the much greater market for antibiotics than antiparasitic drugs, in 2019 many large pharmaceutical companies reduced their efforts to develop new ones. This was because the volume of sales was not enough to recoup the costs involved. For example, a patient typically takes an antibiotic for 7–14 days, whilst they might take drugs for a chronic medical condition over many years. In addition, doctors tend to prescribe older and cheaper antibiotics and keep new antibiotics in reserve. Consequently, it makes commercial sense to concentrate on more profitable areas of research, and there is little incentive to invest in the even less profitable area of developing antiparasitic drugs. The costs of anticancer drugs are notoriously high, and therefore the returns are potentially much greater. For example, ipilimumab, which is used to treat melanoma, reportedly costs US$ 120,000 for a course of therapy in return for increasing the life expectancy of the patient by 4 months (Howard et al. 2015).

Nevertheless, there are economies of scale if novel compounds are first tested for their effectiveness in a range of uses for which there are larger markets. Subsequently, the knowledge gained from a chemical's use in one application can be transferred to its use against parasites. Unfortunately, some of the existing antiparasitic drugs are highly toxic and/or resistance is becoming an increasing problem. For example, the antimonial drugs used for the treatment of leishmaniasis are notoriously toxic, whilst some helminth populations are becoming increasingly resistant to many of the benzimidazole anthelmintics (Rashwan et al. 2017). Therefore, there is a serious need for new drugs to treat many parasites. However, new drugs are not reaching the market fast enough, and many scientists are advocating novel approaches to speed up the process of drug identification. One way of doing this is through establishing product development partnerships (Weng et al. 2018). These involve scientists in pharmaceutical companies, universities, and institutes and international organisations cooperating in the identification of potential new targets and undertaking the initial screening novel compounds. Another approach would be to change the way in which commercial companies are rewarded for producing drugs, so they are not so reliant on their volume of sales. For example, this might be through a subscription approach in which the company is paid 'up front' by a health service or other provider for access to the drug. This is another example of where the identification of an effective drug/vaccine forms only one part of a parasite control strategy – there are also financial matters to consider.

The traditional means of screening compounds for antiparasitic activity is a slow and costly process. This is because most parasites cannot be cultured *in vitro* and therefore activity is assessed by giving the drug to a laboratory animal infected with the parasite or one of its close relatives. For example, antimalarial activity can be assessed by treating mice infected with *Plasmodium berghei*. The use of laboratory animals is always a controversial topic. Apart from the ethical considerations, their use can be criticised on the basis that laboratory models do not always provide a reliable basis for understanding how the intended parasite target would respond to a particular drug in its normal host. For example, there is an ongoing controversy of whether *P. berghei* infections in mice provide a reliable model for the pathophysiological processes associated with cerebral malaria caused by *P. falciparum* (Craig et al. 2012). Nevertheless, up until recently this was virtually the only way in which potential antiparasitic drugs could be screened for activity. A novel alternative that is still in the development stage is to adopt organoid technology and organ-on-a-chip technology (Duque-Correa et al. 2020). This involves culturing parasites *in vitro* but within a physiological environment closer to that of their normal host. Organoid technology involves growing tissues derived from primary tissues, embryonic stem cells, or pluripotent stem cells within a 3D medium. For example, the complete *Cryptosporidium* life cycle can be obtained within organoids grown from donated human small intestine and lung tissue (Heo et al. 2018). Organ-on-a-chip technology involves growing miniature organs within a microfluidic device. These devices can now be purchased commercially and are about the size of a computer memory stick. The cells grow on the walls of hollow microfluidic channels and can be exposed to different physiological, environmental conditions. Therefore, assuming that parasites could be cultured upon or within these cells, they would provide an ideal test bed for studying the effects of therapeutic agents. However, there is an ongoing debate about how accurately organ-on-a-chip technology mimics natural conditions. At the time of writing, it was uncertain which parasites would grow and develop normally using this technology.

Another approach to identifying candidate drugs is to use a combination of bioinformatics and chemo informatics (Saldívar-González et al. 2017). This ensures that only those chemicals with a high chance of success are tested in laboratory animals. The process is based on identifying potential target molecules from computational analysis of genome and protein sequences and then identifying chemical structures that would interact with the target molecule without causing pathology in the host. For example, Ferreira et al. (2015) discuss this in relation to drugs for schistosomiasis. The full or partial sequences of many parasite species are now known, and so it is becoming easier to assign a function to many genes. This makes it possible to identify unique features that could be targeted by either novel drugs or those already known to express structure–activity relationships. Clearly, it is not enough to simply identify a unique parasite protein – it must also be shown that the protein performs an essential function within the parasite. This is relatively easy for protozoan parasites that can be maintained in culture or in laboratory animals by selectively activating or inhibiting the target protein using knockout/knockdown mutants or RNAi (RNA interference). For example, Tewari et al. (2010) analysed the genome of *P. berghei* and discovered that 23 out of 66 genes coding for protein kinases were functionally redundant for asexual development within red blood cells. Protein kinases are essential for many cellular processes and therefore logical drug targets. By identifying the specific protein kinases that are essential at certain life cycle stages, it is possible to reduce the list of genes to be targeted and thereby prioritise further research. Similarly, there are two isoprenoid biosynthesis pathways: the mevalonate pathway that operates in vertebrates, fungi and some bacteria; and the MEP (2-*C*-methyl-D-erythritol 4 phosphate) pathway that occurs in many bacteria, algae, higher plants, and those organisms with a photosynthetic lineage. The latter includes parasites such as *Plasmodium* and *Prototheca*. Consequently, by selectively targeting enzymes involved in the MEP pathway, it may be possible to design safe and selective drugs that will simultaneously kill parasites and some of the accompanying secondary bacterial infections. Unfortunately, similar phenotypic experiments are more difficult to perform in platyhelminth and nematode parasites because of their complex life cycles and protein function often must be inferred from model organisms such as *Caenorhabditis elegans*. This is less than ideal because the model organisms often lack the metabolic pathways that parasites utilise to breakdown host molecules such as haemoglobin and avoid the host immune response.

14.4 DNA/RNA Technology

With every passing year, it is becoming easier and cheaper to sequence whole genomes. Consequently, it is often said that we are living in the 'post-genomic era', which will revolutionise medicine, agriculture, and many other aspects of our lives. We are promised personalised medicine in which we will receive treatment that is tailored to our own unique physiological and genomic idiosyncrasies. For example, our genetic constitution affects the way in which we metabolise drugs and other chemicals, as well as the functioning of our immune system. Individual genomic sequencing would enable us to avoid giving drugs to people in which they would not work or induced harmful side effects. However, the molecular mechanisms controlling our bodies are enormously complicated and transforming laboratory experiments into commercial products will not be easy. It was once believed that a gene coded for a single protein, but we now know that a single gene can code for tens or even thousands of different proteins. Consequently, achieving a desired physiological effect is not necessarily a question of switching on or off a few genes. Furthermore, the new technologies are not cheap and, at least for the time being, there is little likelihood of personalised medicine becoming an important factor for the treatment of parasitic diseases of poor people living in developing countries.

One of the most fundamental discoveries in biology in recent years is that epigenetic mechanisms are responsible for many aspects of cell regulation (Allis and Jenuwein 2016; Klemm et al. 2019). Epigenetic factors are those mechanisms that regulate genetic expression without changing the DNA sequence. Epigenetic regulation is important in all organisms and has relevance for host parasite relationships because it governs the host's immune response and the parasite's life cycle transformations, virulence, ability to overcome the host's immune system, and adapt to drugs (Cheeseman and Weitzman 2015; Villares et al. 2020). It is also a potential target for antiparasite therapies that selectively target unique epigenetic pathways in parasites without harming the host (Coetzee et al. 2020).

Epigenetic factors include DNA methylation, histone modification, and regulatory RNA molecules. DNA methylation occurs through the addition of methyl groups to cytosine to produce 5-methylcytosine. This normally happens at CpG sites (cytosine-phosphate-guanine) and extensive methylation of CpG sites within a gene sequence results in the silencing of the gene. Chromatin consists of DNA wrapped around the large structural protein histone. If the sequence of amino acids that comprise histone is modified, it alters the three-dimensional shape of the molecule, and this affects the expression of gene activity of the associated DNA. Histone modification occurs in several different ways (e.g., acetylation or methylation), and these have different effects on gene expression. It was once thought that RNA was merely a 'messenger molecule' but we now know that there are various single and double-stranded RNA molecules and that small non-coding micro-RNA molecules are involved in the regulation of gene expression at the level of translation through mechanisms such as RNA interference.

RNA interference regulates gene activity and is also part of a cell's natural means of protection against viruses. It is brought about by short sequences of double-stranded RNA, and in many but not all organisms, the RNAi pathway is as follows. Specific double-stranded RNA is cleaved by a ribonuclease enzyme called 'dicer' to form small (short) interfering RNA (siRNA) consisting of about 20–25 nucleotides. The siRNA is then assembled to form an 'RNA-induced silencing complex' (RISC) that includes the antisense strand of the targeted mRNA and an endonuclease enzyme. The silencing complex binds to the mRNA and then the endonuclease enzyme (Argonaute) brings about its degradation. Consequently, the mRNA is not translated and the protein it codes for is not formed. This natural process can be exploited by exposing cells to synthetic double-stranded RNA or siRNA and thereby influencing their gene activity.

Another class of small (~22 nucleotides) non-coding RNA molecules called microRNAs (miRNA) also regulate gene expression. These differ from siRNA in deriving from RNA transcripts that fold back to form hairpin-shaped molecules. Initially discovered in the free-living nematode *C. elegans*, they are now known to operate in many organisms, including us, and some viruses. Most miRNAs interact with mRNA to suppress gene expression although some act as gene promotors. Some parasites release miRNA into their surroundings as do their hosts in response to infection. Consequently, these have potential as diagnostic markers and providing insights into host:parasite relationships (Britton et al. 2014; Cai et al. 2016).

Antisense DNA and RNA

Within a cell, the first step in the production of a protein is when the gene coding for it in the cell's DNA is transcribed into a sequence of messenger RNA (mRNA) oligonucleotides. The single-stranded mRNA molecule then moves to the ribosomes where it is translated into a sequence of amino acids. The mRNA is referred to as a 'sense' strand, whilst its non-coding complementary strand is the 'antisense strand'. For example, if the sense strand had the sequence 5′-AACGAAUUAC-3′, its antisense strand would be 3′-UUGCUUAAUG-5′. If sense and antisense strands came into contact, they would bind together to form a non-functional duplex molecule. Consequently, the sense strand would not be translated and the protein molecule it coded for would not be formed. Natural antisense RNA transcripts occur in both prokaryotic and eukaryotic organisms (including protozoan and helminth parasites) in which they presumably function in the regulation of RNA although their purpose is not always apparent. Engineered antisense RNA inhibition is accomplished by inserting a gene, which codes for the antisense sequence of mRNA into an organism's DNA. This blocks the expression of a specific protein and is a basis to produce transgenic organisms and many studies on gene function. It could also be a potential area of development for therapies.

The DNA helix consists of two complimentary strands of DNA: these strands are known as the 'sense' and 'antisense' strands. The antisense strand is transcribed to produce the sense strand of mRNA. Short strands (oligodeoxynucleotides) of synthetic antisense DNA will bind to mRNA to form an RNA/DNA heteroduplex, which is then a substrate for endogenous cellular RNAases. Consequently, the protein coded for by the mRNA is not formed. This can be used as means of studying gene function and to selectively kill target cells/organisms by preventing them from producing essential molecules. Oligonucleotides are rapidly broken down within cells and therefore modified phosphorothioate oligodeoxynucleotides that are more resistant to nuclease attack are usually employed. Morpholino antisense oligonucleotides are showing promise for the treatment of several genetic diseases, as well as bacterial and viral infections. In addition, they inhibit translation and cis splicing of RNA in *Trypanosoma cruzi* and thereby prevent its growth and infectivity (Hashimoto et al. 2016). They may therefore have activity against other trypanosome parasites. Morpholino-oligomers consist of the same four bases as DNA: adenine, thymine, cytosine guanine, but morpholine provides the links between bases rather than the sugar ribose. Morpholine is a ring-shaped molecule with the empirical formula $O(CH_2CH_2)_2NH$.

Whilst manipulating epigenetic processes offers the prospect of killing pathogens and enhancing the host immune response, the technology to accomplish this is still at the experimental stage. Nevertheless, RNAi therapy for cancer and viral diseases is advancing rapidly and will undoubtedly inform the treatment of parasitic diseases. The first RNAi-based drug was approved by the US Food and Drug Administration in 2018. This drug, patisiran, is intended for the treatment of an extremely rare form of hereditary polyneuropathy. Experimental work indicates that several helminths, especially schistosomes, are susceptible to artificial RNA interference (Da'dara and Skelly 2015). However, the effectiveness of this approach varies between parasite species. For example, several parasitic protozoa, including *Babesia bovis*, *Leishmania major*, *Theileria* spp., and *Trypanosoma cruzi*, lack an active RNA interference pathway (Militello et al. 2008). Consequently, this approach is unlikely to be useful against them.

Despite the potential, it is necessary to proceed with care. Although RNAi therapy is theoretically more specific in its mode of action, it suffers from many of the same concerns that affect conventional pharmaceutical drugs. In particular, the siRNA sequences must be delivered to the target cells and remain there long enough to exert an effect. There also remain safety concerns such as the possibility of siRNA effects on non-target gene sequences and the potential for over-stimulation of RNAi pathways. This latter concern may not be a problem if synthetic siRNA does not act directly on RNAi pathways. However, if they act non-specifically through Toll-like receptor 3 pathways, then this could lead to a more widespread reaction that affected non-target tissues.

14.5 Molecular Chaperones (Heat Shock Proteins)

Nuttall's Standard Dictionary (1921 edition) quaintly describes a chaperone as 'a matron who attends a young lady in public places as a protector'. Molecular chaperones do not ward off predatory suitors, but they do ensure that their charges are able to engage in their duties in a seemly fashion! They are protein molecules that facilitate the non-covalent folding/unfolding and assembly/disassembly of other macromolecules. They do not, however, form part of the functioning macromolecule after it is assembled or correctly folded. Molecular chaperones are found in both prokaryotic and eukaryotic organisms and play an essential role in many basic functions such as responding to environmental cues, signal transduction, differentiation, and development. The best known of the molecular chaperones are the group known as the 'heat shock proteins' (Hsp). Their name derives from their initial discovery as novel proteins expressed in response to heat stress. The term 'heat shock' is a bit of a misnomer since it was subsequently found that they are also expressed in response to numerous stresses, including infection, and as part of normal physiological processes (i.e., they are 'constitutively expressed'). Heat shock proteins are classed into families according to their molecular weight (e.g., Hsp20, Hsp40, Hsp60, Hsp70, Hsp90, and Hsp100), and members of each family tend to show structural and functional similarity. An organism may have several Hsps within each family, and some of them work by interacting with co-chaperones.

Hsp90 is currently the most promising drug target for several pathogens and for the treatment of cancer. Clinical trials targeting Hsp90 for the treatment of breast cancer and other neoplasms have been in progress for many years although at the time of writing their success was limited. All parasites investigated to date express Hsp90 molecules, and these show structural differences from human Hsp90 that may enable them to be specifically targeted (Faya et al. 2015). This is of particular interest for *P. falciparum* in which Hsp90 is essential for both the circulatory and liver stages of the parasite's life cycle. Several candidate drugs targeting parasite Hsp90 have been identified but these remain at the early stages of investigation (Posfai et al. 2018).

The drug 'geldanamycin' and its analogues, such as 17-AAG and 17-DMAG, are effective inhibitors of Hsp90 in various protozoan, trematode, and nematode parasites, including *P. falciparum*, *Trypanosoma evansi*, *Leishmania donovani*, *Schistosoma japonicum*, *Trichinella spiralis*, and the filarial nematodes *Brugia pahangi* and *Brugia malayi*. Geldanamycin is a benzoquinone-containing ansamycin (a group of microbial secondary metabolites) that was originally identified from the actinobacterium *Streptomyces hygroscopicus*. Interestingly, although the Hsp90 found in the nematodes *Brugia* spp. and *C. elegans* share considerable structural similarity, that of the free-living species is not affected by geldanamycin. Indeed, among nematodes, geldanamycin only binds to Hsp90 in those species that lack free-living stages, or the transmission stage is enclosed within an egg. The reason for this is probably because the activity of Hsp90 is regulated through several co-chaperones and post-translational modifications that vary with the task it undertakes. Consequently, the ability of geldanamycin to bind to nematode Hsp90 relates to the nematode's life history strategy (Him et al. 2009). This also emphasises the problems associated with using free-living organisms as models for parasitic species when attempting to identify potential drugs and drug targets.

Other heat shock proteins are also involved in the life cycle of parasites (Chen et al. 2018). Examples include Hsp100, which is important for the infectivity and survival of *Leishmania* within macrophages, whilst Hsp70 enables *Toxoplasma gondii* to avoid the host's inflammatory response. Genomic analysis of *P. falciparum* indicates the presence of six Hsp70s, and these could be targeted by direct inhibition or through preventing them from interacting with their co-chaperones (Zininga et al., 2017). In schistosomes, Hsp70 appears to be involved in triggering a cercaria to begin penetrating the skin of a potential definitive host (Ishida and Jolly 2016). It may also be involved in protecting the adult parasite from the harmful effects of the drug praziquantel (Abou-El-Naga 2020). Several heat shock proteins have been characterised from cestodes, including Hsp 60 and Hsp 70, but their functions currently remain poorly understood.

14.6 Nanotechnology

Nanomaterials are solid colloidal particles 1–100 nm in diameter with applications in computing, clothing, engineering, and pharmaceuticals. Nanomaterials can be manufactured from elements such as gold, silver, and carbon, compounds such as iron oxide, as well as organic polymers such as chitosan. Perhaps surprisingly, there is no accepted legal definition of what constitutes an engineered nanomaterial. For parasitologists, their main interest is as drug/antigen/gene carriers, vaccine adjuvants, and diagnostic tools (Mehlhorn 2016). For example, antisense oligodeoxynucleotides delivered using chitosan nanoparticles are effective at preventing the growth of *P. falciparum* (Föger et al. 2006), whilst the incorporation of praziquantel within liposome-based nanoparticles increases its absorption across the tegument of schistosomes (Adekiya et al. 2020).

Nanoparticles come in various shapes and sizes. Nanocapsules have a polymer membrane and the chemical to be delivered is encapsulated within it or adsorbed onto its outer wall. Nanospheres are solid and the associated chemical is either dispersed within its matrix or adsorbed onto its outer surface. As their name suggests, nanotubes have a tubular profile but may be short, long, solid, hollow, single walled, or multi-walled. Particles that are less than 100 nm in size have an enormous surface area to volume ratio and exhibit very different chemical and physical properties to larger particles. This can prove useful for delivering drugs that are either bound to or enclosed within nanoparticles. By altering the size and shape of the nanoparticle, one can radically alter a drug's characteristics. For example, one can improve a drug's solubility in water. Solubility is an important attribute if a drug is to be absorbed and dispersed around the body. Lack of solubility is the reason that many candidate drugs fail during screening. In addition, by delivering drugs via nanoparticles, one can extend the time they remain in the circulation, adjust the rate at which they are released, target them to a specific organ, and combine two or more drugs where combination therapy is required. For example, andrographolide is a natural product derived from the leaves of *Andrographis paniculata* that combines anti-*Leishmania* activity with low toxicity. However, its therapeutic use has not been exploited because it has a short plasma half-life and does not reach its target in sufficient concentration to exert a strong antiparasitic effect. By delivering the drug using nanoparticles one can target the drug to macrophages and increase its effect on *Leishmania donovani* (Akbari et al. 2017; Das et al. 2018). However, the studies to date involve *in vitro* assays, and it currently remains uncertain whether a curative effect is possible *in vivo*.

Nanoparticles have promise for the treatment of intracellular parasites since the drug-bearing particles can be targeted to specific cell types, which would then internalise them through endocytosis. Nanoparticle delivery could also extend the usefulness of older drugs for which the patent has expired so that they can be reintroduced to the market at comparatively low cost. In the process of reformulating older drugs using nanoparticles, they can become more effective and have fewer side effects than in their original guise. For example, nanoencapsulation of quinine increases its efficacy against *P. berghei* malaria in mice and reduces its toxicity (Gomes et al. 2018).

Gold Nanoparticles for the Diagnosis and Treatment of Parasitic Infections

Gold is chemically stable and does not corrode. However, although gold is inert, gold nanoparticles will adsorb protein and other organic molecules onto their surface and therefore they make good delivery vehicles. Gold nanoparticles can be manufactured in a variety of shapes and sizes depending upon the physical and chemical properties required. They have great potential for the treatment of cancer, and they could prove equally useful for combating parasitic diseases (Sengani et al. 2017). For example, gold nanoparticles can act as carriers of hydrophobic drugs and by conjugating an antibody to the surface of the particles they can target specific cells. One can also make gold nanoparticle nucleic acid conjugates to deliver DNA or RNA sequences across cell membranes.

Raman reporters or 'Raman tags' are molecules that are excited when stimulated by specific wavelengths. If one attaches Raman reporters to gold nanoparticles, these can be visualised after administration using Surface Enhanced Raman Spectroscopy (SERS). Consequently, one can locate a tumour or parasite by using gold nanoparticles bearing the appropriate antibodies and Raman reporters. Similarly, SERS can detect nanoparticle nucleic acid conjugates after they have bound to their target sequence and thereby identify the presence of a particular organism. This holds great potential in molecular diagnostics because unlike PCR, there is no requirement to extract and amplify the target sequence. Furthermore, unlike PCR, the procedure does not use enzymes and therefore is less likely to suffer from problems with inhibitors or produce false positives. For example, Ngo et al. (2018) used SERS to detect *Plasmodium falciparum* RNA directly in infected red blood cell lysates.

Certain types of gold nanoparticles convert absorbed light into near-infrared radiation (heat) and have potential for laser photoablation (photodynamic therapy). The basis of this approach is that one targets the nanoparticles to specific cell types and then directs a laser beam onto them. This results in localised heating that kills the cells to which the gold nanoparticles have bound. Pissuwan et al. (2009) have used this approach to kill the tachyzoites of *Toxoplasma gondii*, and it could presumably also be used for other parasites. Alternatively, a laser can deliver a specific wavelength that stimulates gold nanoparticles that have reached their target to release bound molecules, such as drugs. This ensures that the target cells experience a sudden therapeutic dose of the drug.

Agglutination-based immunoassays are based on the detection of aggregates that develop when an antigen binds to an antibody. Gold nanoparticles can be used as a basis for agglutination assays by detecting the aggregates that form when antigens/antibodies bind to antibodies/antigens cross-linked to nanoparticles. Aggregation of gold nanoparticles within a fluid shifts the absorption peak to a longer wavelength, and this can be detected colorimetrically. Alternatively, the binding of antigens/antibodies with nanoparticles can be measured using piezoelectric biosensors that respond to the mass change that follows binding. Wang et al. (2004) describe how *T. gondii* infections can be identified by detecting anti-*T. gondii* immunoglobulins in serum using gold nanoparticles coated with *T. gondii* antigens.

Although nanoparticles have great promise in human and veterinary medicine, most potential applications are still at the experimental stage. Nanoparticles are not without their problems and risks although these are difficult to predict. For example, nanoparticles are prone to aggregation, and this can make them difficult to handle in both liquid and dry formulations. Substances that are safe in one size range may become poisonous or carcinogenic at another size. In addition, the chemistry of the nanomaterials, their shape, and their porosity are all factors that influence their toxicity. Silver nanoparticles have numerous domestic and industrial applications, including incorporation into socks and sports clothing to kill bacteria and reduce smells, but there is mounting concern about the quantities finding their way into the world's ecosystems. This is because silver is a toxic metal and silver nanoparticles could potentially affect microbial and invertebrate communities. There is also evidence that nanoparticles can accumulate through food chains and pose a risk to both wild animals and humans (Deng et al. 2017; Lopez-Chaves et al. 2018). Nanoparticles may also damage DNA directly or through interacting with hydrogen peroxide (which is naturally present in cells) to produce hydroxyl radicals that then damage nuclear DNA. Particularly worrying is the observation that cobalt-chromium nanoparticles can induce DNA damage in human fibroblast cells without ever crossing the cell membrane. In addition, some scientists consider nanoparticles pose a potential danger to our reproductive health (Brohi et al., 2017).

14.7 Quantum Dots

Quantum dots (Qdots) are nanocrystal semiconductors that are of great interest for their electronic and optical characteristics (Efros 2019). In biology, they have great potential for bioimaging. For example, when attached to molecules or cells, they enable movements to be tracked in real time. Quantum dots could prove useful in parasite diagnosis. For example, one can attach parasite antigens to quantum dots, which are then integrated into a lateral flow assay. If specific antibodies to the parasite are present within a serum sample, then their binding to the antigen induces the quantum dots to fluoresce, and this can be detected using a mobile phone reader. This approach has been used to develop potential point of care tests for several parasites, including for malaria (Kim et al. 2017) and *Taenia solium* (Lee et al. 2019). However, at the time of writing, these were all at the proof-of-principal stages of development.

Quantum dots can be made into delivery vehicles by attaching both specific antibodies and other molecules to them. For example, they can deliver gene-silencing RNA (siRNA) to specific cells. This is of particular interest for the treatment of cancer, but the same approach could be used for pathogens. Quantum dots can also be directly toxic (Lin et al. 2019). For example, carbon dots that are manufactured to have a highly positive charge will bind to and damage the negatively charged membranes of bacteria. By adjusting the shape of the carbon dots, it is also possible to increase their toxicity to microbes. Another approach is to shine specific wavelengths of light onto the photodynamic carbon dots and thereby induce them to cause the formation of free radicals that then kill the microorganisms. Some carbon dots, such as those prepared from graphene are intrinsically photodynamic and do not require any modifications, whilst others are coupled to a photosensitiser. These dots are effective at killing many species of bacteria, but at the time of writing, there were few accounts of their effect on parasites.

14.8 Natural Remedies

Up until the development of the pharmaceutical industry in the late 1800s, the only drugs available for treating parasitic diseases were derived from plants, animals, and mineral substances. For example, in the United Kingdom, gastrointestinal nematode infections in both humans and domestic animals were commonly treated with turpentine, which is a volatile oil distilled from pine resin, male fern extract (*Dryopteris felix-mas*), or areca nut powder (*Areca catechu*). These remedies were less effective than modern pharmaceuticals and often deeply unpleasant to the patient. However, some natural remedies were undoubtedly life savers. For example, bark from South American Cinchona trees contains quinine and was for hundreds of years the basis of the only effective treatment for malaria available. Unfortunately, natural products are often inconsistent in their effectiveness because of varying levels of the active ingredient(s). For example, there are many species of Cinchona tree, and their barks vary in their quinine content. Consequently, once chemists became skilled in manufacturing substances that were both more effective and could be delivered in precise dosages, many people lost interest in natural remedies.

There is a resurgence of interest in traditional remedies in both developing and industrialised countries. In developing countries, this interest results (at least in part) from the high cost of pharmaceutical drugs and limited access to modern medicine at a local level. This means that the role of traditional healers within these countries often remains strong. Furthermore, nationalist commentators often complain that western medicine is a continuation of colonialism and a means of maintaining the reliance of poor people on western industrial countries. In reality, India and China are among the most important producers of pharmaceutical medicines by volume. In industrialised countries, traditional medicine is gaining popularity because some people distrust modern western-style medicine and are wary of (or prejudiced against) major drug companies. Many people are also scared by reports of medicines causing harmful side effects, and they are attracted to cures that are purported to be more 'natural'. In addition, many diseases are difficult or impossible to cure and therefore those who suffer from them feel let down by the medical profession and look for alternatives. Social media is therefore awash with enthusiastic accounts of herbal and ethnic medicines whose efficacy often lacks scientific support. For example, in 2020, the price of elephant dung in Namibia rose dramatically when a rumour spread that it is an effective treatment for SARS-COVID-19. It isn't. Elephant dung is commonly sold in some African markets because there is a traditional belief that it has various beneficial effects, the most common one being to facilitate childbirth. There is no scientific evidence for this either. The faeces of various animals were, and remain, common components of traditional medicines in many societies. For example, the British naturalist Edward Topsell quotes that a cure for a horse afflicted by worms would be to 'ride him, having his bit first anointed with dung coming hot from the man' (Topsell 1658). Another consequence of the SARS-COVID epidemic was the removal of bat faeces from the Chinese pharmacopoeia in 2020. Bat faeces was assumed to help improve eyesight and cure certain eye diseases. Any treatments that involve consuming faeces are likely to facilitate the spread of zoonotic diseases and SARS-COVID19 probably originated in bats.

One should never dismiss natural remedies out of hand, and some have a proven record of antiparasitic activity, particularly against ectoparasites. For example, Tea Tree (*Melaleuca alternifolia*) oil has long been recognised as an effective treatment for head lice (Di Campli et al. 2012). Other essential oils, such as those derived from lavender, *Lavandula angustifolia*, are effective at preventing egg laying by blowflies and thereby the onset of myiasis (Bedini et al. 2019). Some essential oils possess antibacterial and antifungal properties, and this potentially enhances their usefulness in topical applications. The scientific evidence for the efficacy of essential oils for the treatment of internal parasites is more limited. For example, essential oils derived from Mediterranean thyme (*Thymbra capitata*) and Oregano (*Origanum virens*) are effective at inhibiting the *in vitro* growth of *Giardia lamblia* (Machado et al. 2010), but whether this translates into an effective therapeutic effect *in vivo* is uncertain. Similarly, various essential oils prevent the hatching of *Haemonchus contortus* eggs under *in vitro* conditions, but it is uncertain whether they would have any effect against the parasitic adult worms living in sheep (Katiki et al. 2017). *in vitro* and *in vivo* experiments can sometimes yield conflicting results. For example, Amora et al. (2017) found that although that several essential oils were highly toxic to the plant parasitic nematode *Meloidogyne javanica* under *in vitro* conditions, their presence was associated with an increase in nematode reproduction and gall formation in greenhouse experiments.

Pharmaceutical companies are interested in natural products because once the active ingredients are identified, the chemicals can be either manufactured *de novo* or through biologically engineered bacteria or, if the structure is too complicated, extracted from cultured plants or animals. For example, the cysteine proteases contained within the latex of fig trees (*Ficus* spp.) and the papaya (*Carica papaya*) break down the cuticle of nematodes but have a low oral toxicity in mammals. They therefore show promise for the treatment of gastrointestinal nematode infections (Stepek et al. 2006). Sponges (Porifera) contain a group of chemicals called kalihinol*s* that are toxic to several species of human parasitic protozoa, including *Plasmodium* (Anjum et al. 2016). Unfortunately, major problems remain in developing artificial culturing conditions for sponges, and there are further problems in formulating the chemicals such that they reach the sites where the parasites are living in sufficient concentration and without losing their biological activity. The microbes associated with sponges are also potential sources of antiparasitic chemicals. For example, the *Streptomyces* associated with certain Mediterranean sponges produce chemicals with activity against *Leishmania major* and *Trypanosoma brucei brucei* (Pimentel-Elardo et al. 2010). Possibly, these might be more amenable to culturing or the genes responsible to produce antiparasitic chemicals could be inserted into *Streptomyces* that are already cultured commercially.

Artemisinin and the Treatment of Malaria

References to the antimalarial properties of extracts of sweet wormwood ('Qinghao', *Artemisia annua*) can be found in Chinese traditional medicine texts that date back several hundred years. However, it was not until the 1970s that a serious attempt was made to determine the active ingredient and exploit it on a widespread basis. As is so often the case with parasitic diseases, the initial impetus for the research came not so much from humanitarian concerns but to support troops engaged in warfare. In this case, the Chinese government was seeking novel antimalarial drugs that could be used to treat Vietnamese Communist soldiers who were engaged in a prolonged jungle war with the Americans.

The active ingredient of the wormwood extract is a sesquiterpene trioxane lactone called 'qinghaosu' or 'artemisinin', which has the empirical formula $C_{15}H_{22}O_5$. The complicated structure makes it difficult to synthesise artificially, although strains of yeast have been biologically engineered to produce artemisinic acid that can then be converted to artemisinin. The amount of artemisinin contained within sweet wormwood varies between strains of the plant and is also affected by the soil and climate in which it is grown. There are hopes that high-yield hybrid plants currently grown in Madagascar, South Africa, and certain other African countries may boost the production of artemisinin.

Artemisinin is highly effective for the treatment of malaria and in particular that caused by *Plasmodium falciparum*. It is unusual amongst current antimalarial drugs in being active against all the life cycle stages present in humans. The discovery of artemisinin coincided with the development of widespread resistance to chloroquine and sulphadoxine-pyrimethamine. Consequently, artemisinin was soon the treatment of choice. Sweet wormwood therefore became an important cash crop for many farmers in China, Vietnam and elsewhere in Asia. However, by 2007 the supply outstripped demand and the value of sweet wormwood fell from a high of US$1,100 kg^{-1} to a low of US$200 kg^{-1}. Farmers therefore turned their attention to more valuable crops, and this was further exacerbated by a coincidental increase in world food prices. No sooner had this happened than there was an increase in the demand for sweet wormwood when the Global Fund to Fight AIDS, Tuberculosis and Malaria released funds to subsidise cheap drugs under its Affordable Medicines Facility – Malaria. Thus, the availability and price of artemisinin is closely tied to the vagaries of the marketplace and sudden injections of cash from large aid organisations (Van Noorden 2010).

Artemisinin has poor bioavailability and is usually modified to produce semi-synthetic derivatives such as artesunate, artemotil, and artelinic acid. The WHO demands that artemisinin and its derivates are used in combination with other drugs (artemisinin-based combination therapies – ACTSs). This ensures that all the parasites are killed and thereby reduces the chances of resistance developing. For example, in many parts of the world, artesunate is used in combination with mefloquine to treat *P. falciparum* infections. Unfortunately, some manufacturers do not heed the WHO and sell artemisinin derivatives in the absence of other antimalarial drugs. In addition, in common with other antimalarial drugs, there is a widespread problem of counterfeit drugs that contain either very low levels of the active ingredients or none (Yadav and Rawal 2016). Counterfeit drugs containing low concentrations of artemisinin derivatives (or other antimalarial drugs) increase the risk of drug resistance developing. The counterfeit drugs lacking artemisinin derivatives are equally harmful because not only will they not cure the patient's malaria, but they sometimes contain carcinogens such as benzene and safrole. Furthermore, because the patient is not cured, they remain a source of infection to mosquitoes and, therefore, disease transmission. It is therefore not surprising that artemisinin-resistant strains of *P. falciparum* were soon identified. Initially, these were localised between the borders of Thailand and Cambodia, but later they were also discovered in Rwanda in Africa. The Rwandan strains have apparently evolved their resistance independently of the Asian strains (Uwimana et al. 2020). There is therefore concern that the genes conferring resistance might now spread more widely or evolve independently elsewhere.

The mode of action of artemisinin and its derivatives is still uncertain, and there are several competing theories (Feng et al. 2019). There is strong evidence that the drug is first activated by interacting with iron, although whether this is free iron, that present in haem, or both, is uncertain. For example, one suggestion is that the drug interacts with the haem moiety in hemozoin, which is a waste product of *Plasmodium* metabolism, to produce free radicals that then kill the parasite by damaging DNA and cell membranes and/or interfering with the physiological processes necessary for further production of hemozoin. Artemisinin may also induce pathology without interacting with iron, since derivatives accumulate within the membranes of *P. falciparum* where they directly induce oxidative damage. It now seems likely that artemisinin exerts its effects through acting on several physiological process within *Plasmodium* (Wang et al. 2015).

In addition to *Plasmodium*, artemisinin derivatives also demonstrate activity against other protozoan (Loo et al. 2017) and helminth (Lam et al. 2018) parasites. These include *Toxoplasma gondii* (Secrieru et al. 2020), *Leishmania donovani* (Want et al. 2017), and *Schistosoma mansoni* (de A.P. Corrêa et al. 2019). Artemisinin derivatives also show promise in cancer chemotherapy (Bhaw-Luximon and Jhurry 2017). Consequently, artemisinin-based drug therapies show promise for the treatment of malaria co-infections and carcinogenic trematode parasites.

Although traditional remedies are often considered by the public to be safe, they have seldom undergone the stringent efficacy and safety testing required of pharmaceutical products. This is not surprising because those who prepare natural remedies seldom have the resources to comply with the requisite legislation. Furthermore, the products often contain a mixture of chemicals, which vary between batches. Pharmaceutical companies and their proponents complain about this, since they spend enormous sums of money in developing drugs, whilst companies that market natural products seldom conduct clinical trials, and there is limited information on the health risks associated with taking the remedies. In addition, herbal remedies sometimes contain high levels of heavy metals and undeclared prescription drugs that could have seriously harmful consequences (Skalicka-Woźniak et al. 2017).

14.9   Homeopathy

Few therapies are more guaranteed to induce fury among scientists and medical professionals than homeopathy. However, it would be perverse to ignore totally 'remedies' that are routinely used by hundreds of thousands of people around the world. Homeopathy is based upon the concept that a disease can be cured by consuming dilute concentrations of substances that would provoke similar signs and symptoms in a healthy person. The idea was first proposed in the 1790s by the German physician Dr Samuel Hahnemann. He observed that drugs obtained from the bark of *Cinchona* that were used at the time to treat malaria induced malaria-like symptoms in those who were uninfected by the disease. Homeopathy has since grown enormously and supports registered practitioners in human and veterinary homeopathic medicine. The global market for homeopathic remedies is worth millions of pounds. Although there are experimental studies reporting beneficial effects of homeopathic treatments for parasitic diseases, these tend to be published in journals of homeopathy and specialist websites rather than in the mainstream scientific press. For example, Rajan et al. (2017) discuss the use of homeopathy in the treatment of malaria, whilst Giuliotti et al. (2016) consider its use for the treatment of strongyle infections in sheep.

Many scientists argue that the concentration of the purported 'active ingredient' in homeopathic remedies is so low that one might as well drink distilled water, and there is no rational scientific basis for homeopathy. In human medicine, the reported beneficial results of homeopathic remedies have usually been ascribed to the placebo effect. That is, if one believes that something is doing you good, then you often feel better. By contrast, the nocebo effect occurs when a person starts with a viewpoint that a drug or other treatment is harmful and then reports adverse reactions that are either non-existent or arise from some other cause. Basically, if you start thinking that something will be painful or unpleasant, then it probably will be. The placebo and nocebo effects are the bane of all drug trials, and why it is so important to have double-blind studies involving as many people as possible. Anyway, most published evidence suggests that homeopathic medicine provides no more benefit than a placebo. One could argue that if a patient feels better, and the treatment is not harmful, then it doesn't matter how the effect is achieved. However, to knowingly prescribe someone, a placebo is not consistent with informed patient choice and some doctors would consider it unethical. Of more concern, especially with parasitic infections, is the possibility that an infected person who initially resorts to homeopathy might put off consulting a medical doctor until the disease has caused serious pathology and/or he/she could have transmitted the infection to others.

15

Parasite Vaccines

CONTENTS

15.1 Introduction

Drugs may completely cure a disease, but reinfection is inevitable if the pathogen's infectious stage and the host renew their acquaintance. Therefore, wherever exposure to a pathogen is unavoidable, remaining disease free only comes from developing a protective immune response. In the past, the acquisition of immunity was dependent upon surviving the infection. Indeed, in the UK and USA, some parents still organise ill advised 'chicken pox parties' so that their offspring have the opportunity to become infected with the disease. Once a child recovers from chicken pox, they usually have a long-lasting protective immunity to the condition. Unfortunately, natural infections can have serious consequences. Although chicken pox is usually a mild disease in children, the blisters it causes occasionally become infected by bacteria and result in serious illness. The purpose of vaccination is to stimulate a protective immune response without the risks associated with a natural infection.

Parasitology: An Integrated Approach, Second Edition. Alan Gunn and Sarah J. Pitt.
© 2022 John Wiley & Sons Ltd. Published 2022 by John Wiley & Sons Ltd.
Companion website: www.wiley.com/go/gunn/parasitology2

15.2   The Design and Use of Vaccines

The development of vaccines to treat viral infections is one of the major success stories of modern medicine. Indeed, smallpox and rinderpest are now extinct in the wild because highly effective vaccines enabled carefully coordinated eradication programmes. For the same reasons, polio is on the verge of eradication. Vaccines are particularly useful in control programmes because, unlike drugs, pathogens do not develop resistance against them. There are also effective vaccines against many bacterial infections although the level and duration of protection they offer are seldom sufficient to make them the basis for an eradication campaign. The world market for vaccines is huge. As of 2018, it was worth US\$33.7 billion, and the forecast was for this to rise to US\$100 billion by 2028. This forecast was made before the arrival of SARS-COVID-19 and the worldwide demand for SARS-COVID-19 vaccines will fuel an even bigger market. Some estimates suggest that the global malaria vaccine market will be worth US\$156.8 million by 2026. The projected returns for a successful vaccine are driving enormous investment and this includes the search for vaccines against parasitic infections. Nevertheless, despite decades of work, relatively few anti-parasitic vaccines have met the criteria for commercial production.

One normally uses a vaccine prophylactically to protect a person or animal from becoming infected rather than to treat an existing condition. That is, vaccines stimulate the recipient to develop an active acquired immunity to the pathogen through the production of antibodies that neutralise the target antigen or make it more vulnerable to phagocytosis by immune cells. For good practical reasons, experimental studies on vaccines are almost entirely undertaken using mice. For example, it allows large-scale studies that would be very difficult to perform using larger animals, such as the observation that vaccination of mice with OVA257-264 peptide triggers the production of more CD8 T cells during the middle of the day than at other times (Nobis et al. 2019). Circadian rhythms influence many aspects of physiology in all animals, including humans, and therefore the timing of vaccinations may influence their effectiveness. Unfortunately, vaccines that work well in mice do not necessarily provide the same level of protection in other, larger, mammals. It is therefore common for vaccines judged highly promising in the laboratory to fail to transfer into commercial products. It will also remain unknown whether vaccines rejected for limited effectiveness in mice models might have worked very well in humans and other animals.

In addition to protecting the recipient against future infections, some workers suggest designing vaccines to reduce the pathology associated with existing infections. The reasoning is that developing vaccines that protect against parasitic infections is proving exceptionally difficult. However, we know that for many of these, such as malaria and schistosomiasis, most of the pathology associated with the infection is caused by the host's own excessive immune response against the parasite. The suggestion is therefore to develop vaccines that target the host's response to parasite antigens. At the time of writing, these were all at the experimental stage of development. Presumably, this approach would help in the treatment of an individual but would not necessarily remove the parasite or prevent the possibility of disease transmission.

15.3   Herd Immunity

Herd immunity is a term that was frequently, and often incorrectly, used during the SARS-COVID-19 viral pandemic that commenced in 2019–2020. Herd immunity does not occur through natural infections and is a goal one hopes to achieve through a successful vaccination strategy. The basis of herd immunity is that as the proportion of immune or less susceptible people (or animals) in a population increases, there is a decrease in the chances of a non-immune individual encountering the infection. Once the number of immune individuals passes a certain threshold, it becomes impossible for further disease transmission to occur. This is the 'herd immunity threshold'. It is essential to reach this threshold because there will always some people for whom it is not possible to vaccinate: for example, the very young, the very elderly or the immunosuppressed. Numerous factors, such as the nature of transmission and effectiveness of the immunity conferred by natural infection/vaccination, influence the herd immunity threshold and therefore it varies between pathogens. The complexities and vagaries of natural infections means that they cannot by themselves result in the reaching of the herd immunity threshold. Even for diseases that are highly infectious and induce a strong life-long immunity, such as measles, it is essential to continue to vaccinate most children to ensure the population is protected.

15.4 Factors Limiting the Production of Commercial Antiparasitic Vaccines

Protozoan and metazoan parasites are considerably more complex organisms than prokaryotic bacteria and viruses, and they have more complicated relationships with their hosts. Although most parasites generate an immune response in their host, in many cases this is not protective. This is often because the parasite can avoid the immune response by hiding within host cells (e.g., *Leishmania*) and/or because it has evolved sophisticated mechanisms for avoiding, depressing, and confusing the host immune system such as variable surface coat antigens generated by trypanosomes. Consequently, the vaccine must do more than stimulate the normal host immune response to the parasite, which is essentially all that is required of an antiviral vaccine. Instead, it must induce an immune response that reaches the parasite, overcomes the parasite defence mechanisms, and induces a much stronger immune response or exploits targets unreachable by the normal host immune response.

One must also consider the commercial concerns. Some estimates suggest developing any new vaccine typically takes 12–14 years to reach the market and costs ~US$1 billion (Plotkin et al. 2017). By contrast, the development of SARS-COVID-19 vaccines took less than 12 months from identification of the virus to approval of several candidate vaccines for widespread deployment. However, this was very much an exception and one in which unprecedented resources were made available. Considering that most candidate vaccines will fail at some stage in the development process, no commercial company can afford to embark upon vaccine development unless it has an extremely good chance of recouping the costs of its investment.

15.5 Properties of an Ideal Vaccine

The 'ideal vaccine' is, like all ideals, impossible to deliver. This is because it depends upon not one but two biological systems – the pathogen and the host – and the vagaries of human priorities. For example, what we consider acceptable and affordable. That is, there are often big discrepancies between what would be ideal and what is realistic. Anyway, in an ideal world, a vaccine would meet the following criteria:

Provides effective immunity: Obviously, a vaccine must stimulate an effective immune response against the target pathogen. No vaccine can provide 100% immunity because individuals vary in their response to immune challenge depending upon numerous factors, such as age, health, and genetics. Most childhood vaccines, such as MMR (measles, mumps, rubella), typically provide protection for 85–95% of their recipients, and this is quite sufficient to provide 'herd immunity' if sufficient of the population are vaccinated.

Provides life-long immunity: Ideally, vaccination should provide life-long immunity, because this means life-long protection and no requirement for periodic 'booster vaccinations' throughout life. This is an impossibility because our immunological memory fades with time. Vaccines vary considerably in the length of immunity they provide. The polio vaccine provides over 99% protection for at least 18 years, whilst the immunity provided by the pertussis (whooping cough) vaccine starts to wane after about 4 years and ceases after 7 years.

Transmission of effective immunity from mother to young: During development in the womb, we, like all mammals, acquire antibodies from our mother across the placenta. After birth, we also absorb antibodies from colostrum – our mother's first milk. Both these sources provide a level of immunity against pathogens whilst our immune system is developing. This protection seldom lasts long: there is an 'immunity gap' between when the protection gained from our mother wears off, and our own immune system becomes fully functional. Despite this, if a mother receives a vaccine that enables her to pass on her immunity to her offspring, then this would provide protection during a vulnerable stage of life. However, some vaccinations can have negative consequences for the young. For example, cows receiving vaccinations against bovine-viral-diarrhoea-virus produce antibodies in their colostrum that cause a potentially fatal condition, bovine neonatal pancytopenia, in their calves (Kasonta et al. 2014).

Provides protection to all ages from infant to elderly: Numerous factors govern the effectiveness of our immune system and one of these is age. As we become elderly, we experience immunosenescence in which, amongst other things, our T-cells and macrophages begin to function less efficiently. Consequently, we become more vulnerable to infections, and vaccinations do not provide us with the same level as protection as they did when we were younger. Similarly, when we are babies, our immune system is not yet fully developed and therefore vaccines may not stimulate a protective immune response. Obviously, an ideal vaccine would stimulate effective immunity in both the very young and the very old.

Provides equal protection to males and females: Males and females often respond differently to both drugs and infections – and the same is true of vaccinations. Men tend to be more susceptible than women to infectious diseases and show a weaker response to vaccinations. This is because of interactions between our sex hormones and the immune system. At menopause, the decline in oestradiol results in a reduction in immune responsiveness (Giefing-Kröll et al. 2015) and in both sexes the decline in sex hormone titres with age contributes to changes in their response to vaccinations.

Provides immunity against all the pathogen's life cycle stages in the vaccine recipient: Many parasites, such as *Plasmodium* spp. and *Schistosoma* spp., exhibit several life-cycle stages within their host. These transformations are often accompanied by changes in their cell membrane components and therefore in their antigenic profile. Furthermore, many parasite species can express numerous antigenic variants in response to immune attack. Consequently, unless one can develop a vaccine against an antigen shared by all stages of the life cycle in the host, it will have limited effectiveness. Similarly, some parasites invade different tissues or cells at different stages in their life cycle and this can alter the ability of the immune system to locate and neutralise them.

Prevents pathogen transmission: The ideal vaccine should prevent a person or animal from becoming infected and shedding the pathogen into the environment. Inactivated vaccines seldom provide sufficient immunity to prevent infection and they act by reducing pathogen multiplication and the development of clinical disease. That is, the pathogen initially invades the mucosal surfaces and starts to reproduce but then, in the vaccinated host, it stimulates sufficient immune reaction to prevent further damage. In this situation, vaccination has prevented clinical disease and reduced pathogen transmission, but it has not prevented transmission entirely, which would be the ideal.

Induces rapid immunity: Obviously, the more rapid the onset of protective immunity, the sooner an individual gains protection and the sooner one is likely to reach the herd immunity threshold. The speed with which an individual acquires protective immunity varies between vaccines and the nature of the immune response they generate. Usually, it will take 7–21 days for protective immunity to develop although inactivated vaccines often require two or more doses to become fully protective.

Requires only one immunisation to provide protection: To ensure herd immunity, it is essential to vaccinate as near 100% of the population as possible. The actual herd immunity threshold value will depend upon the nature of pathogen, its transmission, and the level and duration of the protection offered by the vaccine. This requires both co-ordination and compliance. As soon as more than one vaccination is required, it is likely that individuals will miss the follow-up vaccination or not receive vaccinations at the correct time intervals.

Does not require injections: Many people dislike injections, and some refuse them even though they offer protection against a potentially fatal disease. If repeated injections are required, there is a further reduction in the likelihood of take up. Injections also present logistical difficulties because trained medical personnel must deliver them. This all reduces the chances of vaccinating sufficient people to ensure herd immunity. Furthermore, in some developing countries, the re-use of needles has resulted in the spread of diseases such as HIV and hepatitis c. There is therefore a lot of work on developing alternative delivery methods such as oral vaccines, nasal sprays, and patches that enable vaccine absorption across the skin.

Do not induce harmful side effects: Vaccines provide protection, but they sometimes cause side effects in a minority of recipients. Most side effects tend to be mild, such as pain at the injection site or headaches. Some side effects result from a reaction to one or more of the vaccine's components. For example, owing to the method of manufacture, some vaccines contain gelatine or egg/chicken proteins, and these can cause allergic reactions in certain people. For some vaccines, there are recognised serious risks. For example, in a very small number of cases, primarily infants, the oral poliovirus vaccine causes vaccine-associated paralytic poliomyelitis.

Stable: Stability is an important consideration for any vaccine intended for use in developing countries. If a vaccine requires constant refrigeration, then this complicates its transport and storage and increases the costs associated with its use. Similarly, a long 'shelf-life' is helpful because this facilitates the maintenance of a ready availability and stockpiling in case of emergencies. It is also common for goods (including vaccines) to be stuck at customs posts for long periods, during which they experience less than ideal conditions.

Cheap: If one intends to use a vaccine in a control programme, it is essential that it is cheap because one will have to deliver it to thousands and possibly even millions of recipients. Even affluent countries cannot consider such undertakings if the vaccine is expensive. Parasitic diseases are primarily a problem for poor people (and their livestock) living in developing countries. Consequently, an effective vaccine is of little use if individuals or their government cannot afford it.

15.6 Types of Vaccine

There are various types of vaccine (Table 15.1), and their composition is typically more complex than that of drugs. In addition to the antigen(s) stimulating the immune response, vaccines also often contain stabilisers, adjuvants, preservatives, and by-products from the production process. This complexity means that vaccines are difficult to manufacture although commercial companies consider this beneficial because it limits the ability of a competitor to manufacture a generic version.

Table 15.1 Types of vaccine.

Type of vaccine	Example
Live attenuated	MMR (measles, mumps, rubella), yellow fever, smallpox
Inactivated	Rabies, Hepatitis A
Subunit/recombinant vaccines	PPV (adult pneumococcal vaccine)/HPV (Human papillomavirus vaccine)
Polysaccharide conjugate vaccines	*Haemophilus* influenza type b (Hib)
Toxoid vaccines	DPT: diphtheria, pertussis (whooping cough), tetanus
Virus-like particles	None in commercial production
Peptide/polypeptide vaccines	None in commercial production
DNA/ RNA vaccines	Pfizer-BioNTech COVID-19 vaccine

After identifying a potential target antigen that is safe and induces protective immunity, a crucial subsequent step is to manufacture it in bulk quantities and then process it for delivery as a commercial vaccine. During these steps, conformational changes to the antigen can occur. For an antigen to induce a protective effect, it is essential for the antigen in the vaccine to retain the same conformation as the intended target antigen on the pathogen. If it does not, the vaccine will generate antibodies that do not attach to the pathogen.

15.6.1   Live Attenuated Vaccines

These vaccines utilise pathogens rendered harmless or less virulent (i.e., 'attenuated') by passaging them through a foreign host, exposing them to chemicals or radiation, or selectively knocking out genes of pathogens reared in culture. Because these vaccines closely resemble the target pathogen and replicate within the vaccine recipient, it means that the recipient's immune system must respond to a variety of pathogen life cycle stages. This is an important consideration for antiparasitic vaccines because many parasites undergo a succession of developmental stages in their host. Consequently, live vaccines elicit an immune response like that of the pathogen but without its associated pathology. Live vaccines are therefore more likely to induce a T-cell-mediated immune response than killed vaccines, and this is important for combating intracellular parasites.

Live vaccines usually stimulate a strong long-lasting immunity, but they need to be stored under refrigerated conditions. This limits their use in tropical countries with basic infrastructures. In addition, they can present a risk to those with a weakened immune system or who are immunocompromised. There is also a risk that the vaccine strain may regain some of its pathogenicity. For example, in areas of poor sanitation and limited vaccine coverage, rare cases of paralysing vaccine-derived poliovirus infection can occur. Similarly, a vaccine developed against *Ancylostoma caninum* in the 1960s utilising X-ray attenuated third stage larvae ceased commercial production partly because of the risk of patent infections. The vaccine gave up to 90% protection in dogs but the combination of patent infections, expensive production costs, and a short shelf life meant that it became commercially unviable.

Another problem with attenuated anti-parasite vaccines is that one cannot grow many parasites in culture. Furthermore, those that can be grown change their phenotype over successive generations so that they become less like the 'wild type' and genetically more homogeneous. This means that they do not engender an immune reaction that is protective against the 'wild' parasite. Nevertheless, commercial live attenuated vaccines against *Theileria parva* and *Theileria annulata* are available and have been for many years. Unfortunately, although they provide effective immunity in cattle, they suffer from practical problems in their delivery and the search for better vaccines against these parasites continues (Nene and Morrison 2016). Live attenuated vaccines prepared using parasites in which genes have been deleted have been trialled for the pre-erythrocytic stages of *Plasmodium* spp. (Mac-Daniel and Ménard 2019), *Leishmania donovani* (Banerjee et al. 2018), and *Trypanosoma cruzi* (Sánchez-Valdéz et al. 2015), but at the time of writing, none of these were in commercial production.

Some workers have undertaken trials using live attenuated organisms as vehicles for recombinant antigens. For example, the yellow fever virus vaccine YF17D is extremely safe and effective, and one can genetically engineer it to express specific antigens that will generate an immune response. By inserting circumsporozoite antigen into the YF17D vaccine, Stoyanov et al. (2010) were able to immunise mice against *Plasmodium yoelii* and thereby protect them from subsequent infection. Similarly, Liu et al. (2018) enabled a live attenuated virus against goatpox virus to express a recombinant antigen against the tapeworm *Echinococcus granulosus*. This holds the prospects of producing a multivalent vaccine that simultaneously vaccinates against two pathogens. Commercially available vaccines against both pathogens are available in China.

15.6.2 Inactivated Vaccines

As the name suggests, these vaccines utilise previously pathogenic organisms that were killed by heat or chemicals. Although they are safer than live attenuated vaccines, they tend to be less immunogenic and therefore require several courses of vaccinations to stimulate a good level of protective immunity. An inactivated vaccine against *Neospora caninum* infection in cattle was once available commercially, but it produced a low-level of immunity (Weston et al. 2012) and was subsequently withdrawn.

15.6.3 Subunit/Recombinant Vaccines

A subunit vaccine utilises a single specific protein as the antigen. The protein may be isolated from the pathogen or manufactured by a vector as a recombinant protein (recombinant vaccine). These vaccines have the benefit of safety since they do not contain any pathogen DNA or RNA. The effectiveness of subunit vaccines depends upon the antigen retaining its three-dimensional shape during its isolation/manufacture. If it does not, the vaccine antigen will still stimulate an immune reaction, but the antibodies that are formed may not 'recognise' the target antigen on the pathogen. This may explain why experimental subunit/recombinant vaccines sometimes yield unpredictable results.

One produces a recombinant vaccine by inserting the gene coding for the antigen of interest from a pathogen into the genome of a vector. The vector then makes large amounts of the protein coded for by the gene, and this is then isolated and used to prepare the vaccine. Many vectors are available including microbes, invertebrate and vertebrate cells, and plants (Table 15.2). Obviously, the recombinant protein must not be harmful to the vector producing it, and one must be able to separate it from the vector. The recombinant protein must also not require any post-translational modification to acquire the correct structure and shape. The vector will lack the physiological mechanisms to do this because it would only contain the gene coding for production and not those for subsequent modification.

Table 15.2   Cell types used to produce vaccines.

Organism type	Example cell type	Product type
Mammalian	VERO (African green monkey [*Chlorocebus* spp.] kidney epithelial cells); Chinese hamster ovary	Live attenuated or inactivated viral products, viral vectors, virus-like protein production, recombinant proteins
Avian	Duck cell lines	Live attenuated or inactivated viral products, viral vectors,
Insect	*Spodoptera frugiperda* (fall armyworm moth)	Live attenuated or inactivated viral products, viral vectors, virus-like protein production, recombinant proteins
Plant	*Nicotiana benthamiana* ('benth' – a close relative of the tobacco plant)	Recombinant proteins
Fungal	*Saccharomyces cerevisiae* (yeast)	Live attenuated or inactivated viral products, viral vectors, recombinant proteins, conjugates
Bacterial	*Escherichia coli* *Streptococcus pneumonia* *Neisseria meningitidis*	Native proteins, recombinant proteins, conjugates, synthetic peptides, poly epitopes expression

Many vector cell types are in use (Table 15.2), and they all have their advantages and disadvantages. Producing vaccines in plants is potentially cost-effective because it is cheaper to set up and maintain a greenhouse than a bioreactor. Typically, one employs a 'plant virus expression system' to introduce the gene coding for the desired antigen into a plant genome (Laere et al. 2016). If food plants are used, it may even be possible to produce an edible vaccine although maintaining dose levels would be difficult. The technology is currently still in its infancy, and the use of transgenic organisms is controversial and strictly regulated. Blagborough et al. (2016) successfully manufactured recombinant P25 protein of *Plasmodium vivax* in tobacco plants. This protein was then isolated and used as the basis for a transmission blocking vaccine. However, at the time of writing, there were no plant-based vaccines approved for human use.

Recombinant subunit vaccines providing a high level of protection are available against a range of cestode parasites, but for various reasons most have not entered commercial production. For example, an effective recombinant antigen vaccine that prevents the development of the cysticerci of *Taenia ovis* in sheep was identified many years ago. After a great deal of research, the vaccine was registered but never released as a commercial product despite 20 million doses being prepared. Rickard et al. (1995) admirably cover the combination of commercial, scientific, and political factors for this. The vaccine aimed to reduce the prevalence of cysticercosis in older lambs before they went to slaughter. However, to mount an eradication campaign, it was necessary to prevent the lambs from becoming infected, and this required them to receive two doses of the vaccine. Antibodies present in the vaccinated ewe's colostrum interfered with the development of a protective response to the first dose of vaccine by the lambs. Consequently, if the maternal antibodies failed to provide sufficient protection, there was a window of opportunity for infection to occur before the lambs received their second vaccination. More work was needed to overcome this problem and rather than spend yet more money, the decision was made to start manufacturing a product that was not entirely suitable for an eradication programme. There was then the question of who should pay for the vaccine since the cost of the infection was predominantly borne by the slaughterhouse through condemned meat rather than the farmer. On top of this was a complex political situation revolving around who was responsible for what, legislation, and how the vaccine would fit into an existing control programme based on treating dogs – which are the definitive hosts of *Taenia ovis*. Arguably, it is cheaper to target a comparatively small number of dogs rather than vaccinate hundreds of sheep (although that depends upon having the means to compel all owners to treat or vaccinate their dogs). This is a classic example of how even if an effective vaccine, drug, or control measure is available, it may not make commercial sense to put it on the market.

A recombinant vaccine against the tapeworm *E. granulosus* has had more success and is currently in commercial production in China and Argentina. It utilises the EG95 antigen derived from the oncosphere. It provides 96–100% protection in sheep and goats and can be adapted for use in cattle (Heath et al. 2012). Recombinant antigens derived from protoscolices of *E. granulosus* can provide a similarly high protection against infection in dogs (Zhang and McManus 2008) although this is not in commercial use. Subunit vaccines have been trialled with promising results against many parasites including protozoa (e.g., *Toxoplasma gondii* [Ching et al. 2016], *Leishmania donovani* [Duthie et al. 2017]), nematodes (e.g., *Trichinella spiralis* [Xu et al. 2020]), and arthropods (e.g., *Psoroptes ovis* [Burgess et al. 2016]).

15.6.4 Peptide/Polypeptide Vaccines

A peptide consists of chains of two or more amino acids linked together by peptide bonds – these are covalent bonds between C1 (carbon atom number one) of one amino acid and N2 (nitrogen atom number 2) of another amino acid. Polypeptides consist of long chains of amino acids – there is no consensus about the point at which the addition of another amino acid turns a peptide into a polypeptide. The peptides used in vaccines are synthesised artificially and usually consist of 20–30 amino acids. Peptides are not immunogenic and must be combined with another substance, called an adjuvant, to stimulate the immune system. A commonly used adjuvant in animal experiments is 'Freund's Complete Adjuvant' (an oil emulsion) although its use in human vaccines is banned owing to its toxicity. Even in animal studies, use of Freund's Complete Adjuvant must follow strict guidelines because of the pain it can cause. Adjuvants licensed for use in human vaccines include alum, virosomes, and squalene. An alternative approach is to conjugate the peptide to a self-adjuvanting 'lipid core peptide system' (LCP): these can subsequently self-assemble into nanoparticles that stimulate sufficient immune response without the need for an additional adjuvant. There is therefore a lower risk of LCP vaccines inducing toxic side effects than those incorporating adjuvants.

At the time of writing, no peptide vaccines were in commercial production although they have the advantages of being safe, cheap, and easy to manufacture. Peptide vaccines have been trialled against several parasitic diseases including leishmaniasis (De Brito et al. 2018), malaria (Skwarczynski et al. 2020), and hookworm infections (Bartlett et al. 2020).

15.6.5 Carbohydrate Vaccines

The first commercially successful carbohydrate vaccines targeted polysaccharides found in the capsule of bacteria such as *Haemophilus influenzae* type b (Hib) and *Neisseria meningitidis*. Carbohydrate vaccines hold promise because all cell membranes include a rich array of carbohydrates attached to proteins and lipids. These molecules fulfil numerous physiological processes that include features essential to parasites, such as cell recognition and attachment.

Carbohydrate antigens are classed as thymus-independent antigens whilst proteins are classed as thymus-dependent antigens. The terms reflect the different ways in which the antigens are processed. Unlike protein antigens, carbohydrate antigens do not form complexes with MHC (major histocompatibility class) class II molecules. Instead, the antigen presenting cells present them directly to carbohydrate antigen-specific B cells. The activated B cells then produce IgM and IgG although in lower amounts than those stimulated by protein antigens and the immune memory is often shorter. Furthermore, babies and children less than 2 years old are unable to develop antibodies in response to carbohydrate antigens. This is therefore a serious consideration for the development of a vaccine aimed at targeting many parasitic infections. A way around these problems is to link the carbohydrate antigen to a protein to form a glycoconjugate. In this manner, it is possible to initiate a thymus-dependent immune response that provides long-lasting immunity.

Parasite carbohydrate vaccines are hard to develop because it is difficult to isolate enough pure material to work with. Unlike bacterial infections, most parasites cannot be cultured under laboratory conditions. Nevertheless, experimental carbohydrate vaccines have shown promise against *Plasmodium* spp., *Leishmania* spp., *Toxoplasma gondii*, and *Schistosoma* spp. (Jaurigue and Seeberger 2017; McWilliam et al. 2012).

15.6.6 Toxoid (Anti-toxin) Vaccines

These vaccines are useful where the toxin(s) produced a pathogen is/are the main virulence factor – this is a feature of a range of bacterial diseases such as anthrax and diphtheria. The vaccine is prepared by isolating the toxin and then inactivating it – for example, using formaldehyde. Because the chemical mimics the toxin biochemically, but is inactive, it is called a 'toxoid'. For example, the DPT vaccine (diphtheria, pertussis, tetanus) includes antigens that elicit an immune response against diphtheria and tetanus toxins.

Parasites seldom manufacture toxins although some of them release waste products of digestion that have harmful effects. For example, malaria parasites release hemozoin, which causes many harmful effects. However, much of the pathology associated with malaria result from the host reaction to glycosylphosphatidylinositols present on the *Plasmodium* surface membranes. These molecules function as anchors for cell surface proteins but they also induce a strong inflammatory response in the host. Some workers consider them to have the characteristics of a toxin and the immune response they cause is responsible for some of the harmful pathology associated with infection. Synthetic glycosylphosphatidylinositol has been used as the basis for an anti-toxin vaccine that provides protection against cerebral malaria in mouse models (Malik et al. 2019). Glycosylphosphatidylinositols are common constituents of cell membranes (including those of humans) and may be responsible for the pathology associated with other parasites. However, a vaccine raised against *Toxoplasma gondii* glycosylphosphatidylinositol was not protective (Götze et al. 2015).

15.6.7 Virus-Like Particles Vaccines

Virus-like particles are produced from natural or synthetic proteins that can self-assemble to form structures that resemble viruses. They contain no DNA or RNA, so they cannot replicate and are not infectious, but they can include proteins that engender a strong immune response. At the time of writing, there were no commercial vaccines using this technology although preclinical trials of vaccines against chikungunya virus and influenza A virus subtype H5N1 were promising.

15.6.8 DNA/RNA Vaccines

These vaccines induce the recipient to manufacture the antigen(s) within their own body. One achieves this by introducing a plasmid coding for the antigen into specific cells in the recipient's body. The body therefore becomes a 'vaccine factory' that efficiently stimulates both T-cell and B-cell responses. Plasmids are unable to replicate within our bodies and therefore to make sufficient antigen they need to be accompanied a strong promoter/terminator.

Preclinical trials of DNA vaccines against *Trypanosoma cruzi*, *Leishmania donovani*, and *Schistosoma* spp. have proved promising (Arce-Fonseca et al. 2020; Kumar and Samant 2016; Tebeje et al. 2016).

15.7 Identification of Antigens for Use in Anti-parasite Vaccines

The physical presence of a parasite within its host's body, the act of feeding upon it, and of releasing secretions and excretions inevitably trigger an immune response. However, this response is not necessarily protective if it is weak or affects an aspect of the parasite's biology that is not crucial for its survival. The search for antigens to use in a vaccine must therefore focus on those that can elicit a strong immune response against an essential physiological process. For example, glutathione S-transferase enzymes (GSTs) are currently considered to hold promise as antigen targets for vaccines against schistosomes. These enzymes perform numerous functions in schistosomes (as they do in humans and other animals) including protection from host-mediated oxidative stress. Experimental vaccines that target schistosome GSTs reduce parasite fecundity; this may be because they make them more vulnerable to the host immune response, but the precise mechanism is currently uncertain (Tang et al. 2019).

Targeting antigens that are normally hidden from the host's immune system: Antigens present in the guts of blood-feeding parasites such as hookworms are normally 'hidden' from their host's immune system. However, by stimulating the host to produce antibodies against these antigens, the parasite will ingest them when it feeds and thereby potentially inflict fatal damage to the parasite's gut. This approach has potential, but there are currently problems in stimulating an immunity that lasts long enough to be of use in field conditions.

Targeting excretory vesicles: Parasitic helminths and nematodes release excretory vesicles into their surroundings, and these are subsequently taken in by host cells. The vesicles contain proteins that are unique to the parasite species releasing them and contain large amounts of miRNA. The miRNA probably modulates the inflammatory cytokine response against the worms and thereby facilitates the establishment and maintenance of the parasites in their host. Vaccines that target components of the excretory vesicles therefore present a potential novel approach and initial results are promising (Drurey et al. 2020; Mekonnen et al. 2018). For example, excretory-secretory products released by *Trichuris muris* stimulate a strong, long-lasting protective immune response in mice (Shears et al. 2018). Similar homologs to the proteins stimulating immunity to *T. muris* occur in the human parasite, *T. trichiura*, and therefore might have potential as vaccine candidates.

Genome-wide screening for antigens: Advances in analytical techniques are enabling the sequencing of increasing numbers of pathogen genomes. These sequences can then be analysed using bioinformatics to identify proteins that would make suitable targets, such as membrane components in a process called 'reverse vaccinology'. This approach is relatively simple for viruses because they contain few proteins but eukaryotes such as protozoa and helminths produce thousands of proteins, and therefore it becomes much harder to narrow the candidate antigens down to the one or two for testing in a vaccination trial. At the time of writing, this approach had been used to identify candidate targets from various parasitic protozoa, helminths, and ticks (Goodswen et al. 2017; Lew-Tabor and Rodriguez Valle 2016), but all remained at the early experimental stage.

15.8 Vaccine Delivery

Vaccination normally involves an injection that may be intravenous, intramuscular, or intradermal. Many people dislike injections, especially when they cause painful or systemic flu-like reactions and if a series of injections are required, it can limit compliance. Ideally, therefore, a vaccination should only require a single injection. In addition, although the health risks associated with re-using needles are well known, the practice still occurs in some countries. A frequently overlooked problem with injections is that they generate lots of waste (plastic syringes and needles) that need to be disposed of safely – and these facilities are often lacking in developing countries. Oral vaccine vaccinations are more 'patient friendly' and are therefore likely to be accepted by most of the population – which is required if 'herd immunity' is to be obtained. Unfortunately, formulating an oral vaccine so that it can survive exposure to the digestive processes of the gut and then be absorbed into the circulation is extremely difficult. Consequently, there are very few commercial oral vaccines (e.g., polio, cholera, typhoid fever, human rotavirus).

Nasal spray vaccines are 'patient friendly' and easy to deliver. As their name indicates, the vaccine is sprayed up the nostrils using a customised device that delivers a set volume and droplet size. The UK has an annual flu vaccination campaign for children that uses a live attenuated strain of the influenza virus delivered as a nasal spray. Consequently, there is a lot of experience in vaccine design and delivery. Nevertheless, there appears to be limited interest delivering antiparasitic vaccines as nasal sprays although Ducournau et al. (2020) report good success in trails of using a nasal spray technique to deliver a nanoparticle-based vaccine against toxoplasmosis in sheep. Carcaboso et al. (2004) demonstrated that it was possible to induce a prolonged rise in antibody levels in mice using an anti-malarial vaccine containing the synthetic peptide SPf66 delivered as a nasal spray. Similarly, Yu et al. (2010) was able to induce immunity in mice to *Cryptosporidium parvum* using a DNA vaccine delivered as a nasal spray. Therefore, nasal spray delivery appears compatible with various types of vaccine.

Needle-free injection devices and drug delivery systems can be used for intradermal, subcutaneous, and intramuscular delivery using accelerated liquids or powder grains (Ravi et al. 2015). The injection impinges on a much smaller area of skin than a conventional injection and can take as little as 40 msec. The injection causes very little damage to the underlying tissues, reduces the risk of needle-borne contamination and is virtually pain-free. Needle-free injections often provide a greater antibody response than conventional injections.

Gene gun immunisation is a needle-free delivery system used to convey DNA or RNA directly into the cells. First, one attaches the DNA/RNA to gold nanoparticles and then accelerates these to supersonic speed using a gene gun to force them into the subcutaneous skin of the recipient (Bergmann-Leitner and Leitner 2015). One can attach more than one DNA-encoded antigen to the gold particles. Because the nanoparticles deliver the DNA/RNA directly to the cytoplasm/nucleus, the procedure requires a lot less material than conventional intramuscular DNA immunisation. At the time of writing, gene guns were still research tools.

15.9 Vaccines Against Malaria

For a vaccine to be effective within a control programme, one must first understand the biology and life cycle of the target parasite. One must also understand how its host mounts an immune response against it. For example, in the case of malaria, an anti-sporozoite vaccine would block new infections. This would be particularly useful for people previously unexposed to the disease who intend to visit a malaria-endemic region. It would also reduce the chances of those already infected from acquiring a more serious infection through repeated challenges. However, an anti-sporozoite vaccine would not help cure an existing infection and a person already infected with *P. vivax* or *Plasmodium ovale* may continue to suffer relapses without appropriate drug treatment since these species have a hypnozoite stage that gives rise to future generations of merozoites. The different species of *Plasmodium* overlap in their geographical distribution, but they are taxonomically very distinct, so any anti-sporozoite vaccine would probably have to contain components specific for each species. By contrast, an anti-gametocyte vaccine would potentially block transmission of the disease within a locality. This approach to control would be particularly effective for parasite species such as *Plasmodium falciparum* in which there are no wild or domestic animal reservoirs of infection. However, an anti-gametocyte vaccine would not directly benefit anyone already suffering from malaria, and it would not prevent someone contracting the disease if bitten by an infectious mosquito. In addition, as a means of control it might run into problems if infected people regularly visited the control programme region (e.g., refugees, border regions) and thereby maintained a level of infection among the local mosquitoes.

Many candidate anti-malaria vaccines have been identified over the years but currently the RTS,S/AS01 vaccine (Mosquirix) that targets the *P. falciparum* circumsporozoite protein is considered the most promising. This vaccine entered Phase 3 field trials in 2009 in seven African countries trials and involved approximately 15,000 children; the trials ended in 2014. The vaccine prevented 39% of the cases of malaria and 29% of cases of severe malaria among children aged 5–17 months, who received four doses of the vaccine. It is essential that for a child to receive all four doses because if they only receive three doses, then the level of protection provided rapidly falls to almost zero within 3 years. Nevertheless, these are the best results to date of any large-scale field trial for an anti-malaria vaccine. The vaccine is as safe as most of the other childhood vaccines, and therefore, the benefits conferred by the vaccine outweigh the risks of a child developing a harmful reaction to it. In 2019, the WHO commenced a phased introduction of the vaccine in further trails in several African countries either as a complementary part of ongoing malaria control measures or as part of the routine childhood vaccinations in areas where there is high transmission of *P. falciparum* (Adepoju, 2019). These further trials will last until 2024.

Development of the RTS,S/AS01 vaccine began in 1987 at the pharmaceutical company GlaxoSmithKline (GSK). It has therefore taken 32 years to arrive at a vaccine that is sufficiently effective to reach the final stages of widespread trials. It has also cost hundreds of millions of dollars and there are questions about who will fund the delivery of vaccinations once the trials are complete in 2024. Furthermore, the level of protection provided is far less than one would wish. Therefore, it is essential that other malaria control measures, such as the use of bed-nets and mosquito control, continue. This emphasises the difficulty of developing vaccines against some parasites and the need to integrate them within other control measures.

Field trials of the PfSPZ vaccine produced by Sanaria also show promising results and early results indicate that it provides a much higher level of protection (>50%) against *P. falciparum* (Epstein et al. 2017). This stronger immune response is probably linked to it being an attenuated vaccine containing irradiated whole sporozoites. It therefore contains numerous antigens rather than a single recombinant protein. The attenuated sporozoites can infect liver cells and form a schizont, thereby generating a CD8+ T cell immune response but cannot develop into merozoites and infect red blood cells. CD8+ T cells have a major role in killing the malaria parasite stages in the liver and therefore stimulating them reduces the chances of further infections. However, one problem with the vaccine is that one must inject it intravenously. Although this is not an insurmountable problem, it does complicate its delivery to large numbers of people.

15.10 Nanobodies (Single Domain Antibodies)

Most mammals produce conventional antibodies that consist of four polypeptide chains: two heavy chains and two light chains. These are large molecules of 150–160 kDa. There are variable regions at the tips of both the heavy (VH) and light (VL) chains at which antigen binding occurs. By contrast, camelids (llamas, alpacas etc.) produce much smaller antibodies (~80 kDa) that are composed solely of two heavy chains and antigen binding is limited to a single variable chain domain (VHH).

The production of nanobodies involves injecting llamas (or alternative suitable animal) with the target antigen. After a llama has mounted a strong enough immune response, its blood cells (lymphocytes) are collected and their RNA isolated and used to prepare a cDNA library. The VHH genes can then be expressed in an expression vector such as the bacterium *Escherichia coli* to produce a monoclonal single domain antibody (sdAb) or VHH nanobody. A VHH nanobody can bind effectively to its target antigen without the necessity for association with the rest of the antibody. It is also much smaller (~15 kDa) and more stable than a conventional antibody. Single domain antibodies can also be produced from certain cartilaginous fish (e.g., nurse sharks) and artificially by splitting and modifying conventional IgG.

Because of their small size, nanobodies can easily penetrate tissues and cross the blood–brain barrier. This coupled with their stability and the ease with which they can be chemically modified, means that they have potential for immunomodulatory therapy, as building blocks for drug design, and for use in diagnostics.

The use of nanobodies in parasitology is still at the experimental stage, but they certainly have potential (Fernandes et al. 2017). For example, it is possible to use nanobodies as drug delivery systems to target *Trypanosoma brucei* with pentamidine (Arias et al. 2015). They can also be employed in the design of highly specific diagnostic tests for *Trypanosoma congolense* (Torres et al. 2018) and *Toxocara canis* (Morales-Yánez et al. 2019).

15.11 Problems with Vaccination Strategies

An effective vaccine offers many advantages and especially so where drug treatment is difficult or expensive or resistance to drugs has developed. Even when cheap and effective drugs are available, such as benzimidazoles for soil-transmitted helminths (e.g., *Ascaris lumbricoides*, *Trichuris trichiura*, and *Necator americanus*), some workers consider developing vaccines against them would be worthwhile. This is because in the absence of improved sanitation and living standards, those treated with drugs rapidly become re-infected. Nevertheless, despite advances in vaccine design and the expenditure of large sums of money, few commercial vaccines are available against the major parasitic diseases. This has led some to question whether the money has been well spent or whether a vaccine is desirable in the first place. For example, Gryseels (2000) has argued that if an effective vaccine for human schistosomiasis was discovered, it would reduce the impetus for providing safe drinking water, sanitation, and health care and thereby compromise the control of other important but less high-profile pathogens. He also points out that human schistosomiasis is a serious disease in only a minority of people – who can already be diagnosed and treated at the primary health care level. These same arguments can also be made for soil-transmitted helminths. Indeed, in parts of Pakistan, vaccination campaigns against diseases such as polio, measles, and bacterial meningitis ran into problems when the public perceived that these were not accompanied by any improvement in health services.

One can expect vaccinations to exert strong selection pressure upon a pathogen, and some workers feel that we should be careful about the type of vaccine we develop. Sylvain Gandon and co-workers (Gandon and Day 2008) suggest that vaccines that reduce the growth rate of pathogens or neutralise the harmful effects of toxins could alter the natural selection within parasite populations in favour of more highly virulent strains. Consequently, those who remain unvaccinated may become exposed to more dangerous varieties of the pathogen. By contrast, they argue that vaccines that block infection may result in selection for less virulent pathogens.

Obviously, one requires a vaccine that meets the qualities of an 'ideal vaccine' as closely as possible. This, however, merely represents a 'good start'. The vaccine then needs delivering to the target population and for them to accept it. One therefore enters the problematic fields of politics, religion, and human behaviour. For example, rumours always spread faster than the truth and mobile phones, and the internet means that they can gain worldwide attention within 24 hours of birth. The reliability of the rumours is immaterial and assurances and reasoned argument seldom assuage irrational fears. For example, some parents continue to refuse to have their children vaccinated with the MRR (measles, mumps, rubella) vaccine because of fears that it causes autism. The fact that the scientific basis for the rumour was comprehensively discredited many years ago makes little difference. Similarly, some vaccines are plagued by rumours that they contain harmful substances such as mercury or traces of pork cells that make them unacceptable to Muslims and Jewish peoples. For some people, the manufacture of vaccines using animal cells is unacceptable, some will not countenance vaccines developed using cells that were derived initially from aborted foetuses even if these cells have been in culture for many years, whilst vaccines derived through genetic engineering would be unacceptable to others. A fundamental difference between a drug and a vaccine is that a drug offers the prospect of relief now whilst a vaccine promises the avoidance of suffering in the future. A person in pain and distress tends to be less concerned about the nature and provenance of their medicine. For the purposes of disease control and eradication, it is essential to include sufficient members of the population to ensure 'herd immunity'. Unfortunately, it does not take many people to fail to engage with the vaccination strategy to compromise its effectiveness. And the higher the herd immunity threshold, the higher the proportion of people that needs to be vaccinated to reach it.

16

Parasite Control

CONTENTS

16.1   Introduction

Attempting to understand parasite life cycles is like learning German grammar: it is complicated and full of exceptions to the rule. However, without the grammar one cannot speak the language correctly. Similarly, knowledge of a parasite's life cycle enables one to understand how it is transmitted and how it causes disease. From the life cycle, one can identify potential weak points and therefore potential targets for control strategies (Table 16.1). For example, in the United Kingdom, rainfall and temperature are the key factors determining the transmission efficiency of the liver fluke *Fasciola hepatica*. This is principally through their influence upon the reproduction and survival of the snail intermediate host. This has enabled the development of a liver fluke forecasting scheme that is operated by the National Animal Disease Information Service (NADIS) (http://www.nadis.org.uk). Farmers use this to determine when their sheep flocks are most at risk of infection and therefore need treating with anthelmintics or, if possible, moved to less risky pasture. Similar liver fluke forecasting schemes exist in other countries. Beltrame et al. (2018) discuss a novel modelling approach to fluke forecasting that could improve forecasting accuracy.

    A simple direct life cycle is one in which the parasite exploits a single host species (or group of related host species) and usually does not reproduce outside its host. For example, the nematode *Ascaris lumbricoides* lives in our intestines, it only infects us (although there remains a debate as to the extent to which transmission to and from pigs takes place), it is transmitted by passive contamination, and one of its eggs gives rise to one adult worm. One might therefore expect that parasites such as this would be easier to control than those with more complicated life cycles

Parasitology: An Integrated Approach, Second Edition. Alan Gunn and Sarah J. Pitt.
© 2022 John Wiley & Sons Ltd. Published 2022 by John Wiley & Sons Ltd.
Companion website: www.wiley.com/go/gunn/parasitology2

Table 16.1 How a parasite's life cycle influences its treatment and control.

Life cycle factor	Importance in treatment/control	Application of life cycle knowledge
Direct life cycle	Sanitation and basic hygiene practices prevent many gastrointestinal parasitic infections	Washing fruit and vegetables in clean water removes protozoan cysts and helminth eggs
Life cycle involves one or more species of vector	Controlling vectors/preventing vector: host contact reduces disease transmission	Bed-nets can prevent mosquitoes transmitting malaria
Life cycle involves one or more species of intermediate hosts	Targeting intermediate host(s) can reduce disease transmission	Drainage removes the habitat of snail intermediate host of *Fasciola hepatica*
Parasite has several definitive hosts	Reservoir hosts are a potential source of re-infection	*Schistosoma japonicum* has numerous reservoir hosts, which can contaminate paddy fields etc. with eggs
Parasite has life cycle stages that are exposed to the environment	Environmental conditions can promote or limit infections	Composting kills the infective stages of many gastrointestinal parasites
Sequence and timing of life cycle stages within a host	Optimal time for diagnosis	Microfilariae of *Wuchereria bancrofti* exhibit periodicity
Location within host	Optimal time for treatment	Cattle should be treated for warble fly infections before the larvae reach their winter resting sites

involving intermediate hosts, reservoir hosts, or vectors. However, this is not always the case. The eggs of *A. lumbricoides* pass out in our faeces, so preventing faecal contamination of the food and water supply limits its transmission. It also simultaneously limits the transmission of many other gastro-intestinal parasites. This, however, is easier said than done. This is because providing even the most basic sanitation demands cooperation from the public and constant maintenance. Anyone who has attended a large open-air event such as a pop concert or rally will be well aware of how there are seldom sufficient toilets. Furthermore, these soon become so noisome that people take to urinating and defecating wherever they can find a sufficiently secluded spot. Absent or poor toilet facilities are a fact of everyday life for residents of slums, those living in remote rural areas in developing countries, and displaced people in refugee camps. Consequently, faecal-orally transmitted diseases can be rife in these communities. Furthermore, once someone in a family or close group becomes infected, the parasite often spreads and infects the others. Although simple and cost-effective measures can prevent gastrointestinal parasite infections, they are often lacking. There are several reasons for this, but they usually relate to money and prioritisation. The provision of sanitation is not photogenic and therefore lacks appeal as a media or research topic: vaccines are far more glamorous and enable photographs of cute babies in the arms of concerned-looking young doctors. In addition, sanitation requires constant maintenance. This means people need employing, materials made available, and checks made to ensure that standards are kept. And this means regular funding. It also requires those using the facilities to act responsibly. Any casual user of university toilets will know how irresponsible many supposedly educated people are. Therefore, one should not expect those living in slums etc. to keep facilities clean out of self-interest. In the United Kingdom, many councils have ceased providing public toilets because of funding issues. If rich countries are unable or unwilling to provide services, one cannot expect those in poor countries to behave more responsibly towards their citizens.

By contrast to *A. lumbricoides*, the filarial nematode *Onchocerca volvulus* (which causes river blindness) has a complex life cycle involving development within and transmission by *Simulium* (blackfly) vectors. Nevertheless, in parts of West Africa this parasite has been successfully controlled solely through the actions of professional teams treating rivers with insecticides to kill off the blackfly larvae. The control, therefore, does not rely upon the cooperation of the local people and they do not need to change the way they live their lives.

16.2   Eradication, Elimination, and Control

There is a world of difference between eradication and control. Eradication indicates that the parasite no longer exists anywhere in the world and, therefore, there is no need to maintain treatment regimens or control programmes. By contrast, if a parasite is merely controlled then it continues to exist, although its incidence and prevalence, morbidity, and mortality in particular areas will be reduced because of the ongoing control measures. In between these two extremes is localised elimination of the parasite from a specific geographical location although this condition can usually only be sustained by control programmes to prevent the parasite from re-invading from surrounding regions.

Although there is evidence that small parasite burdens are useful for the maintenance of a well-regulated immune system (Chapter 9), parasites are usually viewed with such loathing that most people would favour their complete eradication. This, however, is seldom a realistic goal owing to numerous biological, sociological, economic, and geographical reasons. Indeed, very few pathogens have been intentionally eradicated from the world, and these are all special cases. For example, the smallpox virus no longer exists in the wild although two stocks continue to be held in laboratories in USA and Russia. If these stocks are destroyed (which currently seems unlikely), then the virus can finally be declared extinct. Many factors contributed to the success of the smallpox eradication campaign but fundamental to this was the availability of a vaccine that provided a high level of protective immunity (Table 16.2). Since the eradication of smallpox in1980, the rinderpest virus was eradicated in 2011, and there are hopes that the polio virus will be eradicated in the next few years.

At the time of writing (2021), there were no examples of successful parasite eradication campaigns although the Guinea worm *Dracunculus medinensis* was considered extremely close to this

Table 16.2   Factors contributing to the success of the smallpox eradication campaign.

Live attenuated vaccine confers long-lasting protective immunity

Single dose of vaccine required, easily delivered

Natural exposure confers long-lasting protective immunity

Absence of animal reservoir hosts

Fear of disease ensures public acceptance of vaccine

Infected individuals always symptomatic and thus identified

Effective co-ordinated vaccine delivery and education campaign

Government cooperation

Funding

point. A *caveat* is required here because the imminent demise of *D. medinensis* has been forecast for many years. Islands and geographically isolated regions (e.g., cut off by mountain ranges) have a good chance of achieving success in local elimination campaigns provided they are suitably policed, and the parasite is not able to fly in from the nearest endemic region. For example, the warble flies *Hypoderma bovis* and *Hypoderma lineatum* have probably been eradicated from the Great Britain using effective drugs and the stringent imposition of regulations on the treatment and movement of cattle. Neither species can fly long distances across the sea, so the country should remain free provided infected cattle are not imported. Within developed countries, many human parasites have been eradicated because of the rise in living standards, effective waste disposal, and improved hygiene. For example, in the United Kingdom, a combination of enforced legislation and meat hygiene inspection has led to the elimination of *Taenia solium* and *Taenia saginata*. Nevertheless, these parasites remain common in many other countries, and they would undoubtedly return to those in which they are currently eliminated should the opportunity arise. This is one of the reasons why many countries have stringent regulations concerning the movement of live animals within and across their borders and also for the import and sale of meat and meat products. Nevertheless, remarkable numbers of living animals and their meat are traded illegally both within and between countries. For example, livestock rustling is common even in developed countries – in the United Kingdom, £2.5 million worth of livestock was stolen in 2018. This was a rise of 11% from 2016. Similarly, there is a multi-million-pound international trade in illegal meat, and some estimates suggest that around 29,000 tons of meat enter the UK illegally every year from outside the EU. There is also a large illegal trade in living and dead wild animals. For example, in 2018, the Border Forces from 92 countries instigated a 6-week global operation called Operation Thunderstorm. During this operation, they made 276 seizures that included 43 tons of wild meat (comprising that of zebras, bears, elephants, crocodiles, and several other wild animals), as well as 27,000 reptiles and almost 4,000 birds. These figures give a good idea of the potential for zoonotic pathogens to disperse rapidly around the world.

It is seldom feasible to achieve even local elimination of parasitic diseases, and therefore one must contemplate the best means of controlling them. This means that the parasite persists, and the community, farmer, or individual must consider both the cost of the control measures and the level of infection that they consider acceptable. For crop pests, it is relatively easy to calculate the economic costs of control by determining the 'gain threshold': this is derived by dividing the management costs by the market value of the crop. Therefore, if the management costs for the application of a pesticide are €100 ha^{-1} and the market value of the crop is €20 ton^{-1}, the gain threshold would be 5 tons ha^{-1}. That is, the farmer would need to save at least 5 tons ha^{-1} for it to be worth this while applying the pesticide. Although a similar procedure could be applied to domestic animals (e.g., by reference to live weight gain, milk production), the farmer would also have to consider the welfare of his animals. For example, although treatment may not make economic sense, by not treating his animals he may cause them suffering that he would consider unacceptable (although farming is a business most farmers like their animals). In addition, it might lead him to being prosecuted under animal welfare legislation. Also, there may be local legislation compelling farmers to treat their animals should they become infected with a specific disease. For example, in the UK, farmers have a legal requirement to treat their whole flock of sheep should any of them become infected with the sheep scab mite *Psoroptes ovis*. In Scotland, there is a further requirement that the farmer must inform their local Animal and Plant Health agency Office. The farmer must bear in mind that parasitic diseases are generally chronic infections and that an infected animal is a potential source of infection for other animals both in the present and in the future (the transmission stages of some parasites can persist for months or even years). The determination of an 'acceptable level of infection' for domestic livestock is therefore not always a simple procedure. Nevertheless, gain thresholds can be useful when comparing different treatments. For example, if the most effective treatment is also the costliest, it may make economic sense if it still provides an acceptable level of protection. For human parasitic diseases, any attempt at calculating an acceptable level of infection is fraught with ethical issues. However, even in the developed nations, there is never sufficient money to treat all those who require medical attention, and some means of prioritising needs must be developed – such as the DALY calculations referred to in Chapter 1.

16.3 Education

Parasitic infections tend to have their most severe impact upon those who are poorly educated but the risk of infection can often be reduced by simple changes in behaviour. It therefore follows that education is usually a key feature of parasite control programmes. For example, if people understand that the large cysts growing in them result from eating and drinking substances contaminated with dog faeces, then, theoretically, it should be relatively easy to reduce the incidence of hydatid disease among a population by making them re-asses their relationship with dogs. Educational material is usually disseminated via leaflets, posters, radio, and television programmes. However, as the world becomes increasingly commercialised the worthy but often dull and simplistic health information can battle to gain the intended recipient's attention. The production of written health information is (or should be) a skilled job since the target audience is seldom highly literate. The information must be presented clearly and concisely in a language that the target audience understands. For example, in India, and parts of Africa and Asia, there are numerous local languages, and one cannot assume that everyone will be conversant with the country's main official language. Avoiding scientific terminology and employing illustrations and photographs can help. In addition, many people in both developed and developing countries are illiterate and therefore cannot understand even simple written guidance. For these people, radio and television can be important sources of information. China has achieved success with so-called bare-foot doctors who dispense simple treatment and verbal guidance to isolated communities. Health messages are often delivered via teachers in schools since the young are usually more impressionable and are likely to pass on their knowledge to their parents or guardians. Health clinics usually provide an opportunity to talk to the women in the community who may otherwise be very difficult to reach (especially in patriarchal societies), but who can often have a strong influence on the behaviour of all members of the family.

Unfortunately, although education is important, just because someone has information does not mean that they are willing or able to act on it. If life was this simple there would not be problems with alcoholism, drug abuse, and unwanted pregnancies. For example, insecticide-treated bed-nets are effective at preventing malaria if used correctly. Therefore, aid agencies in parts of Africa distribute free bed-nets and instruct people how to use them. However, despite this, the level of infant mortality from malaria sometimes remains high. This is partly because the nets are not always appropriate for the way people live. For a start, the bed-nets that are distributed often must be hung from the ceiling and unless there is sufficient space they need to be erected and disassembled every day. This can quickly become a chore and the responsibility of a single person – and if they are not present the bed-nets are not used. In many countries, people like to sleep outside at night because it is too hot inside – and there is therefore often nowhere to hang the bed-net. Even if the people are sleeping indoors, the bed-net traps heat, which can then make it difficult to sleep. If the only source of light in the room is a naked flame, it would be considered too risky to use a bed-net in proximity. Some people also consider bed-nets to be a potential danger to young unsupervised children and think that they could become trapped within or strangle themselves trying to get in and out. In many African cultures, the young children sleep with their mother until they are weaned and two bodies under a single bed-net can make it uncomfortably hot. In short, people might be well aware of the benefits of a particular behaviour (in this case sleeping under a bed-net) but for a whole variety of reasons or excuses they choose to ignore it. Similarly, although people might know that freezing or cooking can kill the infective stages of most parasites, they may be unable to do either of these things. Poor people seldom have access to freezers, whilst firewood or other fuel may be expensive or unavailable. In addition, local practices handed down over generations are a major determinant of how food is prepared and whether it is cooked at all.

Nevertheless, in some countries, education has proved enormously successful in the control of certain parasitic diseases. For example, during the 1800s in Iceland, hydatid disease was effectively controlled for many years through the provision of educational pamphlets. It probably helped that although the population was literate the only competing reading material in Icelandic at the time was the Bible and some Viking Sagas. Subsequently, legislation and more effective treatments became available, and by the 1960s, the disease was eliminated from the country (Craig and Larrieu 2006). Similarly, a large part of the success of the Guinea worm eradication programme is because of the effectiveness of the education programme that encourages people living in regions where *Dracunculus medinensis* is endemic to only drink water that is passed through a filter to remove infected copepods.

16.4 Environmental Modification and Cultural Control

Up until the development of effective drugs and pesticides, environmental modification and cultural control were the main means of combatting pests, parasites, and diseases. It was seldom understood why certain practices were effective, but experience taught us to do or not to do certain things at certain times of the year. For example, for many years, people thought that malaria was caused by breathing in the air that surrounds marshy land. People therefore avoided living close to wetlands or they drained the land. This also improved the health of domestic animals by reducing their risk of exposure to *F. hepatica*, as well as reducing the survival of eggs and infectious stages of many parasitic nematodes. However, it is a difficult and expensive business to drain land, and we now realise that it can have harmful ecological consequences.

Organic Livestock Farming and Parasitic Diseases

There is an increasing market for organic produce in many parts of the world, and this fetches a premium in the marketplace. However, officially certificated organic farming poses a problem for livestock farmers because they are unable to employ many of the drugs commonly used to control parasitic infections. Consequently, organically farmed animals often contain higher parasite burdens than animals that are farmed conventionally. Not only can parasites reduce productivity to the point at which the losses make farming uneconomic, but they also raise welfare issues.

Organic farmers can utilise certain approved natural products to control parasites. Diatomaceous earths are naturally occurring soils that are comprised largely of the fossilised remains of diatoms. They have a wide variety of applications from a component of cat litter to DNA purification. They have proven action against some pest insects and also some tick and mite infestations (Murillo and Mullens 2016). They act through damaging the cuticle and the target arthropod, then dies of desiccation. Although marketed as a treatment of gastrointestinal helminth infections in us and our domestic animals, the scientific evidence for their effect is very limited. This is hardly surprising because death by desiccation is unlikely for anything living in the gut. Many organic farmers provide their animals with nutritional supplements to offset the losses they suffer from larger worm burdens compared to conventionally reared stock.

Natural products are seldom as effective as pharmaceutical drugs at treating parasite infections, so organic farmers rely heavily on cultural control techniques. Lower stocking rates can reduce the risk of transfer of infections and contamination of the land with parasitic protozoan cysts or helminth eggs. Nevertheless, over time the number of infective stages on a piece of land will steadily increase. Farmers with plenty of land can rest pasture for prolonged periods to allow time for the transmission stages to starve to death. How long this would take would depend upon the pathogen and environmental conditions, but it could take many months or even years. It is therefore seldom a feasible option. The number of transmission stages can also be reduced by alternating grazing with other livestock species or combining grazing with using the land for a hay crop. The influence of plant species composition on parasite transmission remains poorly understood. Work in New Zealand indicates that grass species composition has a major impact on parasite burdens and productivity in lambs. Furthermore, forage crops containing high levels of condensed tannins, such as *Hedysarum coronarium*, could have natural anthelmintic properties (Niezen et al. 1995). Unfortunately, it is difficult to determine the protective effect of forage composition because numerous genetic and environmental factors influence the composition and concentration of their secondary metabolites. Some grasses, such as molasses grass *Melinis minutiflora*, have a deterrent effect on ticks and reduce their survival (Fernández-Ruvalcaba et al. 2004), but they have not been utilised to any great extent. A more robust anti-tick response is generated by the legume *Stylosanthes* that captures and kills ticks on sticky secretions (Sutherst et al. 1982).

Parasites are susceptible to pathogens just like other organisms and biological control could be exploited within organic farming. However, whilst fungi will kill helminth eggs and some fungi specialise in capturing and killing nematodes (de Freitas Soares et al. 2018), it has proved difficult to utilise them as part of a routine control programme. Similarly, although fungi, such as *Amoebophilus simplex* kill free-living amoebae, their use against parasitic species would present considerable logistical difficulties. Currently, most interest is focused upon the nematophagous fungus *Duddingtonia flagrans*. The spores of this fungus can survive passage through the gut of a ruminant and will then germinate and grow rapidly once passed with the faeces. Consequently, the spores can be incorporated into feed blocks or controlled release devices and the fungi kill parasitic nematode larvae that are passed with the faces or subsequently hatch from eggs whilst in the faecal pat. There are many other species of nematophagous fungi and da Silveira et al. (2017) showed that *D. flagrans* used in combination with some of these is effective at reducing the numbers of the free-living stages of gastrointestinal nematode parasites on pastureland. Nematophagous fungi are not specific predators of nematode parasites and will also attack free-living nematodes but there are no reports of them having an adverse effect on the environment (Saumell et al. 2016).

In countries with seasonal climates, the risk of infection is often most pronounced at certain times of year. Therefore, it is sometimes possible to predict when or whether a parasite will be a problem and take appropriate action. For example, the fluke-forecasting scheme can provide warning of when it is appropriate to move sheep to safer pastures. This, of course, assumes that a farmer has alternative less risky pasture on which to move his animals and poor people can seldom move to a less disease-prone environment.

Relatively simple environmental modifications can reduce the risk of exposure to many parasitic diseases. For example, increasing the water flow rate in irrigation channels and removing plant growth can make them unsuitable for the snails that act as the intermediate hosts of schistosomes. Similarly, concrete farmyards and animal pens are easier to keep clean of faeces and solid hut or house walls and ceilings do not provide hiding places for the hemipteran bug vectors of *Trypanosoma cruzi*. Although simple environmental modifications are usually relatively cheap, they may still be too expensive for the poorest people who would benefit most. Their implementation therefore often requires funding by government or aid organisations, and they usually require constant upkeep to remain effective.

The safe disposal of human and animal faeces is central to the control of many parasitic diseases. However, human and animal waste is used as a fertilizer for the growth of crops and in aquaculture in many parts of the world, and this can inadvertently contribute to the spread of pathogens. This is because alternative industrial nitrogen fertilizer is either not available or too expensive. Furthermore, if waste were not used as a fertilizer, it would have to be disposed of in some other manner that would undoubtedly be expensive or pose environmental problems. The infective stages of many parasites are protected by thick cyst walls (e.g., *Entamoeba histolytica*) or eggshells (e.g., *Ascaris lumbricoides*) and can survive for weeks or even years under ideal conditions. However, if the faeces are composted in an appropriate manner for a period of time before it is applied to crops, then all the parasite infective stages will be killed. There are two basic composting processes: aerobic and anaerobic. Aerobic composting is where microorganisms that utilise oxygen break down the organic matter and respire carbon dioxide. The oxidation of organic compounds releases a lot of heat but if the temperature remains below 45 °C, the process of decomposition is slow. The arrival of thermophilic species enables the temperature to rise to above 60 °C and the speed of decomposition increases dramatically. Anaerobic decomposition takes place in the absence of oxygen and the breakdown products are mostly ammonia, hydrogen sulphide, organic acids, and methane although some carbon dioxide is formed. Anaerobic decomposition does not yield much heat and is slower than aerobic decomposition – it also yields many more unpleasant smells. Although anaerobic decomposition kills parasite transmission stages eventually, the low temperature means that it is less effective than aerobic decomposition. Consequently, aerobic composting is generally recommended for the composting of farm animal waste and in villages that lack mains waste disposal facilities. Where anaerobic composting is undertaken, it is often preceded by a period of aerobic composting to destroy the pathogens. There are many ways of undertaking aerobic composting and when it is done effectively, the inner temperature can rise to over 60 °C whilst the outer temperature can reach 50 °C. Exposure to temperatures above 45 °C for five or so days is sufficient to inactivate even parasite transmission stages as robust as *Ascaris* eggs.

16.5 Remote Sensing and GIS Technology

Remote sensing (RS) satellite data and geographic information systems (GIS) technology are proving increasingly useful for monitoring the epidemiology of parasites and forecasting the risk of disease outbreaks (Fletcher-Lartey and Caprarelli 2016). Remote sensing is a means of monitoring the environment without physically contacting it. This is mostly achieved through satellite technology using a combination of passive and active monitoring devices. Passive detectors monitor different wavelengths that are emitted or reflected from the land beneath. Active detectors emit particular wavelengths and measure the time taken for them to return. Remote sensing can monitor temperature, ground cover, forestation, and many other environmental variables on a regular basis in regions in which it would be impossible to gain ground-based measurements owing to the cost and/ or risks involved. Various RS satellite data sets are available including LANDSAT, MODIS, NDVI, and SRTM DEM. Geographic information systems are means of capturing, storing, updating, retrieving, analysing, and displaying any form of geographically referenced digital information. It is not a single entity but a collection of computer hardware, software, and geographical data: a GIS that is suitable for one situation is not necessarily suitable for another application.

GIS is particularly useful for parasite surveillance and simulating the consequences of intervention strategies or changes in the environment (e.g., global warming). Through GIS, it is possible to map simultaneously one or more of the following on either a regional, national, or global scale: the occurrence of the parasite, the disease it causes, its host, vector/intermediate host, co-infections, and environmental variables. For example, disease maps are a quick and simple means of visualizing spatial and temporal 'hotspots' of where disease is clustering, the linkages between parasite distribution and environmental variables and the effectiveness of control measures. Similarly, remote sensing can identify those environmental variables that promote the breeding of vectors and intermediate hosts and therefore where problems are likely to arise.

Although GIS has enormous potential, it has remained a specialist topic rather than a day-to-day tool for workers in human and veterinary science. This is because many of the GIS tools such as ArcGIS are complicated and take months of dedicated practice to master. In addition, the hardware and software can be expensive and does not always integrate well with modelling software that uses spatial statistics. For example, agent-based modelling is popular as a means of simulating pathogen transmission, but it has proved difficult to use it in conjunction with GIS. Nevertheless, GIS does not have to be expensive. Freeware GIS software such as DIVA GIS can be downloaded for free from the internet and is suitable for mapping the vectors and intermediate hosts of parasitic diseases. For example, Ruberanziza et al. (2019) used this approach in Rwanda to identify populations most at risk of contracting gastrointestinal nematode parasites, such as *Ascaris lumbricoides*, *Trichuris trichiura*, and the hookworms *Ancylostoma duodenale* and *Necator americanus*. Furthermore, spatial data sets of parasites, intermediate hosts and vectors are available for download through biodiversity databases such as GBIF (http://www.gbif.org/). However, it is worth noting that for those working with human diseases, there are confidentiality concerns, since location and patient data are often linked together and consequently, it would be possible to reverse engineer maps to identify the home address of individual patients.

16.6 Whether to Treat the Individual or the Population

The traditional basis of disease control programmes has always been to reduce the prevalence of infection – ideally to zero. However, for many parasitic diseases, and particularly human gastrointestinal helminth infections in developing countries, this is an expensive and unrealistic goal. This is because there is little prospect of breaking the transmission cycle of parasites such as *Ascaris lumbricoides* until there is adequate sanitation. In the absence of sanitation, anyone cured of their infection usually becomes quickly re-infected. Therefore, the prevalence of infection among a community remains the same or soon returns to its original level after a brief decline. This is often cited as a reason why a vaccine that provides long-lasting immunity against these diseases would be beneficial. However, there is little prospect of a commercial vaccine against most gastrointestinal parasites becoming available for many years. Therefore, control will continue to depend upon a combination of drugs and improving sanitation.

Low burdens of gastrointestinal helminths seldom cause serious pathology unless the patient is malnourished or suffering from another health problem. Consequently, the aim is to reduce the worm burden within individuals rather than the number of people infected with worms. In particular, one aims to reduce the number of people harbouring high worm burdens. In us, as in other animals, most of the parasites are usually found within a minority of the population. These highly infected individuals suffer the most harmful effects from parasitism and are also a source of infection for the rest of the population. It may be too costly to identify these vulnerable individuals but if they are treated on a regular basis along with everyone else, then it can prevent them from developing a serious infection. Because many people remain infected with parasites, some might argue that this approach is a waste of time and money. However, reducing the worm burden to sub-clinical levels can result in marked increases in health and productivity.

Treating the whole population works well for many gastrointestinal parasites of humans and domestic animals because there are cheap safe drugs and low parasite burdens are seldom pathogenic. It does, however, risk facilitating the arise of drug resistance. This is because wide-spectrum drugs do not affect all the target parasites equally. Consequently, some may experience sub-lethal levels, and this encourages the spread of resistant phenotypes. For example, SNP polymorphisms linked to benzimidazole resistance have been described in *Trichuris trichiura* and *Necator americanus,* and there are concerns about praziquantel resistance developing in schistosomes. In addition, there will always be a small minority of people who react badly to any drug or vaccine. And involving more people means more reports of harmful reactions, and this can result in bad publicity, compromise the control programme, and lead to claims for compensation. Where the only available drugs are expensive or have serious side effects, those who are infected must be identified and treated on an individual basis.

In some situations, identifying exactly which individuals have the disease and targeting their treatment appropriately is not only important for the individual's welfare, but may be the key to effective control. Advances in laboratory diagnostics have proved helpful in this regard. For example, in areas where malaria is endemic, it still relatively common to treat everyone with malaria-like symptoms with antimalarial drugs. Even when laboratory tests are available, there can be a perception among the doctors and nurses that a 'negative' blood slide result might be incorrect, and so the patient should receive antimalarials anyway. Presumptive treatment means that those genuinely infected with *Plasmodium* spp. receive antimalarials, thus relieving their symptoms and reducing the likelihood of transmission to others. However, it does have some risks. One is that the patient may not have malaria, but another serious infection and without the appropriate treatment their condition could become worse or even fatal. Another is that the indiscriminate use of antimalarials, particularly chloroquine, has encouraged the development of drug-resistant strains of *Plasmodium* in many parts of the world and these are therefore harder to control. Drakeley and Reyburn (2009) argue that rapid diagnostic tests (RDTs) for malaria mean that we should abandon or at least re-consider the strategy of treating everyone with antimalarials. RDTs are reliable, sensitive, and provide a quick result and healthcare workers increasingly perceive them as more trustworthy than blood slide observations. In regions of low-level malaria transmission, this targeted approach means that antimalarials are restricted to patients who actually need them. This is a better use of limited resources and should enhance the control efforts.

16.7   Piggy-Backing Control Programmes

Although centrally administered control programmes or those delivered by aid organisations can provide excellent results, they are seldom sustainable. A change in government priorities or the withdrawal of aid can result in a control programme collapsing and consequently the prevalence of the target parasite (or other disease) increases. The return of the disease can be severe since the local population may have lost some of its immunity during the period of reduced parasite exposure. The WHO therefore advocates that, where possible, control measures are 'piggy-backed' onto existing networks. This approach is primarily being used for the control of gastrointestinal parasites in children but can be adapted for other parasites and infectious diseases. Using existing networks reduces the need for trained personnel and involves many people already employed doing other jobs. Consequently, the long-term delivery of the control programme is inherently more stable and less susceptible to the vagaries of external funding. It is also cheap and reduces the burden on the existing local health system that usually lacks staff and resources.

Young children are particularly vulnerable to the harmful effects of gastrointestinal soil-transmitted helminths and schools are ideal venues for contacting them. Many children within a locality will attend school and the school itself is a 'non-threatening' focal point of the community: the same might not be the case with other official buildings. In poor communities, some children do not go to school or attend only sporadically. However, they are almost certainly going to be familiar with the school and can be encouraged to attend for a 'special free event'. In addition, teachers are often respected members of society in poor communities whose views and opinions are valued. Therefore, if the teachers explain the reasons why it is important for all children to receive a pill, it is more likely that the community will participate than if an outside organisation suddenly turned up 'out of the blue'. Because they are educated, teachers can be quickly trained to organise de-worming days during which they administer tablets of albendazole or some other safe anthelmintic for which there is no need for complicated treatment regimes, injections, or worries about potentially serious reactions. Although it could be argued that the good will of the teachers is being exploited, there are clear advantages for them. This is because children who are healthy are more likely to attend regularly and perform better in their lessons than those who are chronically ill. In addition, high-risk children would benefit the most from this approach to parasite control – that is, the children from the poorest families and girls (who are often discriminated against in patriarchal societies). De-worming days can be organised on a regular basis to ensure that those who are most vulnerable continue to receive adequate protection.

In some African countries, aid organisations facilitate Child Health Days with the aim of providing anthelmintics, vitamin A supplements, and other health care to preschool-age children. These events present an opportunity to piggyback treatment opportunities to women of reproductive age who might be otherwise difficult to reach in some societies (Gyorkos et al. 2018). Similarly, other one-off events or feeding programmes, and water and sanitation initiatives provide an opportunity to simultaneously reach thousands of people.

16.8   Disruptions to Control Programmes

The Prussian Field Marshal Helmut von Moltke (1800–1891) made the oft-repeated quote that no plan survives first contact with the enemy. This is as true of parasite control programmes as it is on the battlefield. Even the best planned and a supported control programme can be disrupted by unforeseen circumstances such as famine, flood, or war. These events often lead to mass migration of people from one region to another and in some cases involves them crossing national borders. Hastily erected refugee camps are usually susceptible to faecal–orally transmitted disease. In addition, people take their infections with them. Therefore, when migrants move from an area where a particular parasite is endemic to one where it is not, it can create public health problems for the indigenous population. For example, in regions where malaria is endemic, people infected with *Plasmodium* could still be active enough to travel. This might lead to the re-introduction of malaria into an area in which was recently eliminated, or refugees could carry drug resistance strains or new species to areas where these had not previously occurred. Provided that a suitable vector mosquito is available in the new area, then transmission can resume. This occurred in Tajikistan during the civil war in the 1990s. Because of the fighting, some people sought refuge in neighbouring Afghanistan, where a chloroquine-resistant strain of *Plasmodium falciparum* was circulating. Once Tajikistan became a safer place to live, people started returning home and some were infected with malaria. Although *Plasmodium vivax* was always quite common in Tajikistan, *P. falciparum* had not been reported there for many years. However, a species of *Anopheles* capable of vectoring *P. falciparum* was present and therefore its transmission within the country resumed (Pitt et al. 1998). This case also highlights another important point in effective control. Efforts were initially hampered by a lack of accurate diagnostics. This was because laboratory staff had only received training in detecting *P. vivax* in blood slides and were reporting *P. falciparum* infections as 'negative'. These events occurred before RDTs became widely used.

The sudden appearance of other diseases can throw even the best laid plans into disarray. For example, sporadic outbreaks of Ebola in several West Africa countries continue to severely curtail the delivery of malaria control programmes (Hayden 2014). Although malaria kills far more people than Ebola in the region, the fear of Ebola and the hostility of some communities to those tasked with its control make it extremely difficult to implement malaria control. Furthermore, both Ebola and malaria cause fever, and therefore, it is essential to rapidly distinguish between the two diseases. The WHO has therefore recommended a change in approach and to implement mass administration of artemisinin combination treatment in areas where Ebola outbreaks occur. Without continued distribution of drugs and in the absence of mosquito control, the prevalence of malaria returns to its original level after a few weeks, but this can be beneficial in the control of Ebola (Aregawi et al. 2016; Walker et al. 2015). For example, it reduces the likelihood of a person presenting with a fever to be suffering from malaria. The global SARS-COVID-19 epidemic has impacted upon the control of many parasitic diseases. For example, malaria control has been hampered in many countries through disrupting the supply of bed nets, medicines, and implementing mosquito control (Thornton 2020). In addition, both malaria and COVID-19 cause febrile symptoms, so it is essential for the two diseases to be distinguished rapidly. Unfounded claims that the antimalarial drugs chloroquine and hydroxychloroquine are effective for treating COVID-19 led to sudden shortages of the drugs. In addition to malaria, chloroquine and hydroxychloroquine are also used in the treatment of rheumatoid arthritis and lupus erythematous. Therefore, shortage of these drugs could impact on many patients other than those with malaria. Furthermore, both drugs can have harmful side effects and there are concerns about people obtaining and self-administering them outside medical supervision. In addition, if lots of people in malaria-endemic regions take the drugs because they suspect that they might have COVID-19, or it could prevent them from contracting it, then it could encourage the evolution of resistance genes within the *Plasmodium* populations. A similar problem arose when ivermectin was promoted on social media as a preventative prophylactic for COVID-19 (Molento 2020). This followed the publication of some laboratory *in vitro* studies and the known activity of ivermectin against the very different dengue virus (Ooi 2020). At the time of writing, there were few patient studies on the effectiveness of ivermectin against COVID-19, and its pharmacokinetics suggest that it may not reach a high enough concentration to be effective against the virus (Peña-Silva et al. 2020). Nevertheless, in parts of South America, so many people began self-prescribing ivermectin that it became difficult to conduct meaningful trials. Although, ivermectin has a wide safety margin in humans, it is not without its risks, and many people are accessing ivermectin formulations designed for use in veterinary medicine. There is also the worry that with so many people taking the drug in an uncontrolled manner that it will promote the development of resistance. It will certainly increase the amount of drug entering the environment. Rather ironically, there are also claims that infection with gastrointestinal helminths may protect against COVID-19 engendering a cytokine storm (Bradbury et al. 2020; Hays et al. 2020). There is currently limited clinical evidence for this, but undoubtedly the prevalence of helminth parasites would be expected to fall among populations that are taking ivermectin in the hope of combatting COVID-19.

16.9   Role of Governments, Foundations, and Aid Organisations

The control of many parasitic diseases depends upon a co-ordinated national or even international response, and this can only be achieved through the active cooperation of governments. This requires stability within the country and a suitable infrastructure for distribution of funds, communication, and travel for expert personnel. Organisation of a control programme may be through national government, local government, or some combination of the two depending upon the way the country operates. Governments accumulate revenue from taxes and determine funding priorities. They therefore determine the extent to which human and animal health services are funded and the priority given to particular diseases. Even within the richest countries, there is never enough money to meet all the health needs of the population and decisions as to which diseases receive the most funding can be heavily influenced by political considerations. Only governments have the power to pass and enforce legislation. They can also ensure that different authorities and departments work together to mount a strategic response. Control programmes work best where there is a strong, stable government that is honest and appreciates the importance of the control programme. Governments also give permission for aid organisations or other charitable bodies to undertake work within their country or region: no matter how serious a disease problem or natural disaster might be, aid organisations cannot undertake operations without the agreement of a country's national or regional government. However, all governments tend to think of their own survival first: fighting a rival for power can be far more important to a government than fighting a disease.

Aid organisations and health authorities must work in the real world. Therefore politics, advertising, and media relations all play a part in determining whether a control programme can be undertaken, how it is undertaken, and its ultimate success. Unless carefully managed at both a national and local level, mass vaccination and other health campaigns can become perceived as vehicles for outside agencies, which are almost invariably seen as having an ulterior and nefarious purpose. For example, in Nigeria, a polio eradication campaign ran into problems because of rumours that the vaccine contained substances that made Muslims infertile. Similarly, in 2011, the CIA mounted a fake Hepatitis B vaccination campaign in Pakistan as part of their attempts to identify whereabouts Osama bin Laden, head of the terrorist organisation al-Qaeda. The CIA knew that bin Laden had young children travelling with him and if they could be identified then it would mean that he must be somewhere in the vicinity. Documentation can be forged, but we will always be identifiable through our DNA. The vaccination campaign provided a cover for surreptitiously collecting the DNA from children living in areas where Osama bin Laden might be. The plan failed but his hiding place was subsequently identified by other means, and he was killed. Nevertheless, it was only a matter of a few months before the ruse became common knowledge and thereafter the public perception of the validity and purpose of future vaccination campaigns was clouded (Kennedy 2017). Such rumours are not confined to developing countries, and health scares accompany most mass vaccination campaigns. The well-organised and vociferous groups spreading misinformation about SARS-COVID-19 vaccines (and the use of masks) is an obvious example. Consequently, the discovery of an effective drug or vaccine to treat a parasitic disease is just one small part of a control programme.

Legislation can be a highly effective tool in the control of parasitic diseases. For example, legislation can force local authorities to provide safe water and sanitation, make it an offence to import pets or domestic animals without a certificate indicating that they are free of disease and require farmers to dispose of dead stock appropriately. However, legislation serves little purpose if it is not enforced or where the penalties are so low that people are not put off committing the offence. Similarly, the target audience of the legislation must have a realistic chance of obeying the rules. For example, the owners of livestock will continue to slaughter and butcher animals at home if there is no official slaughterhouse within easy reach or it charges too much for its services. Similarly, farmers may not dispose of dead livestock correctly if the cost/logistics are prohibitive (Figure 16.1a,b). Effective legislation needs backing up with information and education because laws that the public does not understand, deems irrelevant, or considers an unjust imposition are seldom enforceable except in a police state.

Governments in developed countries are a major source of aid, and they also influence how the money is spent within the recipient country. This can cause tensions and conflicts of interest since donor governments often favour the purchase of products and services from their own country. It can be argued that the recipient of aid has a better appreciation of the best ways in which the

(a)

(b)

Figure 16.1 Legislation is not effective unless it is enforceable. (a) The Middle Eastern country where this photograph was taken had regulations requiring sheep to be slaughtered at official slaughterhouses where meat could be inspected and waste disposed of safely. However, numerous, small, unlicensed butchers continued to operate throughout the country. (b) Dead sheep disposed of by the roadside in Derbyshire, UK. The farmer has removed the sheep's head, so there are no ear tags or other means of identifying the owner. Many farmers operate on small profit margins and whilst legislation requires stock dying of natural causes to be disposed of appropriately, safe burial may not be feasible, and paying a renderer to collect the body is considered an expense to be avoided.

donated money could be spent. However, this must be balanced by the disappearance of huge amounts of aid money through corruption and mismanagement.

In addition to governments, there are numerous charities and aid organisations that fund and deliver parasite control programmes. The Bill and Melinda Gates Foundation has put enormous sums of money into the treatment and control of malaria, whilst several public-private partnerships have been established to fund the sequencing of parasite genomes and the development of antiparasitic drugs and vaccines (e.g., Drugs for Neglected Diseases Initiative, Institute for One World Health, Medicines for Malaria Venture). Large donors will inevitably influence policy decisions. Although this should not necessarily be considered a bad thing, it can be a source of concern for those who hold different views. For example, in 2009, following pressure from the Bill and Melinda Gates Foundation, the PATH Malaria Vaccine Initiative (MVI) changed the way in which it identified and funded candidate malaria vaccines. In particular, a much greater emphasis was placed upon identifying transmission-blocking vaccines because these are seen as having greater potential in contributing to the long-term eradication of the disease.

Regardless of the source of funding for a control programme, its delivery is often influenced by the prevailing management ethos. Many industrialised nations have a short-term target-based culture that can be highly effective in business but does not always transfer well to disease control in the developing world. For example, when considering how best to control malaria, the distribution of insecticide-impregnated bed-nets is easily measured, and can therefore become an end in itself. As a result, sometimes, little consideration is given to whether the type of bed-nets is appropriate for the lifestyle of the recipients and whether bed-nets are being used effectively. In addition, the distribution of free bed-nets does not contribute to sustainable malaria control since it can put the local bed-net manufacturers out of business. Consequently, after a couple of years when the donated nets need to be repaired or replaced, there is no one able to do the job. In addition, some scientists worry that the exposure of mosquitoes to sub-lethal levels of insecticides within the bed-nets may encourage the development of insecticide resistance.

In conclusion, to be effective, control programmes aimed at human parasites are almost always long-term projects that require the support of the authorities and the engagement of the local community. They must also be economically feasible and fully costed. Ultimately, many human parasitic infections will never be controlled until underlying problems of poverty, discrimination, and inequality are properly addressed. Similarly, many of the parasites of domestic animals persist because of a combination of poverty and a lack of appreciation of the importance of animal welfare.

References

Abdoli, A. and Ardakani, H.M. (2020). Potential application of helminth therapy for resolution of neuroinflammation in neuropsychiatric disorders. *Metabolic Brain Disease* **35**: 95–110. https://doi.org/10.1007/s11011-019-00466-5.

Abdul-Ghani, R. (2016). Combat against neglected parasitic diseases in Yemen: the need for mapping as a prerequisite for elimination. *Yemeni Journal for Medical Sciences* **10**: 1–5. https://doi.org/10.20428/YJMS.10.1.E.

Abele, L.G. and Gilchrist, S. (1977). Homosexual rape and sexual selection in acanthocephalan worms. *Science* **197**: 81–83. https://doi.org/10.1126/science.867055.

Abhishek, K., Das, S., Kumar, A. et al. (2018). *Leishmania donovani* induced unfolded protein response delays host cell apoptosis in PERK dependent manner. *PLoS Neglected Tropical Diseases* **12**: https://doi.org/10.1371/journal.pntd.0006646.

Abnave, P., Mottola, G., Gimenez, G. et al. (2014). Screening in planarians identifies MORN2 as a key component in LC3-associated phagocytosis and resistance to bacterial infection. *Cell Host & Microbe* **16**: 338–350. https://doi.org/10.1016/j.chom.2014.08.002.

Abner, S.R., Parthasarathy, G., Hill, D.E. et al. (2001). *Trichuris suis*: detection of antibacterial activity in excretory-secretory products from adults. *Experimental Parasitology* **99**: 26–36. PMID: 11708831. doi: 10.1006/expr.2001.4643.

Abonwa, M., Martin, R.J., and Robertson, A.P. (2017). A brief review on the mode of action of antinematodal drugs. *Acta Veterinaria* **67**: 137–152. https://doi.org/10.1515/acve-2017-0013.

Abou-El-Naga, I.F. (2020). Heat shock protein 70 (Hsp70) in *Schistosoma mansoni* and its role in decreased adult worm sensitivity to praziquantel. *Parasitology* **147**: 634–642. https://doi.org/10.1017/S0031182020000347.

Abrahão, J., Silva, L., Oliveira, D. et al. (2018). Lack of evidence of mimivirus replication in human PBMCs. *Microbes and Infection* **20**: 281–283. https://doi.org/10.1016/j.micinf.2018.03.003.

Abruzzi, A., Fried, B., and Alikhan, S.B. (2016). Coinfection of *Schistosoma* species with hepatitis b or hepatitis c viruses. *Advances in Parasitology* **91**: 111–231. https://doi.org/10.1016/bs.apar.2015.12.003.

Acosta, D.B., Ruiz, M., and Sanchez, J.P. (2019). First molecular detection of *Mycoplasma suis* in the pig louse *Haematopinus suis* (Phthiraptera: Anoplura) from Argentina. *Acta Tropica* **194**: 165–168. https://doi.org/10.1016/j.actatropica.2019.04.007.

Adamson, P.B. (1988). Dracontiasis in antiquity. *Medical History* **32**: 204–209. https://doi.org/10.1017/S0025727300048006.

Parasitology: An Integrated Approach, Second Edition. Alan Gunn and Sarah J. Pitt.
© 2022 John Wiley & Sons Ltd. Published 2022 by John Wiley & Sons Ltd.
Companion website: www.wiley.com/go/gunn/parasitology2

Adedeji, S.O., Ogunba, E.O., and Dipeolu, O.O. (1989). Synergistic effect of migrating *Ascaris* larvae and *Escherichia coli* in piglets. *Journal of Helminthology* **63**: 19–24. https://doi.org/10.1017/S0022149X00008671.

Adekiya, T.A., Kondiah, P.P., Choonara, Y.E. et al. (2020). A review of nanotechnology for targeted anti-schistosomal therapy. *Frontiers in Bioengineering and Biotechnology* **8**: 32. https://doi.org/10.3389/fbioe.2020.00032.

Adepoju, P. (2019). RTS,S malaria vaccine pilots in three African countries. *Lancet* **393**: 1685. https://doi.org/10.1016/S0140-6736(19)30937-7.

Afzelius, B.A., Alberti, G., Dallai, R. et al. (1989). Virus-and Rickettsia-infected sperm cells in arthropods. *Journal of Invertebrate Pathology* **53**: 365–377. https://doi.org/10.1016/0022-2011(89)90102-X.

Agha, S., El-Mashad, N., El-Malky, M. et al. (2006). Prevalence of low positive anti-HCV antibodies in blood donors: *Schistosoma mansoni* co-infection and possible role of autoantibodies. *Microbiology and Immunology* **50**: 447–452. https://doi.org/10.1111/j.1348-0421.2006.tb03813.x.

Aguinaldo, A.M.A., Turbeville, J.M., Linford, L.S. et al. (1997). Evidence for a clade of nematodes, arthropods and other moulting animals. *Nature* **387**: 489–493. https://doi.org/10.1038/387489a0.

Ahne, W. (1985). *Argulus foliaceus* L. and *Piscicola geometra* L. as mechanical vectors of spring viraemia of carp virus (SVCV). *Journal of Fish Diseases* **8**: 241–242. https://doi.org/10.1111/j.1365-2761.1985.tb1220.x.

Akama, T., Zhang, Y.K., Freund, Y.R. et al. (2018). Identification of a 4-fluorobenzyl l-valinate amide benzoxaborole (AN11736) as a potential development candidate for the treatment of Animal African Trypanosomiasis (AAT). *Bioorganic and Medicinal Chemistry Letters* **28**: 6–10. https://doi.org/10.1016/j.bmcl.2017.11.028.

Akbari, M., Oryan, A., and Hatam, G. (2017). Application of nanotechnology in treatment of leishmaniasis: a review. *Acta Tropica* **172**: 86–90. https://doi.org/10.1016/j.actatropica.2017.04.029.

Alarcón de Noya, B., Díaz-Bello, Z., Colmenares, C. et al. (2010). Large urban outbreak of orally acquired acute Chagas disease at a school in Caracas, Venezuela. *Journal of Infectious Diseases* **201**: 1308–1315. https://doi.org/10.1086/651608.

Alba, A., Tetreau, G., Chaparro, C. et al. (2019). Natural resistance to *Fasciola hepatica* (Trematoda) in *Pseudosuccinea columella* snails: a review from literature and insights from comparative "omic" analyses. *Developmental and Comparative Immunology* **101**: https://doi.org/10.1016/j.dci.2019.103463.

Al-Kinani, A.T. (2006). Depleted uranium in the food chain at south of Iraq. *Iranian Journal of Radiation Research* **4**: 143–148.

Allan, J.C. and Craig, P.S. (2006). Coproantigens in taeniasis and echinococcosis. *Parasitology International* **55**: S75–S80. https://doi.org/10.1016/j.parint.2005.11.010.

Allen, J.M., Light, J.E., Perotti, M.A. et al. (2009). Mutational meltdown in primary endosymbionts: selection limits Muller's Ratchet. *PLoS One* **4**: https://doi.org/10.1371/journal.pone.0004969.

Allis, C.D. and Jenuwein, T. (2016). The molecular hallmarks of epigenetic control. *Nature Reviews Genetics* **17**: 487–500. https://doi.org/10.1038/nrg.2016.59.

Almeria, S., Cinar, H.N., and Dubey, J.P. (2019). *Cyclospora cayetanensis* and cyclosporiasis: an update. *Microorganisms* **7**: 317. https://doi.org/10.3390/microorganisms7090317.

Alrefaei, A.F., Low, R., Hall, N. et al. (2019). Multilocus analysis resolves the European finch epidemic strain of *Trichomonas gallinae* and suggests introgression from divergent trichomonads. *Genome Biology and Evolution* **11**: 2391–2402. https://doi.org/10.1093/gbe/evz164.

Al-Salem, W.S., Pigott, D.M., Subramaniam, K. et al. (2016). Cutaneous leishmaniasis and conflict in Syria. *Emerging Infectious Diseases* **22**: 931–933. https://doi.org/10.3201/eid2205.160042.

Alves, E.B.D.S., Conceição, M.J., and Leles, D. (2016). *Ascaris lumbricoides, Ascaris suum*, or *"Ascaris lumbrisuum"*? *The Journal of Infectious Diseases* **213**: 1355–1355. https://doi.org/10.1093/infdis/jiw027.

Amanzougaghene, N., Mediannikov, O., Ly, T.D.A. et al. (2020). Molecular investigation and genetic diversity of *Pediculus* and *Pthirus lice* in France. *Parasites & Vectors* **13**: 177. https://doi.org/10.1186/s13071-020-04036-y.

Ambroise-Thomas, P. and Petersen, E. (2013). *Congenital Toxoplasmosis. Scientific Background, Clinical Management, and Control*. Heidelberg, Germany: Springer-Verlag.

Amoah, P., Drechsel, P., Abaidoo, R.C. et al. (2006). Pesticide and pathogen contamination of vegetables in Ghana's urban markets. *Archives of Environmental Contamination and Toxicology* **50**: 1–6. https://doi.org/10.1007/s00244-004-0054-8.

Amora, D.X., de Podesta, G.S., Nasu, E.D.G.C. et al. (2017). effect of essential oils on the root-knot nematode management. *Agri-Environmental Sciences* **3**: 15–23. https://revista.unitins.br/index.php/agri-environmental-sciences/article/view/225.

Anderson, A.L. and Chaney, E. (2009). Pubic lice (*Pthirus pubis*): history, biology and treatment vs. knowledge and beliefs of US college students. *International Journal of Environmental Research and Public Health* **6**: 592–600. https://doi.org/10.3390/ijerph6020592.

Anderson, D., Nathoo, N., Lu, J. et al. (2018). *Sarcocystis* myopathy in a patient with HIV-AIDS. *Journal of Neurovirology* **24**: 376–378. https://doi.org/10.1007/s13365-018-0620-x.

Andreani, G., Lodge, R., Richard, D. et al. (2012). Mechanisms of interaction between protozoan parasites and HIV. *Current Opinion in HIV and AIDS* **7**: 275–281. https://doi.org/10.1097/COH.0b13e32835211e9.

Angulo-Valadez, C.E., Scholl, P.J., Cepeda-Palacios, R. et al. (2010). Nasal bots . . . a fascinating world! *Veterinary Parasitology* **174**: 19–25. https://doi.org/10.1016/j.vetpar.2010.08.011.

Anjum, K., Abbas, S.Q., Shah, S.A.A. et al. (2016). Marine sponges as a drug treasure. *Biomolecules & Therapeutics* **24**: 347–362. https://doi.org/10.4062/biomolther.2016.067.

Antoine-Moussiaux, N., Büscher, P., and Desmecht, D. (2009). Host-parasite interactions in trypanosomiasis: on the way to an anti-disease strategy. *Infection and Immunity* **77**: 1276–1284. https://doi.org/10.1128/IAI.01185-08.

Antunes, S., Rosa, C., Couto, J. et al. (2017). Deciphering *Babesia*-vector interactions. *Frontiers in Cellular and Infection Microbiology* **7**: 429. https://doi.org/10.3389/fcimb.2017.00429.

Anuradha, R., Munisankar, S., Bhootra, Y. et al. (2017). Modulation of *Mycobacterium tuberculosis*-specific humoral immune responses is associated with *Strongyloides stercoralis* co-infection. *PLoS Neglected Tropical Diseases* **11**: https://doi.org/10.1371/journal.pntd.0005569.

Arasu, P. (2001). *in vitro* reactivation of *Ancylostoma caninum* tissue-arrested third-stage larvae by transforming growth factor-β. *Journal of Parasitology* **87**: 733–738. https://doi.org/10.1645/0022-3395(2001)087[0733:IVROAC]2.0.CO;2.

Araujo, A., Reinhard, K.J., Ferreira, L.F. et al. (2008). Parasites as probes for prehistoric human migrations? *Trends in Parasitology* **24**: 112–115. https://doi.org/10.1016/j.pt.2007.11.007.

Aravindhan, V., Mohan, V., Surendar, J. et al. (2010). Decreased prevalence of lymphatic filariasis among subjects with type-1 diabetes. *American Journal of Tropical Medicine and Hygiene* **83**: 1336–1339. https://doi.org/10.4269/ajtmh.2010.10-0410.

Arce-Fonseca, M., Carbajal-Hernández, A.C., Lozano-Camacho, M. et al. (2020). DNA vaccine treatment in dogs experimentally infected with *Trypanosoma cruzi*. *Journal of Immunology Research* **2020**: https://doi.org/10.1155/2020/9794575.

Arce-Fonseca, M., Rios-Castro, M., del Carmen Carrillo-Sánchez, S. et al. (2015). Prophylactic and therapeutic DNA vaccines against Chagas disease. *Parasites & Vectors* **8**: 121. https://doi.org/10.1186/s13071-015-0738-0.

Aregawi, M., Smith, S.J., Sillah-Kanu, M. et al. (2016). Impact of the Mass Drug Administration for malaria in response to the Ebola outbreak in Sierra Leone. *Malaria Journal* **15**: 480. https://doi.org/10.1186/s12936-016-1493-1.

Aregawi, W.G., Agga, G.E., Abdi, R.D. et al. (2019). Systematic review and meta-analysis on the global distribution, host range, and prevalence of *Trypanosoma evansi*. *Parasites & Vectors* **12**: 67. https://doi.org/10.1186/s13071-019-3311-4.

Argy, N., Lariven, S., Rideau, A. et al. (2020). Congenital leishmaniasis in a newborn infant whose mother was coinfected with leishmaniasis and HIV. *Journal of the Pediatric Infectious Diseases Society* **9**: 277–280. https://doi.org/10.1093/jpids/piz055.

Arias, J.L., Unciti-Broceta, J.D., Maceira, J. et al. (2015). Nanobody conjugated PLGA nanoparticles for active targeting of African Trypanosomiasis. *Journal of Controlled Release* **197**: 190–198. https://doi.org/10.1016/j.jconrel.2014.11.002.

Arlian, L.G. and Morgan, M.S. (2017). A review of *Sarcoptes scabiei*: past, present and future. *Parasites & Vectors* **10**: 297. https://doi.org/10.1186/s13071-017-2234-1.

Armoo, S., Doyle, S.R., Osei-Atweneboana, M.Y. et al. (2017). Significant heterogeneity in *Wolbachia* copy number within and between populations of *Onchocerca volvulus*. *Parasites & Vectors* **10**: 188. https://doi.org/10.1186/s13071-017-2126-4.

Arnott, S.A., Barber, I., and Huntingford, F.A. (2000). Parasite-associated growth enhancement in a fish-cestode system. *Proceedings of the Royal Society of London. Series B: Biological Sciences* **267**: 657–663. https://doi.org/10.1098/rspb.2000.1052.

Ashford, R.W., Barnish, G., and Viney, M.E. (1992). *Strongyloides fuelleborni kelly*i: infection and disease in Papua New Guinea. *Parasitology Today* **8**: 314–318. https://doi.org/10.1016/0169-4758(92)90106-C.

Ashour, D.S. (2015). Toll-like receptor signalling in parasitic infections. *Expert Review of Clinical Immunology* **11**: 771–780. https://doi.org/10.1586/1744666X.2015.1037286.

Attarian, S., Serratrice, J., Mazodier, C. et al. (2003). Guillain-Barré syndrome revealing visceral leishmaniasis in an immunocompetent woman. *Revue Neurologique* **159**: 1046–1048. (in French) PMID: 14710025.

Audebert, C., Even, G., Cian, A. et al. (2016). Colonization with the enteric protozoa *Blastocystis* is associated with increased diversity of human gut bacterial microbiota. *Scientific Reports* **6**: 1–11. https://doi.org/10.1038/srep25255.

Auld, S.K., Tinkler, S.K., and Tinsley, M.C. (2016). Sex as a strategy against rapidly evolving parasites. *Proceedings of the Royal Society B: Biological Sciences* **283**: 2226. https://doi.org/10.1098/rspb.2016.2226.

Autier, B., Belaz, S., Razakandrainibe, R. et al. (2018). Comparison of three commercial multiplex PCR assays for the diagnosis of intestinal protozoa. *Parasite (Paris, France)* **25**: 48. https://doi.org/10.1051/parasite/2018049.

Averkhin, A.I., Gapon, N.M., and Strel'chik, V.A. (1974). Pathological changes in experimental *Moniezia* infections in calves. *Nauchnye Trudy-Omskii Veterinarnogo Instituta* **30**: 38–46. (In Russian).

Aydin, Ö. (2007). Incidental parasitic infestations in surgically removed appendices: a retrospective analysis. *Diagnostic Pathology* **2**: 16. https://doi.org/10.1186/1746-1596-2-16.

Ayres, J. and Schneider, D.S. (2009). The role of anorexia in resistance and tolerance to infections in *Drosophila*. *PLoS Biology* **7**: https://doi.org/10.1371/journal.pbio.1000150.

Babal, P., Milcheva, R., Petkova, S. et al. (2011). Apoptosis as the adaptation mechanism in survival of *Trichinella spiralis* in the host. *Parasitology Research* **109**: 997–1002. https://doi.org/10.1007/s00436-011-2343-2.

Badman, S.G., Causer, L.M., Guy, R. et al. (2016). A preliminary evaluation of a new GeneXpert (Gx) molecular point-of-care test for the detection of *Trichomonas vaginalis*. *Sexually Transmitted Infections* **92**: 350–352. https://doi.org/10.1136/sextrans-2015-052384.

Baek, M.H., Kamiya, M., Kushibiki, T. et al. (2016). Lipopolysaccharide-bound structure of the antimicrobial peptide cecropin P1 determined by nuclear magnetic resonance spectroscopy. *Journal of Peptide Science* **22**: 214–221. https://doi.org/10.1002/psc.2865.

Bailey, F., Mondragon-Shem, K., Haines, L.R. et al. (2019). Cutaneous leishmaniasis and co-morbid major depressive disorder: a systematic review with burden estimates. *PLoS Neglected Tropical Diseases* **13**: https://doi.org/10.1371/journal.pntd.0007092.

Bailey, S.L., Price, J., and Llewelyn, M. (2011). Fluke infertility: the late cost of a quick swim. *Journal of Travel Medicine* **18**: 61–62. https://doi.org/10.1016/j.ijpara.2016.02.006.

Baldacchino, F., Bruno, M.C., Visentin, P. et al. (2017). Predation efficiency of copepods against the new invasive mosquito species *Aedes koreicus* (Diptera: Culicidae) in Italy. *European Zoological Journal* **84**: 43–48. https://doi.org/10.1080/11250003.2016.1271028.

Bale, J., van Lenteren, J., and Bigler, F. (2008). Biological control and sustainable food production. *Philosophical Transactions of the Royal Society B: Biological Sciences* **363**: 761–776. https://doi.org/10.1098/rstb.2007.2182.

Balloux, F., Brynildsrud, O.B., Van Dorp, L. et al. (2018). From theory to practice: translating whole-genome sequencing (WGS) into the clinic. *Trends in Microbiology* **26**: 1035–1048. https://doi.org/10.1016/j.tim.2018.08.004.

Banerjee, A., Bhattacharya, P., Dagur, P.K. et al. (2018). Live attenuated *Leishmania donovani* centrin gene–deleted parasites induce IL-23–dependent IL-17–protective immune response against visceral leishmaniasis in a murine model. *Journal of Immunology* **200**: 163–176. https://doi.org/10.4049/jimmunol.1700674.

Bao, M., Pierce, G.J., Strachan, N.J. et al. (2019). Human health, legislative and socioeconomic issues caused by the fish-borne zoonotic parasite *Anisakis*: challenges in risk assessment. *Trends in Food Science & Technology* **86**: 298–310. https://doi.org/10.1016/j.tifs.2019.02.013.

Barash, N.R., Maloney, J.G., Singer, S.M. et al. (2017). *Giardia* alters commensal microbial diversity throughout the murine gut. *Infection and Immunity* **85**: https://doi.org/10.1128/IAI.00948-16.

Barber, I., Hoare, D., and Krause, J. (2000). Effects of parasites on fish behaviour: a review and evolutionary perspective. *Reviews in Fish Biology and Fisheries* **10**: 131–165. https://doi.org/10.1023/A:1016658224470.

Barennes, H., Sayasone, S., Odermatt, P. et al. (2008). A major trichinellosis outbreak suggesting a high endemicity of *Trichinella* infection in northern Laos. *The American Journal of Tropical Medicine and Hygiene* **78**: 40–44. https://doi.org/10.4269/ajtmh.2008.78.40.

Barnard, J.W., Biro, M.G., Lo, S.K. et al. (1995). Neutrophil inhibitory factor prevents neutrophil-dependent lung injury. *Journal of Immunology* **155**: 4876–4881. PMID: 7594491.

Barnes, T.S., Morton, J.M., and Coleman, G.T. (2007). Clustering of hydatid infection in macropodids. *International Journal for Parasitology* **37**: 943–952. https://doi.org/10.1016/j.ijpara.2007.01.014.

Barratt, J., Gough, R., Stark, D. et al. (2016). Bulky trichomonad genomes: encoding a Swiss army knife. *Trends in Parasitology* **32**: 783–797. https://doi.org/10.1016/j.pt.2016.05.014.

Barratt, J.L. and Sapp, S.G. (2020). Machine learning-based analyses support the existence of species complexes for *Strongyloides fuelleborni* and *Strongyloides stercoralis*. *Parasitology* **147**: 1184–1195. https://doi.org/10.1017/S0031182020000979.

Bartlett, S., Eichenberger, R.M., Nevagi, R.J. et al. (2020). Lipopeptide-based oral vaccine against hookworm infection. *Journal of Infectious Diseases* **221**: 934–942. https://doi.org/10.1093/infdis/jiz528.

Barton, D.P., Baker, A., Porter, M. et al. (2020). Verification of rabbits as intermediate hosts for *Linguatula serrata* (Pentastomida) in Australia. *Parasitology Research* **119**: 1553–1562. https://doi.org/10.1007/s00436-020-06670-y.

Basch, P.F. (1991). *Schistosomes: Development, Reproduction and Host Relations*. Oxford, UK: Oxford University Press.

Bathie, P. (2020). Military safety: a systems perspective on Lariam. *RUSI Journal* **165**: 64–76. https://doi.org/10.1080/03071847.2020.1755107.

Bedini, S., Flamini, G., Cosci, F. et al. (2019). Toxicity and oviposition deterrence of essential oils of *Clinopodium nubigenum* and *Lavandula angustifolia* against the myiasis-inducing blowfly *Lucilia sericata*. *PLoS One* **14**: https://doi.org/10.1371/journal.pone.0212576.

Beltrame, L., Dunne, T., Vineer, H.R. et al. (2018). A mechanistic hydro-epidemiological model of liver fluke risk. *Journal of the Royal Society Interface* **15**: https://doi.org/10.1098/rsif.2018.0072.

Beltrame, M.O., Pruzzo, C., Sanabria, R. et al. (2020). First report of pre-Hispanic *Fasciola hepatica* from South America revealed by ancient DNA. *Parasitology* **147**: 371–375. https://doi.org/10.1017/S0031182019001719.

Beltran, S., Cézilly, F., and Boissier, J. (2009). Adult sex ratio affects divorce rate in the monogamous endoparasite *Schistosoma mansoni*. *Behavioral Ecology and Sociobiology* **63**: 1363–1368. https://doi.org/10.1007/s00265-009-0757-y.

Benabdelkader, S., Andreani, J., Gillet, A. et al. (2019). Specific clones of *Trichomonas tenax* are associated with periodontitis. *PloS One* **14**: https://doi.org/10.1371/journal.pone.0213338.

Bennink, S., Kiesow, M.J., and Pradel, G. (2016). The development of malaria parasites in the mosquito midgut. *Cellular Microbiology* **18**: 905–918. https://doi.org/10.1111/cmi.12604.

Bennis, I., Verdonck, K., El Khalfaoui, N. et al. (2018). Accuracy of a rapid diagnostic test based on antigen detection for the diagnosis of cutaneous leishmaniasis in patients with suggestive skin lesions in Morocco. *American Journal of Tropical Medicine and Hygiene* **99**: 716–722. https://doi.org/10.4269/ajtmh.18-0066.

Bentwich, Z., Maartens, G., Torten, D. et al. (2000). Concurrent infections and HIV pathogenesis. *AIDS* **14**: 2071–2081. https://doi.org/10.1097/00002030-200009290-00002.

Berbudi, A., Ajendra, J., Wardani, A.P. et al. (2016). Parasitic helminths and their beneficial impact on type 1 and type 2 diabetes. *Diabetes/Metabolism Research and Reviews* **32**: 238–250. https://doi.org/10.1002/dmrr.2673.

Bergmann-Leitner, E.S. and Leitner, W.W. (2015). Vaccination using gene-gun technology. In: *Malaria Vaccines. Methods in Molecular Biology*, vol. **1325** (ed. A. Vaughan), 289–302. New York, NY: Humana Press https://doi.org/10.1007/978-1-4939-2815-6_22.

Berke, O., Romig, T., and von Keyserlingk, M. (2008). Emergence of *Echinococcus multilocularis* among red foxes in northern Germany, 1991–2005. *Veterinary Parasitology* **155**: 319–322. https://doi.org/10.1016/j.vetpar.2008.05.017.

Berland, B. (1961). Copepod *Ommatokoita elongata* (Grant) in the eyes of the Greenland shark – a possible cause of mutual dependence. *Nature* **191**: 829–830. https://doi.org/10.1038/191829a0.

Bernabeu, M. and Smith, J.D. (2017). EPCR and malaria severity: the center of a perfect storm. *Trends in Parasitology* **33**: 295–308. https://doi.org/10.1016/j.pt.2016.11.004.

Bernhardt, V., Finkelmeier, F., Verhoff, M.A. et al. (2019). Myiasis in humans – a global case report evaluation and literature analysis. *Parasitology Research* **118**: 389–397. https://doi.org/10.1007/s00436-018-6145-7.

Beros, S., Lenhart, A., Scharf, I., *et al.* (2021) Extreme lifespan extension in tapeworm-infected ant workers. *Royal Society Open Science* **8**, http://doi.org/10.1098/rsos.202118.

Berry, A.S., Salazar-Sánchez, R., Castillo-Neyra, R. et al. (2019). Sexual reproduction in a natural *Trypanosoma cruzi* population. *PLoS Neglected Tropical Diseases* **13**: https://doi.org/10.1371/journal.pntd.0007392.

Berto, B.P., Flausino, W., McIntosh, D. et al. (2011). Coccidia of New World passerine birds (Aves: Passeriformes): a review of *Eimeria* Schneider, 1875 and *Isospora* Schneider, 1881 (Apicomplexa: Eimeriidae). *Systematic Parasitology* **80**: 159. https://doi.org/10.1007/s11230-011-9317-8.

Beschin, A., De Baetselier, P., and Van Ginderachter, J.A. (2013). Contribution of myeloid cell subsets to liver fibrosis in parasite infection. *Journal of Pathology* **229**: 186–197. https://doi.org/10.1002/path.4112.

Besuschio, S.A., Llano Murcia, M., Benatar, A.F. et al. (2017). Analytical sensitivity and specificity of a loop-mediated isothermal amplification (LAMP) kit prototype for detection of *Trypanosoma cruzi* DNA in human blood samples. *PLoS Neglected Tropical Diseases* 11. 10.1371/journal.pntd.0005779.

Bethony, J., Brooker, S., Albonico, M. et al. (2006). Soil-transmitted helminth infections: ascariasis, trichuriasis, and hookworm. *Lancet* **367**: 1521–1532. https://doi.org/10.1016/S0140-6736(06)68653-4.

Betts, A., Gray, C., Zelek, M. et al. (2018). High parasite diversity accelerates host adaptation and diversification. *Science* **360**: 907–911. https://doi.org/10.1126/science.aam9974.

Beveridge, I. (2014). A review of the genus *Paramoniezia* Maplestone et Southwell, 1923 (Cestoda: Anoplocephalidae), with a new genus, *Phascolocestus*, from wombats (Marsupialia) and redescriptions of *Moniezia mettami* Baylis, 1934 and *Moniezia phacochoeri* (Baylis, 1927) comb. n. from African warthogs (Artiodactyla). *Folia Parasitologica* **61**: 21–33. https://doi.org/10.14411/fp.2014.008.

Bhattacharya, C., Singh, R.N., Misra, S., and Rathaur, S. (1997). Diethylcarbamazine: effect on lysosomal enzymes and acetylcholine in *Wuchereria bancrofti* infection. *Tropical Medicine International Health* **2**: 686–690. https://doi.org/10.1046/j.1365-3156.1997.d01-353.x.

Bhattacharya, P., Ghosh, S., Ejazi, S.A. et al. (2016). Induction of IL-10 and TGFβ from CD4+ CD25+ FoxP3+ T cells correlates with parasite load in Indian kala-azar patients infected with *Leishmania donovani*. *PLoS Neglected Tropical Diseases* **10**: https://doi.org/10.1371/journal.pntd.0004422.

Bhaw-Luximon, A. and Jhurry, D. (2017). Artemisinin and its derivatives in cancer therapy: status of progress, mechanism of action, and future perspectives. *Cancer Chemotherapy and Pharmacology* **79**: 451–466. https://doi.org/10.1007/s00280-017-3251-7.

Bhutta, Z.A. (2002). Ethics in international health research: a perspective from the developing world. *Bulletin of the World Health Organization* **80**: 114–120. http://www.cmhealth.org/docs/wg2_paper4.pdf).

Bilic, I. and Hess, M. (2020). Interplay between *Histomonus meleagridis* and bacteria: mutualistic or predator–prey? *Trends in Parasitology* **36**: 232–235. https://doi.org/10.1016/j.pt.2019.12.015.

Billet, A.C., Salmon Rousseau, A., Piroth, L. et al. (2019). An underestimated sexually transmitted infection: amoebiasis. *BMJ Case Reports CP* **12**: e228942. https://doi.org/10.1136/bcr-2018-228942.

Bitilinyu-Bangoh, J., Voskuijl, W., Thitiri, J. et al. (2019). Performance of three rapid diagnostic tests for the detection of *Cryptosporidium* spp. and *Giardia duodenalis* in children with severe acute malnutrition and diarrhoea. *Infectious Diseases of Poverty* **8**: 1–8. https://doi.org/10.1186/s40249-019-0609-6.

Bjørnson, S. and Oi, D. (2014). Microsporidia biological control agents and pathogens of beneficial insects. In: *Microsporidia: Pathogens of Opportunity* (ed. L.M. Weiss and J.J. Becnel), 635–670. Chichester, UK: Wiley.

Black, C.K., Mihai, D.M., and Washington, I. (2014). The photosynthetic eukaryote *Nannochloris eukaryotum* as an intracellular machine to control and expand functionality of human cells. *Nano Letters* **14**: 2720–2725.

Blackburn, B.G. and Montoya, J.G. (2019). Parasitic infections in transplant recipients: toxoplasmosis, strongyloidiasis, and other parasites. In: *Principles and Practice of Transplant Infectious Diseases* (ed. A. Safdar), 775–792. New York, NY: Springer.

Blagborough, A.M., Musiychuk, K., Bi, H. et al. (2016). Transmission blocking potency and immunogenicity of a plant-produced Pvs25-based subunit vaccine against *Plasmodium vivax*. *Vaccine* **34**: 3252–3259. https://doi.org/10.1016/j.vaccine.2016.05.007.

Blatner, C., Oppel, E.M., Wieser, A. et al. (2017). The small liver fluke (*Dicrocoelium dendriticum*): an unusual clinical finding in chronic urticaria. *Allergo Journal International* **26**: 165–167. https://doi.org/10.1007/s40629-017-0021-3.

Blaxter, M.L., De Ley, P., Garey, J.R. et al. (1998). A molecular evolutionary framework for the phylum Nematoda. *Nature* **392**: 71–75. https://doi.org/10.1038/32160.

Bleise, A., Danesi, P.R., and Burkart, W. (2003). Properties, use and health effects of depleted uranium (DU): a general overview. *Journal of Environmental Radioactivity* **64**: 93–112. https://doi.org/10.1016/S0265-931X(02)00041-3.

Bloomfield, R.D., Suarez, J.R., and Malangit, A.C. (1978). Transplacental transfer of Bancroftian filariasis. *Journal of the National Medical Association* **70**: 597–598. PMID: 702594; PMCID: PMC2537219.

Blumental, S., Lambermont, M., Heijmans, C. et al. (2015). First documented transmission of *Trypanosoma cruzi* infection through blood transfusion in a child with sickle-cell disease in Belgium. *PLoS Neglected Tropical Diseases* **9**: https://doi.org/10.1371/journal.pntd.0003986.

Boillat, M., Hammoudi, P.M., Dogga, S.K. et al. (2020). Neuroinflammation-associated aspecific manipulation of mouse predator fear by *Toxoplasma gondii*. *Cell Reports* **30**: 320–334. https://doi.org/10.1016/j.celrep.2019.12.019.

Boissière, A., Tchioffo, M.T., Bachar, D. et al. (2012). Midgut microbiota of the malaria mosquito vector *Anopheles gambiae* and interactions with *Plasmodium falciparum* infection. *PLoS Pathogens* **8**: https://doi.org/10.1371/journal.ppat.1002742.

Bones, A.J., Jossé, L., More, C. et al. (2019). Past and future trends of *Cryptosporidium in vitro* research. *Experimental Parasitology* **196**: 28–37. https://doi.org/10.1016/j.exppara.2018.12.001.

Bonner, M., Fresno, M., Gironès, N. et al. (2018). Reassessing the role of *Entamoeba gingivalis* in periodontitis. *Frontiers in Cellular and Infection Microbiology* **8**: 379. https://doi.org/10.3389/fcimb.2018.00379.

Booth, M. (2018). Climate change and the neglected tropical diseases. *Advances in Parasitology* **100**: 39–126. https://doi.org/10.1016/bs.apar.2018.02.001.

Borges, F., Sybrecht, G.W., and von Samson-Himmelstjerna, G. (2019). First reported case of *Hypoderma diana* Brauer, 1985 (Diptera: Oestridae) – associated myiasis in a horse in Germany. *Equine Veterinary Education* **31**: 122–125. https://doi.org/10.1111/eve.12810.

Bossi, P., Caumes, E., Paris, L. et al. (1998). *Toxoplasma gondii*-associated Guillain-Barré syndrome in an immunocompetent patient. *Journal of Clinical Microbiology* **36**: 3724–3725. https://doi.org/10.1128/JCM.36.12.3724-3725.1998.

Boulanger, N., Bulet, P., and Lowenberger, C. (2006). Antimicrobial peptides in the interactions between insects and flagellate parasites. *Trends in Parasitology* **22**: 262–268. https://doi.org/10.1016/j.pt.2006.04.003.

Bourke, C.D., Berkley, J.A., and Prendergast, A.J. (2016). Immune dysfunction as a cause and consequence of malnutrition. *Trends in Immunology* **37**: 386–398. https://doi.org/10.1016/j.it.2016.04.003.

Boyce, M.R. and O'Meara, W.P. (2017). Use of malaria RDTs in various health contexts across sub-Saharan Africa: a systematic review. *BMC Public Health* **17**: 1–15. https://link.springer.com/article/10.1186/s12889-017-4398-1.

Boyce, R., Reyes, R., Matte, M. et al. (2016). Severe flooding and malaria transmission in the Western Ugandan Highlands: implications for disease control in an era of global climate change. *Journal of Infectious Diseases* **214**: 1403–1410. https://doi.org/10.1093/infdis/jiw363.

Boyd, B.M., Allen, J.M., Nguyen, N.P. et al. (2017). Primates, lice and bacteria: speciation and genome evolution in the symbionts of hominid lice. *Molecular Biology and Evolution* **34**: 1743–1757. https://doi.org/10.1093/molbev/msx117.

Boyer, M., Madoui, M.A., and Gimenez, G. (2010). Phylogenetic and phyletic studies of informational genes in genomes highlight existence of a 4th domain of life including giant viruses. *PLoS One* **5**: https://doi.org/10.1371/journal.pone.0015530.

Braae, U.C., Kabululu, M., Nørmark, M.E. et al. (2015). *Taenia hydatigena* cysticercosis in slaughtered pigs, goats, and sheep in Tanzania. *Tropical Animal Health and Production* **47**: 1523–1530. https://doi.org/10.1007/s11250-015-0892-6.

Bradbury, R.S., Piedrafita, D., Greenhill, A. et al. (2020). Will helminth co-infection modulate COVID-19 severity in endemic regions? *Nature Reviews Immunology* **20**: 342. https://doi.org/10.1038/s41577-020-0330-5.

Bradbury, R.S., Roy, S., Ali, I.K. et al. (2019). Case report: cervicovaginal co-colonization with *Entamoeba gingivalis* and *Entamoeba polecki* in association with an intrauterine device. *American Journal of Tropical Medicine and Hygiene* **100**: 311–313. https://doi.org/10.4269/ajtmh.18-0522.

Brahmbhatt, H., Sullivan, D., Kigozi, G. et al. (2008). Association of HIV and malaria with mother-to-child transmission, birth outcomes, and child mortality. *JAIDS Journal of Acquired Immune Deficiency Syndromes* **47**: 472–476. https://doi.org/10.1097/QAI.0b13e318162afe0.

Breeze, R.G., Budowle, B., and Schutzer, S.E. (2005). *Microbial Forensics*. Burlington, MA: Elsevier https://doi.org/10.1016/B978-0-12-088483-4.X5000-8.

Brennan, J.J. and Gilmore, T.D. (2018). Evolutionary origins of toll-like receptor signalling. *Molecular Biology and Evolution* **35**: 1576–1587. https://doi.org/10.1093/molbev/msy050.

Bresciani, K.D.S. and da Costa, A.J. (2018). *Congenital Toxoplasmosis in Humans and Domestic Animals*. Sharjah, UAE: Bentham Science Publishers.

Briggs, N., Weatherhead, J., Sastry, K.J. et al. (2016). The hygiene hypothesis and its inconvenient truths about helminth infections. *PLoS Neglected Tropical Diseases* **10**: https://doi.org/10.1371/journal.pntd.0004944.

Brindley, P.J. and Loukas, A. (2017). Helminth infection–induced malignancy. *PLoS Pathogens* **13**: e1006393. https://doi.org/10.1371/journal.ppat.1006393.

Brinkmann, U.K., Krämer, P., Presthus, G.T. et al. (1976). Transmission *in utero* of microfilariae of *Onchocerca volvulus*. *Bulletin of the World Health Organization* **54**: 708. PMID: 1088518; PMCID: PMC2366585.

Britton, C., Winter, A.D., Gillan, V. et al. (2014). microRNAs of parasitic helminths–Identification, characterization and potential as drug targets. *International Journal for Parasitology: Drugs and Drug Resistance* **4**: 85–94. https://doi.org/10.1016/j.ijpddr.2014.03.001.

Brohi, R.D., Wang, L., Talpur, H.S. et al. (2017). Toxicity of nanoparticles on the reproductive system in animal models: a review. *Frontiers in Pharmacology* **8**: 606. https://doi.org/10.3389/fphar.2017.00606.

Brooker, S.J. and Pullan, R.L. (2013). *Ascaris lumbricoides* and ascariasis: estimating numbers infected and burden of disease. In: *Ascaris: The Neglected Parasite* (ed. C. Holland), 343–362. Amsterdam: Elsevier.

Brown, H.W. and Cort, W.W. (1927). The egg production of *Ascaris lumbricoides*. *Journal of Parasitology* **14**: 88–90. https://www.jstor.org/stable/3271720.

Brown, S.P. and Grenfell, B.T. (2001). An unlikely partnership: parasites, concomitant immunity and host defence. *Proceedings of the Royal Society of London. Series B: Biological Sciences* **268**: 2543–2549. https://doi.org/10.1098/rspb.2001.1821.

Brun, R., Blum, J., Chappuis, F., and Burri, C. (2010). Human African trypanosomiasis. *Lancet* **375**: 148–159. https://doi.org/10.1016/S0140-6736(09)60829-1.

Bryant, A.S. and Hallem, E.A. (2018). Terror in the dirt: sensory determinants of host seeking in soil-transmitted mammalian-parasitic nematodes. *International Journal for Parasitology: Drugs and Drug Resistance* **8**: 496–510. https://doi.org/10.1016/j.ijpddr.2018.10.008.

Buck, J.C., De Leo, G.A., and Sokolow, S.H. (2020). Concomitant immunity and worm senescence may drive schistosomiasis epidemiological patterns: an eco-evolutionary perspective. *Frontiers in Immunology* **11**: 160. https://doi.org/10.3389/fimmu.2020.00160.

Buckingham, L. (2019). *Molecular Diagnostics: Fundamentals, Methods, and Clinical Applications*, 3rde. Philadelphia, PA: F.A. Davis Company.

Bucur, I., Gabriël, S., Van Damme, I. et al. (2019). Survival of *Taenia saginata* eggs under different environmental conditions. *Veterinary Parasitology* **266**: 88–95. https://doi.org/10.1016/j.vetpar.2018.12.011.

Bueno, L.L., Morais, C.G., Araujo, F.F. et al. (2010). *Plasmodium vivax*: induction of CD4+ CD25+ FoxP3+ regulatory T cells during infection are directly associated with level of circulating parasites. *PloS One* **5**: https://doi.org/10.1371/journal.pone.0009623.

Bundy, D.A.P. and Cooper, E.S. (1989). *Trichuris* and trichuriasis in humans. *Advances in Parasitology* **28**: 107–173. https://doi.org/10.1016/S0065-308X(08)60332-2.

Burgess, S.T., Nunn, F., Nath, M. et al. (2016). A recombinant subunit vaccine for the control of ovine psoroptic mange (sheep scab). *Veterinary Research* **47**: 26. https://doi.org/10.1186/s13567-016-0315-3.

Burgos, G., Yanqui-Rivera, F., Mollocana, D. et al. (2017). Multiplex PCR in non-human DNA molecular identification of *Ascaris* spp. in forensic biology. *Forensic Science International: Genetics Supplement Series* **6**: e568–e569. https://doi.org/10.1016/j.fsigss.2017.09.227.

Burke, N.W. and Bonduriansky, R. (2017). Sexual conflict, facultative asexuality, and the true paradox of sex. *Trends in Ecology & Evolution* **32**: 646–652. https://doi.org/10.1016/j.tree.2017.06.002.

Büscher, P., Cecchi, G., Jamonneau, V. et al. (2017). Human African trypanosomiasis. *Lancet* **390**: 2397–2409. https://doi.org/10.1016/S0140-6736(17)31510-6.

Buss, S.N., Leber, A., Chapin, K. et al. (2015). Multicenter evaluation of the BioFire FilmArray gastrointestinal panel for etiologic diagnosis of infectious gastroenteritis. *Journal of Clinical Microbiology* **53**: 915–925. https://doi.org/10.1128/JCM.02674-14.

Bustos, J.A., Ninaquispe, B.E., Rodriguez, S. et al. (2019). Performance of a sandwich antigen-detection ELISA for the diagnosis of porcine *Taenia solium* cysticercosis. *American Journal of Tropical Medicine and Hygiene* **100**: 604–608. https://doi.org/10.4269/ajtmh.18-0697.

Buvé, A., Weiss, H.A., Laga, M. et al. (2001). The epidemiology of trichomoniasis in women in four African cities. *Aids* **15**: S89–S96. PMID: 11686470. https://doi.org/10.1097/00002030-200108004-00010.

Cacciò, S.M., Lalle, M., and Svärd, S.G. (2018). Host specificity in the *Giardia duodenalis* species complex. *Infection, Genetics and Evolution* **66**: 335–345. https://doi.org/10.1016/j.meegid.2017.12.001.

Cadman, E.T. and Lawrence, R.A. (2010). Granulocytes: effector cells or immunomodulators in the immune response to helminth infection? *Parasite Immunology* **32**: 1–19. https://doi.org/10.1111/j.1365-3024.2009.01147.x.

Cai, P., Gobert, G.N., and McManus, D.P. (2016). MicroRNAs in parasitic helminthiases: current status and future perspectives. *Trends in Parasitology* **32**: 71–86. https://doi.org/10.1016/j.pt.2015.09.003.

Calavas, D. and Martin, P.M. (2014). Schistosomiasis in cattle in Corsica, France. *Emerging Infectious Diseases* **20**: 2163–2164. https://doi.org/10.3201/eid2012.141474.

Calegar, D.A., Nunes, B.C., Monteiro, K.J.L. et al. (2016). Frequency and molecular characterisation of *Entamoeba histolytica*, *Entamoeba dispar*, *Entamoeba moshkovskii*, and *Entamoeba hartmanni* in the context of water scarcity in northeastern Brazil. *Memórias do Instituto Oswaldo Cruz* **111**: 114–119. https://doi.org/10.1590/0074-02760150383.

Calero-Bernal, R. and Gennari, S.M. (2019). Clinical toxoplasmosis in dogs and cats: an update. *Frontiers in Veterinary Science* **6**: 54. https://doi.org/10.3389/fvets.2019.00054.

Calero-Bernal, R., Horcajo, P., Hernández, M. et al. (2019). Absence of *Neospora caninum* DNA in human clinical samples, Spain. *Emerging Infectious Diseases* **25**: 1226–1227. https://doi.org/10.3201/eid2506.181431.

Caley, J. (1975). *in vitro* hatching of the tapeworm *Moniezia expansa* (Cestoda: Anoplocephalidae) and some properties of the egg membranes. *Zeitschrift für Parasitenkunde* **45**: 335–346. https://doi.org/10.1007/BF00329823.

Calvo-Barreiro, L., Eixarch, H., Montalban, X. et al. (2018). Combined therapies to treat complex diseases: the role of the gut microbiota in multiple sclerosis. *Autoimmunity Reviews* **17**: 165–174. https://doi.org/10.1016/j.autrev.2017.11.019.

Calvopina, M., Romero-Alvarez, D., Macias, R. et al. (2017). Severe pleuropulmonary paragonimiasis caused by *Paragonimus mexicanus* treated as tuberculosis in Ecuador. *American Journal of Tropical Medicine and Hygiene* **96**: 97–99. https://doi.org/10.4269/ajtmh.16-0351.

Camacho, M., Araújo, A., Morrow, J. et al. (2018). Recovering parasites from mummies and coprolites: an epidemiological approach. *Parasites & Vectors* **11**: 248. https://doi.org/10.1186/s13071-018-2729-4.

Campbell, W.C. and Blair, L.S. (1974). Chemotherapy of *Trichinella spiralis* infections (a review). *Experimental Parasitology* **35**: 304–334. https://doi.org/10.1016/0014-4894(74)90037-X.

Canavan, C., West, J., and Card, T. (2014). The epidemiology of irritable bowel syndrome. *Clinical Epidemiology* **6**: 71–80. https://doi.org/10.2147/CLEP.S40245.

Canning, E.U. (1975). *The Microsporidian Parasites of Platyhelminthes: Their Morphology, Development, Transmission and Pathogenicity*. Buckinghamshire, England: Commonwealth Institute of Helminthology Miscellaneous Publications No. 2. Commonwealth Agricultural Bureaux.

Canning, E.U. and Gunn, A. (1984). *Nosema helminthorum* (Moniez, 1887) (Microspora, Nosematidae): a taxonomic enigma. *Journal of Protozoology* **31**: 525–531. https://doi.org/10.1111/j.1550-7408.1984.tb05496.x.

Cantacessi, C., Dantas-Torres, F., Nolan, M.J. et al. (2015). The past, present, and future of *Leishmania* genomics and transcriptomics. *Trends in Parasitology* **31**: 100–108. https://doi.org/10.1016/j.pt.2014.12.012.

Cantanhêde, L.M., Fernandes, F.G., Ferreira, G.E.M. et al. (2018). New insights into the genetic diversity of *Leishmania RNA Virus 1* and its species-specific relationship with *Leishmania* parasites. *PLoS ONE* **13**: https://doi.org/10.1371/journal.pone.0198727.

Capewell, P., Cren-Travaille, C., Marchesi, F. et al. (2016). The skin is a significant but overlooked anatomical reservoir for vector-borne African trypanosomes. *Elife* **5**: 17716. https://doi.org/10.7554/eLife.17716.

Carcaboso, A.M., Hernandez, R.M., Igartua, M. et al. (2004). Potent, long lasting systemic antibody levels and mixed Th1/Th2 immune response after nasal immunization with malaria antigen loaded PLGA microparticles. *Vaccine* **22**: 1423–1432. https://doi.org/10.1016/j.vaccine.2003.10.020.

Carlson, A.L., Xia, K., Azcarate-Peril, M.A. et al. (2018). Infant gut microbiome associated with cognitive development. *Biological Psychiatry* **83**: 148–159. https://doi.org/10.1016/j.biopsych.2017.06.021.

Carpenter, H.A. (1998). Bacterial and parasitic cholangitis. *Mayo Clinic Proceedings* **73**: 473–478. https://doi.org/10.1016/S0025-6196(11)63734-8.

Carvalho, T., Trindade, S., Pimenta, S. et al. (2018). *Trypanosoma brucei* triggers a marked immune response in male reproductive organs. *PLoS Neglected Tropical Diseases* **12**: https://doi.org/10.1371/journal.pntd.0006690.

Cavalier-Smith, T. (2018). Kingdom Chromista and its eight phyla: a new synthesis emphasising periplastid protein targeting, cytoskeletal and periplastid evolution, and ancient divergences. *Protoplasma* **255**: 297–357. https://doi.org/10.1007/s00709-017-1147-3.

Ceccarelli, F., Agmon-Levin, N., and Perricone, C. (2017). Genetic factors of autoimmune diseases. *Journal of Immunology Research* **2789242**: https://doi.org/10.1155/2017/2789242.

Chacín-Bonilla, L. (2008). Transmission of *Cyclospora cayetanensis* infection: a review focusing on soil-borne cyclosporiasis. *Transactions of the Royal Society of Tropical Medicine and Hygiene* **102**: 215–216. https://doi.org/10.1016/j.trstmh.2007.06.005.

Chagas, A.C., Oliveira, F., Debrabant, A. et al. (2014). Lundep, a sand fly salivary endonuclease increases *Leishmania* parasite survival in neutrophils and inhibits XIIa contact activation in human plasma. *PLoS Pathogens.* **10**: https://doi.org/10.1371/journal.ppat.1003923.

Champion, T.S., Connelly, S., Smith, C.J. et al. (2021). Monitoring schistosomiasis and sanitation interventions – the potential of environmental DNA. *Wiley Interdisciplinary Reviews: Water* **8**: 10.1002/wat2.1491.

Chan, A., Dziedziech, A., Kirkman, L.A. et al. (2020). A histone methyltransferase inhibitor can reverse epigenetically acquired drug resistance in the malaria parasite Plasmodium falciparum. *Antimicrobial Agents and Chemotherapy* **64**: https://doi.org/10.1128/AAC.02021-19.

Chan, F.L., Kennedy, B., and Nelson, R. (2018). Fatal *Strongyloides* hyperinfection syndrome in an immunocompetent adult with review of the literature. *Internal Medicine Journal* **48**: 872–875. https://doi.org/10.1111/imj.13940.

Chan, J.A., Fowkes, F.J., and Beeson, J.G. (2014). Surface antigens of *Plasmodium falciparum*-infected erythrocytes as immune targets and malaria vaccine candidates. *Cellular and Molecular Life Sciences* **71**: 3633–3657. https://doi.org/10.1007/s00018-014-1614-3.

Chandler, A.C. and Read, C.P. (1961). *Introduction to Parasitology*, 10the. Chichester, UK: John Wiley and Sons.

Chapuis, E. (2009). Correlation between parasite prevalence and adult size in a trematode-mollusc system: evidence for evolutionary gigantism in the freshwater snail *Galba truncatula*? *Journal of Molluscan Studies* **75**: 391–396. https://doi.org/10.1093/mollus/eyp035.

Chapuis, E., Arnal, A., and Ferdy, J.B. (2012). Trade-offs shape the evolution of the vector-borne insect pathogen *Xenorhabdus nematophila*. *Proceedings of the Royal Society B: Biological Sciences* **279**: 2672–2680. https://doi.org/10.1098/rspb.2012.0228.

Cheesbrough, M. (2009). *District Laboratory Practice in Tropical Countries. Part 1*, 2nde. Cambridge, UK: Cambridge University Press.

Cheeseman, K. and Weitzman, J.B. (2015). Host–parasite interactions: an intimate epigenetic relationship. *Cellular Microbiology* **17**: 1121–1132. https://doi.org/10.1111/cmi.12471.

Chen, H.Y., Cheng, Y.S., and Shih, H.H. (2018). Heat shock proteins: role, functions and structure in parasitic helminths. In: *Heat Shock Proteins in Veterinary Medicine and Sciences* (ed. A. Asea and P. Kaur), 339–348. Cham, Switzerland: Springer https://doi.org/10.1007/978-3-319-73377-7_12.

Chi, J.F., Lawson, B., Durrant, C. et al. (2013). The finch epidemic strain of *Trichomonas gallinae* is predominant in British non-passerines. *Parasitology* **140**: 1234–1245. https://doi.org/10.1017/S0031182013000930.

Chin, A., Guo, F.C., Bernier, N.J. et al. (2004). Effect of *Cryptobia salmositica*-induced anorexia on feeding behaviour and immune response in juvenile rainbow trout *Oncorhynchus mykiss*. *Diseases of Aquatic Organisms* **58**: 17–26. https://doi.org/10.3354/dao058017.

Ching, X.T., Fong, M.Y., and Lau, Y.L. (2016). Evaluation of immunoprotection conferred by the subunit vaccines of GRA2 and GRA5 against acute toxoplasmosis in BALB/c mice. *Frontiers in Microbiology* **7**: https://doi.org/10.3389/fmicb.2016.00609.

Cifre, S., Gozalbo, M., Ortiz, V. et al. (2018). *Blastocystis* subtypes and their association with irritable bowel syndrome. *Medical Hypotheses* **116**: 4–9. https://doi.org/10.1016/j.mehy.2018.04.006.

Clark, C.G. and Diamond, L.S. (1992). Colonization of the uterus by the oral protozoan *Entamoeba gingivalis*. *American Journal of Tropical Medicine and Hygiene* **46**: 158–160. https://doi.org/10.4269/ajtmh.1992.46.158.

Clark, E. and Serpa, J.A. (2019). Tropical diseases in HIV. *Current Treatment Options in Infectious Diseases* **11**: 215–232. https://doi.org/10.1007/s40506-019-00194-5.

Clark, I.A., Alleva, L.M., Budd, A.C. et al. (2008). Understanding the role of inflammatory cytokines in malaria and related diseases. *Travel Medicine and Infectious Disease* **6**: 67–81. https://doi.org/10.1016/j.tmaid.2007.07.002.

Clayton, D.H. and Tompkins, D.M. (1995). Comparative effects of mites and lice on the reproductive success of rock doves (*Columba livia*). *Parasitology* **110**: 195–206. https://doi.org/10.1017/S0031182000063964.

Cleveland, C.A., Eberhard, M.L., Thompson, A.T. et al. (2017). Possible role of fish as transport hosts for *Dracunculus* spp. larvae. *Emerging Infectious Diseases* **23**: 1590–1592. https://doi.org/10.3201/eid2309.161931.

Cleveland, C.A., Garrett, K.B., Cozad, R.A. et al. (2018). The wild world of Guinea Worms: a review of the genus *Dracunculus* in wildlife. *International Journal for Parasitology: Parasites and Wildlife* **7**: 289–300. https://doi.org/10.1016/j.ijppaw.2018.07.002.

Clough, D., Prykhodko, O., and Råberg, L. (2016). Effects of protein malnutrition on tolerance to helminth infection. *Biology Letters* **12**: https://doi.org/10.1098/rsbl.2016.0189.

Cockburn, I.A., Mackinnon, M.J., O'Donnell, A. et al. (2004). A human complement receptor 1 polymorphism that reduces *Plasmodium falciparum* rosetting confers protection against severe malaria. *Proceedings of the National Academy of Sciences of the United States of America* **101**: 272–277. https://doi.org/10.1073/pnas.0305306101.

Codrington, R.H. (1881). Religious beliefs and practices in Melanesia. *Journal of the Anthropological Institute of Great Britain and Ireland* **10**: 261–316. https://www.jstor.org/stable/2841527.

Coelho, P.N. and Henry, R. (2017). Copepods against *Aedes* mosquitoes: a very risky strategy. *BioScience* **67**. 489 489. https://doi.org/10.1093/biosci/bix006.

Coetzee, N., von Grüning, H., Opperman, D. et al. (2020). Epigenetic inhibitors target multiple stages of *Plasmodium falciparum* parasites. *Scientific Reports* **10**: 2355. https://doi.org/10.1038/s41598-020-59298-4.

Colebunders, R., Abd-Elfarag, G., Carter, J.Y. et al. (2018). Clinical characteristics of onchocerciasis-associated epilepsy in villages in Maridi County, Republic of South Sudan. *Seizure* **62**: 108–115. https://doi.org/10.1016/j.seizure.2018.10.004.

Coleman, W.B. and Tsongalis, G.J. (2017). *Molecular Pathology: The Molecular Basis of Human Disease*, 2nde. London, UK: Academic Press.

Colwell, D.D., Hall, M.J.R., and Scholl, P.J. (2006). *The Oestrid Flies. Biology, Host-Parasite Relationships, Impact and Management*. Wallingford, UK: CABI Publishing.

Coman, B.J. and Rickard, M.D. (1975). The location of *Taenia pisiformis*, *Taenia ovis* and *Taenia hydatigena* in the gut of the dog and its effect on net environmental contamination with ova. *Zeitschrift für Parasitenkunde* **47**: 237–248. https://doi.org/10.1007/BF00389883.

Cong, W., Zhang, N.Z., Hou, J.L. et al. (2017). First detection and genetic characterization of *Toxoplasma gondii* in market-sold oysters in China. *Infection, Genetics and Evolution* **54**: 276–278. https://doi.org/10.1016/j.meegid.2017.07.014.

Conlan, J.V., Vongxay, K., Fenwick, S. et al. (2009). Does interspecific competition have a moderating effect on *Taenia solium* transmission dynamics in Southeast Asia? *Trends in Parasitology* **25**: 398–403. https://doi.org/10.1016/j.pt.2009.06.005.

Connor, B.A., Johnson, E.J., and Soave, R. (2001). Reiter syndrome following protracted symptoms of *Cyclospora* infection. *Emerging Infectious Diseases* **7**: 453–454. https://doi.org/10.3201/eid0703.010317.

Conor, D.H., George, G.H., and Gibson, D.W. (1985). Pathologic changes in human onchocerciasis: implications for future research. *Reviews of Infectious Diseases* **7**: 809–819. https://doi.org/10.1093/clinids/7.6.809.

Conrad, P.A., Miller, M.A., Kreuder, C. et al. (2005). Transmission of *Toxoplasma*: clues from the study of sea otters as sentinels of *Toxoplasma gondii* flow into the marine environment. *International Journal for Parasitology* **35**: 1155–1168. https://doi.org/10.1016/j.ijpara.2005.07.002.

Conway Morris, S. (1981). Parasites and the fossil record. *Parasitology* **82**: 489–509. https://doi.org/10.1017/S0031182000067020.

Cook, G.C. (1994). Enterobius vermicularis infection. *Gut* **35**: 1159–1162. https://doi.org/10.1136/gut.35.9.1159.

Cope, J.R., Landa, J., Nethercut, H. et al. (2019). The epidemiology and clinical features of *Balamuthia mandrillaris* disease in the United States, 1974–2016. *Clinical Infectious Diseases* **68**: 1815–1822. https://doi.org/10.1093/cid/ciy813.

Corcuera, M.T., Rodríguez-Bobada, C., Zuloaga, J. et al. (2018). Exploring tumourigenic potential of the parasite *Anisakis*: a pilot study. *Parasitology Research* **117**: 3127–3136. https://doi.org/10.1007/s00436-018-6008-2.

Corrêa Soares, G.H., Santos da Silva, A.B., de Sousa Ferreira, L.S. et al. (2020). Case report: coinfection by *Leishmania amazonensis* and HIV in a Brazilian diffuse cutaneous leishmaniasis patient. *American Journal of Tropical Medicine and Hygiene* **103**: 1076–1080. https://doi.org/10.4269/ajtmh.20-0131.

Corsini, M., Geissbühler, U., Howard, J. et al. (2015). Clinical presentation, diagnosis, therapy and outcome of alveolar echinococcosis in dogs. *Veterinary Record* **177**: 569–569. https://doi.org/10.1136/vr.103470.

Cortez-Aguirre, G.R., Jiménez-Coello, M., Gutiérrez-Blanco, E. et al. (2018). Stray dog population in a city of Southern Mexico and its impact on the contamination of public areas. *Veterinary Medicine International* **2018**: https://doi.org/10.1155/2018/2381583.

Costain, A.H., MacDonald, A.S., and Smits, H.H. (2018). Schistosome egg migration: mechanisms, pathogenesis and host immune responses. *Frontiers in Immunology* **9**: https://doi.org/10.3389/fimmu.2018.03042.

Costa-Macedo, L.M.D. and Rey, L. (1990). *Ascaris lumbricoides* in neonate: evidence of congenital transmission of intestinal nematodes. *Revista do Instituto de Medicina Tropical de São Paulo* **32**: 351–354. https://doi.org/10.1590/S0036-46651990000500007.

Cotton, J.A., Bennuru, S., Grote, A. et al. (2016). The genome of *Onchocerca volvulus*, agent of river blindness. *Nature Microbiology* **2**: 16216. https://doi.org/10.1038/nmicrobiol.2016.216.

Coulibaly, J.T., N'Goran, E.K., Utzinger, J. et al. (2013). A new rapid diagnostic test for detection of anti-*Schistosoma mansoni* and anti-*Schistosoma haematobium* antibodies. *Parasites & Vectors* **6**: 1–8. https://doi.org/10.1186/1756-3305-6-29.

Craig, A.G., Grau, G.E., Janse, C. et al. (2012). The role of animal models for research on severe malaria. *PLoS Pathogens* **8**: https://doi.org/10.1371/journal.ppat.1002401.

Craig, P.S. and Larrieu, E. (2006). Control of cystic echinococcosis/hydatidosis: 1863–2002. *Advances in Parasitology* **61**: 443–508. https://doi.org/10.1016/S0065-308X(05)61011-1.

Cressey, R. and Boxshall, G. (1989). *Kabatarina pattersoni*, a fossil parasitic copepod (Dichelesthiidae) from a Lower Cretaceous fish. *Micropaleontology* **35**: 150–167. https://doi.org/10.2307/1485466.

Criscione, C.D., Cooper, B., and Blouin, M.S. (2006). Parasite genotypes identify source populations of migratory fish more accurately than fish genotypes. *Ecology* **87**: 823–828. https://doi.org/10.1890/0012-9658(2006)87[823:PGISPO]2.0.CO;2.

Croese, J., O'neil, J., Masson, J. et al. (2006). A proof of concept study establishing *Necator americanus* in Crohn's patients and reservoir donors. *Gut* **55**: 136–137. https://doi.org/10.1136/gut.2005.079129.

Croft, A.M. (2007). A lesson learnt: the rise and fall of Lariam and Halfan. *Journal of the Royal Society of Medicine* **100**: 170–174. https://doi.org/10.1177/014107680710011411.

Croft, A.M., Bager, P., and Garg, S.K. (2012). Helminth therapy (worms) for allergic rhinitis. *Cochrane Database of Systematic Reviews* **4**: 10.1002/14651858.CD009238.pub2.

Cruz-Saavedra, L., Díaz-Roa, A., Gaona, M.A. et al. (2016). The effect of *Lucilia sericata*-and *Sarconesiopsis magellanica*-derived larval therapy on *Leishmania panamensis*. *Acta Tropica* **164**: 280–289. https://doi.org/10.1016/j.actatropica.2016.09.020.

Cunningham, J., Hasker, E., Das, P. et al. (2012). A global comparative evaluation of commercial immunochromatographic rapid diagnostic tests for visceral leishmaniasis. *Clinical Infectious Diseases* **55**: 1312–1319. https://doi.org/10.1093/cid/cis716.

Cunningham, J., Jones, S., Gatton, M.L. et al. (2019). A review of the WHO malaria rapid diagnostic test product testing programme (2008–2018): performance, procurement and policy. *Malaria Journal* **18**: 1–15. https://doi.org/10.1186/s12936-019-3028-z.

Cuypers, B., Van den Broeck, F., Van Reet, N. et al. (2017). Genome-wide SNP analysis reveals distinct origins of *Trypanosoma evansi* and *Trypanosoma equiperdum*. *Genome Biology and Evolution* **9**: 1990–1997. https://doi.org/10.1093/gbe/evx102.

Cytryńska, M., Rahnamaeian, M., Zdybicka-Barabas, A. et al. (2020). Proline-rich antimicrobial peptides in medicinal maggots of *Lucilia sericata* interact with bacterial DnaK but do not inhibit protein synthesis. *Frontiers in Pharmacology* **11**: 532. https://doi.org/10.3389/fphar.2020.00532.

da Silva, D.F., da Silva, R.J., da Silva, M.G. et al. (2007). Parasitic infection of the appendix as a cause of acute appendicitis. *Parasitology Research* **102**: 99–102. https://doi.org/10.1007/s00436-007-0735-0.

da Silva, M.S., Viviescas, M.A., Pavani, R.S. et al. (2017). Nuclear and kinetoplast DNA replication in trypanosomatids. In: *Frontiers in Parasitology: Molecular and Cellular Biology of Pathogenic Trypanosomatids* (ed. M.S. da Silva and M.I.N. Cano), 134–194. Sharjah, UAE: Bentham Books.

da Silveira, W.F., Braga, F.R., de Oliveira Tavela, A. et al. (2017). Nematophagous fungi combinations reduce free-living stages of sheep gastrointestinal nematodes in the field. *Journal of Invertebrate Pathology* **150**: 1–5. https://doi.org/10.1016/j.jip.2017.08.013.

Da'dara, A.A. and Skelly, P.J. (2015). Gene suppression in schistosomes using RNAi. In: *Parasite Genomics Protocols. Methods in Molecular Biology*, vol. **1201** (ed. C. Peacock). New York, NY: Humana Press https://doi.org/10.1007/978-1-4939-1438-8_8.

Dagci, H., Zeyrek, F., Gerzile, Y.K. et al. (2008). A case of myiasis in a patient with psoriasis from Turkey. *Parasitology International* **57**: 239–241. https://doi.org/10.1016/j.parint.2007.12.010.

Dakubo, J.C.B., Naaeder, S.B., and Kumodji, R. (2008). Totemism and the transmission of human pentastomiasis. *Ghana Medical Journal* **42**: 165–168. PMID: 19452026; PMCID: PMC2673832.

Darby, M., Schnoeller, C., Vira, A. et al. (2015). The M3 muscarinic receptor is required for optimal adaptive immunity to helminth and bacterial infection. *PLoS Pathogens* **11**: https://doi.org/10.1371/journal.ppat.1004636.

Das, S., Halder, A., Mandal, S. et al. (2018). Andrographolide engineered gold nanoparticle to overcome drug resistant visceral leishmaniasis. *Artificial Cells, Nanomedicine, and Biotechnology* **46**: 751–762. https://doi.org/10.1080/21691401.2018.1435549.

Das, V.N.R., Pandey, K., Verma, N. et al. (2009). Development of Post–Kala-Azar Dermal Leishmaniasis (PKDL) in Miltefosine-treated visceral leishmaniasis. *The American Journal of Tropical Medicine and Hygiene* **80**: 336–338. https://doi.org/10.4269/ajtmh.2009.80.336.

Davies, I.M., McHenery, J.G., and Rae, G.H. (1997). Environmental risk from dissolved ivermectin to marine organisms. *Aquaculture* **158**: 263–275. https://doi.org/10.1016/S0044-8486(97)00209-3.

de A.P. Corrêa, S., de Oliveira, R.N., Mendes, T.M. et al. (2019). *in vitro* and *in vivo* evaluation of six artemisinin derivatives against *Schistosoma mansoni*. *Parasitology Research* **118**: 505–516. https://doi.org/10.1007/s00436-018-6188-9.

De Bona, E., Lidani, K.C.F., Bavia, L. et al. (2018). Autoimmunity in chronic Chagas disease: a road of multiple pathways to cardiomyopathy? *Frontiers in Immunology* **9**: 1842. https://doi.org/10.3389/fimmu.2018.01842.

De Bont, J. and Vercruysse, J. (1998). Schistosomiasis in cattle. *Advances in Parasitology* **41**: 285–364. https://doi.org/10.1016/S0065-308X(08)60426-1.

De Brito, R.C., Jamille, M.D.O., Reis, L.E. et al. (2018). Peptide vaccines for leishmaniasis. *Frontiers in Immunology* **9**: https://doi.org/10.3389/fimmu.2018.01043.

de Dood, C.J., Hoekstra, P.T., Mngara, J. et al. (2018). Refining diagnosis of *Schistosoma haematobium* infections: antigen and antibody detection in urine. *Frontiers in Immunology* **9**: https://doi.org/10.3389/fimmu.2018.02635.

de Freitas Soares, F.E., Sufiate, B.L., and de Queiroz, J.H. (2018). Nematophagous fungi: far beyond the endoparasite, predator and ovicidal groups. *Agriculture and Natural Resources* **52**: 1–8. https://doi.org/10.1016/j.anres.2018.05.010.

De Kyvon, M.A.L.C., Maakaroun-Vermesse, Z., Lanotte, P. et al. (2016). Congenital trypanosomiasis in child born in France to African mother. *Emerging Infectious Diseases* **22**: 935–937. https://doi.org/10.3201/eid2205.160133.

de León, G.P.P. and Hernández-Mena, D.I. (2019). Testing the higher-level phylogenetic classification of Digenea (Platyhelminthes, Trematoda) based on nuclear rDNA sequences before entering the age of the 'next-generation' Tree of Life. *Journal of Helminthology* **93**: 260–276. https://doi.org/10.1017/S0022149X19000191.

de Oliveira, A.C., Soccol, V.T., and Rogez, H. (2019). Prevention methods of foodborne Chagas disease: disinfection, heat treatment and quality control byRT-PCR. *International Journal of Food Microbiology* **301**: 34–40. https://doi.org/10.1016/j.ijfoodmicro.2019.04.009.

de Queiroz, A. and Alkire, N. (1998). The phylogenetic placement of *Taenia* cestodes that parasitize humans. *Journal of Parasitology* **84**: 379–383. https://www.jstor.org/stable/3284501.

De Ruiter, K., Tahapary, D.L., Sartono, E. et al. (2017). Helminths, hygiene hypothesis and type 2 diabetes. *Parasite Immunology* **39**: https://doi.org/10.1111/pim.12404.

de Souza, J.N., Oliveira, C.D.L., Araújo, W.A. et al. (2020). *Strongyloides stercoralis* in alcoholic patients: implications of alcohol intake in the frequency of infection and parasite load. *Pathogens* **9**: 422. https://doi.org/10.3390/pathogens9060422.

de Souza, T.K., Soares, S.S., Benitez, L.B. et al. (2017). Interaction between Methicillin-Resistant *Staphylococcus aureus* (MRSA) and *Acanthamoeba polyphaga*. *Current Microbiology* **74**: 541–549. https://doi.org/10.1007/s00284-017-1196-z.

Debrah, A.Y., Specht, S., Klarmann-Schulz, U. et al. (2015). Doxycycline leads to sterility and enhanced killing of female *Onchocerca volvulus* worms in an area with persistent microfilaridermia after repeated ivermectin treatment: a randomized, placebo-controlled, double-blind trial. *Clinical Infectious Diseases* **61**: 517–526. https://doi.org/10.1093/cid/civ363.

Deere, J.R., Parsons, M.B., Lonsdorf, E.V. et al. (2019). *Entamoeba histolytica* infection in humans, chimpanzees and baboons in the Greater Gombe Ecosystem, Tanzania. *Parasitology* **146**: 1116–1122. https://doi.org/10.1017/S0031182018001397[Opens in a new window].

Del Castillo, M., Szymanski, A.M., Slovin, A. et al. (2017). Congenital *Plasmodium falciparum* malaria in Washington, DC. *American Journal of Tropical Medicine and Hygiene* **96**: 167–169. https://doi.org/10.4269/ajtmh.15-0630.

Del Grande, C., Galli, L., Schiavi, E. et al. (2017). Is *Toxoplasma gondii* a trigger of bipolar disorder? *Pathogens* **6**: https://doi.org/10.3390/pathogens6010003.

Del Puerto, F., Ozorio, M., Trinidad, B. et al. (2020). Detection and characterization of *Plasmodium* spp. by semi-nested multiplex PCR both in mosquito vectors and in humans residing in historically endemic areas of Paraguay. *Parasite Epidemiology and Control* **11**: https://doi.org/10.1016/j.parepi.2020.e00174.

Del Valle, A., Jones, B.F., Harrison, L.M. et al. (2003). Isolation and molecular cloning of a secreted hookworm platelet inhibitor from adult *Ancylostoma caninum*. *Molecular and Biochemical Parasitology* **129**: 167–177. https://doi.org/10.1016/S0166-6851(03)00121-X.

Delves, P.J., Martin, S.J., Burton, D.R. et al. (2017). *Roitt's Essential Immunology*, 13the. Chichester, UK: Wiley-Blackwell.

Deng, R., Lin, D., Zhu, L. et al. (2017). Nanoparticle interactions with co-existing contaminants: joint toxicity, bioaccumulation and risk. *Nanotoxicology* **11**: 591–612. https://doi.org/10.1080/17435390.2017.1343404.

Deng, Y., Wu, T., Zhai, S.Q. et al. (2019). Recent progress on anti-toxoplasma drugs discovery: design, synthesis and screening. *European Journal of Medicinal Chemistry* **183**: https://doi.org/10.1016/j.ejmech.2019.111711.

Deribe, K., Cano, J., Trueba, M.L. et al. (2018). Global epidemiology of podoconiosis: A systematic review. *PLoS Neglected Tropical Diseases* **12**: https://doi.org/10.1371/journal.pntd.0006324.

Desowitz, R.S. (1987). *New Guinea Tapeworms and Jewish Grandmothers Tales of Parasites and People*. New York, NY: W.W. Norton Press.

Desportes, I., Charpentier, Y.L., Galian, A. et al. (1985). Occurrence of a new microsporidian: *Enterocytozoon bieneusi* ng, n. sp., in the enterocytes of a human patient with AIDS 1. *Journal of Protozoology* **32**: 250–254. https://doi.org/10.1111/j.1550-7408.1985.tb03046.x.

Desportes, I. and Schrével, J. (2013). *Treatise on Zoology-Anatomy, Taxonomy, Biology. The Gregarines*, vol. **2 vols**. Leiden: Brill.

Dever, M.L., Kahn, L.P., and Doyle, E.K. (2015). Removal of tapeworm (*Moniezia* spp.) did not increase growth rates of meat-breed lambs in the Northern Tablelands of NSW. *Veterinary Parasitology* **208**: 190–194. https://doi.org/10.1016/j.vetpar.2015.01.016.

Devleesschauwer, B., Allepuz, A., Dermauw, V. et al. (2017). *Taenia solium* in Europe: still endemic? *Acta Tropica* **165**: 96–99. https://doi.org/10.1016/j.actatropica.2015.08.006.

Dholakia, S., Buckler, J., Jeans, J.P. et al. (2014). Pubic lice: an endangered species? *Sexually Transmitted Diseases* **41**: 388–391. https://doi.org/10.1097/OLQ.0000000000000142.

Di Campli, E., Di Bartolomeo, S., Pizzi, P.D. et al. (2012). Activity of tea tree oil and nerolidol alone or in combination against *Pediculus capitis* (head lice) and its eggs. *Parasitology Research* **111**: 1985–1992. https://doi.org/10.1007/s00436-012-3045-0.

Díaz-Roa, A., Gaona, M.A., Segura, N.A. et al. (2016). Evaluating *Sarconesiopsis magellanica* blowfly-derived larval therapy and comparing it to *Lucilia sericata*-derived therapy in an animal model. *Acta Tropica* **154**: 34–41. https://doi.org/10.1016/j.actatropica.2015.10.024.

DiNardo, A.R., Mace, E.M., Lesteberg, K. et al. (2016). Schistosome soluble egg antigen decreases *Mycobacterium tuberculosis*–specific CD4+ T-cell effector function with concomitant arrest of macrophage phago-lysosome maturation. *Journal of Infectious Diseases* **214**: 479–488. https://doi.org/10.1093/infdis/jiw156.

Diro, E., Lynen, L., Ritmeijer, K. et al. (2014). Visceral leishmaniasis and HIV coinfection in East Africa. *PLoS Neglected Tropical Diseases* **8**: https://doi.org/10.1371/journal.pntd.0002869.

Dixon, B., Mihajlovic, B., Couture, H. et al. (2016). Qualitative risk assessment: *Cyclospora cayetanensis* on fresh raspberries and blackberries imported into Canada. *Food Protection Trends* **36**: 18–32.

Dobrowsky, P.H., Khan, S., Cloete, T.E. et al. (2016). Molecular detection of *Acanthamoeba* spp., *Naegleria fowleri* and *Vermamoeba* (*Hartmanella*) *vermiformis* as vectors for *Legionella* spp. in untreated and solar pasteurized harvested rainwater. *Parasites & Vectors* **9**: 539. https://doi.org/10.1186/s13071-016-1829-2.

Dolgin, E. (2017). Climate change: as the ice melts. *Nature* **543**: S54–S55. https://doi.org/10.1038/543S54a.

Dominey, A., Tschen, J., Rosen, T. et al. (1989). Pityriasis folliculorum revisited. *Journal of the American Academy of Dermatology* **21**: 81–84. https://doi.org/10.1016/S0190-9622(89)70152-3.

Doncaster, C.C. (1981). Observations on relationships between infective juveniles of bovine lungworm, *Dictyocaulus viviparus* (Nematoda: Strongylida) and the fungi, *Pilobolus kleinii* and *P. crystallinus* (Zygomycotina: Zygomycetes). *Parasitology* **82**: 421–428. https://doi.org/10.1017/S0031182000066956.

Dong, Y. and Dimopoulos, G. (2009). *Anopheles* fibrinogen-related proteins provide expanded pattern recognition capacity against bacteria and malaria parasites. *Journal of Biological Chemistry* **284**: 9835–9844. https://doi.org/10.1074/jbc.M807084200.

Donovan, S.K. (2015). A prejudiced review of ancient parasites and their host echinoderms: CSI Fossil Record or just an excuse for speculation? *Advances in Parasitology* **90**: 291–328. https://doi.org/10.1016/bs.apar.2015.05.003.

Doolan, D.L., Dobaño, C., and Baird, J.K. (2009). Acquired immunity to malaria. *Clinical Microbiology Reviews* **22**: 13–36. https://doi.org/10.1128/CMR.00025-08.

Doonan, J., Tarafdar, A., Pineda, M.A. et al. (2019). The parasitic worm product ES-62 normalises the gut microbiota bone marrow axis in inflammatory arthritis. *Nature Communications* **10**: 1–14. https://doi.org/10.1038/s41467-019-09361-0.

Dorchies, P., Duranton, C., and Jacquiet, P.H. (1998). Pathophysiology of *Oestrus ovis* infection in sheep and goats: a review. *Veterinary Record* **142**: 487–489. https://doi.org/10.1136/vr.142.18.487.

Dorchies, P., Wahetra, S., Lepetitcolin, E. et al. (2003). The relationship between nasal myiasis and the prevalence of enzootic nasal tumours and the effects of treatment of *Oestrus ovis* and milk production in dairy ewes of Roquefort cheese area. *Veterinary Parasitology* **113**: 169–174. https://doi.org/10.1016/S0304-4017(03)00032-3.

dos Santos, T.R., Maia, M.O., and da Costa, A.J. (2018). Congenital toxoplasmosis in ewes. In: *Congenital Toxoplasmosis in Humans and Domestic Animals* (ed. K.D.S. Bresciani and A.J. da Costa), 75–81. Sharjah, UAE: Bentham Science Publishers.

Douvres, F.W., Tromba, F.G., and Malakatis, G.M. (1969). Morphogenesis and migration of *Ascaris suum* larvae developing to fourth stage in swine. *Journal of Parasitology* **55**: 689–712. https://www.jstor.org/stable/3277198.

Downs, J.A., Mguta, C., Kaatano, G.M. et al. (2011). Urogenital schistosomiasis in women of reproductive age in Tanzania's Lake Victoria region. *American Journal of Tropical Medicine and Hygiene* **84**: 364–369. https://doi.org/10.4269/ajtmh.2011.10-0585.

Doyle, S.R. and Cotton, J.A. (2019). Genome-wide approaches to investigate anthelmintic resistance. *Trends in Parasitology* **35**: 289–301. https://doi.org/10.1016/j.pt.2019.01.004.

Drakeley, C. and Reyburn, H. (2009). Out with the old, in with the new: the utility of rapid diagnostic tests for malaria diagnosis in Africa. *Transactions of the Royal Society of Tropical Medicine and Hygiene* **103**: 333–337. https://doi.org/10.1016/j.trstmh.2008.10.003.

Dreyer, G., Norões, J., and Addiss, D. (1997). The silent burden of sexual disability associated with lymphatic filariasis. *Acta Tropica* **63**: 57. PMID: 9083585. https://doi.org/10.1016/s0001-706x(96)00604-3.

Dreyer, G., Noroes, J., Figueredo-Silva, J. et al. (2000). Pathogenesis of lymphatic disease in bancroftian filariasis: a clinical perspective. *Parasitology Today* **16**: 544–548. PMID: 11121854. https://doi.org/10.1016/s0169-4758(00)01778-6.

Drezen, J.M., Leobold, M., Bézier, A. et al. (2017). Endogenous viruses of parasitic wasps: variations on a common theme. *Current Opinion in Virology* **25**: 41–48. https://doi.org/10.1016/j.coviro.2017.07.002.

Driscoll, M.S., Rothe, M.J., Grant-Kels, J.M. et al. (1993). Delusional parasitosis: a dermatologic, psychiatric, and pharmacologic approach. *Journal of the American Academy of Dermatology* **29**: 1023–1033. PMID: 7902366. https://doi.org/10.1016/0190-9622(93)70284-z.

Drisdelle, R. (2011). *Parasites: Tales of Humanity's Most Unwelcome Guests*. Berkeley, CA: University of California Press.

Drurey, C., Coakley, G., and Maizels, R.M. (2020). Extracellular vesicles: new targets for vaccines against helminth parasites. *International Journal for Parasitology* **50**: 623–633. https://doi.org/10.1016/j.ijpara.2020.04.011.

Du, R., Hotez, P.J., Al-Salem, W.S. et al. (2016). Old world cutaneous leishmaniasis and refugee crises in the Middle East and North Africa. *PLoS Neglected Tropical Diseases* **10**: https://doi.org/10.1371/journal.pntd.0004545.

Dubey, J.P. (2002). A review of toxoplasmosis in wild birds. *Veterinary Parasitology* **106**: 121–153. https://doi.org/10.1016/S0304-4017(02)00034-1.

Dubey, J.P. and Almeria, S. (2019). *Cystoisospora belli* infections in humans: the past 100 years. *Parasitology* **146**: 1490–1527. https://doi.org/10.1017/S0031182019000957.

Ducournau, C., Moiré, N., Carpentier, R. et al. (2020). Effective nanoparticle-based nasal vaccine against latent and congenital toxoplasmosis in sheep. *Frontiers in Immunology* **11**: https://doi.org/10.3389/fimmu.2020.02183.

Dudek, B.M., Kochert, M.N., Barnes, J.G. et al. (2018). Prevalence and risk factors of *Trichomonas gallinae* and trichomonosis in golden eagle (*Aquila chrysaetos*) nestlings in Western North America. *Journal of Wildlife Diseases* **54**: 755–764. https://doi.org/10.7589/2017-11-271.

Duggal, P., Guo, X., Haque, R. et al. (2011). A mutation in the leptin receptor is associated with *Entamoeba histolytica* infection in children. *Journal of Clinical Investigation* **121**: 1191–1198. PMID: 21393862. https://doi.org/10.1172/JCI45294.

Dunn, A.M., Terry, R.S., and Smith, J.E. (2001). Transovarial transmission in the microsporidia. *Advances in Parasitology* **48**: 57–100. https://doi.org/10.1016/S0065-308X(01)48005-5.

Duque-Correa, M.A., Maizels, R.M., Grencis, R.K. et al. (2020). Organoids–new models for host–helminth interactions. *Trends in Parasitology* **36**: 170–181. https://doi.org/10.1016/j.pt.2019.10.013.

Duthie, M.S., Pereira, L., Favila, M. et al. (2017). A defined subunit vaccine that protects against vector-borne visceral leishmaniasis. *NPJ Vaccines* **2**: 10.1038/s41541-017-0025-5.

Easton, A.V., Quiñones, M., Vujkovic-Cvijin, I. et al. (2019). The impact of anthelmintic treatment on human gut microbiota based on cross-sectional and pre-and post-deworming comparisons in Western Kenya. *MBio* **10**: e00519–e00519. https://doi.org/10.1128/mBio.00519-19.

Efros, A.L. (2019). Quantum dots realize their potential. *Nature* **575**: 604–605. https://doi.org/10.1038/d41586-019-03607-z.

Elgharably, A., Gomaa, A.I., Crossey, M.M. et al. (2016). Hepatitis C in Egypt – past, present, and future. *International Journal of General Medicine* **10**: 1–6. https://doi.org/10.2147/IJGM.S119301.

Elkington, R.A., Humphries, M., Commins, M. et al. (2009). A *Lucilia cuprina* excretory–secretory protein inhibits the early phase of lymphocyte activation and subsequent proliferation. *Parasite Immunology* **31**: 750–765. https://doi.org/10.1111/j.1365-3024.2009.01154.x.

Epstein, J.E., Paolino, K.M., Richie, T.L. et al. (2017). Protection against *Plasmodium falciparum* malaria by PfSPZ Vaccine. *JCI Insight* **2**: https://doi.org/10.1172/jci.insight.89154.

Ergun, U.G., Celik, M., and Ozer, H.T.E. (2007). Reactive arthritis due to zoophilic (canine) sexual intercourse. *International Journal of STD and AIDS* **18**: 285–286. https://doi.org/10.1258/095646207780658890.

Eritja, R., Palmer, J.R., Roiz, D. et al. (2017). Direct evidence of adult *Aedes albopictus* dispersal by car. *Scientific Reports* **7**: 14399. https://doi.org/10.1038/s41598-017-12652-5.

Escalante, A.A. and Ayala, F.J. (1995). Evolutionary origin of *Plasmodium* and other Apicomplexa based on rRNA genes. *Proceedings of National Academy of Sciences of the United States of America* **92**: 5793–5797. https://doi.org/10.1073/pnas.92.13.5793.

Escobedo, A.A., Almirall, P., Alfonso, M. et al. (2014). Sexual transmission of giardiasis: a neglected route of spread? *Acta Tropica* **132**: 106–111. https://doi.org/10.1016/j.actatropica.2013.12.025.

Escolà-Vergé, L., Arando, M., Vall, M. et al. (2017). Outbreak of intestinal amoebiasis among men who have sex with men, Barcelona (Spain), October 2016 and January 2017. *Euro surveillance: bulletin Europeen sur les maladies transmissibles = European Communicable Disease Bulletin* **22**: 30581. 10.2807/1560-7917.ES.2017.22.30.30581.

Escueta-de Cadiz, A., Kobayashi, S., Takeuchi, T. et al. (2010). Identification of an avirulent *Entamoeba histolytica* strain with unique tRNA-linked short tandem repeat markers. *Parasitology International* **59**: 75–81. https://doi.org/10.1016/j.parint.2009.10.010.

Espinoza-Jiménez, A., De Haro, R., and Terrazas, L.I. (2017). *Taenia crassiceps* antigens control experimental type 1 diabetes by inducing alternatively activated macrophages. *Mediators of Inflammation* **2017**: https://doi.org/10.1155/2017/8074329.

Eugenin, E.A., Martiney, J.A., and Berman, J.W. (2019). The malaria toxin hemozoin induces apoptosis in human neurons and astrocytes: potential role in the pathogenesis of cerebral malaria. *Brain Research* **1720**: https://doi.org/10.1016/j.brainres.2019.146317.

Faya, N., Penkler, D.L., and Bishop, Ö.T. (2015). Human, vector and parasite Hsp90 proteins: a comparative bioinformatics analysis. *FEBS Open Bio* **5**: 916–927. https://doi.org/10.1016/j.fob.2015.11.003.

Fayer, R., Esposito, D.H., and Dubey, J.P. (2015). Human infections with *Sarcocystis* species. *Clinical Microbiology Reviews* **28**: 295–311. https://doi.org/10.1128/CMR.00113-14.

Fayer, R., Speer, C.A., and Dubey, J.P. (2018). General biology of *Cryptosporidium*. In: *Cryptosporidiosis of Man and Animals* (ed. J.P. Dubey), 1–30. Boca Raton, FL: CRC Press.

Feary, J.R., Venn, A.J., Mortimer, K. et al. (2010). Experimental hookworm infection: a randomized placebo-controlled trial in asthma. *Clinical and Experimental Allergy* **40**: 299–306. https://doi.org/10.1111/j.1365-2222.2009.03433.x.

Feng, L.U., Xin-Long, H.E., Richard, C. et al. (2019). A brief history of artemisinin: Modes of action and mechanisms of resistance. *Chinese Journal of Natural Medicines* **17**: 331–336. https://doi.org/10.1016/S1875-5364(19)30038-X.

Fernandes, C.F., Pereira, S.D.S., Luiz, M.B. et al. (2017). Camelid single-domain antibodies as an alternative to overcome challenges related to the prevention, detection, and control of neglected tropical diseases. *Frontiers in Immunology* **8**: 653. https://doi.org/10.3389/fimmu.2017.00653.

Fernández, C., Porcel, M.A., Alonso, A. et al. (2011). Semifield assessment of the runoff potential and environmental risk of the parasiticide drug ivermectin under Mediterranean conditions.

Environmental Science and Pollution Research **18**: 1194–1201. https://doi.org/10.1007/s11356-011-0474-8.

Fernández-Huerta, M., Zarzuela, F., Barberá, M.J. et al. (2019). Sexual transmission of intestinal parasites and other enteric pathogens among men who have sex with men presenting gastrointestinal symptoms in an STI Unit in Barcelona, Spain: a cross-sectional study. *American Journal of Tropical Medicine and Hygiene* **101**: 1388–1391. https://doi.org/10.4269/ajtmh.19-0312.

Fernández-Ruvalcaba, M., Preciado-De-La Torre, F., Cruz-Vazquez, C. et al. (2004). Anti-tick effects of *Melinis minutiflora* and *Andropogon gayanus* grasses on plots experimentally infested with *Boophilus microplus* larvae. *Experimental and Applied Acarology* **32**: 293–299. https://doi.org/10.1023/B:APPA.0000023233.63268.cc.

Ferreira, L.G., Oliva, G., and Andricopulo, A.D. (2015). Target-based molecular modeling strategies for schistosomiasis drug discovery. *Future Medicinal Chemistry* **7**: 753–764. https://doi.org/10.4155/fmc.15.21.

Field, N., Clifton, S., Alexander, S. et al. (2018). *Trichomonas vaginalis* infection is uncommon in the British general population: implications for clinical testing and public health screening. *Sexually Transmitted Infections* **94**: 226–229. https://doi.org/10.1136/sextrans-2016-052660.

Figueiredo de Sá, B.S.L., Rezende, A.M., Melo Neto, O.P.D. et al. (2019). Identification of divergent *Leishmania* (*Viannia*) *braziliensis* ecotypes derived from a geographically restricted area through whole genome analysis. *PLoS Neglected Tropical Diseases* **13**: https://doi.org/10.1371/journal.pntd.0007382.

Filho, A.O., Dias, D., Miranda, Á. et al. (2018). Oral myiasis in older adult with severe Alzheimer's disease. *Special Care in Dentistry* **38**: 99–106. https://doi.org/10.1111/scd.12277.

Finlay, C.M., Stefanska, A.M., Coleman, M.M. et al. (2017). Secreted products of *Fasciola hepatica* inhibit the induction of T cell responses that mediate allergy. *Parasite Immunology* **39**: https://doi.org/10.1111/pim.12460.

Fischer, P., Supali, T., and Maizels, R.M. (2004). Lymphatic filariasis and *Brugia timori*: prospects for elimination. *Trends in Parasitology* **20**: 351–355. https://doi.org/10.1016/j.pt.2004.06.001.

Fisher, M.C. and Garner, T.W. (2020). Chytrid fungi and global amphibian declines. *Nature Reviews Microbiology* **18**: 332–343. https://doi.org/10.1038/s41579-020-0335-x.

Flegr, J. and Kaňková, Š. (2020). The effects of toxoplasmosis on sex ratio at birth. *Early Human Development* **141**: https://doi.org/10.1016/j.earlhumdev.2019.104874.

Fleming, J., Hernandez, G., Hartman, L. et al. (2019). Safety and efficacy of helminth treatment in relapsing-remitting multiple sclerosis: results of the HINT 2 clinical trial. *Multiple Sclerosis Journal* **25**: 81–91. https://doi.org/10.1177/1352458517736377.

Fletcher-Lartey, S.M. and Caprarelli, G. (2016). Application of GIS technology in public health: successes and challenges. *Parasitology* **143**: 401–415. https://doi.org/10.1017/S0031182015001869.

Fleury, A., Cardenas, G., Adalid-Peralta, L. et al. (2016). Immunopathology in *Taenia solium* neurocysticercosis. *Parasite Immunology* **38**: 147–157. https://doi.org/10.1111/pim.12299.

Fogelman, R.M., Kuris, A.M., and Grutter, A.S. (2009). Parasitic castration of a vertebrate: effect of the cymothoid isopod, *Anilocra apogonae*, on the five-lined cardinalfish, *Cheilodipterus quinquelineatus*. *International Journal for Parasitology* **39**: 577–583. https://doi.org/10.1016/j.ijpara.2008.10.013.

Föger, F., Noonpakdee, W., Loretz, B. et al. (2006). Inhibition of malarial topoisomerase II in *Plasmodium falciparum* by antisense nanoparticles. *International Journal of Pharmaceutics* **319**: 139–146. https://doi.org/10.1016/j.ijpharm.2006.03.034.

Formenti, F., La Marca, G., Perandin, F. et al. (2019). A diagnostic study comparing conventional and real-time PCR for *Strongyloides stercoralis* on urine and on faecal samples. *Acta Tropica* **190**: 284–287. https://doi.org/10.1016/j.actatropica.2018.12.001.

Forrer, A., Khieu, V., Schär, F. et al. (2017). *Strongyloides stercoralis* is associated with significant morbidity in rural Cambodia, including stunting in children. *PLoS Neglected Tropical Diseases* **11**: https://doi.org/10.1371/journal.pntd.0005685.

Forzán, M.J., Vanderstichel, R., Melekhovets, Y.F. et al. (2010). Trichomoniasis in finches from the Canadian Maritime provinces – an emerging disease. *Canadian Veterinary Journal* **51**: 391–396. PMC2839828. PMID: 20592828.

Foster, J.M., Landmann, F., Ford, L. et al. (2014). Absence of *Wolbachia* endobacteria in the human parasitic nematode *Dracunculus medinensis* and two related *Dracunculus* species infecting wildlife. *Parasites & Vectors* **7**: 140. https://doi.org/10.1186/1756-3305-7-140.

Fried, B., Reddy, A., and Mayer, D. (2011). Helminths in human carcinogenesis. *Cancer Letters* **305**: 239–249. https://doi.org/10.1016/j.canlet.2010.07.008.

Friedman, D.J. and Pollak, M.R. (2016). Apolipoprotein L1 and kidney disease in African Americans. *Trends in Endocrinology and Metabolism: TEM* **27**: 204–215. https://doi.org/10.1016/j.tem.2016.02.002.

Fromstein, S.R., Harthan, J.S., Patel, J. et al. (2018). *Demodex* blepharitis: clinical perspectives. *Clinical Optometry* **10**: 57–63. https://doi.org/10.2147/OPTO.S142708.

Fujioka, H., Phelix, C.F., Friedland, R.P. et al. (2013). Apolipoprotein E4 prevents growth of malaria at the intraerythrocyte stage: implications for differences in racial susceptibility to Alzheimer's disease. *Journal of Health Care for the Poor and Underserved* **24**: 70–78. https://doi.org/10.1353/hpu.2014.0009.

Fuller, G. (1962). How screwworm eradication will affect wildlife: the eradication of the screwworm in the Southwest will result in a larger deer population in the region. *The Cattleman* **48**: 82–84. May.

Fumagalli, M., Pozzoli, U., Cagliani, R. et al. (2009). Parasites represent a major selective force for interleukin genes and shape the genetic predisposition to autoimmune conditions. *Journal of Experimental Medicine* **206**: 1395–1408. https://doi.org/10.1084/jem.20082779.

Furch, B.D., Koethe, J.R., Kayamba, V. et al. (2020). Interactions of *Schistosoma* and HIV in Sub-Saharan Africa: a systematic review. *American Journal of Tropical Medicine and Hygiene* **102**: 711–718. https://doi.org/10.4269/ajtmh.19-0494.

Furuya, K., Nakajima, H., Sasaki, Y. et al. (2016). A scabies outbreak in a diabetic and collagen disease ward: management and prevention. *Experimental and Therapeutic Medicine* **12**: 3711–3715. https://doi.org/10.3892/etm.2016.3845.

Galal, L., Hamidović, A., Dardé, M.L. et al. (2019). Diversity of *Toxoplasma gondii* strains at the global level and its determinants. *Food and Waterborne Parasitology* **15**: https://doi.org/10.1016/j.fawpar.2019.e00052.

Gamble, A., Ramos, R., Parra-Torres, Y. et al. (2019). Exposure of yellow-legged gulls to *Toxoplasma gondii* along the Western Mediterranean coasts: Tales from a sentinel. *International Journal for Parasitology: Parasites and Wildlife* **8**: 221–228. https://doi.org/10.1016/j.ijppaw.2019.01.002.

Gandon, S. and Day, T. (2008). Evidences of parasite evolution after vaccination. *Vaccine* **26**: C4–C7. https://doi.org/10.1016/j.vaccine.2008.02.007.

Gansser, A.W.E. (1956). *Warble Flies and other Oestridae: Biology and Control*. Surrey, UK: The Hide and Allied Trades Improvement Society.

Gao, H.W., Wang, L.P., Liang, S. et al. (2012). Change in rainfall drives malaria re-emergence in Anhui Province. *China. PLoS One* **7**: e43686. https://doi.org/10.1371/journal.pone.0043686.

Gao, L., Li, A., Li, N. et al. (2017). Innate and intrinsic immunity in planarians. *Invertebrate Survival Journal* **14**: 443–452. http://isj-new.dmz-ext.unimore.it/index.php/ISJ/article/view/71.

Garcia, H.H., Castillo, Y., Gonzales, I. et al. (2018). Low sensitivity and frequent cross-reactions in commercially available antibody detection ELISA assays for *Taenia solium* cysticercosis. *Tropical Medicine & International Health* **23**: 101–105. https://doi.org/10.1111/tmi.13010.

García, H.H., Gonzalez, A.E., Evans, C.A. et al. (2003). *Taenia solium* cysticercosis. *Lancet* **362**: 547–556. https://doi.org/10.1016/S0140-6736(03)14117-7.

Garcia, L.S. (2006). *Diagnostic Medical Parasitology*. Washington, DC: American Society for Microbiology Press.

Garcia, L.S. (2021). *Practical Guide to Diagnostic Parasitology*, 3rde. Chichester, UK: John Wiley & Sons.

Garcia, L.S., Arrowood, M., Kokoskin, E. et al. (2018). Practical guidance for clinical microbiology laboratories: laboratory diagnosis of parasites from the gastrointestinal tract. *Clinical Microbiology Reviews* **31**: https://doi.org/10.1128/CMR.00025-17.

García-Bernalt Diego, J., Fernández-Soto, P., Crego-Vicente, B. et al. (2019). Progress in loop-mediated isothermal amplification assay for detection of *Schistosoma mansoni* DNA: towards a ready-to-use test. *Nature Scientific Reports* **9**: 14744. https://doi.org/10.1038/s41598-019-51342-2.

Gargantini, P.R., del Carmen Serradell, M., Ríos, D.N. et al. (2016). Antigenic variation in the intestinal parasite *Giardia lamblia*. *Current Opinion in Microbiology* **32**: 52–58. https://doi.org/10.1016/j.mib.2016.04.017.

Garver, L.S., de Almeida Oliveira, G., and Barillas-Mury, C. (2013). The JNK pathway is a key mediator of *Anopheles gambiae* anti-plasmodial immunity. *PLoS Pathogens* **9**: https://doi.org/10.1371/journal.ppat.1003622.

Gasim, G.I., Bella, A., and Adam, I. (2015). Schistosomiasis, hepatitis B and hepatitis C co-infection. *Virology Journal* **12**: https://doi.org/10.1186/s12985-015-0251-2.

Gaugler, R. (2018). *Entomopathogenic Nematodes in Biological Control*. Boca Raton, FL: CRC Press.

Gaydos, C.A., Klausner, J.D., Pai, N.P. et al. (2017). Rapid and point-of-care tests for the diagnosis of *Trichomonas vaginalis* in women and men. *Sexually Transmitted Infections* **93**: S31–S35. https://doi.org/10.1136/sextrans-2016-053063.

Gazzinelli-Guimaraes, P.H. and Nutman, T.B. (2018). Helminth parasites and immune regulation. *F1000Research* **7**: doi: 10.12688/f1000research.15596.1.

GBD 2015 Mortality and Causes of Death Collaborators (2016). Global, regional, and national life expectancy, all-cause mortality, and cause-specific mortality for 249 causes of death, 1980–2015: a systematic analysis for the Global Burden of Disease Study 2015. *Lancet* **388**: 1459–1544. https://doi.org/10.1016/S0140-6736(16)31012-1.

Geisbert, T.W., Hensley, L.E., Jahrling, P.B. et al. (2003). Treatment of Ebola virus infection with a recombinant inhibitor of factor VIIa/tissue factor: a study in rhesus monkeys. *Lancet* **362**: 1953–1958. https://doi.org/10.1016/S0140-6736(03)15012-X.

Geissler, P.W., Mwaniki, D., Thiong'o, F. et al. (1998). Geophagy as a risk factor for geohelminth infections: a longitudinal study of Kenyan primary schoolchildren. *Transactions of the Royal Society of Tropical Medicine and Hygiene* **92**: 7–11. https://doi.org/10.1016/S0035-9203(98)90934-8.

Gentil, K., Lentz, C.S., Rai, R. et al. (2014). Eotaxin-1 is involved in parasite clearance during chronic filarial infection. *Parasite Immunology* **36**: 60–77. https://doi.org/10.1111/pim.12079.

Getachew, A.M., Innocent, G., Trawford, A.F. et al. (2012). Gasterophilosis: a major cause of rectal prolapse in working donkeys in Ethiopia. *Tropical Animal Health and Production* **44**: 757–762. https://doi.org/10.1007/s11250-011-9961-7.

Ghazanfari, N., Mueller, S.N., and Heath, W.R. (2018). Cerebral malaria in mouse and man. *Frontiers in Immunology* **9**: https://doi.org/10.3389/fimmu.2018.02016.

Gholami, S., Tanzifi, A., Sharif, M. et al. (2018). Demographic aspects of human hydatidosis in Iranian general population based on serology: a systematic review and meta-analysis. *Veterinary World* **11**: 1385–1396. 10.14202/vetworld.2018.1385–1396.

Ghoshal, U.C., Park, H., and Gwee, K.A. (2010). Bugs and irritable bowel syndrome: the good, the bad and the ugly. *Journal of Gastroenterology and Hepatology* **25**: 244–251. https://doi.org/10.1111/j.1440-1746.2009.06133.x.

Giangaspero, A., Traversa, D., Trentini, R. et al. (2011). Traumatic myiasis by *Wohlfahrtia magnifica* in Italy. *Veterinary Parasitology* **175**: 109–112. https://doi.org/10.1016/j.vetpar.2010.09.028.

Gibson, W. and Peacock, L. (2019). Fluorescent proteins reveal what trypanosomes get up to inside the tsetse fly. *Parasites & Vectors* **12**: https://doi.org/10.1186/s13071-018-3204-y.

Gibson, W.C. (2005). The SRA gene: the key to understanding the nature of *Trypanosoma brucei* rhodesiense. *Parasitology* **131**: 143. PMID: 16145931. https://doi.org/10.1017/s0031182005007560.

Giefing-Kröll, C., Berger, P., Lepperdinger, G. et al. (2015). How sex and age affect immune responses, susceptibility to infections, and response to vaccination. *Aging Cell* **14**: 309–321. https://doi.org/10.1111/acel.12326.

Gilbert, B.W. and Slechta, J. (2018). A case of ivermectin-induced warfarin toxicity: first published report. *Hospital Pharmacy* **53**: 393–394. https://doi.org/10.1177/0018578718758972.

Gillespie, D., Frye, F.L., Stockham, S.L. et al. (2000). Blood values in wild and captive Komodo dragons (*Varanus komodoensis*). *Zoo Biology* **19**: 495–509. https://doi.org/10.1002/1098-2361(2000)19:6<495::AID-ZOO2>3.0.CO;2-1.

Gilroy, S.A. and Bennett, N.J. (2011). *Pneumocystis* pneumonia. *Seminars in Respiratory and Critical Care Medicine* **32**: 775–782. https://doi.org/10.1055/s-0031-1295725.

Giraud, E., Svobodová, M., Müller, I. et al. (2019). Promastigote secretory gel from natural and unnatural sand fly vectors exacerbate *Leishmania major* and *Leishmania tropica* cutaneous leishmaniasis in mice. *Parasitology* **146**: 1796–1802. https://doi.org/10.1017/S0031182019001069.

Giribet, G. and Edgecombe, G.D. (2017). Current understanding of Ecdysozoa and its internal phylogenetic relationships. *Integrative and Comparative Biology* **57**: 455–466. https://doi.org/10.1093/icb/icx072.

Giuliotti, L., Pisseri, F., di Sarsina, P.R. et al. (2016). Gastrointestinal strongyles burden monitoring in a flock of Zerasca sheep treated with homeopathy. *European Journal of Integrative Medicine* **8**: 235–238. https://doi.org/10.1016/j.eujim.2015.09.133.

Goddard, J.H.R., Torchin, M.E., Kuris, A.M. et al. (2005). Host specificity of *Sacculina carcini*, a potential biological control agent of the introduced European green crab *Carcinus maenas* in California. *Biological Invasions* **7**: 895–912. https://doi.org/10.1007/s10530-003-2981-0.

Gomes, G.S., Maciel, T.R., Piegas, E.M. et al. (2018). Optimization of curcuma oil/quinine-loaded nanocapsules for malaria treatment. *AAPS PharmSciTech* **19**: 551–564. https://doi.org/10.1208/s12249-017-0854-6.

Gómez, E.A., Giraldo, P., and Orduz, S. (2017). InverPep: a database of invertebrate antimicrobial peptides. *Journal of Global Antimicrobial Resistance* **8**: 13–17. https://doi.org/10.1016/j.jgar.2016.10.003.

Gómez-Arreaza, A., Haenni, A.L., Dunia, I. et al. (2017). Viruses of parasites as actors in the parasite-host relationship: a 'ménage à trois'. *Acta Tropica* **166**: 126–132. https://doi.org/10.1016/j.actatropica.2016.11.028.

Gómez-Morales, M.Á., Pezzotti, P., Ludovisi, A. et al. (2021). Collaborative studies for the detection of *Taenia* spp. Infections in humans within CYSTINET, the European Network on Taeniosis/Cysticercosis. *Microorganisms* **9**: 1173. https://doi.org/10.3390/microorganisms9061173.

Gomez-Puerta, L.A. and Mayor, P. (2017). Congenital filariasis caused by *Setaria bidentata* (Nematoda: Filarioidea) in the red brocket deer (*Mazama americana*). *Journal of Parasitology* **103**: 123–126. https://doi.org/10.1645/16-86.

Gonzalez-Ceron, L., Santillan, F., Rodriguez, M.H. et al. (2003). Bacteria in midguts of field-collected *Anopheles albimanus* block *Plasmodium vivax* sporogonic development. *Journal of Medical Entomology* **40**: 371–374. https://doi.org/10.1603/0022-2585-40.3.371.

Goodswen, S.J., Kennedy, P.J., and Ellis, J.T. (2017). On the application of reverse vaccinology to parasitic diseases: a perspective on feature selection and ranking of vaccine candidates. *International Journal for Parasitology* **47**: 779–790. https://doi.org/10.1016/j.ijpara.2017.08.004.

Götze, S., Reinhardt, A., Geissner, A. et al. (2015). Investigation of the protective properties of glycosylphosphatidylinositol-based vaccine candidates in a *Toxoplasma gondii* mouse challenge model. *Glycobiology* **25**: 984–991. https://doi.org/10.1093/glycob/cwv040.

Goudal, A., Laude, A., and Valot, S. (2019). Rapid diagnostic tests relying on antigen detection from stool as an efficient point of care testing strategy for giardiasis and cryptosporidiosis? Evaluation of a new immunochromatographic duplex assay. *Diagnostic Microbiology and Infectious Disease* **93**: 33–36. https://doi.org/10.1016/j.diagmicrobio.2018.07.012.

Graf, J., Kikuchi, Y., and Rio, R.V. (2006). Leeches and their microbiota: naturally simple symbiosis models. *Trends in Microbiology* **14**: 365–371. https://doi.org/10.1016/j.tim.2006.06.009.

Granroth-Wilding, H.M., Burthe, S.J., Lewis, S. et al. (2015). Indirect effects of parasitism: costs of infection to other individuals can be greater than direct costs borne by the host. *Proceedings of the Royal Society B: Biological Sciences* **282**: https://doi.org/10.1098/rspb.2015.0602.

Grassberger, M. and Frank, C. (2003). Temperature-related development of the parasitoid wasp *Nasonia vitripennis* as forensic indicator. *Medical and Veterinary Entomology* **17**: 257–262. https://doi.org/10.1046/j.1365-2915.2003.00439.x.

Greer, A.W. (2008). Trade-offs and benefits: implications of promoting a strong immunity to gastrointestinal parasites in sheep. *Parasite Immunology* **30**: 123–132. https://doi.org/10.1111/j.1365-3024.2008.00998.x.

Grice, E.A. and Segre, J.A. (2011). The skin microbiome. *Nature Reviews Microbiology* **9**: 244–253. https://doi.org/10.1038/nrmicro2537.

Grote, A., Voronin, D., Ding, T. et al. (2017). Defining *Brugia malayi* and *Wolbachia* symbiosis by stage-specific dual RNA-seq. *PLoS Neglected Tropical Diseases* **11**: e0005357. https://doi.org/10.1371/journal.pntd.0005357.

Grunin, K.Y. (1973). The first discovery of larvae of the mammoth bot fly *Cobboldia* (*Mamontia*, subgen. II.) *russanovi* sp. n. (Diptera, Gasterophilidae). *Entomological Review* **52**: 165–169.

Gryseels, B. (2000). Schistosomiasis vaccines: a devils' advocate view. *Parasitology Today* **16**: 46–48. https://doi.org/10.1016/S0169-4758(99)01597-5.

Guernier, V., Brennan, B., Yakob, L. et al. (2017). Gut microbiota disturbance during helminth infection: can it affect cognition and behaviour of children? *BMC Infectious Diseases* **17**: 58. https://doi.org/10.1186/s12879-016-2146-2.

Guerrant, R.L., Brush, J., Ravdin, J.I. et al. (1981). Interaction between *Entamoeba histolytica* and human polymorphonuclear neutrophils. *Journal of Infectious Diseases* **143**: 83–93. https://doi.org/10.1093/infdis/143.1.83.

Guerrini, V.H. (1997). Excretion of ammonia by *Lucilia cuprina* larvae suppresses immunity in sheep. *Veterinary Immunology and Immunopathology* **56**: 311–317. https://doi.org/10.1016/S0165-2427(96)05744-3.

Guezala, M.C., Rodriguez, S., Zamora, H. et al. (2009). Development of a species-specific coproantigen ELISA for human *Taenia solium* taeniasis. *American Journal of Tropical Medicine and Hygiene* **81**: 433–437.

Guglielmini, J., Woo, A.C., Krupovic, M. et al. (2019). Diversification of giant and large eukaryotic dsDNA viruses predated the origin of modern eukaryotes. *Proceedings of the National Academy of*

Sciences of the United States of America **116**: 19585–19592. https://doi.org/10.1073/pnas.1912006116.

Guimaraes, A.J., Gomes, K.X., Cortines, J.R. et al. (2016). *Acanthamoeba* spp. as a universal host for pathogenic microorganisms: One bridge from environment to host virulence. *Microbiological Research* **193**: 30–38. https://doi.org/10.1016/j.micres.2016.08.001.

Gunderson, E.L., Vogel, I., Chappell, L. et al. (2020). The endosymbiont Wolbachia rebounds following antibiotic treatment. *PLoS pathogens* **16**: https://doi.org/10.1371/journal.ppat.1008623.

Gunn, A. (1980). A case of *Ascaris suum* infection in lambs. *Veterinary Record* **107**: 581.

Gunn, A. (2019). *Essential Forensic Biology*, 3rde. Chichester, UK: Wiley-Blackwell.

Gunn, A. and Probert, A.J. (1983). *Moniezia expansa*: the interproglottidal glands and their secretions. *Journal of Helminthology* **57**: 51–58. https://doi.org/10.1017/S0022149X00007884.

Gyorkos, T.W., Montresor, A., Belizario, V. et al. (2018). The right to deworming: the case for girls and women of reproductive age. *PLoS Neglected Tropical Diseases* **12**: https://doi.org/10.1371/journal.pntd.0006740.

Haas, W., Grabe, K., Geis, C. et al. (2002). Recognition and invasion of human skin by *Schistosoma mansoni* cercariae: the key role of L-arginine. *Parasitology* **124**: 153–167. https://doi.org/10.1017/S0031182001001032.

Habetha, M., Anton-Erxleben, F., Neumann, K. et al. (2003). The *Hydra viridis/Chlorella* symbiosis. Growth and sexual differentiation in polyps without symbionts. *Zoology* **106**: 101–108. https://doi.org/10.1078/0944-2006-00104.

Hagerty, J.R. and Jolly, E.R. (2019). Heads or tails? Differential translational regulation in cercarial heads and tails of schistosome worms. *PloS One* **14**: https://doi.org/10.1371/journal.pone.0224358.

Haider, N., Laaksonen, S., Kjær, L.J. et al. (2018). The annual, temporal and spatial pattern of *Setaria tundra* outbreaks in Finnish reindeer: a mechanistic transmission model approach. *Parasites & Vectors* **11**: 565. https://doi.org/10.1186/s13071-018-3159-z.

Hajek, A.E., Hurley, B.P., Kenis, M. et al. (2016). Exotic biological control agents: a solution or contribution to arthropod invasions? *Biological Invasions* **18**: 953–969. https://doi.org/10.1007/s10530-016-1075-8.

Halliez, M. and Buret, A.G. (2015). Gastrointestinal parasites and the neural control of gut functions. *Frontiers in Cellular Neuroscience* **9**: 452. https://doi.org/10.3389/fncel.2015.00452.

Hall-Mendelin, S., Craig, S.B., Hall, R.A. et al. (2011). Tick paralysis in Australia caused by *Ixodes holocyclus* Neumann. *Annals of Tropical Medicine & Parasitology* **105**: 95–106. https://doi.org/10.1179/136485911X12899838413628.

Hamilton, W.D., Axelrod, R., and Tanese, R. (1990). Sexual reproduction as an adaptation to resist parasites (a review). *Proceedings of the National Academy of Sciences of the United States of America* **87**: 3566–3573. https://doi.org/10.1073/pnas.87.9.3566.

Hanington, P.C. and Zhang, S.M. (2011). The primary role of fibrinogen-related proteins in invertebrates is defense, not coagulation. *Journal of Innate Immunity* **3**: 17–27. https://doi.org/10.1159/000321882.

Happi, A.N., Milner, D.A., and Anti, R.E. (2012). Blood and tissue leukocyte apoptosis in *Trypanosoma brucei* infected rats. *Journal of Neuroparasitology* **3**: 1–10. https://doi.org/10.4303/jnp/N120101.

Harnett, M.W. and Harnett, W. (2017). Can parasitic worms cure modern world's ills? *Trends in Parasitology* **33**: 694–705. https://doi.org/10.1016/j.pt.2017.05.007.

Harris, A.F., Matias-Arnéz, A., and Hill, N. (2006). Biting time of *Anopheles darlingi* in the Bolivian Amazon and implications for control of malaria. *Transactions of the Royal Society of Tropical Medicine and Hygiene* **100**: 45–47. https://doi.org/10.1016/j.trstmh.2005.07.001.

Hashimoto, M., Nara, T., Mita, T. et al. (2016). Morpholino antisense oligo inhibits trans-splicing of pre-inositol 1, 4, 5-trisphosphate receptor mRNA of *Trypanosoma cruzi* and suppresses parasite growth and infectivity. *Parasitology International* **65**: 175–179. https://doi.org/10.1016/j.parint.2015.12.001.

Hasnain, S.Z., Thornton, D.J., and Grencis, R.K. (2011). Changes in the mucosal barrier during acute and chronic *Trichuris muris* infection. *Parasite Immunology* **33**: 45–55. https://doi.org/10.1111/j.1365-3024.2010.01258.x.

Haugen, B., Karinshak, S.E., Mann, V.H. et al. (2018). Granulin secreted by the food-borne liver fluke *Opisthorchis viverrini* promotes angiogenesis in human endothelial cells. *Frontiers in Medicine* **5**: 30. https://doi.org/10.3389/fmed.2018.00030.

Haugerud, R.E. and Nilssen, A.C. (1990). Life history of the reindeer sinus worm, *Linguatula arctica* (Pentastomida), a prevalent parasite in reindeer calves. *Rangifer* **10**: 333–334. https://doi.org/10.7557/2.10.3.875.

Hay, S.I., Abajobir, A.A., Abate, K.H. et al. (2017). Global, regional, and national disability-adjusted life-years (DALYs) for 333 diseases and injuries and healthy life expectancy (HALE) for 195 countries and territories, 1990–2016: a systematic analysis for the Global Burden of Disease Study 2016. *Lancet* **390**: 1260–1344. https://doi.org/10.1016/S0140-6736(17)32130-X.

Hayashida, K., Kajino, K., Hachaambwa, L. et al. (2015). Direct blood dry LAMP: a rapid, stable, and easy diagnostic tool for human African trypanosomiasis. *PLoS Neglected Tropical Diseases* **9**: https://doi.org/10.1371/journal.pntd.0003578.

Hayden, E.C. (2014). Ebola obstructs malaria control: outbreak is shutting down prevention and treatment programmes in West Africa. *Nature* **514**: 15–17. PMID: 25279895. https://doi.org/10.1038/514015a.

Hayes, K.S., Bancroft, A.J., Goldrick, M. et al. (2010). Exploitation of the intestinal microflora by the parasitic nematode *Trichuris muris*. *Science* **328**: 1391–1394. https://doi.org/10.1126/science.1187703.

Hays, R., Pierce, D., Giacomin, P. et al. (2020). Helminth coinfection and COVID-19: an alternate hypothesis. *PLoS Neglected Tropical Diseases* **14**: https://doi.org/10.1371/journal.pntd.0008628.

He, K., Lin, K., Wang, G. et al. (2016). Genome sizes of nine insect species determined by flow cytometry and k-mer analysis. *Frontiers in Physiology* **7**: 569. https://doi.org/10.3389/fphys.2016.00569.

Heath, D.D., Robinson, C., Shakes, T. et al. (2012). Vaccination of bovines against *Echinococcus granulosus* (cystic echinococcosis). *Vaccine* **30**: 3076–3081. https://doi.org/10.1016/j.vaccine.2012.02.073.

Hegertun, I.E.A., Gundersen, K.M.S., Kleppa, E. et al. (2013). *S. haematobium* as a common cause of genital morbidity in girls: a cross-sectional study of children in South Africa. *PLoS Neglected Tropical Diseases* **7**: 364–369. https://doi.org/10.4269/ajtmh.2011.10-0585.

Heimlich, H.J., Chen, X.P., Xiao, B.Q. et al. (1997). Malariotherapy for HIV patients. *Mechanisms of Ageing and Development* **93**: 79–85. https://doi.org/10.1016/S0047-6374(96)01813-1.

Hembrough, T.A., Swartz, G.M., Papathanassiu, A. et al. (2003). Tissue factor/factor VIIa inhibitors block angiogenesis and tumor growth through a nonhemostatic mechanism. *Cancer Research* **63**: 2997–3000. PMID: 12782609.

Hendrickx, E., Thomas, L.F., Dorny, P. et al. (2019). Epidemiology of *Taenia saginata* taeniosis/cysticercosis: a systematic review of the distribution in West and Central Africa. *Parasites & Vectors* **12**: 324. https://doi.org/10.1186/s13071-019-3584-7.

Hendrix, C.M. and Robinson, E. (2016). *Diagnostic Parasitology for Veterinary Technicians*, 5the. St Louis, MO: Elsevier.

Heo, I., Dutta, D., Schaefer, D.A. et al. (2018). Modelling *Cryptosporidium* infection in human small intestinal and lung organoids. *Nature Microbiology* **3**: 814–823. https://doi.org/10.1038/s41564-018-0177-8.

Hertz, M.I., Nana-Djeunga, H., Kamgno, J. et al. (2018). Identification and characterization of *Loa loa* antigens responsible for cross-reactivity with rapid diagnostic tests for lymphatic filariasis. *PLoS Neglected Tropical Diseases* **12**: https://doi.org/10.1371/journal.pntd.0006963.

Heydorn, A.O. and Melhorn, H. (2002). *Neospora caninum* is an invalid species name: an evaluation of facts and statements. *Parasitology Research* **88**: 175–184. https://doi.org/10.1007/s00436-001-0513-3.

Heyworth, M.F. (2016). *Giardia duodenalis* genetic assemblages and hosts. *Parasite (Paris, France)* **23**: 13. https://doi.org/10.1051/parasite/2016013.

Heyworth, M.F. (2017). Genetic aspects and environmental sources of microsporidia that infect the human gastrointestinal tract. *Transactions of the Royal Society of Tropical Medicine and Hygiene* **111**: 18–21. https://doi.org/10.1093/trstmh/trx001.

Higazi, T.B., Filiano, A., Katholi, C.R. et al. (2005). *Wolbachia* endosymbiont levels in severe and mild strains of *Onchocerca volvulus*. *Molecular and Biochemical Parasitology* **141**: 109–112. https://doi.org/10.1016/j.molbiopara.2005.02.006.

Hildebrandt, J.P. and Lemke, S. (2011). Small bite, large impact–saliva and salivary molecules in the medicinal leech, *Hirudo medicinalis*. *Naturwissenschaften* **98**: 995–1008. https://doi.org/10.1007/s00114-011-0859-z.

Him, N.A., Gillan, V., Emes, R.D. et al. (2009). Hsp-90 and the biology of nematodes. *BMC Evolutionary Biology* **9**: 254. https://doi.org/10.1186/1471-2148-9-254.

Hinz, R., Schwarz, N.G., Hahn, A., and Frickmann, H. (2017). Serological approaches for the diagnosis of schistosomiasis – a review. *Molecular and Cellular Probes* **31**: 2–21. https://doi.org/10.1016/j.mcp.2016.12.003.

Hisaeda, H., Yasutomo, K., and Himeno, K. (2005). Malaria: immune evasion by parasites. *The International Journal of Biochemistry & Cell Biology* **37**: 700–706. https://doi.org/10.1016/j.biocel.2004.10.009.

Hite, J.L. and Cressler, C.E. (2019). Parasite-mediated anorexia and nutrition modulate virulence evolution. *Integrative and Comparative Biology* **59**: 1264–1274. https://doi.org/10.1093/icb/icz100.

Hoffman, C.M., Fritz, L., Radebe, O. et al. (2018). Rectal *Trichomonas vaginalis* infection in South African men who have sex with men. *International journal of STD & AIDS* **29**: 1444–1447. https://doi.org/10.1177/0956462418788418.

Hohmann, C.D., Stange, R., Steckhan, N. et al. (2018). The effectiveness of leech therapy in chronic low back pain: a randomized controlled trial. *Deutsches Ärzteblatt International* **115**: 785–792. https://doi.org/10.3238/arztebl.2018.0785.

Hollander, E., Uzunova, G., Taylor, B.P. et al. (2020). Randomized crossover feasibility trial of helminthic *Trichuris suis* ova versus placebo for repetitive behaviors in adult autism spectrum disorder. *World Journal of Biological Psychiatry* **21**: 291–299. https://doi.org/10.1080/15622975.2018.1523561.

Hollis, A.C. (1909). *The Nandi: Their Language and Folklore*. Oxford, UK: Clarendon Press.

Holmes, S.D. and Fairweather, I. (1982). *Hymenolepis diminuta*: the mechanism of egg hatching. *Parasitology* **85**: 237–250. https://doi.org/10.1017/S0031182000055219.

Holterman, M., van der Wurff, A., van den Elsen, S. et al. (2006). Phylum-wide analysis of SSU rDNA reveals deep phylogenetic relationships among nematodes and accelerated evolution toward crown clades. *Molecular Biology and Evolution* **23**: 1792–1800. https://doi.org/10.1093/molbev/msl044.

Horta, M.F., Andrade, L.O., Martins-Duarte, É.S. et al. (2020). Cell invasion by intracellular parasites – the many roads to infection. *Journal of Cell Science* **133**: PMID: 32079731. https://doi.org/10.1242/jcs.232488.

Hosseini, S.A., Amouei, A., Sharif, M. et al. (2019). Human toxoplasmosis: a systematic review for genetic diversity of *Toxoplasma gondii* in clinical samples. *Epidemiology and Infection* **147**: https://doi.org/10.1017/S0950268818002947.

Hotez, P.J. (2016). Southern Europe's coming plagues: vector-borne neglected tropical diseases. *PLoS Neglected Tropical Diseases* **10**: https://doi.org/10.1371/journal.pntd.0004243.

Hotez, P.J., Bethony, J.M., Diemert, D.J. et al. (2010). Development of vaccines to combat hookworm infection and intestinal schistosomiasis. *Nature Reviews Microbiology* **8**: 814–826. https://doi.org/10.1038/nrmicro2438.

Hotterbeekx, A., Perneel, J., Mandro, M. et al. (2020). Comparison of diagnostic tests for *Onchocerca volvulus* in the Democratic Republic of Congo. *Pathogens* **9**: https://doi.org/10.3390/pathogens9060435.

Houpt, E., Barroso, L., Lockhart, L. et al. (2004). Prevention of intestinal amebiasis by vaccination with the *Entamoeba histolytica* Gal/GalNac lectin. *Vaccine* **22**: 611–617. https://doi.org/10.1016/j.vaccine.2003.09.003.

Howard, D.H., Bach, P.B., Berndt, E.R. et al. (2015). Pricing in the market for anticancer drugs. *Journal of Economic Perspectives* **29**: 139–162. https://doi.org/10.1257/jep.29.1.139.

Howard, E.J., Xiong, X., Carlier, Y. et al. (2014). Frequency of the congenital transmission of *Trypanosoma cruzi*: a systematic review and meta-analysis. *BJOG: An International Journal of Obstetrics & Gynaecology* **121**: 22–33. https://doi.org/10.1111/1471-0528.12396.

Howell, A.K., McCann, C.M., Wickstead, F. et al. (2019). Co-infection of cattle with *Fasciola hepatica* or *F. gigantica* and *Mycobacterium bovis*: A systematic review. *PloS One* **14**: https://doi.org/10.1371/journal.pone.0226300.

Huang, L. and Appleton, J.A. (2016). Eosinophils in helminth infection: defenders and dupes. *Trends in Parasitology* **32**: 798–807. https://doi.org/10.1016/j.pt.2016.05.004.

Huang, Y., Abuzeid, A.M., Liu, Y. et al. (2020). Identification and localization of hookworm platelet inhibitor in *Ancylostoma ceylanicum*. *Infection, Genetics and Evolution* **77**: 104102. https://doi.org/10.1016/j.meegid.2019.104102.

Huggett, J.F., Cowen, S., and Foy, C.A. (2015). Considerations for digital PCR as an accurate molecular diagnostic tool. *Clinical Chemistry* **61**: 79–88. https://doi.org/10.1373/clinchem.2014.221366.

Huwe, T., Prusty, B.K., Ray, A. et al. (2019). Interactions between parasitic infections and the human gut microbiome in Odisha, India. *American Journal of Tropical Medicine and Hygiene* **100**: 1486–1489. https://doi.org/10.4269/ajtmh.18-0968.

Huyse, T., Boon, N.A.M., Van den Broek, F. et al. (2018). Evolutionary epidemiology of schistosomiasis: linking parasite genetics with disease phenotypes in humans. *International Journal for Parasitology* **48**: 107–115. PMID: 29154994. https://doi.org/10.1016/j.ijpara.2017.07.010.

Hviid, L. and Jensen, A.T. (2015). PfEMP1 – a parasite protein family of key importance in *Plasmodium falciparum* malaria immunity and pathogenesis. *Advances in Parasitology* **88**: 51–84. https://doi.org/10.1016/bs.apar.2015.02.004.

Hyman, P., Atterbury, R., and Barrow, P. (2013). Fleas and smaller fleas: virotherapy for parasite infections. *Trends in Microbiology* **21**: 215–220. https://doi.org/10.1016/j.tim.2013.02.006.

Ichikawa-Seki, M., Peng, M., Hayashi, K. et al. (2017). Nuclear and mitochondrial DNA analysis reveals that hybridization between *Fasciola hepatica* and *Fasciola gigantica* occurred in China. *Parasitology* **144**: 206–213. https://doi.org/10.1017/S003118201600161X.

Idowu, O.A. and Rowland, S.A. (2006). Oral fecal parasites and personal hygiene of food handlers in Abeokuta, Nigeria. *African Health Sciences* **6**: 160–164. https://doi.org/10.5555/AFHS.2006.6.3.160.

Ikede, B.O., Elhassan, E., and Akpavie, S.O. (1988). Reproductive disorders in African trypanosomiasis: a review. *Acta Tropica* **45**: 5–10. PMID: 2896446.

Ilmonen, P., Taarna, T., and Hasselquist, D. (2000). Experimentally activated immune defence in female pied flycatchers results in reduced breeding success. *Proceedings of the Royal Society of London, Series B – Biological Sciences* **267**: 665–670. https://doi.org/10.1098/rspb.2000.1053.

Imai, K. (1999). The haemoglobin enzyme. *Nature* **401**: 437–439. https://doi.org/10.1038/46707.

Imbert-Establet, D. and Combes, C. (1986). *Schistosoma mansoni:* comparison of a Caribbean and African strain and experimental crossing based on compatibility with intermediate hosts and *Rattus rattus. Experimental Parasitology* **61**: 210–218. https://doi.org/10.1016/0014-4894(86)90154-2.

Inoue, H., Motani-Saitoh, H., Sakurada, K. et al. (2014). Genotypic polymorphisms of hepatitis b virus provide useful information for estimating geographical origin or place of long-term residence of unidentified cadavers. *Journal of Forensic Sciences* **59**: 236–241. https://doi.org/10.1111/1556-4029.12257.

Isaäcson, M. and Frean, J.A. (2001). African malaria vectors in European aircraft. *Lancet* **357**: 235.

Ishida, K. and Hsieh, M.H. (2018). Understanding urogenital schistosomiasis-related bladder cancer: an update. *Frontiers in Medicine* **5**: 223. 10.3389/fmed.2018.00223.

Ishida, K. and Jolly, E.R. (2016). Hsp70 may be a molecular regulator of schistosome host invasion. *PLoS Neglected Tropical Diseases* **10**: https://doi.org/10.1371/journal.pntd.0004986.

Ito, A., Putra, M.I., Subahar, R. et al. (2002). Dogs as alternative intermediate hosts of *Taenia solium* in Papua (Irian Jaya), Indonesia confirmed by highly specific ELISA and immunoblot using native and recombinant antigens and mitochondrial DNA analysis. *Journal of Helminthology* **76**: 311–314. https://doi.org/10.1079/JOH2002128.

Ito, K., Karasawa, M., Kawano, T. et al. (2000). Involvement of decidual Vα14 NKT cells in abortion. *Proceedings of the National Academy of Sciences of the United States of America* **97**: 740–744. https://doi.org/10.1073/pnas.97.2.740.

Ivory, C.P., Prystajecky, M., Jobin, C. et al. (2008). Toll-like receptor 9-dependent macrophage activation by *Entamoeba histolytica* DNA. *Infection and Immunity* **76**: 289–297. https://doi.org/10.1128/IAI.01217-07.

Iwanaga, S. and Lee, B.L. (2005). Recent advances in the innate immunity of invertebrate animals. *Journal of Biochemistry and Molecular Biology* **38**: 128–150. https://doi.org/10.5483/bmbrep.2005.38.2.128.

Jacobs, R.T., Lunde, C.S., Freund, Y.R. et al. (2019). Boron-pleuromutilins as anti-*Wolbachia* agents with potential for treatment of onchocerciasis and lymphatic filariasis. *Journal of Medicinal Chemistry* **62**: 2521–2540. https://doi.org/10.1021/acs.jmedchem.8b1854.

Jacobs, R.T., Nare, B., Wring, S.A. et al. (2011). SCYX-7158, an orally-active benzoxaborole for the treatment of stage 2 human African trypanosomiasis. *PLoS Neglected Tropical Diseases* **5**: https://doi.org/10.1371/journal.pntd.0001151.

Jagielski, T., Krukowski, H., Bochniarz, M. et al. (2019). Prevalence of *Prototheca* spp. on dairy farms in Poland – a cross-country study. *Microbial Biotechnology* **12**: 556–566. https://doi.org/10.1111/1751-7915.13394.

Janssen, M.E., Takagi, Y., Parent, K.N. et al. (2015). Three-dimensional structure of a protozoal double-stranded RNA virus that infects the enteric pathogen *Giardia lamblia. Journal of Virology* **89**: 1182–1194. https://doi.org/10.1128/JVI.02745-14.

Jarquín-Díaz, V.H., Balard, A., Jost, J. et al. (2019). Detection and quantification of house mouse Eimeria at the species level – challenges and solutions for the assessment of Coccidia in wildlife. *International Journal for Parasitology: Parasites and Wildlife* **10**: 29–40. https://doi.org/10.1016/j.ijppaw.2019.07.004.

Jaurigue, J.A. and Seeberger, P.H. (2017). Parasite carbohydrate vaccines. *Frontiers in Cellular and Infection Microbiology* **7**: 248. https://doi.org/10.3389/fcimb.2017.00248.

Jaworowski, A., Fernandes, L.A., Yosaatmadja, F. et al. (2009). Relationship between human immunodeficiency virus type 1 coinfection, anemia, and levels and function of antibodies to variant surface antigens in pregnancy-associated malaria. *Clinical and Vaccine Immunology* **16**: 312–319. https://doi.org/10.1128/CVI.00356-08.

Jayasekara, S., Sissons, J., Tucker, J. et al. (2004). Postmortem culture of *Balamuthia mandrillaris* from the brain and cerebrospinal fluid of a case of granulomatous amoebic encephalitis, using human brain microvascular endothelial cells. *Journal of Medical Microbiology* **53**: 1007–1012. https://doi.org/10.1099/jmm.0.45721-0.

Jenkins, M., Fetterer, R., and Miska, K. (2009). Co-infection of chickens with *Eimeria praecox* and *Eimeria maxima* does not prevent development of immunity to *Eimeria maxima*. *Veterinary Parasitology* **161**: 320–323. https://doi.org/10.1016/j.vetpar.2009.01.012.

Jeudy, J.M. (2010). Surgical repair of a giant scrotal elephantiasis. BJUI Case Reports. BJUI.org.

Jian, B., Kolansky, A.S., Baloach, Z.W. et al. (2008). *Entamoeba gingivalis* pulmonary abscess-diagnosed by fine needle aspiration. *Cytojournal* **5**: 12. https://doi.org/10.4103/1742-6413.43179.

Jimenez, A., Rees-Channer, R.R., Perera, R. et al. (2017). Analytical sensitivity of current best-in-class malaria rapid diagnostic tests. *Malaria Journal* **16**: 1–9. https://doi.org/10.1186/s12936-017-1780-5.

Jin, Z., Akao, N., and Ohta, N. (2008). Prolactin evokes lactational transmission of larvae in mice infected with *Toxocara canis*. *Parasitology International* **57**: 495–498. https://doi.org/10.1016/j.parint.2008.06.006.

Joerger, T., Sulieman, S., Carson, V.J. et al. (2020). Chronic meningitis due to Prototheca zopfii in an adolescent girl. *Journal of the Pediatric Infectious Diseases Society* https://doi.org/10.1093/jpids/piaa049.

Johnigk, S.A. and Ehlers, R.U. (1999). Endotokia matricida in hermaphrodites of *Heterorhabditis* spp. and the effect of the food supply. *Nematology* **1**: 717–726. https://doi.org/10.1163/156854199508748.

Johnson, H.J. and Koshy, A.A. (2020). Latent toxoplasmosis effects on rodents and humans: how much is real and how much is media hype? *Mbio* **11**: https://doi.org/10.1128/mBio.02164-19.

Johnson, P.T.J., Lunde, K.B., Ritchie, E.G. et al. (1999). The effect of trematode infection on amphibian limb development and survivorship. *Science* **284**: 803–804. https://doi.org/10.1126/science.284.5415.802.

Johnson, P.T.J. and Sutherland, D.R. (2003). Amphibian deformities and *Ribeiroia* infection: an emerging hcminthiasis. *Trends in Parasitology* **19**: 332–335. https://doi.org/10.1016/S1471-4922(03)00148-X.

Jones, R.A., Brophy, P.M., Davis, C.N. et al. (2018). Detection of *Galba truncatula, Fasciola hepatica* and *Calicophoron daubneyi* environmental DNA within water sources on pasture land, a future tool for fluke control? *Parasites Vectors* **11**: 342. https://doi.org/10.1186/s13071-018-2928-z.

Jones, R.A., Brophy, P.M., Mitchell, E.S. et al. (2017). Rumen fluke (*Calicophoron daubneyi*) on Welsh farms: prevalence, risk factors and observations on co-infection with *Fasciola hepatica*. *Parasitology* **144**: 237–247. https://doi.org/10.1017/S0031182016001797[Opens in a new window].

Jones-Brando, L., Torrey, E.F., and Yolken, R. (2003). Drugs used in the treatment of schizophrenia and bipolar disorder inhibit the replication of *Toxoplasma gondii*. *Schizophrenia Research* **62**: 237–244. https://doi.org/10.1016/S0920-9964(02)00357-2.

Junqueira, C., Santos, L.I., Galvão-Filho, B. et al. (2011). *Trypanosoma cruzi* as an effective cancer antigen delivery vector. *Proceedings of the National Academy of Sciences of the United States of America* **108**: 19695–19700. https://doi.org/10.1073/pnas.1110030108.

Kafle, P., Peacock, S.J., Grond, S. et al. (2018). Temperature-dependent development and freezing survival of protostrongylid nematodes of Arctic ungulates: implications for transmission. *Parasites & Vectors* **11**: 400. https://doi.org/10.1186/s13071-018-2946-x.

Kalra, S.K., Sharma, P., Shyam, K. et al. (2020). *Acanthamoeba* and its pathogenic role in granulomatous amebic encephalitis. *Experimental Parasitology* **208**: https://doi.org/10.1016/j.exppara.2019.107788.

Kamel, B., Laidemitt, M.R., Lu, L. et al. (2021). Detecting and identifying *Schistosoma* infections in snails and aquatic habitats: a systematic review. *PLoS Neglected Tropical Diseases* **15**: 10.1371/journal.pntd.0009175.

Kaňková, Š., Hlaváčová, J., and Flegr, J. (2020). Oral sex: A new, and possibly the most dangerous, route of toxoplasmosis transmission. *Medical Hypotheses* **141**: https://doi.org/10.1016/j.mehy.2020.109725.

Kaňková, Š., Kodym, P., Frynta, D. et al. (2007b). Influence of latent toxoplasmosis on the secondary sex ratio in mice. *Parasitology* **134**: 1709–1717. https://doi.org/10.1017/S0031182007003253.

Kaňková, Š., Šulc, J., Nouzová, K. et al. (2007a). Women infected with parasite *Toxoplasma* have more sons. *Naturwissenschaften* **94**: 122–127. https://doi.org/10.1007/s00114-006-0166-2.

Kano, R. (2020). Emergence of fungal-like organisms: *Prototheca*. *Mycopathologia* **185**: 747–754. https://doi.org/10.1007/s11046-019-00365-4.

Kasonta, R., Holsteg, M., Duchow, K. et al. (2014). Colostrum from cows immunized with a vaccine associated with bovine neonatal pancytopenia contains allo-antibodies that cross-react with human MHC-I molecules. *PLoS One* **9**: https://doi.org/10.1371/journal.pone.0109239.

Kassa, F., Van Den Ham, K., Rainone, A. et al. (2016). Absence of apolipoprotein E protects mice from cerebral malaria. *Scientific Reports* **6**: https://doi.org/10.1038/srep33615.

Kassa, M., Abdellati, S., Cnops, L. et al. (2020). Diagnostic accuracy of direct agglutination test, rK39 ELISA and six rapid diagnostic tests among visceral leishmaniasis patients with and without HIV coinfection in Ethiopia. *PLoS Neglected Tropical Diseases* **14**: https://doi.org/10.1371/journal.pntd.0008963.

Katiki, L.M., Barbieri, A.M.E., Araujo, R.C. et al. (2017). Synergistic interaction of ten essential oils against *Haemonchus contortus*. *Veterinary Parasitology* **243**: 47–51. https://doi.org/10.1016/j.vetpar.2017.06.008.

Katzer, F., Ngugi, D., Walker, A.R. et al. (2010). Genotypic diversity, a survival strategy for the apicomplexan parasite *Theileria parva*. *Veterinary Parasitology* **167**: 236–243. https://doi.org/10.1016/j.vetpar.2009.09.025.

Kaufer, A., Ellis, J., Stark, D. et al. (2017). The evolution of trypanosomatid taxonomy. *Parasites & Vectors* **10**: 287. https://doi.org/10.1186/s13071-017-2204-7.

Kazumba, L.M., Kaka, J.C.T., Ngoyi, D.M. et al. (2018). Mortality trends and risk factors in advanced stage-2 human African trypanosomiasis: a critical appraisal of 23 years of experience in the Democratic Republic of Congo. *PLoS Neglected Tropical Diseases* **12**: https://doi.org/10.1371/journal.pntd.0006504.

Kean, B.H., Sun, T., and Ellsworth, R.M. (1991). *Color Atlas/Text of Ophthalmic Parasitology*. Tokyo, Japan: Igaku-Shoin.

Kearn, G.C. (1998). *Parasitism and the Platyhelminths*. London, UK: Chapman and Hall.

Kellerová, P. and Tachezy, J. (2017). Zoonotic *Trichomonas tenax* and a new trichomonad species, *Trichomonas brixi* n. sp., from the oral cavities of dogs and cats. *International Journal for Parasitology* **47**: 247–255. https://doi.org/10.1016/j.ijpara.2016.12.006.

Kennedy, J. (2017). How drone strikes and a fake vaccination program have inhibited polio eradication in Pakistan: an analysis of national level data. *International Journal of Health Services* **47**: 807–825. https://doi.org/10.1177/0020731417722888.

Kennedy, P.G. and Rodgers, J. (2019). Clinical and neuropathogenetic aspects of human African trypanosomiasis. *Frontiers in Immunology* **10**: 39. https://doi.org/10.3389/fimmu.2019.00039.

Kern, E., Kim, T., and Park, J.K. (2020). The mitochondrial genome in nematode phylogenetics. *Frontiers in Ecology and Evolution* **8**: 250. https://doi.org/10.3389/fevo.2020.00250.

Kerney, R., Leavitt, J., Hill, E. et al. (2019). Co-cultures of *Oophila amblystomatis* between *Ambystoma maculatum* and *Ambystoma gracile* hosts show host-symbiont fidelity. *Symbiosis* **78**: 73–85. https://doi.org/10.1007/s13199-018-00591-2.

Keusch, G.T. (2003). The history of nutrition: malnutrition, infection and immunity. *Journal of Nutrition* **133**: 336S–340S. https://doi.org/10.1093/jn/133.1.336S.

Khalil, S.H.A., Megallaa, M.H., Rohoma, K.H. et al. (2018). Prevalence of type 2 diabetes mellitus in a sample of the adult population of Alexandria, Egypt. *Diabetes Research and Clinical Practice* **144**: 63–73. PMID: 30056190. https://doi.org/10.1016/j.diabres.2018.07.025.

Khan, I.D., Sahni, A.K., Sen, S. et al. (2018). Outbreak of *Prototheca wickerhamii* algaemia and sepsis in a tertiary care chemotherapy oncology unit. *Medical Journal Armed Forces India* **74**: 358–364. https://doi.org/10.1016/j.mjafi.2017.07.012.

Khorvash, F., Keshteli, A.H., Salehi, H. et al. (2008). Unusual transmission route of *Lymphogranuloma venereum* following sexual contact with a female donkey. *International Journal of STD and AIDS* **19**: 563–564. https://doi.org/10.1258/ijsa.2008.008073.

Kim, C., Hoffmann, G., and Searson, P.C. (2017). Integrated magnetic bead–quantum dot immunoassay for malaria detection. *ACS Sensors* **2**: 766–772. https://doi.org/10.1021/acssensors.7b00119.

Kim, J.S., Choi, T.H., Kim, N.G. et al. (2007). The replantation of an amputated tongue by supermicrosurgery. *Journal of Plastic, Reconstructive & Aesthetic Surgery* **60**: 1152–1155. https://doi.org/10.1016/j.bjps.2007.01.009.

Kim, T.I., Oh, S.-R., Dai, F. et al. (2017). Inactivation of *Paragonimus westermani* metacercariae in soy sauce-marinated and frozen freshwater crabs. *Parasitology Research* **116**: 1003–1006. PMID: 28127717. https://doi.org/10.1007/s00436-017-5380-7.

Kirchgäßner, M., Schmahl, G., Al-Quraishy, S. et al. (2008). What are the infectious larvae in *Ascaris suum* and *Trichuris muris*? *Parasitology Research* **103**: 603–607. https://doi.org/10.1007/s00436-008-1018-0.

Kirkness, E.F., Haas, B.J., Sun, W. et al. (2010). Genome sequences of the human body louse and its primary endosymbiont provide insights into the permanent parasitic lifestyle. *Proceedings of the National Academy of Sciences of the United States of America* **107**: 12168–12173. https://doi.org/10.1073/pnas.1003379107.

Kissinger, P. (2015). *Trichomonas vaginalis*: a review of epidemiologic, clinical and treatment issues. *BMC Infectious Diseases* **15**: 307. https://doi.org/10.1186/s12879-015-1055-0.

Klein, S., Wendt, M., Baumgärtner, W. et al. (2010). Systemic toxoplasmosis and concurrent porcine circovirus-2 infection in a pig. *Journal of Comparative Pathology* **142**: 228–234. https://doi.org/10.1016/j.jcpa.2009.08.155.

Klemm, S.L., Shipony, Z., and Greenleaf, W.J. (2019). Chromatin accessibility and the regulatory epigenome. *Nature Reviews Genetics* **20**: 207–220. https://doi.org/10.1038/s41576-018-0089-8.

Koeppen, D., Aurich, M., Pasalar, M. et al. (2020). Medicinal leech therapy in venous congestion and various ulcer forms: perspectives of Western, Persian and Indian medicine. *Journal of Traditional and Complementary Medicine* **10**: 104–109. https://doi.org/10.1016/j.jtcme.2019.08.003.

Kojour, A.M., Han, Y.S., and Jo, Y.H. (2020). An overview of insect innate immunity. *Entomological Research* **50**: 282–291. 10.1111/1748-5967.12437.

Kokoza, V., Ahmed, A., Shin, S.W. et al. (2010). Blocking of *Plasmodium* transmission by cooperative action of Cecropin A and Defensin A in transgenic *Aedes aegypti* mosquitoes. *Proceedings of the National Academy of Sciences of the United States of America* **107**: 8111–8116. https://doi.org/10.1073/pnas.1003056107.

Kolkhir, P., Balakirski, G., Merk, H.F. et al. (2016). Chronic spontaneous urticaria and internal parasites – a systematic review. *Allergy* **71**: 308–322. PMID: 26648083. https://doi.org/10.1111/all.12818.

König, A., Romig, T., Thoma, D. et al. (2005). Drastic increase in the prevalence of *Echinococcus multilocularis* in foxes (*Vulpes vulpes*) in southern Bavaria, Germany. *European Journal of Wildlife Research* **51**: 277–282. https://doi.org/10.1007/s10344-005-0100-5.

Koonin, E.V. and Starokadomskyy, P. (2016). Are viruses alive? The replicator paradigm sheds decisive light on an old but misguided question. *Studies in History and Philosophy of Science part C: Studies in History and Philosophy of Biological and Biomedical Sciences* **59**: 125–134. https://doi.org/10.1016/j.shpsc.2016.02.016.

Kooyman, F.N., Wagenaar, J.A., and Zomer, A. (2019). Whole-genome sequencing of dog-specific assemblages C and D of *Giardia duodenalis* from single and pooled cysts indicates host-associated genes. *Microbial Genomics* **5**: e000302. https://doi.org/10.1099/mgen.0.000302.

Kordalis, N.G., Arsenopoulos, K., Vasileiou, N.G.C. et al. (2019). Field evidence for association between increased gastrointestinal nematode burden and subclinical mastitis in dairy sheep. *Veterinary Parasitology* **265**: 56–62. https://doi.org/10.1016/j.vetpar.2018.11.010.

Kosintsev, P.A., Lapteva, E.G., Trofimova, S.S. et al. (2010). The intestinal contents of a baby woolly mammoth (*Mammuthus primigenius* Blumenbach, 1799) from the Yuribey River (Yamal Peninsula). *Doklady Biological Sciences* **432**: 209–211. https://doi.org/10.1134/S0012496610030129.

Kozlov, D.P. (1974). The role of birds in the dissemination of infection of Anoplocephalata. *Trudy Gel'mintologicheskoi Laboratorii (Ekologiya i Geografiya Gel'mintov)* **24**: 62–63. (in Russian).

Krementsov, N. (2009). *Trypanosoma cruzi*, câncer e a Guerra Fria. *História, Ciências, Saúde-Manguinhos* **16**: 75–94. https://doi.org/10.1590/S0104-59702009000500005.

Kreuder, C., Miller, M.A., Jessup, D.A. et al. (2003). Patterns of mortality in southern sea otters (*Enhydra lutris nereis*) from 1998–2001. *Journal of Wildlife Diseases* **39**: 495–509. https://doi.org/10.7589/0090-3558-39.3.495.

Kriel, J.D. (2003). Witches, healers, and helminths: Sotho beliefs regarding the utilisation of the latent power of phenomena. *Anthropology Southern Africa* **26**: 167–171.

Krishnamani, R. and Mahaney, W.C. (2000). Geophagy among primates: adaptive significance and ecological consequences. *Animal Behaviour* **59**: 899–915. https://doi.org/10.1006/anbe.1999.1376.

Kroidl, I., Saathoff, E., Maganga, L. et al. (2016). Effect of *Wuchereria bancrofti* infection on HIV incidence in southwest Tanzania: a prospective cohort study. *Lancet* **388**: 1912–1920. https://doi.org/10.1016/S0140-6736(16)31252-1.

Kruger, F.J. and Evans, A.C. (1990). Do all human urinary infections with *Schistosoma mattheei* represent hybridisation between *S. haematobium* and *S. mattheeei*? *Journal of Helminthology* **64**: 330–332. https://doi.org/10.1017/S0022149X00012384.

Kucknoor, A.S., Mundodi, V., and Alderete, J.F. (2009). Genetic identity and differential gene expression between *Trichomonas vaginalis* and *Trichomonas tenax*. *BMC Microbiology* **9**: https://doi.org/10.1186/1471-2180-9-58.

Kudo, R. and Hetherington, D.C. (1922). Notes on a microsporidian parasite of a nematode. *Journal of Parasitology* **8**: 129–132.

Kuhn, K.G., Campbell-Lendrum, D.H., Armstrong, B. et al. (2003). Malaria in Britain: past, present, and future. *Proceedings of the National Academy of Sciences of the United States of America* **100**: 9997–10001. https://doi.org/10.1073/pnas.1233687100.

Kumar, A. and Samant, M. (2016). DNA vaccine against visceral leishmaniasis: a promising approach for prevention and control. *Parasite Immunology* **38**: 273–281. https://doi.org/10.1111/pim.12315.

Kumar, R., Verma, A.K., Shrivas, S. et al. (2020). First successful field evaluation of new, one-minute haemozoin-based malaria diagnostic device. *EClinicalMedicine* **22**: https://doi.org/10.1016/j.eclinm.2020.100347.

Kumar, V., Abbas, A.K., and Aster, J.C. (2020). *Robbins & Cotran Pathologic Basis of Disease*, 10the. Philadelphia, PA: Elsevier.

Kumari, P., Nigam, R., Singh, A. et al. (2017). *Demodex canis* regulates cholinergic system mediated immunosuppressive pathways in canine demodicosis. *Parasitology* **144**: 1412–1416. https://doi.org/10.1017/S0031182017000774.

Kurup, S.P., Anthony, S.M., Hancox, L.S. et al. (2019). Monocyte-derived CD11c+ cells acquire *Plasmodium* from hepatocytes to prime CD8 T cell immunity to liver-stage malaria. *Cell Host & Microbe* **25**: 565–577. https://doi.org/10.1016/j.chom.2019.02.014.

Kuznetsov, M.I. (1968). Length of life of *Moniezia expansa* in lambs. *Papers on Helminthology Presented to Academician K.I. Skryabina on His 90th Birthday*, Moscow, Izdat. Akademii Nauchnykh S.S.R. 220–222. (in Russian).

Kwak, M.L. and Schubert, J. (2019). Utilizing ticks as forensic indicators in a livestock investigation. *Forensic Science, Medicine and Pathology* **15**: 119–121. https://doi.org/10.1007/s12024-018-0067-7.

Kwenti, T.E. (2018). Malaria and HIV coinfection in sub-Saharan Africa: prevalence, impact, and treatment strategies. *Research and Reports in Tropical Medicine* **9**: 123–136. https://doi.org/10.2147/RRTM.S154501.

Kwiatkowski, D. (1990). Tumour necrosis factor, fever and fatality in falciparum malaria. *Immunology Letters* **25**: 213–216. https://doi.org/10.1016/0165-2478(90)90117-9.

Ladle, R.J. (1992). Parasites and sex: catching the red queen. *Trends in Ecology and Evolution* **7**: 405–408. https://doi.org/10.1016/0169-5347(92)90021-3.

Laere, E., Ling, A.P.K., Wong, Y.P. et al. (2016). Plant-based vaccines: production and challenges. *Journal of Botany* https://doi.org/10.1155/2016/4928637.

Lafferty, K.D. (2006). Can the common brain parasite, *Toxoplasma gondii*, influence human culture? *Proceedings of the Royal Society B: Biological Sciences* **273**: 2749–2755. https://doi.org/10.1098/rspb.2006.3641.

Lafferty, K.D. and Kuris, A.M. (2009). Parasitic castration: the evolution and ecology of body snatchers. *Trends in Parasitology* **25**: 564–572. PMID: 19800291. https://doi.org/10.1016/j.pt.2009.09.003.

Lafferty, K.D. and Mordecai, E.A. (2016). The rise and fall of infectious disease in a warmer world. *F1000Research* **5**: doi: 10.12688/f1000research.8766.1.

Lainson, R. and Shaw, J.J. (1987). Evolution, classification and geographical distribution. In: *The Leishmaniasis in Biology and Epidemiology* (ed. W. Peters and R. Killick-Kendrick), 1–120. London, UK: Academic Press.

Laird, M. (1977). *Tsetse: The Future for Biological Methods in Integrated Control*. Ottawa, Canada: International Development Research Centre.

Lam, N.S., Long, X., Su, X.Z. et al. (2018). Artemisinin and its derivatives in treating helminthic infections beyond schistosomiasis. *Pharmacological Research* **133**: 77–100. https://doi.org/10.1016/j.phrs.2018.04.025.

Lamb, T. (2012). *Immunity to Parasitic Infections*. Chichester, UK: Wiley-Blackwell.

Lambrechts, L., Halbert, J., Durand, P. et al. (2005). Host genotype by parasite genotype interactions underlying the resistance of anopheline mosquitoes to *Plasmodium falciparum*. *Malaria Journal* **4**: 3. https://doi.org/10.1186/1475-2875-4-3.

Langley, J.G., Goldsmid, J.M., and Davies, N. (1987). Venereal trichomoniasis: role of men. *Sexually Transmitted Infections* **63**: 264–267. https://doi.org/10.1136/sti.63.4.264.

Larsen, M.N. and Roepstorff, A. (1999). Seasonal variation in development and survival of *Ascaris suum* and *Trichuris suis* eggs on pastures. *Parasitology* **119**: 209–220. https://doi.org/10.1017/S0031182099004503.

Laudisoit, A., Leirs, H., Makundi, R.H. et al. (2007). Plague and the human flea, Tanzania. *Emerging Infectious Diseases* **13**: 687–693. https://doi.org/10.3201/eid1305.061084.

Laurenti, M.D., de Santana Leandro Jr, M.V., Tomokane, T.Y. et al. (2014). Comparative evaluation of the DPP* CVL rapid test for canine serodiagnosis in area of visceral leishmaniasis. *Veterinary Parasitology* **205**: 444–450. https://doi.org/10.1016/j.vetpar.2014.09.002.

Lazar, V., Ditu, L.M., Pircalabioru, G.G. et al. (2018). Aspects of gut microbiota and immune system interactions in infectious diseases, immunopathology, and cancer. *Frontiers in Immunology* **9**: 1830. https://doi.org/10.3389/fimmu.2018.01830.

Leander, B.S. (2004). Did trypanosomatid parasites have photosynthetic ancestors? *Trends in Microbiology* **12**: 251–258. https://doi.org/10.1016/j.tim.2004.04.001.

Ledizet, M., Harrison, L.M., Koski, R.A. et al. (2005). Discovery and pre-clinical development of antithrombotics from hematophagous invertebrates. *Current Medicinal Chemistry – Cardiovascular & Hematological Agents* **3**: 1–10. https://doi.org/10.2174/1568016052773315.

Lee, C., Noh, J., O'Neal, S.E. et al. (2019). Feasibility of a point-of-care test based on quantum dots with a mobile phone reader for detection of antibody responses. *PLoS Neglected Tropical Diseases* **13**: https://doi.org/10.1371/journal.pntd.0007746.

Lee, D.L. (1996). Why do some nematode parasites of the alimentary tract secrete acetylcholinesterase? *International Journal for Parasitology* **26**: 499–508. https://doi.org/10.1016/0020-7519(96)00040-9.

Lee, D.L. (2001). *The Biology of Nematodes*. London: Taylor & Francis.

Lefèvre, T., Vantaux, A., Dabire, K.R. et al. (2013). Non-genetic determinants of mosquito competence for malaria parasites. *PLoS Pathogens* **9**: https://doi.org/10.1371/journal.ppat.1003365.

Lefoulon, E., Bain, O., Bourret, J. et al. (2015). Shaking the tree: multi-locus sequence typing usurps current onchocercid (Filarial Nematode) phylogeny. *PLoS Neglected Tropical Diseases* **9**: e0004233. https://doi.org/10.1371/journal.pntd.0004233.

Lefoulon, E., Bain, O., Makepeace, B.L. et al. (2016). Breakdown of coevolution between symbiotic bacteria *Wolbachia* and their filarial hosts. *PeerJ* **4**: e1840. https://doi.org/10.7717/peerj.1840.

Leger, E. and Webster, J.P. (2017). Hybridizations within the genus *Schistosoma*: implications for evolution, epidemiology and control. *Parasitology* **144**: 65–80. https://doi.org/10.1017/S0031182016001190.

Lehane, M.J. (2005). *The Biology of Blood-Sucking Insects*, 2nde. Cambridge, UK: Cambridge University Press.

Lepage, O.M., Doumbia, A., Perron-Lepage, M.F. et al. (2012). The use of maggot debridement therapy in 41 equids. *Equine Veterinary Journal* **44**: 120–125. https://doi.org/10.1111/j.2042-3306.2012.00609.x.

Lestinova, T., Rohousova, I., Sima, M. et al. (2017). Insights into the sand fly saliva: blood-feeding and immune interactions between sand flies, hosts, and *Leishmania*. *PLoS Neglected Tropical Diseases* **11**: https://doi.org/10.1371/journal.pntd.0005600.

Leung, K., Ras, E., Ferguson, K.B. et al. (2020). Next-generation biological control: the need for integrating genetics and genomics. *Biological Reviews* **95**: 1838–1854. https://doi.org/10.1111/brv.12641.

Leung, T.L. (2017). Fossils of parasites: what can the fossil record tell us about the evolution of parasitism? *Biological Reviews* **92**: 410–430. https://doi.org/10.1111/brv.12238.

Leventhal, R. and Cheadle, R.F. (2019). *Medical Parasitology: A Self-Instructional Text*. Philadelphia, PA: FA Davis Company.

Lewin, R.A. (1999). *Merde Excursions into Scientific, Cultural and Socio-Historical Coprology*. London, UK: Aurum Press.

Lew-Tabor, A.E. and Rodriguez Valle, M. (2016). A review of reverse vaccinology approaches for the development of vaccines against ticks and tick borne diseases. *Ticks and Tick-borne Diseases* **7**: 573–585. https://doi.org/10.1016/j.ttbdis.2015.12.012.

Li, J., Wang, R., Chen, Y. et al. (2020). *Cyclospora cayetanensis* infection in humans: Biological characteristics, clinical features, epidemiology, detection method and treatment. *Parasitology* **147**: 160–170. https://doi.org/10.1017/S0031182019001471.

Light, J.E., Toups, M.A., and Reed, D.L. (2008). What's in a name: the taxonomic status of human head and body lice. *Molecular Phylogenetics and Evolution* **47**: 1203–1216. https://doi.org/10.1016/j.ympev.2008.03.014.

Ligoxygakis, P. (2017). *Insect Immunity*. Cambridge, MA: Academic Press.

Lim, L., Sayers, C.P., Goodman, C.D. et al. (2016). Targeting of a transporter to the outer apicoplast membrane in the human malaria parasite *Plasmodium falciparum*. *PloS One* **11**: https://doi.org/10.1371/journal.pone.0159603.

Lin, F., Bao, Y.W., and Wu, F.G. (2019). Carbon dots for sensing and killing microorganisms. *C – Journal of Carbon Research* **5**: 33. https://doi.org/10.3390/c5020033.

Lincicome, D.R. (1971). The goodness of parasitism: a new hypothesis. In: *Aspects of the Biology of Symbiosis* (ed. T.C. Cheng), 139–227. Baltimore, MD: University Park Press.

Linger, R.J., Belikoff, E.J., Yan, Y. et al. (2016). Towards next generation maggot debridement therapy: transgenic *Lucilia sericata* larvae that produce and secrete a human growth factor. *BMC Biotechnology* **16**: 1–12. https://doi.org/10.1186/s12896-016-0263-z.

Lippitz, B.E. (2013). Cytokine patterns in patients with cancer: a systematic review. *Lancet Oncology* **14**: e218–e228. PMID: 23639322. https://doi.org/10.1016/S1470-2045(12)70582-X.

Little, M.D. (1961). Observations on the possible role of insects as paratenic hosts for *Ancylostoma caninum*. *Journal of Parasitology* **47**: 263–267. https://www.jstor.org/stable/3275302.

Liu, F., Fan, X., Li, L. et al. (2018). Development of recombinant goatpox virus expressing *Echinococcus granulosus* EG95 vaccine antigen. *Journal of Virological Methods* **261**: 28–33. https://doi.org/10.1016/j.jviromet.2018.08.002.

Liu, W., Li, Y., Learn, G.H. et al. (2010). Origin of the human malaria parasite *Plasmodium falciparum* in gorillas. *Nature* **467**: 420–425. https://doi.org/10.1038/nature09442.

Liu, X., Walton, S., and Mounsey, K. (2014). Vaccine against scabies: necessity and possibility. *Parasitology* **141**: 725–732. https://doi.org/10.1017/S0031182013002047.

Liu, Y., Reichel, M.P., and Lo, W.C. (2020). Combined control evaluation for *Neospora caninum* infection in dairy: economic point of view coupled with population dynamics. *Veterinary Parasitology* **277**: https://doi.org/10.1016/j.vetpar.2019.108967.

Lively, C.M. (1996). Host-parasite coevolution and sex. *Bioscience* **46**: 107–114. https://doi.org/10.2307/1312813.

Lively, C.M. and Jokela, J. (2002). Temporal and spatial distributions of parasites and sex in a freshwater snail. *Evolutionary Ecology Research* **4**: 219–226.

Löhr, K.F., Pholpark, S., Siriwan, P. et al. (1986). *Trypanosoma evansi* infection in buffaloes in North-East Thailand. II. Abortions. *Tropical Animal Health and Production* **18**: 103–108. https://doi.org/10.1007/BF02359721.

Loke, P. and Lim, Y.A. (2015). Helminths and the microbiota: parts of the hygiene hypothesis. *Parasite Immunology* **37**: 314–323. https://doi.org/10.1111/pim.12193.

Loker, E.S. (1983). A comparative study of the life-histories of mammalian schistosomes. *Parasitology* **87**: 343–369. https://doi.org/10.1017/S0031182000052689.

Loo, C.S.N., Lam, N.S.K., Yu, D. et al. (2017). Artemisinin and its derivatives in treating protozoan infections beyond malaria. *Pharmacological Research* **117**: 192–217. https://doi.org/10.1016/j.phrs.2016.11.012.

Lopes, L.B., Nicolino, R., Capanema, R.O. et al. (2015a). Economic impacts of parasitic diseases in cattle. *CAB Reviews* **10**: 1–10. https://doi.org/10.1079/PAVSNNR201510051.

Lopes, V.V., dos Santos, H.A., da Silva, A.V.M. et al. (2015b). First case of human infection by *Bertiella studeri* (Blanchard, 1891) Stunkard, 1940 (Cestoda; Anoplocephalidae) in Brazil. *Revista do Instituto de Medicina Tropical de São Paulo* **57**: 447–450. https://doi.org/10.1590/S0036-46652015000500015.

Lopes, W.D.Z., Rodriguez, J.D.A., Souza, F.A. et al. (2013). Sexual transmission of *Toxoplasma gondii* in sheep. *Veterinary Parasitology* **195**: 47–56. https://doi.org/10.1016/j.vetpar.2012.12.056.

Lopez-Chaves, C., Soto-Alvaredo, J., Montes-Bayon, M. et al. (2018). Gold nanoparticles: distribution, bioaccumulation and toxicity. *in vitro* and *in vivo* studies. *Nanomedicine: Nanotechnology. Biology and Medicine* **14**: 1–12. https://doi.org/10.1016/j.nano.2017.08.011.

Lorenz, L.M. and Koella, J.C. (2011). Maternal environment shapes the life history and susceptibility to malaria of *Anopheles gambiae* mosquitoes. *Malaria Journal* **10**: https://doi.org/10.1186/1475-2875-10-382.

Lorenzo-Morales, J., Cabello-Vílchez, A.M., Martín-Navarro, C.M. et al. (2013). Is *Balamuthia mandrillaris* a public health concern worldwide? *Trends in Parasitology* **29**: 483–488. https://doi.org/10.1016/j.pt.2013.07.009.

Lotter, H., González-Roldán, N., Lindner, B. et al. (2009). Natural killer T cells activated by a lipopeptidophosphoglycan from *Entamoeba histolytica* are critically important to control amebic liver abscess. *PLoS Pathogens* **5**: https://doi.org/10.1371/journal.ppat.1000434.

Loy, D.E., Liu, W., Li, Y. et al. (2017). Out of Africa: origins and evolution of the human malaria parasites *Plasmodium falciparum* and *Plasmodium vivax*. *International Journal for Parasitology* **47**: 87–97. https://doi.org/10.1016/j.ijpara.2016.05.008.

Lu, D., Macchietto, M., Chang, D. et al. (2017). Activated entomopathogenic nematode infective juveniles release lethal venom proteins. *PLoS Pathogens* **13**: https://doi.org/10.1371/journal.ppat.1006302.

Lucht, E., Evengård, B., Skott, J. et al. (1998). *Entamoeba gingivalis* in human immunodeficiency virus type 1-infected patients with periodontal disease. *Clinical Infectious Diseases* **27**: 471–473. https://doi.org/10.1086/514709.

Lund, M.E., Greer, J., Dixit, A. et al. (2016). A parasite-derived 68-mer peptide ameliorates autoimmune disease in murine models of Type 1 diabetes and multiple sclerosis. *Scientific Reports* **6**: https://doi.org/10.1038/srep37789.

Lunde, C.S., Stebbins, E.E., Jumani, R.S. et al. (2019). Identification of a potent benzoxaborole drug candidate for treating cryptosporidiosis. *Nature Communications* **10**: 2816. https://doi.org/10.1038/s41467-019-10687-y.

Luo, J., Yu, L.I., Xie, G. et al. (2017). Study on the mitochondrial apoptosis pathways of small cell lung cancer H446 cells induced by *Trichinella spiralis* muscle larvae ESPs. *Parasitology* **144**: 793–800. https://doi.org/10.1017/S0031182016002535.

Luong, L.T., Platzer, E.G., Zuk, M. et al. (2000). Venereal worms: sexually transmitted nematodes in the decorated cricket. *Journal of Parasitology* **86**: 471–477. https://doi.org/10.1645/0022-3395(2000)086[0471:VWSTNI]2.0.CO;2.

Ma, G., Holland, C.V., Wang, T. et al. (2018). Human toxocariasis. *Lancet Infectious Diseases* **18**: e14–e24. https://doi.org/10.1016/S1473-3099(17)30331-6.

Ma, G., Wang, T., Korhonen, P.K. et al. (2019). Comparative bioinformatic analysis suggests that specific dauer-like signalling pathway components regulate *Toxocara canis* development and migration in the mammalian host. *Parasites & Vectors* **12**: 32. https://doi.org/10.1186/s13071-018-3265-y.

Ma, K., Zhang, H., and Baloch, Z. (2016). Pathogenetic and therapeutic applications of tumor necrosis factor-α (TNF-α) in major depressive disorder: a systematic review. *International Journal of Molecular Sciences* **17**: 733. https://doi.org/10.3390/ijms17050733.

Mabbott, N.A. (2018). The influence of parasite infections on host immunity to co-infection with other pathogens. *Frontiers in Immunology* **9**: 2579. https://doi.org/10.3389/fimmu.2018.02579.

Mac-Daniel, L. and Ménard, R. (2019). Live vaccines against *Plasmodium* preerythrocytic stages. In: *Malaria Control and Elimination. Methods in Molecular Biology*, vol. **2013** (ed. F. Ariey, F. Gay and R. Ménard), 189–198. New York, NY: Humana 10.1007/978-1-4939-9550-9_14.

Machado, M., Dinis, A.M., Salgueiro, L. et al. (2010). Anti-*Giardia* activity of phenolic-rich essential oils: effects of *Thymbra capitata*, *Origanum virens*, *Thymus zygis* subsp. *sylvestris*, and *Lippia graveolens* on trophozoites growth, viability, adherence, and ultrastructure. *Parasitology Research* **106**: 1205–1215. https://doi.org/10.1007/s00436-010-1800-7.

Maciver, S.K., Piñero, J.E., and Lorenzo-Morales, J. (2020). Is *Naegleria fowleri* an emerging parasite? *Trends in Parasitology* **36**: 19–28. https://doi.org/10.1016/j.pt.2019.10.008.

Macnab, V. and Barber, I. (2012). Some (worms) like it hot: fish parasites grow faster in warmer water and alter host thermal preferences. *Global Change Biology* **18**: 1540–1548. https://doi.org/10.1111/j.1365-2486.2011.02595.x.

Macnab, V., Katsiadaki, I., Tilley, C.A. et al. (2016). Oestrogenic pollutants promote the growth of a parasite in male sticklebacks. *Aquatic Toxicology* **174**: 92–100. https://doi.org/10.1016/j.aquatox.2016.02.010.

Maffrand, R., Avila-Vazquez, M., Princich, D. et al. (2006). Congenital ocular toxocariasis in a premature neonate. *Anales de Pediatria* **64**: 599–600. (in Spanish) https://doi.org/10.1157/13089931.

Magalhães, L.M.D., Passos, L.S.A., Fujiwara, R.T. et al. (2020). Immunopathology and modulation induced by hookworms: from understanding to intervention. *Parasite Immunology* https://doi.org/10.1111/pim.12798.

Magez, S. and Radwanska, M. (ed.) (2014). *Trypanosomes and Trypanosomiasis*. Vienna, Austria: Springer.

Magro, A.G., Assis, V.P., Silva, L.C. et al. (2017). *Leishmania infantum* is present in vaginal secretions of naturally infected bitches at lower levels in oestrogenized bitches than in non-oestrogenized bitches. *Acta Parasitologica* **62**: 625–629. https://doi.org/10.1515/ap-2017 0076.

Mahinc, C., Flori, P., Delaunay, E. et al. (2017). Evaluation of a new immunochromatography technology test (LDBio diagnostics) to detect *Toxoplasma* IgG and IgM: comparison with the routine architect technique. *Journal of Clinical Microbiology* **55**: 3395–3404. https://doi.org/10.1128/JCM.01106-17.

Mahmood, K. (2015). *Naegleria fowleri* in Pakistan-an emerging catastrophe. *Journal of the College Physicians Surgery Pakistan* **25**: 159–160. PMID: 25772952.

Maia, Z., Lírio, M., Mistro, S. et al. (2012). Comparative study of rK39 *Leishmania* antigen for serodiagnosis of visceral leishmaniasis: systematic review with meta-analysis. *PLoS Neglected Tropical Diseases* **6**: https://doi.org/10.1371/journal.pntd.0001484.

Maizels, R.M. (2016). Parasitic helminth infections and the control of human allergic and autoimmune disorders. *Clinical Microbiology and Infection* **22**: 481–486. https://doi.org/10.1016/j.cmi.2016.04.024.

Maizels, R.M. (2020). Regulation of immunity and allergy by helminth parasites. *Allergy* **75**: 524–534. https://doi.org/10.1016/j.cmi.2016.04.024.

Maizels, R.M. and McSorley, H.J. (2016). Regulation of the host immune system by helminth parasites. *Journal of Allergy and Clinical Immunology* **138**: 666–675. https://doi.org/10.1016/j.jaci.2016.07.007.

Major, R.H. (1954). *A History of Medicine*, vol. **1**. Springfield, MO: Charles Thomas.

Makala, L.H., Baban, B., Lemos, H. et al. (2011). *Leishmania major* attenuates host immunity by stimulating local indoleamine 2, 3-dioxygenase expression. *Journal of Infectious Diseases* **203**: 715–725. https://doi.org/10.1093/infdis/jiq095.

Malik, A., Steinbeis, F., Carillo, M.A. et al. (2019). Immunological evaluation of synthetic glycosylphosphatidylinositol glycoconjugates as vaccine candidates against malaria. *ACS Chemical Biology* **15**: 171–178. https://doi.org/10.1021/acschembio.9b00739.

Maltby, R. (1999). The language of Plautus's parasites. http://www2.open.ac.uk/ClassicalStudies/GreekPlays/Conf99/Maltby.htm (accessed 1 December 2012).

Manguin, S., Bangs, M.J., Pothikasikorn, J. et al. (2010). Review on global co-transmission of human *Plasmodium* species and *Wuchereria bancrofti* by *Anopheles* mosquitoes. *Infection, Genetics and Evolution* **10**: 159–177. https://doi.org/10.1016/j.meegid.2009.11.014.

Mans, B.J., De Klerk, D.G., Pienaar, R. et al. (2014). The host preferences of Nuttalliella namaqua (Ixodoidea: Nuttalliellidae): a generalist approach to surviving multiple host-switches. *Experimental and Applied Acarology* **62**: 33–240. https://doi.org/10.1007/s10493-013-9737-z.

Manzoor, M.U., Faruqui, Z.S., Ahmed, Q. et al. (2008). Aspergilloma complicating newly diagnosed pulmonary echinococcal (hydatid) cyst: a rare occurrence. *British Journal of Radiology* **81**: e279–e281. https://doi.org/10.1259/bjr/23110281.

Marino, A.M.F., Giunta, R.P., Salvaggio, A. et al. (2019). *Toxoplasma gondii* in edible fishes captured in the Mediterranean basin. *Zoonoses and Public Health* **66**: 826–834. https://doi.org/10.1111/zph.12630.

Marris, E. (2007). Linnaeus at 300: the species and the specious. *Nature* **446**: 250–253. https://doi.org/10.1038/446250a.

Martínez-Rojano, H., Noguez, J.C., and Huerta, H. (2018). Nosocomial myiasis caused by *Lucilia sericata* (Diptera: Calliphoridae) and neonatal myiasis by *Sarcophaga* spp. (Diptera: Sarcophagidae) in Mexico. *Case Reports in Infectious Diseases* **2018**: https://doi.org/10.1155/2018/5067569.

Martin-Martin, I., Chagas, A.C., Guimaraes-Costa, A.B. et al. (2018). Immunity to LuloHya and Lundep, the salivary spreading factors from *Lutzomyia longipalpis*, protects against *Leishmania major* infection. *PLoS Pathogens* **14**: https://doi.org/10.1371/journal.ppat.1007006.

Martín-Vega, D., Garbout, A., Ahmed, F. et al. (2018). 3D virtual histology at the host/parasite interface: visualisation of the master manipulator, *Dicrocoelium dendriticum,* in the brain of its ant host. *Scientific Reports* **8**: 1–10. https://doi.org/10.1038/s41598-018-26977-2.

Marty, M., Lemaitre, M., Kémoun, P. et al. (2017). *Trichomonas tenax* and periodontal diseases: a concise review. *Parasitology* **144**: 1417–1425. https://doi.org/10.1017/S0031182017000701.

Marzano, V., Mancinelli, L., Bracaglia, G. et al. (2017). 'Omic' investigations of protozoa and worms for a deeper understanding of the human gut 'parasitome'. *PLoS Neglected Tropical Diseases* **11**: https://doi.org/10.1371/journal.pntd.0005916.

Masha, S.C., Cools, P., Sanders, E.J. et al. (2019). *Trichomonas vaginalis* and HIV infection acquisition: a systematic review and meta-analysis. *Sexually Transmitted Infections* **95**: 36–42. https://doi.org/10.1136/sextrans-2018-053713.

Masiero, F.S., Aguiar, E.S.V., Pereira, D.I.B. et al. (2020). First report on the use of larvae of *Cochliomyia macellaria* (Diptera: Calliphoridae) for wound treatment in veterinary practice. *Journal of Medical Entomology* **57**: 965–968. PMID: 27696133. https://doi.org/10.1007/s13744-016-0444-4.

Mathison, B.A., Bishop, H.S., Sanborn, C.R. et al. (2016). *Macracanthorhynchus ingens* infection in an 18-month-old child in Florida: a case report and review of acanthocephaliasis in humans. *Clinical Infectious Diseases* **63**: 1357–1359. https://doi.org/10.1093/cid/ciw543.

Mathison, B.A., Couturier, M.R., and Pritt, B.S. (2019). Diagnostic identification and differentiation of microfilariae. *Journal of Clinical Microbiology* **57**: https://doi.org/10.1128/JCM.00706-19.

Mathison, B.A. and Pritt, B.S. (2017). Update on malaria diagnostics and test utilization. *Journal of Clinical Microbiology* **55**: https://doi.org/10.1128/JCM.02562-16.

Mathur, V., Kolisko, M., Hehenberger, E. et al. (2019). Multiple independent origins of apicomplexan-like parasites. *Current Biology* **29**: 2936–2941. https://doi.org/10.1016/j.cub.2019.07.019.

Maudlin, I. and Welburn, S.C. (1989). A single trypanosome is sufficient to infect a tsetse fly. *Annals of Tropical Medicine and Parasitology* **83**: 431–433. https://doi.org/10.1080/00034983.1989.11812368.

Mautner, S.I., Cook, K.A., Forbes, M.R. et al. (2007). Evidence for sex ratio distortion by a new microsporidian parasite of a Corophiid amphipod. *Parasitology* **134**: 1567–1573. https://doi.org/10.1017/S0031182007003034.

McAuley, C.F., Webb, C., Makani, J. et al. (2010). High mortality from *Plasmodium falciparum* malaria in children living with sickle cell anemia on the coast of Kenya. *Blood, The Journal of the American Society of Hematology* **116**: 1663–1668. https://doi.org/10.1182/blood-2010-01-265249.

McBurney, S., Kelly-Clark, W.K., Forzán, M.J. et al. (2017). Persistence of *Trichomonas gallinae* in birdseed. *Avian Diseases* **61**: 311–315. https://doi.org/10.1637/11545-113016-RegR1.

McDermott, J.R., Leslie, F.C., D'Amato, M. et al. (2006). Immune control of food intake: enteroendocrine cells are regulated by CD4+ T lymphocytes during small intestinal inflammation. *Gut* **55**: 492–497. https://doi.org/10.1136/gut.2005.081752.

McDonald, R.A., Wilson-Aggarwal, J.K., Swan, G.J. et al. (2020). Ecology of domestic dogs *Canis familiaris* as an emerging reservoir of Guinea worm *Dracunculus medinensis* infection. *PLOS Neglected Tropical Diseases* **14**: https://doi.org/10.1371/journal.pntd.0008170.

McDougald, L.R. and Fuller, L. (2005). Blackhead disease in turkeys: direct transmission of *Histomonas meleagridis* from bird to bird in a laboratory model. *Avian Diseases* **49**: 328–331. https://doi.org/10.1637/7257-081004R.1.

McKee, K.M., Koprivnikar, J., Johnson, P.T. et al. (2020). Parasite infectious stages provide essential fatty acids and lipid-rich resources to freshwater consumers. *Oecologia* **192**: 477–488. https://doi.org/10.1007/s00442-019-04572-0.

McLachlan, R.S. (2010). Julius Caesar's late onset epilepsy: a case of historic proportions. *Canadian Journal of Neurological Science* **37**: 557–561. https://doi.org/10.1017/S0317167100010696.

McMichael, A.J. (2000). The urban environment and health in a world of increasing globalization: issues for developing countries. *Bulletin of the World Health Organization* **78**: 1117–1126.

McNew, S.M. and Clayton, D.H. (2018). Alien invasion: biology of Philornis flies highlighting *Philornis downsi*, an introduced parasite of Galápagos Birds. *Annual Review of Entomology* **63**: 369–387. https://doi.org/10.1146/annurev-ento-020117-043103.

McWilliam, H.E., Driguez, P., Piedrafita, D. et al. (2012). Novel immunomic technologies for schistosome vaccine development. *Parasite Immunology* **34**: 276–284. https://doi.org/10.1111/j.1365-3024.2011.01330.x.

Meckel, L.A., McDaneld, C.P., and Wescott, D.J. (2018). White-tailed deer as a taphonomic agent: photographic evidence of white-tailed deer gnawing on human bone. *Journal of Forensic Sciences* **63**: 292–294. https://doi.org/10.1111/1556-4029.13514.

Mehlhorn, H. (2016). *Animal Parasites: Diagnosis, Treatment, Prevention*. Cham: Springer.

Mehraj, V., Hatcher, J., Akhtar, S. et al. (2008). Prevalence and factors associated with intestinal parasitic infection among children in an urban slum of Karachi. *PloS One* **3**: https://doi.org/10.1371/journal.pone.0003680.

Mekonnen, G.G., Pearson, M., Loukas, A. et al. (2018). Extracellular vesicles from parasitic helminths and their potential utility as vaccines. *Expert Review of Vaccines* **17**: 197–205. https://doi.org/10.1080/14760584.2018.1431125.

Menees, S. and Chey, W. (2018). The gut microbiome and irritable bowel syndrome. *F1000Research* **7**: F1000 Faculty Rev-1029: 10.12688/f1000research.14592.1.

Menendez, C., Ordi, J., Ismail, M.R. et al. (2000). The impact of placental malaria on gestational age and birth weight. *Journal of Infectious Diseases* **181**: 1740–1745. https://doi.org/10.1086/315449.

Menezes, R.G., Kanchan, T., Rai, S. et al. (2010). An autopsy case of sudden unexplained death caused by malaria. *Journal of Forensic Sciences* **55**: 835–838. https://doi.org/10.1111/j.1556-4029.2010.01328.x.

Mercer, F. and Johnson, P.J. (2018). *Trichomonas vaginalis*: Pathogenesis, symbiont interactions, and host cell immune responses. *Trends in Parasitology* **34**: 683–693. https://doi.org/10.1016/j.pt.2018.05.006.

Merola, V.M. and Eubig, P.A. (2012). Toxicology of avermectins and milbemycins (macrocyclic lactones) and the role of *P*-glycoprotein in dogs and cats. *Veterinary Clinics: Small Animal Practice* **42**: 313–333. https://doi.org/10.1016/j.cvsm.2011.12.005.

Midha, A., Janek, K., Niewienda, A. et al. (2018). The intestinal roundworm *Ascaris suum* releases antimicrobial factors which interfere with bacterial growth and biofilm formation. *Frontiers in Cellular and Infection Microbiology* **8**: 271. https://doi.org/10.3389/fcimb.2018.00271.

Mi-ichi, F., Ishikawa, T., Tam, V.K. et al. (2019). Characterization of *Entamoeba histolytica* adenosine 5′-phosphosulfate (APS) kinase; validation as a target and provision of leads for the development of new drugs against amoebiasis. *PLoS Neglected Tropical Diseases* **13**: https://doi.org/10.1371/journal.pntd.0007633.

Militello, K.T., Refour, P., Comeaux, C.A. et al. (2008). Antisense RNA and RNAi in protozoan parasites: working hard or hardly working? *Molecular and Biochemical Parasitology* **157**: 117–126. https://doi.org/10.1016/j.molbiopara.2007.10.004.

Milutinović, B. and Kurz, J. (2016). Immune memory in invertebrates. *Seminars in Immunology* **28**: 328–342. https://doi.org/10.1016/j.smim.2016.05.004.

Minetti, C., Chalmers, R.M., Beeching, N.J. et al. (2016). Giardiasis. *BMJ* **355**: https://doi.org/10.1136/bmj.i5369.

Minetti, C., Lamden, K., Durband, C. et al. (2015). Case-control study of risk factors for sporadic giardiasis and parasite assemblages in North West England. *Journal of Clinical Microbiology* **53**: 3133–3140. https://doi.org/10.1128/JCM.00715-15.

Mitchell, S., Bell, S., Wright, I. et al. (2016). Tongue worm (*Linguatula* species) in stray dogs imported into the UK. *Veterinary Record* **179**: 259–260. https://doi.org/10.1136/vr.i4829.

Moazeni, M. and Ahmadi, A. (2016). Controversial aspects of the life cycle of *Fasciola hepatica*. *Experimental Parasitology* **169**: 81–89. https://doi.org/10.1016/j.exppara.2016.07.010.

Mockler, B.K., Kwong, W.K., Moran, N.A. et al. (2018). Microbiome structure influences infection by the parasite *Crithidia bombi* in bumble bees. *Applied Environmental Microbiology* **84**: https://doi.org/10.1128/AEM.02335-17.

Molento, M.B. (2020). COVID-19 and the rush for self-medication and self-dosing with ivermectin: a word of caution. *One Health* **10**: 100148. https://doi.org/10.1016/j.onehlt.2020.100148.

Molloy, D.P., Vinikour, W.S., and Anderson, R.V. (1999). New North American records of aquatic insects as paratenic hosts of Pheromermis (Nematoda: Mermithidae). *Journal of Invertebrate Pathology* **74**: 89–95. https://doi.org/10.1006/jipa.1999.4860.

Momčilović, S., Cantacessi, C., Arsić-Arsenijević, V. et al. (2019). Rapid diagnosis of parasitic diseases: current scenario and future needs. *Clinical Microbiology and Infection* **25**: 290–309. https://doi.org/10.1016/j.cmi.2018.04.028.

Mongan, A.E., Tuda, J.S.B., and Runtuwene, L.R. (2020). Portable sequencer in the fight against infectious disease. *Journal of Human Genetics* **65**: 35–40. https://doi.org/10.1038/s10038-019-0675-4.

Monge-Maillo, B., Norman, F.F., Cruz, I. et al. (2014). Visceral leishmaniasis and HIV coinfection in the Mediterranean region. *PLoS Neglected Tropical Diseases* **8**: https://doi.org/10.1371/journal.pntd.0003021.

Moody, A. (2002). Rapid diagnostic tests for malaria parasites. *Clinical Microbiology Reviews* **15**: 66–78. https://doi.org/10.1128/CMR.15.1.66-78.2002.

Moore, J. and Brooks, D.R. (1987). Asexual reproduction in cestodes (Cyclophyllidea: Taeniidae): ecological and phylogenetic influences. *Evolution* **41**: 882–891. https://doi.org/10.1111/j.1558-5646.1987.tb05861.x.

Moore, R.B., Oborník, M., Janouškovec, J. et al. (2008). A photosynthetic alveolate closely related to apicomplexan parasites. *Nature* **451**: 959–963. https://doi.org/10.1038/nature06635.

Morales-Hojas, R. (2009). Molecular systematics of filarial parasites, with an emphasis on groups of medical and veterinary importance, and its relevance for epidemiology. *Infection, Genetics and Evolution* **9**: 748–759. https://doi.org/10.1016/j.meegid.2009.06.007.

Morales-Yánez, F., Trashin, S., Hermy, M. et al. (2019). Fast one-step ultrasensitive detection of *Toxocara canis* antigens by a nanobody-based electrochemical magnetosensor. *Analytical Chemistry* **91**: 11582–11588. https://doi.org/10.1021/acs.analchem.9b1687.

Moran, N.A. and Bennett, G.M. (2014). The tiniest tiny genomes. *Annual Review of Microbiology* **68**: 195–215. https://doi.org/10.1146/annurev-micro-091213-112901.

Moreira, D. and López-Garcia, P. (2009). Ten reasons to exclude viruses from the tree of life. *Nature Reviews: Microbiology* **7**: 306–311. https://doi.org/10.1038/nrmicro2108.

Morgan, M.S. and Arlian, L.G. (2010). Response of human skin equivalents to *Sarcoptes scabiei*. *Journal of Medical Entomology* **47**: 877–883. https://doi.org/10.1093/jmedent/47.5.877.

Morley, N.J., Irwin, S.W.B., and Lewis, J.W. (2003). Pollution toxicity to the transmission of larval digeneans through their molluscan hosts. *Parasitology* **126**: S5–S26. https://doi.org/10.1017/s0031182003003755.

Morris, A., Robinson, G., Swain, M.T. et al. (2019). Direct sequencing of cryptosporidium in stool samples for public health. *Frontiers in Public Health* **7**: 360. https://doi.org/10.3389/fpubh.2019.00360.

Morris, S.R., Bristow, C.C., Wierzbicki, M.R. et al. (2021). Performance of a single-use, rapid, point-of-care PCR device for the detection of *Neisseria gonorrhoeae*, *Chlamydia trachomatis*, and *Trichomonas vaginalis*: a cross-sectional study. *Lancet Infectious Diseases* **21**: 668–676. https://doi.org/10.1016/S1473-3099(20)30734-9.

Morris, U. and Aydin-Schmidt, B. (2021). Performance and application of commercially available loop-mediated isothermal amplification (LAMP) kits in malaria endemic and non-endemic settings. *Diagnostics* **11**: 336. https://doi.org/10.3390/diagnostics11020336.

Mostafa, M.H., Sheweita, S.A., and O'Connor, P.J. (1999). Relationship between schistosomiasis and bladder cancer. *Clinical Microbiology Reviews* **12**: 97–111. PMID: 9880476; PMCID: PMC88908.

Mote, S., Schönberg, C.H., Samaai, T. et al. (2019). A new clionaid sponge infests live corals on the west coast of India (Porifera, Demospongiae, Clionaida). *Systematics and Biodiversity* **17**: 190–206. https://doi.org/10.1080/14772000.2018.1513430.

Moure, Z., Zarzuela, F., Espasa, M. et al. (2017). *Dicrocoelium dendriticum*: an unusual parasitological diagnosis in a reference international health unit. *American Journal of Tropical Medicine and Hygiene* **96**: 355–357. https://doi.org/10.4269/ajtmh.16-0549.

Mueller, M.C., Marx, M., Peyerl-Hoffmann, G. et al. (2020). Spatial distribution and incidence trend of human alveolar echinococcosis in southwest Germany: increased incidence and urbanization of the disease? *Infection* **48**: 923–927. https://doi.org/10.1007/s15010-020-01479-4.

Mugnier, M.R., Stebbins, C.E., and Papavasiliou, F.N. (2016). Masters of disguise: antigenic variation and the VSG coat in *Trypanosoma brucei*. *PLoS Pathogens* **12**: e1005784. https://doi.org/10.1371/journal.ppat.1005784.

Muhanguzi, D., Picozzi, K., Hatendorf, J. et al. (2014). Collateral benefits of restricted insecticide application for control of African trypanosomiasis on *Theileria parva* in cattle: a randomized controlled trial. *Parasites & Vectors* **7**: 432. https://doi.org/10.1186/1756-3305-7-432.

Mukhtar, M.M., Eisawi, O.A., Amanfo, S.A. et al. (2019). *Plasmodium vivax* cerebral malaria in an adult patient in Sudan. *Malaria Journal* **18**: 316. https://doi.org/10.1186/s12936-019-2961-1.

Mulcahy, G., O'Neill, S., Fanning, J. et al. (2005). Tissue migration by parasitic helminths–an immunoevasive strategy? *Trends in Parasitology* **21**: 273–277. https://doi.org/10.1016/j.pt.2005.04.003.

Mulder, C. (2017). Pathogenic helminths in the past: much ado about nothing. *F1000Research* **6**: 852. https://doi.org/10.12688/f1000research.11752.3.

Mullen, G. and Durden, L. (2002). *Medical and Veterinary Entomology*. Amsterdam: Academic Press.

Müller, C., Lukas, P., Böhmert, M. et al. (2020). Hirudin or hirudin-like factor-that is the question: insights from the analyses of natural and synthetic HLF variants. *FEBS Letters* **594**: 841–850. https://doi.org/10.1002/1873-3468.13683.

Mumcuoglu, K.Y., Huberman, L., Cohen, R. et al. (2010). Elimination of symbiotic *Aeromonas* spp. from the intestinal tract of the medicinal leech, *Hirudo medicinalis*, using ciprofloxacin feeding. *Clinical Microbiology and Infection* **16**: 563–567. https://doi.org/10.1111/j.1469-0691.2009.02868.x.

Munro, H.M.C. and Thrushfield, M.V. (2001). 'Battered pets': sexual abuse. *Journal of Small Animal Practice* **42**: 333–337. https://doi.org/10.1111/j.1748-5827.2001.tb02468.x.

Munson, L., Terio, K.A., Kock, R. et al. (2008). Climate extremes promote fatal co-infections during canine distemper epidemics in African lions. *PloS One* **3**: https://doi.org/10.1371/journal.pone.0002545.

Murdoch, M.E. (2018). Onchodermatitis: where are we now? *Tropical Medicine and Infectious Disease* **3**: 94. https://doi.org/10.3390/tropicalmed3030094.

Murillo, A.C. and Mullens, B.A. (2016). Timing diatomaceous earth-filled dustbox use for management of northern fowl mites (Acari: Macronyssidae) in cage-free poultry systems. *Journal of Economic Entomology* **109**: 2572–2579. https://doi.org/10.1093/jee/tow165.

Murrell, K.D. (2016). The dynamics of *Trichinella spiralis*: out to pasture? *Veterinary Parasitology* **231**: 92–96. https://doi.org/10.1016/j.vetpar.2016.03.020.

Musharrafieh, U., Hamadeh, G., Touma, A. et al. (2018). Nasopharyngeal linguatulosis or halzoun syndrome: clinical diagnosis and treatment. *Revista da Associação Médica Brasileira* **64**: 1081–1084. http://orcid.org/0000-0002-5368-8764.

Muturi, E.J., Mbogo, C.M., Mwangangi, J.M. et al. (2006). Concomitant infections of *Plasmodium falciparum* and *Wuchereria bancrofti* on the Kenyan coast. *Filaria Journal* **5**: 8. https://doi.org/10.1186/1475-2883-5-8.

Muzny, C.A., Blackburn, R.J., Sinsky, R.J. et al. (2014). Added benefit of nucleic acid amplification testing for the diagnosis of *Trichomonas vaginalis* among men and women attending a sexually transmitted diseases clinic. *Clinical Infectious Diseases* **59**: 834–841. https://doi.org/10.1093/cid/ciu446.

Na, B.K., Pak, J.H., and Hong, S.J. (2020). *Clonorchis sinensis* and clonorchiasis. *Acta Tropica* **203**: https://doi.org/10.1016/j.actatropica.2019.105309.

Nader, J.L., Mathers, T.C., Ward, B.J. et al. (2019). Evolutionary genomics of anthroponosis in *Cryptosporidium*. *Nature Microbiology* **4**: 826–836. https://doi.org/10.1038/s41564-019-0377-x.

Naessens, J. (2006). Bovine trypanotolerance: a natural ability to prevent severe anaemia and haemophagocytic syndrome? *International Journal for Parasitology* **36**: 521–528. https://doi.org/10.1016/j.ijpara.2006.02.012.

Nagayasu, E., Htwe, M.P.P.T.H., Hortiwakul, T. et al. (2017). A possible origin population of pathogenic intestinal nematodes, *Strongyloides stercoralis*, unveiled by molecular phylogeny. *Scientific Reports* **7**: 1–13. https://doi.org/10.1038/s41598-017-05049-x.

Nakada-Tsukui, K. and Nozaki, T. (2016). Immune response of amebiasis and immune evasion by *Entamoeba histolytica*. *Frontiers in Immunology* **7**: 175. https://doi.org/10.3389/fimmu.2016.00175.

Nakao, J.H., Collier, S.A., and Gargano, J.W. (2017). Giardiasis and subsequent irritable bowel syndrome: a longitudinal cohort study using health insurance data. *Journal of Infectious Diseases* **215**: 798–805. https://doi.org/10.1093/infdis/jiw621.

Namroodi, S., Shirazi, A.S., Khaleghi, S.R. et al. (2018). Frequency of exposure of endangered Caspian seals to canine distemper virus, *Leptospira interrogans*, and *Toxoplasma gondii*. *PloS One* **13**: https://doi.org/10.1371/journal.pone.0196070.

Nash, T.E., Bustos, J.A., Garcia, H.H. et al. (2017). Disease centered around calcified *Taenia solium* granuloma. *Trends in Parasitology* **33**: 65–73. https://doi.org/10.1016/j.pt.2016.09.003.

Nausch, N., Dawson, E.M., Midzi, N. et al. (2014). Field evaluation of a new antibody-based diagnostic for *Schistosoma haematobium* and *S. mansoni* at the point-of-care in northeast Zimbabwe. *BMC Infectious Diseases* **14**: 1–9. https://doi.org/10.1186/1471-2334-14-165.

Ndimubanzi, P.C., Carabin, H., Budke, C.M. et al. (2010). A systematic review of the frequency of neurocyticercosis with a focus on people with epilepsy. *PLoS Neglected Tropical Diseases* **4**: https://doi.org/10.1371/journal.pntd.0000870.

Neelakanta, G., Sultana, H., Fish, D. et al. (2010). *Anaplasma phagocytophilum* induces *Ixodes scapularis* ticks to express an antifreeze glycoprotein gene that enhances their survival in the cold. *Journal of Clinical Investigation* **120**: 3179–3190. https://doi.org/10.1172/JCI42868.

Nemati, M., Malla, N., Yadav, M. et al. (2018). Humoral and T cell-mediated immune response against trichomoniasis. *Parasite Immunology* **40**: https://doi.org/10.1111/pim.12510.

Nene, V., Kiara, H., Lacasta, A. et al. (2016). The biology of *Theileria parva* and control of East Coast fever – current status and future trends. *Ticks and Tick-borne Diseases* **7**: 549–564. https://doi.org/10.1016/j.ttbdis.2016.02.001.

Nene, V. and Morrison, W.I. (2016). Approaches to vaccination against *Theileria parva* and *Theileria annulata*. *Parasite Immunology* **38**: 724–734. https://doi.org/10.1111/pim.12388.

Newton, R.B. (1908). Fossil pearl-growths. *Proceedings of the Malacological Society of London* **8**: 128–139. https://doi.org/10.1093/oxfordjournals.mollus.a066243.

Ngo, H.T., Freedman, E., Odion, R.A. et al. (2018). Direct detection of unamplified pathogen RNA in blood lysate using an integrated lab-in-a-stick device and ultrabright SERS Nanorattles. *Scientific Reports* **8**: 4075. https://doi.org/10.1038/s41598-018-21615-3.

Niezen, J.H., Waghorn, T.S., Charleston, W.A.G. et al. (1995). Growth and gastrointestinal nematode parasitism in lambs grazing either lucerne (*Medicago sativa*) or sulla (*Hedysarum coronarium*) which contains condensed tannins. *Journal of Agricultural Science* **125**: 281–289. https://doi.org/10.1017/S0021859600084422.

Nigam, Y. and Morgan, C. (2016). Does maggot therapy promote wound healing? The clinical and cellular evidence. *Journal of the European Academy of Dermatology and Venereology* **30**: 776–782. https://doi.org/10.1111/jdv.13534.

Nobis, C.C., Laramée, G.D., Kervezee, L. et al. (2019). The circadian clock of CD8 T cells modulates their early response to vaccination and the rhythmicity of related signalling pathways. *Proceedings of the National Academy of Sciences* **116**: 20077–20086. https://doi.org/10.1073/pnas.1905080116.

Nogami, Y., Fujii-Nishimura, Y., Banno, K. et al. (2016). Anisakiasis mimics cancer recurrence: two cases of extragastrointestinal anisakiasis suspected to be recurrence of gynaecological cancer on

PET-CT and molecular biological investigation. *BMC Medical Imaging* **16**: 31. https://doi.
org/10.1186/s12880-016-0134-z.

Noireau, F. (1992). Infestation by *Auchmeromyia senegalensis* as a consequence of the adoption of
non-nomadic life by the Pygmies in the Congo. *Transactions of the Royal Society of Tropical Medicine
and Hygiene* **86**: 329. https://doi.org/10.1016/0035-9203(92)90334-9.

Norris, D.E. (1971). The migratory behavior of the infective-stage larvae of Ancylostoma braziliense
and *Ancylostoma tubaeforme* in rodent paratenic hosts. *Journal of Parasitology* **57**: 998–1009. https://
www.jstor.org/stable/3277855.

Notomi, T., Mori, Y., Tomita, N. et al. (2015). Loop-mediated isothermal amplification (LAMP):
principle, features, and future prospects. *Journal of Microbiology* **53**: 1–5. https://doi.org/10.1007/
s12275-015-4656-9.

Nour, N.M. (2010). Schistosomiasis: health effects on women. *Reviews in Obstetrics and Gynecology*
3: 28–32. PMID: 20508780; PMCID: PMC2876318.

Noyes, H.A., Alimohammadian, M.H., Agaba, M. et al. (2009). Mechanisms controlling anaemia in
Trypanosoma congolense infected mice. *PLoS One* **4**: https://doi.org/10.1371/journal.pone.0005170.

Noyes, H.A., Morrison, D.A., Chance, M.L. et al. (2000). Evidence for a neotropical origin of
Leishmania. *Memórias do Instituto Oswaldo Cruz* **95**: 575–578. https://doi.org/10.1590/
S0074-02762000000400021.

Nozaki, T. and Bhattacharya, A. (2015). *Amebiasis. Biology and Pathogenesis of Entamoeba*. Tokyo,
Japan: Springer.

Nutman, T.B. (2017). Human infection with *Strongyloides stercoralis* and other related *Strongyloides*
species. *Parasitology* **144**: 263–273. https://doi.org/10.1017/S0031182016000834.

Obame-Nkoghe, J., Rahola, N., Bourgarel, M. et al. (2016). Bat flies (Diptera: Nycteribiidae and
Streblidae) infesting cave-dwelling bats in Gabon: diversity, dynamics and potential role in
Polychromophilus melanipherus transmission. *Parasites & Vectors* **9**: 333. https://doi.org/10.1186/
s13071-016-1625-z.

Obieglo, K., Schuijs, M.J., Ozir-Fazalalikhan, A. et al. (2018). Isolated *Schistosoma mansoni* eggs
prevent allergic airway inflammation. *Parasite Immunology* **40**: e12579. https://doi.org/10.1111/
pim.12579.

O'Brien, M.D. and Jagathesan, T. (2016). Lesson of the month 1: post-malaria neurological syndromes.
Clinical Medicine **16**: 292–293. https://doi.org/10.7861/clinmedicine.16-3-292.

Ochodo, E.A., Gopalakrishna, G., Spek, B. et al. (2015). Circulating antigen tests and urine reagent
strips for diagnosis of active schistosomiasis in endemic areas. *Cochrane Database of Systematic
Reviews* **3**: https://doi.org/10.1002/14651858.CD009579.pub2.

Ocholi, R.A., Kalejaiye, J.O., and Okewole, P.A. (1989). Acute disseminated toxoplasmosis in two
captive lions (*Panthera leo*) in Nigeria. *Veterinary Record* **124**: 515–516. PMID: 2756626. https://doi.
org/10.1136/vr.124.19.515.

Odum, E.P. (1959). *Fundamentals of Ecology*, 2nde. Philadelphia, PA: W.B. Saunders.

Ogawa, A., Streit, A., Antebi, A. et al. (2009). A conserved endocrine mechanism controls the
formation of dauer and infective larvae in nematodes. *Current Biology* **19**: 67–71. https://doi.org/
10.1016/j.cub.2008.11.063.

Ogier, J., Pagès, S., Frayssinet, M., and Gaudriault, S. (2020). Entomopathogenic nematode-associated
microbiota: from monoxenic paradigm to pathobiome. *Microbiome* **8**: 25. https://doi.org/10.1186/
s40168-020-00800-5.

Okaka, F.O. and Odhiambo, B. (2018). Relationship between flooding and outbreak of infectious
diseases in Kenya: a review of the literature. *Journal of Environmental and Public Health* https://doi.
org/10.1155/2018/5452938.

Okamura, B. and Gruhl, A. (2016). Myxozoa+ *Polypodium*: a common route to endoparasitism. *Trends in Parasitology* **32**: 268–271. https://doi.org/10.1016/j.pt.2016.01.007.

Okoye, A.A. and Picker, L.J. (2013). CD4(+) T-cell depletion in HIV infection: mechanisms of immunological failure. *Immunological Reviews* **254**: 54–64. https://doi.org/10.1111/imr.12066.

Okoye, I.C. and Onwuliri, C.O. (2007). Epidemiology and psycho-social aspects of onchocercal skin diseases in northeastern Nigeria. *Filaria Journal* **6**: 15. https://doi.org/10.1186/1475-2883-6-15.

Okwor, I. and Uzonna, J.E. (2013). The immunology of Leishmania/HIV co-infection. *Immunologic Research* **56**: 163–171. https://doi.org/10.1007/s12026-013-8389-8.

Oldroyd, H. (1964). *The Natural History of Flies*. New York, NY: Norton Library.

Oliveira, F.M., Neumann, E., Gomes, M.A. et al. (2015). *Entamoeba dispar*: Could it be pathogenic. *Tropical Parasitology* **5**: 9–14. https://doi.org/10.4103/2229-5070.149887.

Oliveira, R.C., Oi, C.A., Vollet-Neto, A. et al. (2016). Intraspecific worker parasitism in the common wasp, *Vespula vulgaris*. *Animal Behaviour* **113**: 79–85. https://doi.org/10.1016/j.anbehav.2015.12.025.

Olivier, M., Atayde, V.D., Isnard, A. et al. (2012). *Leishmania* virulence factors: focus on the metalloprotease GP63. *Microbes and Infection* **14**: 1377–1389. https://doi.org/10.1016/j.micinf.2012.05.014.

Olivier, M. and Zamboni, D.S. (2020). *Leishmania Viannia guyanensis*, LRV1 virus and extracellular vesicles: a dangerous trio influencing the faith of immune response during muco-cutaneous leishmaniasis. *Current Opinion in Immunology* **66**: 108–113. https://doi.org/10.1016/j.coi.2020.08.004.

Olsen, A., Permin, A., and Roepstorff, A. (2001). Chickens and pigs as transport hosts for *Ascaris*, *Trichuris* and *Oesophagostomum* eggs. *Parasitology* **123**: 325–330. https://doi.org/10.1017/S0031182001008435.

Olson, P.D., Cribb, T.H., Tkach, V.V. et al. (2003). Phylogeny and classification of the Digenea (Platyhelminthes: Trematoda). *International Journal for Parasitology* **33**: 733–755. https://doi.org/10.1016/S0020-7519(03)00049-3.

Olson, P.D., Poddubnaya, L.G., Littlewood, D.T.J. et al. (2008). On the position of *Archigetes* and its bearing on the early evolution of the tapeworms. *Journal of Parasitology* **94**: 898–904. https://doi.org/10.1645/GE-1456.1.

Olupot-Olupot, P., Engoru, C., Uyoga, S. et al. (2017). High frequency of blackwater fever among children presenting to hospital with severe febrile illnesses in eastern Uganda. *Clinical Infectious Diseases* **64**: 939–946. https://doi.org/10.1093/cid/cix003.

Onukwugha, E., McRae, J., Kravetz, A. et al. (2016). Cost-of-illness studies: an updated review of current methods. *PharmacoEconomics* **34**: 43–58. PMID: 26385101. https://doi.org/10.1007/s40273-015-0325-4.

Onyilagha, C. and Uzonna, J.E. (2019). Host immune responses and immune evasion strategies in African trypanosomiasis. *Frontiers in Immunology* **10**: 2738. https://doi.org/10.3389/fimmu.2019.02738.

Ooi, E.E. (2020). Repurposing ivermectin as an anti-dengue drug. *Clinical Infectious Diseases*. **72**: e594–e595. https://doi.org/10.1093/cid/ciaa1341.

Orem, J., Mbidde, E.K., Lambert, B. et al. (2007). Burkitt's lymphoma in Africa, a review of the epidemiology and etiology. *African Health Sciences* **7**: 166–175. https://doi.org/10.5555/afhs.2007.7.3.166.

Orosz, F. (2018). Does apicortin, a characteristic protein of apicomplexan parasites and placozoa, occur in Eumetazoa? *Acta Parasitologica* **63**: 617–633. https://doi.org/10.1515/ap-2018-0071.

Ortega-Vargas, S., Espitia, C., Sahagún-Ruiz, A. et al. (2019). Moderate protection is induced by a chimeric protein composed of leucine aminopeptidase and cathepsin L1 against *Fasciola hepatica* challenge in sheep. *Vaccine* **37**: 3234–3240. https://doi.org/10.1016/j.vaccine.2019.04.067.

Oryan, A., Sadjjadi, S.M., Mehrabani, D. et al. (2008). The status of *Linguatula serrata* infection of stray dogs in Shiraz, Iran. *Comparative Clinical Pathology* **17**: 55–60. https://doi.org/10.1007/s00580-007-0707-x.

Ostfeld, R.S. and Brunner, J.L. (2015). Climate change and *Ixodes* tick-borne diseases of humans. *Philosophical Transactions of the Royal Society B: Biological Sciences* **370**: https://doi.org/10.1098/rstb.2014.0051.

Özdemir, M.H., Aksoy, U., and Akisu, Ç. (2003). Investigating demodex in forensic autopsy cases. *Forensic Science International* **135**: 226–231. https://doi.org/10.1016/S0379-0738(03)00216-0.

Özen, D., Böhning, D., and Gürcan, İ.S. (2016). Estimation of stray dog and cat populations in metropolitan Ankara, Turkey. *Turkish Journal of Veterinary and Animal Sciences* **40**: 7–12. https://doi.org/10.3906/vet-1505-70.

Özüm, Ü., Karadağ, Ö., Eği slmez, R. et al. (2008). A case of brain abscess due to Entamoeba species, *Eikenella corrodens* and *Prevotella* species. *British Journal of Neurosurgery* **22**: 596–598. https://doi.org/10.1080/02688690801894646.

Page, L.K., Swihart, R.K., and Kazacos, K.R. (1998). Raccoon latrine structure and its potential role in transmission of *Baylisascaris procyonis* to vertebrates. *American Midland Naturalist* **140**: 180–185. https://doi.org/10.1674/0003-0031(1998)140[0180:RLSAIP]2.0.CO;2.

Pakharukova, M.Y. and Mordvinov, V.A. (2016). The liver fluke *Opisthorchis felineus*: biology, epidemiology and carcinogenic potential. *Transactions of the Royal Society of Tropical Medicine and Hygiene* **110**: 28–36. PMID: 26740360. https://doi.org/10.1093/trstmh/trv085.

Papaiakovou, M., Gasser, R.B., and Littlewood, D.T.J. (2019). Quantitative PCR-based diagnosis of soil-transmitted helminth infections: faecal or fickle? *Trends in Parasitology* **35**: 491–500. 10.1016/j.pt.2019.04.006.

Paparau, C., Brincus, C.G., and Ghita, A.E. (2018). Mechanical asphyxia with *Ascaris lumbricoides*-a forensic case report. *Journal of Forensic Research* **9**: 1–4. https://doi.org/10.4172/2157-7145.1000415.

Park, S., Hitchcock, M.M., Gomez, C.A. et al. (2017). Is follow-up testing with the FilmArray gastrointestinal multiplex PCR panel necessary? *Journal of Clinical Microbiology* **55**: 1154–1161. https://doi.org/10.1128/JCM.02354-16.

Parker, D. and Booth, A.J. (2013). The tongue-replacing isopod *Cymothoa borbonica* reduces the growth of largespot pompano *Trachinotus botla*. *Marine Biology* **160**: 2943–2950. https://doi.org/10.1007/s00227-013-2284-7.

Parks, R. (2014). The rise, critique and persistence of the DALY in global health. *The Journal of Global Health* **4**: 28–32. doi: 10.7916/thejgh.v4i1.4893.

Parra-Henao, G., Oliveros, H., Hotez, P.J. et al. (2019). In search of congenital chagas disease in the Sierra Nevada de Santa Marta, Colombia. *American Journal of Tropical Medicine and Hygiene* **101**: 482–483. https://doi.org/10.4269/ajtmh.19-0110.

Parshad, S., Grover, P.S., Sharma, A. et al. (2002). Primary cutaneous amoebiasis: case report with review of the literature. *International Journal of Dermatology* **41**: 676–680. https://doi.org/10.1046/j.1365-4362.2002.01569.x.

Passos, M.R.L., Barreto, N.A., Varella, R.Q. et al. (2004). Penile myiasis: a case report. *Sexually Transmitted Infections* **80**: 183–184. https://doi.org/10.1136/sti.2003.008235.

Patel, C., Keller, L., Welsche, S. et al. (2021). Assessment of fecal calprotectin and fecal occult blood as point-of-care markers for soil-transmitted helminth attributable intestinal morbidity in a case-control substudy conducted in Côte d'Ivoire, Lao PDR and Pemba Island, Tanzania. *EClinicalMedicine* **32**: https://doi.org/10.1016/j.eclinm.2021.100724.

Patel, R. (2015). MALDI-TOF MS for the diagnosis of infectious diseases. *Clinical Chemistry* **61**: 100–111. https://doi.org/10.1373/clinchem.2014.221770.

Patz, J.A. and Olson, S.H. (2006). Malaria risk and temperature: influences from global climate change and local land use practices. *Proceedings of the National Academy of Sciences of the United States of America* **103**: 5635–5636. https://doi.org/10.1073/pnas.0601493103.

Paulo, D.F., Williamson, M.E., Arp, A.P. et al. (2019). Specific gene disruption in the major livestock pests *Cochliomyia hominivorax* and *Lucilia cuprina* using CRISPR/Cas9. *G3: Genes Genomes, Genetics* **9**: 3045–3055. https://doi.org/10.1534/g3.119.400544.

Pays, E. and Vanhollebeke, B. (2009). Human innate immunity against African trypanosomes. *Current Opinion in Immunology* **21**: 493–498. https://doi.org/10.1016/j.coi.2009.05.024.

Peacock, L., Bailey, M., and Gibson, W. (2016). Dynamics of gamete production and mating in the parasitic protist *Trypanosoma brucei*. *Parasites Vectors* **9**: 404. https://doi.org/10.1186/s13071-016-1689-9.

Peña-Silva, R., Duffull, S.B., Steer, A.C. et al. (2020). Pharmacokinetic considerations on the repurposing of ivermectin for treatment of COVID-19. *British Journal of Clinical Pharmacology* **87**: 1589–1590. 10.1111/bcp.14476.

Pennino, M.G., Bachiller, E., Lloret-Lloret, E. et al. (2020). Ingestion of microplastics and occurrence of parasite association in Mediterranean anchovy and sardine. *Marine Pollution Bulletin* **158**: https://doi.org/10.1016/j.marpolbul.2020.111399.

Pennisi, E. (2018). Hybridization may give some parasites a leg up. *Science* **361**: 832–833. https://doi.org/10.1126/science.361.6405.832.

Perotti, M.A. and Braig, H.R. (2011). Eukaryotic ectosymbionts of Acari. *Journal of Applied Entomology* **135**: 514–523. https://doi.org/10.1111/j.1439-0418.2011.01639.x.

Perotti, M.A., Kirkness, E.F., Reed, D.L. et al. (2008). Endosymbionts of lice. In: *Insect Symbiosis*, vol. **3** (ed. K. Bourtzis and T.A. Miller), 205–220. Boca Raton, FL: CRC Press.

Peters, A., Das, S., and Raidal, S.R. (2020). Diverse *Trichomonas* lineages in Australasian pigeons and doves support a columbid origin for the genus *Trichomonas*. *Molecular Phylogenetics and Evolution* **143**: https://doi.org/10.1016/j.ympev.2019.106674.

Phillips, B.P., Wolfe, P.A., Rees, C.W. et al. (1955). Studies on the ameba-bacteria relationship in amebiasis: comparative results of the intracecal inoculation of germfree, monocontaminated, and conventional guinea pigs with *Entamoeba histolytica*. *American Journal of Tropical Medicine and Hygiene* **4**: 675–692. https://doi.org/10.4269/ajtmh.1955.4.675.

Phosuk, I., Sanpool, O., Thanchomnang, T. et al. (2018). Molecular identification of *Trichuris suis* and *Trichuris trichiura* eggs in human populations from Thailand, Lao PDR, and Myanmar. *American Journal of Tropical Medicine and Hygiene* **98**: 39–44. https://doi.org/10.4269/ajtmh.17-0651.

Pichakacheri, S.K. (2019). Guinea-worm (*Dracunculus medinensis*) infection presenting as a diabetic foot abscess: a case report from Kerala. *National Medical Journal of India* **32**: 22–23. https://doi.org/10.4103/0970-258X.272111.

Picot, S., Cucherat, M., and Bienvenu, A.-L. (2020). Systematic review and meta-analysis of diagnostic accuracy of loop-mediated isothermal amplification (LAMP) methods compared with microscopy, polymerase chain reaction and rapid diagnostic tests for malaria diagnosis. *International Journal of Infectious Diseases* **98**: 408–419. https://doi.org/10.1016/j.ijid.2020.07.009.

Pidone, F.A.M. and dos Santos, D.P. (2020). Modulation of apoptotic pathways by *Trypanosoma cruzi* and its relationship with the progression of heart disease in the host. *World Journal of Advanced Research and Reviews* **7**: 146–154. 10.30574/wjarr.2020.7.2.0292.

Pier, A.C., Smith, J.M.B., Alexiou, H. et al. (1994). Animal ringworm – its aetiology, public health significance and control. *Journal of Medical and Veterinary Mycology* **32**: 133–150. https://doi.org/10.1080/02681219480000791.

Pietsch, T.W. (2005). Dimorphism, parasitism, and sex revisited: modes of reproduction among deep-sea ceratioid anglerfish (Teleostei: Lophiiformes). *Ichthyological Research* **52**: 207–236. https://doi.org/10.1007/s10228-005-0286-2.

Pike, A.W. (1989). Sea lice – major pathogens of farmed Atlantic salmon. *Parasitology Today* **5**: 291–297. https://doi.org/10.1016/0169-4758(89)90020-3.

Pillai, A., Ueno, S., Zhang, H. et al. (2005). Cecropin P1 and novel nematode cecropins: a bacteria-inducible antimicrobial peptide family in the nematode *Ascaris suum*. *Biochemical Journal* **390**: 207–214. https://doi.org/10.1042/BJ20050218.

Pimentel, L.A., Dantas, A.F.M., Uzal, F. et al. (2012). Meningoencephalitis caused by *Naegleria fowleri* in cattle of northeast Brazil. *Research in Veterinary Science* **93**: 811–812. https://doi.org/10.1016/j.rvsc.2012.01.002.

Pimentel-Elardo, S.M., Kozytska, S., Bugni, T.S. et al. (2010). Anti-parasitic compounds from *Streptomyces* sp. strains isolated from Mediterranean sponges. *Marine Drugs* **8**: 373–380. https://doi.org/10.3390/md8020373.

Pinaud, S., Portet, A., Allienne, J.F. et al. (2019). Molecular characterisation of immunological memory following homologous or heterologous challenges in the schistosomiasis vector snail, *Biomphalaria glabrata*. *Developmental & Comparative Immunology* **92**: 238–252. https://doi.org/10.1016/j.dci.2018.12.001.

Pink, R., Hudson, A., Mouriès, M.A. et al. (2005). Opportunities and challenges in antiparasitic drug discovery. *Nature Reviews Drug Discovery* **4**: 727–740. https://doi.org/10.1038/nrd1824.

Pisano, S., Ryser-Degiorgis, M.P., Rossi, L. et al. (2019). Sarcoptic mange of fox origin in multiple farm animals and scabies in humans, Switzerland, 2018. *Emerging Infectious Diseases* **25**: 1235–1238. https://doi.org/10.3201/eid2506.181891.

Pissuwan, D., Valenzuela, S.M., Miller, C.M. et al. (2009). Destruction and control of *Toxoplasma gondii* tachyzoites using gold nanosphere/antibody conjugates. *Small* **5**: 1030–1034. https://doi.org/10.1002/smll.200801018.

Pitt, S., Pearcy, B.E., Stevens, R.H. et al. (1998). War in Tajikistan and re- emergence of *Plasmodium falciparum*. *Lancet* **352**: 1279. https://doi.org/10.1016/S0140-6736(98)00040-3.

Plotkin, S., Robinson, J.M., Cunningham, G. et al. (2017). The complexity and cost of vaccine manufacturing: an overview. *Vaccine* **35**: 4064–4071. https://doi.org/10.1016/j.vaccine.2017.06.003.

Poinar, G. (1984). Fossil evidence of nematode parasitism. *Revue Nématogia* **7**: 201–203.

Poinar, G. and Poinar, R. (2008). *What Bugged the Dinosaurs? Insects, Disease, and Death in the Cretaceous*. Princeton, NJ: Princeton University Press.

Pomajbíková, K.J., Jirků, M., Levá, J. et al. (2018). The benign helminth *Hymenolepis diminuta* ameliorates chemically induced colitis in a rat model system. *Parasitology* **145**: 1324–1335. https://doi.org/10.1017/S0031182018000896.

Pomari, E., Piubelli, C., Perandin, F. et al. (2019). Digital PCR: a new technology for diagnosis of parasitic infections. *Clinical Microbiology and Infection* **25**: 1510–1516. https://doi.org/10.1016/j.cmi.2019.06.009.

Ponnudurai, G. and Rani, N. (2017). *Veterinary Clinical and Diagnostic Parasitology*. Delhi, India: Narendra Publishing House.

Portes, J., Barrias, E., Travassos, R. et al. (2020). *Toxoplasma gondii* mechanisms of entry into host cells. *Frontiers in Cellular and Infection Microbiology* **10**: 294. https://doi.org/10.3389/fcimb.2020.00294.

Poser, C.M. and Bruyn, G.W. (1999). *An Illustrated History of Malaria*. New York, NY: Parthenon Publishing Group.

Posfai, D., Eubanks, A.L., Keim, A.I. et al. (2018). Identification of Hsp90 inhibitors with anti-Plasmodium activity. *Antimicrobial Agents and Chemotherapy* **62**: https://doi.org/10.1128/AAC.01799-17.

Post, C.F. and Juhlin, E. (1963). *Demodex folliculorum* and blepharitis. *Archives of Dermatology* **88**: 298–302. https://doi.org/10.1001/archderm.1963.01590210056008.

Pouillevet, H., Dibakou, S.-E., Ngoubangoye, B. et al. (2017). A comparative study of four methods for the detection of nematode eggs and large protozoan cysts in Mandrill faecal material. *Folia Primatologica* **88**: 344–357. https://doi.org/10.1159/000480233.

Poulin, R. (2000). Manipulation of host behaviour by parasites: a weakening paradigm? *Proceedings of the Royal Society B* **267**: 787–792. https://doi.org/10.1098/rspb.2000.1072.

Power, G., Moore, Z., and O'Connor, T. (2017). Measurement of pH, exudate composition and temperature in wound healing: a systematic review. *Journal of Wound Care* **26**: 381–397. 10.12968/jowc.2017.26.7.381.

Pozio, E. (2016). Adaptation of *Trichinella* spp. for survival in cold climates. *Food and Waterborne Parasitology* **4**: 4–12. https://doi.org/10.1016/j.fawpar.2016.07.001.

Pozio, E., Armignacco, O., Ferri, F. et al. (2013). *Opisthorchis felineus*, an emerging infection in Italy and its implication for the European Union. *Acta Tropica* **126**: 54–62. https://doi.org/10.1016/j.actatropica.2013.01.005.

Praet, N., Verweij, J.J., and Mwape, K.E. (2013). Bayesian modelling to estimate the test characteristics of coprology, coproantigen ELISA and a novel real-time PCR for the diagnosis of taeniasis. *Tropical Medicine & International Health* **18**: 608–614. 10.1111/tmi.12089.

Prasertbun, R., Mori, H., Pintong, A.R. et al. (2017). Zoonotic potential of *Enterocytozoon* genotypes in humans and pigs in Thailand. *Veterinary Parasitology* **233**: 73–79. https://doi.org/10.1016/j.vetpar.2016.12.002.

Prokop, P., Usak, M., and Fančovičová, J. (2010). Risk of parasite transmission influences perceived vulnerability to disease and perceived danger of disease-relevant animals. *Behavioural Processes* **85**: 52–57. https://doi.org/10.1016/j.beproc.2010.06.006.

Proto, W., Coombs, G., and Mottram, J. (2013). Cell death in parasitic protozoa: regulated or incidental? *Nature Reviews Microbiology* **11**: 58–66. https://doi.org/10.1038/nrmicro2929.

Qiao, T., Ma, R.H., Luo, Z.L. et al. (2014). *Clonorchis sinensis* eggs are associated with calcium carbonate gallbladder stones. *Acta Tropica* **138**: 28–37. https://doi.org/10.1016/j.actatropica.2014.06.004.

Queiroz, R., Benz, C., Fellenberg, K. et al. (2009). Transcriptome analysis of differentiating trypanosomes reveals the existence of multiple post-transcriptional regulons. *BMC Genomics* **10**: 1–19. https://doi.org/10.1186/1471-2164-10-495.

Rae, R.G., Tourna, M., and Wilson, M.J. (2010). The slug parasitic nematode *Phasmarhabditis hermaphrodita* associates with complex and variable bacterial assemblages that do not affect its virulence. *Journal of Invertebrate Pathology* **104**: 222–226. https://doi.org/10.1016/j.jip.2010.04.008.

Raina, A.K., Kumar, R., Sridhar, V.R. et al. (1985). Oral transmission of *Trypanosoma evansi* infection in dogs and mice. *Veterinary Parasitology* **18**: 67–69. https://doi.org/10.1016/0304-4017(85)90009-3.

Rajamanickam, A., Munisankar, S., Dolla, C. et al. (2020). Helminth infection modulates systemic pro-inflammatory cytokines and chemokines implicated in type 2 diabetes mellitus pathogenesis. *PLoS Neglected Tropical Diseases* **14**: https://doi.org/10.1371/journal.pntd.0008101.

Rajan, A., Ravichandran, R., and Bagai, U. (2017). Homeopathy against malaria: it's potential as a third millennium drug. *Alternative and Integrative Medicine* **6**: 2. https://doi.org/10.4172/2327-5162.1000232.

Rallis, T., Day, M.J., Saridomichelakis, M.N. et al. (2005). Chronic hepatitis associated with canine leishmaniosis (*Leishmania infantum*): a clinicopathological study of 26 cases. *Journal of Comparative Pathology* **132**: 145–152. https://doi.org/10.1016/j.jcpa.2004.09.004.

Ramaiah, K.D., Das, P.K., Michael, E. et al. (2000). The economic burden of lymphatic filariasis in India. *Parasitology Today* **16**: 251–253. https://doi.org/10.1016/S0169-4758(00)01643-4.

Ramírez-Romero, R., Rodríguez-Tovar, L.E., Nevárez-Garza, A.M. et al. (2010). *Chlorella* infection in a sheep in Mexico and minireview of published reports from humans and domestic animals. *Mycopathologia* **169**: 461–466. https://doi.org/10.1007/s11046-010-9287-4.

Ramírez-Toloza, G., Sosoniuk-Roche, E., Valck, C. et al. (2020). *Trypanosoma cruzi* calreticulin: immune evasion, infectivity, and tumorigenesis. *Trends in Parasitology* **36**: 368–381. https://doi. org/10.1016/j.pt.2020.01.007.

Ramos, F., Marques, C.B., Reginato, C.Z. et al. (2020). Field and molecular evaluation of anthelmintic resistance of nematode populations from cattle and sheep naturally infected pastured on mixed grazing areas at Rio Grande do Sul, Brazil. *Acta Parasitologica* **65**: 118–127. https://doi.org/10.2478/ s11686-019-00137-6.

Ranasinghe, S.L. and McManus, D.P. (2018). *Echinococcus granulosus*: cure for cancer revisited. *Frontiers in Medicine* **5**: 60. https://doi.org/10.3389/fmed.2018.00060.

Randolph, S.E. (1991). The effect of *Babesia microti* on feeding and survival in its tick vector, *Ixodes trianguliceps*. *Parasitology* **102**: 9–16. https://doi.org/10.1017/S0031182000060285.

Rao, S., Tsai, H., Tsai, E. et al. (2019). *Strongyloides stercoralis* hyperinfection syndrome as a cause of fatal gastrointestinal hemorrhage. *ACG Case Reports Journal* **6**: 1–3. 10.14309/crj.0000000000000018.

Raoult, D., Dutour, O., Houhamdi, L. et al. (2006). Evidence for louse-transmitted diseases in soldiers of Napoleon's Grand Army in Vilnius. *Journal of Infectious Diseases* **193**: 112–120. https://doi. org/10.1086/498534.

Rashid, M., Akbar, H., Rashid, I. et al. (2018). Economic significance of tropical theileriosis on a Holstein Friesian dairy farm in Pakistan. *Journal of Parasitology* **104**: 310–313. https://doi. org/10.1645/16-179.

Rashwan, N., Scott, M., and Prichard, R. (2017). Rapid genotyping of β-tubulin polymorphisms in *Trichuris trichiura* and *Ascaris lumbricoides*. *PLoS Neglected Tropical Diseases* **11**: https://doi. org/10.1371/journal.pntd.0005205.

Rausch, S., Midha, A., Kuhring, M. et al. (2018). Parasitic nematodes exert antimicrobial activity and benefit from microbiota-driven support for host immune regulation. *Frontiers in Immunology* **9**: https://doi.org/10.3389/fimmu.2018.02282.

Ravi, A.D., Sadhna, D., Nagpaal, D. et al. (2015). Needle free injection technology: a complete insight. *International Journal of Pharmaceutical Investigation* **5**: 192–199. https://doi.org/10.4103/ 2230-973X.167662.

Ravida, A., Aldridge, A.M., Driessen, N.N. et al. (2016). *Fasciola hepatica* surface coat glycoproteins contain mannosylated and phosphorylated *N*-glycans and exhibit immune modulatory properties independent of the mannose receptor. *PLoS Neglected Tropical Diseases* **10**: https://doi.org/10.1074/ mcp.M116.059774.

Read, A.F. and Sharping, A. (1995). The evolution of tissue migration by parasitic nematode larvae. *Parasitology* **111**: 359–371. https://doi.org/10.1017/S0031182000081919.

Recht, J., Ashley, E.A., and White, N.J. (2018). Use of primaquine and glucose-6-phosphate dehydrogenase deficiency testing: divergent policies and practices in malaria endemic countries. *PLoS Neglected Tropical Diseases* **12**: https://doi.org/10.1371/journal.pntd.0006230.

Reddy, A. and Fried, B. (2007). The use of *Trichuris suis* and other helminth therapies to treat Crohn's disease. *Parasitology Research* **100**: 921–927. https://doi.org/10.1007/s00436-006-0416-4.

Reddy, A. and Fried, B. (2009). An update on the use of helminths to treat Crohn's and other autoimmune diseases. *Parasitology Research* **104**: 217–221. https://doi.org/10.1007/s00436-008-1297-5.

Redpath, S.A., Fonseca, N.M., and Perona-Wright, G. (2014). Protection and pathology during parasite infection: IL-10 strikes the balance. *Parasite Immunology* **36**: 233–252. https://doi.org/10.1111/pim.12113.

Reichel, M.P., Ayanegui-Alcérreca, A.M., Gondim, L.F. et al. (2013). What is the global economic impact of *Neospora caninum* in cattle – the billion dollar question. *International Journal of Parasitology* **43**: 133–142. https://doi.org/10.1016/j.ijpara.2012.10.022.

Reiling, S.J., Measures, L., Feng, S. et al. (2019). *Toxoplasma gondii, Sarcocystis* sp. and *Neospora caninum*-like parasites in seals from northern and eastern Canada: potential risk to consumers. *Food and Waterborne Parasitology* **17**: https://doi.org/10.1016/j.fawpar.2019.e00067.

Reilly, D.F. (2020). The things they carry: diphyllobothriasis at sea, a case report. *Military Medicine* **185**: e510–e512. https://doi.org/10.1093/milmed/usz462.

Reimão, J.Q., Coser, E.M., Lee, M.R. et al. (2020). Laboratory diagnosis of cutaneous and visceral leishmaniasis: current and future methods. *Microorganisms* **8**: https://doi.org/10.3390/microorganisms8111632.

Reiter, P. (2000). From Shakespeare to Defoe: malaria in England in the Little Ice Age. *Emerging Infectious Diseases* **6**: 1–11. https://doi.org/10.3201/eid0601.000101.

Renslo, A.R. and McKerrow, J.H. (2006). Drug discovery and development for neglected parasitic diseases. *Nature Chemical Biology* **2**: 701–710. https://doi.org/10.1038/nchembio837.

Resetarits, E.J., Torchin, M.E., and Hechinger, R.F. (2020). Social trematode parasites increase standing army size in areas of greater invasion threat. *Biology Letters* **16**: https://doi.org/10.1098/rsbl.2019.0765.

Resnik, D.B. (2018). Ethics of community engagement in field trials of genetically modified mosquitoes. *Developing World Bioethics* **18**: 135–143. https://doi.org/10.1111/dewb.12147.

Retief, F.P. and Cilliers, J.F.G. (2006). The illnesses of Herod the Great. *Acta Theologica* **26**: 278–293. https://doi.org/10.4314/actat.v26i2.52580.

Ricci, C. (2017). Bdelloid rotifers: 'sleeping beauties' and 'evolutionary scandals', but not only. *Hydrobiologia* **796**: 277–285. https://doi.org/10.1007/s10750-016-2919.

Richards, L., Erko, B., Ponpetch, K. et al. (2019). Assessing the nonhuman primate reservoir of *Schistosoma mansoni* in Africa: a systematic review. *Infectious Diseases of Poverty* **8**: 32, https://doi.org/10.1186/s40249-019-0543-7.

Richardson, R.F. Jr., Remler, B.F., Katirji, B. et al. (1998). Guillain–Barré syndrome after *Cyclospora* infection. *Muscle & Nerve* **21**: 669–671. PMID: 9572253. 10.1002/(sici)1097-4598(199805)21:5<669::aid-mus20>3.0.co;2-p.

Richens, J. (2004). Genital manifestations of tropical diseases. *Sexually Transmitted Infections* **80**: 12–17. https://doi.org/10.1136/sti.2003.004093.

Rickard, M.D., Harrison, G.B.L., Heath, D.D. et al. (1995). *Taenia ovis* recombinant vaccine-'quo vadit'. *Parasitology* **110**: S5–S10. https://doi.org/10.1017/S0031182000001438.

Ries, J., Komarek, A., Gottschalk, J. et al. (2016). A case of possible Chagas transmission by blood transfusion in Switzerland. *Transfusion Medicine and Hemotherapy* **43**: 415–417. https://doi.org/10.1159/000446264.

Riesgo, A., Burke, E.A., Laumer, C. et al. (2017). Genetic variation and geographic differentiation in the marine triclad *Bdelloura candida* (Platyhelminthes, Tricladida, Maricola), ectocommensal on the American horseshoe crab *Limulus polyphemus*. *Marine Biology* **164**: 111. https://doi.org/10.1007/s00227-017-3132-y.

Rijo-Ferreira, F., Carvalho, T., Afonso, C. et al. (2018). Sleeping sickness is a circadian disorder. *Nature Communications* **9**: 62. https://doi.org/10.1038/s41467-017-02484-2.

Rio, R.V., Attardo, G.M., and Weiss, B.L. (2016). Grandeur alliances: symbiont metabolic integration and obligate arthropod hematophagy. *Trends in Parasitology* **32**: 739–749. https://doi.org/10.1016/j.pt.2016.05.002.

Rivera-Correa, J. and Rodriguez, A. (2017). A role for autoimmunity in the immune response against malaria. In: *Malaria* (ed. M. Mota and A. Rodriguez), 81–95. Cham: Springer 10.1007/978-3-319-45210-4_5.

Robert-Gangneux, F., Meroni, V., Dupont, D. et al. (2018). Toxoplasmosis in transplant recipients, Europe, 2010–2014. *Emerging Infectious Diseases* **24**: 1497–1504. https://doi.org/10.3201/eid2408.180045.

Roberts, F.H.S. (1934). The large roundworm of pigs, *Ascaris lumbricoides* L., 1758, its life history in Queensland, economic importance and control. *Queensland Department for Agriculture and Stocking, Animal Health Station Yeerongpilly Bulletin* **1**: 1–81.

Roberts, L.J., Huffam, S.E., Walton, S.F. et al. (2005). Crusted scabies: clinical and immunological findings in seventy-eight patients and a review of the literature. *Journal of Infection* **50**: 375–381. https://doi.org/10.1016/j.jinf.2004.08.033.

Robertson, L.J., Devleesschauwer, B., de Noya, B.A. et al. (2016). *Trypanosoma cruzi*: time for international recognition as a foodborne parasite. *PLoS Neglected Tropical Diseases* **10**: https://doi.org/10.1371/journal.pntd.0004656.

Robinson, W. (1940). Ammonium bicarbonate secreted by surgical maggots stimulates healing in purulent wounds. *American Journal of Surgery* **47**: 111–115. https://doi.org/10.1016/S0002-9610(40)90125-8.

Rodgers, J. (2010). Trypanosomiasis and the brain. *Parasitology* **137**: 1995–2006. https://doi.org/10.1017/S0031182009991806.

Roelke-Parker, M.E., Munson, L., Packer, C. et al. (1996). A canine distemper virus epidemic in Serengeti lions (*Panthera leo*). *Nature* **379**: 441–445. https://doi.org/10.1038/379441a0.

Rohr, J.R., Schotthoefer, A.M., Raffel, T.R. et al. (2008). Agrochemicals increase trematode infections in a declining amphibian species. *Nature* **455**: 1235–1239. https://doi.org/10.1038/nature07281.

Rojas, F. and Matthews, K.R. (2019). Quorum sensing in African trypanosomes. *Current Opinion in Microbiology* **52**: 124–129. https://doi.org/10.1016/j.mib.2019.07.001.

Romig, T., Ebi, D., and Wassermann, M. (2015). Taxonomy and molecular epidemiology of *Echinococcus granulosus sensu lato*. *Veterinary Parasitology* **213**: 76–84. https://doi.org/10.1016/j.vetpar.2015.07.035.

Roozbahani, M., Hammersmith, K.M., Rapuano, C.J. et al. (2018). *Acanthamoeba* keratitis: are recent cases more severe? *Cornea* **37**: 1381–1387. https://doi.org/10.1097/ICO.0000000000001640.

Rosa, B.A., Choi, Y.J., McNulty, S.N. et al. (2020). Comparative genomics and transcriptomics of 4 *Paragonimus* species provide insights into lung fluke parasitism and pathogenesis. *GigaScience* **9**: https://doi.org/10.1093/gigascience/giaa073.

Rossi, M. and Fasel, N. (2018). The criminal association of *Leishmania* parasites and viruses. *Current Opinion in Microbiology* **46**: 65–72. https://doi.org/10.1016/j.mib.2018.07.005.

Ruberanziza, E., Owada, K., Clark, N.J. et al. (2019). Mapping soil-transmitted helminth parasite infection in Rwanda: estimating endemicity and identifying at-risk populations. *Tropical Medicine and Infectious Disease* **4**: 93. https://doi.org/10.3390/tropicalmed4020093.

Ruggiero, M.A., Gordon, D.P., Orrell, T.M. et al. (2015). A higher level classification of all living organisms. *PLoS One* **10**: https://doi.org/10.1371/journal.pone.0130114.

Rutkowski, N., Dong, Y., and Dimopoulos, G. (2020). Field-deployable molecular diagnostic platform for arbovirus detection in *Aedes aegypti*. *Parasites & Vectors* **13**: 1–14. https://doi.org/10.1186/s13071-020-04357-y.

Ryan, U. and Hijjawi, N. (2015). New developments in *Cryptosporidium* research. *International Journal for Parasitology* **45**: 367–373. https://doi.org/10.1016/j.ijpara.2015.01.009.

Ryan, U., Paparini, A., Monis, P. et al. (2016). It's official – *Cryptosporidium* is a gregarine: What are the implications for the water industry? *Water Research* **105**: 305–313. https://doi.org/10.1016/j.watres.2016.09.013.

Saadi, H., Pagnier, I., Colson, P. et al. (2013). First isolation of Mimivirus in a patient with pneumonia. *Clinical Infectious Diseases* **57**: e127–e134. https://doi.org/10.1093/cid/cit354.

Sabourin, E., Alda, P., Vázquez, A. et al. (2018). Impact of human activities on fasciolosis transmission. *Trends in Parasitology* **34**: 891–903. https://doi.org/10.1016/j.pt.2018.08.004.

Sadeghi, M., Riahi, S.M., Mohammadi, M. et al. (2019). An updated meta-analysis of the association between *Toxoplasma gondii* infection and risk of epilepsy. *Transactions of the Royal Society of Tropical Medicine and Hygiene* **113**: 453–462. PMID: 31034025. https://doi.org/10.1093/trstmh/trz025.

Saidin, S., Othman, N., and Noordin, R. (2019). Update on laboratory diagnosis of amoebiasis. *European Journal of Clinical Microbiology & Infectious Diseases* **38**: 15–38. https://doi.org/10.1007/s10096-018-3379-3.

Salao, K., Spofford, E.M., Price, C. et al. (2020). Enhanced neutrophil functions during *Opisthorchis viverrini* infections and correlation with advanced periductal fibrosis. *International Journal for Parasitology* **50**: 145–152. PMID: 32006550. 10.1016/j.ijpara.2019.11.007.

Saldívar-González, F.I., Naveja, J.J., Palomino-Hernández, O. et al. (2017). Getting SMARt in drug discovery: chemoinformatics approaches for mining structure–multiple activity relationships. *RSC Advances* **7**: 632–641. https://doi.org/10.1039/C6RA26230A.

Salgame, P., Yap, G.S., and Gause, W.C. (2013). Effect of helminth-induced immunity on infections with microbial pathogens. *Nature Immunology* **14**: 1118–1126. https://doi.org/10.1038/ni.2736.

Samie, A., Barrett, L.J., Bessong, P.O. et al. (2010). Seroprevalence of *Entamoeba histolytica* in the context of HIV and AIDS: the case of Vhembe district, in South Africa's Limpopo province. *Annals of Tropical Medicine & Parasitology* **104**: 55–63. https://doi.org/10.1179/136485910X12607012373911.

Sánchez-Valdéz, F.J., Pérez Brandán, C., Ferreira, A. et al. (2015). Gene-deleted live-attenuated *Trypanosoma cruzi* parasites as vaccines to protect against Chagas disease. *Expert Review of Vaccines* **14**: 681–697. https://doi.org/10.1586/14760584.2015.989989.

Santi-Rocca, J., Rigothier, M.C., and Guillén, N. (2009). Host-microbe interactions and defense mechanisms in the development of amoebic liver abscesses. *Clinical Microbiology Reviews* **22**: 65–75. https://doi.org/10.1128/CMR.00029-08.

Santi-Rocca, J., Weber, C., Guigon, G. et al. (2008). The lysine- and glutamic acid-rich protein KERP1 plays a role in *Entamoeba histolytica* liver abscess pathogenesis. *Cellular Microbiology* **10**: 202–217. https://doi.org/10.1111/j.1462-5822.2007.01030.x.

Saracino, M.P., Calcagno, M.A., Beauche, E.B. et al. (2016). *Trichinella spiralis* infection and transplacental passage in human pregnancy. *Veterinary Parasitology* **231**: 2–7. https://doi.org/10.1016/j.vetpar.2016.06.019.

Sargison, N.D., Baird, G.J., Sotiraki, S. et al. (2012). Hepatogenous photosensitisation in Scottish sheep caused by *Dicrocoelium dendriticum*. *Veterinary Parasitology* **189**: 233–237. https://doi.org/10.1016/j.vetpar.2012.04.018.

Sasai, M., Pradipta, A., and Yamamoto, M. (2018). Host immune responses to *Toxoplasma gondii*. *International Immunology* **30**: 113–119. https://doi.org/10.1093/intimm/dxy004.

Satoh, K., Ooe, K., Nagayama, H. et al. (2010). *Prototheca cutis* sp. nov., a newly discovered pathogen of protothecosis isolated from inflamed human skin. *International Journal of Systematic and Evolutionary Microbiology* **60**: 1236–1240. https://doi.org/10.1099/ijs.0.016402-0.

Saumell, C.A., Fernández, A.S., Echevarria, F. et al. (2016). Lack of negative effects of the biological control agent *Duddingtonia flagrans* on soil nematodes and other nematophagous fungi. *Journal of Helminthology* **90**: 706–711. https://doi.org/10.1017/S0022149X1500098X.

Saunders, N., Wilson, R.A., and Coulson, P.S. (1987). The outer bilayer of the adult schistosome tegument surface has a low turnover rate in vitro and in vivo. *Molecular and Biochemical Parasitology* **25**: 123–131. https://doi.org/10.1016/0166-6851(87)90001-6.

Schallig, H.D., Hu, R.V., Kent, A.D. et al. (2019). Evaluation of point of care tests for the diagnosis of cutaneous leishmaniasis in Suriname. *BMC Infectious Diseases* **19**: 1–6. https://doi.org/10.1186/s12879-018-3634-3.

Schantz-Dunn, J. and Nour, N.M. (2009). Malaria and pregnancy: a global health perspective. *Reviews in Obstetrics and Gynecology* **2**: 186–192. PMID: 19826576; PMCID: PMC2760896.

Schär, F., Giardina, F., Khieu, V. et al. (2016). Occurrence of and risk factors for *Strongyloides stercoralis* infection in South-East Asia. *Acta Tropica* **159**: 227–238. https://doi.org/10.1016/j.actatropica.2015.03.008.

Schetters, T.P.M. and Eling, W.M.C. (1999). Can *Babesia* infections be used as a model for cerebral malaria. *Parasitology Today* **15**: 492–497. https://doi.org/10.1016/S0169-4758(99)01566-5.

Schmid-Hempel, P. (2008). Parasite immune evasion: a momentous molecular war. *Trends in Ecology & Evolution* **23**: 318–326. https://doi.org/10.1016/j.tree.2008.02.011.

Schneeberger, P.H.H., Coulibaly, J.T., Panic, G. et al. (2018). Investigations on the interplays between *Schistosoma mansoni*, praziquantel, and the gut microbiome. *Parasites & Vectors* **11**: https://doi.org/10.1186/s13071-018-2739-2.

Scholz, T., Garcia, H.H., Kuchta, R. et al. (2009). Update on the human broad tapeworm (genus Diphyllobothrium), including clinical relevance. *Clinical Microbiology Reviews* **22**: 146–160. https://doi.org/10.1128/CMR.00033-08.

Scholz, T. and Kuchta, R. (2016). Fish-borne, zoonotic cestodes (*Diphyllobothrium* and relatives) in cold climates: a never-ending story of neglected and (re)-emergent parasites. *Food and Waterborne Parasitology* **4**: 23–38. https://doi.org/10.1016/j.fawpar.2016.07.002.

Schönian, G., Lukeš, J., Stark, O. et al. (2018). Molecular Evolution and Phylogeny of *Leishmania*. In: *Drug Resistance in Leishmania Parasites* (ed. A. Ponte-Sucre and M. Padrón-Nieves), 19–57. Cham: Springer https://doi.org/10.1007/978-3-319-74186-4_2.

Schwede, A., Kramer, S., and Carrington, M. (2012). How do trypanosomes change gene expression in response to the environment? *Protoplasma* **249**: 223–238. https://doi.org/10.1007/s00709-011-0282-5.

Scott, D.M., Baker, R., Tomlinson, A. et al. (2020). Spatial distribution of sarcoptic mange (*Sarcoptes scabiei*) in urban foxes (*Vulpes vulpes*) in Great Britain as determined by citizen science. *Urban Ecosystems* **23**: 1127–1140. https://doi.org/10.1007/s11252-020-00985-5.

Scudellari, M. (2017). News feature: cleaning up the hygiene hypothesis. *Proceedings of the National Academy of Sciences of the United States of America* **114**: 1433–1436. https://doi.org/10.1073/pnas.1700688114.

Secrieru, A., Costa, I.C., O'Neill, P.M. et al. (2020). Antimalarial agents as therapeutic tools against toxoplasmosis – a short bridge between two distant illnesses. *Molecules* **25**: 1574. https://doi.org/10.3390/molecules25071574.

Seddon, H.R. (1931). The development in sheep of immunity to *Moniezia expansa*. *Annals of Tropical Medicine and Parasitology* **25**: 431–435. https://doi.org/10.1080/00034983.1931.11684692.

Selzer, P.M. and Epe, C. (2020). Antiparasitics in animal health: quo vadis? *Trends in Parasitology* **37**: 77–89. https://doi.org/10.1016/j.pt.2020.09.004.

Sene-Wade, M., Marchand, B., Rollinson, D. et al. (2018). Urogenital schistosomiasis and hybridization between *Schistosoma haematobium* and *Schistosoma bovis* in adults living in Richard-Toll, Senegal. *Parasitology* **145**: 1723–1726. https://doi.org/10.1017/S0031182018001415.

Sengani, M., Grumezescu, A.M., and Rajeswari, V.D. (2017). Recent trends and methodologies in gold nanoparticle synthesis – a prospective review on drug delivery aspect. *OpenNano* **2**: 37–46. https://doi.org/10.1016/j.onano.2017.07.001.

Sessions, S.K. and Ruth, S.B. (1990). Explanation for naturally occurring supernumerary limbs in amphibians. *Journal of Experimental Zoology* **254**: 38–47. https://doi.org/10.1002/jez.1402540107.

Sesterhenn, A.M., Pfützner, W., Braulke, D.M. et al. (2009). Cutaneous manifestation of myiasis in malignant wounds of the head and neck. *European Journal of Dermatology* **19**: 64–68. https://doi.org/10.1684/ejd.2008.0568.

Severance, E.G. and Yolken, R.H. (2020). Deciphering microbiome and neuroactive immune gene interactions in schizophrenia. *Neurobiology of Disease* **135**: 104331. https://doi.org/10.1016/j.nbd.2018.11.016.

Shadrach, W.S., Rydzewski, K., Laube, U. et al. (2005). *Balamuthia mandrillaris*, free-living ameba and opportunistic agent of encephalitis, is a potential host for *Legionella pneumonophila* bacteria. *Applied and Environmental Microbiology* **71**: 2244–2249. https://doi.org/10.1128/AEM.71.5.2244-2249.2005.

Shanks, G.D. (2017). The multifactorial epidemiology of Blackwater Fever. *American Journal of Tropical Medicine and Hygiene* **97**: 1804–1807. https://doi.org/10.4269/ajtmh.17-0533.

Shao, C.C., Xu, M.J., Alasaad, S. et al. (2014). Comparative analysis of microRNA profiles between adult *Ascaris lumbricoides* and *Ascaris suum*. *BMC Veterinary Research* **10**: 99. https://doi.org/10.1186/1746-6148-10-99.

Shapiro, K., Bahia-Oliveira, L., Dixon, B. et al. (2019). Environmental transmission of *Toxoplasma gondii*: oocysts in water, soil and food. *Food and Waterborne Parasitology* **15**: https://doi.org/10.1016/j.fawpar.2019.e00049.

Sharara, S.L. and Kanj, S.S. (2014). War and infectious diseases: challenges of the Syrian Civil War. *PLoS Pathogens* **10**: https://doi.org/10.1371/journal.ppat.1004438.

Shariati, A., Fallah, F., Pormohammad, A. et al. (2019). The possible role of bacteria, viruses, and parasites in initiation and exacerbation of irritable bowel syndrome. *Journal of Cellular Physiology* **234**: 8550–8569. https://doi.org/10.1002/jcp.27828.

Sharma, L. and Shukla, G. (2017). Placental malaria: a new insight into the pathophysiology. *Frontiers in Medicine* **4**: https://doi.org/10.3389/fmed.2017.00117.

Sharma, S. and Tripathi, P. (2019). Gut microbiome and type 2 diabetes: where we are and where to go? *Journal of Nutritional Biochemistry* **63**: 101–108. https://doi.org/10.1016/j.jnutbio.2018.10.003.

Shears, R.K., Bancroft, A.J., Sharpe, C. et al. (2018). Vaccination against whipworm: identification of potential immunogenic proteins in *Trichuris muris* excretory/secretory material. *Scientific Reports* **8**: https://doi.org/10.1038/s41598-018-22783-y.

Shetty, J.B., Kini, S., Phulpagar, M. et al. (2018). Coinfection of malaria and filaria with unusual crisis forms. *Tropical Parasitology* **8**: 44–46. https://doi.org/10.4103/tp.TP_17_16.

Shimokawa, C., Kabir, M., Taniuchi, M. et al. (2012). *Entamoeba moshkovskii* is associated with diarrhea in infants and causes diarrhea and colitis in mice. *Journal of Infectious Diseases* **206**: 744–751. https://doi.org/10.1093/infdis/jis414.

Shirley, M.W., Smith, A.L., and Blake, D.P. (2007). Challenges in the successful control of the avian coccidia. *Vaccine* **25**: 5540–5547. https://doi.org/10.1016/j.vaccine.2006.12.030.

Shobha, M., Bithika, D., and Bhavesh, S. (2013). The prevalence of intestinal parasitic infections in the urban slums of a city in Western India. *Journal of Infection and Public Health* **6**: 142–149. https://doi.org/10.1016/j.jiph.2012.11.004.

Short, E.E., Caminade, C., and Thomas, B.N. (2017). Climate change contribution to the emergence or re-emergence of parasitic diseases. *Infectious Diseases: Research and Treatment* **10**: https://doi.org/10.1177/1178633617732296.

Shrestha, A., Abd-Elfattah, A., Freudenschuss, B. et al. (2015). *Cystoisospora suis* – a model of mammalian cystoisosporosis. *Frontiers in Veterinary Science* **2**: 68. https://doi.org/10.3389/fvets.2015.00068.

Shukla, G., Bhatia, R., and Sharma, A. (2016). Prebiotic inulin supplementation modulates the immune response and restores gut morphology in *Giardia duodenalis*-infected malnourished mice. *Parasitology Research* **115**: 4189–4198. https://doi.org/10.1007/s00436-016-5196-x.

Siddiqui, R., Ali, I.K.M., Cope, J.R. et al. (2016). Biology and pathogenesis of *Naegleria fowleri*. *Acta Tropica* **164**: 375–394. https://doi.org/10.1186/1756-3305-5-6.

Sig, A.K., Guney, M., Guclu, A.U. et al. (2017). Medicinal leech therapy – an overall perspective. *Integrative Medicine Research* **6**: 337–343. https://doi.org/10.1016/j.imr.2017.08.001.

Siles-Lucas, M., González-Miguel, J., Geller, R. et al. (2020). Potential influence of helminth molecules on COVID-19 pathology. *Trends in Parasitology* **37**: 11–14. https://doi.org/10.1016/j.pt.2020.10.002.

Singh, B. and Daneshvar, C. (2013). Human infections and detection of *Plasmodium knowlesi*. *Clinical Microbiology Reviews* **26**: 165–184. https://doi.org/10.1128/CMR.00079-12.

Singhal, N., Kumar, M., and Virdi, J.S. (2016). MALDI-TOF MS in clinical parasitology: applications, constraints and prospects. *Parasitology* **143**: 1491–1500. https://doi.org/10.1017/S0031182016001189.

Skalicka-Woźniak, K., Georgiev, M.I., and Orhan, I.E. (2017). Adulteration of herbal sexual enhancers and slimers: the wish for better sexual well-being and perfect body can be risky. *Food and Chemical Toxicology* **108**: 355–364. https://doi.org/10.1016/j.fct.2016.06.018.

Skelly, P.J. and Wilson, R.A. (2006). Making sense of the schistosome surface. *Advances in Parasitology* **63**: 185–284. https://doi.org/10.1016/S0065-308X(06)63003-0.

Skrjabin, A.S. (1967). A gigantic diphyllobothriid, *Polygonoporus giganticus* n.g., n.sp. parasite of the cachalot (In Russian). *Zhurnal Parazitologica* **1**: 131–136.

Skwarczynski, M., Chandrudu, S., Rigau-Planella, B. et al. (2020). Progress in the development of subunit vaccines against malaria. *Vaccines* **8**: 373. https://doi.org/10.3390/vaccines8030373.

Slavinsky, V.S., Chugunov, K.V., Tsybankov, A.A. et al. (2018). *Trichuris trichiura* in mummified remains of southern Siberian nomads. *Antiquity* **92**: 410–420. 10.15184/aqy.2018.12.

Slesak, G., Inthalad, S., Strobel, M. et al. (2011). Chromoblastomycosis after a leech bite complicated by myiasis: a case report. *BMC Infectious Diseases* **11**: 14. https://doi.org/10.1186/1471-2334-11-14.

Smout, M.J., Sotillo, J., Laha, T. et al. (2015). Carcinogenic parasite secretes growth factor that accelerates wound healing and potentially promotes neoplasia. *PLoS Pathogens* **11**: https://doi.org/10.1371/journal.ppat.1005209.

Snowdon, F.M. (2006). *The Conquest of Malaria. Italy, 1900–1962*. New Haven, CT: Yale University Press.

Sobczak, N. and Kantyka, M. (2014). Hirudotherapy in veterinary medicine. *Annals of Parasitology* **60**: 89–92. PMID: 25115059.

Sobotková, K., Parker, W., Levá, J. et al. (2019). Helminth therapy – from the parasite perspective. *Trends in Parasitology* **35**: 501–515. https://doi.org/10.1016/j.pt.2019.04.009.

Söderhäll, K. (ed.) (2011). *Invertebrate Immunity*. Heidelberg: Landes Bioscience and Springer Science+Business Media.

Sokolova, Y.Y. and Overstreet, R.M. (2020). Hyperparasitic spore-forming eukaryotes (Microsporidia, Haplosporidia, and Myxozoa) parasitizing trematodes (Platyhelminthes). *Invertebrate Zoology* **17**: 93–117. https://aquila.usm.edu/fac_pubs/18273.

Solaymani-Mohammadi, S. and Singer, S.M. (2010). *Giardia duodenalis*: the double-edged sword of immune responses in giardiasis. *Experimental Parasitology* **126**: 292–297. https://doi.org/10.1016/j.exppara.2010.06.014.

Soldanova, M., Selbach, C., and Sures, B. (2016). The early worm catches the bird? Productivity and patterns of *Trichobilharzia szidati* cercarial emission from *Lymnaea stagnalis*. *PloS One* **11**: https://doi.org/10.1371/journal.pone.0149678.

Sonoiki, E., Ng, C., Lee, M. et al. (2017). A potent antimalarial benzoxaborole targets a *Plasmodium falciparum* cleavage and polyadenylation specificity factor homologue. *Nature Communications* **8**: https://doi.org/10.1038/ncomms14574.

Sorobetea, D., Svensson-Frej, M., and Grencis, R. (2018). Immunity to gastrointestinal nematode infections. *Mucosal Immunology* **11**: 304–315. https://doi.org/10.1038/mi.2017.113.

Spraker, T.R., DeLong, R.L., Lyons, E.T. et al. (2007). Hookworm enteritis with bacteremia in California sea lion pups on San Miguel Island. *Journal of Wildlife Diseases* **43**: 179–188. https://doi.org/10.7589/0090-3558-43.2.179.

Sriamporn, S., Pisani, P., Pipitgool, V. et al. (2004). Prevalence of *Opisthorchis viverrini* infection and incidence of cholangiocarcinoma in Khon Kaen, Northeast Thailand. *Tropical Medicine & International Health* **9**: 588–594. https://doi.org/10.1111/j.1365-3156.2004.01234.x.

Stadler, F. (2020). The maggot therapy supply chain: a review of the literature and practice. *Medical and Veterinary Entomology* **34**: 1–9. https://doi.org/10.1111/mve.12397.

Stahlman, S., Williams, V.F., and Taubman, S.B. (2017). Incident diagnoses of leishmaniasis, active and reserve components, US Armed Forces, 2001–2016. *MSMR* **24**: 2–7. PMID: 28234494.

Stark, D., Barratt, J.L.N., Van Hal, S. et al. (2009). Clinical significance of enteric protozoa in the immunosuppressed human population. *Clinical Microbiology Reviews* **22**: 634–650. https://doi.org/10.1128/CMR.00017-09.

Stark, D., Van Hal, S., Fotedar, R. et al. (2008). Comparison of stool antigen detection kits to PCR for diagnosis of amebiasis. *Journal of Clinical Microbiology* **46**: 1678–1681. https://doi.org/10.1128/JCM.02261-07.

Steenvoorde, P., Jacobi, C.E., and Oskam, J. (2005). Maggot debridement therapy: free-range or contained? An *in-vivo* study. *Advances in Skin and Wound Care* **18**: 430–435. https://doi.org/10.1097/00129334-200510000-00010.

Stensgaard, A.S., Vounatsou, P., Sengupta, M.E. et al. (2019). Schistosomes, snails and climate change: Current trends and future expectations. *Acta Tropica* **190**: 257–268. https://doi.org/10.1016/j.actatropica.2018.09.013.

Stentiford, G.D., Bass, D., and Williams, B.A.P. (2019). Ultimate opportunists – the emergent *Enterocytozoon* group Microsporidia. *PLoS Pathogens* **15**: https://doi.org/10.1371/journal.ppat.1007668.

Stepek, G., Lowe, A.E., Buttle, D.J. et al. (2006). *in vitro* and *in vivo* anthelmintic efficacy of plant cysteine proteinases against the rodent gastrointestinal nematode, *Trichuris muris*. *Parasitology* **132**: 681–689. PMID: 16448585. https://doi.org/10.1017/S003118200500973X.

Stoyanov, C.T., Boscardin, S.B., Deroubaix, S. et al. (2010). Immunogenicity and protective efficacy of a recombinant yellow fever vaccine against the murine malarial parasite *Plasmodium yoelii*. *Vaccine* **28**: 4644–4652. https://doi.org/10.1016/j.vaccine.2010.04.071.

Stracke, K., Clarke, N., Awburn, C.V. et al. (2019). Development and validation of a multiplexed-tandem qPCR tool for diagnostics of human soil-transmitted helminth infections. *PLoS Neglected Tropical Diseases* **13**: https://doi.org/10.1371/journal.pntd.0007363.

Strand, M.R. and Burke, G.R. (2019). Polydnaviruses: evolution and function. In: *Insect Molecular Virology. Advances and Emerging Trends* (ed. B.C. Bonning), 163–181. Poole, UK: Caister Academic Press.

Streit, A. (2008). Reproduction in *Strongyloides* (Nematoda): a life between sex and parthenogenesis. *Parasitology* **135**: 285–294. https://doi.org/10.1017/S003118200700399X.

Strickland, G.T. (2006). Liver disease in Egypt: hepatitis C superseded schistosomiasis as a result of iatrogenic and biological factors. *Hepatology* **43**: 915–922. https://doi.org/10.1002/hep.21173.

Stromberg, B.E. (1997). Environmental factors influencing transmission. *Veterinary Parasitology* **72**: 247–264. https://doi.org/10.1016/S0304-4017(97)00100-3.

Sturgess-Osborne, C., Burgess, S., Mitchell, S. et al. (2019). Multiple resistance to macrocyclic lactones in the sheep scab mite *Psoroptes ovis*. *Veterinary Parasitology* **272**: 79–82. https://doi.org/10.1016/j.vetpar.2019.07.007.

Suarez, C., Lentini, G., Ramaswamy, R. et al. (2019). A lipid-binding protein mediates rhoptry discharge and invasion in *Plasmodium falciparum* and *Toxoplasma gondii* parasites. *Nature Communications* **10**: 1–14. 10.1038/s41467-019-11979-z.

Sugden, K., Moffitt, T.E., Pinto, L. et al. (2016). Is *Toxoplasma gondii* infection related to brain and behavior impairments in humans? Evidence from a population-representative birth cohort. *PLoS One* **11**: https://doi.org/10.1371/journal.pone.0148435.

Summers, R.W., Elliott, D.E., Urban, J.F. et al. (2005). *Trichuris suis* therapy in Crohn's disease. *Gut* **54**: 87–90. https://doi.org/10.1136/gut.2004.041749.

Sundermann, C.A. and Estridge, B.H. (1999). Growth and competition between *Neospora caninum* and *Toxoplasma gondii* in vitro. *International Journal for Parasitology* **29**: 1725–1732. https://doi.org/10.1016/S0020-7519(99)00114-9.

Sures, B., Nachev, M., Selbach, C. et al. (2017). Parasite responses to pollution: what we know and where we go in 'Environmental Parasitology'. *Parasites & Vectors* **10**: 65. https://doi.org/10.1186/s13071-017-2001-3.

Sutherst, R.W., Jones, R.J., and Schnitzerling, H.J. (1982). Tropical legumes of the genus *Stylosanthes* immobilize and kill cattle ticks. *Nature* **295**: 320–321. PMID: 7057894. https://doi.org/10.1038/295320a0.

Sutterland, A.L., Kuin, A., Kuiper, B. et al. (2019). Driving us mad: the association of *Toxoplasma gondii* with suicide attempts and traffic accidents – a systematic review and meta-analysis. *Psychological Medicine* **49**: 1608–1623. https://doi.org/10.1017/S0033291719000813.

Svartman, M., Finklea, J., Potter, E. et al. (1972). Epidemic scabies and acute glomerulonephritis in Trinidad. *Lancet* **299**: 249–251. PMID: 4109709. https://doi.org/10.1016/s0140-6736(72)90634-4.

Syed, A.A. and Mehta, A. (2018). Target specific anticoagulant peptides: a review. *International Journal of Peptide Research and Therapeutics* **24**: 1–12. https://doi.org/10.1007/s10989-018-9682-0.

Székely, C., Cech, G., Atkinson, S.D. et al. (2015). A novel myxozoan parasite of terrestrial mammals: description of *Soricimyxum minuti* sp. n. (Myxosporea) in pygmy shrew *Sorex minutus* from Hungary. *Folia Parasitologica* **62**: 45–49. https://doi.org/10.14411/fp.2015.045.

Tachezy, J. (2019). *Hydrogenosomes and Mitosomes: Mitochondria of Anaerobic Eukaryotes*, 2nde (Microbiology Monographs, vol. **9**. Cham: Springer.

Takken, W. and Lindsay, S. (2019). Increased threat of urban malaria from *Anopheles stephensi* mosquitoes, Africa. *Emerging Infectious Diseases* **25**: 1431–1433. https://doi.org/10.3201/eid2507.190301.

Tan, D. and Vyas, A. (2016). *Toxoplasma gondii* infection and testosterone congruently increase tolerance of male rats for risk of reward forfeiture. *Hormones and Behavior* **79**: 37–44. https://doi.org/10.1016/j.yhbeh.2016.01.003.

Tan, K.S. (2008). New insights on classification, identification, and clinical relevance of *Blastocystis* spp. *Clinical Microbiology Reviews* **21**: 639–665. https://doi.org/10.1128/CMR.00022-08.

Tang, C.L., Zhou, H.H., Zhu, Y.W. et al. (2019). Glutathione *S*-transferase influences the fecundity of *Schistosoma japonicum*. *Acta Tropica* **191**: 8–12. https://doi.org/10.1016/j.actatropica.2018.12.027.

Tantawi, T.I., Williams, K.A., and Villet, M.H. (2010). An accidental but safe and effective use of *Lucilia cuprina* (Diptera: Calliphoridae) in maggot debridement therapy in Alexandria. *Egypt. Journal of Medical Entomology* **47**: 491–494. https://doi.org/10.1093/jmedent/47.3.491.

Tappe, D. and Büttner, D.W. (2009). Diagnosis of human visceral pentastomiasis. *PLoS Neglected Tropical Diseases* **3**: e320. https://doi.org/10.1371/journal.pntd.0000320.

Tasiemski, A., Massol, F., Cuvillier-Hot, V. et al. (2015). Reciprocal immune benefit based on complementary production of antibiotics by the leech *Hirudo verbana* and its gut symbiont *Aeromonas veronii*. *Scientific Reports* **5**: https://doi.org/10.1038/srep17498.

Tasiemski, A., Massol, F., Cuvillier-Hot, V. et al. (2015). Reciprocal immune benefit based on complementary production of antibiotics by the leech *Hirudo verbana* and its gut symbiont *Aeromonas veronii*. *Nature Scientific Reports* **5**: https://doi.org/10.1038/srep17498.

Tatham, P., Stadler, F., Murray, A. et al. (2017). Flying maggots: a smart logistic solution to an enduring medical challenge. *Journal of Humanitarian Logistics and Supply Chain Management* **7**: 172–193. https://doi.org/10.1108/JHLSCM-02-2017-0003.

Tatiya-aphiradee, N., Chatuphonprasert, W., and Jarukamjorn, K. (2018). Immune response and inflammatory pathway of ulcerative colitis. *Journal of Basic and Clinical Physiology and Pharmacology* **30**: 1–10. https://doi.org/10.1515/jbcpp-2018-0036.

Taylor, A.W. (1930). Experiments on the mechanical transmission of West African strains of *Trypanosoma brucei* and *T. gambiense* by *Glossina* and other biting flies. *Transactions of the Royal Society of Tropical Medicine and Hygiene* **24**: 289–303. https://doi.org/10.1016/S0035-9203(30)92062-7.

Taylor, M.A., Coop, L., and Wall, R. (2007). *Veterinary Parasitology*, 3rde. Oxford, UK: Blackwell Publishing.

Taylor, M.J., Cross, H.F., Ford, L. et al. (2001). *Wolbachia* bacteria in filarial immunity and disease. *Parasite Immunology* **23**: 401–409. https://doi.org/10.1046/j.1365-3024.2001.00400.x.

Taylor, M.J., Hoerauf, A., and Bockarie, M. (2010). Lymphatic filariasis and onchocerciasis. *Lancet* **376**: 1175–1185. https://doi.org/10.1016/S0140-6736(10)60586-7.

Taylor, M.J., von Geldern, T.W., Ford, L. et al. (2019). Preclinical development of an oral anti-*Wolbachia* macrolide drug for the treatment of lymphatic filariasis and onchocerciasis. *Science Translational Medicine* **11**: https://doi.org/10.1126/scitranslmed.aau2086.

Tebeje, B.M., Harvie, M., You, H. et al. (2016). Schistosomiasis vaccines: where do we stand? *Parasites & Vectors* **9**: 528–543. https://doi.org/10.1186/s13071-016-1799-4.

Teichmann, C.E., Da Silva, A.S., Monteiro, S.G. et al. (2011). Evidence of venereal and transplacental transmission of canine visceral leishmaniasis in southern Brazil. *Acta Scientiae Veterinariae* **39**: 1–4. https://www.redalyc.org/articulo.oa?id=289022118015.

Teixeira, W.F.P., Tozato, M.E.G., Pierucci, J.C. et al. (2017). Investigation of *Toxoplasma gondii* in semen, testicle and epididymis tissues of primo-infected cats (*Felis catus*). *Veterinary Parasitology* **238**: 90–93. https://doi.org/10.1016/j.vetpar.2017.04.003.

Tekle, S.G., Corpolongo, A., D'Abramo, A. et al. (2018). Case report: delayed diagnosis of congenital malaria by *Plasmodium vivax* in a newborn of an Eritrean woman with varicella infection. *American Journal of Tropical Medicine and Hygiene* **99**: 620–622. https://doi.org/10.4269/ajtmh.18-0091.

Tenter, A.M., Heckeroth, A.R., and Weiss, L.M. (2000). *Toxoplasma gondii*: from animals to humans. *International Journal for Parasitology* **30**: 1217–1258. https://doi.org/10.1016/S0020-7519(00)00124-7.

Terrazas, C., de Dios Ruiz-Rosado, J., Amici, S. et al. (2017). Helminth-induced Ly6C^{hi} monocyte-derived alternatively activated macrophages suppress experimental autoimmune encephalomyelitis. *Scientific Reports* **7**: 40814. https://doi.org/10.1038/srep40814.

Tetreau, G.H., Dhinaut, J., Gourbal, B. et al. (2019). Trans-generational immune priming in invertebrates: current knowledge and future prospects. *Frontiers in Immunology* **10**: 1938. https://doi.org/10.3389/fimmu.2019.01938.

Tewari, R., Straschil, U., Bateman, A. et al. (2010). The systematic functional analysis of *Plasmodium* protein kinases identifies essential regulators of mosquito transmission. *Cell Host & Microbe* **8**: 377–387. https://doi.org/10.1016/j.chom.2010.09.006.

Thakur, L., Kushwaha, H.R., Negi, A. et al. (2020). *Leptomonas seymouri* co-infection in Cutaneous Leishmaniasis cases caused by *Leishmania donovani* from Himachal Pradesh, India. *Frontiers in Cellular and Infection Microbiology* **10**: 345. https://doi.org/10.3389/fcimb.2020.00345.

Thamsborg, S.M., Ketzis, J., Horii, Y. et al. (2017). *Strongyloides* spp. infections of veterinary importance. *Parasitology* **144**: 274–284. https://doi.org/10.1017/S0031182016001116.

Thomas, F., Madsen, T., Giraudeau, M. et al. (2019). Transmissible cancer and the evolution of sex. *PLoS Biology* **17**: https://doi.org/10.1371/journal.pbio.3000275.

Thompson, R.C.A. (2004). The zoonotic significance and molecular epidemiology of *Giardia* and giardiasis. *Veterinary Parasitology* **126**: 15–35. https://doi.org/10.1016/j.vetpar.2004.09.008.

Thompson, R.C.A. and Ash, A. (2019). Molecular epidemiology of *Giardia* and *Cryptosporidium* infections – what's new? *Infection, Genetics and Evolution* **75**: https://doi.org/10.1016/j.meegid.2019.103951.

Thompson, R.C.A., Lymbery, A.J., and Smith, A. (2010). Parasites, emerging disease and wildlife conservation. *International Journal for Parasitology* **40**: 1163–1170. https://doi.org/10.1016/j.ijpara.2010.04.009.

Thomson, R., Genovese, G., Canon, C. et al. (2014). Evolution of the primate trypanolytic factor APOL1. *Proceedings of the National Academy of Sciences of the United States of America* **111**: E2130–E2139. https://doi.org/10.1073/pnas.1400699111.

Thorley-Lawson, D., Deitsch, K.W., Duca, K.A. et al. (2016). The link between *Plasmodium falciparum* malaria and endemic Burkitt's Lymphoma – new insight into a 50-year-old enigma. *PLoS Pathogens* **12**: https://doi.org/10.1371/journal.ppat.1005331.

Thornton, J. (2020). Covid-19: keep essential malaria services going during pandemic, urges WHO. *British Medical Journal* **369**: https://doi.org/10.1136/bmj.m1637.

Thresher, R.E., Werner, M., Høeg, J.T. et al. (2000). Developing the options for managing marine pests: specificity trials on the parasitic castrator, *Sacculina carcini*, against the European crab, *Carcinus maenas*, and related species. *Journal of Experimental Marine Biology and Ecology* **254**: 37–51. https://doi.org/10.1016/S0022-0981(00)00260-4.

Tindih, H.S., Marcotty, T., Naessens, J. et al. (2010). Demonstration of differences in virulence between two *Theileria parva* isolates. *Veterinary Parasitology* **168**: 223–230. https://doi.org/10.1016/j.vetpar.2009.11.006.

Tizard, I.R. (2017). *Veterinary Immunology*, 10the. Amsterdam: Elsevier.

Tomás, G., Martín-Gálvez, D., Ruiz-Rodríguez, M. et al. (2017). Intraspecific avian brood parasites avoid host nests infested by ectoparasites. *Journal of Ornithology* **158**: 561–567. https://doi.org/10.1007/s10336-016-1409-4.

Tomczuk, K., Kostro, K., Grzybek, M. et al. (2015). Seasonal changes of diagnostic potential in the detection of *Anoplocephala perfoliata* equine infections in the climate of Central Europe. *Parasitology Research* **114**: 767–772. https://doi.org/10.1007/s00436-014-4279-9.

Tompkins, D.M., Jones, T., and Clayton, D.H. (1996). Effect of vertically transmitted parasites on the reproductive success of swifts (*Apus apus*). *Functional Ecology* **10**: 733–740. https://doi.org/10.2307/2390508.

Topsell, E. (1658). *The History of Four-footed Beasts and Serpents*. London, UK: E. Coates.

Torgerson, P.R., Robertson, L.J., Enemark, H.L. et al. (2020). Source attribution of human echinococcosis: a systematic review and meta-analysis. *PLoS Neglected Tropical Diseases* **14**: https://doi.org/10.1371/journal.pntd.0008382.

Torres, J.E.P., Goossens, J., Ding, J. et al. (2018). Development of a nanobody-based lateral flow assay to detect active *Trypanosoma congolense* infections. *Nature Scientific Reports* **8**: 9019. https://doi.org/10.1038/s41598-018-26732-7.

Torres-Estrada, J.L., Rodríguez, M.H., Cruz-López, L. et al. (2001). Selective oviposition by *Aedes aegypti* (Diptera: Culicidae) in response to *Mesocyclops longisetus* (Copepoda: Cyclopoidea) under laboratory and field conditions. *Journal of Medical Entomology* **38**: 188–192. https://doi.org/10.1603/0022-2585-38.2.188.

Tracz, E.S., Al-Jubury, A., Buchmann, K. et al. (2019). Outbreak of swimmer's itch in Denmark. *Acta Dermato-Venereologica* **99**: 1116–1120. https://doi.org/10.2340/00015555-3309.

Tram, N.T., Hoang, L.M., Cam, P.D. et al. (2008). *Cyclospora* spp. in herbs and water samples collected from markets and farms in Hanoi, Vietnam. *Tropical Medicine & International Health* **13**: 1415–1420. https://doi.org/10.1111/j.1365-3156.2008.02158.x.

Tretina, K., Gotia, H.T., Mann, D.J. et al. (2015). *Theileria*-transformed bovine leukocytes have cancer hallmarks. *Trends in Parasitology* **31**: 306–314. https://doi.org/10.1016/j.pt.2015.04.001.

Trivers, R.L. and Willard, D.E. (1973). Natural selection of parental ability to vary the sex ratio of offspring. *Science* **179**: 90–92. https://doi.org/10.1126/science.179.4068.90.

Troeger, C., Blacker, B.F., Khalil, I.A. et al. (2018). Estimates of the global, regional, and national morbidity, mortality, and aetiologies of diarrhoea in 195 countries: a systematic analysis for the Global Burden of Disease Study 2016. *Lancet Infectious Diseases* **18**: 1211–1228. https://doi.org/10.1016/S1473-3099(17)30276-1.

Truc, P., Büscher, P., Cuny, G. et al. (2013). Atypical human infections by animal trypanosomes. *PLoS Neglected Tropical Diseases* **7**: https://doi.org/10.1371/journal.pntd.0002256.

Tsui, C.K.M., Miller, R., Uyaguari-Diaz, M. et al. (2018). Beaver fever: whole-genome characterization of waterborne outbreak and sporadic isolates to study the zoonotic transmission of giardiasis. *mSphere* **3**: https://doi.org/10.1128/mSphere.00090-18.

Tu, H.L.C., Nugraheni, Y.R., Tiawsirisup, S. et al. (2021). Development of a novel multiplex PCR assay for the detection and differentiation of *Plasmodium caprae* from *Theileria luwenshuni* and *Babesia* spp. in goats. *Acta Tropica* **220**: https://doi.org/10.1016/j.actatropica.2021.105957.

Tweet, J., Chin, K., and Ekdale, A.A. (2016). Trace fossils of possible parasites inside the gut contents of a hadrosaurid dinosaur, Upper Cretaceous Judith River Formation, Montana. *Journal of Palaeontology* **90**: 279–287. https://doi.org/10.1017/jpa.2016.43.

Uneke, C.J. (2007). Congenital *Plasmodium falciparum* malaria in sub-Saharan Africa: a rarity or frequent occurrence? *Parasitology Research* **101**: 835–842. https://doi.org/10.1007/s00436-007-0577-9.

Uribe-Querol, E. and Rosales, C. (2020). Immune response to the enteric parasite *Entamoeba histolytica*. *Physiology* **35**: 244–260. https://doi.org/10.1152/physiol.00038.2019.

Utzinger, J. and Tanner, M. (2013). Socioeconomic development to fight malaria, and beyond. *Lancet* **382**: 920–922. https://doi.org/10.1016/S0140-6736(13)61211-8.

Uwimana, A., Legrand, E., Stokes, B.H. et al. (2020). Emergence and clonal expansion of in vitro artemisinin-resistant *Plasmodium falciparum* kelch13 R561H mutant parasites in Rwanda. *Nature Medicine* **26**: 1602–1608. https://doi.org/10.1038/s41591-020-1005-2.

Vadlamudi, R.S., Chi, D.S., and Krishnaswamy, G. (2006). Intestinal strongyloidiasis and hyperinfection syndrome. *Clinical and Molecular Allergy* **4**: https://doi.org/10.1186/1476-7961-4-8.

Valenzuela-Moreno, L.F., Rico-Torres, C.P., Cedillo-Peláez, C. et al. (2020). Stray dogs in the tropical state of Chiapas, Mexico, harbour atypical and novel genotypes of *Toxoplasma gondii*. *International Journal for Parasitology* **50**: 85–90. https://doi.org/10.1016/j.ijpara.2019.12.001.

Van Damme, I., Trevisan, C., Mwape, K.E. et al. (2021). Trial design for a diagnostic accuracy study of a point-of-care test for the detection of *Taenia solium* taeniosis and (neuro) cysticercosis in community settings of highly endemic, resource-poor areas in Zambia: challenges and rationale. *Diagnostics* **11**: 1138. https://doi.org/10.3390/diagnostics11071138.

van de Velde, N., Devleesschauwer, B., Leopold, M. et al. (2016). *Toxoplasma gondii* in stranded marine mammals from the North Sea and Eastern Atlantic Ocean: findings and diagnostic difficulties. *Veterinary Parasitology* **230**: 25–32. https://doi.org/10.1016/j.vetpar.2016.10.021.

Van Den Abbeele, J., Caljon, G., and De Ridder, K. (2010). *Trypanosoma brucei* modifies the tsetse salivary composition, altering the fly feeding behavior that favors parasite transmission. *PLoS Pathogens* **6**: https://doi.org/10.1371/journal.ppat.1000926.

van den Biggelaar, A.H., van Ree, R., Rodrigues, L.C. et al. (2000). Decreased atopy in children infected with *Schistosoma haematobium*: a role for parasite-induced interleukin-10. *Lancet* **356**: 1723–1727. https://doi.org/10.1016/S0140-6736(00)03206-2.

Van den Bossche, D., Cnops, L., Verschueren, J. et al. (2015). Comparison of four rapid diagnostic tests, ELISA, microscopy and PCR for the detection of *Giardia lamblia*, *Cryptosporidium* spp. and *Entamoeba histolytica* in feces. *Journal of Microbiological Methods* **110**: 78–84. https://doi.org/10.1016/j.mimet.2015.01.016.

Van Der Plas, M.J., Andersen, A.S., Nazir, S. et al. (2014). A novel serine protease secreted by medicinal maggots enhances plasminogen activator-induced fibrinolysis. *PLoS One* **9**: https://doi.org/10.1371/journal.pone.0092096.

Van Kruiningen, H.J. and West, A.B. (2005). Potential danger in the medical use of *Trichuris suis* for the treatment of inflammatory bowel disease. *Inflammatory Bowel Diseases* **11**: 515–515. https://doi.org/10.1097/01.mib.0000160369.47671.a2.

Van Noorden, R. (2010). Demand for malaria drug soars: farmers and scientists struggle to keep up with needs of ambitious medicine-subsidy programme. *Nature* **466**: 672–674. PMID: 20686539. https://doi.org/10.1038/466672a.

van Paridon, B.J., Colwell, D.D., Goater, C.P. et al. (2017). Population genetic analysis informs the invasion history of the emerging trematode *Dicrocoelium dendriticum* into Canada. *International Journal for Parasitology* **47**: 845–856. https://doi.org/10.1016/j.ijpara.2017.04.006.

Van Xong, H., De Baetselier, P., Pays, E. et al. (2002). Selective pressure can influence the resistance of *Trypanosoma congolense* to normal human serum. *Experimental Parasitology* **102**: 61–65. https://doi.org/10.1016/S0014-4894(03)00032-8.

Varcasia, A., Jia, W.Z., Yan, H.B. et al. (2012). Molecular characterization of subcutaneous and muscular coenurosis of goats in United Arab Emirates. *Veterinary Parasitology* **190**: 604–607. https://doi.org/10.1016/j.vetpar.2012.07.012.

Varcasia, A., Tamponi, C., Tosciri, G. et al. (2015). Is the red fox (*Vulpes vulpes*) a competent definitive host for *Taenia multiceps*? *Parasites & Vectors* **8**: 491. https://doi.org/10.1186/s13071-015-1096-7.

Vejzagić, N., Adelfio, R., Keiser, J. et al. (2015). Bacteria-induced egg hatching differs for *Trichuris muris* and *Trichuris suis*. *Parasites & Vectors* **8**: 1–7. https://doi.org/10.1186/s13071-015-0986-z.

Verdon, R., Mirelman, D., and Sansonetti, P.J. (1992). A model of interaction between *Entamoeba histolytica* and *Shigella flexneri*. *Research in Microbiology* **143**: 67–74. https://doi.org/10.1016/0923-2508(92)90035-M.

Verma, A.K., Bharti, P.K., and Das, A. (2018). HRP-2 deletion: a hole in the ship of malaria elimination. *Lancet Infectious Diseases* **18**: 826–827. https://doi.org/10.1016/S1473-3099(18)30420-1.

Verma, S.K., Knowles, S., Cerqueira-Cézar, C.K. et al. (2018). An update on *Toxoplasma gondii* infections in northern sea otters (*Enhydra lutris kenyoni*) from Washington State, USA. *Veterinary Parasitology* **258**: 133–137. https://doi.org/10.1016/j.vetpar.2018.05.011.

Vigani, A., Schnoke, A., and Pozzi, A. (2011). Maggot debridement and leech therapy as treatment of a partial digital amputation injury in a dog. *Wounds* **23**: E9–E15. Corpus ID: 35118260.

Vila, C.C., Saracino, M.P., Falduto, G.H. et al. (2019). Protein malnutrition impairs the immune control of *Trichinella spiralis* infection. *Nutrition* **60**: 161–169. https://doi.org/10.1016/j.nut.2018.10.024.

Villaescusa, J.M., Angulo, I., Pontón, A. et al. (2016). Infestation of a diabetic foot by *Wohlfahrtia magnifica*. *Journal of Vascular Surgery Cases and Innovative Techniques* **2**: 119–122. https://doi.org/10.1016/j.jvscit.2016.04.007.

Villares, M., Berthelet, J., and Weitzman, J.B. (2020). The clever strategies used by intracellular parasites to hijack host gene expression. *Seminars in Immunopathology* **42**: 215–226. https://doi.org/10.1007/s00281-020-00779-z.

Vink, M.M., Nahzat, S.M., Rahimi, H. et al. (2018). Evaluation of point-of-care tests for cutaneous leishmaniasis diagnosis in Kabul, Afghanistan. *EBioMedicine* **37**: 453–460. https://doi.org/10.1016/j.ebiom.2018.10.063.

Visvesvara, G.S., Moura, H., and Schuster, F.L. (2007). Pathogenic and opportunistic free-living amoebae: *Acanthamoeba* spp., *Balamuthia mandrillaris*, *Naegleria fowleri*, and *Sappinia diploidea*. *FEMS Immunology and Medical Microbiology* **50**: 1–26. https://doi.org/10.1111/j.1574-695X.2007.00232.x.

Volf, P., Benkova, I., Myskova, J. et al. (2007). Increased transmission potential of *Leishmania major*/*Leishmania infantum* hybrids. *International Journal for Parasitology* **37**: 589–593. https://doi.org/10.1016/j.ijpara.2007.02.002.

Von Allmen, N., Christen, S., Forster, U. et al. (2006). Acute trichinellosis increases susceptibility to *Giardia lamblia* infection in the mouse model. *Parasitology* **133**: 139–149. https://doi.org/10.1017/S0031182006000230.

von Sinner, W.N. and Stridbeck, H. (1992). Hydatid disease of the spleen: ultrasonography, CT and MR imaging. *Acta Radiologica* **33**: 459–461. PMID: 1389656. 10.1177/028418519203300517.

Voslářvá, E. and Passantino, A. (2012). Stray dog and cat laws and enforcement in Czech Republic and in Italy. *Annali dell'Istituto superiore di sanitÃ* **48**: 97–104. https://doi.org/10.4415/ANN_12_01_16.

Vukman, K.V., Adams, P.N., Metz, M. et al. (2013). *Fasciola hepatica* tegumental coat impairs mast cells' ability to drive Th1 immune responses. *Journal of Immunology* **190**: 2873–2879. https://doi.org/10.4049/jimmunol.1203011.

Wait, L.F., Dobson, A.P., and Graham, A.L. (2020). Do parasite infections interfere with immunisation? A review and meta-analysis. *Vaccine* **38**: 5582–5590. https://doi.org/10.1016/j.vaccine.2020.06.064.

Walker, P.G., White, M.T., Griffin, J.T. et al. (2015). Malaria morbidity and mortality in Ebola-affected countries caused by decreased health-care capacity, and the potential effect of mitigation strategies: a modelling analysis. *Lancet Infectious Diseases* **15**: 825–832. https://doi.org/10.1016/S1473-3099(15)70124-6.

Wallace, L.A., Gwynne, L., and Jenkins, T. (2019). Challenges and opportunities of pH in chronic wounds. *Therapeutic Delivery* **10**: 719–735. https://doi.org/10.4155/tde-2019-0066.

Wambua, S., Mwangi, T.W., Kortok, M. et al. (2006). The effect of α+-thalassaemia on the incidence of malaria and other diseases in children living on the coast of Kenya. *PLoS Medicine* **3**: https://doi.org/10.1371/journal.pmed.0030158.

Wang, A., Huen, S.C., Luan, H.H. et al. (2018). Glucose metabolism mediates disease tolerance in cerebral malaria. *Proceedings of the National Academy of Sciences of the United States of America* **115**: 11042–11047. https://doi.org/10.1073/pnas.1806376115.

Wang, H., Lei, C., Li, J. et al. (2004). A piezoelectric immunoagglutination assay for *Toxoplasma gondii* antibodies using gold nanoparticles. *Biosensors and Bioelectronics* **19**: 701–709. https://doi.org/10.1016/S0956-5663(03)00265-3.

Wang, H., Naghavi, M., Allen, C. et al. (2016). Global, regional, and national life expectancy, all-cause mortality, and cause-specific mortality for 249 causes of death, 1980–2015: a systematic analysis for the Global Burden of Disease Study 2015. *Lancet* **388**: 1459–1544. https://doi.org/10.1016/S0140-6736(16)31012-1.

Wang, J., Zhang, C.J., Chia, W.N. et al. (2015). Haem-activated promiscuous targeting of artemisinin in *Plasmodium falciparum*. *Nature Communications* **6**: 10111. https://doi.org/10.1038/ncomms10111.

Wang, T., Zhou, J., and Gan, et al. (2014). *Toxoplasma gondii* induce apoptosis of neural stem cells via endoplasmic reticulum stress pathway. *Parasitology* **141**: 988–995. PMID: 24612639. https://doi.org/10.1017/S0031182014000183.

Wang, X., Zhang, Y., Zhang, R. et al. (2019). The diversity of pattern recognition receptors (PRRs) involved with insect defence against pathogens. *Current Opinion in Insect Science* **33**: 105–110. https://doi.org/10.1016/j.cois.2019.05.004.

Wang, Z.D., Wang, S.C., Liu, H.H. et al. (2017). Prevalence and burden of *Toxoplasma gondii* infection in HIV-infected people: a systematic review and meta-analysis. *Lancet HIV* **4**: e177–e188. https://doi.org/10.1016/S2352-3018(17)30005-X.

Want, M.Y., Islammudin, M., Chouhan, G. et al. (2017). Nanoliposomal artemisinin for the treatment of murine visceral leishmaniasis. *International Journal of Nanomedicine* **12**: 2189–2204. https://doi.org/10.2147/IJN.S106548.

Watanabe, N., Bruschi, F., and Korenaga, M. (2005). IgE: a question of protective immunity in *Trichinella spiralis* infection. *Trends in Parasitology* **21**: 175–178. https://doi.org/10.1016/j.pt.2005.02.010.

Waykar, V., Wourms, K., Tang, M. et al. (2020). Delusional infestation: an interface with psychiatry. *BJPsych Advances* **1–6**: https://doi.org/10.1192/bja.2020.69.

Weatherall, D.J. (2018). The evolving spectrum of the epidemiology of thalassemia. *Hematology/Oncology Clinics* **32**: 165–175. https://doi.org/10.1016/j.hoc.2017.11.008.

Weedall, G.D. (2020). The *Entamoeba* lysine and glutamic acid rich protein (KERP1) virulence factor gene is present in the genomes of *Entamoeba nuttalli*, *Entamoeba dispar* and *Entamoeba moshkovskii*. *Molecular and Biochemical Parasitology* **238**: https://doi.org/10.1016/j.molbiopara.2020.111293.

Weinstock, J.V. and Elliott, D.E. (2009). Helminths and the IBD hygiene hypothesis. *Inflammatory Bowel Diseases* **15**: 128–133. https://doi.org/10.1002/ibd.20633.

Weirather, J.L., Jeronimo, S.M., Gautam, S. et al. (2011). Serial quantitative PCR assay for detection, species discrimination, and quantification of *Leishmania* spp. in human samples. *Journal of Clinical Microbiology* **49**: 3892–3904. https://doi.org/10.1128/JCM.r00764-11.

wen Su, C., Chen, C.Y., Li, Y. et al. (2018). Helminth infection protects against high fat diet-induced obesity via induction of alternatively activated macrophages. *Scientific Reports* **8**: 1–14. https://doi.org/10.1038/s41598-018-22920-7.

Weng, H., Chen, H., and Wang, M. (2018). Innovation in neglected tropical disease drug discovery and development. *Infectious Diseases of Poverty* **7**: 67. https://doi.org/10.1186/s40249-018-0444-1.

Weston, J.F., Heuer, C., and Williamson, N.B. (2012). Efficacy of a *Neospora caninum* killed tachyzoite vaccine in preventing abortion and vertical transmission in dairy cattle. *Preventative Veterinary Medicine* **103**: 136–144. https://doi.org/10.1016/j.prevetmed.2011.08.010.

Whitrow, M. (1990). Wagner-Jauregg and fever therapy. *Medical History* **34**: 294–310. https://doi. org/10.1017/S0025727300052431.

Whittaker, C., Walker, M., Pion, S.D. et al. (2018). The population biology and transmission dynamics of *Loa loa*. *Trends in Parasitology* **34**: 335–350. https://doi.org/10.1016/j.pt.2017.12.003.

WHO (2020). Schistosomiasis. Fact Sheet. https://www.who.int/news-room/fact-sheets/detail/ schistosomiasis.

WHO (2021). Global technical strategy for malaria 2016–2030. 2021 update (who.int).

Wiedemann, M. and Voehringer, D. (2020). Immunomodulation and immune escape strategies of gastrointestinal helminths and schistosomes. *Frontiers in Immunology* **11**: 572865. https://doi. org/10.3389/fimmu.2020.572865.

Wijayawardena, B.K., Minchella, D.J., and DeWoody, J.A. (2013). Hosts, parasites, and horizontal gene transfer. *Trends in Parasitology* **29**: 329–338. https://doi.org/10.1016/j.pt.2013.05.001.

Wijesinghe, R., Wickremasinghe, A., Ekanayake, S. et al. (2007). Physical disability and psychosocial impact due to chronic filarial lymphoedema in Sri Lanka. *Filaria Journal* **6**: https://doi. org/10.1186/1475-2883-6-4.

Wilcox, C.M. (1997). Chronic unexplained diarrhea in AIDS: approach to diagnosis and management. *AIDS Patient Care and STDs* **11**: 13–17. PMID: 11361724. https://doi.org/10.1089/apc.1997.11.13.

Willcocks, L.C., Carr, E.J., Niederer, H.A. et al. (2010). A defunctioning polymorphism in FCGR2B is associated with protection against malaria but susceptibility to systemic lupus erythematosus. *Proceedings of the National Academy of Sciences of the United States of America* **107**: 7881–7885. https://doi.org/10.1073/pnas.0915133107.

Williams, A.R. (2011). Immune-mediated pathology of nematode infection in sheep – is immunity beneficial to the animal? *Parasitology* **138**: 547–556. https://doi.org/10.1017/S0031182010001654.

Williams-Blangero, S., VandeBerg, J.L., Subedi, J. et al. (2008). Localization of multiple quantitative trait loci influencing susceptibility to infection with *Ascaris lumbricoides*. *Journal of Infectious Diseases* **197**: 66–71. https://doi.org/10.1086/524060.

Wilson, C.S., Jenkins, D.J., Brookes, V.J. et al. (2020). Assessment of the direct economic losses associated with hydatid disease (*Echinococcus granulosus sensu stricto*) in beef cattle slaughtered at an Australian abattoir. *Preventive Veterinary Medicine* **176**: https://doi.org/10.1016/j. prevetmed.2020.104900.

Wilson, M.R., Nigam, Y., Knight, J. et al. (2019). What is the optimal treatment time for larval therapy? A study on incubation time and tissue debridement by bagged maggots of the greenbottle fly, *Lucilia sericata*. *International Wound Journal* **16**: 219–225. https://doi.org/10.1111/iwj.13015.

Winter, M.D., Wright, C., and Lee, D.L. (1997). The mast cell and eosinophil response of young lambs to a primary infection with *Nematodirus battus*. *Parasitology* **114**: 189–193. https://doi.org/10.1017/ S0031182096008311.

Woesner, M.E. (2019). What is old is new again: the use of whole-body hyperthermia for depression recalls the medicinal uses of hyperthermia, fever therapy, and hydrotherapy. *Current Neurobiology* **10**: 56–66.

Wolff, E.D., Salisbury, S.W., Horner, J.R. et al. (2009). Common avian infection plagued the tyrant dinosaurs. *PLoS One* **4**: e7288. doi10.1371/journal.pone.0007288.

Wollina, U., Karte, K., Herold, C. et al. (2000). Biosurgery in wound healing–the renaissance of maggot therapy. *Journal of the European Academy of Dermatology and Venereology* **14**: 285–289. https://doi. org/10.1046/j.1468-3083.2000.00105.x.

Woods, D.J. and Knauer, C.S. (2010). Discovery of veterinary antiparasitic agents in the 21st century: a view from industry. *International Journal for Parasitology* **40**: 1177–1181. https://doi.org/10.1016/ j.ijpara.2010.04.005.

World Health Organization (2015). *Guidelines for the Treatment of Malaria*. Geneva: World Health Organization.

Wu, X., Wang, W., Li, Q. et al. (2020). Case report: surgical intervention for *Fasciolopsis buski* infection: a literature review. *The American Journal of Tropical Medicine and Hygiene* **103**: 2282–2287. https://doi.org/10.4269/ajtmh.20-0572.

Wu, Z., Nagano, I., and Takahashi, Y. (2013). *Trichinella*: what is going on during nurse cell formation? *Veterinary Parasitology* **194**: 155–159. https://doi.org/10.1016/j.vetpar.2013.01.044.

Xia, S., Ma, J.X., Wang, D.Q. et al. (2016). Economic cost analysis of malaria case management at the household level during the malaria elimination phase in the People's Republic of China. *Infectious Diseases of Poverty* **5**: 50. https://doi.org/10.1186/s40249-016-0141-x.

Xu, D., Tang, B., Yang, Y. et al. (2020). Vaccination with a DNase II recombinant protein against *Trichinella spiralis* infection in pigs. *Veterinary Parasitology* https://doi.org/10.1016/j.vetpar.2020.109069.

Yadav, S. and Rawal, G. (2016). The menace due to fake antimalarial drugs. *International Journal of Pharmaceutical Chemistry and Analysis* **3**: 53–55. https://doi.org/10.5958/2394-2797.2016.00007.1.

Yao, C. and Köster, L.S. (2015). *Tritrichomonas foetus* infection, a cause of chronic diarrhoea in the domestic cat. *Veterinary Research* **46**: 35. https://doi.org/10.1186/s13567-015-0169-0.

Yaw, T.J., O'Neil, P., Gary, J.M. et al. (2019). Primary amebic meningoencephalomyelitis caused by *Naegleria fowleri* in a south-central black rhinoceros (*Diceros bicornis minor*). *Journal of the American Veterinary Medical Association* **255**: 219–223. https://doi.org/10.2460/javma.255.2.219.

Yolken, R., Torrey, E.F., and Dickerson, F. (2017). Evidence of increased exposure to *Toxoplasma gondii* in individuals with recent onset psychosis but not with established schizophrenia. *PLoS Neglected Tropical Diseases* **11**: https://doi.org/10.1371/journal.pntd.0006040.

Yorimitsu, N., Hiraoka, A., Utsunomiya, H. et al. (2013). Colonic intussusception caused by anisakiasis: a case report and review of the literature. *Internal Medicine* **52**: 223–226. https://doi.org/10.2169/internalmedicine.52.8629.

Yoshida, A., Doanh, P.N., and Maruyama, H. (2019). *Paragonimus* and paragonimiasis in Asia: an update. *Acta Tropica* **199**: https://doi.org/10.1016/j.actatropica.2019.105074.

Yoshida, S. (1920). On the resistance of Ascaris eggs. *Journal of Parasitology* **6**: 132–139. https://www.jstor.org/stable/3271066.

Young, S.L., Goodman, D., Farag, T.H. et al. (2007). Geophagia is not associated with *Trichuris* or hookworm transmission in Zanzibar, Tanzania. *Transactions of the Royal Society of Tropical Medicine and Hygiene* **101**: 766–772. https://doi.org/10.1016/j.trstmh.2007.04.016.

Yssouf, A., Almeras, L., Raoult, D. et al. (2016). Emerging tools for identification of arthropod vectors. *Future Microbiology* **11**: 549–566. https://doi.org/10.2217/fmb.16.5.

Yu, Q., Li, J., Zhang, X. et al. (2010). Induction of immune responses in mice by a DNA vaccine encoding *Cryptosporidium parvum* Cp12 and Cp21 and its effect against homologous oocyst challenge. *Veterinary Parasitology* **172**: 1–7.

Zaccone, P., Burton, O.T., and Cooke, A. (2008). Interplay of parasite-driven immune responses and autoimmunity. *Trends in Parasitology* **24**: 35–42. https://doi.org/10.1016/j.pt.2007.10.006.

Zachary, J.F. (2016). *Pathologic Basis of Veterinary Medicine*, 6the. St Louis, MO: Elsevier.

Zaidi, F. and Chen, X.X. (2011). A preliminary survey of carrion breeding insects associated with the Eid ul Azha festival in remote Pakistan. *Forensic Science International* **209**: 186–194. https://doi.org/10.1016/j.forsciint.2011.01.027.

Zaiss, M.M. and Harris, N.L. (2016). Interactions between the intestinal microbiome and helminth parasites. *Parasite Immunology* **38**: 5–11. https://doi.org/10.1111/pim.12274.

Zakeri, A. (2017). Helminth-induced apoptosis: a silent strategy for immunosuppression. *Parasitology* **144**: 1663–1676. https://doi.org/10.1017/S0031182017000841.

Zakovic, S. and Levashina, E.A. (2017). NF-κB-like signaling pathway REL2 in immune defenses of the malaria vector *Anopheles gambiae*. *Frontiers in Cellular and Infection Microbiology* **7**: https://doi.org/10.3389/fcimb.2017.00258.

Zaky, W.I., Tomaino, F.R., Pilotte, N. et al. (2018). Backpack PCR: a point-of-collection diagnostic platform for the rapid detection of *Brugia* parasites in mosquitoes. *PLoS Neglected Tropical Diseases* **12**: https://doi.org/10.1371/journal.pntd.0006962.

Zarlenga, D., Thompson, P., and Pozio, E. (2020). *Trichinella* species and genotypes. *Research in Veterinary Science* **133**: https://doi.org/10.1016/j.rvsc.2020.08.012.

Zarowiecki, M.Z., Huyse, T., and Littlewood, D.T.J. (2007). Making the most of mitochondrial genomes – markers for phylogeny, molecular ecology and barcodes in *Schistosoma* (Platyhelminthes: Digenea). *International Journal for Parasitology* **37**: 1401–1418. https://doi.org/10.1016/j.ijpara.2007.04.014.

Zavala, G.A., Garcia, O.P., Campos-Ponce, M. et al. (2016). Children with moderate-high infection with *Entamoeba coli* have higher percentage of body and abdominal fat than non-infected children. *Pediatric Obesity* **11**: 443–449. https://doi.org/10.1111/ijpo.12085.

Zepeda, N., Copitin, N., Solano, S. et al. (2011). *Taenia crassiceps*: infections of male mice lead to severe disruption of seminiferous tubule cells and increased apoptosis. *Experimental Parasitology* **127**: 153–159. https://doi.org/10.1016/j.exppara.2010.07.008.

Zhang, G., Sachse, M., Prevost, M.C. et al. (2016). A large collection of novel nematode-infecting microsporidia and their diverse interactions with *Caenorhabditis elegans* and other related nematodes. *PLoS Pathogens* **12**: https://doi.org/10.1371/journal.ppat.1006093.

Zhang, M., Sun, K., Wu, Y. et al. (2017). Interactions between intestinal microbiota and host immune response in inflammatory bowel disease. *Frontiers in Immunology* **8**: 942. https://doi.org/10.3389/fimmu.2017.00942.

Zhang, N., Zhang, H., Yu, Y. et al. (2019). High prevalence of *Pentatrichomonas hominis* infection in gastrointestinal cancer patients. *Parasites & Vectors* **12**: 423. https://doi.org/10.1186/s13071-019-3684-4.

Zhang, W. and McManus, D.P. (2008). Vaccination of dogs against *Echinococcus granulosus*: a means to control hydatid disease? *Trends in Parasitology* **24**: 419–424. https://doi.org/10.1016/j.pt.2008.05.008.

Zhou, X., Kambalame, D.M., Zhou, S. et al. (2019). Human *Chrysomya bezziana* myiasis: a systematic review. *PLoS Neglected Tropical Diseases* **13**: https://doi.org/10.1371/journal.pntd.0007391.

Zhou, X., Kaya, H.K., Heungens, K. et al. (2002). Response of ants to a deterrent factor (s) produced by the symbiotic bacteria of entomopathogenic nematodes. *Applied and Environmental Microbiology* **68**: 6202–6209. https://doi.org/10.1128/AEM.68.12.6202-6209.2002.

Zhu, N., Zhang, D., Wang, W. et al. (2020). A novel coronavirus from patients with pneumonia in China, 2019. *New England Journal of Medicine*. https://doi.org/10.1056/NEJMoa2001017.

Zhu, Q., Hastriter, M.W., Whiting, M.F. et al. (2015). Fleas (Siphonaptera) are Cretaceous and evolved with Theria. *Molecular Phylogenetics and Evolution* **90**: 129–139. https://doi.org/10.1016/j.ympev.2015.04.027.

Zininga, T., Pooe, O.J., Makhado, P.B. et al. (2017). Polymyxin B inhibits the chaperone activity of *Plasmodium falciparum* Hsp70. *Cell Stress and Chaperones* **22**: 707–715. https://doi.org/10.1007/s12192-017-0797-6.

Index

Parasitology: An Integrated Approach, Second Edition. Alan Gunn and Sarah J. Pitt.
© 2022 John Wiley & Sons Ltd. Published 2022 by John Wiley & Sons Ltd.
Companion website: www.wiley.com/go/gunn/parasitology2